High-Energy Physics and Nuclear Structure

High-Energy Physics and Nuclear Structure

Proceedings of the Third International Conference on High Energy
Physics and Nuclear Structure sponsored by the International Union
of Pure and Applied Physics, held at Columbia University, New York
City, September 8-12, 1969

Edited by Samuel Devons

Department of Physics
Columbia University
New York City

 PLENUM PRESS • NEW YORK-LONDON • 1970

Library of Congress Catalog Card Number 72-112272

SBN 306-30473-2

© 1970 Plenum Press, New York
A Division of Plenum Publishing Corporation
227 West 17th Street, New York, N. Y. 10011

United Kingdom edition published by Plenum Press, London
A Division of Plenum Publishing Corporation, Ltd.
Donington House, 30 Norfolk Street, London W.C. 2, England

Printed in the United States of America

THIRD INTERNATIONAL CONFERENCE ON HIGH-ENERGY PHYSICS AND NUCLEAR STRUCTURE

Sponsored by the International Union of Pure and Applied Physics

INTERNATIONAL ADVISORY COMMITTEE:

E. Amaldi
Rome

M. Danysz
Warsaw

A. de-Shalit
Weizmann Institute

V. P. Dzhelepov
Dubna

B. P. Gregory
CERN

E. P. I lincks
Nat'l. R. C. Canada

P. Huber
Basel

J. H. D. Jensen
Heidelberg

A. M. L. Messiah
Saolay

I. Nonaka
Tokyo

R. E. Peierls
Oxford

I. S. Shapiro
Moscow

V. F. Weisskopf
MIT

ORGANIZING COMMITTEE:

R. J. Glauber
Harvard University

V. W, Hughes
Yale University

H. Primakoff
University of Pennsylvania

J. Rainwater
Columbia University

R. T. Siegel
College of William and Mary

C. S. Wu
Columbia University

S. Devons
Columbia University

FINANCIAL SPONSORS:

U.S. Atomic Energy Commission
National Science Foundation
University of Pennsylvania
Yale University
Columbia University

PREFACE

In preparing the program for this Conference, the third in the series, it soon became evident that it was not possible to include in a conference of reasonable duration all the topics that might be subsumed under the broad title, "High Energy Physics and Nuclear Structure." From their initiation, in 1963, it has been as much the aim of these Conferences to provide some bridges between the steadily separating domains of particle and nuclear physics, as to explore thoroughly the borderline territory between the two -- the sort of no-man's-land that lies unclaimed, or claimed by both sides.

The past few years have witnessed the rapid development of many new routes connecting the two major areas of 'elementary particles' and 'nuclear structure', and these now spread over a great expanse of physics, logically perhaps including the whole of both subjects. (As recently as 1954, an International Conference on 'Nuclear and Meson Physics' did, in fact, embrace both fields!) Since it is not now possible to traverse, in one Conference, this whole network of connections, still less to explore the entire territory it covers, the choice of topics has to be in some degree arbitrary. It is hoped that ours has served the purpose of fairly exemplifying many areas where physicists, normally separated by their diverse interests, can find interesting and important topics which bring them together.

But nothing exclusive is claimed for our particular choice; nor are there profound reasons for the many omissions. For example, there were no substantial deliberations, at this Conference, on the subject of the 'elementary' particle interactions themselves -- nucleon-nucleon, pion-nucleon, etc. On the other hand, there is included a session on the properties of pions and muons, and another on the development of accelerators, etc., of the 'intermediate energy' class. This is not so much because the scope of this Conference is regarded as synonymous with "Intermediate Energy Physics" as a recognition of the fact that it is in this energy range -- and the techniques associated with it -- that much of the relevant experimental work is done. No doubt, the next Conference in this series will span the same or cover an even wider range of physics, but will do so by selecting and emphasizing a different set of topics.

The contributions included in the proceedings are of two sorts:
i) Records of papers presented at the Conference, with brief synopses of the ensuing discussions (where the record was decipherable!),
ii) Contributed papers -- of a summary nature -- which were submitted to and made available at the Conference, but not presented orally.

In preparing the record some compromise had to be struck between the conflicting demands of speed and accuracy. In addition the exigencies of space have imposed the need for some pruning and editing.* It has not been practicable, in the time available, to send to each contributor proofs of the final version of his contribution to the discussion. Although the editorial work has been a cooperative effort of the Editor and the Scientific Secretaries of the sessions (fortified by the moral support of the Organizing Committee) it is the Editor who must bear the responsibility for all errors, omissions, and misinterpretations.

A few days before the Conference began we learned the tragic news about Professor Amos de-Shalit, whose efforts were in so large a measure responsible for the existence of this series of Conferences. At the final session, the Conference as a whole decided to dedicate these proceedings to his memory.

I would like to record here thanks to all those who helped with the arrangements for the Conference and preparation of these proceedings -- especially Mrs. Jane Orr, who helped in the early stages, Miss Marjorie Tillary in the latter ones, and Mr. Barry Barr for his invaluable efforts throughout.

SAMUEL DEVONS

New York, N.Y.
November, 1969

* Due to some misunderstanding, one or two of the contributions have been presented in much lengthier form than was intended. At the time they were submitted, the only alternatives were to omit or include them as they stood. The latter was felt to be in the best interest of the Conference participants and others.

LIST OF PARTICIPANTS

J.M. Abillon, Laboratoire de Physique Atomique et Nucleaire,
Paris, France.
R.J. Adler, Virginia Polytechnic Institute, Blacksburg, Virginia.
S.L. Adler, School of Natural Sciences, Princeton, New Jersey.
J. Alster, University of Tel-Aviv, Israel.
R.D. Amado, University of Pennsylvania, Philadelphia, Pennsylvania.
H.L. Anderson, University of Chicago, Chicago, Illinois.
N. Auerbach, M.I.T., Cambridge, Massachusetts.
S.M. Austin, Michigan State University, East Lansing, Michigan.
G. Backenstoss, CERN, Geneva Switzerland.
M. Baranger, Carnegie-Mellon University, Pittsburgh, Pennsylvania.
V.S. Barashenkov, Dubna, Joint Institute for Nuclear Research,
Moscow, U.S.S.R.
C. Barber, M.I.T., Cambridge, Massachusetts.
M. Bardon, Columbia University, Irvington, New York.
P. Barnes, Carnegie-Mellon University, Pittsburgh, Pennsylvania.
R.C. Barrett, University of Surrey, Surrey, England.
C.J. Batty, Rutherford High Energy Laboratory, Berkshire, England.
M. Bazin, Rutgers State University, New Brunswick, New Jersey.
F. Beck, Institut fur Theoretische Kernphysik, Darmstadt, Germany.
F. Becker, University of Strasbourg, Strasbourg, France.
R.L. Becker, Oak Ridge National Laboratories, Oak Ridge, Tennessee.
D. Bender, University of British Columbia, Vancouver, Canada.
J.B. Bellicard, CEN-Saclay, France.
A. Bernstein, M.I.T., Cambridge, Massachusetts.
W. Bertozzi, M.I.T., Cambridge, Massachusetts.
J.P. Blaser, Laboratory of High Energy Physics, Zurich, Switzerland.
M. Blecher, Virginia Polytechnic Institute, Newport News, Virginia.
C. Bloch, CEN-Saclay, France.
A.R. Bodmer, University of Illinois, Chicago, Illinois.
E.F. Borie, U.S. Department of Commerce, National Bureau of
Standards, Washington, D.C.
E.T. Boschitz, NASA Lewis Research Center, Cleveland, Ohio.
C.C. Bouchiat, Laboratoire de Physique Theorique, Orsay, France.
I. Brissaud, Institut de Physique Nucleaire, Orsay, France.
G.E. Brown, S.U.N.Y., Stony Brook, New York.

R. Burman, LASAL, Los Alamos, New Mexico.

B. Budick, Jerusalem, Israel.

A. Bussiere, CEN-Saclay, France.

N. Byers, University of California, Los Angeles, California.

V.G. Campos, Instituto Superiore di Sanita, Rome, Italy.

J. Carroll, College of William and Mary, Williamsburg, Virginia.

P. Catillon, Commissariat a l'Energie Atomique, CEN-Saclay, France.

C. Cernigoi, Instituto Fisica Universita, Trieste, Italy.

M. Chemtob, Centre D'Etudes Nucleaires de Saclay, France.

M.Y. Chen, Columbia University, New York, New York.

W.K. Cheng, Stevens Institute of Technology, Hoboken, New Jersey.

B. Chertok, American University, Washington, D.C.

D.C. Choudhury, Polytechnic Institute of Brooklyn, Brooklyn,
                                                              New York.

W.E. Cleland, University of Massachusetts, Amherst, Massachusetts.

S. Cohen, Hebrew University of Jerusalem, Jerusalem, Israel

C.L. Critchfield, Los Alamos Scientific Laboratory, Los Alamos,
                                                              New Mexico.

A.H. Cromer, Niels-Bohr Institute, Copenhagen, Denmark.

K.M. Crowe, University of California, Berkeley, California.

W. Czyż, Instytut Fizyki Jadrowej w Krakowie, Krakow, Poland.

M. Danos, National Bureau of Standards, Washington, D.C.

A. Dar, Technion-Israel Institute of Technology, Haifa, Israel.

P. Darriulat, CERN, Geneva, Switzerland.

N. DeBotton, Service de Haute Energie, CEN-Saclay, France.

M. Demeur, Universite Libre de Bruxelles, Bruxelles, Belgium.

J. Deutsch, CERN, Geneva, Switzerland.

S. Devons, Columbia University, New York, New York.

L. Dick, CERN, Geneva, Switzerland.

L. DiLella, Columbia University, New York, New York.

J. Duclos, Service de Haute Energie, CEN-Saclay, France.

V.P. Dzhelepov, Joint Institute for Nuclear Research, Moscow, U.S.S.R.

M. Eckhause, College of William and Mary, Williamsburg, Virginia.

R.M. Edelstein, Carnegie-Mellon University, Pittsburgh, Pennsylvania.

R. Engfer, Institute für Technische Kernphysik der Technischen
            Hochschule, Darmstadt, West Germany.

I. Eramzhyan, Joint Institute for Nuclear Research, Dubna, Moscow,
                                                              U.S.S.R.

T.E.O. Ericson, M.I.T., Cambridge, Massachusetts.

V.S. Evseev, Joint Institute for Nuclear Research, Moscow, U.S.S.R.

S.T. Emerson, Rice University, Houston, Texas.

G. Faldt, CERN, Geneva Switzerland.

S. Fallieros, Brown University, Providence, Rhode Island.

H.A. Ferwerda, Universiteitscomplex Paddepoel, Groningen,
                                                     The Netherlands.

H. Feshbach, M.I.T., Cambridge, Massachusetts.

G. Finocchiaro, SUNY, Stony Brook, New York.

E. Fiorini, Instituto Nazionale di Fisica Nucleare, Milano, Italy.

E. Fischbach, Institute for Theoretical Physics, Stony Brook,
                                                              New York.

W. Fischer, SIN-ETH, Zurich, Switzerland.
V. Flaminio, Brookhaven National Laboratories, Upton, New York.
L.L. Foldy, Case Western Reserve University, Cleveland, Ohio.
K. Foley, Brookhaven National Laboratories, Upton, New York.
K.W. Ford, University of California, Irvine, California.
W. Ford, NASA Lewis Research Center, Cleveland, Ohio.
V. Franco, Los Alamos Sciencific Laboratory, Los Alamos, New Mexico.
C. Franzinetti, Universita di Torino, Torino, Italy.
J.L. Friedes, Brookhaven National Laboratories, Upton, New York.
W.A. Friedman, Princeton University, Princeton, New Jersey.
J. Friar, University of Washington, Seattle, Washington.
B.W. Fricke, Institut f. Theoretische Physik der Universität,
                                        Frankfurt, Germany.
A. Fujii, Sophia University, Tokyo, Japan.
J.I. Fujita, University of Tokyo, Tokyo, Japan.
H.O. Funsten, College of· William and Mary, Williamsburg, Virginia.
A. Gal, M.I.T., Cambridge, Massachusetts.
G.T. Garvey, Princeton, University, Princeton, New Jersey.
J. Gillespie, CEN-Saclay, France.
V. Gillet, CEN-Saclay, France.
A.E. Glassgold, CEN-Saclay, France.
R.J. Glauber, Harvard University, Cambridge, Massachusetts.
A.S. Goldhaber, State University of New York, Stony Brook, New York.
G. Scharff-Goldhaber, Brookhaven National Laboratories, Upton,
                                        New York.
M. Goldhaber, Brookhaven National Laboratory, Upton, New York.
K. Gotow, Virginia Polytechnic Institute, Newport News, Virginia.
K. Gottfried, M.I.T., Cambridge, Massachusetts.
B. Goulard, Laval University, Quebec, Canada.
L. Grodzins, M.I.T., Cambridge, Massachusetts.
P.C. Gugelot, Virginia Associated Research Center, Newport News,
                                        Virginia.
R.P. Haddock, University of California, Los Angeles, California.
D. Hamilton, The University of Sussex, Sussex, England.
R. Hammerstein, Michigan State University, East Lansing, Michigan.
R.C. Hanna, Rutherford High Energy Laboratory, Berkshire, England.
C.K. Hargrove, National Research Council of Canada, Ottawa, Ontario,
                                        Canada.
D.R. Harrington, Rutgers University, New Brunswick, New Jersey.
K. Hartt, University of Rhode Island, Kingston, Rhode Island.
J.H. Heisenberg, Max-Planck-Institut Fur Physik und Astrophysik,
                                        Munchen, Germany.
R. Hess, University of Geneva, Geneva, Switzerland.
E.P. Hincks, National Research Council of Canada, Ottawa, Canada.
W. Hirt, CERN, Geneva, Switzerland.
R. Hofstadter, Stanford University, Stanford, California.
H.D. Holmgren, University of Maryland, College Park, Maryland.
M.G. Huber, Max Planck Institut f. Kernphysik, Heidelberg, Germany.
P. Huber, Physikalisches Institut der Universität Basel, Basel,
                                        Switzerland.

J. Hüfner, M.I.T., Cambridge, Massachusetts.
E.B. Hughes, Stanford University, Stanford, California.
V.W. Hughes, Yale University New Haven, Connecticut.
H. Hultzsch, Universitat Mainz, Mainz, Germany.
D.P. Hutchinson, Princeton-Penn Accelerator, Princeton, New Jersey.
D. Isabelle, Universite de Clermont, Clermont Ferrand, France.
G. Jacob, U. do Rio Grande do Sul, Porto Alegre, Brazil.
A.N. James, The University of Liverpool, Liverpool, England.
H. Jehle, George Washington University, Washington, D.C.
D.A. Jenkins, Virginia Polytechnic Institute at VARC, Newport News,
                                                              Virginia.
G. Jones, University of British Columbia, Vancouver, Canada.
J. Julien, Service de Haute Energie, CEN-Saclay, France.
E. Kankeleit, Institut Fur Tech. Kernphysik, Darmstadt, West Germany.
S.N. Kaplan, University of California, Berkeley, California.
H.A. Kastrup, Sektion Physik der Universitat-Theoretische, Munchen,
                                                               Germany.
J. Keren, Northwestern University, Evanston, Illinois.
D. Kessler, Carleton University, Ontario, Canada.
F.C. Khanna, Chalk River Nuclear Laboratories, Ontario, Canada.
C.W. Kim, John Hopkins University, Baltimore, Maryland.
L.S. Kisslinger, Case Western Reserve, Cleveland, Ohio.
A. Klein, University of Pennsylvania, Philadelphia, Pennsylvania.
R.H. Klein, Cleveland State University, Cleveland, Ohio.
T.M. Knasel, DESY, Hamburg, Germany.
T. Kohmura, The University of Rochester, Rochest , New York.
D. Koltun, The University of Rochester, Rochester, New York.
E.H. Kone, American Institute of Physics, New York, N.Y.
T. Kozlowski, Columbia University, New York, New York.
J. Kossler, College of William and Mary, Williamsburg, Virginia.
S. Kowalski, M.I.T., Cambridge, Massachusetts.
D. Kurath, Argonne National Laboratory, Argonne, Illinois.
A.A. Kuznetsov, Joint Institute for Nuclear Research, Moscow, U.S.S.R.
L.M. Langer, Indiana University, Bloomington, Indiana.
C. Lazard, Institut de Physique Nucleaire d'Orsay, Orsay, France.
K.J. LeCouteur, The Australian National University, Canberra,
                                                              Australia.
L. Lederman, Columbia University, New York, New York.
W.Y. Lee, Columbia University, New York, New York.
D.W.G.S. Leith, Stanford University, Stanford, California.
J.S. Levinger, Rensselaer Polytechnic Institute, Troy, New York.
D.B. Lichtenberg, Indiana University, Bloomington, Indiana.
S.J. Lindenbaum, Brookhaven National Laboratory, Upton, New York.
V.M. Lobashov, A.F. Ioffe Physico-Technical Institute, Leningrad,
                                                                U.S.S.R.
M.P. Locher, CERN, Geneva, Switzerland.
R. Lombard, Institut de Physique Nucleaire, Orsay, France.
D.C.J. Lu, Yale University, New Haven, Connecticut.
H.J. Lubatti, M.I.T., Cambridge, Massachusetts.

S.W. MacDowell, Yale University, New Haven, Connecticut.
B.A. MacDonald, Virginia Polytechnic Institute, Newport News,
                                                    Virginia.
J. McCarthy, Stanford University, Stanford, California.
R.J. McKee, National Research Council of Canada. Ottawa, Canada.
B.H.J. McKellar, University of Sydney, Sydney, Australia.
J.M. McKinley, Oakland University, Rochester, Michigan.
E. Macagno, Columbia University, New York, New York.
R.J. Macek, Los Alamos Scientific Laboratory, Los Alamos, New Mexico.
P. Macq, Universite de Louvain, Heverle, Belgium.
A. Magnon, Service de Haute Energie, CEN-Saclay, France.
R. Malvano, INFN, Torino, Italy.
J. Manassah, Columbia University, New York, New York.
B. Margolis, McGill University, Quebec, Canada.
N. Marty, Institut de Physique Nucleaire d'Orsay, Orsay, France.
L.C. Maximon, SUNY, Stony Brook, New York.
D.F. Measday, CERN, Geneva, Switzerland.
A. Messiah, CEN-Saclay, France.
R.C. Minehart, University of Virginia, Charlottesville, Virginia.
P. Minkowski, Centre de Physique Nucleaire, Heverle - Louvain, Belgium.
L. Mo, Stanford University, Stanford, California.
R.B. Moore, McGill University, Montreal, Canada.
J. Morgenstern, Service de Haute Energie, CEN-Saclay, France.
R.J. Morrison, University of California, Santa Barbara, California.
G. Myatt, Brookhaven National Laboratories, Upton, New York.
K. Nagatani, Brookhaven National Laboratories, Upton, New York.
D.E. Nagel, Los Alamos Scientific Laboratories, Upton, New York.
B. Nefkens, University of California, Los Angeles, California.
P. Nemethy, Service de Haute Energie, CEN-Saclay, France.
C. Noack, Institut fur Theoretische Physik, Heidelberg, Germany.
J.V. Noble, University of Pennsylvania, Philadelphia, Pennsylvania.
Y. Nogami, McMaster University, Ontario, Canada.
T.B. Novey, Argonne National Laboratory, Argonne, Illinois.
S. Ozaki, Brookhaven National Laboratory, Upton, New York.
R.S. Panvini, Setauket, New York.
R.E. Peierls, University of Oxford, Oxford, England.
C.F. Perdrisat, College of William and Mary, Williamsburg, Virginia.
V. Perez-Mendez, University of California, Berkeley, California.
G.A. Peterson, The University of Massachusetts, Amherst, Massachusetts.
J. Picard, Center d'etudes Nucleaires de Saclay, France.
E. Picasso, CERN, Geneva, Switzerland.
H. Pietschmann, Institute of Theoretical Physics, Vienna, Austria.
C.A. Piketty, Laval University, Quebec, Canada.
H. Pilkuhn, University of Karlsruhe, Kaisertsr., Germany.
H.S. Plendl, The Florida State University, Tallahassee, Florida.
A.M. Poskanzer, Lawrence Radiation Laboratory, Berkeley, California.
R. Powers, Virginia Associated Research Center, Newport News,
                                                    Virginia.
E. Predazzi, Indiana University, Bloomington, Indiana.
H. Primakoff, The University of Pennsylvania, Philadelphia,
                                                    Pennsylvania.

M. Priou, Service de Haute Energie, CEN-Saclay, France.
I.I. Rabi, Columbia University, New York, New York.
P. Radvanyi, Institut de Physique Nucleaire d'Orsay, Orsay, France.
J. Rainwater, Columbia University, New York, New York.
R. Raphael, Catholic University of America, Washington, D.C.
A. Reitan, Technical University of Norway, Trondheim, Norway.
E.A. Remler, College of William and Mary, Williamsburg, Virginia.
L.P. Remsberg, Brookhaven National Laboratories, Upton, New York.
M. Rho, Service de Physique Theorique, CEN-Saclay, France.
A. Richter, Argonne National Laboratory, Argonne, Illinois.
M.E. Rickey, Indiana University, Bloomington, Indiana.
A. Roberts, National Accelerator Laboratory, Batavia, Illinois.
L.P. Robertson, University of Victoria, Victoria, B.C., Canada.
G.L. Rogosa, U.S.A.E.C., Washington, D.C.
D.R. Rote, University of Illinois, Chicago, Illinois.
K. Runge, CERN, Geneva, Switzerland.
J.E. Russell, University of Cincinnati, Cincinnati, Ohio.
A. Sachs, Columbia University, New York, New York.
J. Saudinos, Service de Moyenne Energie, CEN-Saclay, France.
C. Schaerf, Laboratori Nazionali de Frascati, Frascati, Italy.
F. Scheck, CERN, Geneva, Switzerland.
P. Schlein, University of California, Los Angeles, California.
H. Schmitt, CERN, Geneva, Switzerland.
R. Schneider, Columbia University, New York, New York.
E. Schwarz, Universite de Neuchatel, Neuchatel, Switzerland.
R.E. Segel, Northwestern University, Evanston, Illinois.
E.G. Segré, University of California, Berkeley, California.
F. Selleri, Bari University, Bari, Italy.
A. Sherwood, CEN-Saclay, France.
I. Sick, Stanford University, Stanford, California.
R.T. Siegel, College of William and Mary, Williamsburg, Virginia.
R. Silbar, University of California, Los Alamos, New Mexico.
W.D. Simpson, Brookhaven National Laboratory, Upton, New York.
S.E. Sobottka, University of Virginia, Charlottesville, Virginia.
R.L. Stearns, Vassar College, Poughkeepsie, New York.
P. Stein, Cornell University, Ithaca, New York.
M.M. Sternheim, University of Massachusetts, Amherst, Massachusetts.
T. Stovall, University of Paris, Orsay, France.
C.R. Sun, State University of New York, Albany, New York.
M.K. Sundaresan, Carleton University, Ontario, Canada.
R. Sundelin, Cornell University, Ithaca, New York.
W. Tanner, University of Oxford, Oxford, England.
V. Telegdi, University of Chicago, Chicago, Illinois.
E.D. Theriot, University of California, Los Alamos, New Mexico.
H.A. Thiessen, Los Alamos Scientific Laboratory, Los Alamos, New Mexico.
P.A. Thompson, Yale University, New Haven, Connecticut.
A. G. Tibell, University of Uppsala, Uppsala, Sweden.
J. Tiomno, Universidade de Sao Paulo, Brazil.
K.K. Trabert, University of Pennsylvania, Philadelphia, Pennsylvania.

C.C. Trail, Brooklyn College, Brooklyn, New York.
T.V. Tran, Laboratoire de Physique Theorique, Orsay, France.
J. Trefil, University of Illinois, Urbana, Illinois.
P. Truoel, Lawrence Radiation Laboratory, Berkeley, California.
W.P. Trower, Virginia Polytechnic Institute, Blacksburg, Virginia.
W. Turchinetz, M.I.T., Cambridge, Massachusetts.
H. Überall, Catholic University of America, Washington, D.C.
H.B. Van Der Raay, University of Birmingham, Birmingham, England.
J.J. Veillet, Laboratoire de l'Accelerateur Lineaire, Orsay, France.
N. Vinh Mau, Institut de Physique Nucleaire, Orsay, France.
H. Wacksmuth,
J.D. Walecka, Stanford University, Stanford, California.
N.S. Wall, University of Maryland, College Park, Maryland.
A.H. Wapstra, Institut Voor Kernphysisch Onderzoek, Amsterdam,
                                                    Netherlands.
J.B. Warren, University of British Columbia, Vancouver, Canada.
A.P. Wattenberg, University of Illinois, Urbana, Illinois.
R.E. Welsh, College of William and Mary, Williamsburg, Virginia.
M.G. White, Princeton-Pennsylvania Accelerator, Princeton,
                                                    New Jersey.
C.E. Wiegand, University of California, Berkeley, California.
L. Wilets, University of Washington, Seattle, Washington.
H.J. Williams, Chatham, New Jersey.
C. Williamson, M.I.T., Cambridge, Massachusetts.
R. Wilson, Harvard University, Cambridge, Massachusetts.
L. Wolfenstein, Carnegie-Mellon University, Pittsburgh, Pennsylvania.
A.M. Wolsky, University of Pennsylvania, Philadelphia, Pennsylvania.
L. Wright, Duke University, Durham, North Carolina.
C.S. Wu, Columbia University, New York, New York.
C.N. Yang, State University of New York, Stony Brook, New York.
A.I. Yavin, University of Illinois, Urbana, Illinois.
G. Yekutieli, SLAC, Stanford, California.
P.F. Yergin, Rensselaer Polytechnic Institute, Troy, New York.

CONTENTS

II - ELECTROMAGNETIC AND WEAK INTERACTION PROBES

Chairman: G. Backenstoss
Scien. Sec.: E. Macagno

**Paper not received in time for inclusion in the proceedings.

III - HADRONIC INTERACTIONS: HIGH-ENERGY

COLLISION PROCESSES

Chairman: C. Bloch
Scien. Sec.:   R. Hess

***The contents of these reports is included in the paper pre-
   sented by V. P. Dzhelepov (page 278).

IV - HADRONIC INTERACTIONS: COLLISION PROCESSES

WITH PIONS AND UNSTABLE PARTICLES

Chairman: V. P. Dzhelepov
Scien. Sec.: J. Manassah

VI – NEW ACCELERATORS AND OTHER EXPERIMENTAL

DEVELOPMENTS

Chairman: J. P. Blaser
Scien. Sec.: R. Schneider

VII - PROPERTIES OF PIONS AND MUONS

Chairman: H. Primakoff
Scien. Sec.: L. DiLella

VIII - FUNDAMENTAL SYMMETRY PROPERTIES

Chairman: C. S. Wu
Scien. Sec.: K. Trabert

IX - SOME THEORETICAL QUESTIONS IN ELEMENTARY

PARTICLE AND NUCLEAR PHYSICS

Chairman: R. E. Peierls
Scien. Sec.:   A. Wolsky

DEDICATION

All his activities gave him deep pleasure -- and
indeed it was this pleasure that was one of the elements
of his great personal charm. We all knew the flashing
twinkle in his eyes when he was learning, teaching,
talking and making physics. But this infectious
enthusiasm was not restricted to physics; it could
emerge in the course of a conversation with a student,
or a world-famous political leader.

II.

It is entirely fitting that the proceedings of
this the Third International Conference on High Energy
Physics and Nuclear Structure be dedicated to Amos
de-Shalit. For he played an essential role in starting
this series of conferences, first of which was held in
CERN in 1963. At that time it took foresight to recognize that studying the interplay between nuclear and
high energy physics would be of great value to both
fields. The first conference, its sequel at Rehovot in
1967, and the course on the same topic at the Enrico
Fermi School in Varenna in the summer of 1966, were all
very successful, and in all of them de-Shalit took the
initiative and played an important part.

III.

Amos de-Shalit was born in Jerusalem September 29,
1926. When he graduated from high school in 1944,
there were only 800 students at the Hebrew University.
The Weizmann Institute of Science did not yet exist.
The times were exciting and dangerous, and the group
at the Hebrew University shared that uniquely intense cam-
araderie which comes from high hopes for a common goal.
It was from this group that many of Amos's life-long
friends and colleagues were drawn.

In 1946 while still at the University he published
his first work, a text based on the lectures of Leo
Motzkin on Analytical and Projective Geometry, a subject
in which he always maintained an interest. The war of
independence broke out sporadically in the fall of
1947, and de-Shalit became a member of an ordnance group.
During this difficult period they studied Heitler's
book on the Quantum Theory of Radiation, and learned
from Racah the application of group theory to quantum
mechanics. After the armistice in 1949 de-Shalit received his M.Sc. degree from the Hebrew University for
an investigation on the self-energy of the electron.

At that point the new Israeli government with remarkable foresight agreed to send some of these bright
young men to the centers of learning in Europe and

the United States for further education. The group tried
to cover a broad area of nuclear physics, and studied
under leading figures such as Pauli and Fermi. de-Shalit
went into experimental physics; he studied under
Scherrer at Zurich, and received his doctorate in 1951.

The following two years were spent at Princeton,
M.I.T., Brookhaven National Laboratory and Saclay.

In 1954 he returned to the Weizmann Institute to
be the first head of the newly created nuclear physics
department. In the period 1954-66 during which de-Shalit
was the head it became one of the world's leading centers
for the study of nuclear physics, rivalled by only a few
other institutions. A most significant achievement of
this period was the establishment of an ambitious program of graduate eduation in physics. Its success
exceeded all reasonble expectations: Everyone is keen-
ly aware of the great impact young Israelis have made
on nuclear and particle physics in recent years. These
accomplishments would have been impossible without the
outstanding group of physicists at the Weizmann Institute. But it did require wise leadership, and this
de-Shalit provided. He was a superb administrator. He
was quick to grasp new ideas. He had good taste in
physics and in people. And above all he was willing to
make bold decisions, and to institute new ventures with
verve and dispatch.

These qualities served him well in the many causes
in which he was enrolled. He had much to do with the
development and expansion of the whole Weizmann Institute, serving as Scientific Director in 1961-63, and
as Director General in the period 1966-68. Starting
in the early sixties he became concerned with the quality of science teaching at the secondary school level.
He organized summer science camps for the young. He
initiated a program at the Institute in science education which led to the creation of a Department of Science
Education of which he was the head at the time of his
death. More than 10,000 Israeli children are now taking science courses first developed at the Institute.

He was very much involved with improving the
communication between nations, and saw their common
interest in science as a possible path towards better
relations. A member of IUPAP, he helped to organize
many international conferences. In addition to his
participation in the organization of this conference,
he served as summary speaker at the Heidelberg Conference
on heavy ions in July, and was to participate in a
round-table discussion at the Montreal Conference on
Nuclear Structure in August.

He was deeply interested in the problems of developing states, serving on United Nations advisory committ-
ees working in this area. At the time of his death he
was to act as host for a Rehovot conference on "Science
and Education in the Developing States". We can gain
some appreciation for his breadth of view by quoting
from the address he had prepared for this conference, an
address he was never to deliver:

> "If the quest for equipartition of rights can
> characterize the period of the French Revolution,
> and if that of the equipartition of property can
> characterize the social revolutions at the
> beginning of this century in both the socialist
> and the capitalist countries, it is the quest
> for the equipartition of knowledge among all
> people that is going to characterize the end of
> this century, and the beginning of the next one,
> since the possession of knowledge is a prerequisite
> for power in its best sense."

And where most of us would be overwhelmed by the enormous education problems faced by the developing nations,
de-Shalit characteristically, saw a golden opportunity:

> "But I do want to stress that those of us who
> are in the fortunate position of planning our
> educational systems essentially from scratch
> may want to consider very seriously the intro-
> duction of such new educational technologies
> in an early stage in our new developments, rather
> than going through the whole procedure of develop-
> ing first a classical educational system, and then
> going through the pains of changing it into a
> more modern one."

IV.

de-Shalit was a brilliant physicist, one of the
very few adept at both experiment and theory. His experimental work, begun at Zurich, was mainly concerned with
radioactive nuclei. He expanded the range of energy
levels and nuclear species that could be examined spec-
troscopically, and greatly improved the precision of
these measurements. These researches set new standards
in the field.

Although he made many contributions to the theory
and understanding of nuclear reactions, his main interests
lay in the shell model. de-Shalit was quick to grasp
the significance of this model when it was proposed by
Mayer and Jensen. His paper on this subject, written

while still a student at Zurich, was the first of a
series in which he applied concepts such as fractional
parentage and seniority, and other group theoretical
methods of Racah and Wigner, to nuclear problems.
Simple interaction models were used to explain empirical
results like the Nordheim rule. He demonstrated the
need for correcting the description of a nuclear state
as given by the simplest form of the shell model. He
showed that in the neighborhood of closed shells the
necessary modifications could be made by mixing in
additional configurations, and evaluated the consequent
effects on magnetic moments, beta and gamma ray transition probabilities, and the effective charge of neutrons.
He contributed to the fundamental understanding of the
shell model and the independent pair theory in an important paper on nuclear matter. His book with Igal Talmi,
entitled "Nuclear Shell Theory", is the standard reference work in this area. In more recent years he had
become interested in the use of unstable particles and
high energy projectiles as nuclear probes. At the time
of his death he had just completed the first volume of
a two-volume book on nuclear theory.

     This bare recital of some of his interests and
accomplishments fails to describe his true role as a
scientist. For he was a great source of ideas, both experimental and theoretical, which he freely shared.
This was true from the beginning. While a graduate student at Zurich he not only proposed his own thesis topic,
but he also suggested many of the problems which other
students undertook. A visit by Amos was looked forward
to with great anticipation. It was sure to be an exciting time. His presence made a discussion more fruitful,
a seminar more instructive, a theory or experiment more
significant. He was delighted by new insights, and in
the give and take involved in exploring them with his
friends. And when he would leave there would remain
behind new ideas which were ready to be developed into
theories and experiments.

                         V.

     Amos de-Shalit was a man of many parts: He was
both gentle and strong, proud and humble, idealistic and
pragmatic. He became powerful and influential, but he
never lost his concern for his fellowmen, or his ability
to laugh at himself. He was profoundly devoted to his
country, but never chauvinistic or parochial. He was
loved by many: over 5,000 people, including virtually
the entire cabinet, came to bid him a last farewell.
Many are the scientists from all over the world who will

miss him deeply, and who will always remember the warm
friendship and hospitality he and his lovely wife
Nechama showed to them.

> F. Bloch
> S. Devons
> S. D. Drell
> H. Feshbach
> M. Goldhaber
> K. Gottfried
> U. Haber-Schaim
> J. D. Walecka
> V. F. Weisskopf

Publications

Decay of Hg^{197}, with H. Fraunfelder, O. Huber and W. Zunti, Phys. Rev. 79 1029 (1950).

Magnetic Moments of Odd Nuclei, Phys. Rev. 80 103 (1950).

Zerfall der Quecksilber-Isomere Hg^{197}, with O. Huber, F. Humbel, H. Schneider and W. Zunti, Helv. Phys. Acta 24 127 (1951).

On the Deviations of Magnetic Moments from the Schmidt Lines, Helv. Phys. Acta. 24 296 (1951).

Spektrometrische Messung von β-β-Koinzidenzen, with O. Huber, F. Humbel and H. Schneider, Helv. Phys. Acta 25 4 (1951).

The Decay of Pt^{195}, Au^{195}, Pt^{197} and Au^{199}, with O. Huber, F. Humbel and H. Schneider, Helv. Phys. Acta 25 629 (1951).

Elektronenspektren zwischen 1 und 10 KeV, with H. Schneider, O. Huber, F. Humbel and W. Zunti, Helv. Phys. Acta 25 258 (1951).

On the Decay of some Odd Isotopes of Pt, Au and Hg (Thesis), Helv. Phys. Acta 25 279 (1952).

On the Measurement of the K/L Capture Ratio, Bull Res. Coun. of Israel 2 64 (1953).

Cosine Interaction between Nucleons, Phys. Rev. 87 843 (1952).

Effects of Departures from the Single Particle Model on Nuclear Magnetic Moments, Phys. Rev. 90 83 (1953).

The Decay of Au^{197m}, with J. W. Mihelich, Phys. Rev. 91 78 (1953).

Neutron Deficient Isotopes of Hg, with J. W. Mihelich, L. P. Gillon, and K. Gopalkrishnan, Phys. Rev. 91 498 (1953).

The Energy Levels of Odd-Odd Nuclei, Phys. Rev. 91 1479 (1953).

Mixed Configurations in Nuclei, with M. Goldhaber, Phys. Rev. 92 1211 (1953).

Nuclear Spectroscopy of Neutron-Deficient Hg Isotopes, with L. P. Gillon, K. Gopalkrishnan and J. W. Mihelich, Phys. Rev. 93 124 (1954).

Some Regularities in the Nuclear Level Spacings of Hg, Au, and Pt, with J. W. Mihelich, Phys. Rev. 93 135 (1954).

Many Particle Configurations in a Central Field, with C. Schwartz, Phys. Rev. 94 1257 (1954).

Angular Momentum in non-Spherical Fields, Bull Res. Council of Israel 3 359 (1954).

Statistical Weights in Many-Particle Systems, with Y. Yeivin, Nuovo Cimento 1 1146 (1955).

On the Description of Collective Motion by the Use of Superfluous Coordinates, with H. J. Lipkin and I. Talmi, Nuovo Cimento Serie X 2 773 (1955).

The Average Field in Many-Particle Systems, Bull. of the Res. Council of Israel 5A 78 (1955).

Electric Quadrupole Transitions in Nuclei, Bull. Res. Council of Israel 5A 212 (1956).

Magnetic Moments of Nuclei, Supp. Nuovo Cimento 4 1195 (1956).

Moments of Inertia of Freely Rotating Systems, with H. J. Lipkin and I. Talmi, Phys. Rev. 103 1773 (1956).

M1 Transition in Tl^{208}, Phys. Rev. 105 1531 (1957).

Interpretation of Regularities in Neutron and Proton Separation Energies, Phys. Rev. 105 1528 (1957).

The 3^- Level in A^{40}, with R. Thieberger, Nucl. Physics 4 469 (1957).

Binding Energies of Heavy Nuclei, with R. Thieberger, Phys. Rev. 108 378 (1957).

Detection of Electron Polarization by Double Scattering, with S. Kuperman, H. J. Lipkin and T. Rothem, Phys. Rev. 107 1459 (1957).

Energy Levels in Odd-Mass and Even Nuclei, Nucl. Phys. 7 225 (1958).

The Wave Function of Nuclear Matter, with V. F. Weisskopf, Ann. of Physics 5 282 (1958).

Measurement of Beta Ray Polarization of Au^{198} by Double Coulomb Scattering, with H. J. Lipkin, S. Cuperman, T. Rothem, Phys. Rev. 109 223 (1958).

Effective Charge of Neutrons in Nuclei, Phys. Rev. 113 547 (1959).

Nuclear Systematics, Supp. Helv. Phys. Acta V 147 (1960).

The Superconducting State in the Behte-Goldstone Approximation, Nuovo Cimento 16 485 (1960).

Ground States of Odd-Odd Nuclei, with J. D. Walecka, Phys. Rev. 120 1790 (1960).

Effective Moments in K^{39} and K^{40}, Nucl. Phys. 22 677 (1961).

Spectra of Odd Nuclei, with J. D. Walecka, Nucl. Phys. 22 184 (1961).

Core Excitations in Nondeformed, Odd-A Nuclei, Phys. Rev. 122 1530 (1961).

Excitations of Two Phonon Surface Vibrations in Nuclei, with R. H. Lemmer and N. S. Wall, Phys. Rev. 124 1155 (1961).

New Evidence for Core Excitation in Au^{197}, with A. Braunstein, Phys. Lett. 1 254 (1962).

Contribution to the Theory of Coulomb Stripping, with A. Dar and S. Reiner, Phys. Rev. 131 1732 (1963).

Core Excitation: Au^{197} Revisited, Phys. Lett. 15 170 (1965).

On the Polarization in Elastic Scattering, with J. Hufner, Phys. Lett. 15 52 (1965).

Polarization and Zeros of the Scattering Amplitude, Preludes in Theoretical Physics, p. 35 North Holland Publishing Co. (1966).

Stripping Via Core Excitation, with B. Kozlovsky, Nuclear Physics, 77, 215 (1966).

Nuclear Spectroscopy in the Alpha-particle, with J. D. Walecka, Phys. Rev. 147 763 (1966).

Conference Reports, Review Articles, etc.

Effects of Configuration Mixing on Electromagnetic Transitions, Proc. Rehovoth Conf. Nucl. Structure p. 202 (1957).

Nuclear Magnetic Moments, Invited Paper, Brookhaven Conference on Nuclear Moments (1960).

Models of Finite Nuclei, Proc. International Conference on Nuclear Structure, Kingston, 1960.

Effective Moments in Nuclei, Proc. of the Gatlinburg Conference 1961 (Nat'l Academy of Science, U.S.A. Publication 974 p. 15 (1962)).

Nuclear Moments, Varenna Summer School XXIII p. 48 (1961).

Electromagnetic Properties of Atomic Nuclei, 4 Selected Topics in Nuclear Structure, p. 209, International AEC, Vienna (1963).

Some Problems in Nuclear Structure, International Conference at Stanford, 24-27, June 1963.

Nuclear Models and Electromagnetic Properties of Nuclei, Scottish Universities Summer School in Physics on Nuclear Structure and Electromagnetic Interactions (1964).

Recent Developments in the Core Excitation Model, USSR Academy of Science (1965).

Diffraction Phenomena in Nuclear Reactions, USSR Academy of Science (1965).

Remarks on Nuclear Structure, Science, 153, 1063-1067 (1966).

Some Open Problems in Nuclear Physics, International School of Physics "Enrico Fermi", Summer Courses 1966.

Recent Developments in Theoretical Nuclear Physics. To be published in a special issue of the Supplement to Il Nuovo Cimento.

Nuclear Structure from High Energy Phenomena.   International Conference on Nuclear Structure.   XXI A. p. 453 Tokyo (1967).

The Nuclear Shell Model, Contemporary Physics II 307-328 Trieste Symposium (1968).

Summary Talk at Heidelberg Conference on Heavy Ions (1969).

Science Education

Objectives and Contents of Science Curricula, Paris Unesco 1969.

The Scientific Methods, Rehovot Conference, 1969.

Scientific Research in Small Countries, the Case of Israel, Lecture delivered in Dusseldorf (Sondersitzung AM 13 July, 1966 in Dusseldorf).   Westdeutscher Verlag Koln and Opladen.

Books

Analytical and Projective Geometry (University text-book, based on lectures of L. Motzkin, Jerusalem 1946).

Nuclear Shell Theory, with I. Talmi - Academic Press, New York 1963.

Preludes in Theoretical Physics, Editor with H. Feshbach and L. van Hove, North Holland Publishing Co. (1966).

Spectroscopic and Group Theoretical Methods in Physics, Editor with F. Bloch, S. G. Cohen, S. Sambursky and I. Talmi (1968).

Nuclear Theory with H. Feshbach, to be published by John Wiley and Sons Inc.

NUCLEAR STRUCTURE WITH ELECTROMAGNETIC PROBES

J. D. Walecka

Institute of Theoretical Physics, Department of Physics

Stanford University, Stanford, California

The electromagnetic interaction is ideal for studying nuclear structure. An electromagnetic probe sees the local current density in the target, so the interaction is known and the target structure can be directly examined. In addition, the interaction is weak (of order α) so the probe does not alter the structure of the target. Finally, electromagnetic sources are readily available in the form of electromagnetic radiation, charged leptons, and heavy ions.

If the target is unpolarized and unobserved, then to lowest order the electromagnetic response (Fig. 1) can be summarized in two tensors whose coefficients $W_{1,2}$ are functions of the Lorentz scalars q_λ^2 and $q \cdot P$. These surfaces $W_{1,2}$ contain all the information on the strong interaction structure which can be obtained with an electromagnetic probe. Since inelastic electron scattering can map out both surfaces for all $q_\lambda^2 \geq 0$, I shall

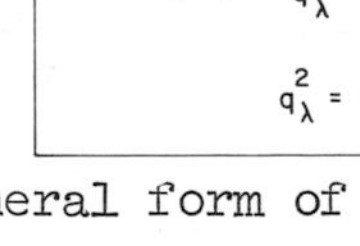

$$W_{\mu\nu} = \frac{(2\pi)^3 \Omega}{Z^2} \sum_{\text{initial state}} \sum_{\text{final state}} \delta^{(4)}(P - P' \cdot q)\langle P | \hat{J}_\nu(o) | P' \rangle \langle P' | \hat{J}_\mu(o) | P \rangle (E)$$

$$= W_1 \left(\delta_{\mu\nu} - \frac{q_\mu q_\nu}{q_\lambda^2} \right) + W_2 \frac{1}{M_T^2} \left(P_\mu - \frac{P \cdot q}{q_\lambda^2} q_\mu \right) \left(P_\nu - \frac{P \cdot q}{q_\lambda^2} q_\nu \right)$$

$$\frac{d^2\sigma}{d\Omega_2 d\epsilon_2} = \frac{4 Z^2 \alpha^2}{q_\lambda^4} \frac{\epsilon_2^2}{M_T^2} \cos^2\frac{\theta}{2} \left[W_2(q_\lambda^2, q \cdot P) + 2 W_1(q_\lambda^2, q \cdot P) \tan^2\frac{\theta}{2} \right]$$

$$q_\lambda^2 = 4 \epsilon_1 \epsilon_2 \sin^2\frac{\theta}{2}$$

Fig. 1. General form of the cross section with one-photon exchange.

1

concentrate on this topic; the cross section is given in Fig. 1.
If only the final electron is detected, there are 3 free variables,
(k_1, k_2, θ) or equivalently $(q^2 = (k_1 - k_2)^2, \omega = k_1 - k_2, \theta)$ or $(q_\lambda^2 \equiv q^2 - \omega^2, q \cdot P \equiv \omega M_T, \theta)$. The two surfaces can be separated by keeping q^2
and ω constant and making a straight line plot against $\tan^2 \frac{1}{2}\theta$ or
by working at $\theta = 180^\circ$ where only W_1 contributes. The photoabsorption cross section for an incoming photon of 4-momentum k
is obtained from the limiting value of W_1 at $k_\lambda^2 = 0$

$$\sigma_\gamma = \frac{(2\pi)^2 \alpha}{[(k \cdot P)^2]^{\frac{1}{2}}} \ W_1(k_\lambda^2 = 0, -k \cdot P)$$

A typical nuclear response surface is sketched in Fig. 2. At $\omega \approx 0$
there is an elastic peak, then excitations to discrete levels
including the giant resonances. If the nucleus were a collection
of nucleons at rest, the surface would consist of a single spike
at $\omega = q^2/2m$. This feature occurs as the quasi-elastic peak and
is spread out by the Fermi motion of the target particles. Finally,
above $\omega = m_\pi$, there is pion production and eventually production
of <u>nucleon</u> resonances.

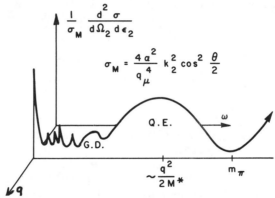

Fig. 2. A typical nuclear electron scattering cross section.

     For transitions between discrete levels, it is appropriate to
make a multipole analysis of the electromagnetic field, assuming a
well-localized target density (see Fig. 3). Some general features
of this analysis are:
1. Because of current conservation, only the matrix elements of
the transverse current and charge density are independent. The
multipole analysis of the transverse interaction is the same as in
real photon processes; however, for a given energy transfer in
inelastic electron scattering, we can vary the 3-momentum transferred
to the target, the only restriction being $q \geq \omega$. Thus we can map
out the complete Fourier transform of the transition current density,
and by inverting, we have a microscope for locating the density
itself.

$$\frac{d\sigma}{d\Omega} = \frac{\pi\alpha^2\cos^2\frac{\theta}{2}}{\epsilon_1^2\sin^4\frac{\theta}{2}}\ \frac{1}{(1+\frac{2\epsilon_1}{M_T}\sin^2\frac{\theta}{2})}\left\{\left(\frac{q_\mu^4}{q^4}\right)\frac{1}{2J_i+1}\sum_{J=0}^{\infty}\left|\langle J_f\|\hat{M}_J^{coul}(q)\|J_i\rangle\right|^2 + \right.$$
$$\left.\left(\frac{q_\mu^2}{2\underset{\sim}{q}^2}+\tan^2\frac{\theta}{2}\right)\frac{1}{2J_i+1}\sum_{J=1}^{\infty}\left(\left|\langle J_f\|\hat{T}_J^{el}(q)\|J_i\rangle\right|^2+\left|\langle J_f\|\hat{T}_J^{MAG}(q)\|J_i\rangle\right|^2\right)\right\}$$

$$\hat{M}_{JM}^{coul}(q) = \int j_J(qx)\, Y_{JM}(\Omega_x)\,\hat{\rho}_N(\underset{\sim}{x})\,d\underset{\sim}{x}$$

$$\hat{T}_{JM}^{el}(q) = \frac{1}{q}\int \underset{\sim}{\nabla}_\Lambda\left[j_J(qx)\,\underset{\sim}{\mathcal{Y}}_{JJ1}^M(\Omega_x)\right]\cdot\hat{\underset{\sim}{J}}_N(\underset{\sim}{x})\,d\underset{\sim}{x}$$

$$\hat{T}_{JM}^{MAG}(q) = \int j_J(qx)\,\underset{\sim}{\mathcal{Y}}_{JJ1}^M(\Omega_x)\cdot\hat{\underset{\sim}{J}}_N(\underset{\sim}{x})\,d\underset{\sim}{x}$$

Fig. 3. Multipole analysis for scattering to discrete levels.

2. The selection rules on the multipole operators are the same as for real photons.

3. The Coulomb and transverse multipoles cannot interfere if only the electron is observed, and their relative contribution can be separated by the methods previously discussed.

4. The transverse multipoles start with $J=1$ since a transverse photon carries unit helicity; in contrast, there is a $J=0$ Coulomb multipole.

5. The long-wavelength behavior of the multipole operators follows from general considerations. By increasing q, we can escape the long-wavelength limit and bring out levels of high multipolarity or of a magnetic character.

6. There is an additional long-wavelength relation between the transverse electric and Coulomb multipoles, which follows from current conservation

$$\langle f|\hat{T}_{JM}^{el}(q)|i\rangle \underset{q\to 0}{\to} \left(\frac{E-E'}{q}\right)\left(\frac{J+1}{J}\right)^{\frac{1}{2}}\langle f|\hat{M}_{JM}^{coul}(q)|i\rangle$$

These results are all quite general. To test our understanding of nuclear structure, however, we must eventually relate the nuclear current operator to the properties of the nucleon. The usual prescription of nuclear physics is to take the matrix elements of the interaction with <u>free</u> nucleons in plane wave states $\langle \underset{\sim}{k}'\lambda'|J_\mu|\underset{\sim}{k}\lambda\rangle$ and write the nuclear operator in second quantization as

$$\hat{J}_\mu = \Sigma\, a_{\underset{\sim}{k}'\lambda'}^\dagger\ \langle \underset{\sim}{k}'\lambda'|J_\mu|\underset{\sim}{k}\lambda\rangle\ a_{\underset{\sim}{k}\lambda}$$

This expression assumes the energy relation $E_k = (\underset{\sim}{k}^2+m^2)^{\frac{1}{2}}$ appropriate to free nucleons and neglects exchange currents. In most applications the operator has been expanded in powers of $1/m$. This approach has proven very successful; however, it must be remembered that this is a basic approximation, and in particular, if we are to interpret data at very high momentum transfer and energy loss in terms of nuclear models, it is essential to have the correct form of the current in these regions. It is also essential to have a better treatment of nuclear recoil, which so far has been handled adequately only within the framework of the harmonic oscillator shell model.

As a first application consider elastic scattering from a spin-zero target, where only the monopole Coulomb moment contributes and we measure

$$F_{el}(q) \equiv \frac{1}{Z} \int \frac{\sin qx}{qx} \rho_{oo}(x) dx^3$$

(For heavy nuclei, this Born approximation is inadequate, and it is necessary to use electron waves distorted by the Coulomb potential of the nucleus. This is in principle straightforward, and done by solving the partial wave Dirac equation.[1]) As an example of the very precise experimental work now being carried out on elastic scattering, Fig. 4 shows some data from Hofstadter's group at Stanford on Ca^{40} and Ca^{48}.[2] These new experiments are revealing more details of the charge density, and it is apparently necessary to include ripples on the charge density, in qualitative accord with the shell-model picture of this quantity. To illustrate these freatures, Fig. 5 shows some recent data of McCarthy and Sick at

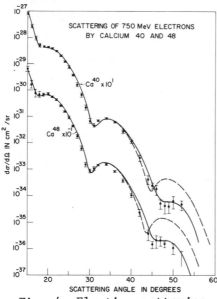

Fig. 4 –Elastic scattering cross-section for Ca^{40} and Ca^{48}.[2]

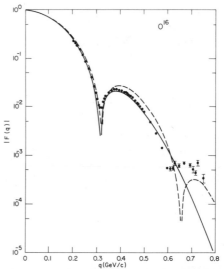

Fig. 5 – Elastic form factor for O^{16}[3]. Compared with harmonic oscillator (solid curves) and Woods-Saxon results(dashed curves).

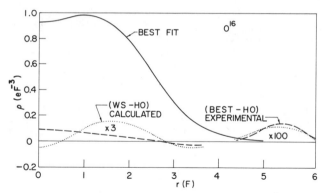

Fig. 6. Charge distributions for O^{16}.[4]

Stanford on O^{16}, compared with both the harmonic oscillator
predictions and the theoretical results of Donnelly and Walker who
use a finite-well Wood-Saxon nuclear potential to generate the
charge density.[3] Of particular interest is the prediction of a
second diffraction minimum in the more realistic calculation. The
charge densities from the harmonic oscillator, finite well, and
best phenomenological fit of Sick are compared in Fig. 6.[4] It is
interesting that the exponential fall off at large r is well
confirmed by the data. This figure clearly illustrates how the
higher Fourier components of the charge density are measured by
the scattering at high momentum transfers.

As an example of the experimental work now being done on inelastic scattering to discrete collective levels, Fig. 7 shows results
of Heisenberg and Sick on the first excited state of Pb^{208}.[5] This
work complements the accurate low-q experiments of the Yale group.[6]
Note particularly the inelastic diffraction minima which have only
recently been observed. So far, the phenomenological fits have used
an inelastic charge distribution obtained by differentiation from the
elastic with the width, location, and strength as adjustable parameters. This transition charge density is of direct interest to
theorists who have models to explain the low-lying collective excitations. Instead of just one transition moment, there is now a whole
function which must be understood. It will also be of great
interest to see when these high-q experiments begin to show up
ripples on the inelastic densities, as one would again expect if
these states are described in the particle-hole model.

I would like to spend some time discussing the particle-hole
model in detail, with particular emphasis on the giant resonance
and other negative parity oscillations of closed-shell nuclei.
The basic idea of the particle-hole model is as follows: If we
make a canonical transformation to particles and holes $c_\alpha \equiv$
$\Theta(\alpha-F)a_\alpha + \Theta(F-\alpha)b_\alpha^\dagger$ where α is a complete set of single-particle
quantum numbers, then the Hamiltonian consists of 3 contributions

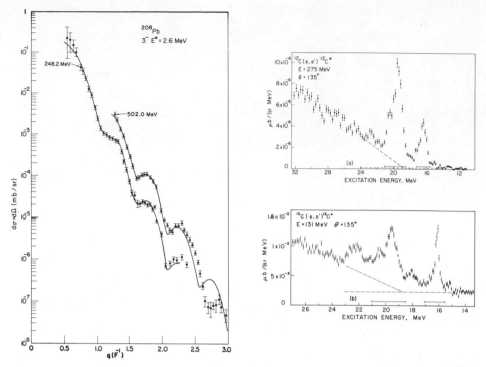

Fig. 7. Inelastic cross section for first-excited state of Pb^{208}.^{5}
Fig. 8. Inelastic cross section at 135° from C^{12}.^{9}

$$H_1 = \Sigma_{\alpha<F} \left(t_\alpha + \tfrac{1}{2}v_\alpha\right)$$

$$\hat{H}_2 = \Sigma_{\alpha>F} \epsilon_\alpha a_\alpha^\dagger a_\alpha - \Sigma_{\alpha<F} \epsilon_\alpha b_\alpha^\dagger b_\alpha$$

$$\hat{H}_3 = \tfrac{1}{2} \Sigma_{\alpha\beta\gamma\delta} \langle\alpha\beta|V|\gamma\delta\rangle N(c_\alpha^\dagger c_\beta^\dagger c_\delta c_\gamma)$$

where $\epsilon_\alpha = t_\alpha + v_\alpha$ and

$$v_\alpha = \Sigma_{\beta<F} \left[\langle\alpha\beta|V|\alpha\beta\rangle - \langle\alpha\beta|V|\beta\alpha\rangle\right]$$

This is an <u>exact</u> transformation provided that the single-particle
wave functions satisfy the Hartree-Fock equations, and ϵ_α is
evidently the Hartree-Fock energy. The simplest description of
the excitations of the system is obtained by assuming the ground
state corresponds to closed shells, $a_\alpha|0\rangle = b_\alpha|0\rangle = 0$, and the
excited state is some linear combination of particle-hole states
$|\Psi\rangle = \Sigma_{\alpha\beta}\psi_{\alpha\beta}^* a_\alpha^\dagger b_\beta^\dagger|0\rangle$. By examining the expression $\langle\Psi|[\hat{H}, a_\alpha^\dagger b_\beta^\dagger]|0\rangle = \omega\psi_{\alpha\beta}$, we are lead to the following equations of motion

$$(\epsilon_\alpha - \epsilon_\beta - \omega)\psi_{\alpha\beta} + \Sigma_{\gamma\delta}\,[\,\langle\gamma\beta\,|\,V\,|\,\delta\alpha\rangle - \langle\beta\gamma\,|\,V\,|\,\delta\alpha\rangle\,]\psi_{\gamma\delta} = 0$$

This is a linear, homogeneous set of equations for the coefficients $\psi_{\alpha\beta}$, and the vanishing of the determinant yields the
eigenvalues ω. The basis for these equations can be reduced by
using angular momentum, isotopic spin, or any other symmetry of the
problem. The transition matrix element of any multipole operator
is given in terms of the coefficients $\psi_{\alpha\beta}$ by

$$\langle\Psi\,|\,\hat{T}_{kq}\,|\,0\rangle = \Sigma_{\alpha\beta}\,\langle\alpha\,|\,T_{kq}\,|\,\beta\rangle\psi_{\alpha\beta}$$

The particle-hole model of Brown[7] assumes that the Hartree-Fock
energies ϵ_α do not vary rapidly as a function of A and takes
the particle and hole energies from the neighboring $A\pm 1$ nuclei.
The shift due to the particle-hole interaction is then computed from
the nucleon-nucleon interaction as indicated above. We have used
a non-singular potential fit to low-energy scattering in our
calculations. It is usually assumed in computing matrix elements
that harmonic oscillator wave functions, with an oscillator parameter determined from elastic electron scattering, give a good
approximation to the Hartree-Fock wave functions (although some
calculations have been done using true continuum wave functions in
a finite well[8]).

Many of the resulting particle-hole states lie in the continuum
above particle threshold and can be grouped together into well-
defined complexes. We interpret these levels as "doorway states"
and compare with experimental inelastic electron scattering data
summed over an interval 0.5-1.0 MeV. With this resolution, the
data also shown well-defined peaks. Since high-spin states can
show up strongly at large q, it is essential to keep all the
states which can be made from a given configuration. At small
scattering angle, collective $J^\pi = 1^-,2^+,3^-,4^+,\cdots(T=0,1)$ levels
are excited through the Coulomb interaction; at large angles and
momentum transfers, magnetic excitation of $J^\pi = 1^+,2^-,3^+,4^-,\cdots$
(T=1) levels through the large isovector magnetic moment of the
nucleon is also important.

I will discuss two detailed comparisons between theory and experiment, C^{12} and O^{16}. The calculations are due to Donnelly et al. and
the new experimental results at $\Theta = 135^\circ$ and momentum transfers up
to 700 MeV/c are due to Sick and Hughes.[9,10,11] Figure 8 shows
data at two different incident energies. The ground state of C^{12}
is assumed to be a closed $p_{3/2}$ shell, and all negative parity states
of the $(p_{3/2})^{-1}d_{5/2}$, $(p_{3/2})^{-1}d_{3/2}$, $(p_{3/2})^{-1}s_{1/2}$, and $(s_{1/2})^{-1}p_{1/2}$
configurations as well as the positive parity $(p_{3/2})^{-1}p_{1/2}$ doublet
are retained. The resulting groupings (with the observed energies)
are 1^+(15 MeV); $2^+,2^-$ (16 MeV); 1^-(18 MeV); $2^-,3^-,4^-$(19 MeV); and
$1^-,1^-,2^-$(21-27 MeV). (A high-lying 3^- and 1^- are not seen in

electron scattering). The form factors of the various complexes
are shown in Figs. 9-12. Note the diffraction minimum for the

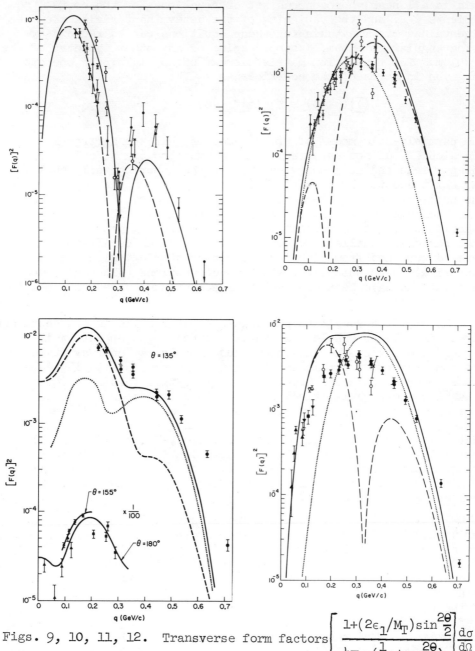

Figs. 9, 10, 11, 12. Transverse form factors $\left[\dfrac{1+(2\epsilon_1/M_T)\sin^2\frac{\theta}{2}}{4\pi\sigma_M(\frac{1}{2}+\tan^2\frac{\theta}{2})}\right]\dfrac{d\sigma}{d\Omega}$

for 15,16, and 19 MeV peaks and Giant Resonance in C^{12}.[9,10]
(clockwise from upper left)

15.1 MeV level (only the high-q data is shown in the figure). The
theoretical amplitude for the 1^+-2^+ doublet has been reduced by a
factor of 2, indicating that the particle-hole model cannot be
correct in detail here. The 19 MeV complex contains the giant
magnetic quadrupole resonance together with the 4^- calculated to
lie very close to it. Note how the role of these two levels is
interchanged as q increases. The quasi-elastic background in the
giant-dipole resonance region (21-27 MeV) has been computed for a
finite square-well potential and is included in the theoretical
curve. The maximum in the form factor is due to the contribution
of the spin-flip electric dipole state.

The ground state of O^{16} is assumed to form a closed p-shell.
All states of the negative-parity configurations $(p_{3/2})^{-1}d_{5/2}$,
$(p_{3/2})^{-1}d_{3/2}$, $(p_{3/2})^{-1}s_{1/2}$, $(p_{1/2})^{-1}d_{5/2}$, $(p_{1/2})^{-1}d_{3/2}$, and $(p_{1/2})^{-1}$
$s_{1/2}$ have been kept. The resulting states fall into the complexes
$1^-,2^-,3^-$(13 MeV); $1^-,2^-$(18 MeV); $2^-,3^-,4^-$(19 MeV); $1^-,2^-$(20.4 MeV);
$1^-,1^-,2^-,3^-$(20.8-26 MeV). The inelastic form factors are shown in
Figs. 13, 14 and 15. The amplitudes for the 13, 18, and 20.4 MeV
peaks have been reduced by $\approx 2/3$. Note the diffraction minimum in
the giant magnetic quadrupole resonance. The 21-26 MeV curve again
includes the calculated quasi-elastic background. The conclusions
I would draw from this work are the following:
1. The particle-hole structure in the region 10-30 MeV seems to
become clearer at large q, where well-defined groupings of states
show up strongly. In particular, high-spin states can dominate the
spectrum at large q.
2. The q^2 dependence of the peaks serves as a valuable tool in
identifying the levels, and some of the complexes still show
diffraction minima.
3. The effects of the spin-flip 1^- and 2^- states, which are
believed to play a dominant role in μ-capture, are clearly seen in
inelastic electron scattering. These states can be grouped with
the electric dipole resonance into a [15]-dimensional SU(4)
supermultiplet of giant resonances rather well in these light
nuclei.
4. The particle-hole model is extremely successful in predicting
the location of the states, generally to better than 1 MeV, and in
predicting the inelastic form factors out to $q \approx 700$ MeV/c.
5. The amplitudes of the particle-hole model can be too high by
up to a factor of 2 (for the 1^+-2^+ doublet in C^{12}). Calculations
of Kurath and Walker, mixing in other configurations, help explain
how the particle-hole model can be wrong in amplitude by factors
of this order, but still do very well on the q^2 dependence.[11,12]

The theoretical calculations have recently been extended to
Si^{28}, S^{32}, and Ca^{40} by Donnelly and Walker,[13] and well-defined
complexes seem to persist in these heavier nuclei. In addition,
high-spin magnetic levels, up to 6^-, are predicted to show up
strongly in electron scattering. The supermultiplet structure of

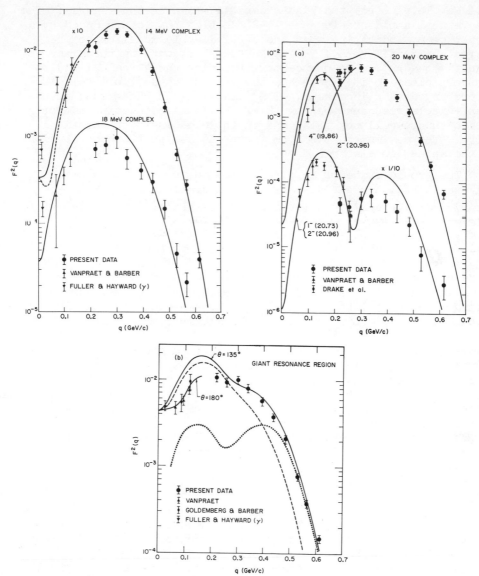

Figs. 13, 14, and 15. Same as Fig. 9. For 13, 18, 19 + 20.4, and 20.4 MeV peaks, and Giant Resonance in O^{16}.[11] (Energies indicated in figures are calculated values.)

the giant resonances, however, appears to be badly fragmented by spin-dependent forces as one goes to higher A.

Let me now go to the quasi-elastic peak. This presents our most direct measurement of the single-particle structure of nuclei and is the dominant feature of the spectrum. Figure 16 shows a

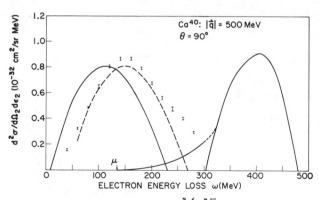

Fig. 16. Quasi-elastic peak in Ca^{40}[14,15] and calculated $N^*(1236)$ production cross section.[14]

comparison between experimental results of Zimmerman and calculations of Moniz for Ca^{40}.[14,15] The calculations are based on a
simple local Fermi gas model of the nucleus with k_F determined
by $N/V = 2k_F^3/3\pi^2$. Moniz kept the full relativistic electromagnetic
nucleon vertex and the correct relativistic kinematics for the
ejected nucleon. To obtain the observed location of the peak, an
average single-particle binding energy ≈ 35 MeV must be included
in the energy-conserving δ-function. The shape and magnitude are
remarkably good. In fact, one can use this result to obtain the
world's best measurement of the Fermi momentum in nuclei, $k_F =
248 \pm 5$ MeV/c. This provides an _independent_ measurement of k_F, and
the agreement with the value obtained from the ground-state density
confirms the essential validity of this simple picture of the
nucleus. To improve these calculations, the best single-particle
model of the nucleus should be used, where the nucleons are bound
in a finite potential and make transitions into the continuum.
This calculation has been carried out by Czyż[16] and recently
extended by deForest[17] using plane waves, and by Donnelly who used
true continuum wave functions generated from a square-well single-
particle potential.[18] Figure 17 compares Donnelly's results for
C^{12} with recent experimental data of Isabelle's group at $\theta = 60°$
and $q = 190$ MeV/c.[19] The giant resonance structure, which lies
below particle threshold in Donnelly's calculation, has been
sketched in with the calculated strength and an empirical width.
It is evident that the finite-well calculations can explain the
high-energy-loss tail of the quasi-elastic peak and locate the
maximum in the correct position; nevertheless, a great deal more
work needs to be done before we really understand the single-
particle structure of nuclei.

Figure 16 includes a calculation of $N^*(1236)$ production by
Moniz whose idea is to use the knowledge of the quasi-elastic peak
to remove the nuclear physics and then to learn something about

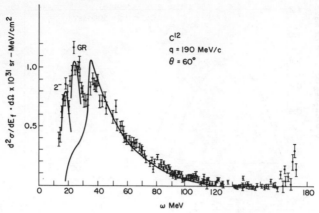

Fig. 17. Quasi-elastic peak in C^{12}.[18,19]

how a nucleon isobar propagates in the nucleus.[15] There are many
interesting questions here -- the modification of the width by
nuclear matter, the extraction of the N-N* cross section, and so
on. I urge experimentalists to get some data in this region.

We can use our knowledge of the nuclear response to an electromagnetic probe as a tool in unravelling the structure of other
processes. I shall very briefly discuss two applications: The
first is the presence of weak magnetism in μ-capture.[20] Consider
electron excitation of the T=1, 1^+, 15.1 MeV level in C^{12}. The
M1 form factor has been measured very well by several groups.
The transition depends primarily on the matrix elements of the
operator $\lambda^\nu \sigma \tau_3 e^{i\underline{q}\cdot\underline{x}}$ as is evident both theoretically and by
comparing the fτ value for β-decay with the M1 lifetime. The
τ_- component of this operator makes the dominant contribution to
the allowed μ-capture transition to the ground-state of B^{12}.
therefore, by taking the form factor from electron scattering,
normalizing to the fτ value for β-decay, and calculating the
small correction terms theoretically

$$\frac{\left|GF_A^\beta\right|^2}{2\pi^2} \sum_{M_f} \left|\int \underline{\sigma}\right|^2 =$$

$$\frac{3\pi}{m_e^5} \frac{\ell n\, 2}{f\tau_{\frac{1}{2}}} \frac{\left|(1/q)\langle 1^+||T_1^{mag}(q)||0^+\rangle\right|^2_{q\,=\,91.67\,MeV}}{\left|(1/q)\langle 1^+||T_1^{mag}(q)||0^+\rangle\right|^2_{q\,=\,0}} \times [.95 \pm 3\%]$$

the square of the dominant matrix element in μ-capture can be
determined to about 5%. The measured μ-capture rate can then be
used to determine the μ-capture coupling constants. This rate is
completely insensitive to the exact value of the induced

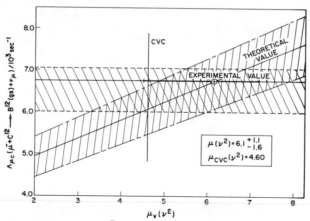

Fig. 18. $C^{12}(\mu^-\nu_\mu)B^{12}$ rate vs weak magnetism moment.[20]

pseudoscalar coupling constant. If we assume $F_A{}^\mu/F_A{}^\beta = 1$ as
indicated by the $\pi \to e+\nu/\pi \to \mu+\nu$ branching ratio, the weak
magnetism contribution to μ-capture can be determined as shown in
Fig. 18. The presence of the weak magnetism term with correct
sign and about the correct magnitude is clearly seen. The accuracy
here is comparable to that obtained in the famous experiments with
this system showing weak magnetism in β-decay.

As a second brief application, consider the nuclear polarization
correction in μ-mesic X-rays. The virtual excitation and de-
excitation of the nucleus by the electromagnetic field of the muon
in an atomic orbit, shifts the energy of the levels. This process
depends on the structure of the target, and at present is the
largest theoretical uncertainty in the analysis of μ-mesic X-rays.
Cole[21] observed that the nuclear vertices are exactly those measured
in inelastic electron scattering. In fact, the dispersion correction
of any discrete nuclear level can be evaluated in a model independent way in terms of experimental inelastic cross sections
(these same observations apply to the dispersion corrections to
electron scattering itself, but so far no one has pursued this
point). Cole tried to estimate the systematic shift as a function
of Z by including Coulomb excitation of the giant dipole resonance
and quasi-elastic peak. He used a model for these regions, but
demanded that it first fit inelastic electron scattering. He also
used closure on the muon but was careful in trying to estimate the
muon closure energy correctly. The resulting downward level shifts
are shown in Fig. 19. Cole estimates his results should be correct
to within a factor of 2.

Finally, I would like to return to the response surfaces and
say a few words about electron excitation of the nucleon itself.
The general analysis for the excitation of discrete levels can be
carried over directly. The one additional feature is that the

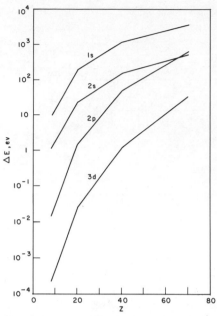

Fig. 19. Total level shifts caused by nuclear polarization.[21]

momentum transfers become large compared to the nucleon mass; thus it is crucial to keep explicit relativistic covariance throughout. A detailed experimental investigation of the resonance region has been carried out at SLAC.[22] I would like to show some of the results together with the predictions of a very simple model. The model assumes that resonance production can be viewed as an

$$W \to \quad \text{(Exc)} \dashrightarrow \text{(J}^{\pi}\text{)} $$

excitation process into the right channel, followed by a final-state enhancement which builds up the resonance. Pritchett, Zucker, and I[23] have carried out a coupled-channel calculation under the assumption that a single eigenphase shift is resonant. The amplitude for production of the resonant channel $|e\rangle = \cos\epsilon|1\rangle + \sin\epsilon|2\rangle$ is then

$$A(k^2,W) = (A_1^{\ell hs}(k^2,W)\cos\epsilon + A_2^{\ell hs}(k^2,W)\sin\epsilon)/D(W)$$

$$D(W) = \exp\left[-\frac{W-M_s}{\pi}\int_{W_0}^{\infty}\frac{\xi(W')dW'}{(W'-M_s)(W'-W-i\eta)}\right]$$

In this expression $A_{1,2}^{\ell hs}(k^2,W)$ are the multipole projections of a set of Feynman amplitudes believed to play an important role in the excitation process, and $D(W)$ is a final-state enhancement factor.

The eigenphase shift ξ and mixing angle ϵ can be obtained from $|\pi N\rangle$ scattering. This model keeps all the general properties of the theory such as relativistic invariance, gauge invariance, analytic properties threshold behaviors, and the final-state theorem. In our calculations, we have kept the $|\pi N\rangle$ and $|\pi N^*(1236)\rangle$ channels and π, N, N*(1236), and ω exchange graphs in the excitation mechanism, using experimental coupling constants and choosing any unknown form factors to make this sum of Feynman graphs explicitly gauge invariant. The ratio $d\sigma_{in}/d\sigma_{el}$ at 6° for the first 3 nucleon resonances is shown in Figs. 20-22. There is one unknown parameter, M_s, for each resonance and its value is obtained by normalizing the resonance contributions to the data at one k^2 (for example, at photoproduction). The curves corresponding to pure threshold behavior are also shown. Models I and II compare the

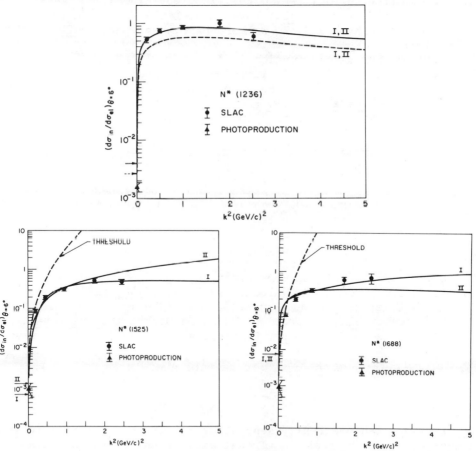

Figs. 20-22. $(d\sigma_{in}/d\sigma_{el})$ at 6° for the $1236(3/2^+,1/2)$, $1525(3/2^-, 1/2)$, and 1688 MeV $(5/2^+,1/2)$ levels of the nucleon.[23]

single $|\pi N\rangle$ channel with the coupled-channel calculations. The main conclusions from these results are:
1. The ratio $d\sigma_{in}/d\sigma_{el} \sim 1$ as $k^2 \to \infty$ for all these resonances.
2. The region of k^2 where the deviation from the threshold result occurs, and the large k^2 behavior, are given correctly by this simple model.
There are many unanswered questions here; for example, the multipolarity of the resonance regions has not been experimentally determined, and the separate contributions of the various amplitudes, which are theoretically predicted to show interesting diffraction structure, has not yet been disentangled.

In summary, electromagnetic probes provide a unique and powerful tool for understanding the structure of hadronic matter.

REFERENCES

1. T. A. Griffy et al., Phys. Rev. 128, 833 (1962).
2. K. J. VanOostrum, et al., Phys. Rev. Letters 16, 528 (1966).
3. T. W. Donnelly and G. E. Walker, Phys. Rev. Letters 22, 1121 (1969).
4. I. Sick, (private communication).
5. J. Heisenberg and I. Sick, (to be published).
6. J. F. Ziegler and G. A. Peterson, Phys. Rev. 165, 1337 (1968).
7. G. E. Brown et al., Nucl. Phys. 22, 1 (1961).
8. J. Friar, (to be published in Phys. Rev.).
9. T. W. Donnelly et al., Phys. Rev. Letters 21, 1196 (1968).
10. T. W. Donnelly, Stanford preprint ITP-332 (1969).
11. I. Sick et al., Stanford preprint ITP-335, HEPL-606 (1969).
12. D. Kurath, Phys. Rev. 134, B1025 (1964).
13. T. W. Donnelly and G. E. Walker, (to be published).
14. P. D. Zimmerman, Ph.D. Thesis, Stanford University (1969).
15. E. J. Moniz, (to be published in Phys. Rev.).
16. W. Czyż, Phys. Rev. 131, 2141 (1963).
17. T. deForest, Orsay preprint, LAL-1208 (1969).
18. T. W. Donnelly, (private communication).
19. D. Isabelle, (private communication).
20. L. L. Foldy and J. D. Walecka, Phys. Rev. 140, B1339 (1965).
21. R. K. Cole, Jr., Phys. Rev. 177, 164 (1969).
22. E. Bloom et al., reported by W. K. Panofsky, International Conference on High-Energy Physics, Vienna 1968, (CERN, Geneva, 1968) p. 23.
23. P. L. Pritchett, et al., (to be published in Phys. Rev.).

DISCUSSION

K. Gottfried: Can one really separate the effects of ripples in the charge density from those of virtual excitations of the nucleus (nuclear dispersion)?   J.D.W.: In principle, no.   The calculations that have been made indicate that the only places where dispersion

corrections show up strongly are in the diffraction minima.   In
fact, Sick found that the data of ^{40}Ca fit very well with a
realistic charge density everywhere except in the diffraction
minima.   R. Hofstadter: Since the same kind of rippling in the
charge distribution shows up for several different nuclei, it would
be surprising if dispersion corrections were responsible in all
cases.

F. Beck: You add the quasi-elastic peak independently to the (more
or less) discrete transitions to resonances.   In the finite-well
shell-model calculations, is not this separation unjustified, so
one should not add them independently?   J.D.W.: Yes, it is all
one process and one should calculate the whole thing with continuum
wavefunctions.   This has been done out to very low momentum transfer and no one has tried to compute the whole surface.   Donnelly's
less complete procedure is not completely self-consistent.

B. Chertok: In the discussion of the ^{12}C 15.1 MeV state, in
connection with the weak magnetism moment, the radiative width of
this state is assumed to be well known.   In fact the experimental
value has decreased from 54 eV (10 years ago) to 32-33 eV (from our
recent low energy inelastic scattering data).   J.D.W.: The data
we used corresponds to a width of 36 ± 1 volts.   B.C.:   From the
inelastic scattering data the width could be as low as 29 eV.   A
really precise comparison here is not yet possible.

M. Goldhaber: Do measurements of the Fermi momentum such as you
showed for ^{40}Ca, exist for other nuclei?   J.D.W.: Only for ^{48}Ca,
where the distribution is the same as for ^{40}Ca, within experimental
error.

EXPERIMENTALIST COMMENTS ON ELECTRON SCATTERING

D.B. Isabelle

Lab. de Physique Nucléaire F-63.   Clermont-Ferrand and

Lab. de l'Accélérateur Linéaire F-91.   Orsay

I am always very embarassed to prepare a review paper on experimental electron scattering for a Conference in which a theoretical review paper covering the same field is also presented. The reason is that there are very few experimental problems and that most of the work of people who have done electron scattering experiments over the last two years have been a comparison of results with various models. So, taking the advantage of the fact that I am not supposed to act as rapporteur, I decided to organize my presentation about a few problems in this field that I believe are worth discussing.

It is interesting to note that despite the fact that many authors emphasize the importance of electron scattering as a tool to study nuclear structure none of the two latest international conference on nuclear physics includes a special session devoted to this topic. It seems to imply that electron scattering is loosing its glamour and has to return to the ranks of the various other techniques used to study experimental nuclear physics.

Advantages of electron scattering are well known, and there is no reason to discuss them again here. I will first present a brief summary of the present trends in experimental facilities, then I will devote some time to the discussion of the radiative correction problems. These two subjects are really the only ones belonging to the official title of my talk. Anyway I will try to make a few comments in fields already covered by Walecka, i.e. elastic, inelastic and quasi-elastic scattering, because I believe that a few points have not been given the appropriate approach up to now.

EXPERIMENTAL EQUIPMENT

There are now more than fifteen electron linacs in the world
used to study nuclear structure by means of electron scattering
and/or of photoreactions.   I will not list them explicitly here,
but the oldest is the Stanford Mark II now rebuilt at Sao Paulo
(Brazil) and the youngest are the Saclay one which was dedicated
last February and the MIT one which should become operational
by the end of 1970.   It is difficult to caracterize those machines
by one single criterion.   If you choose maximum available energy
it ranges from 30 MeV to 2.3 GeV.   Now, if you compare the average
beam power it ranges from a few watts to 200 kW.

To the experimentalists, the most important parameters are
the duty cycle and the characteristics of the beam reaching the
target.   Poor duty cycle has always been the main disadvantage
of linacs as compared to synchrotrons.   A lot of skill and money
has been invested in machine design and we have now to our dis-
posal electron linacs with a 2% duty cycle (Saclay and MIT).
Efforts are now underway to build c-w linacs using cryogenic techniques (Stanford and Illinois).   The results obtained recently at
Stanford seem very encouraging.[1]   For the classical linacs using
cryogenic techniques (Stanford and Illinois).   The results obtained
recently at Stanford seem very encouraging (1).   For the classical
linacs which have recently started operation the intensity of the
beam onto the target is of the order of 5 μA average current for a
momentum spread of 0.1%.   To close those remarks I must mention
that some efforts are also devoted to the production of polarized
electrons.   A recent paper by Jost and Zeman (2) gives a complete
survey of this topic.   The sources available to-day deliver a
very low current [∿ $10-9$ A] but 80 to 90% polarization can be
achieved.

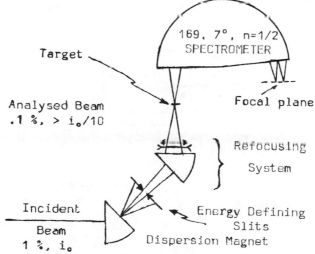

Figure 1 - Classical electron scattering set-up

Up to now all electron scattering facilities followed the same general design Figure 1): An achromatic beam handling systems provides an analysed beam onto the target, then the scattered electrons are momentum analysed by means of a double focusing spectrometer. The best energy resolution is achieved if the spectrometer mean bending angle is 169.7° (the so-called Sakai-Penner magic angle)[3] : a resolution $\Delta\rho/\rho = 4.10^{-4}$ FWHM has been obtained for the NBS spectrometer[4] (radius: .75m) while the Saclay group hope to reach 2.10^{-4} FWHM (radius: 1.80m). Actually, an overall energy resolution of 10^{-3} can be reached for thin target (< 50 mg/cm^2) and a 5.10^{-4} spread in momentum for the incident beam for a solid angle of 5×10^{-3} Sr.

This system does not allow to make a full use of intense beam available with modern linacs as the use of such spectrometers implies a very small beam spot and a small solid angle. This is why the MIT group suggested about two years ago to extend to electron scattering a technique already used at some cyclotron facilities and for pion experiment at Frascati. Such a system generally called an "energy loss spectrometer" or the "dispersed beam technique" is described by Kowalsky et al.[5] "ɛs a typical non-dispersive beam handling system with the spectrometer taking the place of the second magnet" (Figure 2). It allows the use of the total beam even with a total incertainty of 1 % in momentum while an energy resolution of less than 2×10^{-4} FWHM can be achieved. It will provide a mean of measuring very small cross-sections (10^{-35} cm^2/sr) in a reasonable time and then to use fruitfully the large beam power available with the new accelerators.

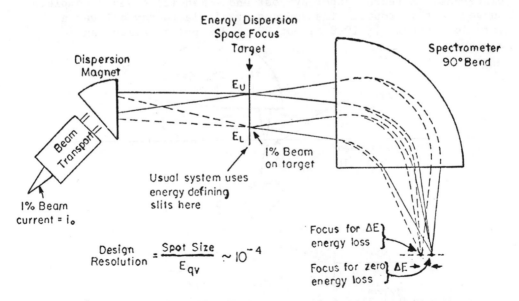

Figure 2 - Schematic diagram of the energy loss spectrometers

The scattered electrons are detected with a ladder counter
located in the spectrometer focal plane. The first generation
used thin scintillation counters in coincidence with a Cerenkov
counter. As energy resolution improved the channel width had to
be decreased, experimentalists then decided to use semi-conductor
detectors (typical size being 1 x 1 x 80 mm^3). But the disadvantage of such a technique is its prohibitive cost due to the large
amount of associated electronics which is needed. So more recently
spark chambers of various types have been used. They provide a
spatial resolution equivalent to what one can reach with counters,
and they have the big advantage of avoiding the need for channel
intercalibrations.

Most electron scattering facilities are now using on-line
computers for data taking and data handling. They directly
control the whole experimental set-up and they perform various
corrections and normalization then provide a display of the
measured spectrum.

To close this section we would like to recall that special
facilities have been built for 180° scattering experiments at
Orsay, Stanford and Amsterdam. They all follow the original design
of Barber at Stanford and they have already been described elsewhere.
I would like to mention a new set-up described by Bergstrom from
MIT[6]. It is based on a theorem by L.A.C. Koerts according to
which a particle passing through a cylindrically symmetric
magnetic field which satisfies the relation :

$$\int_0^\infty B_z \, r \, dr = 0$$

will have the same z component of angular momentum in a field free
region near the symmetry axis as it has in a field free region outside the magnet. Therefore a particle approaching radially will
pass pass through the axis of symmetry and reciprocally (see Fig. 3).
Such a device has been studied by the MIT group and is now under
construction. It will also allow scattering at angles smaller
than 180° which are difficult to reach with a classical set-up.

RADIATIVE CORRECTION PROBLEMS

The main difficulty in the analysis of electron scattering
data is due to the fact that such a process is always perturbated
by the simultaneous emission of photons. It is beyond the scope of
this conference to discuss in detail the calculations of radiative
corrections and radiative tails. All the useful formulas have
been recently summarized by L.C. Maximon[7] in a very complete paper.
It contains a discussion of the validity of those formulas and the
author comments on certain aspects of the subject that need
further theoretical investigations. Two problems are of particular

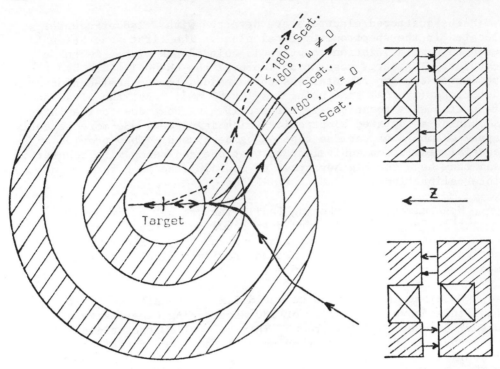

Figure 3 — Schematic diagram of the MIT proposal for 180° Scattering

importance to the experimentalist. The first one deals with the
fact in the present calculations the nuclear recoil effect is not
taken into account while this effect becomes important at high
energy for light nuclei. The second topic is even more fundamental
as it is related to the basic formalism used to obtain the useful
formulas. All of them have been derived using the first Born
approximation and we have very little knowledge of what will occur
when higher Born radiative corrections will be included. I would
like to mention the recent work of Brodsky and Gillespie[8] who computed
the wide angle pair production and bremsstrahlung in the second
Born approximation. Their results can be extended to electron
scattering.

Another important theoretical work must also be emphasized.
It was done by Maximon and Tzara[9] who have shown that the diminution
of the radiative problems in the case of muon scattering as compared
to the electron case is not as large as was generally stated. It
turned out that the advantage of muons is only true for momentum
transfers smaller than 50 MeV/c which are of little interest for
nuclear structure studies.

For the practical use of the formulas some computer programs
have been written either to compute the shape of elastic peaks[10]

or to unfold the experimental spectra[11]. For this last problem,
all the programs are based on the original idea of H.W. Kendall[12].
They may differ in the mathematical treatment but they all take
into account properly the nuclear part of the interaction using
experimental data without any nuclear model assumptions.

Various tests have been done to check the validity of those
calculations and they seem to confirm its success at least for
light nuclei. Nevertheless we must point out that although it has
been possible to check the validity of the electron-radiation field
interaction treatment by means of electron-electron scattering
experiments (either with colliding beams or with atomic electron
targets)[13] we do not have any direct check on the validity of the
formalism used to calculate the electron-nuclear potential interaction.

ELASTIC SCATTERING

The use of elastic scattering to determine charge distribution
as well as its comparison with other techniques are discussed by
other authors at this Conference. Apart from the Wood-Saxon single
particle potential reported by Walecka other models can be used
to analyse elastic electron scattering. Vinciguerra and Stovall[14]
have performed an improved Nilsson model calculation taking into
account matrix elements with principal quantum number different by
two units or more. By comparing their results with the experimental
form factors for deformed p-shell nuclei they obtain information,
in particular, about the quadrupole moment.

All the information derived from high energy elastic electron
scattering are obviously model dependent, and I believe that a
wrong philosophical approach has been taken by some authors when
they want to demonstrate that such measurements prove the existence
of nuclear phenomena such as short range correlations among nucleons
inside the nuclei. One must keep in mind that an electron scattering
experiment is a diffraction process from which one can derive
information about the size and shape of the diffracting object, but
not about the way it is built. We must choose a model to interpret
the observed data and then I will say that on a general basis any
model is as good as any other one. We have to rely on other types
of measurements to decide which one is the closest to the truth.
This is why we should be very cautious if we want to demonstrate
the importance of phenomena such as dispersive corrections,
correlations or proton halo. If those effects exist they will all
be mixed and it will be very difficult to separate them.

The dispersive corrections are due to the fact that during the
scattering process the nucleus is not rigid but can be deformed and
polarized. Most of the calculations[15] done in this field are incomplete as their authors do not include all the possible excited states.

But recently De Forest[16] calculated the second Born approximation correction to the first Born approximation cross section using for the nucleus the harmonic oscillator model with an energy dependent potential but no center of mass corrections[17]. His results seem to indicate that dispersive corrections have an opposite sign to the Coulomb correction, and that their magnitude is small (less than 7 % at 420 MeV for ^{16}O). This confirms the previous publications and indicates that more accurate experiments should be done to see such an effect. In fact, I believe that a comparison between negaton and positon scattering could bring some light on this problem. This is due to the fact that the cross term in the calculation of the cross-section taking into account one and two photon exchange have a sign related to the charge of the scattered particle. I am sure that a cooperative effort between theoreticians and experimentalists in this field will be very fruitful. In fact it will be interesting to study the importance of two-photon terms with a process which can only occur through such a term: a good candidate will be $0^+ \rightarrow 0^-$ transition between the ^{16}O ground state and its excited state at 12.78 MeV.

The inclusion of short range correlations in the analysis of electron scattering data is very much in fashion presently. It is based on two facts : first it was shown by Cioffi degli Atti[19] that they should be taken into account to explain the (e, e'p) results; second following the success of the Glauber theory to analyse the high energy scattering of nuclear probes (protons, deutons, pions) by nuclei it was suggested by Czyz and Lezniak[18] to follow a similar approach to analyse elastic electron scattering data. Many calculations have been performed recently in this direction[19]. I would like to report about some recent calculations by a group at Saclay[20]. Instead of using Green type correlation in the fundamental they calculated the elastic cross-section for a charge density to which they added the corrections to Hartree-Fock due to the long range correlation computed in an approximate way within the frame work of RPA (Table 1). They also performed the

$q_{fm}-1$	1.5	2	2.5	3	3.5	4
^{16}O	-14	76	-26	-17	-16	27
^{40}Ca	3	-3	31	-42	-29	-13

Table I - Effects of RPA ground state correlations for elastic scattering 750 MeV electrons expressed as the ratio $(\frac{\sigma_o - \sigma}{\sigma_o + \sigma} \times 100)$ where σ_o and σ are the values of the cross-section without and with correlation.

same calculation including short range correlations using Jastrow
wave functions. The most striking result of those calculations
is that the RPA correlation is as important as the short range one
in the domain of momentum transfer ($g \sim 2\text{-}3$ fm^{-1}) where the effects
are maximum (Table 11).

q fm^{-1}	1.5	2	2.5
with RPA correlations (Gillet-Lumbroso)	4.4	1.4	80
with short range correlations (Ripka)	26	13.6	55.1

Table II — RPA versus short range effects in ^{40}Ca for
elastic scattering of 250 MeV electrons expressed as
$100 \times \dfrac{|\sigma_0 - \sigma|}{\sigma_0 + \sigma}$.

In my opinion the only thing that one knows for sure is that
a good fit to the experimental data at large momentum transfers
is only achieved if one includes short range correlation. But I do
not believe that it can be taken as a direct proof of the existence
of this effect. Such a proof can only be obtained if one correlates
information from various sources such as hadron scattering, electron
elastic scattering, (e, e'p), (p, 2p) and pion interactions. The
real advantage of electron elastic scattering is the absence of
effects due to final state interactions.

Another correction effect has been recently raised, it has to
do with the possible existence of a proton halo. Van Niftrick
et al.[21] have done a careful study of the importance of this effect
in the elastic scattering of electrons by nuclei. They found that
measurements at both low and high momentum transfers are needed
to provide information. They discussed very carefully all the
causes of uncertainty in the experiment, and they finally have a
word of caution about the fact that the influence of this effect
will be folded with those I have already mentioned.

As a last word on elastic scattering: I would like to say
how surprised I am to see that very little work has been done in
the field of 180° scattering. To my knowledge apart from the
works done at Stanford and Orsay a few years ago the only team
working on this problem is the NRL in Washington[22]. These 180°
measurements provide information about the static magnetic moments
as well as the form factors of nuclear magnetization densities.
The knowledge of this last quantity is of particular importance
in computing the Bohr-Weisskopf magnetic hfs, needed in the analysis
of muonic X-rays experiments[23].

A detail study of this type of measurements has been done by Barber[24]. A difficulty arises from the fact that the finite apperture of the spectrometer implies that both magnetic and Coulomb scattering are simultaneously detected. But this difficulty can be overcome by performing measurements on a neighboring spherical nucleus (^{208}pb for ^{209}Bi, for example) to measure the importance of the Coulomb cross-section. The usefulness of the information derived from such experiment must induce the experimentalists to tackle this problem. With the high intensity linac (100 µA) and large solid angle which will be available in the near future such measurements can be performed in a reasonable time.

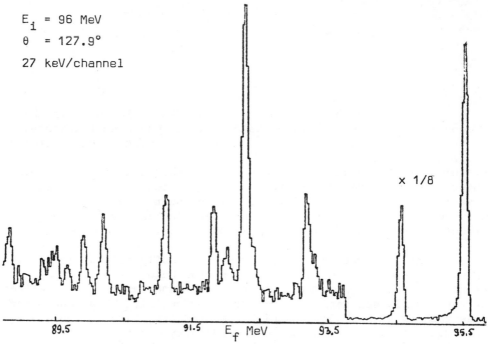

E_i = 96 MeV

θ = 127.9°

27 keV/channel

× 1/8

89.5   91.5 E_f MeV   93.5   95.5

Figure 4 - Spectrum of electrons scattered on ^{48}Ti

INELASTIC SCATTERING

Two points must be emphasized to summarize the significant situation in this field. The first one has to do with the energy resolution presently obtained, which is of the order of 60 keV for an incident energy of 100 MeV. This fact is dramatically displayed in Figure 4 which presents some recent measurements of the NBS group on ^{48}Ti [24]. The same group has been able to separate the 6.02 MeV and 6.13 MeV levels in ^{16}O [25]. The second point is surely as important as the first one; various laboratories scattered around the world have performed measurements for the same nuclei and it is striking to see that a perfect agreement is obtained among all the available data.

There is no point to discuss here the various nuclear models used to analyse the data. But I must emphasize the fact that inelastic electron scattering at large momentum transfer will provide new nuclear information such as the influence of short range correlations. It seems to me more appropriate to discuss here the way one proceeds to compare experiment and theory in this field. The problem is relatively easy for bound states as long as the corresponding scattering peaks can be experimentally observed. Then determination of the experimental cross-section is trivial.

But in the case of excitation region corresponding to unbound states the advantage of electron scattering in allowing excitation with all kinds of multipoles (and not only dipole) becomes a disadvantage, because it is practically impossible to separate the contributions due to the various states. As a matter of fact the old fashion analysis of the giant resonance region by making a decomposition in peaks cannot be used anymore. We know from various work[26] that the peaks are less interesting than the valleys. A typical example of the situation is shown on Figure 5 which displays the inelastic cross-section in ^{12}C [27] in the region of the giant resonance. The observed structure is far more complex than the one expected from particle-hole calculations.

Then what can we do? I see two alternatives. The first one is to use information from other experiments as a guideline to the analysis. Photoabsorbtion, (p, γ), (d, γ), (^{3}He, γ) experiments have seen their accuracy improve due to be the development of various experimental techniques. They must be compared to low momentum transfer electron scattering experiments in which only dipole states are strongly excited. Then by slowly increasing the momentum transfer one could try to determine in a model independent way the multipolarities existing in the parts of the spectrum which do not correspond to dipole states.

But is the result really worth the effort? I am not sure. Presently theoreticians seem to be more interested in obtaining experimental values of the reduced matrix element, than in information about the behavior of the form factors at large momentum transfers. The second alternative will be to try to compare sum rules. It is well known that there is a discrepancy of a factor of 2 between the experimental and theoretical values of the dipole sum rule for ^{16}O. But here I must also remark that this information could be obtained from photoabsorbtion experiments. The development of high energy and high intensity linacs provides us with new facilities of intense monoenergetic γ-ray beams. Then I wonder if it will not be worth while to spend some efforts in

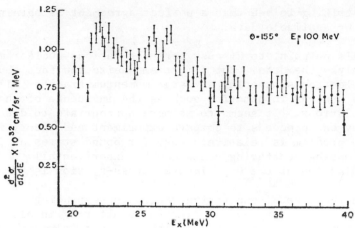

Figure 5 - Inelastic scattering cross-section in ^{12}C (Ex = excitation energy).

this direction before accumulating a lot of electron scattering data that no one will be able to use properly.

If the emitted particle is detected[28] then one will gain additional information. Such a program is now under way at the NBS by Dodge et al.[29] who use simultaneously a magnetic spectrometer and energy-range semi-conductors to detect protons, deutons, tritons, ... The same technique has been used by Shoda et al.[30] to study analogue states in ^{90}Zr and ^{140}Ce.

In a paper presented at this Conference Drechsel and Uberall[31] derive the general angular correlation formulas for a coincidence experiment in which the scattered electron is detected simultaneously with a heavy particle or a photon emitted by the nucleus subsequently to its electroexcitation. The advantages of such an experiment over simple electron scattering are that the radiative tail problem is eliminated (but not the radiative corrections) and that spin and parities of resonance levels may be determined in a model-independent way. Such an experiment should be possible with the high duty cycle linacs, and I do believe that such a program should be started as soon as those linacs come to operation.

Electron inelastic scattering is a perfect tool to study few nucleon systems. This is why some efforts have been recently devoted to the measurement of the electrodisintegration cross-sections of ^2H and ^3He near threshold. As we are interested in the M 1 components of those quantities the experiment must be performed at backward angles and preferably at 180°. ^3He has been studied by Chertok et al.[32] while deuteron experiments were performed by Katz et al. at Saskatoon[33] at 155° for momentum transfers 0.35, 0.66 and 0.84 fm^{-1} and by Ganichot et al.[34] at Orsay at 180° for

momentum transfers between 0.8 and 2.0 fm^{-1}. In the deuteron the
main contribution in the low energy excitation region is due to
the M 1 transitions to the ^1So and $^3S_1$ unbound states. Previous
measurements[35] of this effect seem to indicate that one needed to
include meson exchange currents to explain the differences observed
between experimental and theoretical cross-sections. The new
data (Figure 5) seems to indicate that the cross-sections obtained
agree with a potential model using a hard core in both initial
and final states without the inclusion of exchange currents. In
fact one will expect that the effect of such currents will be
small in the deuterium as in this nucleus the nucleons are relatively
far away from each other. This may explain why the effect cannot
be seen in those experiments. On the other hand it has been
recently demonstrated by Bosco et al.[36] using the dispersive
approach that in the effective range approximation the dominating
effect is the ^1So virtual resonance of the n-p system.

On the other hand the results of Chertok et al. demonstrate
that M 1 contribution is certainly present in the two-body and
three-body break up of ^3He. But more calculations are needed to
determine if the observed M 1 is attributable to meson exchange
currents or to other effects.

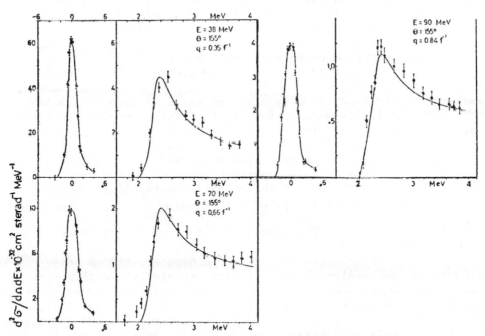

Figure 6 - Elastic and inelastic cross-section on deuterium. The
solid line in the inelastic region is the theoretical prediction
folded with the experimental resolution[33].

Related to this topic I must mention the very interesting work presented by Chemtob and Lumbroso at this Conference[37] on the influence of the meson exchange effects on the inelastic form factors for ^{12}C level at excitation energies above 14 MeV. They retain in their calculations as mechanisms that could produce two-body currents those associated to the one pion exchange diagrams. They choose ^{12}C as for this nucleus the situation is favorable since Coulomb and dispersive corrections are expected to be small. They found that the exchange corrections to the form factor is generally small i.e.a 10 to 40 % increase to the squared impulse form factor for q < 200 MeV/c, but the increase is more substantial for higher q. This is in good agreement with the presently available data but measurements at larger q are needed.

QUASI ELASTIC SCATTERING

Here we are in a field where the advantage of electron scattering may be more obvious, as the weakness of electromagnetic interaction causes the electron to be an excellent probe to study the nucleon-nucleon forces. Two kinds of experiments are performed in this field either one measures the spectrum of quasi-elastically scattered electrons or one detects in coincidence the scattered electron and the high energy emitted proton (or neutron, seuteron or α). Up to now this last kind of measurements have only be performed by the Istituto di Sanita Group at Frascati[38], but no new results have been obtained during the last two years as they had to modify their experimental equipment. Measurements should be resumed soon and they will hopefully provide a better energy resolution. Another experiment of this type is also scheduled at Saclay[39]. They hope to achieve an energy resolution of 1.5 MeV. They are planning to measure for ^{12}C the missing energy spectrum for values of the emerging proton kinetic energy ranging from 40 to 230 MeV, then to obtain for the l s and l p levels the angular distribution of the emerging proton up to very large values of its momentum. Analysis of this experiment should be eventually reconsidered in the light of the recent results obtained from (p, 2p) experiments by the Liverpool Group.

The measurements of electron spectra have been pursued at Orsay on ^{12}C while ^{40}Ca and ^{48}Ca were studied at Stanford[42]. Those two series of experiments show that the high excitation energy side of the quasi-elastic peak present "shoulders" which may be due to nucleon-nucleon correlation. This topic has already been discussed by Walecka. I would like to add to his report that it will be interesting to extend those measurements to 180° scattering as at this angle only transverse terms will contribute and the determination of the sum rules will provide information about the values of magnetic moments for nucleons bound in nuclei[13].

CONCLUSIONS

As stated in the introduction I did not intend in this paper to present a complete picture of the field and I must apologize to those who have not been mentioned. This is due to the fact that I believe that electron scattering results should be discussed in conjunction with the information provided by other experiments. By now everybody is aware of the advantages and limitations of this technique and we must consider that the infant state is finished at least if one considers the simple electron scattering.

In the future some efforts should be devoted either to extend already proved techniques such as 180° scattering and positron scattering or to develop new ways of investigations such as coincidence measurements in electrodisintegration experiments.

ACKNOWLEDGMENTS

It is my pleasure to acknowledge the very interesting and fruitful discussions I have had during the preparation of this paper with J.B. Bellicard, W. Bertozzi, M. Danos, T. De Forest, V. Gillet, M.G. Huber, J.F. Leiss, S. Penner and G.H. Rawitscher.

REFERENCES

[1] E. Jones, M.S. Mc Ashan and L.R. Suelze - Proceeding of Particle Acc. Conf., Washington, D.C.

[2] K. Jost and H.D. Zeman - Submitted to ReV. Mod. Phys.

[3] M. Sakai - Nucl. Instr. and Methods, $\underline{8}$, 61 (1960); S. Penner, Internal Report National Bureau of Standards, June 5, (1961) (Unpublished)

[4] S. Penner and J.W. Lightbody, Jr. - IEE Proceedings on Magnet Technology (SLAC 1965)

[5] S. Kowalsky, W. Bertozzi and C.P. Sargent - MIT 1967 Summer Study P.39 - Report TID 24667

[6] J. Bergstrom - MIT 1967 Summer Study, P. 207 - Report TID 24667

[7] L.C. Maximon - Rev. Mod. Phys., $\underline{41}$, 193 (1969)

[8] S.J. Brodsky and J.R. Gillespie - Phys. Rev. $\underline{173}$, 1011 (1968)

[9] L.C. Maximon and C. Tzara - Phys. Lett. $\underline{26B}$, 201 (1968)

[10] J. Bergstrom - MIT 1967 Summer Study, p. 251 - Report TID 24667

[11] H. Crannel - Nucl. Instr. and Methods, $\underline{71}$, 208 (1969); J. Berthot- Thesis University of Clermont-Ferrand (1968) unpublished

[12] D.B. Isabelle - Rendiconti della Scuola Internazionale "E. Fermi", Corso XXXVIII p. 302 (1967)

[13] A. Browman, B. Grossetete and D. Yount - Phys. Rev. $\underline{151}$, 1094 (1967)

[14] D. Vinciguerra and T. Stovall - Nucl. Phys. $\underline{A132}$, 410 (1969)

[15] See Ref (7) for a list of references on this topic

[16] T. De Forest, Jr. - Private communication

[17] T. De Forest, Jr. - Nucl. Phys. $\underline{A132}$, 305 (1969)

[18]W. Cayz and L. Lesniak - Phys. Lett. 25B, 319 (1967)
[19]T. Stovall and D. Vinciguerra - Nuovo Cimento Lett. 1, 100(1969)
  C. Cioffi degli Atti - Nucl. Phys. A129, 350 (1969)
  M. Malecki and P. Picchi - Nuovo Cimento Lett. 1, 81 (1969)
  S.T. Tuan et al. - Phys. Rev. Lett. 23, 174 (1969)
  C. Cioffi degli Atti and N.M. Kabachnik - This Conference
[20]J.B. Bellicard, V. Gillet, A. Lumbroso et R. Ripka - Private
  communication
[21]G.J.C. Van Niftrick et al. - Phys. Rev. 177, 1797 (1969)
[22]W.C. Barber - MIT 1967 Summer Study, p. 212 - Report TID 24667
[23]R. Baader et al. - Phys. Lett. 27B, 428 (1968)
[24]D. Blum and S. Penner - Private Communication
[25]MIT-NBS Collaboration - Private Communication
[26]V. Gillet - in "Nuclear Structure" p. 271, IAEA Vienna (1968)
[27]G. Ricco, H.S. Caplan, R.M. Hutcheon and R. Malvano - Nucl. Phys.
  A114, 685 (1968)
[28]W.R. Dodge and W.C. Barber - Phys. Rev. 127, 1746 (1962)
[29]W.R. Dodge - Private Communication
[30]K. Shoda et al. - Phys. Lett. 22B, 30 (1968)
[31]D. Drechsel and H. Uberall - This Conference
[32]B.T. Chertok et al. - Phys. Rev. Lett. 23, 34 (1969) and this
  Conference
[33]L. Katz et al. - Phys. Lett. 28B, 114 (1968)
[34]D. Ganichot et al. - Private communication
[35]R.E. Rand et al. - Phys. Rev. Lett. 18, 469 (1967)
[36]B. Bosco - Private Communication
[37]M. Chemtob and A. Lumbroso - This Conference
[38]G. Cortellessa - MIT 1967 Study Group, p. 427 - Report TID 24667
[39]J. Mougey - Private Communication
[40]A.N. James - This Conference
[41]J. Berthot et al. - Private Communication
[42]M.R. Yearian et al. - Bull. Am. Phys. Soc. 14, 105 (1969)

DISCUSSION

V. Telegdi: Is there not only one concrete piece of evidence for
exchange current, and that is in the thermal neutron-proton capture,
where,according to the calculation by Sachs and Alston, there is an
effect of some 6% according as one includes or excludes the exchange
currents?   Is this consistent with electron-scattering?
B. Chertok: This is still the best evidence for exchange currents
in the two-nucleon problem.   In electron scattering on deuteron
below 200 MeV, the typical meson exchange currents give an effect of
2-4%, while the experimental data is only good to 10%.   J.D. Walecka:
There is a direct way of measuring the correlation in the e,e'p n
experiment, one can directly measure the two-nucleon wavefunction
inside the target nucleus.   D.B.I.: There is no inconsistency with
electron scattering data.   Up to momentum transfers of 4 or 5 fm^{-1}
there seems to be no need to include meson exchange currents.

FINE STRUCTURE IN NUCLEAR CHARGE DISTRIBUTION

J. Heisenberg[*], R. Hofstadter[*], J. S. McCarthy[*], I. Sick[*],

B. C. Clark[†], R. Herman[†], and D. G. Ravenhall[‡]

*High Energy Physics Lab, Stanford Univ., Stanford, Calif.

†General Motors Research Labs, Warren, Michigan 48090

‡Physics Dept., Univ. of Illinois, Urbana, Illinois 61801

(Presented by J. S. McCarthy)

The original elastic electron scattering experiments on nuclei produced charge distributions that were characterized by two basic parameters; the half-density radius $r_{0.5}$ and the skin thickness t . It was possible using a phenomenological charge distribution, containing only these two parameters, to fit the experimental data from a wide range of nuclei. This situation has changed considerably in recent years. The amount of experimental cross sections in existence has increased and the data has been extended to higher values of the recoil momentum q . The ability to measure cross sections at larger values of q together with the better accuracy of the experimental data allows details of the charge distribution to be seen that were previously unobservable.

The experimental results for the two doubly-magic nuclei Ca^{40} and Ca^{48} are shown in Fig. 1.[1] These results are typical of the type of data that are being obtained for other nuclei. The dashed curves shown in the figure are a theoretical phase shift calculation of the cross section using a parabolic Fermi shape for the charge distribution. The three parameters were determined from best fits to earlier 250 MeV data. There is excellent agreement up to a scattering angle of 35°, beyond which the predicted curves are considerably higher than the measured points.

There are several possible explanations for this effect, among which is the possible failure of the basic description of the scattering process in terms of a static charge distribution $\rho(r)$. Another alternative is to slightly change $\rho_o(r)$ so that reasonable agreement with experimental data can be maintained. This has been the approach used to analyze the Ca isotopes and

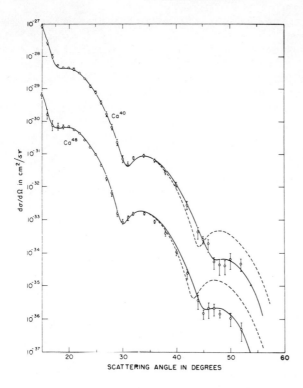

Fig. 1. Experimental and theoretical differential cross sections at 757.5 MeV. The cross section for Ca^{40} is times 10 and that for Ca^{48} times 10^{-1} .

has subsequently been used in the analysis of other nuclei where similar effects are seen.

   What is required to fit the data is a small oscillatory addition to the basic charge. This is seen in Fig. 2 where the experimentally determined undulation is compared to a typical

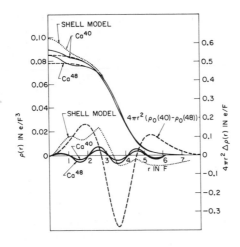

Fig. 2. Charge distributions for the previous 250 MeV fits (dashed line) and for the new data (solid lines). The right ordinate scale refers to $4\pi r^2$ times the change $\Delta\rho(r)$ between the new and old distributions.

shell model calculation using a Woods-Saxon potential.[2] The
theoretical prediction is qualitatively similar in amplitude and
wavelength to the one determined from the data. However the
amplitude is considerably reduced in magnitude from that seen and
is in itself an interesting feature to investigate. This reduction
in magnitude occurs for all the nuclei we have analyzed using this
method. Some typical charge distributions, which also include small
oscillations, have been determined for other nuclei and are shown
in Fig. 3.[11]

If the data are to be used to obtain such fine details of the
charge distribution, the presence of nuclear polarization, i.e.,
virtual excitation in the electron scattering process, requires
further investigation. Calculations of this process have been
done by Rawitscher[4] and Onley.[5] A generalized conclusion concerning these calculations would predict that the nuclear polarization
gives a small, smooth addition to the oscillating form factor, therefore affecting cross sections predominantly in the diffraction
minima.

To gain further information on this energy-dependent effect,
very precise measurements were taken on Ca^{40} at two different
energies; 249.3 MeV and 496.8 MeV . To test the model dependence of the analysis of the data, different functional forms
have been used. A least-squares fit to the cross section produces
almost indistinguishable charge distributions for the functional

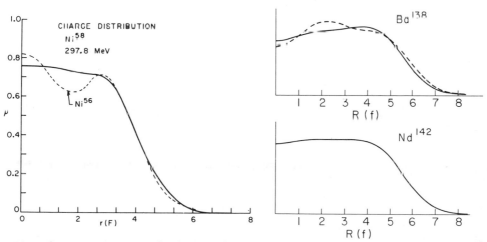

Fig. 3. Experimental charge distribution compared to Hartree-Fock
calculations.[3] (The calculations do not include the finite size of
the proton, which when included slightly reduces the magnitude of
the charge oscillation.)

forms used. These different distributions also predict to the level of accuracy of the data nearly the same muonic X-ray levels, showing to this extent model independent results.

A possible energy dependence in the method of analysis was checked by using the charge distribution obtained at 249.3 MeV to predict the energy of the 496.8 MeV data. To within the ± 0.1% uncertainty in the experimental energy there was good agreement. This allows one to use with increased confidence the charge distribution as determined from electron scattering to predict the muonic X-ray energy. In calculating the muon energy levels, the Lamb shift, nuclear polarization, and all other corrections which are expected to produce effects smaller than the accuracy of our comparison are neglected.

A preliminary analysis of the 249.6 MeV data gives for the value of the $2p_{3/2} \rightarrow 1s_{1/2}$ transition energy.

$$E(K_{\alpha 1}) = (783.65 \pm 0.50)\ keV \quad .$$

This is to be compared with the experimental muonic atom results of one of the Chicago groups,

$$E(K_{\alpha 1}) = (784.05 \pm 0.16)\ keV \quad .$$

If we assume that any discrepancy between the two energies can be attributed to a finite size difference between the muon and electron, then it is possible to set upper limits on the muon size.[6] Assuming zero "size" for the electron we obtain a result

$$\langle r^2 \rangle^{1/2}_{muon} \leq 0.18\ F \quad ,$$

and this compares favorably with other measurements.

The proposed proton "halo" would also be another source of any discrepancy, since the spreading of 1% of the proton charge over a much larger volume has a significant effect on the muon 1s state. This state is raised in energy by 2.4 keV while the 2p state at a considerably larger distance from the nucleus is unaffected by the halo. The total effect is to decrease the muon $K_{\alpha 1}$ X-ray energy from the value predicted from electron scattering by 2.4 keV. This result compared to the agreement reached between the measurements suggests that any possible halo must be decreased either in magnitude by a factor of 4 or have a much smaller rms radius than the proposed 8 F .

Measurements on the nucleus Pb^{208} have been improved in accuracy and extended to higher values of the recoil momentum than

the earlier measurements of Bellicard and van Oostrum.[7]   Some of
the conclusions reached concerning the nuclear charge distribution
have been noted before, particularly the characteristic central
depression.   However, the precise details of the inner part of
$\rho(r)$ could only be determined by a combination of electron and
muon experiments.   The new data on Pb^{208} shows that the accessi-
bility of this type of information from electron scattering alone
depends on the accuracy and extent of the data.   More information
is obtained, of course, if the very precisely measured muonic X-ray
energies[8] are included in the analysis.

The experimental facilities at the Stanford Mark III accelerator
have been continually improved and are responsible to a large extent
for the increased accuracy of the measurements.   Most significant
is a new energy calibration of the accelerator so that the energy is
known to better than $\pm 0.1\%$ and is reproducible to about $+ 0.05\%$.
A spectrum of scattered electrons analyzed in momentum is shown in
Fig. 4.   This is not the original experimental spectrum since the
data points have been corrected for relative efficiencies of the
counters and the important radiative effects.   The elastic peak is
visible together with the most strongly excited level, the 3^-
state at 2.6 McV .

The method used in analyzing the data is very similar to that
employed with the calcium isotopes.   A variety of functional forms
are assumed for the charge distribution and a search is made for
characteristics which are independent of the method of fitting and
of the assumed functional form.   If we combine the results from
muonic atom data, the two energies from the K and L lines allow
us to fix two of our adjustable parameters.

In Fig. 5 are shown the experimental data and the fit to them.
The data is shown at two energies, 248 MeV and 502 MeV.   To

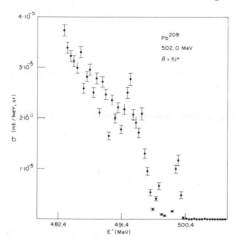

Pb^{208}

502.0 MeV

$\theta = 51°$

Fig. 4.   Spectrum of scattered
electrons from Pb^{208} .

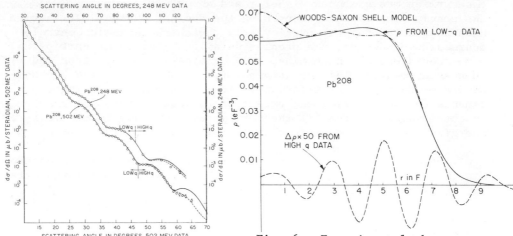

Fig. 5 – Elastic scattering cross-sections for Pb^{208}.

Fig. 6 – Experimental charge distribution for Pb^{208} (solid line) compared to shell model calculation (dashed line).

within the precision of the measurements, there is no detectable energy dependence in the charge distribution as determined for these two fits over the same range of q .

The 502 MeV data shows a similar deviation from the low-q predictions as in the calcium isotopes. The dashed curve in Fig. 5 is the best-fit obtained by the addition of a small undu-lation $\Delta\rho(r)$ which is added to the basic smooth charge $\rho(r)$. The dashed curve in Fig. 6 represents the undulation. The actual magnitude of the oscillation is so small that $\rho(r) + \Delta\rho(r)$ is indistinguishable from $\rho(r)$ as drawn in the figure. The details of the analysis of this wiggle are still preliminary. The various functional forms used to analyze the low-q part of the data are identical to the accuracy of the drawing.

The resulting nuclear charge distribution is characterized by the following: (1) The fit to the low-q part, which is a smooth average of the actual charge distribution, has a central depression of about 7%. (2) The extreme edge of the distribution falls off more rapidly than the usual Fermi shape but is in agreement with the calculation of an independent-particle shell model.[9] The value for the rms radius determined from electron scattering alone, which is not a directly measured quantity, centers around r_{rms} = 5.54 F . This is to be compared to the value obtained from muonic atoms of r_{rms} = (5.4978 ± 0.0030) F .[8]

The final conclusion that can be drawn concerning both shell model and Hartree-Fock calculations of nuclear charge distributions is that they over-estimate the magnitude of the charge oscillation. Both these calculations are independent-particle models and ignore correlations between nucleons. This reduction of the theoretical magnitude of the oscillation attributed in Ca^{40} and Ca^{48} to the effect of nucleon correlations,[10] now appears to be much more pronounced in the heavier nuclei. It is then possible to conjecture that the central depression and the much diminished oscillation are both pieces of evidence for strong correlations in the heavier nuclei such as Pb^{208}

REFERENCES

1. J. B. Bellicard et al, Phys. Rev. Letters, 19, 527 (1967).

2. L. R. Mather, J. M. McKinly, and D. G. Ravenhall, unpublished calculations.

3. R. M. Tarbutton and K. T. R. Davies, Nucl. Phys. A120, 1 (1968).

4. G. H. Rawitscher, Phys. Rev. 151, 846 (1966).

5. D. S. Onley, Nucl. Phys. A118, 436 (1968).

6. R. F. Frosch et al, Phys. Rev. 174, 1380 (1968).

7. J. B. Bellicard and K. J. van Oostrum, Phys. Rev. Letters 19, 242, (1967).

8. H. L. Anderson et al, Phys. Rev. Letters 22, 221 (1969).

9. L. R. Mather, R. H. Landau, and D. G. Ravenhall, unpublished calculations.

10. F. C. Khanna, Phys. Rev. Letters 20, 871 (1968).

11. J. Heisenberg, J. S. McCarthy, I. Sick, unpublished.

DISCUSSION

R. Hofstadter: De Shalit first suggested that we compare the ^{40}Ca and ^{48}Ca isotopes.

R. C. Barrett: One final way of testing the existence of a proton 'halo' would be to examine the relative positions of the 2s and 2p levels in the muonic H-atom.

H. L. Anderson: Have you, in the case of Pb, reversed the procedure, i.e., calculated the X-ray transition energies from your charge distributions?   J.S.M.: Yes – but we required our charge distributions to fit the X-ray transition energies.   If we use only electron scattering data there is a difference of some 30 keV, but we don't expect precise agreement, because there are significant dispersion corrections, and we do not measure the r.m.s. radius well.

LOW MOMENTUM TRANSFER ELECTRON SCATTERING

F. Beck

Institut für Theoretische Kernphysik

Technische Hochschule Darmstadt, Germany

As we have heard in the previous reports, elastic and inelastic electron scattering has become an important tool in nuclear studies. Let me once more stress the basic advantage of using electrons as electromagnetic probes, namely the decoupling of energy- and momentum-transfer to the nucleus.

Having this in mind one may very well ask if going to the limit of low momentum transfer, or phrased differently, approaching the real photon point $\tilde{q}^2 \rightarrow k^2$ ($k = \Delta E/hc$), one is not giving away the very advantage of that specific probe. The answer to this question depends of course on what property of the nucleus one is looking for. Let me, quite arbitrarily, define the low q range by, say, $q \lesssim 0.5$ fm^{-1} (this corresponds to electron accelerators with maximum energy <60 MeV). Then, with the nuclear size parameter R we have $qR \sim 1$, and with the nucleon size parameter, the hard core radius r_c, we obtain $qr_c \sim 0.2$. From a nuclear structure point of view, long range correlations are very well within our low q while everything connected with short range correlations clearly is not. Low momentum transfer electron scattering sees the smooth parts of the nucleus and is thus connected with conventional nuclear spectroscopy. It is quite clear from this, that in order to compete with other spectroscopic methods, e.g. γ ray spectroscopy, one must reach a high level of precision, and this is just where modern development is aiming.

In the following I shall present some recent results

from this viewpoint. I leave out completely elastic
scattering since this is connected with nuclear ground
state properties which is discussed in another context during this conference.

Let me very briefly remind you how to extract spectroscopic data from measured cross-sections. I shall do
this in the language of first-order Born approximation
for ease of discussion. I should like to emphasize, however, that precision analysis requires even for light
nuclei distorted-wave calculations though they are numerically cumbersome. The nuclear physics enters the
measured cross sections through the form factors $|F_\lambda(q)|^2 \sim$
$q^{2\lambda} B(\lambda,q)$ where λ is the multipole order of the partial
cross section[1]. There are three of them, longitudinal
(from Coulomb scattering), transverse electric and transverse magnetic. At low q one prefers to work with the
transition rates $B(\lambda,q)$ rather than with the form factors
itself. The transversal transition rates $B(E\lambda,q)$, $B(M\lambda,q)$
equal the reduced radiative transition probabilities at
the photon point q=k.

As it has been demonstrated in many practical applications[2], a combination of constant q (varying scattering
angle θ) and constant θ (varying momentum transfer q)
measurements can frequently disentangle the different
contributions to the measured cross-section. This is so,
because low q enhances strongly the lowest possible multipoles. Thus the multipolarities λ, the parities and the
transition rates $B(\lambda,q)$ are extracted. What then are we
doing with this information? In the first instant the answer is once more conventional spectroscopy. Extrapolation
of the measured transition rates to the photon point
gives us the absolute value of the ground to excited
state transition probabilities, a quantity which is not
easy to obtain in γ ray spectroscopy. In this way the
spectroscopic material of light nuclei has been amended
considerably during the past years.

We have, however, more information from the measured
cross sections. There is on one hand the longitudinal contribution, and on the other the functional dependence on
q. Does this contain additional nuclear structure information, besides serving as a technical tool to obtain
the spectroscopic data proper? Clearly, near the photon
point we have normally kR<<1 and thus we can work out
everything in the long wave length limit. This leads to
the connection[1]

$$B(E\lambda,q) \xrightarrow[q \to 0]{} (k/q)^2 B(C\lambda,q) \tag{1}$$

Thus as long as this relation holds the longitudinal
and transverse electric cross–sections give us essentially
the same information. Because of the extra $1/q^2$ dependence of the transverse part, for increasing q Coulomb
scattering will dominate, and it is the longitudinal
transition rate which should be extrapolated to yield the
radiative width of electric transitions. Expanding the
Bessel function in the nuclear matrix element one obtains[2,3]

$$B(C\lambda,q) = B(C\lambda,0) \ \{1- \frac{1}{2\lambda+3}(qR_t)^2+O(q^4)\};R_t^2=\frac{<r^{\lambda+2}>}{<r^\lambda>} \quad (2)$$

The expansion coefficient R_t^2 is certainly a model dependent quantity, and it has been argued therefore[4] that extrapolation to the photon point is essentially model dependent. However, with the improved measuring techniques,
R_t can be determined experimentally, leading to one extra
piece of information on the transition density. In addition,
because of the very rapid convergence of the expansion
(2) in the low q range, a practically model independent
extrapolation to the photon point is possible. Fig. 1
shows you an example how this works.

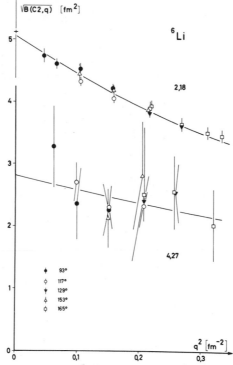

Fig. 1: Example of the determination of R_t and extrapolation
to the photon point (F.Eigenbrod,DALINAC).+ is from spectroscopic measurement. Upper curve: E_x = 2.18 MeV (3^+,T=0) level.
Lower curve: E_x = 4.27 MeV level. Both transitions are C2.

We have seen before that the low q range extends well
up to qR∿1 and then the magnetization and convection current
terms in the transverse electric matrix element

$$M_{fi}(E\lambda,q) \sim \{- \frac{k}{q}<\rho_N> + \frac{q}{\lambda+1} (<\vec{\mu}_N \cdot \vec{L}> + i <\vec{j}_N \cdot \vec{r}>)\} \qquad (3)$$

are by no means neglegible as in the long wave length limit. It has been pointed out by Walecka and others[1] that
the characteristic q-dependence which results from squaring
(3) can give us interesting information on these terms not
obtainable with real photons. The effect is to be expected most strongly for non collective transitions in light
nuclei and has been observed for the less collective component of the giant dipole resonance in ^{12}C and ^{16}O [5].
Another example, where the magnetization term sets in at
even lower q, namely the E1 transitions to positive parity states of ^{13}C and ^9Be, has been submitted to this conference[6]. Fig. 1 of this contribution shows the importance
of the magnetic moment term ($\mu=0$ corresponds still to the
limit (1)).

There is a serious obstacle for obtaining precise
inelastic electron scattering data at low momentum transfer which is connected with the radiative tail of the
elastic line. To extract cross sections from measured excitation curves, one has to subtract background. Now it
is not difficult to calculate the bremsstrahlung tail of
elastic scattering, but the observed background was always about 30 % higher than the calculated tail. It is
now clear that this is due to instrumental scattering,
mainly from the spectrometer backside. Recently improved
spectrometers have been installed in several places which
reduce instrumental scattering to a minimum. Another method
employed in Darmstadt[7] consists in measuring the instrumental background and to subtract it, together with the
calculated radiative tail, from the measured spectra. An
example of this procedure is shown in fig. 2 for inelastic scattering from ^{12}C. The continuous contribution
above 7 MeV excitation can be explained by the (e,e'3α)
reaction whose threshold is indicated in the figure. This
success in handling the background makes it possible now,
to extend low q inelastic scattering to the continuum
region where resonances overlap. I should mention, however,
that for low primary energies, E<70 MeV, say, some discrepancies still seem to persist[7] which may very well indicate a breakdown of the Bethe-Heitler formula for higher
Z. Thus, improved bremsstrahlung calculations, including
Coulomb distortion, would be desirable.

In closing I should like to present some recent
results of low q inelastic scattering work which may have

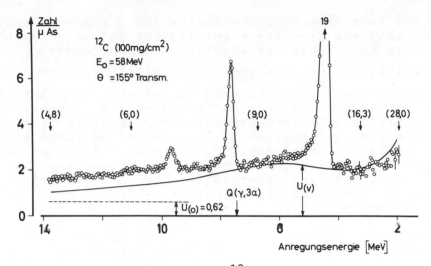

Fig. 2: Inelastic spectrum of ^{12}C after subtraction of
the calculated radiative tail. Counting rates
before subtraction are indicated by the figures
above the arrows. $U_{(v)}$ is the measured instrumental background (A. Goldmann[7]).

some general interest in nuclear structure. There are
several regularities in the excitation spectra of different
nuclei which reflect common features of the nuclear interactions and thus are especially challenging to nuclear
models. These regularities are best known in heavier nuclei where they are a domain of collective models, but can
also be found in light nuclei. Here the chance of a
"microscopic" interpretation is greater. Examples are the
low lying $J^{\pi}=0^+$, T=0 states of 4n-nuclei, and the giant
resonance.

0^+-States. The lowest 0^+,T=0 excited states of 4n-
nuclei show a striking regularity of excitation energy,
starting with about 8 MeV in ^{12}C and degrading to about
3 MeV in ^{40}Ca. The occurence of these states is presumably
connected with the coexistence of "spherical" and "deformed" states[8,9] in these nuclei.

Monopole transitions in light nuclei have been studied
systematically by the Darmstadt group[10,11,12]. The results
for the monopole matrix elements and the transition radii
to the excited 0^+ states are given in table 1. Note the
difference in the data for closed shell nuclei and the
other ones.

	E_x (MeV)	ME (fm^2)	R_t/R_{ms}
^{12}C	7.65	5.37 $\pm$ 0.22	1.77 $\pm$ 0.10
^{16}O 0^+(1)	6.05	3.80 $\pm$ 0.19[a]	2.3 $\pm$ 0.6
0^+(2)	12.05	4.40 $\pm$ 0.44	2.7 $\pm$ 0.4
0^+(3)	14.00	3.3 $\pm$ 0.7	2.9 $\pm$ 0.7
^{24}Mg	6.44	6.33 $\pm$ 0.29	1.90 $\pm$ 0.12
^{40}Ca	3.35	2.91 $\pm$ 0.06[a]	
		(3.6 $\pm$ 1.1)	

Table 1: Monopole matrix elements and transition radii in units of ground state rms-radius for low lying 0^+ states.([a]) from Devons et al., Proc. Phys. Soc. 67 (1954) 134)

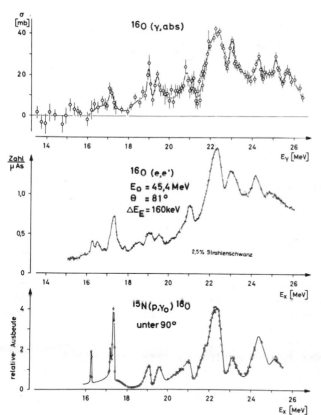

Fig. 3 – Electron scattering data[7] in the giant resonance of ^{16}O in comparison with γ absorption[16] and (p,γ)[17] results.

Giant Resonance

Using the technique described before to subtract background, the giant resonance region of ^{12}C, ^{16}O, ^{26}Mg and ^{40}Ca has recently been investigated by the Darmstadt group[7,13]. Fig. 3 shows a spectrum at forward angle compared with photon work, indicating about the same resolution in both techniques.

Fig. 4 shows what we can learn from a study of the q-dependence. The levels at 19 and 20 MeV clearly show a much faster growth with increasing q than the rest of the region, and a detailed analysis leads to an M2 assignment, going to theoretically predicted[14] 2$^-$ states.

There is still the quartet splitting of the main part of
the giant resonance between 21 and 26 MeV, not yet understood theoretically, except in a very frivolous model
employing particle hole excitations upon a slightly deformed ground state[15]. At 23.5 MeV, however, a yet unresolved level of different multipolarity, or at least
with a markedly different transverse behaviour grows
up.

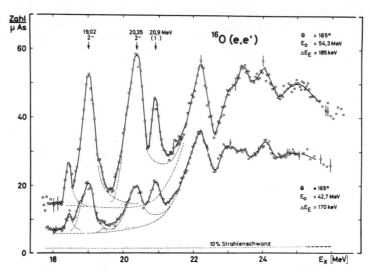

Fig. 4 - Electron scattering spectra in ^{16}O at
constant angle but different momentum transfers.
The dashed line at the bottom indicates how
absolute values change if the calculated radiative tail is increased by 10% (from 7).

I hope to have shown to you that low momentum transfer electron
scattering has grown to a useful tool of nuclear spectroscopy. It
is also clear that low momentum transfer is really a restriction and
has to go hand by hand with results from higher energies. Being
pushed, however, from both sides, experimenters with low energy
machines may have succeeded in converting poverty into virtue.

REFERENCES

1) A derivation can be found in many articles, e.g. T.
   de Forest, Jr. and J.D. Walecka, Adv. in Physics 15
   (1966) 1
2) E. Spamer, Z. Physik 191 (1966) 24

3) D.R. Isabelle and G.R. Bishop, Nuclear Phys. 45 (1963) 209

4) R.S. Willey, Nuclear Phys. 40 (1963) 529

5) F.H. Lewis, Jr., J.D. Walecka, I. Goldemberg and W.C. Barber, Phys. Rev. Letters 10 (1963) 493

6) F. Beck, L. Grünbaum and M. Tomaselli, Contribution to this Conference

7) A. Goldmann, to be published and private communication

8) G.E. Brown and A.M. Green, Nuclear Phys. 75 (1966) 401

9) G. Kluge and P. Manakos, Phys. Letters 29 B (1969) 277

10) M. Stroetzel, Z. Physik 214 (1968) 357

11) P. Strehl and Th. H. Schucan, Phys. Letters 27 B (1968) 641, and private communication

12) M. Stroetzel and A. Goldmann, Phys. Letters 29 B (1969) 306

13) A. Goldmann and E. Spamer, to be published

14) T. de Forest, Jr., Phys. Rev. 139 (1965) B 1217

15) G. Kluge, Z. Physik 197 (1966) 288

16) D. S. Dolbilkin, V. I. Korin, L. E. Lazareva and F. A. Nikolaev, JETP Letters 1, 1481 (1965)

17) E. D. Earle and N. W. Tanner, Nuclear Phys. A95, 241 (1967)

ANALOGS OF GIANT RESONANCES FROM PHOTO-PION PRODUCTION AND FROM MUON AND PION CAPTURE

H. Überall

Catholic University of America* and

Naval Research Laboratory, Washington, D. C.

The giant resonances have originally been seen in photonuclear reactions such as

$$\gamma \;+\; {}^{16}O \;\longrightarrow\; {}^{16}O^{*}_{g.res.} \;\longrightarrow\; {}^{15}O + n \;({}^{15}N + p), \tag{1}$$

and as was pointed out in the preceding reports, they may also be excited by inelastic electron scattering. In $T = 0$ nuclei (e.g. ${}^{12}C$ or ${}^{16}O$), they form a $T = 1$ isotriplet whose $T_3 = \pm 1$ components exist in the neighboring nuclei (Fig. 1) but whose $T_3 = 0$ component only may be excited by photons or electrons. The photonuclear giant resonance was explained by Goldhaber and Teller [1] as a collective $\Delta T = 1$ dipole vibration of protons vs. neutrons (isospin or i-waves) see Fig. 2a. This picture has been extended in several respects:

(1) Spin-isospin (si) and spin (s) waves were introduced [2] as additional collective dipole modes (Fig. 2a). The spin waves cannot easily be excited and shall be disregarded here. The si-waves ($S = 1$) in the dipole case ($L = 1^{-}$) form states $J = 0^{-}$, 1^{-} and 2^{-} whereas the i-waves ($S = 0$) appear as only one state $J = 1^{-}$; all three modes together constitute a 15-dimensional SU4 supermultiplet.

In photoabsorption, only i-waves are strongly excited, because of the fixed momentum transfer ($q = \omega$), whereas in electroexcitation where q ranges from ω (forward) to $2E_1 - \omega$ (backward), si-waves appear strongly at large values of q [3], demonstrating the superiority of electro- over photoexcitation for studies of nuclear structure.

* Supported by the National Science Foundation

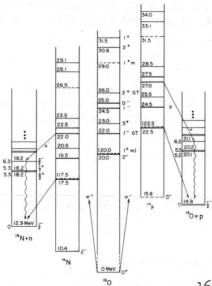

Fig. 1. The Giant Resonance isotriplet in ^{16}O, and decay channels
of the T_3 = ± 1components

(2) Giant quadrupole vibrations ($L = 2^+$) may take place [4],
see Fig. 2b, leading to 2^+ (i) and 1^+, 2^+ and 3^+ (si) states. There
is some evidence for their existence from electron scattering [5].

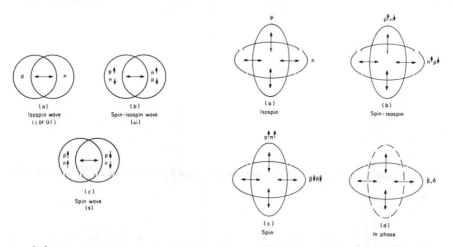

Fig. 2.(a) left: Collective dipole vibrations of nuclear matter;
(b) right: Collective quadrupole vibrations of nuclear matter
(modes of the generalized Goldhaber-Teller model)

(3) The $T_3 = \pm 1$ analogs of the giant resonances may be searched for. They may be excited in reactions containing $\tau_\pm$, viz.

(a) Muon capture, e.g.

$$\mu^- + {}^{16}O \longrightarrow {}^{16}N^*_{g.res.} + \nu .\tag{2}$$

The momentum transfer $q = m_\mu - \omega$ is fixed (but large), and the Fermi transition (τ_+) excites the i-waves, the Gamow-Teller transition $(\tau_+ \vec{\sigma})$ the si-waves. Neutron spectra may be observed as in (γ, n) reactions.

(b) Radiative pion capture, e.g.

$$\pi^- + {}^{16}O \longrightarrow {}^{16}N^*_{g.res.} + \gamma .\tag{3}$$

Again $q = m_\pi - \omega$ is fixed (large); the matrix element excites si-waves only, and besides neutron spectra, one may measure photon spectra.

(c) Pion photoproduction, e.g.

$$\gamma + {}^{16}O \longrightarrow {}^{16}N^*_{g. res.} + \pi^+ \cdot ({}^{16}F^*_{g.res.} + \pi^-) .\tag{4}$$

Here, q varies depending on ϑ_π, rendering (4) more informative than (2) and (3). Only si-waves are excited near threshold, and pion spectra (if γ is monochromatic) or n (p) spectra may be measured.

The following calculations and experiments have been done:

(1) <u>Muon Capture</u>. Fig. 3a shows calculated neutron spectra

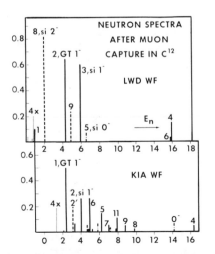

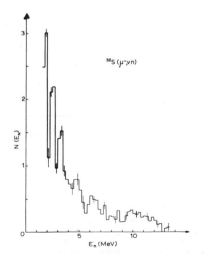

Fig. 3. Neutron spectra following muon capture (a),left, in ^{12}C, calculated with p-h models, and (b), right, measured in ^{32}S.

following muon capture [6] in ^{12}C obtained with particle-hole wave
functions of Lewis - Walecka - deForest (top) and Arima (bottom),
and Fig. 3b presents recently measured neutron spectra after muon
capture [7] in ^{32}S which seem to confirm the existence of the resonance capture mechanism.

An interesting variant is the observation of de-excitation photons (see Fig. 1) after muon capture. In the non-analog case, these
have been observed in ^{16}O after photoexcitation [8], see Fig. 4a.
Kaplan [9] also observed such photons after muon capture in ^{16}O
(Fig. 4b), with rates in good agreement with calculations based on
the resonance mechanism.

(2) Radiative Pion Capture. It has been shown [10] that over
80% of the captures take place from the 2p rather the 1s orbit, in
which case giant quadrupole states should appear strongly [11].
Fig. 5a shows calculated neutron spectra [12] following radiative
pion capture in ^{16}O which bear this out. Fig. 5b presents early
measured photon spectra [13] in ^{12}C and ^{32}S (solid curves), compa-

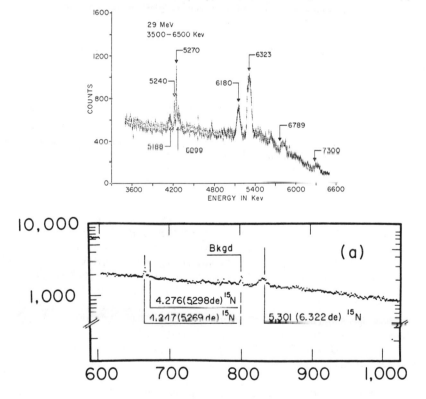

Fig. 4(a) top: Decay photon spectra following ^{16}O photoexcitation
(graph courtesy K. Murray); (b) bottom: decay photon spectra following muon capture in ^{16}O.

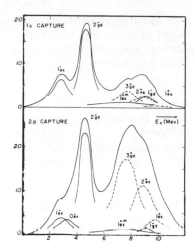
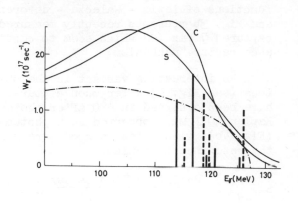

Fig. 5(a) left: Calculated neutron spectra following 1s and 2p radiative pion capture in ^{16}O; (b) right: photon spectra of radiative pion capture: experiments in ^{12}C, ^{32}S (solid curves), and particle-hole calculation in ^{16}O (verticle lines).

red with calculated spectra (1s capture considered only) for ^{16}O [14]. Better resolution spectra are now being obtained by K. Crowe (private communication) using a pair spectrometer. The phase space curve (dashed) gives too little peaking, whereas a Fermi gas, direct-capture mechanism fit (not shown) gives too much, besides failing in absolute magnitude. An alternative mechanism (proton exchange "pole model") was proposed recently [15] to explain the broad peaking of the photon spectra. It must be left up to better experiments to show if such alternate capture modes compete with the resonance mechanism.

The best evidence so far for giant resonance excitation in radiative pion capture comes from an experiment in ^{12}C by Lam et al [16] which measured neutron spectra at various angles with respect to the photon, by the time-of-flight method using the photon as a coincidence signal. Fig. 6 clearly shows a 4 MeV peak (top) as predicted by the particle-hole calculations [6] (center). The direct-emission (Fermi gas) background calculated by D.K. Anderson (bottom) is too small to explain the observed rate, but it provides an observable peaking the the $\theta_{\pi\gamma}$ angular distribution at 180°. Similar experiments in ^{16}O and ^{28}Si [17] indicated however little structure in the neutron spectra.

The appearance of a high-lying photon peak in the Liverpool experiment [13] for nuclei from Li to Cu suggests the presence of (analog) si-vibrations in nuclei through the periodic table, similar to the photonuclear (i) giant resonance.

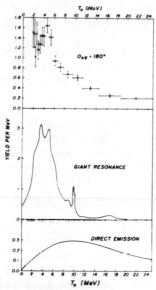

Fig. 6. Neutron spectra after radiative pion capture in ^{12}C: experiment (top), resonance calculation (center), and Fermi gas calculation (bottom). Graph courtesy B. MacDonald.

(3) <u>Inelastic Pion Photoproduction</u>. This was recently calculated [18] on the resonance model, and in Fig. 7, we present pion

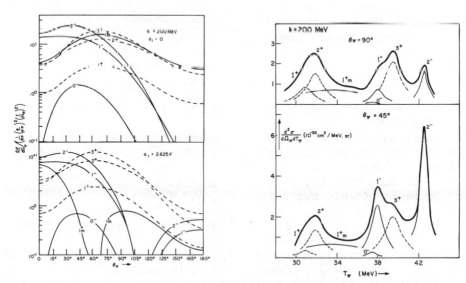

Fig. 7. Photo-pion angular distributions, left (a), and spectra, right (b) in inelastic photoproduction from ^{16}O.

angular distributions and spectra for ^{16}O near threshold. Fig. 7a shows the angular distributions with excitation of the states shown in Fig. 1 assuming volume production (top) and surface production (bottom); in either case, it is seen that giant quadrupoles (dashed) dominate over the dipoles (solid curves) for $\Theta_\pi \gtrsim 60°$. Fig. 7b shows the corresponding (volume production) pion spectra at $\Theta_\pi = 45°$ and $90°$.

To measure pion spectra, monochromatic 200 Mev photons would be desirable. These are now becoming available using annihilation radiation or crystal bremsstrahlung (Saclay, Livermore linacs). In any case, spectra of decay nucleons may also be measured, and will likewise show resonance effects, similarly to the Dodge-Barber (e,e'p) spectra.

It should be repeated that the variability of q in inelastic pion photoproduction, with e.g. a corresponding enhancement of quadrupole over dipole states, should render this process as superior as compared to the capture process for a study of analog giant resonances, as inelastic electron scattering has become compared to photoabsorption for a study of the non-analog giant resonances.

REFERENCES

1 M. Goldhaber and E. Teller, Phys. Rev. 74, 1046 (1948)
2 W. Wild, Bayr. Akad. Wiss. Math.-Nat. Kl. 18, 371 (1956); see also H. Überall, Phys. Rev. 137, B502 (1965)
3 H. Überall, Nuovo Cimento 41B, 25 (1966)
4 M. Danos and H. Steinwedel, Z. Naturf. 6a, 217 (1951)
5 R. Raphael, H. Überall, and C. Werntz, Phys. Rev. 152, 899 (1966)
6 F. J. Kelly and H. Überall, Nucl. Phys. A118, 302 (1968)
7 V. Evseyev et al, Phys. Letters 28B, 553 (1969)
8 K. Murray and J. C. Ritter, Bull. Am. Phys. Soc. Ser. II 13, 717 (1968)
9 S. N. Kaplan et al, Phys. Rev. Letters 22, 795 (1969)
10 Y. Eisenberg and D. Kessler, Phys. Rev. 123, 1472 (1961)
11 D. K. Anderson and J. M Eisenberg, Phys. Letters 22, 164 (1966)
12 J. D. Murphy et al, Phys. Rev. Letters 19, 714 (1967)
13 H. Davies et al, Nucl. Phys. 78, 673 (1966)
14 M. Kawaguchi et al, Phys. Letters 27B, 358 (1968)
15 L. G. Dakhno and Yu. D. Prokoshkin, Sov. J. Nucl. Phys. 7, 351 (1968)
16 W. C. Lam et al, Bull. Am. Phys. Soc. Ser. II 13, 715 (1968)
17 M. Holland et al, Bull. Am. Phys. Soc. Ser. II 14, 538 (1969)
18 F. J. Kelly, L. J. McDonald, and H. Überall, contribution to this conference, and to be published.

DISCUSSION

M. Rho: Do you consider final-state interaction in photo-production
of pions?   H.U.: This is done in a very rough way – by the surface
production model: the wavefunctions are cut off at the nuclear
radius.   An improved treatment is needed, but I think the salient
features are present in this model.   M.R.: If one does not deal
with the pion production at threshold, the operator becomes very
complicated.   How can one make direct connection with the giant
resonance?   H.U.: The pion production to the ground state and to
the low excited states has been calculated taking into account the
effect of final state interaction.   This did not seem to be a very
big effect.   M.R.:   That depends upon which nucleus you are looking at.

T. Ericson: For all these considerations it is essential that the
interaction is known.   The basic assumption of Dr. Uberall is that
the effective interaction Hamiltonian essentially has Gamow-Teller
matrix-elements.   There are only three pieces of experimental
evidence: the total radiative absorption rate of pions by nuclei;
the Panofsky ratio for ^{3}He; and Π^- capture in ^{6}Li going into ^{6}He
ground state.   This last is the only one which can be considered
quantitative.   It seems most desirable to establish that this interaction is really the right one, as well as to investigate the
detailed consequences of this assumption.   H.U.: Agreed; but at
the low-momentum transfer limit the G-T interaction should be
correct; and for the rough calculations here it should be adequate.

R. Eramzhyan: The gross structure of decay of the giant resonance
seems to be clear, but there remains the problem of fine-structure.
Kaplan's (Ref. 9) results indicate that 15% of the decay of the
giant resonance goes to the positive parity states of ^{15}N.   Theoretical calculations (particle-hole model) give only 2%.   The situation is similar for photo-disintegration.   It seems necessary to
include more complicated configurations in the ground and excited
states (of ^{16}O).

ELECTRON SCATTERING EXPERIMENTS AT THE MAINZ 300 MeV LINEAR ACCELERATOR

H.G. Andresen, H. Averdung, H. Ehrenberg, G. Fricke,
J. Friedrich, H. Hultzsch, R.M. Hutcheon, P. Junk,
G. Luhrs, G. Mulhaupt, R. Neuhausen, H.D. Wohlfahrt

University of Mainz, Germany

The Mainz electron scattering facility provides energies between 80 and 300 MeV and scattering angles up to 158° which correspond to a useful momentum transfer region $0.4 \leq q \leq 3.0$ fm^{-1}. The accelerator and analyzing system produce approximately 1 µA average current with the overall resolution in the final spectra of usually 0.15% to 0.30%.

Measurements on the nuclei ^6Li, ^{12}C, ^{24}Mg, ^{28}Si, ^{32}S, ^{208}Pb have been performed or are under way.

$\underline{^6Li}$ Three narrow levels have been seen in the spectrum up to 10 MeV. The cross-section for the transition to the first level (2.18 MeV, 3^+)

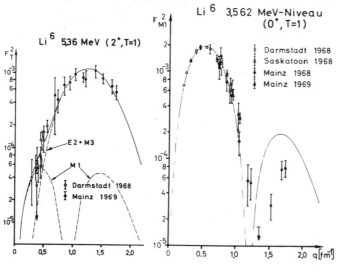

Fig. 1 – Form Factors for electron inelastic scattering in ^6Li.

seems to be pure longitudinal ($F_T^2 < 0.02\ F_L^2$) (Z. Physik 220 (1969) 456). The form factors for the other two levels (Fig. 1) are transverse; the first, being of pure M1 character, displays a clearly defined diffraction minimum and shows a discrepancy to the simple shell model; the second is mixed M1, E2, M3. The calculations for this level use intermediate coupling shell model wavefunctions.

A broad quasi-continuous peak is seen centered at 4.2 MeV with a width of about 2 MeV and starting at the alpha-deuteron breakup threshold. Finally, in a spectrum at 100 MeV, 60°, out to an excitation energy of 32 MeV no structure was observed (statistical accuracy $\sim$ 5%), which is in contrast to the Orsay spectrum (Phys. Lett. 5 (1963) 140).

^{12}C Both the longitudinal and transverse form factors were determined for the two 4.44 and 16.11 MeV electric quadrupole transitions by measurements at constant momentum transfer (q $\sim$ 1.0 fm^{-1}) for different angles. Checks were carried out by evaluating the pure 7.65 and 15.11 MeV transitions.

For the collective 4.44 MeV quadrupole transition F_T^2/F_L^2 is 3.1×10^{-3} in agreement with the liquid drop model (Z. Physik 181 (1964) 542). The same ratio for the single particle 16.11 MeV quadrupole transition is two orders of magnitude larger ($F_T^2/F_L^2 = 0.75$) but is, however, a factor of three lower than particle-hole model calculations (Phys. Lett. 10 (1964) 308). A broad continuous cross-section is clearly seen underlying the 7.65, 9.64 and 10.84 MeV levels (^{12}C (e, e'α) ^{8}Be). (Fig. 2)

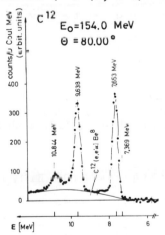

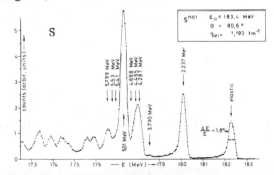

Fig. 2 – Spectrum of scattered electrons: from sulfur (above) and detail of spectrum from ^{12}C (left).

Longitudinal and Transverse form factor of ^{12}C

MeV	I	T	q fm^{-1}	F_L^2	F_T^2
4.439	2^+	0	0.985	$(1.37 \pm 0.03)10^{-2}$	$(4.20 \pm 3.10)10^{-5}$
7.653	0^+	0	0.970	$(2.85 \pm 0.11)10^{-3}$	$(0.10 \pm 1.00)10^{-5}$
15.109	1^+	0	0.939	$(0.90 \pm 5.10)10^{-5}$	$(1.32 \pm 0.12)10^{-4}$
16.106	2^+	1	0.931	$(3.58 \pm 0.49)10^{-4}$	$(2.78 \pm 0.18)10^{-4}$

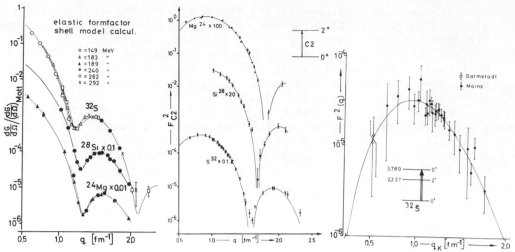

Fig. 3 - Elastic & inelastic scattering from ^{24}Mg, ^{28}Si & ^{32}S

^{24}Mg, ^{28}Si, ^{32}S   (0^+ ground states) A S^{32}-spectrum is shown in
Fig. 2. The elastic cross-sections (Fig. 3) were taken up to a
momentum transfer of q $\sim$ 2.35 fm^{-1} and normalized to the ^{12}C data.
We can describe our data by using phase shift analysis with a shell
model or Fermi-3-parameter charge distribution.   In ^{32}S, the
presence of two 2s-protons enlarges the 1s-shell-model-parameter
enormously. We explain this effect through short range correlations.

shell model	a_{1S}	a_{1P}	a_{1D}	a_{2S}	fermi	c	t	w
^{24}Mg	1.62	1.88	1.89	---		3.06	2.56	-0.12
^{32}S	2.80	1.77	1.87	1.89		3.53	2.69	-0.26

     In all three nuclei the form factor for the excitation of the
first 2^+-levels displays a well defined minimum (Fig. 3). A first
Born approximation analysis using either a liquid drop or Helm model
yields the following parameters:

^{24}Mg (1.37 MeV) R = 3.31 fm g = 0.91 fm β = 0.30 H.M.
^{24}Mg (4.23 MeV) R = 3.03 fm g = 1.15 fm β = 0.03 H.M.
^{28}Si (1.78 MeV) R = 3.50 fm λ^2 = 0.12   L.D.M.
^{32}S (2.24 MeV) R = 3.67 fm g = 0.81 fm β = 0.06 H.M.

     These parameters result in preliminary values for the level
width of 0.71 ± 0.14 meV for Si^{28} and 2.6 ± 0.5 meV for S^{32}.   The
form factor of the 3.78 MeV monopole transition (Fig. 3) of ^{32}S, we

find to be smaller than given in the literature (Nucl. Phys. 59 (1964) 398).   Our preliminary Helm-model parameters are R = 3.31 fm, g = 1.06 fm, β = 0.89 × 10^{-4}.

Measurements in the giant resonance region of ^{28}Si show a peak at 14.2 MeV with a large transverse component and transitions to as yet unidentified states at 13.2, 13.5 and 13.8 MeV.

180° ELECTRON SCATTERING FROM ^{3}He AND ^{4}He

B. T. Cheltok[*] and E. C. Jones[+]
The American University, Washington, D.C. 20016
and
W. L. Bendel, L. W. Fagg, H. F. Kaiser and S. K. Numrich
U.S. Naval Research Laboratory, Washington, D.C. 20390

I.   HELIUM 3

The photomagnetic dipole disintegrations of the trinucleon doublets, ^{3}H and ^{3}He, are forbidden processes which, accordingly, are difficult to observe in competition with the E1 disintegration produced by photons. The extent of this orthogonality is evident by comparing the radiative capture cross sections of the proton and of the deuteron by thermal neutrons. The measured values are 0.5 mb for ^{2}H(n,γ)^{3}H and 334 mb for ^{1}H(n,γ)^{2}H where the latter is a M1 allowed radiative transition. It is by now widely appreciated that inelastic electron scattering at 180° is an experimental tool to sizably enhance M1 in competition with E1 processes.[1] This is because the longitudinal part of the virtual photon spectrum vanishes as $\theta \rightarrow \pi$, while the transverse part of the equivalent spectrum is dominated at 180° by M1 compared to E1 as q^2/k^2 where $\vec{q} = \vec{p_o} - \vec{p}$ and $k = \Delta E/\hbar c$. Furthermore the photomagnetic orthogonality in ^{3}He may be broken by inelastic electron scattering.[2]

The M1 selection rule in ^{3}H photodisintegration or the equivalent inverse (p,γ) reaction arises from two factors. The magnetic moment operator conserves the symmetric part of the ground state, $^2S_{\frac{1}{2}}$, which is 92% of the ^{3}He ground state. Furthermore the radial part of the continuum wave functions for the S state are largely in the same potential as the ground state. These considerations rule out $^2S \rightarrow {}^4S$ and $^2S \rightarrow {}^2S$ photomagnetic dipole transitions. The finite D(n,γ)^{3}H cross section at threshold has been interpreted as arising from the mixed symmetry part of the ground state (<2% S') and from meson exchange currents whose operators project out the orthogonal parts of $^2S \rightarrow {}^2S$ and $^2S \rightarrow {}^4S$.[3] For electron induced processes such as ^{3}He(e,e')d,p the magnetic spin interaction

60

Hamiltonian is no longer $\sum\limits_{j=1}^{A} \mu_j \vec{\sigma}_j \cdot \vec{e} \times \vec{k} e^{i\vec{k}\cdot\vec{r}_j}$ but $\sum\limits_{j=1}^{A} \mu_j \vec{\sigma}_j \times \vec{q} e^{i\vec{q}\cdot\vec{r}_j}$

and the orthogonality may be broken for qr ~ 1.2

    Having made these remarks on photon and electron induced re-
actions in the trinucleon system, we report on an extension of our
measurements at the Naval Research Laboratory linear accelerator on
180° scattering of 56 MeV electrons from ^{3}He, ^{4}He and ^{1}H.[4] The figure displays electron scattering data from ^{3}He and ^{4}He up to 13 MeV
excitation in the ^{3}He system. The ^{4}He data, which is taken under
identical conditions, is the background spectrum since ^{4}He has no
known nuclear excitation below 20 MeV. The peak in the spectra at
56 MeV is due to the .00025" Havar windows covering each end of the
2.0" long gas chamber. The ^{3}He and ^{4}He difference between 49 and
54 MeV are electrons produced largely by magnetic bremsstrahlung
from the static magnetic moment of ^{3}He. The difference spectrum in
the region 41 to 48 MeV is the nuclear excitation of ^{3}He. Preliminary analysis of the data indicates three significant features:

    a) the threshold for ^{3}He(e,e')d,p occurs at 6.0 MeV not 5.5
MeV as expected from binding energy considerations alone,

    b) the 180° electrodisintegration cross section appears as a
superposition of two smooth excitation functions with the second one
appearing at about 9.6 MeV so that the 3-body breakup appears at ~ 2
MeV above its binding energy threshold, and

    c) the cross section for electrodisintegration is about
2.7×10^{-33} cm^2/sr-MeV at 9 MeV excitation.

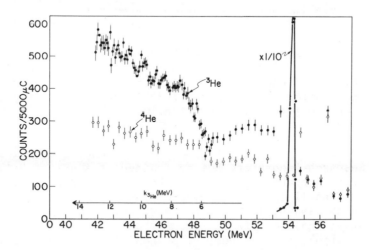

The electron scattering spectra are presented for ^{3}He and
^{4}He at 56.5 MeV, $\theta = 178.9°$ and a resolution $\Delta p/p = 0.5\%$.

The ^{3}He excitation spectrum is essentially pure M1 since it can be shown that the E1 contributions to the 180° electron scattering spectrum in this instance are negligible. Known measurements of photon and electron disintegration of ^{3}He have been used in making the E1 estimates.

The theoretical question in understanding the significance of the M1 continuum in ^{3}He is whether the cross section reported above results primarily from breaking the M1 orthogonality in the trinucleon system by direct nucleon spin currents or by meson exchange currents? A calculation of the former using a non-relativistic reduction of the electron-nucleon interaction[5] together with zero range wave functions is in progress.[6]

II. HELIUM 4

It was evident in our early measurements of 180° electron scattering from ^{3}He and ^{4}He, that ^{4}He was no longer a good background target at k>20 MeV. This was not unexpected since ^{4}He has a rich continuum of negative parity states for k > 21 MeV. Hydrogen spectra taken under the same conditions as the helium, i.e. 56.5 Mev and θ = 178.9°, have been used for the background. Except for the region near the proton's elastic scattering peak, the ^{4}He - ^{1}H is statistically zero up to 21 MeV in the ^{4}He system. The ^{4}He electroexcitation is observed as a smooth continuum rising from 21 MeV to a plateau between 24-27 MeV excitation. The cross section for 24 < k < 27 MeV is $d^2\sigma/d\Omega dk \simeq 3.5 \times 10^{-33}$ cm^2/sr-MeV.

These preliminary data provide a good fit to a calculation by C. Werntz of the ^{4}He continuum as expected from 180° inelastic electron scattering. The nuclear scattering states, in this case the T = 1,1$^-$ & 2$^-$ states, are described by R-matrix theory.[7] The 1$^-$ state contributes about 80% to the theoretical excitation. An independent estimate of the E1 electrodisintegration cross section from recent photodisintegration measurements of ^{4}He supports the predominant E1 nature of the observed cross section.

_____

* Work supported in part by the National Science Foundation
+ Naval Research Laboratory Edison Trainee at American University
1. G. A. Peterson and W. C. Barber, Phys. Rev. 128, 812 (1962).
2. J. S. O'Connell (private communication).
3. T. K. Radha and N. T. Meister, Phys. Rev. 136, B388 (1964); 138, AB7 (1965).
4. B. T. Chertok, E. C. Jones, W. L. Bendel and L. W. Fagg, Phys. Rev. Lett. 23, 34 (1969).
5. K. W. McVoy and L. Van Hove, Phys. Rev. 125, 1934 (1962).
6. J. S. O'Connell and B. F. Gibson (private communication).
7. L. Crone and C. Werntz, Nucl. Phys. (in press)

ANOMALOUS q-DEPENDANCE OF TRANSVERSE ELECTROMAGNETIC FORM FACTORS IN ^{13}C AND ^{9}Be

F. Beck, L. Grünbaum and M. Tomaselli

Institut für Theoretische Kernphysik

Technische Hochschule Darmstadt, Germany

Recent inelastic electron scattering work[1,2] has revealed large transverse cross sections for some odd parity transitions in p-shell nuclei at momentum transfers $q < 0.6$ fm^{-1}. Thus the otherwise in that q-range still established[3] low q-limit

$$B(E\lambda,q) \to (k^2/q^2) \, B(C\lambda,q) \qquad (1)$$

does not hold for those transitions. This indicates (a) a "non-collective" transition, and (b) the importance of transverse and/or magnetization current terms in the transverse electric matrix elements. We have used the particle-core coupling model to describe low lying positive parity states of ^{13}C and ^{9}Be, and to study the q-dependance of inelastic form factors.

In this model[4] the total Hamiltonian is approximated by: (I) A single particle operator describing the odd nucleon, restricted to the s-d-shell. (II) A core contribution representing the energies of the ground (0^+) and first excited (2^+) states of ^{12}C and ^{8}Be, resp. (III) A particle-core interaction which was taken as the effective two-body interaction, derived from the Yale potential[5].

The interaction matrix elements have been calculated by taking the core states of ^{12}C and ^{8}Be in the L-S-coupling limit and employing the usual fractional parentage expansion. This same L-S-limit has also been used to describe the $1/2^-$ and $3/2^-$ ground states of ^{13}C and ^{9}Be, resp.

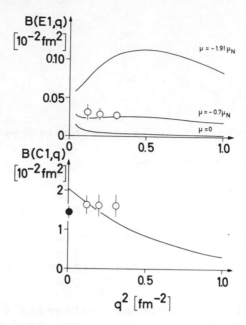

Fig. 1: Longitudinal and transverse transition rates for the dipole transition to the first excited state in ^{13}C. Results are taken from ref. [2].

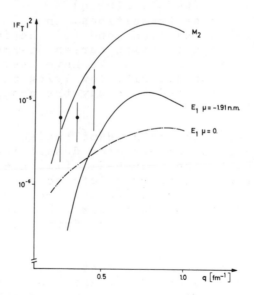

Fig. 2: Square of transverse form factors for the E1 and M2 transitions to the 1/2$^+$ state at 1.66 MeV in ^{9}Be. Experimental points from [1].

Table 1:   Comparison of experimental and theoretical results for the first three positive parity states in ^{13}C. (a) see ref.[6]; (b) see ref.[7].

Energy exp.	MeV calc.	Spin, parity	B.E. calculated Transitions to G.S.	B.E. experimental Transitions to G.S.	
3.08	3.08	$\frac{1}{2}^+$	E_1: 2.05 x 10^{-2} Fm^2	1.44 x 10^{-2} Fm^2	(a)
3.85	3.82	$\frac{5}{2}^+$	M_2: 4.7 x 10^{-2} Fm^4	4.9 x 10^{-2} Fm^4	(b)
			E_3: 1.8 x 10^{-1} Fm^6	109 Fm^6	(b)
6.86	6.44	$\frac{5}{2}^+$	M_2: 0.038 Fm^4		
			E_3: 0.13 Fm^6		

Table 2:   The same as Table 1 for the second and third positive parity states in ^{9}Be. Experimental values are from ref.[1].

Energy exp.	MeV calc.	Spin,Parity	q^2fm^{-2}	$\lvert F_T(E_1)\rvert^2$	$\lvert F_T(M_2)\rvert^2$	$\lvert F_T(E_3)\rvert^2$	Sum x10^{-5}	Exp. x10^{-5}	$\lvert F_L(C_1)\rvert^2$ x10^{-5}	Exp. x10^{-5}
3.04	3.02	$\frac{5}{2}^+$	0.057	6.14x10^{-7}	3.02x10^{-7}	1.41x10^{-10}	0.09	2.92±.87	1.48	23.4±10
			0.143	1.25x10^{-6}	1.01x10^{-6}	8.36x10^{-10}	0.22	4.45±.84	2.45	18.2±9.1
			0.208	2.20x10^{-6}	2.28x10^{-6}	3.46x10^{-9}	0.45	5.503±.56	3.19	14. ±7.
4.7	5.7	$\frac{3}{2}^+$	0.228	1.55x10^{-5}	3.91x10^{-6}	3.46x10^{-9}	1.94	0.7 ±0.7	1.87	35.1±7.

Fig. 1 shows the results for the transition to the $1/2^+$-state at 3.08 MeV in ^{13}C. The transverse transition rates are given for three different values of the intrinsic magnetic moment of the odd neutron. The curve $\mu=0$ corresponds very closely to the limit of equ. (1). Thus it is the magnetization current that leads to the large transverse contribution. The value $\mu=-0.7$ n.m. is about the same "effective" moment as needed for the ground state magnetic moment of ^{13}C in the L-S-coupling limit. Table 1 contains the radiative transition rates for the first three positive parity states. All results have been obtained with an effective charge 0.4 for the odd neutron and a harmonic oscillator parameter of 1.5 fm. No q-dependance has been measured for the other levels yet.

Fig. 2 and table 2 give the results for ^9Be calculated with an effective charge of 0.4 and oscillator parameters 1.76 fm. Nuclear form factors are quoted since all the transitions are mixed and can not be separated experimentally. The calculated longitudinal contributions are one order of magnitude below the experimental results. The disagreement of the transverse part for the $5/2^+$-level favours the suggestion[1] that this transition goes mainly to a not resolved negative parity state.

REFERENCES
1) H.G. Clerc, K.J. Wetzel and E. Spamer, Nucl. Phys. A 120 (1968) 441
2) M. Tomaselli et al., Phys. Lett., in print
3) cf. e.g. P. Strehl and Th.H. Schucan, Phys. Lett. 27 B (1968) 641
4) F.C. Barker, Nucl. Phys. 28 (1961) 96
5) C.M. Shakin et al., Phys. Rev. 161 (1967) 1006 and 1015
6) S.W. Robinson, C.P. Swann and V.K. Rasmussen, Phys. Lett. 26 B (1968) 298
7) F. Riess, P. Paul, J.-B. Thomas and S.S. Hanna, Phys. Rev. 176 (1968) 1140

SINGLE PARTICLE WAVEFUNCTIONS IN A NON-LOCAL POTENTIAL

L. R. B. Elton, S. J. Webb, and R. C. Barrett

Department of Physics, University of Surrey

Guildford, Surrey

The analysis of elastic electron scattering cross-sections and of shell separation energies, as obtained from the (p,2p) reaction, in terms of single-particle wavefunctions in lowest order in a potential well was at first (1) carried out in terms of an energy independent local potential. When this work was first reported, at the Paris Conference in 1964, Brueckner (2) pointed out that "there are impeccable theoretical reasons for believing that the potential must depend on the state of the particle". For that reason, and because it soon became apparent that the shell separation energies could in fact not be fitted by a state-independent potential, subsequent analyses (3,4) used an energy dependent local potential. Other work (5,6), which has used an energy independent local potential and been content to fit the separation energy of the least bound proton, cannot fit all the data.

The "impeccable theoretical reasons" come, of course, from the relation of the potential to the Hartree-Fock self-consistent potential (4). This is known to be non-local, and for that reason we have now analyzed the data in terms of an energy independent non-local potential, first used by Perey and Buck (7) for an analysis of low energy neutron scattering,

$$V(\underset{\sim}{r},\underset{\sim}{r}') = \pi^{-3/2} \beta^{-3} U\left(\left|\frac{\underset{\sim}{r}+\underset{\sim}{r}'}{2}\right|\right) \exp\left| -\left(\frac{\underset{\sim}{r}-\underset{\sim}{r}'}{\beta}\right)^2\right|$$

where

$$U(s) = V_o f_o(s) + V_{so}\left(\frac{\hbar}{m_\pi c}\right)^2 \frac{1}{s}\frac{d f_{so}}{ds} \underset{\sim}{\ell}\cdot\underset{\sim}{\sigma} ,$$

$$f_{o,so}(s) = \left[1 + \exp \frac{s - r_{o,so}A^{1/3}}{a}\right]^{-1}.$$

We have used this potential to fit individual nuclei from C^{12} to Pb^{208} and have obtained excellent fits with the parameters shown in Table 1.   These are close to those used by Perey and Buck, and it is most encouraging to see how little they vary over the whole nuclear range.   A typical fit to electron scattering is shown in Figure 1. The potential also yields the experimental separation energies, and in particular gives a value of only 47 MeV for the 1s-shell energy in Ca^{40}, in agreement with the recent (p,2p) results of James (8).

Table 1

Non-local potential parameters and r.m.s. radii for charge distributions

	V_0 (MeV)	r_0 (fm)	a (fm)	V_{SO} (MeV)	r_{SO} (fm)	β (fm)	$\langle r^2 \rangle^{1/2}$ (fm)
C^{12}	72	1.20	0.35	10.0	1.0	0.9	2.52
O^{16}	72	1.25	0.36	10.0	1.0	0.9	2.67
P^{31}	78	1.20	0.70	10.5	0.8	0.9	3.29
S^{32}	78	1.20	0.70	10.5	0.8	0.9	3.34
Ca^{40}	78	1.18	0.65	10.5	0.8	0.9	3.45
Ca^{42}	78	1.183	0.68	10.5	0.8	0.9	3.48
Ca^{44}	78	1.18	0.65	10.5	0.8	0.9	3.45
Ca^{48}	78	1.187	0.45	10.0	0.8	0.9	3.42
Ni^{58}	78	1.18	0.80	10.5	0.8	0.9	3.89
Ni^{64}	78	1.18	0.85	10.5	0.8	0.9	3.95
Pb^{208}	83	1.185	0.70	10.0	1.0	0.9	5.41

Table 2

Potential parameters and shell occupation numbers for Ca^{40} and Ca^{48} with the inclusion of 2p-1f admixture

	Ca^{40}	Ca^{48}
V_0(MeV)	72	78
r_0 (fm)	1.22	1.18
a (fm)	0.45	0.40
V_{SO} (MeV)	10.5	10.5
r_{SO} (fm)	0.8	0.8
β (fm)	0.9	0.9
2s (occ. nr.)	1.7	1.7
$1d_{5/2}$ (occ.nr.)	4.8	4.8
$1f_{7/2}$ (occ.nr.)	0.5	0.5
$2p_{3/2}$ (occ.nr.)	1.0	1.0

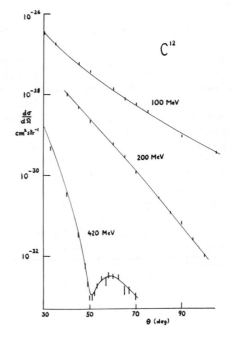

Fig. 1.   Elastic electron scattering from C^{12} at 100, 200 and 420 MeV.

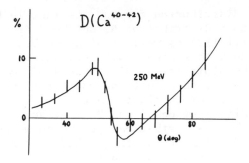

Fig. 2.   Difference in
electron scattering for
$Ca^{40,42}$ at 250 MeV.

$$D(Ca^{40-42}) = \frac{100(\sigma_{40} - \sigma_{42})}{(\sigma_{40} + \sigma_{42})}$$

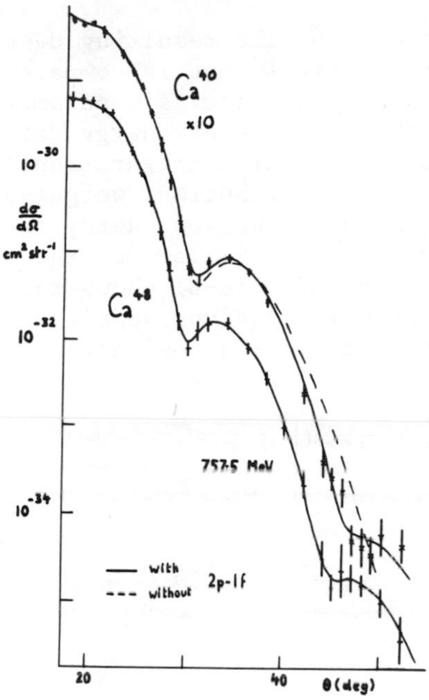

Fig. 3.   Elastic electron scattering
from Ca^{40} and Ca^{48} at 757.5 MeV.

We next turn to a more detailed study of electron scattering by
the $Ca^{40,42,44,48}$ and $Ni^{58,64}$ isotopes.   For each set of isotopes it
is possible to take essentially one set of parameters, as is shown in
Table 1, except that the surface parameter a is unusually small for
Ca^{48}.   A typical fit is shown in Figure 2.   The fact that the half-
way radii of the potential are proportional to $A^{1/3}$ does not, however,
have any fundamental significance.   In particular, a calculation of
the r.m.s. radii of the charge distributions of isotopes in terms of
a potential in which the half-way radius is assumed to be strictly
proportional to $A^{1/3}$ does not constitute an "explanation" of the so-
called anomalous increase in the r.m.s. radius.   The correct explanation, due to de Shalit (9), lies in the increase in the binding
energy of the least bound protons for heavier isotopes.

For Ca^{40} and Ca^{48} there are also electron scattering results at
757.5 MeV, which it proved impossible to fit with the simple model
used so far.   However, it is well known (10,11) that the ground-
state of Ca^{40} has an admixture of about one particle in the 2p-1f
shell, due to particle-hole interaction.   Using this fact, we have
been able to obtain excellent agreement (see Figure 3) with parameters
as given in Table 2.   It will be noticed that these parameters differ
considerably from those used for wavefunctions in lowest configuration.
In particular, there is now no longer a strict $A^{1/3}$-dependence for the
half-way radius of the potential.   (It should also be noted that a
deeper well has to be used for the 2p level, which was otherwise

unbound.)   The resulting density distribution for Ca^{40} is labelled
A in Figure 4, and is compared with a distribution B, obtained from
a phenomenological six-parameter fit (12) as well as with the best
fit C to the lower energy data made with a single particle distribution in lowest configuration.   The plots of the differences of the
density distribution, weighted with r^2, show that the modification
required to the low energy fit C in order to fit the high energy
data is very similar for the two distributions A and B.   Although
the plot of r^2(A-B) shows that the two distributions are by no means
identical, it gives hope that the details of the fit really give significant information on the structure of the Ca^{40} nucleus.

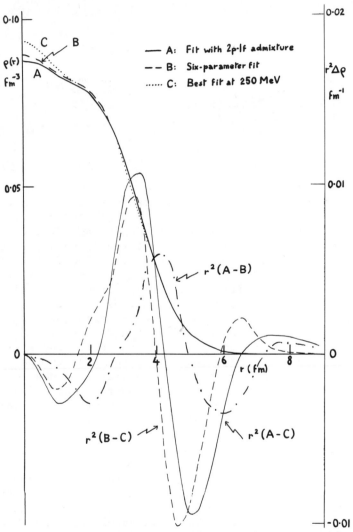

Fig. 4.   Charge density distributions for Ca^{40}.

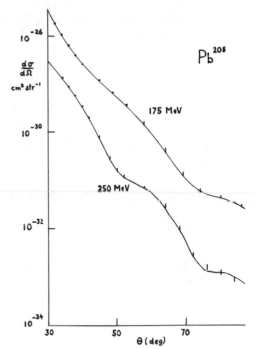

Fig. 5.   Elastic electron
   scattering from Pb^{208}
   at 175 and 250 MeV.

The fit for Pb^{208}, which is shown in Figure 5, is of some interest.   With the energy dependent local potential, the r.m.s. radius
had come to 5.37 fm, as against 5.51 fm for the fit to muonic X-rays
(13).   The so-called Perey effect (14) for non-local potentials
increases the electron scattering value to 5.41 ± 0.02 fm, while the
latest measurement of muonic X-rays leads to a value of 5.498 ± 0.003
fm (15).   Figure 6 shows that the density distribution, as based on
the single particle wavefunctions has a definite maxumum at the
centre, (17) as opposed to the best fit for a parametrized winebottle
distribution, which has a minimum (16).   This minimum is difficult
to reconcile with any density distribution derived from a Hartree-
Fock approach.   On the other hand, the best value obtained from low

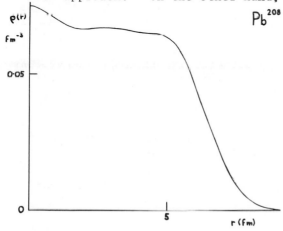

Fig. 6.   Charge density
   distribution for Pb^{208}.

energy electron scattering (17) is 5.48 ± 0.07 fm, which is in better
agreement with the muonic X-ray value than the high energy electron
value obtained by us.

We have now reached the stage of sophistication in theoretical
techniques where we ought to be able to reconcile results from different experiments.   It is no longer satisfactory to use a model
for high energy scattering which gives an r.m.s. radius that conflicts with the muonic X-ray value.   In the case of purely phenomenological distributions this conflict has been removed by using a
winebottle distribution.   We have not quite reached this stage in
the single particle description, although we feel we are coming
close.   The r.m.s. radius of Pb^{208} will certainly be raised when we
take into account two-body correlations.   The fact that it is necessary to raise it further is brought home forcefully by doing an
X-ray calculation with our single particle distribution: the 2p-1s
transitions are too high by 80 keV.   However, a recent Hartree-Fock
calculation by Vautherin and Veneroni gave an r.m.s. radius of
5.50 F and fitted the electron scattering moderately well (seeing
there were no adjustable parameters).

The difficulty of doing these single particle fits is such that
we shall obviously see many more analyses in terms of phenomenological
shapes.   We feel that these should be chosen differently so as to
resemble single particle and Hartree-Fock distributions more closely.
In particular, it is very unsatisfying to us to see a winebottle distribution used for Pb^{208} when we are convinced that there is a hump
in the middle.   We would be interested to see phenomenological fits
using different formulae, e.g.

$$(1 + w(1 - \frac{r}{c})^2) \times \text{Fermi} \quad .$$

To conclude, we would like to draw attention to the fact that
single particle wavefunctions and density distributions which fit
elastic electron scattering have been used successfully in the analyses of many nuclear reactions.   Recent work includes an analysis of
the (p,2p) reaction at 156 MeV (18), pick-up at 155 MeV (19), the
(p,n) reaction (20), low energy proton and neutron scattering (21),
high energy nucleon and pion elastic (22) and inelastic (23) scattering, alpha-scattering (24), and the effect on Pauli correlations
of the correct choice of wavefunction (25).   It would appear that
now that wavefunctions which fit the fairly stringent conditions of
elastic electron scattering are available, at least for spherical
and near spherical nuclei, these should be used in analyses of
nuclear reactions.   To the extent that, with the exception of Ca^{40},
all fits were made in lowest configuration, even these wavefunctions
may be inadequate for reactions that are very sensitive to the tail
of the wavefunctions.

Our thanks are due to Dr. D. F. Jackson for a most helpful
discussion.

(1)   R. R. Shaw, A. Swift, and L. R. B. Elton, Proc. Phys.
      Soc. 86, 513 (1965).
(2)   K. A. Brueckner, Congrès Internationale de Physique
      Nucléaire, Paris 1964, Vol. I, p.462.
(3)   A. Swift, and L. R. B. Elton, Phys. Rev. Lett. 17,
      484 (1966).
(4)   L. R. B. Elton, and A. Swift, Nucl. Phys. A94, 52
      (1966).
(5)   F. G. Perey, and J. P. Schiffer, Phys. Rev. Lett.
      17, 324 (1966).
(6)   B. F. Gibson, and K. J. van Oostrum, Nucl. Phys. A90,
      159 (1967).
(7)   F. G. Perey, and B. Buck, Nucl. Phys. 32, 353 (1962).
(8)   A. N. James et al, preprint.
(9)   A. de Shalit, private communication, 1965.
(10) V. Gillet, and E. A. Sanderson, Nucl. Phys. A91,
      292 (1967).
(11) W. J. Gerace, and A. M. Green, Nucl. Phys. A93, 110
      (1967).
(12) J. B. Bellicard et al, Phys. Rev. Lett. 19, 527
      (1967).
(13) L. R. B. Elton, International Conference on Electro-
      magnetic Sizes of Nuclei, Ottawa 1967.
(14) F. G. Perey, Direct Reactions and Nuclear Reaction
      Mechanisms, Gordon and Breach, New York 1962,
      p.125.
(15) H. L. Anderson et al, preprint.
(16) J. Bellicard, and K. J. van Oostrum, Phys. Rev. Lett.
      19, 242 (1967).
(17) G. J. C. van Niftrik, and R. Engfer, Phys. Lett. 22,
      940 (1966).
(18) D. F. Jackson, and B. K. Jain, Phys. Lett. 27B, 147
      (1968).
(19) I. S. Towner, Nucl. Phys. A126, 97 (1969).
(20) C. J. Batty, E. Friedman, and G. W. Greenlees, Nucl.
      Phys. A129, 368 (1969).
(21) G. W. Greenlees, G. J. Pyle, and Y. C. Tang, Phys.
      Rev. 171, 1115 (1968).
(22) V. K. Kembhavi, and D. F. Jackson, contribution to
      this conference.
(23) D. F. Jackson, Nuovo Cim., to be published, and
      contribution to this conference.
(24) C. G. Morgan, and D. F. Jackson, preprint.
(25) D. F. Jackson, and S. Murugesu, private communi-
      cation.

SHORT-RANGE CORRELATIONS AND HIGH-ENERGY ELECTRON SCATTERING

C. Ciofi degli Atti and N. M. Kabachnik*

Physics Laboratory, Instituto Superiore di Sanità

Rome, Italy

In recent calculations[1] it has been shown that the high momentum part of the nuclear form factors is very sensitive to the effects of the short-range dynamical correlations. Actually, the very recent experimental data[2] on elastic electron scattering by ^6Li and ^{16}O can be explained (using harmonic oscillator wave functions) only by taking into account the effect of these correlations. However, Donnelly and Walker[3] showed that the form factors are also very sensitive to which nuclear potential is used to generate the single particle wave functions. In this connection we have calculated the charge form factor of ^{16}O describing this nucleus firstly with a Slater determinant composed of single particle orbitals generated in a Wood Saxon well, $V(r)=V_O/\{1+\exp[(r-R)/a]\}$, with spin-orbit and Coulomb terms included, and secondly with the correlated Jastrow-type density of Ref.(1). In the first model, the C.M. motion effect has been carefully taken into account by performing the Gartenhaus-Schwartz transformation, i.e. using

$$F(q) = \frac{1}{Z}\langle\Psi| \sum_{k=1}^{A} \mathcal{E}_k\, e^{i\vec{q}(\vec{r}_k - \frac{1}{A}\sum_{\ell=1}^{A}\vec{r}_\ell)} |\Psi\rangle \qquad (1)$$

The two-body correlations have been taken into account using the Iwamoto-Yamada cluster expansion, i.e.

$$F(q)=\frac{1}{Z}\left\{\sum_\alpha\langle\alpha|e^{i\vec{q}\vec{r}_1}|\alpha\rangle-\sum_{\alpha\beta}[\langle\alpha\beta|e^{i\vec{q}\vec{r}_1}f(r_{12})|\alpha\beta-\beta\alpha\rangle - \langle\alpha|e^{i\vec{q}\vec{r}_1}|\alpha\rangle\langle\alpha\beta|f(r_{12})|\alpha\beta-\beta\alpha\rangle]\right\} \qquad (2)$$

---

* On leave from Institute of Nuclear Physics of Moscow State University USSR

where α and β are harmonic oscillator occupied states and $f=\exp(-r^2/b^2)$.

The results of calculations using the correlated model are shown in Fig.1 where the full line corresponds to the uncorrelated shell model with $a=\hbar/\sqrt{M\hbar\omega} = 1.77$ fm and the dashed and dot-dashed lines correspond to the correlated model with $a = 1.7$ fm, $b = 0.75$ fm and $a = 1.64$ fm, $b = 0.89$ fm, respectively; the values of the healing distance (as defined in Ref.1) are $r_h=1.3\sim1.5$ fm and $r_h = 1.5\sim1.8$ fm, respectively.

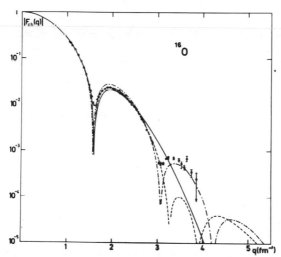

Fig. 1.   The ^{16}O charge form factor.   Full line: Harmonic oscillator without correlations;   dashed and dot-dashed lines: harmonic oscillator with Jastrow correlations.   The values of the healing distance are (in fm) 1.3~1.5 for the dashed line and 1.5~1.8 for the dot-dashed line.   The experimental points represent the new 375 and 500 MeV data[2].

The predictions by the Wood-Saxon well are shown in Fig.2 where the full line corresponds to the best fitting parameters $V_0=51.7$ MeV, $R= 3.27$ fm, $a=0.7$ fm. It can be seen that the high momentum part cannot be satisfactorily reproduced. It should be mentioned, moreover, that the best fitting parameters place the 1s level at $E_{1s} = 27.2$ MeV, while the experimental value of the separation energy given by the (p,2p) reactions is ~40-44 MeV. Using a Wood-Saxon well which reasonably gives the nuclear sizes and the values of the separation energies[4] the dashed line of Fig.2 is obtained.

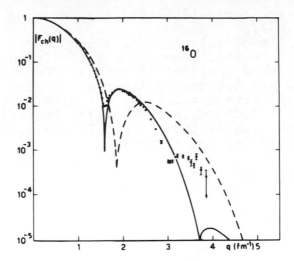

Fig. 2.   The ^{16}O charge form factor calculated with Wood-Saxon wave functions.

It seems therefore difficult to explain the charge form factor of ^{16}O (as well as of ^{4}He and ^{6}Li) simply by using the Wood - Saxon well instead of the harmonic oscillator one. Probably, other types of central wells could reproduce the experimental data. However we preferred to look at possible deviations from the independent - particle shell model and to this end we considered the effect of the short-range dynamical correlations. It can be seen from Fig.1 that this effect is very large and, moreover, in agreement with the experimental data.

References

1) C. Ciofi degli Atti, Phys. Rev. 175, 1256 (1968); Nucl. Phys. A129, 350 (1969)
2) J. G. McCarthy, I. Sick and R. R. Whitney, private communications.
3) T. W. Donnelly and G. E. Walker, Phys. Rev. Lett., to be published
4) D. H. Wilkinson and M. E. Mafethe, Nucl. Phys, 85, 97 (1966)

DIPOLE CHARGE-TRANSFER EXCITATIONS IN NUCLEI

B. COULARD*   and   S. FALLIEROS†

Université Laval   Brown University

In this paper, we study some properties of nuclear states that can be excited from the ground state of a nucleus via an interaction operator of the form[1]:

$$V_\lambda = \sum_{i=1}^{A} z_i t_\lambda (i) \tag{1}$$

In this expression, z_i is the third component of the relative position operator of the ith nucleon and t_λ ($\lambda=0$, ± 1) are the spherical components of the isospin operator. The $\lambda=0$ component describes electric dipole absorption, $\lambda=-1$ is related to μ^- capture (or n,p) reactions and $\lambda= 1$ corresponds to neutrino induced (or p,n) reactions. The total strengths associated with these processes can be written in the form

$$< 0| V_{-1}^{+} V_{-1} |0> = |d|^2 \lfloor 1-\alpha_1 + \alpha_2 \rfloor \tag{2a}$$

$$<0| V_{o}^{+} V_{o} |0> = |d|^2 \tag{2b}$$

$$<0| V_{+1}^{+} V_{+1} |0> = |d|^2 [1 + \alpha_1 + \alpha_2] \tag{2c}$$

α_1 represents the isovector component in $V_\lambda^{+} V_\lambda$; it is clearly absent for $\lambda=0$ and has opposite signs in the other two cases. The difference due to the isotensor component of $V_\lambda^{+} V_\lambda$ between $\lambda=0$ and $|\lambda|=1$ is given by α_2. In the following, we give estimates of these parameters[2].

An alternative way of writing eqs (2a, b and c) follows from the observation that V_{-1} acting on the ground state $|0>$ of a nucleus with isospin $T = N-Z/2$ leads uniquely to states with isospin T+1. V_o leads to states with isospin T+1 and T, and V_{+1} can excite states with T+1, T or T-1. The strengths associated with each isospin are defined by the equations:

$$<0|V_{-1}^+ V_{-1}|0> = |b_{T+1}|^2 \tag{3a}$$

$$<0|V_o^+ V_o|0> = \frac{1}{T+1}|b_{T+1}|^2 + \frac{T}{T+1}|b_T|^2 \tag{3b}$$

$$<0|V_{+1}^+ V_{+1}|0> = \frac{1}{(T+1)(2T+1)}|b_{T+1}|^2 \frac{1}{T+1}|b_T|^2 + \frac{2T-1}{2T+1}|b_{T-1}|^2 \tag{3c}$$

The intensities $|b|^2$ are essentially reduced matrix elements and can be expressed in terms of $|d|^2$, α_1 and α_2. In order to estimate α_1, we made use of the commutator,

$$[V_{+1}V_{-1}] = -\sum_{i=1}^{A} z_i^2 t_3(i) \tag{4}$$

From eq. (2a,c) we immediately obtain

$$2|d|^2 \alpha_1 = \frac{T}{3}\int d^3 \vec{r} \, r^2 \rho(r) \tag{5}$$

where $\rho(r)$ is the isovector density defined by

$$\rho(r) = \frac{1}{Z-N} <0|\sum_i \delta(\vec{r}-\vec{r}_i)\tau_3(i)|0> . \tag{6}$$

For the evaluation of α_2, it is useful to consider the analog of the state $|0>$ defined by

$$|a> = (\sqrt{2T})^{-1} T_+|0> . \tag{7}$$

A straightforward calculation gives

$$2|d|^2\alpha_2 = T[<a|V_o^+V_o|a> - <0|V_o^+V_o|0>] \tag{8}$$

We can obtain an explicit expression for α_2 if we make use of a fimiliar result which follows immediately from a simple shell model picture (with harmonic oscillator wave functions) for $|0>$ and $|a>$. The result is:

$$<0|V_o^+V_o|0> \equiv |d|^2 = \frac{NZ}{2\gamma A} \tag{9a}$$

and

$$<a|V_o^+V_o|a> = \frac{(N-1)(Z+1)}{2\gamma A} \tag{9b}$$

where γ is the radial parameter related to the oscillator frequency by $\hbar\gamma = M\omega$. From eqs (8), (9a) et (9b) we then obtain

$$\alpha_2 = \frac{T(T-\frac{1}{2})}{NZ} \qquad (10)$$

which clearly vanishes for $T=0$ or $\frac{1}{2}$.

For the evaluation of α_1 we need again the result of eq. (9a) and an explicit expression for the isovector mean-square radius. The simplified expression

$$\gamma \int r^2 \rho(r) d^3\vec{r} = \frac{9}{8} A^{1/3}$$

$$(3)$$

was used in previous work. From this, we then obtain

$$\alpha_1 = T \frac{A}{NZ} \frac{3}{8} A^{1/3} \approx T\frac{3}{2} A^{-2/3} . \qquad (11)$$

It is clear from eq. (10) and (11) that α_2 is smaller than α_1 by at least an order of magnitude. If we neglect this term altogether, we arrive at the following simplified expressions for isospin intensities:

$$\left| b_{T+1} \right|^2 = \left| d \right|^2 (1-\alpha_1) \qquad (12a)$$

$$\left| b_T \right|^2 = \left| d \right|^2 (1+\frac{1}{T}\alpha_1) \qquad (12b)$$

$$\left| b_{T-1} \right|^2 = \left| d \right|^2 (1+ \frac{T+1}{T} \alpha_1) \qquad (12c)$$

----

*Supported in part by the National Research Council
                                    of Canada
†Supported in part by the US Atomic Energy Commission.

(1) A more general discussion including higher order multipoles, spin dependence and retardation effect will be published elsewhere.

(2) Discussions related to the present work are mentioned in reference 7 of B. Goulard, T. Hughes and S. Fallieros, Phys. Rev. 176-4, 1345 (1968).

(3) B. Goulard and S. Fallieros, Can. J. Phys. 45, 3221 (1967).

ELECTRON SCATTERING FROM LIGHT DEFORMED ORIENTED NUCLEI

J. Langworthy and H. Überall*

Naval Research Laboratory, Washington, D. C.

20390

Elastic and inelastic electron scattering from several orientations of completely aligned ^{10}B, represented by harmonic oscillator and Tassie model fits of unoriented experimental data, is calculated and compared with scattering from the unoriented target. Results show large effects due to orientation.

The Coulomb part of the first Born approximation cross section for electron scattering from an oriented nucleus[1] requires, besides kinematical quantities, specification of multipole form factors and orientation parameters. The multipole form factors have been redefined by applying the multipole expansion directly to the densities.[2] Multipole order is restricted to 0 and 2 since inclusion of the quadrupole is sufficient to exhibit major orientation effects. Choice of a low charge nucleus gives two advantages over calculations in recent heavy nucleus work:[3] electron wave distortion can be neglected as shown in an accurate phase shift calculation;[4] furthermore, elastic scattering may be separated from inelastic scattering experimentally. For elastic scattering the forms of charge and quadrupole densities used, have been suggested by shell model calculations using harmonic oscillator wave functions. The charge radius was obtained from a fit of 1 p shell data[5] and the quadrupole moment from a

---

*Permanent Address: Department of Physics, The Catholic University of America, Washington, D. C.   20017

fit of unoriented ^{10}B data.[4] The transforms of these
multipole moment densities then give the forms of the
multipole form factors. Their specification is completed by inserting the two fit parameters, a_0 and Q. Complete alignment is assumed, thus specifying orientation
parameters. The oriented cross section was calculated
for two beam energies, 50 and 200 MeV, and for four
orientation directions, one longitudinal to the beam (L)
and three transverse, one in the scattering plane (T1),
one perpendicular (T3) and the other between these two
(T2). Comparison is made with the randomly oriented
cross section. Since both cross sections contained the
Mott cross section as a factor, it is divided out and
the results are obtained in terms of the charge form
factor. Maximum orientation effect amounts to 12% for
the 50 MeV beam and 160% for the 200 MeV beam. The
latter is shown in Fig. 1 where the label R denotes
random orientation. The dip in the longitudinal result
is in about the right place for a diffraction dip and
it does drop to zero if spherical symmetry is imposed by
removing the quadrupole moment. Other results explicitly exhibit dependence on the sign of the quadrupole
moment.

The same calculation was made for inelastic
scattering from the 6.02 MeV (4^+) level. (The ground
state is 3^+.) Form factors for this level have been

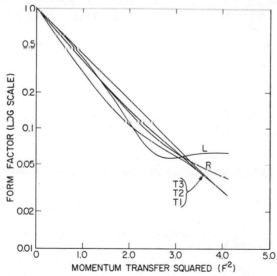

Fig. 1. Elastic electron scattering from four orientations of ^{10}B compared to random orientation. (See text
for orientation labels)

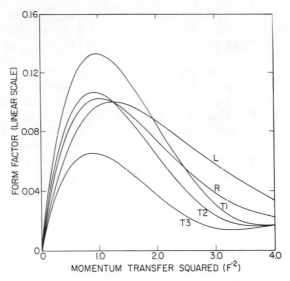

Fig. 2.   Inelastic scattering from 6.02 MeV (4^+) level.

developed from the ground state form factors by applying the Tassie model, and a fit of these has been made[2]
to unoriented scattering data[6] to obtain the vibration
amplitude. The results show scattering effects up to
80% for the 50 MeV beam and over 150% for a 200 MeV
beam. The latter is shown in Fig. 2. Since the data
in the inelastic fit showed error flags around 20-30%,
rising to 50% at the highest momentum transfer (q^2 =
3.61 F^{-2}), a considerable margin may be conceded towards
a more realistic alignment, without challenging the
detectability of the orientation effect.

References

1. L.H. Weigert and M.E. Rose, Nucl. Phys. 51 (1964)
    529; T. deForest, Jr., and D. Walecka, Adv. Phys.
    15 (1966) 1.
2. H. Überall and P. Uginčius, Phys. Rev. 178 (1969)
    1565.
3. R. D. Safrata, J. S. McCarthy, W. S. Little, M. R.
    Yearian and R. Hofstadter, Phys. Rev. Letters 18
    (1967) 667
4. T. Stovall, J. Goldenberg and D. B. Isabelle, Nucl.
    Phys. 86 (1966) 225
5. R. Hofstadter, Ann. Rev. Nucl. Sci. 7 (1957) 231
6. G. Fricke, G. R. Bishop and D. B. Isabelle, Nucl.
    Phys. 67 (1965) 187.

EFFECTS OF LONG RANGE CORRELATIONS ON ELASTIC ELECTRON SCATTERING

D. Vinciguerra, D. Blum, and T. Stovall

Ecole Normale Supérieure

Laboratoire de l'Accélérateur Linéaire 91 - Orsay, France

Because one finds experimental deviations from the shell model calculations of the elastic electron scattering cross sections from nuclei at high values of the momentum transfer ($q^2 \sim 3$ fm^{-2}), one has been prone to interpret these deviations as being caused by the short range correlations between nucleons. Here we would like to stress the fact that effects, just as large in the same range of q, are caused by long range correlations. The particular long range correlations we shall examine are those that are manifests by inter-orbital mixing in the particle-hole model.

In order to emphasize this point, we have chosen to examine the elastic electron scattering of ^{40}Ca of Croissiaux et al.[1], using the particle-hole model with harmonic oscillator basis states. The procedure is to calculate elastic scattering cross section using a phase shift analysis of the spherical part of the nuclear charge density $\rho(\vec{r})$:

$$\rho(\vec{r}) = \int \rho_{PN}(\vec{r}') \, \rho_{proton}(|\vec{r} - \vec{r}'|) \, d\vec{r}'$$

where ρ_{proton} is the charge density of the proton and ρ_{PN} is is the point nucleon nuclear charge density, defined in terms of the nuclear wave function $\psi(\vec{r}, \ldots \vec{r}_A)$ by

$$\rho_{PN} = \sum_{J=1}^{A} \varepsilon_J \int \delta(\vec{r} - \vec{r}_J) \, |\psi(\vec{r}, \ldots \vec{r}_A)|^2 \, d\vec{r}, \ldots d\vec{r}_A$$

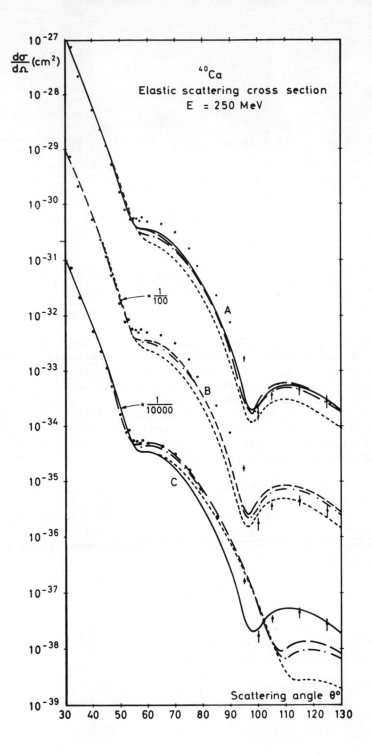

where ε_J = 1 unity for proton and zero for neutrons. In these calculations ρ_{proton} is assumed to have a gaussian form.

An approximate center of mass correction was made by multiplying $\rho(\vec{r})$ by $(A/\pi^3 a_o)^{-\frac{1}{4}} \exp(A R^2/2 a_o^2)$, which is rigorously correct if the center of mass wave function is in an s-state. a_o, A and R are the harmonic oscillator radial parameter, the atomic number and the center of mass coordinate respectively.

We have tried wave functions of the form

$$\psi(^{40}Ca) = a(0p - 0h) + b(2p - 2h)$$

Some typical results are shown in the figure. The solid lines in the sets of curves A and C are identical and represent the shell model calculations. Other lines in the sets of curves A, B and C, show the effects of a 50 % admixtures of two holes states, with the holes in the 1d, 1p and 2s shells respectively. In each set of curves, the broken line, dashed line and dot-dashed line correspond to putting the missing particles in the 1f, 2p and 1g orbital respectively. So the dashed curve in set B, represents the cross section calculated with a wave function that is half no particle no holes state and half two particle two holes state with the holes in the 1p orbital and the particles in the 2p orbital.

Note that the deviations from the shell model are large enough to account for the discrepancies with the data. Also, note the importance of holes in the 2-s shell for achieving a fit in the neighborhood of the first minimum. Holes in the s shells are equivalent to reducing the charge density at the center of the nucleus. This is reminiscent of the work of Gibson and Von Oostrum[2] who found that starting with Saxon-Woods orbitals, there had to truncate the central hump of the charge distribution in order to get a fit.

A particular case of a ^{40}Ca wave function is that reported by Cerace and Green [3] :

$$\psi_{GG}(^{40}Ca) = .904(0p - 0h) + .414(2p-2h) + .111(4p-4h)$$

This wave function yield results that are barely distinguishable from the solid lines curves of the figure.

In closing, we would like to say that in ^{4}He, where the deviations from the shell model are much more pronounced, one may be able to distinguish short range from long range effects.

REFERENCES

(1) M.Croissiaux et al. Physical Review 137 (1965) B 865.
(2) B.F.Gibson & K.J. Von Oostrum, Nuclear Physics A90(1967) 159.
(3) W.J.Gerace & A.M. Green, Nuclear Physics A93 (1967) 110.

MESON EXCHANGE EFFECTS ON ELECTRON SCATTERING FORM FACTORS IN ^{12}C

Marc CHEMTOB and Albert LUMBROSO

Service de Physique Théorique - CEN.Saclay

BP. n° 2 - 91 - GIF-sur-YVETTE - France

The role of meson exchange currents in electromagnetic (and weak) interactions of nuclei has been repeatedly inquired into. Interest was generally on static moments, reactions with photons or sum rules. In the past years, however, excitation of nuclear discrete levels by electron scattering has developed into a powerful tool for the investigation of nuclear structure. Now that measurements are being improved towards better resolution and higher momentum transfers, there is an urgent need to get a feeling of this role here. This is the aim of the work being reported, which also consists in a systematic study of form factors of levels in ^{12}C at excitation energy above 14 MeV .

Our theoretical model for the interaction currents consists in the non-covariant perturbation approach for extracting non-relativistic "equivalent" Hamiltonian operators from the field-theoretic S-matrix. Among the mechanisms that could produce two-body currents were retained those associated to the one pion exchange (OPE) diagrams where the current is induced either by the exchanged pion or by the excitation of a nucleon-antinucleon pair. The corresponding operators, denoted respectively by $\vec{j}^{\pi}(\vec{\xi})$ and $\vec{j}^{\pi N}(\vec{\xi})$, were evaluated in renormalized perturbation theory with point couplings[1]. The interaction current coming from nucleon recoil was not considered, since it is reduced with respect to the others by a factor equal to the ratio of pion to nucleon masses. An attempt was also made to include the vertex (or rescattering) correction to the radiative vertices, by using the expression of the pion photoproduction amplitude that obtains in the static Chew-Low theory (in the magnetic dipole approximation)[1]. This correction brought, however, minor changes on the results and was therefore dropped. Among other still smaller effects that were disregarded we may mention those due to wave

86

function normalization and those due to heavy-meson exchange diagrams. Estimates made for the static magnetic moments suggest these
effects to be small and strongly reduced by the short-range correlations.

For the study of electron scattering form factors of discrete
levels it is necessary to project out the transverse multipoles of
the interaction currents[2]. In the case of the electric transverse
multipoles, approximate verification of the continuity equation (to
which we come back below) justified the neglect of the part proportional to the divergence of the interaction currents. The further
decomposition of the multipole operators in irreducible tensors requires a large number of algebraic manipulations, but does not present, in principle, an insurmountable task. What presents, however,
serious problems is the numerical evaluation of the basic two-body
matrix elements. To make their calculation manageable an approximation was devised which can be given physical support. It consisted
in an expansion of the two-body current density at point ξ around
the centre-of-mass of the pair of nucleons $\vec{R}$; or, equivalently, in
an expansion in powers of $\vec{q}.(\vec{\xi} - R)$, where $\vec{q}$ is the momentum
transfer. Terms of second order in the expansion were dropped. Physically, this amounts to neglect the sharing of electric charge between the two nucleons by attributing it entirely to their centre-
of-mass. An important aspect of this approximation concerns the
condition of charge conservation referred to above. Thus, although
the currents $\vec{j}^{\pi}(\vec{\xi})$ and $\vec{j}^{\pi N}(\vec{\xi})$ do not exhaust all the OPE diagrams
that may appear, their sum still satisfies within this limit the
continuity equation :

$$\vec{\nabla}_{\vec{\xi}} \cdot \left(\vec{j}^{\pi}(\vec{\xi}) + \vec{j}^{\pi N}(\vec{\xi}) \right) = - i \left[V, Q(\vec{\xi}) \right] \quad ,$$

where Q stands for the one-body charge density and V for the OPEP.
Since, on the other hand, in each case that was examined an important cancellation occurred between these two currents, one is temp-
ted to conclude that charge conservation conspires to inhibit the
interaction effect : a result reminiscent of the Siegert theorem.

The model sketched above was then applied to the case of ^{12}C
where the situation is favorable since Coulomb and dispersive effects are expected to be small. Since interaction currents for the
OPE process have only an isovector component, application was made
to the isospin one levels, strongly excited by electron scattering.
Apart from the isolated 1^+ level at 15.1 MeV and the region of the
giant resonance above 21 MeV, the electron excitation spectrum shows
essentially two peaks at 16 and 19 MeV. The description of the corresponding levels in the shell model is both reliable and simple :
in fact, many of these levels were assigned pure one particle-one
hole configurations in j-j coupling. When mixing of configurations
was important, the particle-hole wave functions of Gillet-Vinh Mau[3]
were used. In discussing the numerical results we followed the lead

of Ref. 4 , where a compilation of previous experimental results is also done.

Our results confirm that the exchange corrections to the form factors are generally small, causing invariably 10 to 40% <u>increase</u> of the impulse form factor squared for momentum transfers $q \leq 200$ MeV/c. The increase is, however, more substantial for higher q (the region up to 700 MeV/c was studied) where it also depends more sensitively on the range parameter of the potential well (harmonic oscillator, with $b = 1.6$ fm, in our calculation) and on the wave functions. The approximation above probably overestimates the interaction effect there. It is remarkable that the behaviour of the interaction form factor with q follows closely that of the impulse form factor. Positions of maxima and minima are, in fact, slightly displaced towards higher q. For the spin-flip 1^+ level, the form factor is raised near the photon point and at the second maximum as required by observation. Nice agreement with data is achieved in this case but with the introduction of a normalization constant as suggested by the intermediate coupling model. The situation for the other complexes of levels presents also several interesting features. With the exception of the high momentum transfer points, there is no however, at this moment, any definite experimental evidence for an increase of the form factors there. Longitudinal form factors had also to be included in order to compare with the data at scattering angle $\theta = 135^\circ$. For low q, their contribution was found to be substantial in all the complexes.

REFERENCES

[1] - M. CHEMTOB, Nucl. Phys. <u>A213</u>, 449 (1969)

[2] - T. DEFOREST Jr. and J.D. WALECKA, Adv. in Phys. <u>15</u>, 1 (1966)

[3] - V. GILLET and N. VINH MAU, Nucl. Phys. <u>54</u>, 321 (1964)

[4] - T.W. DONNELLY, J.D. WALECKA, I. SICK and E.B. HUGHES, Phys. Rev. Letters <u>21</u>, 1196 (1968)

GIANT AND PYGMY RESONANCES IN A = 3, 5, 11, 13, 15 AND 17 NUCLEI*

D. J. Albert, R.F. Wagner, H. Überall,[+] and C. Werntz[++]

The Catholic University of America, Washington, D.C. 20017

We have carried out a two particle-one hole shell model calculation for the giant and pygmy resonance region of ^{5}He, ^{13}C and ^{17}O, and a corresponding two hole-one particle calculation for ^{3}He, ^{11}B and ^{15}N, and have also calculated the electron scattering form factors for the resulting states.

The residual interaction for this work was taken as the smooth, separable potential due to Tabakin.[1] The appreciable second-order contributions of this force were calculated using oscillator states as the intermediate set. It is now known that this method gives the same order of corrections as using plane waves as the intermediate set. Spurious shell model states were constructed and eliminated by the Schmidt method. The eigenstates were finally found by diagonalizing the energy matrix in the non-spurious representation.

The electron scattering form factors for the resulting states were calculated and photo dipole strengths found from these by considering the case $q = E$ for $TE1(q)$.

---

*Supported by National Science Foundation and Office of Naval Research

[+]Also at Naval Research Laboratory

[++]Also at Naval Ordnance Laboratory

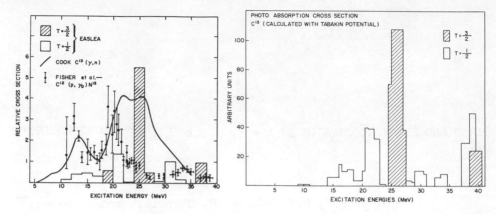

Dipole strength distribution for ^{13}C: Fig. 1 (left): with Soper force (Easlea); Fig. 2 (right): with Tabakin force (this work).

Figs. 1 & 2 show respectively the dipole strength distribution obtained by Easlea[2] for ^{13}C using a zero range force, and our results with the "realistic" force of Tabakin. They agree in identifying the lower two peaks in Cook's[3] ^{13}C photo cross section as T = 1/2 so-called "pygmy" states and the upper peak as a T = 3/2 "giant" resonance. These assignments were confirmed by the ^{12}C(p, γ)^{13}N experiment of Fisher et al.[4] in which only T = 1/2 states are excited. However, whereas Easlea found it necessary to strengthen the J^{π}, T = 2$^+$, O particle-hole interaction to reproduce the lower pygmy, the result came out of our calculation without any adjustment.

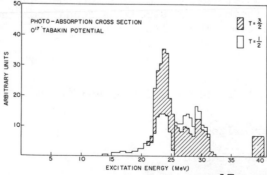

Fig. 3. Dipole Strength distribution for ^{17}O with Tabakin Potential

In Fig. 3 we show our results for the distribution of the dipole states in ^{17}O, for which no data seem to exist. In this case the pygmy T = 1/2 resonance is not separated from the T = 3/2 "giant". A photo-capture reaction such as protons or neutrons on ^{16}O can be used to distinguish the T = 1/2 strength.

In Figs. 4 & 5 we give examples of our form factor results for the case of ^{13}C. For the transverse electric form factor we find that the low momentum transfer region is dominated by a state at 25.4 Mev which is an isospin or Goldhaber-Teller oscillation. At higher momentum transfers the 30.2 Mev state predominates. The strength comes mainly from the spin-flip part of the dipole operator, that is, in the generalized Goldhaber-Teller scheme[2] it is a spin-isospin oscillation.

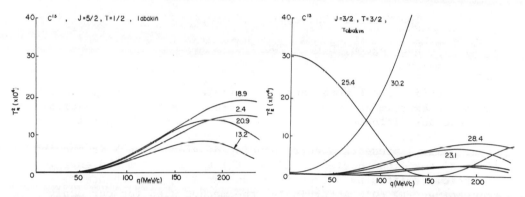

Electron scattering form factors for ^{13}C.
Fig. 4 (left): T_M^2 vs. q; Fig. 5 (right): T_E^2 vs. q.

Similar results have been obtained for the other nuclei listed above.

REFERENCES

1   F. Tabakin, Ann. Phys. 30, 51 (1964)
2   B. R. Easlea, Phys. Letters 1, 163 (1962)
3   B. C. Cook, Phys. Rev. 106, 300 (1957)
4   P. S. Fisher, D. F. Measday, F. A. Nikolaev, A. Kalmykov and A. B. Clegg, Nuclear Physics 45, 113 (1963)
5   H. Überall, Nuovo Cimento 41B, 25 (1966)

PARTICLE AND PHOTON DECAY OF NUCLEI

FOLLOWING ELECTROEXCITATION

Dieter Drechsel[*] and H. Überall[*+]

The Catholic University of America

Washington, D.C.   20017

We derive the general angular correlation formulas for a co-incidence experiment in which an electron, scattered inelastically by a nucleus, is detected simultaneously with a heavy particle or a photon emitted by the nucleus subsequently to its electroexcitation. Fig. 1 presents a corresponding diagram where N is the initial, N' the final nucleus, and (p, E_p) the momentum and energy of a particle emitted by the excited nucleus N^* from a resonance state of energy ω. Accordingly, the method used for our analysis is that of a two-step resonance model in which the nucleus is first excited by the passing electron (with single-photon exchange), and subsequently decays in a manner independent of the

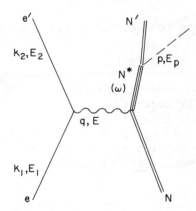

Diagram for coincidence experiment in two-step resonance model.

[*]Supported by a grant of the National Science Foundation
[+]Also at U.S. Naval Research Laboratory, Washington, D.C.   20390

mode of excitation. The case of overlapping nuclear levels is included, and the emitted particles may have spin 0 (e.g., α), $\frac{1}{2}$ (e.g., nucleon), or 1 (e.g., photon).

The general form of the coincidence cross section was shown by de Forest[1] to be given by

$$\frac{d^3\sigma}{d\Omega\, d\Omega_p dE_2} = \frac{\alpha^2}{\Delta^4}\frac{k_2}{k_1}\, p\, E_p \left\{ V_C(\theta)W_C + V_T(\theta)W_T \right.$$

$$\left. + V_I(\theta,\phi_p)W_I + V_S(\theta,\phi_p)W_S \right\} \qquad (1)$$

with four known kinematic functions $V_C \ldots V_S$; here $\Delta^2 = q^2 - \omega^2$, $\vec{q} = \vec{k}_1 - \vec{k}_2$, $\vec{k}_{1,2}$ = initial, final electron momenta (energies $E_{1,2}$); θ = scattering angle, (θ_p,ϕ_p) = emission angles referring to $\vec{q}$ as z axis. The four form factors $W_C \ldots W_S$ are model dependent and have been evaluated by us, with the result

$$W_n = \frac{1}{\pi^2(2J_i+1)} \sum_{J_oJ_o'} \sum_{L\,L'} \frac{(-1)^L \sqrt{2J_o+1}\sqrt{2J_o'+1}\ \sqrt{2L+1}\ \sqrt{2L'+1}}{(E-\omega_{J_o} + \frac{i}{2}\Gamma_{J_o}^{tot})\,(E-\omega_{J_o'} + \frac{i}{2}\Gamma_{J_o}^{tot})}$$

$$\times \sum_I \frac{1}{2}\left[1 + (-1)^I \pi_o\pi_o'\right] W(IJ_o'L\, J_i\,; J_oL')\, S_{J_oJ_o'}^{IJ_f}\, Z_{LL'}^I(n) \qquad (2)$$

where π_o is the parity of the intermediate state J_o; the excitation functions

$$Z_{LL'}^I(C) = (L0,\, L'0\ I0)\, M_L^*\, M_{L'}\, P_I(\cos\theta_p) \qquad (3)$$

etc. are given in terms of standard[2] multipole Coulomb (M_L) and transverse ($T_L^{e,m}$) nuclear matrix elements. The decay properties are contained in our expressions (e.g., for emitted nucleons)

$$S_{J_oJ_o'}^{IJ_f} = (4\pi)^{-1}(-1)^{J_f-J_o-\frac{1}{2}} \sum_{jlj'l'} \sqrt{2j+1}\sqrt{2j'+1}\ (j\,\tfrac{1}{2},\, j'-\tfrac{1}{2}\mid I0)$$

$$\times\, W(J_oJ_o'jj';\, IJ_f)\, \langle J_f\| jl\| J_o\rangle^*\, \langle J_f\| j'l'\| J_o'\rangle, \qquad (4)$$

in which the reduced nuclear matrix elements[3] may be treated as phenomenological parameters. Fig. 2 shows nucleon emissions in the giant resonance region of ^{12}C as an example (θ_p',ϕ_p' refer to $\vec{k}_1$ as z axis).

Advantages of coincidence experiments over an observation of only the scattered electrons are: (a) the radiative tail is eliminated ; (b) interference terms determine relative phases of matrix elements; (c) spins and parities of resonance levels may be determined in a model-independent way; (d) electron scattering may excite levels that cannot be reached in photoexcitation. For these reasons, a performance of coincidence experiments should be

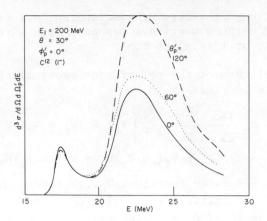

Coincidence cross section as a function of excitation energy E for overlapping giant resonance states in ^{12}C and subsequent decay by proton emission

considered at the new high-duty cycle ($\geq$ some %) electron linacs (Saclay, MIT, Los Alamos).

[1]T. de Forest, Ann. Phys. (N.Y.) <u>45</u>, 365 (1967)

[2]T. de Forest and J.D. Walecka, Adv. Phys. <u>15</u>, 1 (1966)

[3]S. Devons and L.J.B. Goldfarb, in Encyclopedia of Physics, edited by S. Flügge (Springer-Verlag, Berlin 1957)

INELASTIC PHOTOPION PRODUCTION WITH EXCITATION OF GIANT RESONANCE ANALOGS: APPLICATION TO ^{16}O

F. J. Kelly, L. J. McDonald, H. Überall

U. S. Naval Research Laboratory, and The Catholic

University of America*, Washington, D. C.

Using a generalized Goldhaber-Teller model of the giant multi-pole resonances in ^{16}O, we have evaluated the differential cross section for the photoproduction of positive pions near threshold leading to giant resonance spin-flip states of various multipolarities[1] in ^{16}N, and have obtained the corresponding pion spectra assuming monochromatic incident photons. A possible subsequent neutron emission was considered also. We find that at forward angles of the emitted pion the spectra are dominated by the dipole resonances, but that at larger angles the quadrupole resonances appear strongly, making it possible to experimentally identify the two types of resonances. Owing to the variability of its momentum transfer, inelastic nuclear pion photoproduction is potentially as powerful for a study of analog excitations, as inelastic electron scattering has become for $\Delta T_3 = 0$ nuclear excitations.

In the capture reactions, the momentum transfer is fixed, so that the various resonances get excited with a fixed strength only. Moreover in radiative pion capture, the matrix element is dominated by the amplitude[2]

$$\mathcal{F} = \sqrt{2} \, \frac{e}{4\pi} \, \frac{f}{m_\pi} \, i \, \vec{\sigma} \cdot \vec{\varepsilon} \, \tau_-$$

(1)

($e^2 = 4\pi/137$, $f^2/4\pi = 0.08$, $\vec{\varepsilon}$ = photon polarization vector), and thus leads mainly to spin-isospin resonances only. For the inverse of radiative pion capture, namely pion photoproduction,

$$\gamma + N(A, Z) \rightarrow \pi^{\pm} + N^*(A, Z \mp 1),$$

(2)

* Work supported in part by a grant from the National Science Foundation

95

the momentum transfer q may be varied by varied by varying the
angle at which the pion is observed. Near threshold, the amplitude
of (2) is also given by Eq. (1) and hence known. The variability
of q may render pion photoproduction a tool as superior or the
study of analog giant resonances compared[3] to muon or radiative pion
capture, as inelastic electron scattering[3] has become compared to
photoabsorption in the study of the ΔT_3 = O giant resonances.
Resonance peaks may be observed in the pion spectra if the incident
photons are monochromatic. Such photons are now becoming available
in the energy range 150[4]450 MeV from positron annihilation in flight
at the new Saclay linac[4], or they may be produced in crystal
bremsstrahlung,[5] preferably using the DeWire-Mozley technique.[6] If
no such photons are available, one may still observe peaks in the
neutron or proton emission subsequent to reaction (2), e.g.

$$N^{*}_{g.res.} (A, Z-1) \rightarrow N^{(*)}(A-1, Z-1) + n.\tag{3}$$

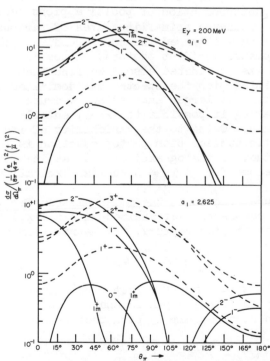

Fig. 1. Photopion angular distribution from ^{16}O for 200 MeV photon
energy, with excitation of giant resonance states; a_1 = O (top);
a_2 = 2.625F (bottom).

We have calculated the differential cross section of reaction (2) with π^+ mesons assuming monochromatic photons, as applied to ^{16}O on the basis of the generalized Goldhaber-Teller model, following our previous procedure[1,7]. Fig. 1 shows the pion angular distribution for excitation of the individual states both for a radial cutoff $a_1 = 0$ (volume production) and $a_1 = 2.625F$ (surface production). The most significant change occurs in the 1^+ monopole state. One notices that for $\Theta_\pi \gtrsim 60°$, the 2^+ and 3^+ quadrupole states exceed the 1^- and 2^- dipole states, so that varying $\Theta_\pi = \sphericalangle(\vec{p}_\pi, \vec{k})$ allows to differentiate between the giant multipole resonances. We have also calculated the energy spectra of the created pions and of the neutrons which are emitted in reaction (3). We find that these spectra also change markedly as the laboratory angle of the observed pions is varied.

REFERENCES

1   R. Raphael, H. Überall, and C. Werntz, Phys. Rev. 152, 899 (1966). The isospin, $\Delta T_3 = 0$ giant quadrupole resonance was first mentioned by D. Drechsel, Nucl. Phys. 78, 465 (1966), and by L. Ligensa, W. Greiner, and M. Danos, Phys. Rev. Letters 16, 364 (1966).

2   D. K. Anderson and J. M. Eisenberg, Phys. Letters 22, 164 (1966); G. F. Chew, M. L. Goldberger, F. E. Low, and Y. Nambu, Phys. Rev. 106, 1345 (1957).

3   T. de Forest and T. D. Walecka, Adv. Phys. 15, 1 (1966).

4   G. Audit, N. deBotton, G. Le Poittevin, C. Schuhl and G. Tamas, preprint, CEN Saclay (unpublished).

5   H. Überall, Phys. Rev. 103, 1055 (1956), Z. Naturforsch. 17a, 332 (1962); G. Diambrini, Revs. Mod. Phys. 40, 611 (1968).

6   J. DeWire and F. R. Mozley, Nuovo Cimento 27, 1281 (1963).

7   J. D. Murphy, R. Raphael, H. Überall, R. F. Wagner, D. K. Anderson, and C. Werntz, Phys. Rev. Letters 19, 714 (1967).

THE PHOTOPRODUCTION OF LEPTON PAIRS FROM NUCLEI

J. S. Trefil

Department of Physics

University of Illinois

It has been pointed out[1,2] that in the photoproduction of electron pairs from hydrogen, a dramatic interference effect should be seen in the region where the invariant mass of the lepton pair is about equal to the mass of the omega mesons. This effect corresponds to interference between those processes which proceed via the photoproduction of a ρ meson which subsequently decays into the lepton pair, and those which proceed via the photoproduction of the ω.

Preliminary data of the photoproduction of electron-positron pairs from Carbon[3] shows no evidence for such an effect. Since the prediction of the interference effect depends only on the assumption of vector dominance, this is a serious problem. The purpose of this contribution is to show that the nuclear effects involved in this process do not change the predictions of the theory in any significant way, so that the absence of ρ-ω interference in nuclear production has the same significance as its absence in production from protons.

To show this, we make use of the multiple scattering theory for nuclear photoproduction which has been developed.[4] The amplitude to produce a lepton pair via either ρ or ω photoproduction is then given by[1,2]

$$
A_{e^+e^-} = m_\rho \Gamma_\rho \left[\frac{A_{\rho N}}{(m^2 - m_\rho^2)^2 + \Gamma_\rho^2 m_\rho^2} + \left(\frac{m_\omega}{m_\rho} \right)^4 \left(\frac{\gamma_\rho}{\gamma_\omega} \right)^2 e^{i\varphi} \right.
$$

$$
\left. \frac{A_{\omega N}}{(m^2 - m_\omega^2)^2 + \Gamma_\omega^2 m_\omega^2} \right] \cdot Q
$$

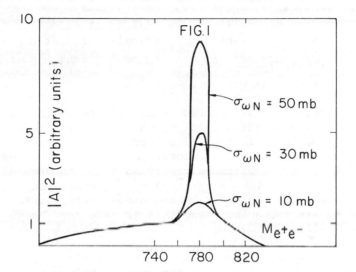

The expected spectrum for the reaction $\gamma + {}^{12}C \rightarrow e^+ e^- + X$. We have set $\varphi = 0$ and $\sigma_{\rho N} = 30$ mb, and neglected energy resolution effects.

142039

where $A_{\rho N}$ and $A_{\omega N}$ are the (suitably normalized) amplitudes for the production of vector mesons on nuclei, and Q describes the decay of the meson into lepton pairs. For production on hydrogen, $A_{\rho N}=A_{\omega N}=1$, and φ is an unknown phase angle which should be near zero.

The quantity $A_{\rho N}$ depends to a very good approximation, only on $\sigma_{\rho N}$, the total ρ-nucleon cross section, which essentially measures the mean free path of the ρ in nuclear matter. This quantity can be extracted from nuclear photoproduction experiments.[4] Similarly, $A_{\omega N}$ depends only on $\sigma_{\omega N}$, which is, in principle, measureable in the same way as $\sigma_{\rho N}$, but which is at present unknown. Therefore, Eq. (1) has one unknown parameter, which is $\sigma_{\omega N}$. As $\sigma_{\omega N}$ is varied, two competing processes occur. First, from vector dominance arguments, a small value of $\sigma_{\omega N}$ corresponds to a low cross section for production on an individual nucleon. On the other hand, once an ω is produced, a small value of $\sigma_{\omega N}$ will mean that it has a higher probability of emerging from the nucleus. Only explicit calculations will be able to determine which of these two effects will win out.

There are, in addition, two other effects which could wipe out the interference effect. If $A_{\rho N}$ had an appreciably different phase from $A_{\omega N}$, the effect would be reduced.[1] For a reasonable value of the phases of the elementary amplitudes, however, the phases of these two amplitudes are found to be about equal, and the phase difference very small. Also, the experimental resolution, if it were on the order of the omega width, would make it difficult to see the peak,[2] and this is true for the nuclear process as well.

In Fig. 1 we show some numerical calculations of the expected e^+e^- mass spectrum from Eq. (1) for typical values of $\sigma_{\omega N}$. We see that the absence of an interference peak near the ω mass can only be explained on the basis of an anomalously small value of $\sigma_{\omega N}$, and hence, from standard vector dominance arguments, a very small diffractive contribution to the photoproduction of the ω on hydrogen (the pion exchange contribution is suppressed, since carbon is a isotopic spin zero nucleus). This would be extremely difficult to understand on the basis of any symmetry model, so the presence or absence of the ρ-ω interference peak in lepton pair production from nuclei remains a good test of our ideas of the basic processes of the photoproduction of vector mesons.

REFERENCES

1) R. G. Parsons and Roy Weinstein, Phys. Rev. Letters 20, 1314 (1968).
2) M. Davier, Phys. Letters 27B, 27 (1968).
3) J. G. Asbury, W. K. Bertram, U. Becker, P. Joos, M. Rhode, A. J.S. Smith, S.C.C. Jordan, and S.C.C. Ting, Phys. Rev. Letters 19, 869 (1967).
4) J. S. Trefil, Phys. Rev. 180, 1379 (1969).

PHOTODISINTEGRATION OF HELIUM NUCLEI AT HIGH ENERGY

P.Picozza[o], C.Schaerf[o], R.Scrimaglio[o], G.Goggi[x]
A.Piazzoli[x], D.Scannicchio[x]
(o) Laboratori Nazionali del C.N.E.N. - Frascati,
(x) Istituto Nazionale di Fisica Nucleare - Pavia,
     Italy

In this paper we report some comparative results on the two-body photodisintegration of ^{3}He and ^{4}He nuclei at a c.m. energy comparable with the first pion-nucleon isobar. We have measured the differential cross section at 90° in the c.m.s. for the reactions :

(1) γ + ^{4}He $\rightarrow$ p + t
(2) γ + ^{3}He $\rightarrow$ p + d

in the gamma ray energy region between 200 and 500 MeV (fig.1).
The experimental apparatuses used for the two experiments were similar and have been described elsewhere[1].
They consisted of a liquid helium target and two telescopes of spark chambers and scintillation counters.
The quantities measured in these experiments were the angles and ranges of the two outgoing charged particles. The selection of events due to reactions (1) and (2) was made by means of a kinematical reconstruction with two constraints.

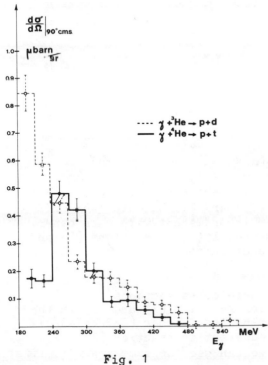

Fig. 1

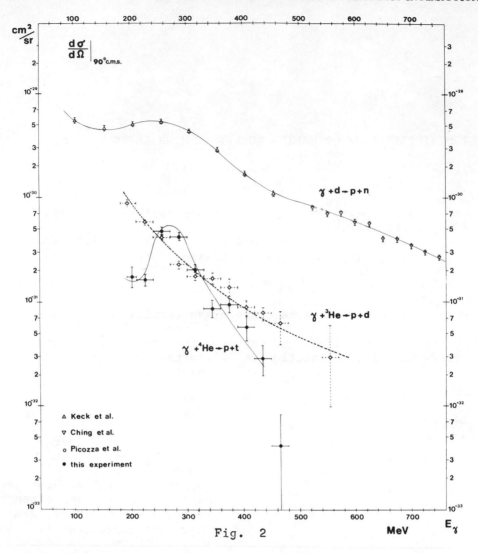

Fig. 2

A comparison of our experimental results with those of previous experiments are indicated in ref.1 and 2. The data from different laboratories do not appear inconsistent with each other. In fig.2 we have indicated our results for reactions (1) and (2) and some results for the reaction[3,4] :

(3) γ + d $\longrightarrow$ p + n

The most relevant feature of these cross sections is a pronounced resonant behaviour for reaction (1) and (3) in the energy region of the first pion-nucleon isobar $(J=3/2; I=3/2)$. No such resonant behaviour is apparent in reaction (2). To understand this phenomenon we can use the same argument first

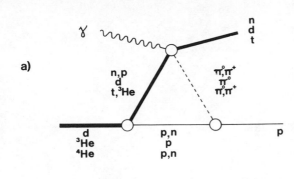

a)

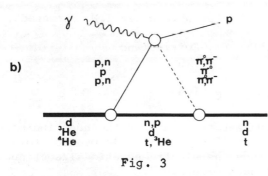

b)

Fig. 3

introduced by Wilson[5] to explain the deuteron data. In fig.3 we have indicated the relevant diagrams for a possible resonant contribution to these reactions. Diagram a) gives a large contribution to the photo- disintegration of deuteron. Instead it is not very strong in the other two cases because the coherent photoproduction of γ mesons in light nuclei at resonance (upper vertex) is known[6] to be depres- sed by multiple pion nuclear scattering. On the other side diagram b) is forbidden in the case of photodisintegra- tion of ^{3}He by isotopic spin conservation at the lower right vertex. This is obviously not true for the other two processes.

With these arguments a relevant resonant behaviour is expected between 250 and 350 MeV in deuteron and ^{4}He but not in ^{3}He photo- disintegration. Our experiments confirm this simple phenomenolo- gical prediction.

References
(1) P.Picozza,C.Schaerf,R.Scrimaglio,G.Goggi,A.Piazzoli,D. Scannicchio: Nuovo Cimento 55A,206(1968).
(2) Idem : submitted to Lettere al Nuovo Cimento.
(3) J.C.Keck and A.V.Tollestrup: Phys.Rev. 101,360(1956).
(4) R.Ching and C.Schaerf: Phys.Rev. 141,1320(1966).
(5) R.R.Wilson: Phys.Rev. 104,218(1956).
(6) M.Davier,D.Benaksas,D.Drickey and P.Lehman: Phys.Rev. 137, B119(1965).

MUONIC ATOMS AND NUCLEAR STRUCTURE

Roland Engfer

Institut für Technische Kernphysik der T.H. Darmstadt,

Germany, and CERN, Geneva, Switzerland

1. INTRODUCTION

A classical problem in nuclear structure is the question of the size of nuclei.   During the last 15 years muons in muonic atoms have been used as probes to study the electromagnetic size of nuclei. The following properties have been determined:

  i) $\rho(r)$ -- the size and shape of the charge distribution -- from K-, L- and M-X-rays yielding $<r^2>$ for light and medium heavy nuclei and two parameters for heavy nuclei;

 ii) $\Delta\rho(r)$ -- differences of charge distributions of isotopes (isotope shift) -- from K- and L- X-rays;

iii) $\delta\rho(r)$ -- differences of charge distributions of nuclear ground and excited states (isomer shift) -- from energy shifts of nuclear γ-rays in muonic atoms;

 iv) Q_0 -- the intrinsic quadrupole moment of deformed nuclei -- from the quadrupole hyperfine structure of K- and L-X-rays;

  v) $\mu(r)$ -- the distribution of the magnetic dipole moment of spherical nuclei -- from a broadening of $2p^{1/2}-1s^{1/2}$ transitions and from the hyperfine structure of nuclear γ-rays in muonic atoms.

Besides these nuclear parameters quantum electrodynamics (vacuum polarization) was tested by accurate energy measurements of muonic X-rays.   It is not possible in the time allotted to go into all details of measurements and analysis yielding this information. Therefore, I have selected several recent papers with a rather subjective viewpoint to explain experimental techniques and results from muonic atom studies. I recommend two recent reviews on muonic atoms by S. Devons and I. Duerdoth and by C.S. Wu and L. Wilets as very instructive references books and summaries [1].

2. RECENT EXPERIMENTAL IMPROVEMENTS

The principal experimental set-up, which was used in the first "classical" experiment by Fitch and Rainwater [2], has not changed very much. A telescope plastic scintillator system is used for the detection of the stopped muon; the NaI(Tl) crystal is replaced by high resolving Ge(Li)-detectors. To achieve an accurate calibration, γ-lines from radioactive sources are measured simultaneously with μ-X lines and stored in different parts of a multichannel analyser. Two different systems are used for this purpose:

i)   A coincidence of β-rays or a second γ-ray characterizes a calibration event, which needs a fairly high source intensity (e.g. Ref. [3]);

ii) All events including background are accepted if not vetoed by an incoming muon; this enables one to use weak radioactive sources but requires a low background. In addition, these events are gated with the time structure of the muon intensity providing an identical load for stored muonic and calibration events (e.g. Refs. [4-6]).

These calibration techniques have obtained an accuracy of $\sim$ 30 eV neglecting uncertainties of the calibration energies.

A NaI(Tl) counter system surrounding the Ge-detector was used effectively in coincidence as a pair spectrometer or in anticoincidence to reduce low-energy Compton background (Fig. 1) [6].

A serious problem in detecting μ-X-rays or nuclear γ-rays in the 100-200 keV region are X-rays from μ-carbon and μ-oxygen emitted from the scintillators and surrounding materials. To reduce this background a NaI crystal can be installed in coincidence with the Ge-detector accepting only γ events of more than $\sim$ 80 keV. Since a number of high-energy X-rays are emitted in heavy μ-atoms the coincidence efficiency is rather high ($\sim$ 50%). On the other hand the disturbing μ-K series of C and O are not in coincidence with high-energy (< 80 keV) X-rays; thus they are reduced by a factor of 120-140. An example of such a measurement is given in Fig. 2 [7]; a fluorine target is attached to the Ta-target to measure this reduction factor.

With the use of on-line computers the energy dependence of the Ge-detector timing was corrected on-line [6]. Spectra taken in

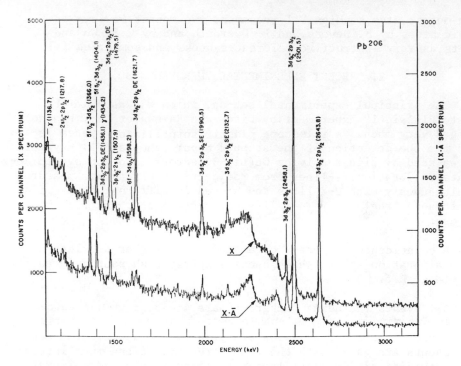

Fig. 1   Prompt μ spectrum in ^{206}Pb [6]. The X spectrum contains all
prompt events, the X • A spectrum the prompt events in anti-coincidence
with the NaI annulus.

different delay time intervals between μ-stop and γ-detection were
used to measure isotope effects in the capture rates of muons in
^{151}Eu and ^{153}Eu [8]. Typical time resolutions are 80 nsec at
100 keV up to 10 nsec at 6 MeV for a 17 ccm coaxial detector [6]
and 10 nsec at 100 keV up to 3 nsec above 300 keV for a 3.5 ccm
planar detector with leading edge timing [7].

3.   CORRECTIONS TO THE BINDING ENERGIES OF MUONIC ATOMS

Radiative correction, nuclear polarization and electron
screening influence the binding energy of a muon in muonic atoms.
For all three effects experimental evidence has been found or
valuable tests were performed recently. Talks on the vacuum
polarization will be given by Anderson and Picasso. Experimental
evidence of nuclear polarization [6,23] will be discussed in Sections 5.1 and 6.1. In the following, recent results on the electron screening [10] are given in some more detail.

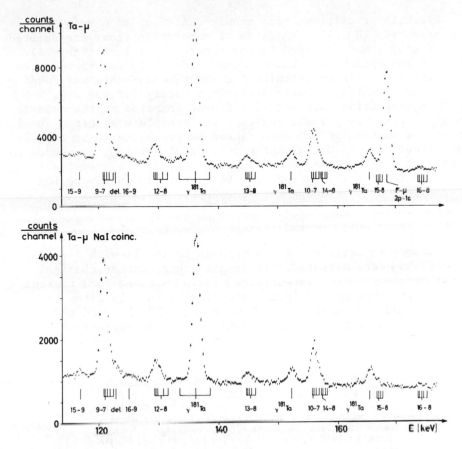

Fig. 2 Prompt μ spectrum in ^{181}Ta [7]. μ-X-rays and nuclear γ-rays are identified. Upper spectrum: Ta and fluorine target without and lower spectrum with NaI coincidence. The 2p-1s μ-X-line of ^{16}O (133.6 keV) coincides with a weak magnetic hf-component of a nuclear γ-transition in μ-^{181}Ta. In the lower spectrum the μ-O line does not appear, but the hf-component was not observed (< 1% of the main hf-component) beyond the statistical errors.

Up to now electron screening of muonic levels has been neglected in the interpretation of muonic X-ray data. This is justified for the lowest muonic states since the electronic charge gives rise only to a constant potential in the region of the lowest muonic orbits. The remaining effect of the non-constant potential term was calculated to be 4.6 eV for the 1s state and 190 eV for the 5g state in Bi [11].

Recently transitions from muonic states up to n = 16 have been measured [10]. The energies of some transitions are shifted by more than 1 keV compared to the unscreened values (Table 1). With the assumption of a Thomas-Fermi model [12] for the electron potential a good agreement with the experimental data was found (Table 1). Small systematic deviations mainly for the 6-5, 9-6 and 10-6 transitions are not significant compared to the experimental errors; they could indicate an incomplete electron shell during the cascade of the muon caused by preceding Auger transitions (Section 4). But higher experimental accuracy is needed to support this effect. An accurate knowledge of the energies is necessary to identify nuclear γ-rays in the low-energy region (50-200 keV) of muonic spectra (see Section 7).

4.   INTENSITIES OF MUONIC X-RAYS

Data on intensities of μ-X-rays up to levels with n = 16 [6,10,13-15] were measured. It is not clear whether chemical shifts of the K-series intensities [13,16] are due to different initial distributions or incomplete electron shells after the μ-capture and the first Auger transitions. For low and medium Z elements it was found that a modifying factor $\exp(a \cdot \ell)$ of the

Table 1

Energies of μ-X-rays in natural osmium [10]. E_{exp}: measured energies; up to 10 fine structure components are taken into account, the energy of the strongest component is given. E_{th}: calculated energies including vacuum polarization [Eq. (2.2) in Ref. 11] and finite size. $\Delta E = E_{exp} - E_{th}$. ΔW: calculated electron screening with Thomas-Fermi model [12].

Transition n-n' Δn	E_{exp} [keV]	E_{th} [keV]	ΔE [keV]	ΔW [keV]
9-8 1	53.48 ± 0.11	53.49	-	-
7-6 1	120.28 ± 0.11	120.410	-0.13 ± 0.11	-0.22
6-5 1	200.11 ± 0.11	200.186	-0.08 ± 0.11	-0.19
9-7 2	131.02 ± 0.11	131.477	-0.46 ± 0.11	-0.53
8-6 2	197.77 ± 0.13	198.222	-0.45 ± 0.13	-0.52
11-8 3	119.02 ± 0.15	119.889	-0.87 ± 0.15	-0.81
9-6 3	251.13 ± 0.11	251.595	-0.47 ± 0.11	-0.81
12-8 4	140.29 ± 0.12	141.360	-1.07 ± 0.12	-1.09
11-7 4	196.64 ± 0.16	197.765	-1.13 ± 0.16	-1.10
10-6 4	289.02 ± 0.15	289.788	-0.77 ± 0.15	-1.10
13-8 5	156.93 ± 0.15	158.070	-1.14 ± 0.15	-1.37
12-7 5	218.55 ± 0.13	219.567	-1.02 ± 0.13	-1.38

initial statistical distribution can be compensated by incomplete
Auger transitions (see Section 3) [17]. The absolute intensities of
higher transitions in heavy μ-atoms are in agreement with a statistical initial population at n = 14 [10,15]. A distribution over n
seems to be necessary to improve the agreement mainly for transitions from levels with n ≥ 12.

5. DETERMINATION OF NUCLEAR CHARGE RADII

5.1 Comparison of μ-Atom Data and Elastic Electron
Scattering Results

For a large number of elements, data on nuclear charge radii
are available and they are collected in Ref. [18]. One interesting
aspect of this information is the comparison with elastic electron
scattering data. Deviations could indicate the influence of dynamical effects (nuclear polarization, dispersion effects) or prove
the validity of radiation corrections. As explained elsewhere
[19,20] a model-independent comparison can be performed with electron scattering data measured at an "equivalent momentum transfer"
$q^2 \simeq 6Z/(R \cdot a_\mu) \simeq 0.1$ fm^{-2} ($R^2 = \frac{5}{3} <r^2>$, $a_\mu = 256$ fm) at which
elastic electron scattering cross-section and muonic atom data show
the same sensitivity to higher moments of the charge distribution.
A comparison of such low momentum data to muonic atom results is
given in Fig. 3 for nuclei with A = 20 to 30 [21]. An excellent
agreement is observed for both methods.

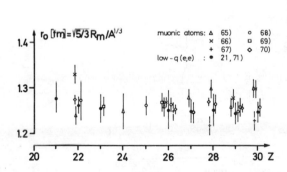

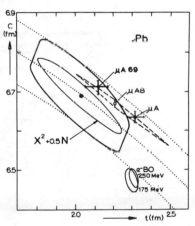

Fig. 3 Comparison of nuclear charge
radii obtained from (e,e) experiments at
∿ 60 MeV [21] and μ-atoms.

Fig. 4 Comparison of Fermi
charge distribution parameters
of natural lead from (e,e) experiments at 40-60 MeV [22] and
μ-atoms. μA: [72], μAB: [63],
μA69: [6], BO: (e,e) at 175 and
250 MeV [73]. The error el-
lipse is the result of the (e,e)
experiment [22], the egg-shaped
area includes systematic errors.

Results of higher Z are sensitive to two parameters of the charge distribution.   This is shown in Fig. 4 for natural Pb [22]. The recent result on μ-atoms of Anderson et al. [6,23] shows an interesting aspect in this comparison.   In the evaluation of the c,t parameters the $1s\frac{1}{2}$ state was omitted, but the $2s\frac{1}{2}$ state was taken into account.   The energy of the $1s\frac{1}{2}$ level calculated with these c,t values was found to be too high by 6.8 ± 2.3 keV which the authors interpret as being due to nuclear polarization.   If the 1s state is included in the fit, only c and t but not $<r^2>$ change. The better agreement of the new value with the low momentum trans- fer data of (e,e) scattering support but do not prove the polariza- tion effect.   Cole [57] and Chen [58] have calculated an increase of the binding of 5.7 keV and 6.0 ± 0.6 keV, respectively, for the 1s level in lead due to nuclear polarization, in excellent agree- ment with the experimental value.   It is unexplained whether the discrepancy of the high-energy (e,e) scattering data [73] is due to the model dependence or dispersion effects.

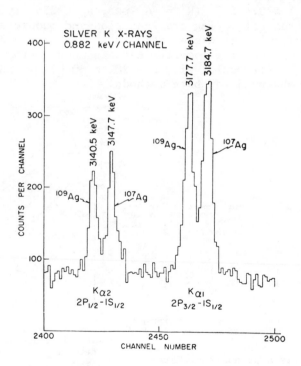

Fig. 5   2p-1s transition in muonic Ag [24].   Natural element with 51.35% [107]Ag and 48.65% [109]Ag.   The double escape peaks are shown but their full energies, E, are indicated.

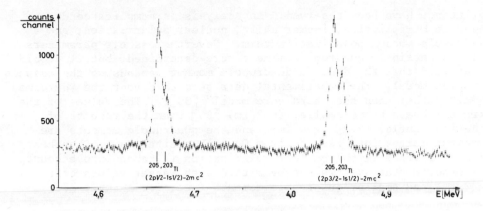

Fig. 6   2p-1s transition in muonic Tl [7].   Natural element with 29.5%
^{203}Tl and 70.5% ^{205}Tl.

5.2 Muonic Isotope Shift

With separated isotopes many data on the isotope shift are
measured with μ-atoms (c.f. [18]).   A detailed discussion is given
by Wu and Wilets [1].   Two recent spectra of natural elements demonstrate the improvement of the resolution of Ge-detectors.
Figure 5 shows the 2p-1s transition in natural Ag [24].   The shift
of the isotopes ^{107,109}Ag clearly is resolved and yields $\delta<r^2>1/2=$
$(2.14 \pm 0.10) \times 10^{-2}$ fm. The intensity ratio of the fine structure
components agree roughly with the expected value of ∿ 2 : 1.   The
same is shown in Fig. 6 for the Thallium isotopes ^{203,205}Tl [7].
The isotope shift was analysed to (9.2 ± 0.3)keV for the 2p1/2-
1s1/2 transition and (9.6 ± 0.2)keV for the 2p3/2-1s1/2 transition,
and yields $\delta<r^2>1/2 = (1.06 \pm 0.02) \times 10^{-2}$ fm at $\Delta t = 0$.   Here the
fine structure intensity ratios are 1.6 ± 0.1 and 1.03 ± 0.07 in
disagreement with the expected value 1.92.   The anomalies are due
to a strong excitation of nuclear levels (see Section 7.2).

6.   HYPERFINE STRUCTURE OF MUONIC X-RAYS

6.1   Deformed Nuclei - Quadrupole hf-Splitting

Muonic X-ray spectra of deformed nuclei have been measured
for a large number of nuclei (Z = 62 to 92) by several groups
[25-33].   These spectra have been analysed in terms of a Fermi-
type charge distribution modified by a deformation of the nucleus
[34].   As shown by Wilets [35] and Jacobson [36] the strong quadrupole interaction of a deformed nucleus with the muon mixes
nuclear and muonic states resulting in an excitation of rotational
levels.   Thus also in even-even nuclei strong quadrupole hyperfine

splittings have been observed. An analysis is complicated due to
magnetic hf-splitting, isomer shift, nuclear polarization, and a
quadrupole vacuum polarization term. Nevertheless c,t parameters
and the intrinsic quadrupole moment were determined, but it should
be noticed that the deduced quadrupole moment depends on the assumed
rotation model. The experimental data seem to favour the "deformed
model" rather than the "hard core model" [30]. The values of the
skin thickness t are smaller (< 2 fm) [30] than the results of
spherical nuclei (2.2 - 2.4 fm), and the quadrupole moments are
somewhat larger than results from Coulomb excitation. Chen has
shown [58] that if the nuclear polarization is taken into account,
the experimental spectra can be fitted with larger values of t
(> 2 fm) and smaller quadrupole moments (see, for example [79]).

Figure 7 shows an example for the 2p-1s transition in the
muonic Eu isotopes [32,33]. These nuclei are at the beginning of
the deformed region. ^{151}Eu has a small quadrupole moment and no
rotational band. The dynamic excitation of the first excited state
at 21.7 keV has to be included in the analysis to explain the
$2p\frac{3}{2}-1s\frac{1}{2}$ structure whereas the broadening of the $2p\frac{1}{2}-1s\frac{1}{2}$ transition is explained by the magnetic hf-splitting (Section 6.2)
(μ = 3.46 nm). ^{153}Eu has a large quadrupole moment Q_0 and because
of this a complicated hf-structure. $Q_0(^{153}$Eu) = (8.02 ± 0.12)b
was deduced from the data.

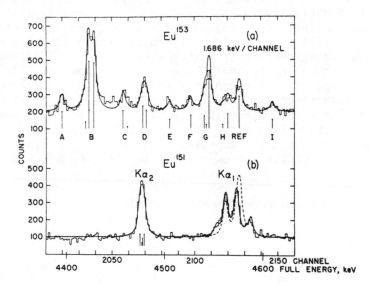

Fig. 7 The 2p-1s muonic transition in ^{151,153}Eu [32,33]. The solid
curves represent the best fit to the data. The dotted line in (b) is
the fit to ^{151}Eu without inclusion of excitation to the 21.7 keV state.
The vertical lines give the positions and intensities of the hf-components.

6.2   Spherical Nuclei --Magnetic Hyperfine-Splitting

Magnetic hyperfine structures have been observed in the $2p\frac{1}{2}$-$1s\frac{1}{2}$ transitions of muonic ^{141}Pr, ^{151}Eu and ^{209}Bi (see Table 2). Since the magnetic splitting is proportional to m^2 and the electric quadrupole splitting proportional to m^3 the magnetic splitting normally is covered by the quadrupole splitting and thus can be analysed separately only from the $2p\frac{1}{2}$-$1s\frac{1}{2}$ transition of spherical nuclei. On the other hand the energy resolution of the Ge-detectors at the corresponding high energies is comparable with or less than the splitting, therefore, only a broadening of the line can be observed (see Kα_2 of ^{151}Eu in Fig. 7).

The displacement of the magnetic substates is given by

$$\Delta W_F = A_1(n\ell j)\left[F(F+1) - I(I+1) - j(j+1)\right]/2Ij .$$

I is the nuclear spin, j the angular momentum of the muonic state and F the total angular momentum of the sublevel. Experimental values of the hyperfine constant A_1 are collected in Table 2. These values are strongly reduced ($\sim 40\%$) compared to the value expected for a point magnetic moment A_1^{point} because of the spatial distribution of the nuclear magnetization density. This finite size effect was calcultated by Bohr and Weisskopf [44] for electrons and for muons by Le Bellac [45]. Results of calculations based on the shell model with configuration mixing as presented in Table 2 are in fair agreement with the experimental results. Since in odd nuclei the magnetic moment is mainly produced by the unpaired nucleon (g_ℓ and g_s factors) the properties of this particle are tested.

Table 2

Magnetic hf-constants in μ-atoms. A_1 is given for the $1s\frac{1}{2}$ state deduced from $2p\frac{1}{2}$-$1s\frac{1}{2}$ transitions. For Tl the values were derived from hf-splittings of nuclear γ-rays.

Isotope	A_1^{exp}[keV]	Ref.	A_1^{theor}[keV]	Ref.	A_1^{point}[keV]
^{115}In	1.65 ± 0.07	80	1.56 a)	81	2.53
^{133}Cs	1.11 ± 0.18	80	1.21 a)	81	1.34
^{141}Pr	1.52 ± 0.06	37	1.47 ± 0.03 a)	37	2.41 ± 0.06
^{141}Pr	1.52 ± 0.15	80	1.47 a)	81	2.38
^{151}Eu	0.80 ± 0.27	32,33	1.04 b)	32	2.04
^{203}Tl	0.66 ± 0.04	5,38	0.72 a)	39,40	1.16
^{205}Tl	0.57 ± 0.01	5,38	0.67 a)	39,40	1.16
^{209}Bi	2.5 ± 0.5	41	2.0 a)	41,42	3.1
	2.1 ± 0.5 c)	41			
	2.1 ± 0.2	43			

a) shell model with configuration mixing;
b) $g_R = Z/A$;
c) an isomer shift of 3 keV is considered.

7.   STUDY OF NUCLEAR γ-RAYS IN MUONIC ATOMS

7.1 Muonic Isomer Shift in Deformed Nuclei

An excitation of rotational states by the cascading muon (Section 6.1) leads to a system with the nucleus in an excited state and the muon in its $1s\frac{1}{2}$ ground state. The lifetime of the $1s\frac{1}{2}$ muon (μ-capture) is $\sim 7 \times 10^{-8}$ sec for heavy nuclei, whereas the nuclear transitions are much faster, 10^{-9}-10^{-11} sec; therefore, the nuclear γ-ray is emitted in the presence of the $1s\frac{1}{2}$ muon. As explained by Devons [46] and Hüfner [47] the excitation energies are changed due to nuclear polarization and the magnetic hf-interaction.   In addition, the energies are shifted by the monopole interaction between the muon and the nucleus in its ground and excited state (isomer shift) [3,4]. These small effects $\lesssim 1$ keV influence the muonic X-rays too, but because of the high X-ray energies they are difficult to observe, whereas the low energy of nuclear γ-rays (~ 100 keV) renders the observation feasible.

For a large number of deformed nuclei, nuclear γ-rays in a μ-atom have been measured.   Energy shifts ΔE_{exp} compared to the corresponding unshifted γ-line from a radioactive source are shown in Table 3.   The isomer shift is given by

$$\Delta E_{is} = e \int |\psi_{1s_{\frac{1}{2}}}(r)|^2 \; \Delta V(r) \; d\tau \; ;$$

$\Delta V(r)$ is the difference in electrostatic potential of the excited and ground state corresponding to $\Delta\rho(r)$, the difference in the charge distribution.   Essentially ΔE_{is} is the same shift as measured in Mössbauer experiments. To deduce an isomer shift from the measured energy shift, the following effects, already mentioned above, have to be considered:

The nuclear polarization is in the order of several keV, as estimated by several authors [52-59] and experimentally supported by Anderson and co-workers [23]. For a nuclear transition, only the difference of the ground state and excited state is relevant; therefore, this is expected to be an order of magnitude less. For a rotational band it can be shown to be zero but it may contribute in cases of a "non-ideal" rotational band. Since no accurate calculations or experimental evidence exist at the moment this effect is neglected.

The magnetic hf-interaction splits the excited and ground state (for $I \geq \frac{1}{2}$) into a doublet (see Fig. 8). This splitting is strongly reduced by the Bohr-Weisskopf effect [44] (cf. Table 2). Only in ^{203,205}Tl the splitting was resolved [5,38]. For the deformed nuclei the magnetic distribution was assumed to be proportional to the charge distribution; this is not confirmed experimentally.

Table 3

Measured energy shifts ΔE_{exp} and isomer shifts ΔE_{is} of nuclear γ-rays in muonic atoms

Isotope	Transition E[keV] I-I'	ΔE_{exp} [eV]	ΔE_{is} [eV]	$\left(\dfrac{\Delta\langle r^2\rangle}{\langle r^2\rangle}\right)\times 10^4$ $\Delta t=0$	Ref.
^{150}Nd	130.2 2^+-0^+	+ 570 ± 120	+ 840 ± 120	5.8 ± 0.8	48
^{152}Sm	121.8 2^+-0^+	+ 560 ± 60	+ 920 ± 70	5.9 ± 0.4	48
		+ 500 ± 40	+ 770	4.8	49
^{154}Gd	123.1 2^+-0^+	+ 670 ± 150	+ 980 ± 150	5.9 ± 0.8	48
^{156}Gd	88.9 2^+-0^+	− 512 ± 200			a)
^{158}Gd	79.5 2^+-0^+	− 499 ± 150			a)
^{160}Gd	75.3 2^+-0^+	− 431 ± 150			a)
^{166}Er	80.6 2^+-0^+	− 350 + 150	− 30 ± 150	−0.16 ± 0.8	48
^{182}W	100.1 2^+-0^+	− 320 ± 100	− 30 ± 100	−0.13 ± 0.5	48
		− 290 ± 90	− 60	−0.2	49
^{184}W	111.2 2^+-0^+	− 340 ± 100	− 25 ± 100	−0.11 ± 0.5	48
		− 350 ± 50	− 100	−0.4	49
^{186}W	122.6 2^+-0^+	− 350 ± 100	− 10 ± 100	−0.04 ± 0.5	48
		− 400 ± 40	− 150	−0.6	49
^{188}Os	155.0 2^+-0^+	− 400 ± 40	− 190	−0.8	49
^{190}Os	186.7 2^+-0^+	− 470 ± 40	− 360	−1.4	49
^{192}Os	205.8 2^+-0^+	− 610 ± 50	− 514	−2.0	49
^{194}Pt	328.5 2^+-0^+	+ 262 ± 45	+ 321	+1.3	50
^{194}Pt	293.6 2^+-2^+	+ 3 ± 50	− 144	−0.6	50
^{196}Pt	355.7 2^+-0^+	+ 390 + 280			50
^{169}Tm	118.2 $9/2-1/2$	− 410 ± 65	− 457	−2.0	50
^{169}Tm	109.8 $5/2-3/2$	+ 57 ± 40	+ 68	+0.3	50
^{169}Tm	130.5 $7/2-3/2$	− 800 ± 700	− 480	−2.1	50
^{181}Ta	136.2 $9/2-7/2$	+ 134 ± 60	+ 110	+0.5	4
^{185}Re	125.3 $7/2-5/2$	+ 175 ± 35			50
^{187}Re	134.3 $7/2-5/2$	+ 174 ± 35			50
^{191}Ir	129.4 $5/2-3/2$	− 108 ± 45	+ 103	+0.4	50
^{193}Ir	139.3 $5/2-3/2$	− 266 ± 45	− 121	−0.5	50
^{203}Tl	279.1 $3/2-1/2$	−	+ 260 ± 180	+0.9	5,50
^{205}Tl	203.8 $3/2-1/2$	−	− 330 ± 150	−1.1	5,50
^{209}Bi	2741.4 $15/2-9/2^-$	+5600 ± 600	+6200		51
^{209}Bi	1132.5 $15/2-13/2$	+2700 ± 800	+3000		51
^{209}Bi	1608.9 $13/2-9/2^-$	+1900 ± 700	+2200		51
^{209}Bi	2563.9 $9/2^+-9/2^-$	+6600 ± 1100	+6600		51

a) preliminary result of the Darmstadt group at CERN.

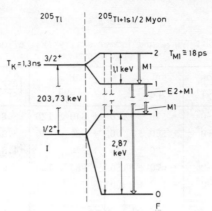

Fig. 8 Level scheme of ^{205}Tl without (left side) and with (right side)
a muon in its $1s^{1/2}$ state [5]. The magnetic hf-splitting is calculated
from shell model with configuration mixing [40].

The excitation mechanism does not populate the hf-levels
statistically [49,60]. In addition a fast M1-interdoublet
transition which is highly converted ($\alpha \sim 10^4$-10^5), depopulates
the higher component (Fig. 8) in competition with the nuclear
transition. This M1-transition first was observed in muonic capture rates by Winston and Telegdi [74,75].

The corrected values of the isomer shift ΔE_{is} are collected
in Table 3. In this correction the initial population, the lifetime and the multipole mixing of the nuclear transition, the M1-
interdoublet transition and the magnetic hf-splitting are taken
into account. Because of uncertainties in the M1-interdoublet
lifetime and the hf-splitting the derived isomer shifts may be incorrect besides the statistical errors up to 100 eV for the even
nuclei and up to 150 eV for the odd nuclei.

The muonic isomer shift can be compared to Mössbauer data.
But whilst a Mössbauer experiment measures a shift proportional to
$\Delta<r^2>$, the muonic isomer shift depends strongly on the shape of
$\Delta\rho(r)$. Figure 9 shows an interpretation of ^{152}Sm and ^{182}W results
in terms of changes in $\Delta<r^2>$ and Δt assuming a two parameter Fermi
charge distribution [49]. Obviously the sign of $\Delta<r^2>$ does not
necessarily determine the sign of ΔE_{is}. However, if $\Delta<r^2>$ is known
from a Mössbauer experiment [76,77], Δt can be derived from the
muonic isomer shift. Corresponding to Fig. 9 the following $\Delta t/t$
values can be deduced: -0.49%(^{152}Sm), +0.23%(^{186}W), +0.05%(^{191}Ir),
+0.1%(^{193}Ir) and +1.6%($\frac{3}{2}^+$ level in ^{169}Tm). Such an interpretation
gives a valuable information on the properties of nuclear states,
but at the moment it is very questionable because of uncertainties
of the Mössbauer data and the neglected muonic nuclear polarization.
Nevertheless the large negative isomer shifts in W, Os, Tm and
^{205}Tl are confirmed. For ^{203,205}Tl a possible explanation of the

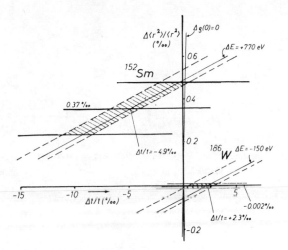

Fig. 9 Analysis of the ^{152}Sm and ^{186}W results in terms of the nuclear
charge parameters $\Delta\langle r^2\rangle$ and Δt [49]. Hatched areas include experimental
errors and the limit of complete de-excitation of the upper hf-level.
Horizontal lines are Mössbauer data [76,77].

different sign is given in [5]. For deformed nuclei no theoretical
interpretation is known. Even the self-consistent cranking model
[78] which contains a negative contribution to the isomer shift
gives results which are too large and exclusively positive.

7.2 Spherical Nuclei

The nuclear excitation in μ-atoms is not restricted to deformed nuclei. An appreciable mixing of nuclear and muonic states
can produce a nuclear excitation if the following condition is
fulfilled [61]:

$$|(E_A - E_B)/M| \lesssim 1 \ .$$

E_A and E_B are muonic states coupled to the nuclear ground state
and excited state, respectively. $M = \langle \psi_A |H_1| \psi_B \rangle$ is the matrix
element due to the perturbation H_1. In addition total angular
momentum and parity of the states A and B have to be equal. The
following resonances have been predicted or observed: E0 in Kr,
Zn [62], E1 in Pb isotopes [6] and Bi [9], E2 and M1 in Tl [5,38,
61] and I [63], E3 in Bi [51,64]. As an example the Bi and Tl
spectra are shown in Figs. 10 and 11. In ^{209}Bi four nuclear
γ-rays have been found [51]; their intensities are consistent
with the anomalous intensity ratio of the muonic 3d-2p fine structure components. Isomer shifts of these high energetic levels

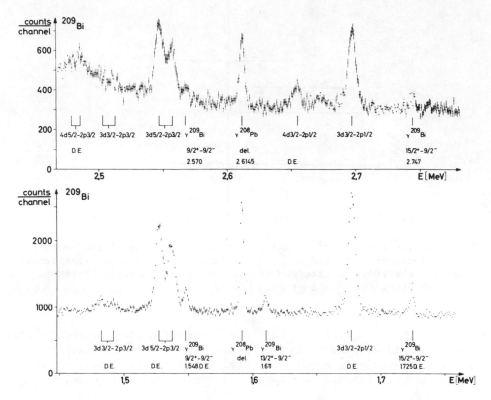

Fig. 10   3d-2p transition in μ-²⁰⁹Bi [10].   D.E.: double escape peak;
γ: nuclear γ-rays; del.: delayed γ-rays after μ-capture.

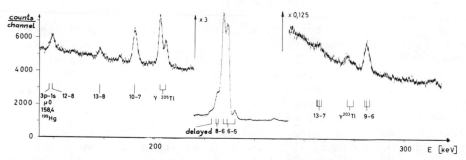

Fig. 11   Nuclear γ-rays in μ-Tl [5].   μ-X transitions are labelled with
n-n'.

(Table 3) of several keV have been determined, which can be explained by the deformation of the $3^- {}^{208}$Pb core of the $^{15}/_2+$ and $^9/_2+$ septuplet states in ^{209}Bi [51].

In ^{203,205}Tl (fig. 11) clearly resolved hf-components have been observed [5,38] which are due only to the hf-splitting of the ground state (c.f. Fig. 8). From this splitting very accurate hf-constants were deduced (Table 2). A shell-model calculation with configuration mixing [39,40] is in a fair agreement with the experimental results. Although the magnetic moments of ^{207,203}Tl are nearly equal, an isotope effect is observed in the hf-constants which can be explained by a different nuclear radii of both isotopes [39]. However, the observed isomer shifts (Table 3) and isotope shift (Fig. 6) cannot be explained simultaneously by the same shell model parameters. Intensities of the nuclear γ-rays in muonic Tl and Bi are discussed in ref. [50].

REFERENCES

1. S. Devons and I. Duerdoth, Adv. in Nucl. Phys. 2, 295 (1969).
    C.S. Wu and L. Wilets, Ann Rev. of Nucl. Science, to be published.
2. V.L. Fitch and J. Rainwater, Phys.Rev. 92, 789 (1953).
3. S. Bernow et al., Phys.Rev.Letters 18, 787 (1967).
4. H. Backe et al., Hyperfine Structure and Nuclear Radiations, North-Holland, Amsterdam (1968), p.65.
5. R. Engfer, Habilitationsschrift, T.H. Darmstadt (1969).
6. H.L. Anderson, C.K. Hargrove, E.P. Hincks, J.D. McAndrew, R.J. McKee, R.D. Barton and D. Kessler, to be published.
7. H. Backe et al., unpublished results of the Darmstadt group.
8. H. Backe, R. Engfer, U. Jahnke, E. Kankeleit, K.H. Lindenberger, C. Petitjean, W.U. Schröder, H.K. Walter and K. Wien, Contributed paper to this conference.
9. C.K. Hargrove et al., Phys.Letters 23, 215 (1969).
10. H. Backe, Thesis, Inst. f. Techn. Kernphysik, T.H. Darmstadt (1969).
11. R.C. Barrett et al., Phys.Rev. 166, 1589 (1968).
12. P. Gombas, Hdb. der Physik XXXVI, p. 109, Springer Verlag Berlin-Göttingen-Heidelberg (1956).
13. D. Kessler et al., Phys.Rev.Letters 18, 1179 (1967).
14. H. Koch et al., Phys.Letters 28B, 280 (1968).
15. H. Backe, R. Engfer, E. Kankeleit, W.V. Schröder, H.K. Walter and K. Wien, contributed paper to this conference.
16. L. Tauscher, Phys.Letters 27A, 581 (1968).
17. A. Suzuki, Phys.Rev.Letters 19, 1005 (1967).
    G. Backenstoss, private communication.
18. L.R.B. Elton, Landolt-Börnstein tables, Springer-Verlag, Berlin (1967), New Series, group I, vol. 2.
19. R. Engfer, Proc. of Int. School of Physics "Enrico Fermi", Varenna (June 1966), p.64.

20. R. Engfer et al., Proc. of the Int. Conf. on Electromagnetic
     Sizes of Nuclei, Ottawa (1967), p.184.
21. H. Theissen, H. Fink and H.A. Bentz, Z. Physik, to be published.
22. G.J.C. van Niftrik, Nucl.Phys., to be published.
23. H.L. Anderson et al., Phys.Rev.Letters 22, 221 (1969).
24. R.A. Carrigan Jr. et al., Phys.Letters 27B, 622 (1968).
25. R.D. Ehrlich et al., Phys.Rev.Letters 13, 550 (1964).
26. G. Backenstoss et al., Nucl.Phys. 62, 449 (1965).
27. H.L. Acker et al., Nucl.Phys. 62, 477 (1965).
28. S. Raboy et al., Nucl.Phys. 73, 353 (1965).
29. R.E. Coté et al., Phys.Letters 19, 18 (1965).
30. S.A. de Wit et al., Nucl.Phys. 87, 657 (1967).
31. D. Hitlin, Proc. of the International Conference on Electro-
     magnetic Sizes of Nuclei, Ottawa (1967), p.254.
32. M.N. Suzuki, Thesis, Carnegie-Mellon University, (Jan. 1968).
33. R.A. Carrigan et al., Phys.Rev.Letters 20, 874 (1968).
34. H.L. Acker, Nucl.Phys. 87, 153 (1966).
35. L. Wilets, Kgl. Dan. Vid. Selsk. Mat.-Fys. Medd. 29, No.3
     (1954).
36. B.A. Jacobsohn, Phys.Rev. 96, 1637 (1957).
37. J.A. Johnson et al., Phys.Letters 29B, 420 (1969).
38. R. Baader et al., Phys.Letters 27B, 428 (1968).
39. F. Scheck, contributed paper to this conference.
40. R. Engfer and F. Scheck, Z. Phys. 216, 274 (1968).
41. R.J. Powers, Phys.Rev. 169, 1 (1968).
42. J. Johnson and R.A. Sorensen, Phys.Letters 26B, 700 (1968).
43. R.A. Carrigan et al., Bull.Am.Phys.Soc. 13, 65 (1968).
44. A. Bohr and V.F. Weisskopf, Phys.Rev. 77, 94 (1950).
45. M. le Bellac, Nucl.Phys. 40, 645 (1963).
46. S. Devons, Proc. of the Rutherford Jubilee Int. Conf.,
     Manchester 1961, Academic Press N.Y., p.624.
47. J. Hüfner, Nucl.Phys. 60, 427 (1964).
48. S. Bernow et al., Phys.Rev.Letters 21, 457 (1968).
49. R. Baader et al., Phys.Letters 27B, 425 (1968).
50. H. Backe, R. Engfer, E. Kankeleit, W.U. Schröder, H.K. Walter
     and K. Wien, contributed paper to this conference.
51. H. Backe, R. Engfer, E. Kankeleit, W.U. Schröder, H.K. Walter
     and K. Wien, contributed paper to this conference.
52. L.N. Cooper and E.M. Henley, Phys.Rev. 92, 801 (1953).
53. E. Nuding, Z. Naturforschung 12a, 187 (1957).
54. W. Greiner, Z. Physik 164, 374 (1961).
55. W. Greiner and H. Marschall, Z. Physik 165, 171 (1961).
56. F. Scheck, Z. Physik 172, 239 (1963).
57. R.K. Cole Jr., Phys.Letters 25B, 178 (1967); Phys.Rev. 177,
     164 (1968).
58. M.Y. Chen, Thesis, Princeton Univ. (1968); Phys.Rev. to be pub.
59. J.M. McKinley, Phys.Rev. to be published.
60. A. Gal et al., Phys.Rev.Letters 21, 453 (1968).
61. J. Hüfner, Z. Physik 190, 81 (1966).
62. E.M. Henley and L. Wilets, Phys.Rev.Letters 20, 1389 (1968).
63. H.L. Acker et al., Nucl.Phys. 87, 1 (1966).

64. J. Hüfner, Phys.Letters 25B, 189 (1967).
65. C.S. Johnson et al., Phys.Rev. 125, 2102 (1962).
66. W. Frati and J. Rainwater, Phys.Rev. 128, 2360 (1962).
67. H.L. Anderson and C.S. Johnson, Phys.Rev. 130, 2468 (1963).
68. D. Quitmann et al., Nucl.Phys. 51, 609 (1964).
69. J.A. Bjorkland et al., Nucl.Phys. 69, 161 (1965).
70. T.T. Bardin et al., Progress Report, Columbia University
     (1966), unpublished.
71. R. Engfer, Z. Physik 192, 29 (1966).
72. H.L. Anderson, Proc. of Int. Conf. on Electromagnetic Sizes
     of Nuclei, Ottawa (1967), p.53.
73. J.B. Bellicard and K.J. van Oostrum, Phys.Rev.Letters 19,
     242 (1967).
74. R. Winston and V.L. Telegdi, Phys.Rev.Letters 7, 104 (1961).
75. R. Winston, Phys.Rev. 129, 2766 (1963).
76. W. Henning et al., Verhandlungen der Deutschen Physikalischen
     Gesellschaft 4, 324 (1968).
77. P. Kienle et al., Hyperfine Structure and Nuclear Radiations,
     North-Holland, Amsterdam (1968), p.971.
78. E.R. Marshalek, Phys.Rev.Letters 20, 217 (1968).
79. D. Hitlin, S. Bernow, S. Devons, I. Duerdoth, J.W. Kast, E.R.
     Macagno, J. Rainwater, C.S. Wu and R.C. Barrett, to be pub-
     lished.
80. W.Y. Lee, S. Bernow, M.Y. Chen, S.C. Cheng, D. Hitlin, J.W.
     Kast, E.R. Macagno, A.M. Rushton, C.S. Wu and B. Budick,
     contributed paper to this conference.
81. J. Johnson, Bull. Am. Phys. Soc. 14, 538 (1969).

DISCUSSION

K. Ford: The isomer-shift, for heavy nuclei, may measure, approximately, the change in $\langle r \rangle$.   Is it not possible that the changes
in $\langle r^2 \rangle$, $\langle r \rangle$ may be different, even in sign?   R.E.: With sufficient
precision, one could determine both changes, the one from Mossbauer
and the other from muonic-atom measurements.   At present this is not
possible; moreover the indications are that changes in surface thickness are small.   R.C. Barrett: The Ford & Wills result - proportionality to change in $\langle r \rangle$ - appears, from checks made with single particle distributions, to be model dependent.

V. Telegdi: The Tℓ measurement is the only case where the H.F.S. can
be readily interpreted in terms of the magnetic dipole distribution.
In Bi the interpretation depends intimately on nuclear excitations.
R.E.: Only for the ground state of Tℓ is the magnetic H.F.S.
observed.   There are no corresponding results for the excited state.
J.D. Walecka: Electron elastic scattering at 180° can also measure
form-factors of old magnetic multipoles.

WEAK INTERACTION AND NUCLEAR STRUCTURE*

M. RHO

Service de Physique Théorique
Centre d'Etudes Nucléaires de Saclay
BP. n° 2 - 91 - GIF-sur-YVETTE

The subjects of μ-mesic atoms and related dynamical effects will be discussed by those actively working in the field. The aim of this talk is to discuss on the subject of nuclear weak interaction which may not be dealt with by others. Since 1966 when J.D. Walecka** gave a talk on the subject at the Williamsburg Conference, there has been little development on the theoretical side of the weak interaction <u>reaction</u> mechanism. It is probably safe to say that but for the many-body complication, the nuclear weak processes are well understood. However the many-body complication is precisely the nuclear physics, and is discussed below in connection with β-decay and μ-capture with the help of theoretical tools recently developed in both nuclear and particle physics.

If one assumes that the fundamental weak interaction is better known than nuclear structure (which is not unreasonable), then one can make interesting statements about the latter via weak interaction probes. For example, one can learn something about correlations in nuclei, i.e., the structure of nuclear wave functions. On a more fundamental level, the weak interaction can be a probe for meson

----
* Review talk given at the Third International Conference on High Energy Physics and Nuclear Structure, Columbia University (10 September, 1969).

** For developments up to 1966, see J.D. Walecka, Proceedings of Williamsburg Conference on Intermediate Energy Physics, (February 1966)

exchange effects, thus leading to informations regarding nuclear force. This is an important consequence of an interplay between particle and nuclear physics.

1 - TRANSITIONS TO BOUND STATES

(a) Consider the process $\mu^- + O^{16}(0^+) \rightarrow \nu_\mu + N^{16}$ (bound states) where the final states are $T = 1$ ground state quartets $0^-, 1^-, 2^-, 3^-$. Both initial and final nuclei are perhaps the best studied objects in nuclear physics and presumably the best known. Also they are the only ones where individual transitions in capture processes have been measured. A simple theoretical picture would be the following : 1) O^{16} ground state considered as doubly closed, 2) N^{16} states taken as $T = 1$, p-h (particle-hole) states with a particle in (sd) shell and a hole in p-shell.

One knows now that this simple picture does not work. The discrepancy seems to do with fundamental properties of the O^{16} ground state and N^{16} quartets. An attractive remedy for the short-coming is the concept of "coexistence" of the deformed $T = 0$ even parity states and spherical $T = 1$ odd-parity states advocated by Brown and Green[1]. In this picture the O^{16} ground state contains (2p-2h), (4p-4h) etc. admixtures in quite different manner from the random phase approximation (RPA) whereas N^{16} quartets are of spherical 1p-1h configuration (with the renormalization due to (2p-2h) states with $T = 1$, odd-parity taken into account). Green and myself[2] argued that on both theoretical and empirical grounds, the (3p-3h) configurations are insignificant in the N^{16} quarters.

This point of view, however, is not favored by everyone in the field. There are some who argue on the basis of "large" shell-model calculations, and some[3] who claim from a method of generating the N^{16} states from deformed ground state, that the 3p-3h mixing can be as large as the 2p-2h mixing of the ground state. In this picture, it is clear that the dependence of capture rates on the ground state admixture would be rather insensitive.

Only an experiment measuring the overlap of these excited states with another state of 4p-4h type (for example 6MeV 0^+ state) can determine which of the two pictures is correct. In the absence of such a direct evidence, we can only rely on indirect evidence , such as energies and theoretically estimated matrix elements. Both indicate that the 3p-3h components can be neglected. If we take this view, then the process can become a useful probe for 2p-2h components in the ground state. This can be seen in Table I. The results are obtained with the N^{16} wave functions obtained by T. Kuo using his screened matrix elements and limiting to 1p-1h space. The renormalization correction referred to above was made by Green, Kallio and Dahlblom[2] and is found to be small, not more than 4% in matrix

element. This is neglected in the results. We have used for the
ground state two models designated I and II. Those can be compared
with the "experimental" wave functions obtained by Purser et al[4]
with (d,t) and (d,He^3) reactions on O^{16}. They give different mixtures for the 2p-2h components. Corresponding to this difference,
one can observe the effects on the individual capture rates. This
can be understood by looking at the correction factors introduced
by the ground state mixtures : roughly they go like (see Table I.A)

$$(\alpha \pm \gamma/\sqrt{12})^2 \quad \text{for} \quad \begin{matrix} 1^- \\ 0^- \end{matrix}$$

$$(\alpha \pm \beta/6)^2 \quad \text{for} \quad \begin{matrix} 3^- \\ 2^- \end{matrix} \quad . \tag{1}$$

Thus, an accurate measurements of capture rates could complement
other experiments such as that of Purser et al to determine the
amplitudes β and γ.

  (b)   Another case where 2p-2h components show up is the decay
of Ni^{56} by K capture to Co^{56}. Here the effect is spectacular. If one
considers Ni^{56} as doubly closed, then 1^+ state in Co^{56} would be essentially a p-h configuration ($f_{5/2}\ f\bar{7}_{/2}$). For this, the Gamow-Teller
transition has $\log_{10} ft = 2.5$ which is much smaller than the experimental value $\log_{10} ft = 5$. This disease was cured by Goode and Zamick[5]

<u>TABLE I.A</u>

O^{16} ground state

$$\psi(0^+) = \alpha |0p-0h\rangle + \beta |d_{5/2}^2\ p_{1/2}^{-2}\rangle + \gamma |s_{1/2}^2\ p_{1/2}^{-2}\rangle$$

	α	β	γ
Model I	0.88	0.44	0.18
II	0.88	0.29	0.38
Exp.			
(d,t)[a]	0.87	0.27	0.26
(d,He^3)[a]	0.82	0.54	0.20

a.   From Purser et al., Ref. 4.

by noticing that the 1^+ state is not predominantly 1p-1h, but rather
2p-2h with small admixture of 1p-1h. The β-decay transition occurs
only via the 1p-1h component, therefore the matrix element is cut
down by the small mixing probability. The theoretical value turns
out to be $\log_{10} ft = 5.8$ in agreement with experiment.

TABLE I.B

Theoretical[b] and Experimental[c] Rates in 10^3 Sec^{-1}

	0^-	1^-	2^-	3^-	Sum
Model I	1.01	1.45	6.87	0.12	9.45
II	0.84	1.74	7.44	0.11	10.1
Exp. C	1.1 $\pm$ 0.2	1.9 $\pm$ 0.1	6.3 $\pm$ 0.7		9.3 $\pm$ 1.0
B	1.6 $\pm$ 0.2	1.4 $\pm$ 0.2			
L	0.85 $\pm$ 0.06	1.85 $\pm$ 0.17		$\leq$ 0.08	

b. N^{16} state wave functions obtained by Kuo (private communication)
were used.

c. C = Columbia, B = Berkeley, L = Louvain experiments. See Ref. 2 for
complete references, and also J.P. Deutsch et al, Phys. Letters
29B, 66 (1969).

2 - TRANSITIONS WITH NEUTRON EMISSION

Let us turn now to another kind of correlation which may have
to do with "short-range" correlation. The process of interest is :

$$\mu^- + Ca^{40} \to K^{39} + n + \nu \tag{2}$$

and we look at neutrons at sufficiently high energy (i.e. >10 MeV)
so that a direct emission mechanism is applicable. Usually two
quantities are measured here, one, neutron spectrum $N_0(E_n)$ and the
other, asymmetry parameter $\alpha(E_n)$ where E_n is neutron energy. If one
denotes P as the residual spin polarization of muon at the time of
capture, Φ as the angle made by μ-spin and neutron momentum, then
the total number of neutrons emitted per capture per MeV per ste-
radian is :

$$N(E_n, \Phi) = (4\pi)^{-1} N_0(E_n) [1 + P\alpha(E_n) \cos \phi] . \tag{3}$$

This definies N_0 and α. If one works out the algebra with the standard weak interaction Hamiltonian, one finds that α has the form

$$\alpha = \frac{A}{B}$$

$$\tag{4}$$

$$B \propto N_0(E_n) ,$$

where both A and B contain nuclear matrix elements of vector, axial-vector, pseudoscalar etc. operators multiplied by appropriate coupling constants. The expressions for them are rather complicated, and are not essential for discussion. Let us just note that B is, apart from constants, just the neutron spectrum N_0 and is dominated by coherent sum of non-relativistic terms. The term A on the other hand comes significantly from relativistic corrections due to the fact that the dominant vector and axial vector matrix elements cancel each other.

The experimental situations are rather confusing at this moment. One may summarize them in three ways :

1. $\alpha \simeq - 1$ for $E_n > 15$ MeV (Russian group[6])
2. α small positive and smoothly varying (Carnegie-Mellon[7])
3. Some structures in α (Columbia[8], Carnegie-Mellon[7]).

Theoretically, $\alpha = - 1$ at high energy is impossible to explain unless one modifies the weak interaction theory[9]. We may accept the weak interaction theory but question the validity of the experiment. The third possibility -that structure exists at high neutron energy- would be equally difficult to understand, though not impossible. Since there seems to be no structure in the denominator, it must be in A, if any. At low neutron energy, one may observe structures coming from giant resonances. The supermultiplets of giant resonances may be split, giving rise to different spectra for the vector and axial-vector transitions. The interference between those two transitions can manifest as structures in α. This was demonstrated by Balashov et al[10] in a contribution to this conference. Fujii et al[11] conjecture that this resonance mechanism can be extended to higher neutron energies, $E_n \gtrsim 15$ MeV, giving rise to the "peak" observed at between 15 and 25 MeV. The resolution of this question will depend upon separate measurements of the excitation functions of the vector and axial-vector transitions (i.e., (γ, p) and radiative capture experiments). At this moment, the situation is not clear.

It is the consensus of the Columbia and Carnegie-Mellon groups that α is small and positive. This also can be explained theoretically without difficulty. As Bogan[12] pointed out, it is the relativistic correction which raises α from negative to positive value for large E_n. The model is simple : a proton in a filled shell $(1d_{3/2}, 2s_{1/2}, \ldots, 1s_{1/2})$ captures a muon making a direct transition to a neutron with E_n in the continuum. Bogan neglects initial correlation and uses nuclear matter value for the wave number of the outgoing neutron to simulate partially the final state interaction. Bogan's calculation is compared with the most recent Carnegie-Mellon data for Ca^{40} in Fig. 1. Although α does become positive at large E_n, the magnitude seems too small and there is a definite disagreement in the energy region of 15 to 25 MeV.

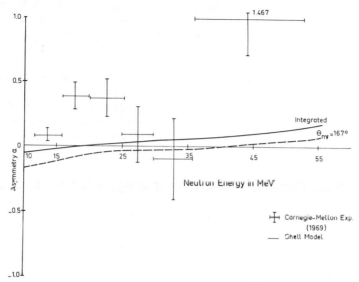

Fig. 1 - Asymmetry parameter for Ca^{40}. Data are from Carnegie-Mellon.

Since the model is based on the assumption that the neutron passes through nucleus without changing its direction, asymmetry parameters smaller than experimental values signify definite disagreements.

An independent check of some matrix elements in the asymmetry parameter can be made by examining the neutron spectrum as emphasized before. The simple model works fairly well for Si, S and Ca as can be seen in Fig. 2 . The small discrepancy which seems to exist even at $E_n \gtrsim 20$ can be removed if the effective wave number is varied a little.

We now argue that this agreement can be a mere accident. This can be seen by considering the photon experiment (γ,p) for photon energy $E_\gamma \gtrsim 30$ MeV. At this energy or above, the kinematical condition is very similar to $(\mu,n\nu)$ process and in fact the matrix elements look essentially the same. Manuzio et al[13] show that the differential cross-section at $45°$ of the reaction $^{12}C(\gamma,p)B^{11}$ at $E_\gamma > 30$ can be understood equally well either by taking into account both the final state interaction and the initial state short-range correlation or by neglecting both entirely. If one neglects the short-range correlation while taking into account the final state interaction, the theoretical cross-section undershoots experiments by a factor of 10 or more. This is probably the case also in N$_o$.

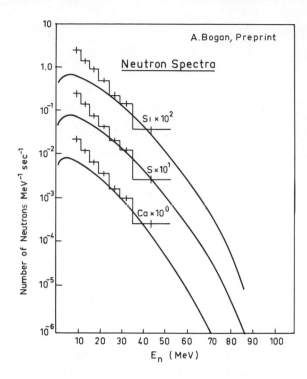

Fig. 2 – Neutron spectra
for Ca, S, Si. Solid is
calculated with the sim-
ple model, compared with
the Carnegie-Mellon data.

It is quite likely that such an accidental cancellation does
not occur in the numerator A and therefore in the asymmetry α. This
is because relativistic terms play important role in the term A.
A direct calculation showing the effect of the short-range correlations has not been completed yet ; it is a delicate calculation involving careful treatments of the final state interaction (first)
and the initial correlation (second) since both effects which are
not prominent at momentum transfers involved, are intermixed in an
intricate way. As an indirect argument for such a treatment, let
us consider α as a function of neutron-neutrino angle $\Theta_{n\nu}$; i.e.
$\alpha(E_n, \Theta_{n\nu})$. This was already studied sometime ago by Klein, Neal and
Wolfenstein[14] in Fermi gas model. We reexamine it in the shell-model
we are using[15]. Since muon is captured practically at rest, the neutron and neutrino would prefer to come off with $\Theta_{n\nu} \simeq 180°$. In nuclei
they can also come off at other angles ($\Theta_{n\nu} < 180°$) with varying
probabilities, and the smaller angle ejection contribution to the
total depends upon the momentum distribution of the bound wave function. The short-range correlation can supply higher momentum components and hence affect the probabilities of smaller angle ejection.
The calculation[15] (Table II) shows that the smaller $\Theta_{n\nu}$, the larger
the asymmetry, the difference being large at energies at which siza-
ble positive asymmetry is observed. Whether or not a consideration
of the correlation discussed here can also explain the possible
structure is not clear.

In summary, at present no consistent calculation exists to explain any of the three alternative observations made by experimentalists. Both initial correlation and final state interaction should be included before one can make any statement about $N(E_n)$ as well as $\alpha(E_n)$. If the structure in α is real, then the reaction mechanism must be revised in the line of Balashov et al and Fujii et al. The asymmetry parameter appears to be an observable in which all the nuclear complications merge together, thus fascinating to nuclear physicists though annoying to weak interaction physicists.

TABLE II

Asymmetry vs. Neutrino-Neutron Angle $\Theta_{n\nu}$ in Ca^{40}

E_n Θ	0°	50°	100°	150°	180°
10 MeV	0.93	0.70	0.22	- 0.10	- 0.17
35	0.93	0.70	0.30	0.01	- 0.09
95	0.52	0.48	0.38	0.28	0.25

3 - MESON EXCHANGE EFFECTS

A correlation of much more subtle nature are effects which require explicitly the meson degrees of freedom ; these are the meson exchange effects in electromagnetic and weak processes and also possibly the correlations at very short distances. Recently there has been an attractive proposal[16] that both effects be simultaneously incorporated by putting baryon resonances (about 1%) into the wave functions. Kisslinger discusses the problem of short-range correlation in the high energy backward proton scattering off deuteron[17]. Danos proposes[18] that the resonance-mixing model can eliminate the double counting which may occur in calculating the exchange current effects in magnetic moments and β-decay. Such schemes hope to go beyond the usual non-relativistic N-N force concepts used in nuclear physics. We shall not discuss this "fundamental" and perhaps "dream" problem, but suggest that presently available (and also powerful) techniques developed in particle physics enable the weak interaction (and the electromagnetic interaction) to provide insight into the ever evasive problem of meson exchange in nuclei. Brown and Green[19] confront directly this problem by considering the nuclear force, in particular, the three-body force in terms of nucleon virtual-pion scattering amplitudes. Here we take the alternative route[20] and use the weak and electromagnetic probes (theoretical) to examine the same thing. Unlike the resonance-mixing models,

here the approach makes use of the low-energy theorems which turn
out to be quite suitable for nuclear properties. In order not to
deviate from the subject   I am supposed to review, we shall consi-
der β-decay, in particular, the process $H^3 \rightarrow He^3 + e^- + \bar{\nu}_e$. We
shall also comment on the magnetic moments of H^3 and He^3 which can
be treated in the same way.

It is well-known that the non-conserved axial current in weak
interaction entails in nuclei many-body currents even at zero momentum transfer at which the β-decay occurs. We shall consider here
only the two-body current. If one denotes the axial current by A_λ ,
then the relevant graph is

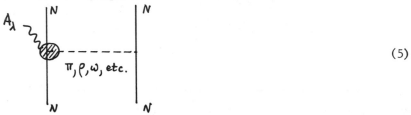

(5)

The object of the game is to obtain an equivalent two-body operator
corresponding to the graph (5). Up to now, there have been three ways
to do it :

a) Phenomenological method. Blin-Stoyle and Tint[21] suggested
that by means of the PCAC, one can extract the operator of (5) from
the reaction $p + p \rightarrow \pi^+ + d$. Their analysis was criticized by Cheng
and Fischbach[22] on the ground that $p + p \rightarrow \pi^+ + d$ gets its dominant
contribution from $N^*_{3,3}$ intermediate state whereas in triton β-decay
$N^*_{3,3}$ should not contribute to the major term. However, if done properly, $N^*_{3,3}$ terms should cancel out, although in practice they may
not because of approximations made.

b) Lagrangian models. Cheng[23] has used Lagrangian models and
PCAC to evaluate the Feyman graphs :

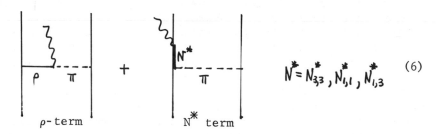

$$N^* = N^*_{3,3}, N^*_{1,1}, N^*_{1,3}$$

(6)

He finds about 10% contribution to the Gamow-Teller matrix element
which is about the correction needed ; however the ρ term which is
large is incorrectly treated, and the coupling constants for πNN^*
seem to be unreasonably large.

c) Low-energy theorem for the axial current. This approach
of Chemtob and Rho[20] is based on the realization that in one-pion
exchange approximation, the blob in (5) is just the vertex for the
weak axial production of soft pions to which Adler and Dothan's
low-energy theorem[24] can be applied. If we denote the matrix element as

$$M_\lambda = \langle \pi(q) N_f | A_\lambda | N_i \rangle$$

and the PCAC relation as

$$k_\lambda M_\lambda = C D \quad , \tag{7}$$

where D is the matrix element for $N_i + \pi(k) \to N_f + \pi(q)$, C is a
known constant, and k and q are pion four-momenta, then the low-
energy theorem gives the non-Born part of M_λ in terms of derivative
of π-N scattering amplitudes evaluated at $k_\lambda = 0$. Those amplitudes
are in principle functions of q^2. They are known at $q^2 = 0$ from
the work of Adler and need only be extrapolated to q^2 relevant to
the mass in nuclei. As pointed out by Brown and Green, pions in nu-
clei are soft which means that $|q^2|$ does not deviate much from m_π^2.
In this case, the extrapolation amounts to q^2/M^2 correction where
M is nucleon mass. This may be safely neglected. Thus to the order
considered, the low-energy theorem gives exact answer for the OPE
term. It also tells us how to handle the ρ term in connection with
the ranges of force involved. The statement made about the Lagrangian
models follows also from this theorem. If the Adler's calculation
of the amplitudes is reliable, then our estimate of the exchange
contribution should be reliable. We find the contribution from non-
Born terms to be about 5% to the axial vector matrix element which
is fairly independent of nuclear models. If one were to decompose
the contributions into the ρ term plus the "rest", the "rest" containing N^* contributions, one finds that the ρ term amounts to about
half of the total non-Born value.

An obvious application of the theorem is to compute the meson
exchange contribution to the elastic neutrino process $\nu + A_i \to A_f + \mu$
(A_i, A_f being initial and final nuclei) at zero momentum transfer.
Due to the Pauli exclusion principle, the cross-section is suppressed at low momentum transfer if one calculates it with any nuclear
model taking only the one-body operator. The experiments[25] do not
show such suppression. If this is real, it may be an ideal place
to "see" the exchange current effect[26].

Now consider the electromagnetic current $J_\lambda^{e.m.}$ which consists
of isoscalar and isovector currents. The meson exchange corrections
to magnetic moments are obtained from the graph (5) by replacing
A_λ by $J_\lambda^{e.m.}$. The isovector current in H^3 and He^3 has already been
computed by Padgett et al[27]. The exchange moment for the isovector
part was mostly accounted for with only the pion-exchange term (7a).

This can be easily understood if a soft-photon theorem is applied. The amplitude $\langle \pi N | J_\lambda^{e.m.} | N \rangle$ for isovector part is given by the Born graphs alone if one confines to order k (photon four-momentum). This is called Kroll-Ruderman theorem. Turning things around, one may apply the soft-pion theorem in which case one gets a M^{-2} correction to the K-R theorem. In either case, the conclusion is that the rescattering correction (N^* contribution) is small. Let us denote the Born term by

$$\text{Born} = \quad\quad\quad + \quad\quad\quad \tag{8}$$

and the rest of contributions such as vector meson exchange, N^*, etc.. as "others". Then we can summarize the situation by saying that the isovector exchange moment can be described by Born terms, while the isocalar moment requires the "others". We have evaluated the "others" with Lagrangian models. Despite the ambiguities in the Lagrangians and coupling constants taken, the results are reasonable. For S-state and a core radius of 0.4 fermi, the meson exchange corrections have the following contents ;

$$\Delta\mu_{\text{mesons}} = \overset{\text{Born}}{(-0.170} \overset{\text{Others}}{- 0.072)}\tau + \overset{\text{Born}}{(0.} \overset{\text{Others}}{+ 0.006)} = -0.242\tau + 0.006$$

where $\tau = 1(-1)$ for He^3 (H^3) .

This accounts satisfactorily for discrepancies remaining with one-body operator in both the isovector and isoscalar moments.

From the discussions given here, one sees that except possibly for the isoscalar exchange moment, N^* alone cannot account for discrepancies at low momentum transfer. This seems to go against the hope of the resonance-mixing models. Further studies along this line will help clarify this new field of nuclear physics.

4 - NUCLEI AS ELEMENTARY PARTICLES

There has been an attempt initiated by Primakoff and Kim[28] to bypass all the complications discussed above by treating nuclei as elementary. The essential idea is to study directly nuclear form factors applying or testing various hypotheses invoked in particle physics. The main application has been directed to β-decay, μ-capture, neutrino-induced processes and π-capture.

Here let me just mention some recent developments which are quite significant.

a) Kim and Frazier[29] found that in nuclei, the Nambu version and Gell-Mann-Lévy version of the PCAC yield different results for the pseudoscalar form factor. They differ by 11% in hydrogen and by 40% in He^3.

b) Ericson and her collaborators[30] estimated the corrections to the soft-pion theorem in π-nucleus charge exchange and threshold pion production by photon and found them to be significant in nuclei in contrast to nucleons where such corrections are small. A systematic study of this kind may lead to a clarification of π-nucleus optical potential commonly accepted in π-mesic atom works. It emerges from these works that both a) and b) are nuclear many-body phenomena and are due to the fact that zero mass pion is a bad approximation in a scheme where nuclei are considered elementary.

5 - ACKNOWLEDGEMENTS

I am grateful to Dr. B. GOULARD for discussions and for the hospitality at the Physics Department of Université Laval, Québec, where a part of this report was written.

REFERENCES

1. G.E. Brown and A.M. Green, Nuclear Physics 75, 401 (1966)
2. A.M. Green and M. Rho, Nuclear Physics A130, 112 (1969)
   A.M. Green, A. Kallio, T. Dahlblom, to be published
3. A. Laverne, B. Goulard and Do Dang Giu, Private communication
   see also G.E. Walker, Phys. Rev. 175, 1290 (1968)
4. K.H. Purser, W.P. Alford, D. Cline, H.W. Fulbright, H.E. Gove
   and M.S. Krick, Nuclear Physics A132, 75 (1969)
5. P. Goode and L. Zamick, Phys. Rev. Letters 22, 958 (1969)
6. V.S. Evseev, V.S. Roganov, V.A. Chernogoroya, Chang Jun-Wa and
   M. Szymczak, Soviet Journal of Nuclear Physics 4, 245 (1967)
7. R.M. Sundelin, R.M. Edelstein, A. Suzuki and K. Takhashi, Phys.
   Rev. Letters 20, 1201 (1968) and private communication (1969)
8. L. Lederman, Private communication
9. However N. Vinh-Mau and A. Bouyssy (unpublished) find that with
   a suitable optical potential for the neutron, and complicated
   ground state wave function, one obtains $\alpha \sim -0.5$ at $E_n = 20$ MeV
   This result seems doubtful in view of reference 12.
10. V.V. Balashov et al, Contribution to session 2
11. A. Fujii et al, Contribution to session 2
12. A. Bogan, Phys. Rev. Letters 22, 71 (1969)
13. G. Manuzio, G. Ricco, M. Sauzone and L. Ferrero, preprint (1969)
14. R. Klein, T. Neal and L. Wolfstein, Phys. Rev. 138, B86 (1965)

15. M. Rho, unpublished
16. A.K. Kerman and L. Kisslinger, Phys. Rev. (to be published)
    H. Arenhövel and M. Danos, Phys. Letters 28B, 299 (1968)
    L. Kisslinger, Phys. Letters 29B, 211 (1969)
17. L. Kisslinger, Contribution to session 9
18. M. Danos, Contribution to session 9
19. G.E. Brown and A.M. Green, Nucl. Physics (to be published)
20. M. Chemtob and M. Rho, Phys. Letters 29B, 540 (1969) and to be
    published
21. R. Blin-Stoyle and M. Tint, Phys. Rev. 160, 803 (1967)
22. W.K. Cheng and E. Fischbach, preprint (1969)
23. W.K. Cheng, Univ. of Pennsylvania Thesis (1966)
24. S.L. Adler, Annals of Physics (N.Y.) 50, 189 (1968)
25. T.B. Novey, Invited talk in session 2
26. H. Primakoff and C.W. Kim, Comments in the session 2 and private
    communication
27. D.W. Padgett, W.M. Frank and J.G. Brennan, Nucl. Phys. 73, 424
    (1965)
28. C.W. Kim and H. Primakoff, Phys. Rev. 139 B1447 (1965)
    140, B566 (1965) and further references given there
29. J. Frazier and C.W. Kim, Phys. Rev. 177, 2568 (1969)
30. M. Ericson and A. Figureau, MIT preprint (1969) ; M. Ericson,
    A. Figureau and A. Molinari, MIT preprint (January 1969).
    See also paper by M. Ericson, Session IX this Conference.

DISCUSSION

V. Evseev: I would like, using as an example Mu-capture in ^{40}Ca, to
show that there are no irreconcilable contradictions between all

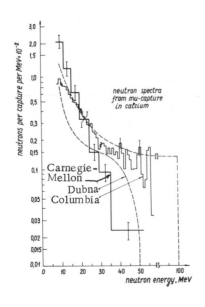

neutron spectra
from mu-capture
in calcium

Carnegie-
Mellon
Dubna
Columbia

neutron energy, MeV

experimental results on the angular
distribution of neutrons from nuclear
Mu-capture.   All these data have
been obtained under different experimental conditions, and direct comparison of these results is not possible.   Indeed, in the Columbia experiment the differential asymmetry is
measured directly for various neutron
energy intervals; in the Carnegie-
Mellon for various energy intervals of
recoil protons; in the Dubna work it is
averaged over a wide region of neutron
energies - those higher than the neutron detection threshold.   For comparison these results should be presented in the same form, but to do
this one must, of course, know the
neutron spectrum (as well as detector
efficiencies, etc.).   Unfortunately

there is not good agreement here between the Columbia and Dubna
measurements and those of Carnegie-Mellon (see figure).   If one
takes an average of the former two for the energy-spectrum, it is
possible to get a moderate fit to the observed asymmetries by
assuming a <u>differential</u> asymmetry coefficient $\alpha(E)$, which is 0,
for $E \lesssim 40$ MeV and -1.0 for $E > 40$ MeV.   This does not reproduce
the Carnegie-Mellon results well, although the agreement is somewhat better if this same asymmetry coefficient is used with <u>their</u>
<u>own</u> energy spectra.

REMARKS ON MUON CAPTURE IN COMPLEX NUCLEI

J.P. Deutsch

Centre de Physique Nucléaire, Université de Louvain

Belgium and CERN, Genéve, Switzerland

This contribution is <u>not</u> a review paper. It is intended to highlight only some significant developments in the field. The author can but hope, that obviously interesting topics he had to bypass because of space- and time- limitation will receive due recognition in other contributions and/or discussions.

1., Muon capture measurements in complex nuclei are aimed: a., to understand the basic interaction, i.e. to fit muon capture in the general framework of the Universal Fermi Interaction (UFI) accounting also for the induced terms important here because of the large momentum-transfer involved, b., to understand the excitation mechanism of nuclear matter by the lepton-field (collective resonances versus direct reaction) and its subsequent evolution (boil-off or direct nucleon emission) and, finally, c., to probe the structure and characteristics of individual nuclei. The tools used to achieve these objectives are: a., measurement of capture rates (total or partial) and b., measurement of correlations between two or more characteristics of the initial state (muon spin, nuclear spin) and the final state (neutrino momentum, nuclear spin and - in the case of unbound final states - momentum and spin of the emitted nucleon). We shall discuss now some relevant experiments confining ourselves mostly to the ones which imply the observation of nuclear gamma-rays.

2., In the field of <u>total capture rate</u> measurements, let us call attention only (without discussing its implications) to an elegant method deviced by the Darmstadt-group to study the isotope-effect even with mixed targets through the observation of characteristic nuclear gamma-rays[1].

3., The detection and identification of nuclear gamma-rays produced in muon-capture yield information complementary to those given by the neutron-multiplicity measurements[2]: it allows − in particular − to pin down simultaneous emission of neutrons and protons. This process − which may be of high yeild in some cases[3]− should be considered also if one uses the neutron-yield and -multiplicity measurements as a probe of the momentum-distribution in nuclei[2]. It should be noted also in this connection, that a great wealth of data begins to accumulate (generally as by-product of other measurements) on the yield of nuclear gamma-rays emitted subsequent to muon-capture[3,4]: a critical examination of these data may help to understand the disagreement found by the authors of ref. 2 between the neutron-multiplicity distributions they obtained and the ones predicted by plausible nucleon momentum-distributions.

4., As for the <u>partial capture rates</u> to individual nuclear levels, let us note first a new success of the "elementary particle" treatment initiated by the authors of ref. 5 for transitions having an observable inverse beta-decay: the agreement between the experimental rate of the capture transition from ^{6}Li to the ground state of ^{6}He $(1600^{+300}_{-120}$ s$^{-1})^6$ and the newly computed rather precise theoretical value $(1350\pm150$ s$^{-1})^7$.

Let us turn now to other partial muon capture rate measurements which may be used for precise coupling constants of UFI. New results became available on the rate of muon capture processes in ^{16}O producing the excited bound levels of ^{16}N [8]. They yeild an accurate measurement of the ratio of capture rates to the (0−) and (1−) levels: 0.46±0.04. This value is in good agreement with the ratio extracted from ref. 9: 0.59±0.11, but disagrees with a less precise value taken from ref. 10: 1.14±0.21. We may combine the first two values disregarding the third one. Using the methods of A.B. Migdal as applied to muon capture by M. Rho[11], one can extract then from this result[8], in first order <u>independently</u> of g_A/g_V, the value: $g_P/g_A = 10.5\pm1.0$. This value is also rather insensitive to the choice of nuclear models: considering the variation implied by the use of different plausible models[12], one obtains: $g_P/g_A = 10\pm3$. We attempted to use this information and the result of a recent muon-capture measurement in gasous hydrogen[13] to distinguish between two g_A/g_V-values extracted from conflicting neutron lifetime measurements[14,15] (assuming − of course − UFI and the hypothesis that g_P/g_A is the same for a free proton and for one bound in ^{16}O [16]. One can see from Fig. 1 that the more recent value $g_A/g_V = 1.23\pm0.01$ is favoured by the comparison.

5., Having considered some measurements of total and partial muon capture rates, let us turn now to the <u>correlation functions</u> which are present in muon capture. We shall but mention the spin-spin correlations in the initial state (hyperfine effect)[17,18]

and the historically more controversial correlation measurements
involving the muon spin and the neutron-momentum which are discussed in detail elsewhere at this Conference. As for the experimental aspect of the other correlation measurements let us note
that a., correlations involving the spin of the final state can be
observed - in principle - through the polarization and/or distribution of the subsequent decay-product and that with the advent of
high-resolution GeLi-detectors, the nuclear recoil and its correlations can be observed through the Doppler-effect they induce on
a gamma-ray emitted shortly after capture[19]. These experiments
seem - however - to lie still on the borderline of actual possibilities and may have to wait - for a more methodic exploitation -

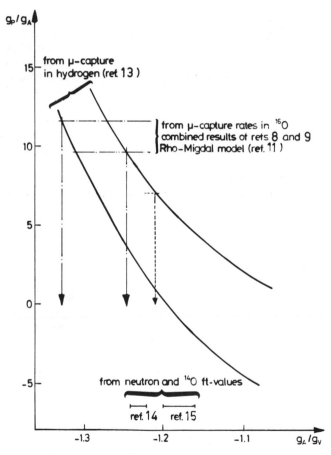

Fig.1:   Comparison of g_A/g_V as extracted indirectly from suitable
         muon-capture measurements and directly from neutron life-time.
         (The thin arrow is representative of the model-dependence).

the advent of higher muon fluxes. Let us discuss - nevertheless -
the information one may hope to gather from these correlation
measurements and the limited amount of experimental information
which is already at hand. For convenience we shall discuss first
capture to the bound and later to the unbound levels of the targets
isobar.

6., The correlation functions involving the <u>bound</u> levels is of
much theoretical interest[20],[21]; it was shown - in particular - that
in the case of "unique" muon capture transitions their sensitivity
to the basic interaction (induced pseudoscalar, G-parity violation,
neutrino helicity) was rather high and - in first order - independent of nuclear structure. Despite these interesting issues, to our
knowledge - no such correlation measurement was reported as yet.

7., Let us turn now to the case of <u>unbound</u>, particle unstable,
final states noting briefly the information the correlation measurements may yeild on the capture-mechanism itself.

In order to account for the muon capture rates, it was proposed
by different authors[22],[23] to consider this process as exciting
mainly the isobar analogue states of the giant resonance levels
excited by photoreactions in the target nucleus. The lepton field
would induce in the nuclear matter (mostly dipole and possible
quadrupole) isospin- and spin-isospin-oscillations. This model does
not consider some features of muon capture such as emission of high-
energy neutrons (probably due to direct reactions) and "boil-off"
phenomena, but probably describes correctly a dominant aspect of
muon capture as shown by the resonance-structure observed in the
neutron spectra[24] and the very nice parallelism found in the gamma
spectra of ^{15}N excited in $^{16}O(\gamma,p)$ ^{15}N and ^{16}O $(\mu,\nu n)$ ^{15}N reactions
[25],[26]. It should be noted also that the measured transition rate
to the 6322 keV ^{15}N-level $(2.50\pm0.23\times10^4$ $s^{-1})[26]$ was predicted to be
3×10^4 s^{-1} by the giant resonance capture hypothesis: an impressive
agreement, indeed.

It is of obvious interest to seek for supplementary and -
possibly - more stringent tests of this capture mechanism. The
neutrino-neutron correlation and the one between the de-excitation
gamma-ray and the recoil of the emitting nucleus could be such
tests: they can be predicted in the framework of the giant capture
mechanism and would be accordingly related to corresponding radiative
pion capture and photonuclear processes. In the following we would
like to show that some limited experimental information exists already along these lines and urge a theoretical computation of the
corresponding correlations.

Two Doppler-broadened gamma-rays were observed from nuclei
produced after neutron-emission form an unbound state of the targets

insobar: from [15]N [25] and from [31]P [27]; let us consider the
information they yield.

    Fig. 2 shows the smeared line-shape of the 6322 keV gamma-ray
observed after muon-capture in [16]O [25]; the life-time of the emitting
state is hsort enough to ascertain complete Doppler-broadening. We
show also - for comparison - the line-shape expected on the basis
of the giant resonance capture mechanism[26] assuming all the involved
directional correlations to be isotropic. The agreement is qualitatively good: the same conclusion was reached also by the authors
of ref. 26 in a contribution to this conference. They note moreover, that the accuracy of the data is not sufficient to draw
quantitative conclusions from possible deviations between the two
curves.

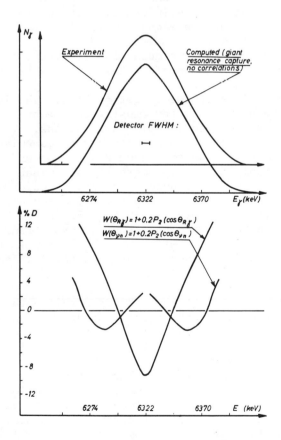

Fig. 2:   Comparison of the experimental and computed line shapes of
          the 6322 keV gamma-ray observed capturing muons in [16]O.
          The experimental resolution was folded in. D. is the
          fractional deviation from "isotropic" line shape.

In order to illustrate the sensitivity of the line-shape to the directional correlations, we indicate in Fig. 2 also the fractional deviation $D = (N_\alpha - N_{\alpha=0})/N_{\alpha=0}$ of the line-shape from the "isotropic" value for reasonable $[1 + \alpha P_2 (\cos\Theta)]$ correlations. (A moderate P_2-contribution to the correlations is expected on the base of photo-reaction-measurements[29]. The gamma-neutron correlation of the corresponding radiative pion capture reaction, however, is nearly isotropic[30]. As we stressed already, we have unfortunately no information as yet on the correlations expected from the hypothesis of giant resonance capture).

A Doppler-broadened line – having an intrinsic FWHM of 23±2 keV was observed also in ^{31}P formed in muon capture ^{32}S. In Fig.3 we compare this line-width to the expected value as a function of neutron-energy (the neutino-neutron correlation was assumed isotropic, the isotropy of the recoil-gamma correlation is guaranteed by the 1/2 spin of the emitting state).

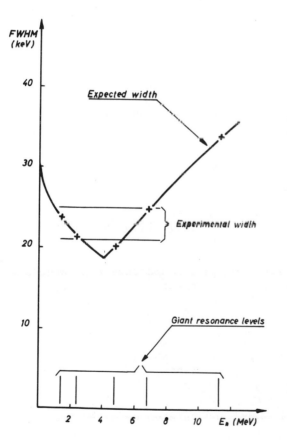

Fig. 3 – Experimental and computed intrinsic line-width (expressed as full-width at half-maximum : FWHM) of the 3135 keV gamma-ray observed capturing muons in ^{32}S.

Fig. 3 shows also the expected position of the neutron-groups
assuming giant resonance capture. Here also, the experimental
broadening agrees reasonably with the one predicted on the basis
of the giant resonance capture mechanism.

In conclusion the line shape of the Doppler-broadened gamma-
rays observed in muon-capture to unbound levels is in qualitative
agreement with the gross features predicted by the giant-resonance
capture mechanism. These line-shapes are sensitive to the correlations involved, correlations which could yeild – if computed – a
supplementary test of the capture mechanism. More experimental
work would render this test – of course – more critical. Along
these lines it would seem reasonable not only to repeat the measurements we quoted with higher accuracy, but also to measure the
neutron energy-spectra yielding the gamma-emitting nuclear states
(as suggested already by the authors of ref. 27) and to observe
the gamma line-shape in coincidence with a detector which defines
the direction of the neutron-momentum. One should note also that
isospin and spin-isospin giant resonance levels yeild opposite
(and large) muon spin-neutrino momentum correlations[19]; a measurement of this correlation is possible by the techniques we discussed
and would also be useful to test details of the capture mechanism.

Even pending these new measurements, it would be interesting,
however, to investigate theoretically the directional correlations
expected in muon capture to unbound nuclear states.

The author is indebted to Drs. L. Grenacs and H. Uberall for
illuminating discussions on the correlation respectively giant
resonance problem in muon capture.

REFERENCES

[1]B. Backe et al., submitted to this Conference.
[2]B. Macdonald et al., Phys. Rev. $\underline{139}$, (1965) B1253 and references
cited herein.
[3]T.A.E.C. Pratt, Nuovo Cim. $\underline{61B}$, (1969) 119.
[4]G. Backenstoss et al., submitted to this Conference H.L. Anderson
et al., EFINS preprint, Chicago, 1969 private communication from
Drs. R. Engfer, D. Kessler and R. Welsh.
[5]C.W. Kim and H. Primakoff, Phys. Rev. $\underline{140}$, (1965)B566.
[6]J.P. Deutsch et al., Phys. Letters $\underline{26B}$, (1968)315.
[7]H. Primakoff, private communication.
[8]J.P. Deutsch et al., Phys. Letters $\underline{29B}$, (1969)66.
[9]R.C. Cohen et al., Nucl. Phys. $\underline{57}$, (1964)255.
[10]A.Astbury et al., Nuovo Cim. $\underline{33}$, (1964)1020.
[11]M.Rho, Phys. Rev. Letters $\underline{18}$, (1967)671 and Phys. Rev. $\underline{161}$.
[12]R.A. Eramzhyan, Dubna Rep. E4-4092 and references cited therin
A. Fujii et al., Progr. Theoret. Phys.(Japan), Suppl. extra

n°(1968)303 and references cited therin.

[13] A. Alberigi Quaranta et al., Phys. Rev. 177, (1969)2118.

[14] C.J. Christensen et al., Phys. Letters 26B, (1967)11.

[15] A.N. Sosnovsky et al. Nucl. Phys. 10, (1959)395.

[16] For a discussion of this assumption, see: M.Rho in CEA Rep. DPh-T/DOC/68-44/EC,19

[17] G. Culligan et al., Phys. Rev. Letters 7, (1969)458.

[18] J.P. Deutsch et al., Phys Letters 28B, (1968)128.

[19] L. Grenacs et al., Nucl. Inst. Methods 58, (1968)1964
T.A.E.C. Pratt Nucl. Methods 70, (1969)213.

[20] M. Morita Phys. Rev. 161, (1967)1028 and references cited therin.

[21] A.P. Bukhvostov and N.P. Popov Sov. Journ. Nucl. Phys. 6(1968)589 and references cited therein.

[22] L.L. Foldy and J.D. Walecka Nuovo Cim. 34, (1964)1026
V.V. Balashov et al., Phys. Letters 9, (1964)168.

[23] II. Uberall, Suppl. Nuovo Cim. 4, (1966)781 and references cited therein.

[24] V. Evseyev et al., Phys. Letters 28B, (1969)553 and submitted to this conference.

[25] R.O. Owens and J.E.E. Baglin, Phys. Rev. Letters 17, (1966)1268.

[26] S.N. Kaplan et al., Phys. Rev. Letters 22, (1969)795.

[27] R.Raphael et al., Phys. Letters 24B, (1967)15.

[28] Louvain group, to be published.

[29] E.D. Earle and N.W. Tanner, Nucl. Phys. A95, (1967)241
J.M. Maison, Thèse, Orsay, 1969.
R.J.J. Stewart et al., Phys. Rev. Letters 23, (1969)323.

[30] K. Gotow, contribution to this conference.

DISCUSSION

S.N. Kaplan: I would like to make a brief remark apropos our published neutron-multiplicity results, about the fraction of μ^- nuclear captures leading to no neutron emission. The manner in which our data are presented has led to some misinterpretation and this seems an appropriate opportunity to set the record straight. We do not give explicitly the probabilities for no neutron emission. These can be calculated from our published numbers by correcting for our neutron detection efficiency of 54.5%. The results (in percent of no-neutron emission per μ^- capture) are:

Al	Si	Ca	Fe	Ag	I	Au	Pb
8±4	36±6	37±3	19±3	7±3	4±3	11±2	0±3

NON-RADIATIVE MUON TRANSITIONS IN HEAVY ELEMENTS

D. Kessler, V. Chan, C.K. Hargrove, E.P. Hincks,
G.R. Mason, R.J. McKee, and S. Ricci
National Research Council-Carleton University
Collaboration, Ottawa, Canada

(Presented by D. Kessler)

We have recently investigated a new type of process which takes place in heavy muonic atoms where it competes with the ordinary electromagnetic (radiative and Auger) transitions which make up most of the mu-atomic cascade process. The new process consists in the radiationless de-excitation of a muonic atom accompanied by an excitation of the nucleus. In many of the cases observed, but not always, the excitation energy is high enough so that neutron emission can take place from the excited nucleus. We still lack a complete understanding of this process, and I will, therefore, concern myself primarily with a presentation of the experimental facts. Before starting, I should mention, however, that one particular type of radiationless transition, namely that leading to fission, was predicted by Wheeler[1] as early as 1949. This process was later studied experimentally as well as theoretically, mainly by Russian groups, and is rather well understood today[2].

Our first observation was the occurrence of the so-called prompt γ rays, first reported by Hargrove[3] at the Ottawa Conference. These are lines which appear in the prompt spectrum together with the muonic x rays, but correspond to well known nuclear transitions. They differ from the capture γ rays in two respects. First, as their name indicates, they appear mainly in the prompt spectrum as opposed to the delayed spectrum which contains most of the capture γ rays. Second, they can be identified as belonging to isotopes of the target element, and not, like capture γ rays, to isotopes of the element with one charge unit less. Thus, all the identified capture γ rays from lead targets belong to thallium isotopes, whereas the prompt γ appear to be associated with transitions in lead. Table I shows

the intensities of prompt γ rays we observe with targets of
separated isotopes of lead.

E_γ Isotope	Pb^{206}	Pb^{207}	Pb^{208}
	Intensity per stopped muon (percent)		
803 keV Pb^{206}	5.3 ± 0.2	5.1 ± 0.3	—
570 keV Pb^{207}	—	4.2 ± 0.2	3.8 ± 0.3
897 keV Pb^{207}	—	1.2 ± 0.3	2.2 ± 0.5

Table I: Observed prompt γ ray intensities from separated isotope
lead targets.

Table I shows that the 803 keV line from the first excited state
of Pb^{206} is observed in 5.3% of all muon stops in a Pb^{206} target.
In Pb^{207}, we observe not only prompt γ rays from the first two
excited states in Pb^{207} at 570 and 897 keV, but also the 803 keV
line from Pb^{206}. Finally, in Pb^{208} we observe again the two
transitions from the first two excited states of Pb^{207}. So far,
we have only weak evidence for the presence of the 2.615 MeV
transition from the first excited state to the ground state of
Pb^{208}. Thus we observe two types of processes: one corresponds
to the excitation of the target nucleus itself, whereas the second
process apparently involves the emission of a neutron followed
by the de-excitation of the isotope with one neutron less.

Following the observation of the prompt γ rays we started
a series of experiments to study the neutrons which are apparently
emitted in many of the radiationless excitation processes. A
schematic drawing of the experimental setup is shown in Fig. 1.
It consists essentially of a muon telescope which signals muon
stops in the chosen target. The neutrons are detected by a neutron
scintillation counter connected to a pulse shape discriminator
which was used to reduce the considerable background of x and γ
rays. The neutron time of flight from the target to the neutron
counter was measured and recorded in a multi-channel analyzer.
The prompt neutrons which we are interested in must be measured
in the presence of a much higher yield of capture neutrons. In
order to eliminate these unwanted neutrons, we have surrounded the
target with a delayed coincidence (DC) counter and require that
this counter be triggered by a capture γ ray or neutron after the

time set aside for measuring the prompt neutron time-of-flight.
The basic timing sequence in Fig. 1b shows how this works: neutrons
are detected during the 50 ns which follow immediately a muon stop,
but they are recorded only if they are followed by a signal from
the delayed coincidence counter during the time interval 60-150 ns.
Experience showed that this requirement was essential to the success
of the experiment.

Of the elements studies so far (gold, separated lead isotopes,
and bismuth) bismuth turned out to have the highest yield of prompt
neutrons. The neutron time-of-flight spectrum obtained using a
bismuth target is shown in Fig. 2a. Due to the comparatively long
time windows used in such a time-of-flight experiment, one can
expect a rather important accidental background. A detailed analysis
of the background shows that it is contributed mainly by real delayed
events combined with accidental neutron counter triggers or by real
neutron events followed by an accidental coincidence in the DC
counter. In order to get a pure background spectrum we use a
silver target, because it is energetically impossible for prompt
neutrons to be emitted from silver, but otherwise the conditions
for silver are very similar to those in bismuth. In particular,
the muon-capture half-life is roughly the same in the two elements,
namely about 50 ns, which is an important consideration in this
experiment which requires the detection of a delayed capture event
within a given time window. The silver spectrum is shown in Fig. 2b
and corresponds to roughly the same number of muon stops as the
bismuth spectrum in Fig. 2a. The silver spectrum has no apparent
structure, except for the prompt x-ray peak, which is due to muonic
x rays detected by the neutron detector and which have not been
rejected by the pulse shape discriminator. When the background
estimated from the silver spectrum is subtracted from the bismuth
spectrum, we are left with a strong signal. This signal has two
characteristic features which are the steep rise at an energy
corresponding to the series limit of the muonic atom with the
daughter nucleus left in the ground state and a large peak corresponding to a $3 \rightarrow 1$ transition of the muonic atom. Other transitions
must contribute to the neutron spectrum to explain the width of
the large peak which is wider than expected from the resolution of
the apparatus and the occurence of a bump around 1.8 MeV neutron
energy.

There are two overlapping particle hole multiplets of Bi^{208}
in the energy interval 930-1150 keV, as described by Alford, Schiffer
and Schwartz[4]. The neutron energies corresponding to the situation
where the daughter nucleus Bi^{208} is left in one of those closely
spaced excited states, is indicated in the figure. It is likely
that these particle-hole states play a significant role in the de-
excitation process. There is another particle-hole multiplet
around 635 keV energy, not indicated in the figure, which may

conceivably also contribute. The total number of events above
background leads to an estimate of 7±2% for the number of prompt
neutrons in bismuth.

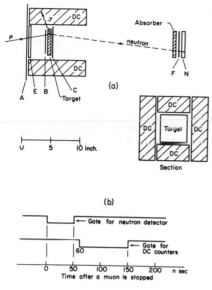

Fig. 1. (a) Experimental set-
up. A,B,C,DC, and F are
plastic scintillation counters
E is a lucite Cerenkov counter
and N is the neutron detector.
(b) Basic timing sequence.

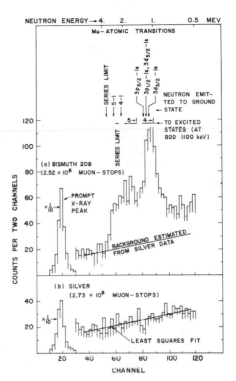

Fig. 2. (a) Neutron time-of-flight
spectrum obtained with a bismuth
target. Transitions likely to
contribute to this spectrum are
indicated at the top of the figure.
(b) Neutron time-of-flight spectrum
obtained with a silver target.

Fig. 3 shows preliminary data obtained recently with a 88%
pure pb207 target. The neutron time-of-flight spectrum shows here
again a marked peak corresponding to the 3-1 transition in muonic
lead with the daughter nucleus left in the ground state. Higher
muonic transitions do not seem to contribute significantly, except
close to the series limit. The steep rise at the series limit is
less pronounced here than in bismuth. The shape of the curve
suggests, however, that those series where the daughter nucleus is
left in an excited state contribute significantly. The fraction
of prompt neutrons emitted in Pb207 is only 73% that obtained in
bismuth, or about 5% per muon stop.

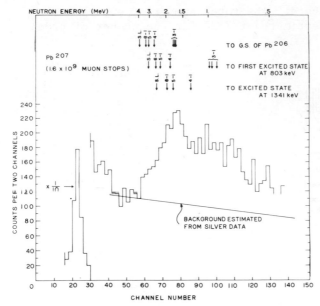

Fig. 3 - Neutron time-of-flight spectrum obtained with Pb^{207}.

Radiationless transitions leading to neutron emission can probably be interpreted with the help of a Zaretsky-type mechanisme.[5] The partial width, Γ_{nr}, for this effect seems to be of the order of the width for radiative transitions, namely of the order of 1 keV. This is much larger than the level spacing $1/\rho$ at the excitation energy involved which should be around 1 eV, so that we have the case of $\rho\Gamma \gg 1$ with $\Gamma_{nr} \simeq \Gamma$, treated by Zaretsky in his second paper. In bismuth, the $2\to 1$ transition does not have enough energy for neutron emission, but the $3\to 1$ transition does. This is therefore the most favorable case and it explains the presence of a strong peak at the neutron energy corresponding to this transition. Higher transitions, such as $4\to 1$, $5\to 1$ etc., are progressively weaker because those states which are significantly populated according to cascade calculations are the circular orbits and transitions from such states to the ground state of the muonic atom involve high multipolarities. It may therefore seem surprising that we see a relatively strong signal close to the series limit. However, it should be remembered that such high orbits are likely to be molecular orbits in which the field is no longer centrally symmetric so that the selection rules with respect to orbital momentum are not necessarily obeyed.

The situation is less clear-cut in the case where no neutron is emitted. A resonance mechanism has been invoked by Hüfner[6] to explain the anomalous intensity ration in bismuth. Backe et al.[7] use this mechanism to explain some of the prompt γ rays in bismuth. While such a resonance mechanism is perfectly legitimate, it seems that the phenomenon of radiationless transitions (with or without

neutron emission) is rather wide-spread and one would like to see
a mechanism which would operate in all heavy atoms. One may argue
that the nuclear level density at an excitation energy corresponding to the 2p-1s muonic transition, which is below the neutron
threshold, is probably not down by a large factor from the level
densities which occur slightly above the neutron separation energy.
If this is the case, there is no reason why the Zaretsky mechanism
should not be operative for, say, the 2p-1s transition in some of
the heavy atoms. This is, in fact, the transition Zaretsky has considered for the radiationless fission of uranium. A different possibility is that the muonic transition which induces the nuclear
excitation occurs from a closely spaced and possibly overlapping
band of mu-molecular orbits. Calculations based on such models are
presently being carried out by Srinivasan and Sundaresan. From the
experimental point of view it will be interesting to see whether
the prompt γ rays are restricted to relatively low excitations or
whether one can find evidence of high excitations up to the particle
threshold. Whatever the mechanism, however, we may expect that
the prompt γ rays will provide us with a rich source of cases for
the study of the isomer shift. They may also be used to study nuclear level densities if the Zaretsky mechanism holds.

REFERENCES

[1] J.S. Wheeler, Rev. Mod. Phys. 21, 133 (1949).
[2] A list of references is given in C.K. Hargrove et al., Phys. Rev.
Letters 23, 215 (1969).
[3] C.K. Hargrove, in Proceedings of the International Conference on
Electromagnetic Sizes of Nuclei, Ottawa, Canada (1967), p. 233.
[4] W.P. Alford, J.P. Schiffer, and J.J. Schwartz, Phys. Rev. Letters
21, 156 (1968).
[5] D.F. Zaretsky and V.M. Novikov, Nucl. Phys. 14, 540 (1959/60) and
Nucl. Phys. 28, 177 (1961).
[6] J. Hüfner, Phys. Letters 25B, 189 (1967).
[7] H. Backe, R. Engfer, E. Kankeleit, W.U. Schröder, H.K. Walter,
and K. Wien, paper presented to this Conference.

NEUTRON SPECTRA AND ASYMMETRIES IN MUON CAPTURE*

Ronald M. Sundelin** and Richard M. Edelstein

Carnegie-Mellon University, Pittsburgh, Pennsylvania

A single large discrepancy has plagued muon physics for the
past decade. This discrepancy is in the asymmetry of neutrons
emitted following muon capture. The experiment which is discussed
in this paper, combined with other recent experimental and
theoretical work, is very close to eliminating this discrepancy.
Early theoretical work had predicted a small negative asymmetry,
and two experiments had obtained very large negative asymmetries
at high neutron energies. We now obtain moderate positive asymmetries, and recent theoretical work predicts positive asymmetries.

The asymmetry in the angular distribution of neutrons is
measured with respect to the direction of the muon spin. This
asymmetry is particularly sensitive to the induced pseudoscalar
coupling constant because of the large momentum transfer provided
by the disappearance of the muon rest mass, and because the vector
and axial vector contributions to the asymmetry are expected to
approximately cancel.

The earliest theoretical calculations predicted a negative
asymmetry with magnitude less than 0.4 [1]. A more recent calculation by Klein, Neal, and Wolfenstein [2], which used the Fermi gas
model with inclusion of terms proportional to the capturing proton
momentum, predicted a negative asymmetry with magnitude less than 0.1.

A number of experiments have been performed to measure the
neutron asymmetry in spin-0 nuclei, including calcium and lighter
elements. These experiments have produced a variety of results.
The first experiment in which background from capture gamma rays was

---

*Work supported by the U. S. Atomic Energy Commission.
**Present address: Cornell University, Ithaca, New York 14850.

eliminated was performed by Baker and Rubbia [3]. Interestingly, their three points are all positive. The next two experiments which were performed used low neutron energy thresholds [4, 5]. These experiments obtained negative asymmetries around -0.3. These results were also surprising in view of the expected dominance of evaporation neutrons at these low energies. Since evaporation neutrons involve an intermediate nuclear state, they cannot exhibit an asymmetry. The next experiment, done at Dubna by Evseev and others [6], measured the asymmetry at a number of different thresholds. This experiment obtained very large negative asymmetries at high neutron energies, with values around -1.0 being obtained for thresholds above 17 MeV. This result was also unexplainable in the existing theory using coupling constants close to the expected ones. The next experiment, done at Columbia by Anderson and others [7], used two different techniques. One method used an oscilloscope and film to record the time between the muon stop and the observation of the neutron event, and to record the pulse height produced by the recoil from the neutron. This method produced large negative asymmetries, near -1.0 at high energies, in good agreement with the Dubna result. The second method used a digital timing system to measure the time between the muon stop and the neutron event. This method produced predominantly small positive asymmetries, with 5 of the 6 points reported being positive, although none of these points was more than 2 standard deviations from zero.

At this point, a group at Carnegie consisting of Edelstein, Suzuki, Takahashi, and myself, decided to do a final experiment to verify the Dubna result, and to close the experimental question once and for all. We performed such an experiment [8], using great care in our experimental technique and accumulating more than three times the data obtained in all previous experiments combined. However, we obtained a very surprising result, and completely reopened the field, instead of closing it. We found small positive asymmetries at low energies, and positive asymmetries between 0.3 and 0.4 at higher energies in Si, S, and Ca. The most significantly positive points were around 15 MeV. The asymmetries remained generally positive up to the highest energy measured, around 50 MeV, but the errors increased with energy. Our technique consisted of precessing stopped muons in a magnetic field, and observing recoils from the neutrons in a liquid scintillation counter. The time between the muon stop and the observation of the neutron event was measured by a digital timing system. This experiment was completed in 1967.

An experiment at Columbia by Sculli and others [9], completed in 1968, used thin plate spark chambers to detect the neutrons. This experiment had the advantage that the neutron energy could be determined directly from the direction and range of the recoil proton, whereas all previous experiments determined neutron energy thresholds from recoil pulse heights. This experiment obtained asymmetries which were generally positive at high energies, in essential agreement with our earlier result.

Edelstein and I have now repeated our original experiment. This experiment was performed at the Carnegie-Mellon University Synchrocyclotron. A number of improvements were made over our first experiment. The most significant of these was the use of a new muon channel, which permitted a considerable reduction in background and an increase in beam intensity. We measured the neutron asymmetry using Si and Ca targets, and accumulated 4 times as much data for each as in our first experiment.

The target was placed at 45° to the beam line. The muons were stopped in the target and were precessed by a vertical magnetic field. A counter which used NE-213 liquid scintillator was placed at right angles to the beam line. This counter was used to detect neutron events by observing the recoils from the neutrons. Events were grouped into bins according to the recoil pulse height. This counter was also used to observe electrons from muon decay. Two thin counters between the target and the neutron detector were used in anticoincidence for neutrons and in coincidence for decay electrons. Neutrons were distinguished from nuclear gamma rays by pulse shape discrimination.

Figure 1 shows our neutron asymmetry results for Si and Ca as a function of the recoil pulse height (expressed in terms of mean proton energy).   Note that the asymmetry is near zero for low energy neutrons, and is around 0.3 at higher energies.   The asymmetries for both Si and Ca agree well with the results of our previous results.

It is possible to unfold the asymmetry as a function of neutron energy from the asymmetry as a function of recoil pulse height. The unfolding can be done with good accuracy because the pulse height spectrum falls off very rapidly with increasing energy, and the

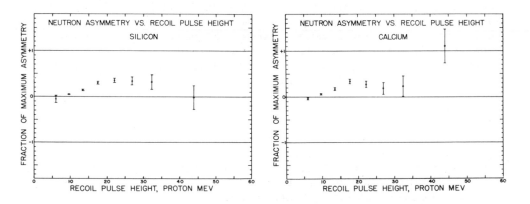

Fig. 1 - Neutron asymmetry vs. recoil pulse height in Si and Ca.

effect of higher energy neutrons on the asymmetry in a given energy
interval is small.   The asymmetry as a function of neutron energy is
therefore quite similar to the asymmetry as a function of pulse
height.   The only additional information required to unfold the asymmetry is the pulse height spectrum and the response of the detector
to monoenergetic neutrons.   A Monte Carlo program was used to
compute the response of the detector to neutrons of various energies.
This program used published neutron scattering cross sections.   The
Monte Carlo results were verified by measuring the response of the
detector to monoenergetic neutrons over the energy spectrum of
interest.

        The (differential) neutron asymmetry as a function of neutron
energy for Si and Ca is shown in Fig. 2.   The most significant
departure from zero asymmetry is still at 17.6 MeV, for both Si and
Ca.

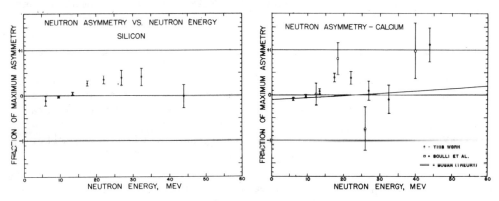

Fig. 2 - Neutron asymmetry vs. neutron energy in Si and Ca.

        Fig. 2 also shows the results of Sculli [9], who obtains the
differential asymmetry directly as a function of neutron energy.
Also shown is a recent theoretical calculation by Bogan [10].

        Bogan's calculation uses the shell model with the inclusion of
terms proportional to the momentum of the capturing proton.   Previous
shell model calculations, which predicted negative asymmetries, did
not include these terms.   Bogan does the calculation with and without
the inclusion of these terms, and obtains negative asymmetries when
they are not included, and obtains the curve shown when they are included.   As mentioned earlier, inclusion of such terms in the Fermi
gas model calculation made the asymmetry less negative.

        The energy spectrum of the neutrons emitted following muon
capture is also of considerable interest.   The energy spectrum is

much more sensitive to the nuclear model used than is the asymmetry, so the spectrum may be of use in determining which model is most appropriate for computing the asymmetry. In addition, the spectrum may be of use in distinguishing the three sources of emitted neutrons: direct, giant dipole resonance, and evaporation. Since the giant dipole resonance and evaporation neutrons involve an intermediate nuclear state, they cannot exhibit an asymmetry.

A number of theoretical calculations of the direct spectrum have been made. A Fermi gas model calculation by Überall [11] falls linearly to zero at 17 MeV. A shell model calculation by Lubkin [12] goes to zero at 11 MeV for Ca^{40}. A shell model calculation by Dolinskii and Blokhintsev [13] goes to zero around 23 MeV. Bogan also has computed the direct neutron spectrum using the shell model. He assumes plane wave states for the emitted neutrons while inside the nucleus, and uses the effective mass of 0.6 characteristic of the Brueckner-Goldstone theory of nuclear matter. The spectrum he calculates rises to a peak around 5 MeV, then falls in a roughly exponential manner to zero around 100 MeV, with an exponential decay constant of 6.8 MeV at 40 MeV.

In our experiment, the same Monte Carlo calculations which were used to unfold the asymmetry were used to unfold the energy spectrum. The energy spectrum was normalized by using the number of decay electrons observed to determine the number of muons which stopped in the target. Fig. 3 shows our neutron energy spectra for Si and Ca. Note that they both have an exponential shape with a decay constant of about 6.5 MeV, in agreement with our earlier measurements.

Two experiments to measure the neutron energy spectrum had been performed prior to our first experiment. One was a bubble chamber experiment by Turner [14] and the other a recoil counter experiment by Hagge [15]. Both of these experiments used calcium as one of their targets, and obtained results which are in good agreement with ours.

A recent measurement of the spectrum in Ca, S, and Pb has been performed by Krieger [16]. This was the same experiment in which the asymmetry was measured by Sculli. Krieger's spectrum for calcium also falls off exponentially, but the decay constant which he obtains is 14 MeV, which is not in good agreement with our results. However, the integrals of the two spectra above 7 MeV are in disagreement by less than 1.6 standard deviations.

Prior to Bogan's calculation, all of the theoretical calculations were in poor agreement with the experiments, as they did not predict the large high energy tails observed. Above 20 MeV, Bogan's calculation is in excellent agreement with our measurement. Below 20 MeV, our spectrum is higher than Bogan's calculated spectrum. This behavior is expected, since our measurement includes all types of neutrons, and Bogan's spectrum includes only directly emitted neutrons.

In conclusion, it is interesting to note that the results of experiments to measure the neutron asymmetry have made a complete cycle.   The first results were moderately positive, followed by several results which were highly negative. The three most recent experiments have yielded moderately positive results. Meanwhile, refinements in the theoretical calculations have caused the predicted asymmetries to change from moderately negative to slightly positive at high energies. The agreement between the most recent experiments and theory is still not extremely good, but it is an order of magnitude better than it was a few years ago.

All measurements of the neutron spectrum are in order of magnitude agreement, and refinements in the theoretical calculations have brought the predictions into excellent agreement with the most recent measurement.

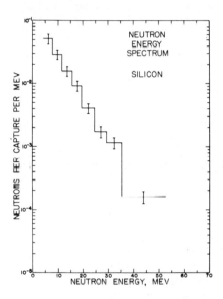 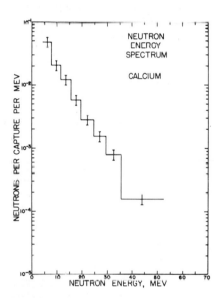

Fig. 3.   Neutron energy spectrum in Si and Ca.

It would be nice if   this major discrepancy in the study of muon physics could be completely removed. We hope that more work will be done on the asymmetry theory to bring it into even better agreement with the experiment. We find Bogan's calculation very interesting because it produces positive asymmetries at high energies. However, more theoretical work is required to obtain asymmetries of even higher precision and to estimate the residual uncertainties in these asymmetries.

The participation of A. Suzuki and K. Takahashi in the first experiment, and the assistance of P. D. Gupta in the second is gratefully acknowledged.

REFERENCES

(1) H. Primakoff, Rev. Mod. Phys. 31, 802 (1959).
(2) R. Klein, T. Neal, and L. Wolfenstein, Phys. Rev. 138B, 86 (1965).
(3) W. F. Baker and C. Rubbia, Phys. Rev. Letters 3, 179 (1959).
(4) A. Astbury, I. M. Blair, M. Hussain, M. A. R. Kemp, H. Muirhead, and R. G. P. Voss, Phys. Rev. Letters 3, 476 (1959), and A. Astbury, J. H. Bartley, I. M. Blair, M. A. R. Kemp, H. Muirhead, and T. Woodhead, Proc. Phys. Soc. 79, 1011 (1962).
(5) V. L. Telegdi, Proceedings of the Rochester Conference on High Energy Physics, 713 (1960).
(6) V. S. Evseev, V. I. Komarov, V. Z. Kush, V. S. Roganov, V. A. Chernogorova, and M. M. Szymczak, JETP 14, 217 (1962), V. S. Evseev, V. S. Roganov, V. A. Chernogorova, M. M. Szymczak, and Chang Run-Hwa, Conference on High Energy Physics at CERN, 425 (1962), V. S. Evseev, V. I. Komarov, V. Kush, V. S. Roganov, V. A. Chernogorova, and M. Szymczak, Acta Physica Polonica 21, 313 (1962), V. S. Evseev, V. S. Roganov, V. A. Chernogorova, Chang Run-Hwa, and M. Szymczak, Physics Letters 6, 332 (1963), V. S. Evseev, V. A. Chernogorova, F. Kilbinger, V. S. Roganov, and M. Szymczak, Joint Institute for Nuclear Research, Dubna, Report E-2516 (1965), and V. S. Evseev, et al., Joint Institute for Nuclear Research, Dubna, Report E-2517 (1965).
(7) E. W. Anderson and J. E. Rothberg, BAPS 10, 80 (1965), and E. W. Anderson, Columbia University, New York, Report NEVIS-136 (1965).
(8) R. M. Sundelin, Carnegie Institute of Technology, Pittsburgh, Report CAR-882-22 (1967), and R. M. Sundelin, R. M. Edelstein, A. Suzuki, and K. Takahashi, Phys. Rev. Letters 20, 1198 and 1201 (1968).
(9) J. Sculli, Columbia University, New York, Report NEVIS-168 (1969).
(10) A. Bogan, Phys. Rev. Letters 22, 71 (1969).
(11) H. Überall, Nuovo Cimento 6, 533 (1957).
(12) E. Lubkin, Annals of Phys. 11, 414 (1960).
(13) E. Dolinskii and L. Blokhintsev, JETP 8, 1040 (1959).
(14) L. Turner, Carnegie Institute of Technology, Pittsburgh, Report CAR-882-5 (1964).
(15) D. Hagge, University of California, Berkeley, Report UCRL-10516 (1963).
(16) M. Krieger, Columbia University, New York, Report NEVIS-172 (1969).

NEUTRON ENERGY SPECTRA FROM NEGATIVE MUON ABSORPTION ON COMPLEX NUCLEI

V. Evseev, T. Kozlowski, V. Roganov, J. Woitkowska

Joint Institute for Nuclear Research

Dubna, U.S.S.R.

As has been suggested in /1-3/ the basic role in the muon capture belongs to the mechanism according to which the collective quasi-bound states of the intermediate nucleus are excited. These states are the isotopic analog ones shaping the "giant dipole resonance" in photonuclear reactions and inelastic electron scattering.

The exact decay mechanism of such states occurring in mu capture has been suggested in refs. /1,3/. According to these two mechanisms neutron energy spectra from nuclear mu capture have been calculated in /1,3,4/. These spectra are of characteristic linear shape. According to these calculations the width of quasi-bound excited states (determining the line width in the neutron spectrum) should be , in some cases , about 10^5eV/4/.

The results of measurements of the neutron energy spectra in the energy range from 1.5 to 13 MeV from mu capture in 0, S, Ca and Pb have been described. Earlier we published some preliminary results for the case of S and Ca /5/. The experiment was performed by using 158 MeV/c pure muon beam obtained by means of the meson channel of the 680 MeV Dubna synchrocyclotron (Laboratory of Nuclear Problems). The apparatus lay-out differs from that shown in ref./5/ only by the fact that the neutron spectrometer is placed at an angle of 90° to the meson beam. Water, melted sulphur, metallic calcium and metallic lead were 2 g/cm^2, 4 g/cm^2, 4 g/cm^2 and 6 g/cm^2 thick, respectively, in the direction of the neutron spectrometer. A stilbene crystal 30 mm in diameter and 20 mm thick with a photomultiplier 56-AVP was used as a neutron spectrometer. In order to separate neutrons and gamma-quanta the method of pulse shape discrimination

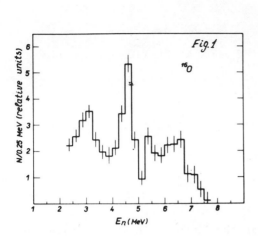

Fig.1

^{16}O

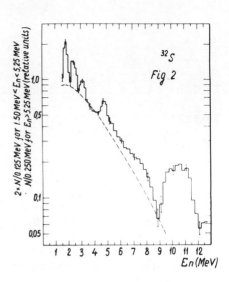

^{32}S

Fig.2

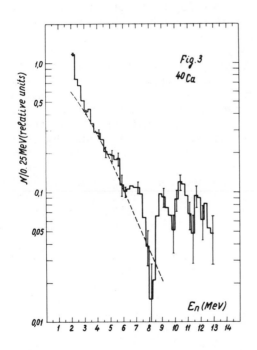

Fig.3

^{40}Ca

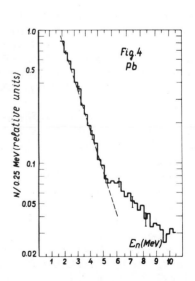

Fig.4

Pb

was applied.   The energy calibration of the neutron spectrometer
was performed by employing standard gamma-sources and a neutron
source (Po-Be).

The recoil-proton amplitude spectrum,detected with an amplitude
analyzer,was separated into 0.25 MeV intervals (or 0.125 MeV intervals within $\leqslant$ 5.25 MeV for the case of sulphur).   Neutron spectra
were obtained by the differentiation of proton spectra.   The final
neutron energy resolution (FWHM) was about 0.5 MeV for E_n = 2.0 MeV,
0.6 MeV for E_n = 5.0 MeV and 1.0 MeV for E_n = 12 MeV for a long
period of operation.   The absolute accuracy of energy scale determination was about ± 5%.

Figs. 1-4 show neutron spectra for 0, S, Ca and Pb, respectively.   As is seen from Figs. 1-4, the structure of neutron
spectra from mu capture in 0, S and Ca evidences for peculiarities
in the form of separate lines or wider bands.   For S, Ca and Pb
a smooth spectrum of "evaporational" character is observed.   The
dashed line of Figs. 2-4 is the "evaporational" spectrum calculated
by assuming temperature values of the final nucleus to be 1.7 MeV,
1.5 MeV and 1.15 MeV /6/ for mu-capture in S, Ca and Pb, respectively.   The relative contribution of this part of the spectrum
increases with increasing the atomic weight.

The observation of the line structure in the neutron energy
spectra from mu-capture is the first direct evidence for the
existence of mu-capture mechanism suggested in refs. /1-3/ and the
decay mechanism /1,3/ of quasi-bound states of the intermediate
nucleus.

References

1.   V. V. Balashov, V. B. Belyaev, N. M. Kabachnik, E. A. Eramzhian,
     Phys. Lett. 9 (1964) 168.

2.   J. Barlow, J. C. Sens, P. Duke and M. R. Kemp, Phys. Lett. 9
     (1964) 84.

3.   H. Uberall, Nuovo Cim. (Suppl.) 4 (1966) 781. Acta Physics
     (1969) to be published.

4.   V. V. Balashov, E. A. Eramzhian. Preprint P2-3258 (1967).

5.   V. Evseyev, T. Kozlowski, V. Roganov, J. Woitkowska.   Phys.
     Lett. 28B (1969) 553.

6.   A. Gilbert, A. G. Cameron. Can.J. Phys. 43 (1965) 1446.

NUCLEAR γ-RAYS FOLLOWING μ-CAPTURE

G. Backenstoss, S. Charalambus, H. Daniel, U. Lynen,
Ch. v.d. Malsburg, G. Poelz, H. Povel, H. Schmitt, ·
K. Springer and L. Tauscher

CERN, Geneva, Switzerland
Max-Planck-Institut für Kernphysik, Heidelberg, Germany
Institut für Experimentelle Kernphysik, Karlsruhe,
Germany
Physik-Department der Technischen Hochschule, München,
Germany

After the cascade process in muonic atoms, the muon decays or is captured by the nucleus from the 1s level through the weak interaction $\mu^- + (Z,A) \rightarrow (Z-1,A) + \nu_\mu$, where a proton is changed into a neutron. The greater part of the liberated energy, of 107 MeV, is carried off by the neutrino, and depends on the momentum of the initial interacting proton; the energy left for the neutron varies from ~ 6 MeV for the proton at rest to some tens of MeV. This energy is distributed among the other nucleons of the nucleus (Z-1,A) which is de-excited predominantly by emission of one or more neutrons. After neutron emission, the resulting nucleus may be in an excited state and its de-excitation takes place by γ-ray emission. We have measured the nuclear γ-rays present in the μ-mesic spectra and assigned them to excited states of nuclei (Z-1,A-X) where X = 0, 1, 2, ... is the number of emitted neutrons. Here we report on γ-ray intensities per muon capture present in the muonic spectra of Mn and Co.

The experiments were performed at the CERN SC muon channel. The usual telescope techniques were applied in conjunction with Ge-detectors. An energy range from 0-1.7 MeV was investigated. The spectra were analysed with a computer programme, which gives the position and the area under the lines. Corrections for detector efficiency and self-absorption in the target have been applied. The yields of γ-rays formed per muonic atom have been determined; this can be done by comparison of the γ-ray intensities with those

Table 1

^{55}Cr		^{54}Cr		^{53}Cr		^{52}Cr	
Energy keV	Yield %	Energy keV	Yield %	Energy keV	Yield %	Energy keV	Yield %
		834.8	56.5 ± 5	563.9	4.2 ± 1.5	1433.5	9.7 ± 3.0
		988.8	22 ± 3	1006.2	9.2 ± 2.5		

Level schemes:

- ^{55}Cr: 574 —— ; 248 —— ; 0
- ^{54}Cr: 1830 —— 4+ ; 835 —— 2+ ; 0
- ^{53}Cr: 1000 —— $\frac{5}{2}-$; 563 —— $\frac{1}{2}-$; 0 $\frac{3}{2}-$
- ^{52}Cr: 2369 —— 4+ ; 1433.6 —— 2+ ; 0 8+

	57 ± 5		13.4 ± 2.9		9.7 ± 3.0	Σ 80.1

Table 2

^{59}Fe		^{58}Fe		^{57}Fe		^{56}Fe		^{55}Fe	
Energy keV	Yield %	Energy keV	Yield %	Energy keV	Yield %	Energy keV	Yield %	Energy keV	Yield %
473.3	2.45±0.5	810.9	60 ±6	122	3.83±0.5	846.9	6.7±1.5	936.6	6.9±1.6
		864.7	3.1±1	136.4	1.35±0.3				
		1677	6.5±2.5	347	3.83±0.5				
				692.3	3.13±0.9				

Level schemes:

- ^{59}Fe: 473 —— $\frac{5}{2}-$; 287 —— $\frac{1}{2}-$; 0 $\frac{3}{2}-$
- ^{58}Fe: 1675 —— 2+ ; 810.5 —— 2+ ; 0
- ^{57}Fe: 706 —— $\frac{5}{2}-$; 367 —— $\frac{3}{2}-$; 136 —— $\frac{5}{2}-$; 14.4 —— $\frac{3}{2}-$; 0 $\frac{1}{2}-$
- ^{56}Fe: 2085 —— 4+ ; 847 —— 2+ ; 0
- ^{55}Fe: 933 —— $\frac{5}{2}-$; 413 —— $\left(\frac{1}{2}\right)-$; 0 $\frac{3}{2}-$

2.5±0.5		66.5±7		12.3±1.3		6.7±1.5		6.9±1.6	Σ 94.9

Yields of nuclear γ-rays per captured muon in Mn (Table 1) and in Co (Table 2). The excited levels are taken from the Table of isotopes by Lederer, Hollander, and Perlman, 6th edition (1967) except for ^{55}Cr which is taken from the Decay scheme of radioactive nuclei by Dželepov and Peker (1966) and ^{55}Fe from the Nuclear Data Sheets.

of the muonic K-series, the sum of which should add up to 100%,
except for a small calculable correction due to the feeding of the
2s level. The fact that the γ-lines are delayed with respect to
the telescope signal was taken into account by using a sufficiently
long coincidence gate or by an appropriate correction. In order to
obtain the γ-ray yield per muon captured by the nucleus, a correction ($\sim$10%) was applied for the decay of muons in the 1s level.

   In Tables 1 and 2 the γ-ray yields per muon capture are given.
Using the decay schemes as indicated in the tables, one obtains the
number of nuclei (Z-1,A-X) per captured muon, which were formed via
their excited states, as shown in the last lines of the tables.
Where necessary, a correction due to internal conversion was applied
in order to obtain the total yield of the excited state. When adding up these numbers one finds 80% and 95% nuclei (Z-1,A-X) per
muon capture for Mn and Co, respectively. This means that the
majority of these nuclei must be formed via their observed excited
states, which implies that the neutron emission probabilities are
not substantially different from the numbers given in the tables.
Nuclei (Z-1,A-1) amount to about 70%. Particular attention was
given to γ-rays resulting from zero neutron emission. If a 248 keV
γ-line from ^{55}Cr exists, this weak line coincides with the strong
3d-2p muonic X-ray line, and it would be difficult to separate it.
The same problem exists in the case of ^{59}Fe; the 287 keV line coincides with the muonic 3d-2p transition. However, a γ-line at
473.3 keV was observed with a yield of 2.5% per muon capture which
can be assigned to the second excited level of ^{59}Fe. Further
studies of nuclear gammas following μ-capture is in progress for
the case of Nb, I, and Bi.

DOPPLER BROADENING OF THE 6.322-MeV γ-RAY FROM ^{15}N FOLLOWING μ^- CAPTURE IN ^{16}O

S. N. Kaplan, R. V. Pyle, L. E. Temple, and
G. F. Valby

Lawrence Radiation Laboratory

University of California, Berkeley, California

The very broad, 6.322-MeV ^{15}N line observed[1] following μ^- capture in ^{16}O can be fitted with a simple kinematic model by using the neutron energy spectra predicted in the resonance-capture calculation of Raphael, Uberall, and Werntz.[2] If we assume complete isotropy in the reaction sequence

$$\mu^- + {}^{16}O \rightarrow {}^{16}N^* + \nu$$

$$^{16}N^* \rightarrow {}^{15}N^* + n$$

$$^{15}N^* \rightarrow {}^{15}N + \gamma$$

and define $\vec{\beta}_1$ as the recoil velocity of the $^{16}N^*$ from ^{16}O, $\vec{\beta}_2$ as the recoil velocity of the $^{15}N^*$ from ^{16}N, and E_0 as the γ-ray energy in the $^{15}N^*$ rest frame, we can easily show that for mono-energetic neutron emission the γ ray will have the trapezoidal line shape shown in Fig. 1.

Summing over the neutron line spectrum of Ref. 2 and folding in detector resolution (12 keV or 2 PHA channels), we obtain the curve in Fig. 2.

The experimental data points can also be fitted by assuming the recoil of a monoenergetic neutron. The quality of this fit as a function of neutron energy (E_n) is indicated in the table.

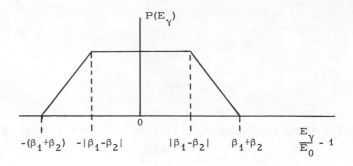

Fig. 1. Calculated line shape for monoenergetic neutron emission.

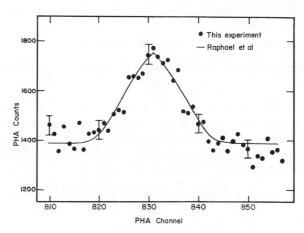

Fig. 2. Experimental data fitted with neutron spectrum of Ref. 2.

E_n(MeV)	χ^2/d	Confidence level (CL) (%)	E_n(MeV)	χ^2/d	Confidence level (CL) (%)
1	70/30	< 0.01	5	25.5/30	> 50
2	47/30	6	6	31.5/30	40
3	28/30	> 50	7	40.8/30	10
4	23.6/30	> 50	8	55.5/30	0.4

The data can, therefore, be well fitted (CL $\gtrsim$ 50%) with any spectrum of neutrons principally in the energy range of 3 to 6 MeV. However, a significant fraction of lower- or higher-energy neutrons would spoil this fit.

These limits imposed on neutron recoil energies are consistent both with the predictions of the resonance model and the μ^- capture neutron spectrum measurements reported by Evseev et al.[3]

It should be pointed out that the assumed isotropy does constitute an oversimplification because of correlation between the nuclear spin axis and the direction of neutrino emission. Therefore some modification of the above result can be expected if detailed account is taken of possible angular correlations.

References:

1. S. N. Kaplan, R. V. Pyle, L. E. Temple, and G. F. Valby, Phys. Rev. Letters 22, 795 (1969).

2. R. Raphael, H. Uberall and C. Werntz, Physics Letters 24B, 15 (1967).

3. Y. Evseev, T. Kozlowski, V. Roganov, and J. Woitkowska (these Proceedings).

MEASUREMENT OF NUCLEAR γ RAYS AND μ X RAYS IN MUONIC ^{209}Bi

H.Backe, R.Engfer[+], E.Kankeleit, W.U.Schröder,
H.K.Walter and K. Wien
Institut für Technische Kernphysik, TH Darmstadt,
Germany
[+] visiting scientist, CERN, Geneva, Switzerland

The anomalous intensity ratios observed in ^{209}Bi
were explained by Hüfner[1] by a resonance mechanism, according to which the nucleus is excited into the $15/2^+$
(2.74 MeV) and $9/2^+$ (2.56 MeV) state. In this paper we
report on the observation of nuclear γ rays following this
excitation. Spectra were taken with a Ge-detector at the
CERN muon channel. Four γ lines were identified in the
prompt (1 2 3 4̄ Ge) spectrum as belonging to transitions
in ^{209}Bi. The energies E_μ and intensities Y_{exp} of these
lines are presented in table 1. They turn out to be shifted in energy by ΔE_{exp} (table 1) compared to energies E_e
as derived from Coulomb excitation[2].

None of the lines showed structure arising from hyperfine interaction. To separate out the isomeric shifts
(which may include a nuclear polarization part) from the
observed shifts the corrections due to the magnetic split-

Transition	E_μ	E_e	ΔE_{exp}	ΔE_{cor}	Y_γ
$\frac{15^+}{2} - \frac{9^-}{2}$	2747.0 ± 0.3	2741.4 ± 0.5	5.6 ± 0.6	6.2	0.025 ± 0.005
$\frac{15^+}{2} - \frac{13^+}{2}$	1135.2 ± 0.3	1132.5 ± 0.7	2.7 ± 0.8	3.0	0.016 ± 0.004
$\frac{13^+}{2} - \frac{9^-}{2}$	1610.8 ± 0.4	1608.9 ± 0.5	1.9 ± 0.7	2.2	0.033 ± 0.008
$\frac{9^+}{2} - \frac{9^-}{2}$	2570.5 ± 0.3	2563.9 ± 1.0	6.6 ± 1.1	6.3...6.9	0.032 ± 0.007

Table 1: Isomer shift and intensities of nuclear γ rays
in ^{209}Bi, energies in keV.

tings in the $15/2^+$, $9/2^+$, $13/2^+$ excited and $9/2^-$ ground state
have to be estimated. The magnetic doublet splitting in the $9/2^-$
state is known to be $\Delta E_m(9/2^-)=4.8$ keV[3]. The $9/2^+$ and $15/2^+$ states
we consider as members of a septuplet arising from a weak coupling
of the h $9/2$-proton to a 3^- vibration of the ^{208}Pb core. For the
latter state the moment has been determined $\mu=1.74\pm0.45$ nm[4] and the
splitting can then be estimated taking the Bohr-Weisskopf effect
into account. We obtain $\Delta E_m (9/2^-=4.2$ keV and $\Delta E_m (15/2^+)=6.4$ keV.
A correction ($\sim$10%) due to the mixing between hf-components of
neighbouring septuplet states is neglected. For the $13/2^+$ level
described by an i $13/2$ proton Scheck found $\Delta E_m-5.5$ keV [5]. We
furtheron evaluate the effect of the M1 interdoublet transitions
and derive the corrected shifts ΔE_{cor} in table 1.

Neglecting for a first discussion the unknown polarization
effects the large positive and about equal shift for the two septuplet states $15/2^+$ and $9/2^+$ can be ascribed to a deformed 3^-
core. An equivalent deformation parameter of $\beta\approx0.06$ can be derived from this shift which is in agreement with $\beta=0.1\pm0.05$ as
obtained from Coulomb excitation in ^{208}Pb 6. For the $13/2$ state

Transition	Δ n	E [MeV]	Yexp.		Yth.
11-8 7-6	1	0.144	0.47	$\pm$ 0.06	0.447
11-7 8-6 6-5	1	0.241	0.62	$\pm$ 0.08	0.616
5-4	1	0.44	0.58	$\pm$ 0.08	0.627
4-3	1	0.98	0.67	$\pm$ 0.06	0.739
3-2	1	2.6	0.63	$\mp$ 0.06	0.863
9-7	2	0.156	0.071	$\pm$ 0.009	0.0831
11-6 7-5	2	0.38	0.087	$\pm$ 0.010	0.104
6-4	2	0.68	0.0713	+ 0.0006	0.0892
5-3	2	1.4	0.052	$\mp$ 0.008	0.0761
10-7	3	0.202	0.031	+ 0.004	0.0328
9-6	3	0.30	0.0225	$\mp$ 0.0023	0.0327
8-5	3	0.48	0.0208	$\mp$ 0.0009	0.0282
7-4	3	0.83	0.015	$\mp$ 0.003	0.0274
13-0	4	0.167	0.012	+ 0.002	0.0203
10-6	4	0.35	0.0137	$\mp$ 0.0023	0.0164
13-8	5	0.187	0.0073	$\mp$ 0.0017	0.0167
12-7	5	0.260	0.0079	$\mp$ 0.0008	0.0137
(14-8	6	0.205	0.0024	+ 0.0006)	0.0210
13-7	6	0.28	0.0064	$\mp$ 0.0012	0.0133
12-6	6	0.40	0.0044	$\mp$ 0.0006	0.00922
11-5	6	0.62	0.0057	$\mp$ 0.0011	0.00609
15-8	7	0.215	0.0041	+ 0.0007	-
13-6	7	0.42	0.0037	$\mp$ 0.0008	0.0105
16-8	8	0.225	0.0018	$\pm$ 0.0005	-

Table 2: Intensities per stopped muon of μ x-rays in ^{209}Bi

shell model calculation[5] gives too large a value for the isomer shift E_{is} (13/2 7 keV. The known vibrational admixture should increase this value even further.

Population probabilities of 4.1% and 3.2% of the $15/2^+$ and $9/2^+$ levels, respectively, were deduced from the measured intensities (table 1). They are consistent with the anomalous relative intensities of the 3d-2p X rays which can be explained by a nuclear excitation of (6.3 ± 1.8)%. The branching ratio $I(15/2^+-13/2^+)/I$ $(15/2^+-9/2^-) = 0.64\pm0.21$ of the $15/2^+$ decay agrees roughly with the value 0.89 0.09 of ref.[2]. The $13/2^+-9/2^-$ intensity is twice the $15/2^+-13/2^+$ intensity indicating an additional excitation of the $13/2^+$ state or an unknown background line.

Intensities of μ X rays, Y_{exp}, per stopped muon measured in the same experiment are presented in table 2. Only the 5-4 intensity was determined absolutely, while all others were obtained relative to this one and normalized correspondingly. The intensities are compared to a cascade calculation, Y_{th}, starting at n=14 with a population proportional to $(2l+1)$ [7]. The low experimental absolute 3d-2p intensity can be explained by the above mentioned nuclear excitation and a radiationless transition of (7 ± 2)%[8] in addition.

This work was supported by the "Bundesministerium für wissenschaftliche Forschung".

REFERENCES

1) J. Hüfner, Phys. Lett. 25B (1967) 189
2) J.W.Hertel, D.G. Fleming, J.P. Schiffer and H.E. Gove, to be published
3) Average of R.A. Carrigan et al., Bull. Am. Phys. Soc. 13 (1968) 65 and R.J. Powers, Phys. Rev. 169 (1968) 1
4) J.D. Bowman, F.C. Zawislak and E.N. Kaufmann, Phys. Lett. 29B (1959) 226
5) F. Scheck, to be published. It was assumed $g_{s eff} = 2.03$ $\mu = 7.8$ nm.
6) A.R. Barnett and W.R. Phillips, IUPAP Conf. Heidelberg 1969
7) The program was kindly supplied by J. Hüfner
8) C.K. Hargrove et al., Phys. Lett. 23 (1969) 215

ISOTOPIC EFFECT OF μ CAPTURE RATES IN ^{151}Eu–^{153}Eu

H.Backe, R.Engfer*, U.Jahnke**,E.Kankeleit,
K.H.Lindenberger**, C.Petitjean***, W.U.Schroder,
H.K.Walter and K. Wien

Institut für Technische Kernphysik, TH Darmstadt,
Germany

Isotopic effects of μ capture rates have been predicted, e.g. by Primakoff[1,2,3] and confirmed by experiments[4], in which enriched isotopes were used and the time distribution of μ capture products (neutrons, γ rays) were measured.

This paper presents experimental results of μ capture rates in Eu isotopes obtained with a new technique.

The experiment was performed at the CERN μ channel. A counter telescope detected μ stops in a natural Eu_2O_3 target. Energy spectra of μ capture γ rays were measured from 50 to 500 keV with a high resolution Ge(Li) counter. Corresponding to three succeeding delay time intervals (10.6, 54.1, 148.1 nsec) between μ stop and γ detection the spectra were stored into three memory blocks of an on-line computer. Fig. 1 shows a part of the measured γ spectra following μ capture in ^{151}Eu and ^{153}Eu. Most lines can be identified as de-excitation γ rays from ^{149}Sm, ^{150}Sm, ^{151}Sm and ^{152}Sm. The strong abundance of ^{152}Sm and ^{150}Sm lines suggest that the most probable μ capture process is ^{A}Eu$(\mu,\nu_\mu$ n$)^{A-1}$Sm. A two neutron emission $(\mu,\nu_\mu 2n)$ seems to take place at a considerably smaller rate (∼20%). γ rays below 500 keV following μ capture without neutron emission neither were observed in Eu nor in a number of other heavy nuclei, which seems to be in contradiction to results of ref. [5].

From a fit to the exponential time distribution of the line intensities the life times $\tau(^{151}$Eu$) = 73.2$ nsec, $\tau(^{153}$Eu$) = 77.9$ nsec and more precisely the difference $\Delta\tau = 4.7 \pm 1.7$ nsec were obtained.

*Visiting scientist, CERN, Switzerland; ** Hahn-Meitner-Institut f. Kernforschung, Berlin, Germany; *** SIN, Zurich, Switzerland.

The main complication in this technique arises from the fact that a certain nuclear excitation may be reached from different capturing isotopes due to multiple neutron emission.   In this case the emission of three neutrons after capture in [153]Eu leads to the same states in [150]Sm as the capture in [151]Eu with emission of one neutron.   The first process though has about a factor of ten smaller probability[5] and can be neglected in the analysis.

The change in capture rates for the two isotopes is well explained by the Primakoff theory[1].   About two thirds of the change is due to the nucleon–nucleon correlation term and one third due to a change in the mean charge distribution or Z_{eff}.

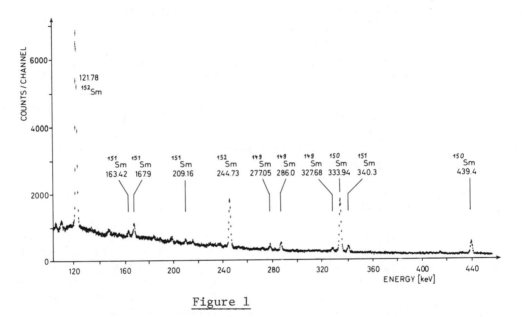

Figure 1

This work was supported by "Bundesministerium fur wissenschaftliche Forschung".

REFERENCES

1)   H. Primakoff, Rev. Mod. Phys. 31 (1959) 802
2)   J.M. Kennedy, Phys. Rev. 87 (1952) 953
3)   H.A. Tolhoek and J.R. Luyten, Nucl. Phys. 3 (1957) 679
4)   W.A. Cramer, V.L. Telegdi, R. Winston and R.A. Lundy,
     Nuov. Cim. 24 (1962) 546
5)   B. MacDonald et al., Phys. Rev. 139 (1965) B 1253

AN INTERPRETATION OF NEUTRON ASYMMETRY IN POLARIZED MUON CAPTURE

A. Fujii*, R. Leonardi**, and M. Rosa-Clot***

*Institut de Physique Nucleaire, Orsay, France & Sophia
University, Tokyo, Japan; **Istituto di Fisica, Univer-
sita di Bologna, Bologna, Italy; ***Dept. of Physics,
Columbia University & Scoula Normale Superiore, Pisa,
Italy

The recent experiments[1,2] on the neutron asymmetry parameter
α in the polarized muon capture by Ca^{40} show the inversion of the
sign (negative, against positive) in the neutron energy region of
about 20 - 30 MeV.

The sign and the magnitude of α depend on the relative strength
of the Fermi (F) and Gamow-Teller (G.T.) transition. If the distribution of these strengths, as a function of the continuous excitation energy, have different shapes, the relative strength at
different energy will have a different value. So that even the
sign of α depends on the excitation energy, or consequently on the
energy of the emitted neutron[3].

In Ca^{40} the allowed F and G.T. matrix elements vanish in the
supermultiplet (SU_4) limit. We define the strength of the first
forbidden F and G.T. transition $|0> \rightarrow |n>$ by

$$f^F_{no}(w_n) = w_n \mid <n \mid \sum_j x_j \frac{\tau}{2j}(+) \mid 0> \mid^2,$$

$$f^{GT}_{no}(w_n) = w_n \mid <n \mid \sum_j x_j \frac{\tau}{2j}(+) \sigma_j \mid 0> \mid^2$$

where $w_n = E_n - E_o$ is the excitation energy. In Primakoff approximation we notice

$$\alpha \rightarrow -0.7 \quad \text{if} \quad f^{GT}_{no} >> f^F_{no},$$

$$\alpha \rightarrow +1 \quad \text{if} \quad f^F_{no} >> f^{GT}_{no}.$$

We can expect that both $f^F_{no}(w)$ and $f^{GT}_{no}(w)$, as functions of continuous variable w, have a giant resonance type shape. In SU_4 limit they coincide. The SU_4 breaking interaction, i.e. the spin-orbit, Bartlett and Heisenberg forces, will remove this degeneracy.

We estimate the separation of the peak by sum rules. With the definition

$$R_F = \sum_n \frac{f_{no}^F}{w_n}, \quad R_{GT} = \sum_n \frac{f_{no}^{GT}}{w_n}, \quad A_F = \sum_n f_{no}^F, \quad A_{GT} = \sum_n f_{no}^{GT}$$

the mean energy (approximately the location of the peak) is given by

$$w_F = \frac{A_F}{R_F} \equiv \frac{1}{R_F} (A_F(SU_4) + A_F(LS) + A_F(B) + A_F(H))$$

$$w_{GT} = \frac{A_{GT}}{R_F} \equiv \frac{1}{R_F} (A_{GT}(SU_4) + A_{GT}(LS) + A_{GT}(B) + A_{GT}(H)),$$

where Su_4, LS, B, H refers to the contribution from the nuclear potential of these specific characteristics. We notice immediately

$$\frac{A_{GT}(SU_4)}{R_{GT}} = \frac{A_F(SU_4)}{R_F} \qquad 4)$$

$$A_F(LS) = A_F(B) = 0, \qquad A_{GT}(LS) \sim 0.$$

Further in the long range approximation the nuclear potential can be factored out, yielding

$$\frac{A_F(H)}{R_F} = \frac{A_{GT}(H)}{R_{GT}} = 2\bar{V}_H, \qquad \frac{A_{GT}(B)}{R_{GT}} = 2\bar{V}_B,$$

where $\bar{V}^H$, $\bar{V}^B$ are the average of the two-body Heisenberg and Bartlett potential, respectively. The separation of the peaks for the F and GT distribution is thus simply given by

$$w_{GT} - w_F = 2\bar{V}_B.$$

A quick estimate may be provided by the experimental information. The electric giant resonance (analogue of the F transition) in doubly closed nuclei has two peaks,[5-7] one corresponding to (T = 1, S = 0), the other (T = 1, S = 1) states of the LS-coupling shell model.

This separation of S = 0 and S = 1 peaks comes essentially from the Bartlett force and equals $2\bar{V}_B$ in the long range approximation. For Ca^{40} this separation is of the order of $5 \sim 7$ MeV experimentally[7] which is large enough as compared with the expected half-width of the S = 0 peak (~ 2 MeV).

Therefore the GT strength exceeds the F strength in some energy region around the peak of the former, changing the sign of α.

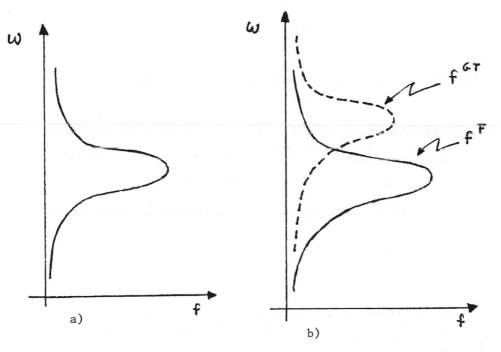

Fig. 1 – Schematic plot of the strength distributions, a) in the SU_4 limit, b) with symmetry breaking.

References

1) J. Sculli, Nevis Cyclotron Report 168 (March 1969).
2) R.F. Sundelin, R.M. Edelstein, A. Suzuki and K. Takahashi, Phys. Rev. Lett. 20, 1201 (1968).
3) J. Fujita and A. Fujii, Phys. Rev., in press.
4) L.L. Foldy and J.D. Walecka, Nuovo Cimento, 34, 1026 (1964).
5) N.A. Burgov et al. J.E.P.T. 18, 1159 (1964).
6) N.W. Tanner et al. N.P. 52, 45 (1962).
7) B.S. Ishkhanou et al. J.E.P.T. 19, 1003 (1965).

RESONANCE MECHANISM OF MUON CAPTURE BY COMPLEX NUCLEI

V.V. Balashov*, R.A. Eramzhyan, N.M. Kabchnik*,

G.Ya. Korenman*, V.L. Korotkih*

Joint Institute for Nuclear Research, Dubna, USSR

Investigating muon capture on ^{32}S and ^{40}Ca, Evseev et al.[1] have observed resonances (peaks) in the neutron spectrum. They conclude that this result is the first direct experimental Confirmation of the idea of collective nuclear excitation in muon capture. The idea of the resonance mechanism in muon capture (giant resonance) gives a general basis for further investigations of various aspects of the muon nuclear interaction.

1. <u>Neutron spectrum and asymmetry in muon capture</u>[2]. As a result of muon capture a proton hole is created in one of the nuclear shells. The interaction of the outgoing neutron with the proton hole causes the excitation of the residual nucleus and as a result the nuclear system goes over to other channels. Thus this interaction leads to the coupling of some various channels. As a result the neutron spectrum has a very nonuniform structure.

In the framework of the approximate method of the unified theory of nuclear reaction, which takes into account both the direct and the resonance processes the amplitude with a neutron in j-channel is given by

$$M_{0 \to j}(E) = \langle \varphi^{(-)}_{jE} | H_\mu | 0 \rangle + \sum_\lambda \frac{\langle \varphi^{(-)}_{jE} | V | \psi_\lambda \rangle \langle \psi_\lambda | H_\mu | 0 \rangle}{E - \mathcal{E}_\lambda}$$

The first term is the direct transition amplitude into the particle hole channel j. The second one is the amplitude of the resonance transition into the same channel. This method was used for muon capture calculation in ^{40}Ca. There were considered allowed, first and second forbidden transitions. The results of the neutron

* Nuclear Research Institute, Moscow State University, Moscow, USSR.

spectrum and asymmetry calculations are given on Fig. 1 and 2.

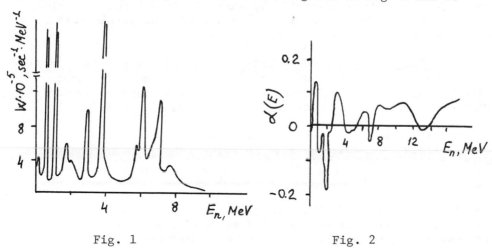

Fig. 1   Fig. 2

2. _Emission of charged particles in muon capture by complex nuclei._

Emission of charged particles in muon capture testifies directly the presence of the correlations between nucleons in nuclei. Using the resonance mechanism idea in muon capture, we are able to give some predictions about the charged particles emission.

After muon capture the number of protons and neutrons changes. In the case of the light nuclei, in an intermediate nucleus the thresholds for the decays with charged particle emission are much higher than for neutrons. Therefore in muon capture the neutron channel must be predominant one. After the intermediate nucleus has emitted the neutron it restores the neutron-proton balance. The charged particle threshold in the daughter nucleus lies at the same energy. Consequently, according to the resonance model the yield of the charged particle χ in muon capture must be necessarily associated not only with the $(\mu^-, \nu \chi')$ channel, but also (and, probably mainly) with the $(\mu^-, \nu n \chi)$ channel. Therefore the total charged particle yield must increase as the nucleons occupy the shell, achieve the maximum when the shell is semifilled (between neighbouring magic nuclei) and decrease with the approach to the nearest magic nuclei. The main part of the spectrum of charged particles must be soft. With increasing atomic number, the charged particle yield decreases. For nuclei with A>40 the charged particle yield will be small.

Such a mechanism will give a negligible yield in the high energy part of the spectrum. One may expect, that high energy particles are due to the other mechanism of muon capture. At such energies the role of the short-range correlations becomes important. Muons may be captured directly by some clusters. However, the total yeild of the charged particles due to cluster absorption is not expected to be large.

3. <u>Transitions to the excited states of the residual nucleus</u>. The results by Kaplan et al.[3] on ^{15}N yield in the $3/2^-$ state after muon capture by ^{16}O were considered also as a confirmation of the idea of the resonance mechanism in muon capture.

It is clear that if in coincidence experiments one defined the γ-lines, one would be able to bring into correspondence each peak of the neutron spectrum with the peaks in excitation spectrum of the intermediate nucleus. Such information would give the possibility for investigating, together with photonuclear reactions and inelastic electron scattering, the analog states in nuclei. Up to now the analogy between photonuclear process and muon capture is used either qualitatively or for the comparison of the total capture rate with the photoabsorption cross sections.

Another problem, which can be investigated by means of the coincidence method, is associated with the study of the neutron asymmetry mechanism in the resonance region. It is clear that the neutron yield corresponding to the definite state of residual nucleus (measured in coincidence with γ-quanta) would be less sensitive to a possible spread of the particle-hole states.

REFERENCES

[1]V. Evseyev et al. Phys. Letters <u>28B</u>, 553 (1969).
[2]V.V. Balashov et al. Preprint E4-4601, 1969 JINR.
[3]S. Kaplan et al. Phys. Rev. Letters <u>22</u>, 795 (1969).

MAGNETIC HYPERFINE STRUCTURE IN MUONIC THALLIUM

Florian Scheck

CERN - Geneva

Switzerland

Recently, a magnetic hyperfine structure in muonic $^{203}_{1}$Tl and ^{205}Tl has been measured by direct observation of nuclear γ-rays. During its cascade to the 1s orbit the muon excites the nuclear $3/2^+$ state at 279 keV and at 204 keV in ^{203}Tl and ^{205}Tl respectively, which then decays while the muon is still present in its 1 s1/2 orbit. Due to the strong magnetic coupling of the muon and the nucleus the excited state splits into a $F_e = 2$ and a $F_e = 1$ level, (F being the total angular momentum), the ground state (with nuclear spin $\frac{1}{2}^+$) into a $F_g = 1$ and a $F_g = 0$ level. The upper excited level $F_e = 2$ is almost entirely depopulated in favour of the lower member with $F_e = 1$ of the excited doublet by strong M1-con= version[2]. Therefore only two nuclear γ-rays, $F_e = 1 \longrightarrow F_g = 1$ and $F_e = 1 \longrightarrow F_g = 0$, are observed, exhibiting the magnetic coupling of the muon in the 1s orbit and the nuclear ground state. This magnetic hfs is well known to be very sensitive to details of the nuclear magnetisation density.

The experimental results[3] for both isotopes are shown in the first line of the Table. The theoretical value for ^{203}Tl in the second line of the Table has been calculated assuming a simple configuration mixing model for the nuclear ground state[4]. The ground state is assumed to be predominantly a $3s\frac{1}{2}$ proton-hole state plus some admixtures of nearby particle-hole states, calculated along the lines of the work of Arima et al.[5] For comparison, the third line of the Table shows what the magnetic hfs would be if the nucleus were pointlike. In both cases the same model gives the correct magnetic moment of Thallium (fourth line).

In the radial integrals which have to be known very accurately, the muonic and nucleonic wave functions were obtained by numerical integration. The muonic wave functions were taken from a code developed

by Acker[6]; the nucleon wave functions were integrated from a realistic
Saxon-Woods potential with spin-orbit interaction[7]. With a standard
set of parameters for the nucleon potential taken from the work of
Blomqvist and Wahlborn[7], the configuration mixing model is seen to be
in fair agreement with the experimental splitting. The remaining difference between the theoretical and the experimental values can be further
reduced by variation of the parameters of the shell model potential.
As long as other information on these is lacking, we do not consider
such a fit to be very meaningful, however. (The set of parameters proposed more recently by Rost[8] give essentially the same result as the
ones of ref. 7).

The experimental results seem to indicate a somewhat __smaller__ splitting in ^{205}Tl than in ^{203}Tl. This result, if it is confirmed, is somewhat puzzling since from configuration mixing alone one would expect the magnetic h fs to follow the trend of the magnetic moments which
__increase__ slightly from ^{203}Tl to ^{205}Tl (see fourth line of Table), as
long as the __spatial__ structure of the two isotopes is not very different.
On the other hand an increase in volume of the nucleus __reduces__ the magnetic hfs even further. Hence the experimental result, if it is maintained, seems to indicate that the finite extension effects dominate the
small change in configuration mixing when going from ^{203}Tl to ^{205}Tl.

Any change in the proton distribution has to reproduce the experimentally measured isotope shift of about -9 keV of the muonic 2p - 1s
transitions[9]. A possible, though certainly not unique, solution consists,
for the proton well, in increasing the diffuseness parameter a (ref.7)
by about 0.1 fm while decreasing the radius parameter by about 0.13 fm
(in order to obtain the correct isotope shift); for the neutron well,
in choosing it somewhat larger but "steeper" at the surface (closeness
to the doubly-magic ^{208}Pb). This leads to the value for ^{205}Tl in the second line of the Table, hence to a reduction of the magnetic hyperfine e
splitting of about 200 eV.

	^{203}Tl	^{205}Tl
δE_{hfs} (Exp)	2.55±0.17	2.27±0.03
δE_{hfs} (Theor)	2.87	2.67
δE_{hfs} (point)	4.65	
magn. moment	1.61169	1.62754

Magnetic hfs (in keV) and magnetic moments in Thallium

REFERENCES

1. R. Baader, H. Backe, R. Engfer, E. Kankeleit, W.U. Schroeder, H. Walter and K. Wien; Phys. Letters 27 B, 428 (1968)

2. For further references on these effects see refs. 3 and 4

3. R. Engfer; Habilitationsschrift. Darmstadt Technische Hochschule preprint, February 1969

4. R. Engfer and F. Scheck; Z. Physik 216, 274 (1968)

5. A. Arima, H. Horie and H. Noya; Progr. Theor. Phys. (Kyoto) 8, 33 (1958); Supplement

6. H. L. Acker, G. Backenstoss, C. Daum, J.C. Sens and S.A. de Witt; Nucl. Phys. 87, 1 (1966)

7. J. Blomqvist and S. Wahlborn; Arkiv Fysik 16, 545 (1960)

8. E. Rost; Phys. Letters 26 B, 184 (1968)

9. R. Engfer; rapporteurs talk at this conference

MUON CAPTURE AND NUCLEAR STRUCTURE

M. Morita, H. Ohtsubo, and A. Fujii*

Osaka University, Toyonaka, Osaka, Japan

*Sophia University, Tokyo, Japan

We have studied the fundamental muon-nucleon interaction
from the muon capture in gaseous and molecular hydrogen and the
nuclear structure problem upon the knowledge of the former from
muon captures in C^{12} and O^{16}.

The phenomenological V - A theory of leptonic interactions
is adopted for the fundamental process $\mu^- + p \longrightarrow n + \nu_\mu$. The
behavior of the form factors is determined under the assumptions
of 1) conserved vector current, 2) isotriplet vector current, 3)
partially conserved axial vector current, 4) definite G-parity
for the current, and 5) muon-electron universality. Calculated
muon capture rates are in agreement with the recent data on
gaseous hydrogen, and also with the data on molecular hydrogen
if the molecular factor $2\gamma = 1.00$ is accepted.

The muon capture rate is calculated for the reaction $\mu^- + C^{12}$
$\longrightarrow B^{12}$ (ground state) $+ \nu_\mu$, with general p-shell wave functions
which were obtained by considering the dynamical properties in
the A = 12 system. The experimental ft value of the beta decay
of B^{12}, which in all previous theories is used to normalize the
muon capture rate, has not been adopted in this analysis. An
excellent fit to the experimental data for the capture rate as
well as for the ft value has been obtained with the B^{12} wave
function proposed by Kurath to explain the magnetic moment of B^{12}
and a C^{12} wave function introduced here. The oscillator strength
parameter is chosen to be b = 1.64 fm: this is in conformity with
the elastic and most of the inelastic electron scattering data.

The muon capture rates are also calculated for the reactions

$\mu^- + O^{16} \longrightarrow N^{16} \ (0^-, 1^-, 2^-) + \nu_\mu$, with the RPA and Tamm–Dancoff models. The calculated capture rates agree well with experimental data, if 1) $C_p/C_A = 5 \sim 14$ and the Gillet wave functions are adopted in the reaction $0^+ \longrightarrow 0^-$, 2) a configuration mixing is taken into account in the reaction $0^+ \longrightarrow 1^-$, and 3) a normalization by the experimental beta decay rate of N^{16} is introduced in the reaction $0^+ \longrightarrow 2^-$.

EXPERIMENTAL DETERMINATION OF THE PROBABILITY OF THE

PROCESS (μ^-,ν) ON ^{27}Al, ^{28}Si, ^{51}V NUCLEI

G.Bunatian,V.Evseev,L.Nikityuk,V.Pokrovsky,
V.Ribakov, I.Yutlandov

Joint Institute for Nuclear Research
Dubna, U S S R

When calculating the total muon capture rates in
complex nuclei allowance for the nucleon residual interactions is to be made to achieve agreement with experimental data. If this effect is taken into account
by means of the methods of finite Fermi-system theory
rather a good agreement can be obtained for about 30
spherical nuclei /1/. It is reasonable to compare
theoretical and experimental values of partial probabilities of muon capture without nucleon emission,
i.e., the (μ^-,ν) process, with a view to further
looking into possibilities of these methods.

The relative probabilities of the (μ^-,ν) process
on ^{27}Al, ^{28}Si and ^{51}V nuclei as well as on ^{56}Fe nucleus
/2/ have been determined by the activation method. For
muon beams of constant intensity the probability is
simply $W_{\mu^-,\nu} = A_\infty/N_\mu$, where A_∞ is the activity of
the reaction product at saturation, N_μ is the number
of muons captured by the target nuclei per unit time.

The targets were irradiated at the muon channel
of the Dubna synchrocyclotron. The numbers of incident
and stopped muons were recorded, the N_μ values being
calculated. The A_∞ values were found by measuring
the gamma-quanta spectra of the induced activity with
a scintillation counter (NaJ(Tl) 100x100 mm), the reaction product being positively identified through the
values of E_γ and $T\frac{1}{2}$. In the case of the relative long-
lived ^{58}Mn radiochemical separation was made.

The results of relevant experiments and calculations are listed in the following Table. Targets are indicated in the first column.

Target	10^5 sec^{-1}, Λ		$W_{\mu^-,\nu} = \dfrac{\Lambda'}{\Lambda}$, %		$W_{\nu 0n}^{\mu^-}$, %/4/
	exp.	theor.	exp.	theor.	
^{27}Al	7.00±0.05	8.0	10±1	6	8±4
^{28}Si	8.49±0.03	9.16	28±4	4.5 or 38	36±6
^{51}V	31.9 ±0.6	27.2	10±1	15	–
^{56}Fe	44.1 ±0.3	46.1	16±3	24	19±3

Column 2 presents total muon capture rates, the experimental ones being taken from ref./3/. The accuracy of theoretical estimates is about 20% .

Column 3 gives the experimental and theoretical values of the (μ^-,ν) process probabilities. The experimental values are the mean ones from 3-5 determinations, the errors being total standard deviations.

For comparison column 4 gives the probabilities of muon capture without neutron emission (but with possible emission of charged particles) evaluated from the data of ref. /4/.

The theoretical estimates of $W_{\mu^-,\nu}$ were determined by calculating the rates Λ' of muon capture on the levels of the product nucleus lying below the emission threshold for the least bound particle. In the case of ^{28}Si the most enhanced transition occurs at the 1^+ level whose energy is about this threshold. The theoretical value $W_{\mu,\nu}$ = 38% corresponds to the energy of this level being smaller than the binding energy of the neutron, while $W_{\mu,\nu}$ = 4.5% refers to the opposite case.

The agreement between experiment and theory for $W_{\mu^-,\nu}$ is seen to be poorer than for the total muon capture rates. This appears to be associated with the necessity of taking into account the effect of mixing configuration and pairing in relatively light nuclei which have a small number of bound states with some of them favoured.

References

1. G.Bunatjan. Yad.Fiz. 3 (1966) 833.

2. G.Bunatjan, V.Evseyev, L.Nikitjuk, A.Nikolina, V.Pokrovsky, V.Roganov, L.Smirnova, I.Yutlandov. Preprint P15-4008, Dubna, 1968.

3. M.Eckhause, R.T.Siegel, R.E.Welsh, T.A.Fillippas. Nucl.Phys. 81 (1966). 575.

4. B.MacDonald, J.A.Diaz, S.N.Kaplan, R.V.Pyle. Phys.Rev. 139 (1965) B1253 .

GAMMA - GAMMA CORRELATION IN RADIATIVE NUCLEAR MUON CAPTURE

Z. Oziewicz and N.P. Popov

A.F. Ioffe Physico-Technical Institute Academy of
Sciences of the USSR, Leningrad

Parallel to normal muon capture investigation, the radiative
nuclear muon capture process is of its own interest. The photon
spectrum after radiative muon capture, which is very sensitive to
the contribution of pseudoscalar induced interaction[1], was studied
experimentally in ^{40}Ca[2].

This paper summarizes our calculations on the angular gamma -
gamma correlation after radiative partial muon capture with daughter
nucleus excitation

$$\mu^- + {}^AZ \rightarrow {}^A(Z-1)^* + \nu + \gamma_r$$
$$^A(Z-1) + \gamma_n$$

There is clear distinction of nuclear gamma rays from radiative
ones by their energy. The energy of nuclear gamma quantum does not
exceed a few MeV whereas radiative photon spectrum has a maximum
about $k \simeq 30$ MeV. Sure enough, the radiative muon capture investigation provides significant experimental difficulties. However
possible information on weak interaction form factors and first of
all on pseudoscalar one (and especially in the case of the partial
radiative muon capture) makes this investigation very desirable. The
radiative muon capture makes it possible to study the dependence of
form factors on momentum transfer (within limits $- m_\mu^2 \lesssim q^2 \lesssim + m_\mu^2$
which the normal muon capture does not allow.

The formula of angular gamma - gamma correlation for unique
allowed $j_0 \xrightarrow{\mu} j_1 \xrightarrow{\gamma_n} j_2$ transitions looks as follows:

$$W = 1 + \frac{a_2(x)}{a_0(x)} \sqrt{6} \Lambda_2^{11} P_2 (k_n \cdot k_r)$$

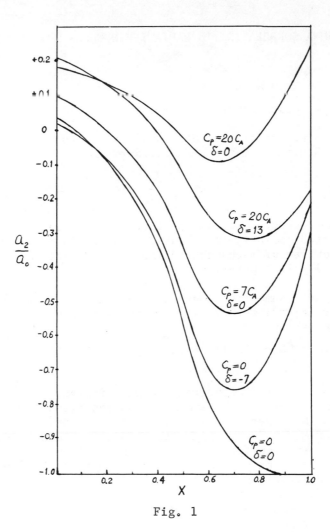

Fig. 1

where k_n and k_r are the unit vectors of momenta of nuclear and radiative gamma quanta respectively.

$$\Lambda_2^{11} = (2j_1+1)\sqrt{5(2L+1)} \ C_{L120}^{L1} \ W(j_0 j_1 12;1j_1) \ W(j_2 L j_1 2;j_1 L)$$

The table of Λ_2^{11} values for certain transitions is given in Ref. [3]. The radiative photon spectrum of this partial transition is determined by $X(1-X)^2 \ a_0(X)$, where $X = k/k_{max}$, k being the energy of radiative photon. One must keep in mind that the dependence of the correlation coefficient on the photon energy is determined by $a_2(X)$ only, because $a_0(X)$ depends on X insignificantly (the integration over the neutrino emission angle has been done). The quantities $a_i(X)$ depend also on weak interaction form factors and on nuclear structure. However in the ratio a_2/a_0 which determined the correlation coefficient the dependence on nuclear structure practically cancels for above mentioned unique transitions. Thus one can interpret the correlation experiment results for partial radiative muon capture more reliably than the total radiative to normal muon capture ratio. Fig. 1 shows the results for a_2/a_0, calculated for the same set of Feynman diagrams as in Ref.[1] The calculation has been done in the UFI theory with various assumptions about the pseudoscalar form factor. We can write the latter as

$$C_p(q^2) = C_p \ (m_\mu^2) \frac{m_\pi^2 + m_\mu^2}{m_\pi^2 + q^2} + C_A \frac{q^2 - m_\mu^2}{m_\pi^2 + q^2} \ \delta$$

In Fig. 1 we denote $C_p \equiv C_p(m_\mu^2)$. δ describes the contributions from the diagrams more massive than the one-pion. There is a string dependence of correlation coefficient on the form of $C_p(q^2)$ for hard photons which are of most interest (c.f. Fig. 1). The curves correspond to the statistical population of h.f. levels of mesoatom or $j_0 = 0$. The dependence on $C_p(q^2)$ slightly diminishes for the capture from the lowest h.f. level only, as would result from conversion between the h.f. levels of the K-shell.

As a conclusion let us add that in the case of radiative muon capture the expansion parameter in exp $(-iS\tau)$ is better (in fact here $S_{av} \simeq m_\mu/2$) than for normal muon capture where $S=m_\mu$. This favours studying heavier nuclei where the probability of radiative muon capture is higher.

References

(1) H.P.C. Rood, On the Theory of Muon Capture in Nuclei (Doctoral Thesis, Groningen Univ., Groningen, Netherlands, 1964); H.P.C. Rood and H.A. Tolhoek, Nucl. Phys. 70, 658, 1965.

(2) M. Conversi, R. Diebold and D. di Lella, Phys. Rev. 136, B 1077, 1964.

(3) Z. Oziewicz and A. Pikulski, Acta Phys. Polonica, 32, 873, 1967.

PARTICLE CORRELATIONS AND HYPERFINE EFFECTS IN MUON CAPTURE

N.P. Popov

A.F. Ioffe Physico-Technical Institute Academy of
Sciences of the USSR, Leningrad

The main objective of the study of nuclear muon capture is to
obtain information concerning both the weak interaction form factors and nuclear structure. Here we touch upon the first aspect of
the problem[1].

In studying muon capture one has often to do with non-zero spin
targets I. In this case the muon may be captured from two mesoa-
tomic hyperfine states with the total angular momenta $F_{\pm} = I \pm 1/2$.

It is possible in principle that the conversion within the
hyperfine doublet of the K-shell occurs. The conversion was experimentally proved by studying the time dependence of the capture
rate for the nuclei ^{11}B and ^{19}F[2]. Here we consider the time dependence of the particle correlations in muon capture. The calculations demonstrate that the dependence may be rather considerable
at the time interval of the order of 1/R, where R is conversion rate.

The study of the polarization and angular correlations is of
interest first of all for obtaining information on the weak coupling
constants in muon capture. The theoretical results for the correlations turn out to depend only slightly on the nuclear structure
(the emission of the nucleons from the nucleus is not considered).

The possibilities for studying particle correlations are the
best in the muon capture with the excitation of the nucleus

$$\mu^- + {}^A Z \to {}^A(Z-1)^* + \nu_\mu \to {}^A(Z-1) + \gamma + \nu_\mu$$

This reaction is an advantage because of a great number of
possible correlations formed by the momentum vector of γ-quantum $\underline{K}$
with that of recoil nucleus $\underline{q}$ and with the muon polarization vector

$\underline{\xi}$. Besides, the experimental method of studying correlations in muon capture based on the Doppler-broadening of γ-spectrum requires the γ-ray to be emitted[3].

In the case of allowed partial muon capture ($\Delta I = 0,1,$no) we have[4] the following most interesting correlations:

$$W\,(\underline{qk}) = 1 + b_2\,P_2\,(\underline{qk})$$

$$W\,(\underline{\xi q}) = 1 - C_o\,\underline{\xi q}$$

$$P_\gamma\,W\,(\underline{qk}) = -b_1\,\underline{qk} - b_3\,P_3\,(\underline{qk})$$

where $P_s\,(\underline{qk})$ are Legendre polinomials; P_γ is circular polarization of the γ-quantum. The correlation coefficients b_i and C_o depend on the process kinematics, weak coupling constants and nuclear corrections.

Let the magnetic moment of the target nucleus be positive. In this case the F+ level is higher than the F_ and time dependence of the coefficients b_i and C_o is of the following general form:

$$b_i(t) = \frac{b_i^+\,W_+\,P_o e^{-Rt} + b_i^-\,W_-\,(1 - P_o e^{-Rt})}{W_+\,P_o e^{-Rt} + W_-\,(1 - P_o e^{-Rt})}$$

where $W\pm$ and $b_i^\pm$ are capture rates and correlation coefficients for muon capture from $F_\pm$ levels respectively. P_o is the initial population probability of the k-shell hyperfine levels ($t = 0$). For statistical population: $P_o = (I + 1)/(2I + 1)$. It should be noted that C_o in the analogous formula exhibit a certain time dependence, unlike $b_i\pm$.

It is known that the hyperfine effect is large for muon capture by hydrogen ($W_- >> W+$). In the case of nuclear capture we consider the pure Gamov-Teller transition with nuclear spin increase ($I \overset{\mu}{\rightarrow} I+1$). For muon capture from F_ state the angular momentum conservation forbids the emission of a neutrino with the total angular momentum $J = 1/2$ in this transition. But the emission of a neutrino with $J \geq 3/2$ is due to correction form factors (e.g. pseudoscalar with the factor m/2M, where m and M is muon and nucleon mass respectively), whose contribution is less than that of the main (axial vector) form factor, determining the muon capture from F+ level in this transition. Hence W_ << W_+ and the total capture rate is strongly decreasing for a time interval of the order of 1/R. As regards the correlation coefficients they should make a jump at some moment t_o. The width of this jump is several 1/R. In fact, at the initial time interval the correlations will be determined mainly by capture from the higher level F_+ (as $W_+ >> W_-$). But later all muons turn out to be in the lower level and the correlations will be determined by capture by capture from F_ level and so again nearly constant in time. The experimental registration of the moment t_o for which the curves $b_i(t)$ make a jump from value about

$b_i(o)$ nearly to $b_i(\infty)$ would give information on the weak coupling constants (then R should be measured previously with enough accuracy from the time dependence of muon capture rate).

At the same time in the $I \xrightarrow{\mu} I - 1$ transition $W_- \gg W_+$ and the correlations are determined by the capture probability from the lower state at any moment. However the correlations in this transition will depend strongly on time for negative magnetic moment of the target nucleus.

Most interesting is the gamma-neutrino angular correlation P_2 (qk) which is the only possible one for unpolarized muon capture without measurement of the γ-quantum polarization. The coefficient b_2 may be highly dependent on pseudoscalar, as this correlation is possible only if the neutrino with $J = 3/2$ is emitted. Apart from studying the time averaged correlation[4] it would be useful to take into consideration its possible time dependence. The curves $b_2(t)$ calculated for some values of pseudoscalar constant in universal Fermi interaction theory with conserved vector current are given in Fig. 1 for muon capture by ^{14}N nucleus.

I am obliged to A.P. Bukhvostov and Z. Oziewicz for useful discussion.

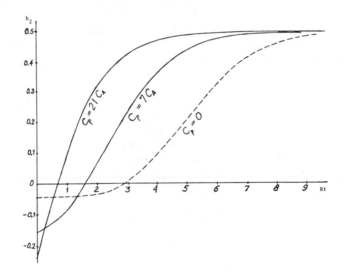

Fig. 1 - The time dependence of the correlation coefficient b_2 in reaction $^{14}N(1^+) \xrightarrow{\mu} {}^{14}C(2^+) \xrightarrow{\gamma} {}^{14}C(0^+)$. ($E\gamma = 7.01$ MeV) calculated for some values of pseudoscalar coupling constant.

REFERENCES

[1] L. Wolfenstein, Proc. 1960 Int. Conf. on High Energy Physics at Rochester, p. 528; H. Primakoff, Proc. 1966 Int. School of Phys. "Enrico Fermi," Course 32, p. 96.

[2] G. Culligan, I.F. Lathrop, V.L. Telegdi, R. Winston, R.A. Lundy, Phys. Rev. Lett. 7, 458, 1961; R. Winston, Phys. Rev. 129, 2766, 1963; I.P. Deutsch, L. Grenacs, I. Lehmann, P. Lipnik, P.C. Macq, Lett., 28B, 178, 1968.

[3] L. Grenacs, I.P. Deutsch, P. Lipnik, P.C. Macq, Nucl. Instr. and Methods, 58, 164, 1968.

[4] N.P. Popov, Proc. 1968 Conf. on Electron Capture and Higher Order Processes in Nuclear Decays, Debrecen, p. 321; A.P. Bukhvostov, N.P. Popov, Phys. Lett. 24B, 497, 1967; Yadernaya Fizika, 6, 809, 1241, 1967; Z. Oziwicz, A. Pkulski, Acta Phys. Polonica 34, 291, 1968.

MEASUREMENTS OF EXCITATION PROBABILITIES AND ISOMER

SHIFTS OF NUCLEAR LEVELS IN μ-ATOMS

H.Backe, R.Engfer[+], E.Kankeleit, W.U.Schröder,
H.K.Walter, and K. Wien
Institut für Technische Kernphysik, TH Darmstadt,
Germany
[+] visiting scientist, CERN, Switzerland

The present paper gives a compilation of recent investigations on nuclear γ rays from heavy μ atoms in extension to previous results on even-even nuclei[1]. The experiments were performed at the CERN muon channel.

1. Excitation probabilities of nuclear levels in μ atoms. Excitation probabilities of nuclear states were deduced from absolute intensities of γ rays measured with a Ge(Li) detector in coincidence with muons stopped in the target (c.f. ref.[1]). The γ intensities were referred to the simultaneously detected muonic K-X ray serie of Aluminium, the emission probability of which was assumed to be unity per stopped muon. Selfabsorption in the targets, effects due to the spatial distribution of the stopped muons in the target and the conversion of the nuclear γ transitions were taken into account. In table 1 the experimental results are compared to theoretical calculations. Apart from ^{188}Os and the spherical Tl-nuclei the quadrupole mixing of nuclear and muonic states[2] describes the experimental excitation probabilities well. The discrepancy for ^{188}Os is unexplained. In muonic Thallium higher nuclear γ rays and anomalies of the 3d-2p structure were observed which could indicate a further excitation besides the one considered by Hüfner[3].

2. Muonic isomer shifts in heavy odd nuclei. As in preceding experiments[1] the muonic nuclear γ lines were compared to corresponding γ lines of radioactive sources. The results are given in table 2. With the exception of the Tl isotopes[4] the magnetic hf-splittings of the ground

Isotope	I^{π}	E_{γ} [KeV]	$P_{exp.}$	$P_{theor.}$	Ref.
^{182}W	(2^+)	100.1	$.40 \pm .11$	.37	2
^{184}W	(2^+)	111.2	$.42 \pm .11$	.37	2
^{186}W	(2^+)	122.6	$.35 \pm .09$	.36	2
^{188}Os	(2^+)	155.0	$.58 \pm .12$	.35	2
^{190}Os	(2^+)	186.7	$.32 \pm .06$	.29	2
^{192}Os	(2^+)	205.8	$.25 \pm .05$	.22	2
^{191}Ir	$(5/2^+)$	129.4	$.18 \pm .04$	.17	2
^{193}Ir	$(5/2^+)$	139.3	$.20 \pm .04$	.18	2
^{203}Tl	$(3/2^+)$	279.2	$.04 \pm .01$	.008	3
^{205}Tl	$(3/2^+)$	203.8	$.14 \pm .03$	.068	3

Table 1: Excitation probabilities of nuclear levels in
μ atoms

Isotope	E_{γ} [KeV]	ΔE_{exp} [eV]	ΔE_{cor} [eV]	$\left(\dfrac{\Delta \langle r^2 \rangle}{\langle r^2 \rangle}\right)_{\Delta t=0} \times 10^4$
^{169}Tm	118.2	-410 ± 65	$- 457$	$- 2.0$
^{169}Tm	109.8	$+ 57 \pm 40$	$+ 68$	$+ 0.3$
^{169}Tm	130.5	-800 ± 700	$- 480$	$- 2.1$
^{185}Re	125.3	$+175 \pm 35$	$-$	$-$
^{187}Re	134.3	$+174 \pm 35$	$-$	$-$
^{191}Ir	129.4	-108 ± 45	$+ 103$	$+ 0.4$
^{193}Ir	139.3	-266 ± 45	$- 121$	$- 0.5$
^{194}Pt	328.5	$+262 \pm 45$	$+ 321$	$+ 1.3$
^{194}Pt	293.6	$+ 3 \pm 50$	$- 144$	$- 0.6$
^{196}Pt	355.7	$+390 \pm 280$	$-$	$-$
^{203}Tl	279.1	$-$	$+260 \pm 180$	$+ 0.9$
^{205}Tl	203.8	$-$	-330 ± 150	$- 1.1$

Table 2: Muonic isomer shifts

and excited states were not resolved. Nevertheless the observed shifts were corrected for the hf-splittings taking into account the Bohr-Weisskopf effect and the M1-interdoublet transition (c.f.[1]). No errors are quoted for the corrected values ΔE_{corr}, the uncertainties of the corrections, however, were estimated to be less than 150 eV.

Neglecting nuclear polarization the isomer shift can be expressed in terms of $\Delta<r^2>$ and Δt with the assumption of the Fermi charge distribution. Values of $\Delta<r^2>/<r^2>$ at $\Delta t=0$ are presented in table 2. $\Delta<r^2>$ values from Mössbauer experiments[5] for ^{191,193}Ir show the same change in sign. Combining Mössbauer and the present ^{191,193}Ir data values of $\Delta t/t=0.5$ $^o/oo$ and 1 $^o/oo$ respectively, can be deduced similarily to fig. 2 (ref.[1]).

It is interesting to observe the large negative shifts of the 3/2, 5/2 and 7/2 excited rotational states in ^{169}Tm. For the 8.4 keV level $(\Delta<r^2>/<r^2>)\Delta t=0 = -2.3\cdot10^{-4}$ was derived whereas the Mössbauer experiment yields $-0.09\cdot10^{-4}$. This difference could be interpreted by a large change in skin thickness $\Delta t/t=16$ $^o/oo$. It should be pointed out, however, that such an interpretation is questionable in view of the uncertainties of the Mössbauer data and the neglected muonic nuclear polarization.

This work was supported by the "Bundesministerium für wissenschaftliche Forschung".

REFERENCES

1) R. Baader, H.Backe, R.Engfer, K.Hesse, E.Kankeleit, U.Schröder, H.K.Walter and K. Wien, Phys. Letters 27B (1968) 425
2) S.A. De Wit, G. Backenstoss, C. Daum, J.C. Sens and H.L. Acker, Nucl. Phys. 87 (1967) 657
3) J. Hüfner, Z. Physik 190 (1966) 81
4) R. Baader, H. Backe, R.Engfer, E.Kankeleit, W.U. Schröder, H.Walter and K. Wien, Phys. Letters 27B (1968) 428
5) P.Kienle, G.M. Kalvius and S.L. Ruby, Hyperfine Structure and Nuclear Radiations; North-Holland, Amsterdam 1968, p. 971

NUCLEAR POLARIZATION IN DEFORMED NUCLEI

Min-yi Chen

Columbia University

New York, New York

The analysis of muonic X-ray spectra in heavy deformed nuclei has been carried out in the past by diagonalizing the electric quadrupole interaction H_Q between the muonic spin doublets $(2P_{3/2}, 2P_{1/2})$, $(3d_{5/2}, 3d_{3/2})$ and the ground state rotational band of the nucleus. To take into account the muonic and nuclear states not included in the diagonalization, we introduce the effective quadrupole interaction H_{eff} by virtual excitations into those states, as shown in the following diagram.

where I, μ are nuclear and muonic intermediate states respectively, and H_ℓ is the electrostatic interaction. Graph 1 represents the quadrupole interaction which has been taken into consideration in the past, and graph 2 represents the corrections. We find the off-diagonal matrix elements of graph 2 are always proportional to the corresponding elements of graph 1; therefore, the inclusion of those elements is equivalent to a renormalization of the intrinsic quadrupole moment of the nucleus. The renormalization constant is calculated to be 1.03 - 1.05. The diagonal matrix elements of graph 2 are always negative, which increases the binding of the 1s and 2p states.

In calculating the intensity ratios of the hyperfine multiplats, we introduce the effective El transition element M_{eff} (El) by the following diagram

Graph 1 represents the El transition matrix elements which have been used in the past. Graphs 2 and 3 take into account the higher muonic levels which have previously been neglected. Graphs 4 and 5 represent the mixing of the giant dipole states. The inclusion of the above corrections in the analysis of the experimental spectra has a profound influence on the results.

PARAMETERS OF THE CHARGE DISTRIBUTION OF DEFORMED NUCLEI*

D. Hitlin[+], S. Bernow, S. Devons, I. Duerdoth, J.W. Kast,
E.R. Macagno, J. Rainwater, and C.S. Wu
      Columbia Univeristy
            and
R.C. Barrett
      University of Surrey

* Supported by the U.S.A.E.C.
+ Now at SLAC, Stanford, Calif.

Precise measurements, using a stable high resolution Ge(Li) spectrometer have been made of the K (2P-1S) and L (3d-2P) muonic X-ray spectra of eight even-even rare-earth nuclei: ^{150}Nd, ^{152}Sm, ^{162}Dy, ^{164}Dy, ^{168}Er, ^{170}Er, ^{182}W, ^{184}W, and ^{186}W (1). From these measurements values of the parameters of the nuclear charge density

$$\rho(r) = \rho_0 \left[1 + \exp \frac{r-c[1+\beta Y_{20}(\theta)]}{a} \right]^{-1}$$

have been determined.

In several different attempts to interpret our experimental results, we have come to appreciate how sensitive the conclusions are to the precision of the theoretical corrections. These include the Lamb Shift and, more important, the nuclear polarization corrections. Nuclear polarization is the effect of all inter-actions which induce mixing of different muon and nuclear states. The dynamic E2 hyperfine spectra (2) can also be considered as a part of the nuclear polarization due to the quadrupole interaction. Recently M. Y. Chen (3) has extended nuclear polarization calcul-ations in deformed nuclei to include states other than the lowest muonic and nuclear states. Before this improved nuclear polariz-ation correction was made, the most unsatisfactory feature of the results was the very small value of the skin thickness obtained. In addition the value of the quadrupole moment was larger than that derived from Coulomb excitation measurements, and it was not possible to obtain a reasonable fit to the relative intensities of the K and L X-rays. In order to bring the quadrupole moment obtained from the h.f.s. spectrum into agreement with that obtained from Coulomb excitation, the skin thickness was assumed to be a

TABLE 1
Parameters of the Charge-Distribution of Deformed Nuclei

Without Nuclear Polarization and Lamb Shift Corrections

a) Columbia (1)

	c, fm.	t, fm.	β	β'	Q_0
^{150}Nd	6.083	1.68±.02	0.274	0	5.27
^{152}Sm	6.093	1.77±.02	0.302	-0.22	5.85
^{162}Dy	6.26	1.59±.02	0.337	-0.28	7.38
^{164}Dy	6.34	1.30±.03	0.329	-0.12	7.53
^{168}Er	6.34	1.51±.02	0.354	-0.92	7.77
^{170}Er	6.42	1.26±.03	0.341	-0.87	7.80
^{182}W	6.47	1.85±.02	0.272	-0.51	6.57
^{184}W	6.49	1.84±.03	0.269	-0.78	6.19
^{186}W	6.55	1.75±.02	0.243	-0.54	6.01

With Nuclear Polarization and Lamb Shift Corrections, $\beta' = 0$.

a') Columbia (1)

c, fm.	t, fm.	β	Q_0	$(Q_0)_{C.E.}$
5.863	2.35	0.278	5.14	5.17 ± .12
5.900	2.37	0.296	5.76	5.85 ± .15
6.00	2.41	0.338	7.34	7.12 ± .12
6.02	2.41	0.340	7.44	7.50 ± .20
6.17	2.17	0.332	7.75	7.66 ± .15
6.27	1.95	0.325	7.71	7.45 ± .13
6.41	2.12	0.248	6.56	6.58 ± .06
6.39	2.23	0.238	6.27	6.21 ± .06
6.46	2.10	0.222	5.88	5.93 ± .05

b) CERN (4)

^{159}Tb	6.2	1.5 ± 0.4	0.30	7.6 ± 0.4
^{165}Ho	6.27	1.5 ± 0.4	0.30	7.9 ± 0.5
^{181}Ta	6.51	1.5 ± 0.4	0.25	7.5 ± 0.4
^{232}Th	7.10	1.49 ± 0.14	0.23	9.8 ± 0.3
^{233}U	7.11	1.50 ± 0.5	0.24	10.3 ± 0.3
^{235}U	7.14	1.44 ± 0.17	0.241	10.6 ± 0.2
^{238}U	7.15	1.46 ± 0.12	0.253	11.25 ± 0.15
^{239}Pu	7.18	1.35 ± 0.3	0.26	12.0 ± 0.3

c) Carnegie-Mellon-Argonne-Binghamton (5)

^{151}Eu	6.27	1.3 ± 0.4	2.75
^{153}Eu	6.06	1.78 ± 0.4	7.01 ± 0.10
^{158}Gd	6.33	0.714	7.15
^{160}Gd	6.35	0.57	7.38
^{175}Lu	6.24	2.07 ± 0.3	8.02 ± 0.12
^{232}Th	7.07	1.75	11.3
^{238}U	7.10	1.75	11.3

Remarks: Argument of the Fermi charge distribution function:

In a)   a')   In b) and c)

$$\frac{r - c\,[1 + \beta\,Y_{20}(\theta)]}{a\,[1 + \beta'\,Y_{20}(\theta)]} \; ; \qquad \frac{r - c\,[1 + \beta\,Y_{20}(\theta)]}{a} \qquad \frac{r\,[1 + \beta\,Y_{20}(\theta)] - c}{a}$$

function of the polar angle: $a' \to a(1+\beta'Y_{20})$.   For the nuclei
studied a best fit to the data was obtained with rather large
negative values of the β' parameter, implying that the nuclear
charge distribution was more diffuse at the "equator" ($\theta=90°$) than
at the poles ($\theta=0°$).   With this four-parameter model the average
skin thickness was still unusually small, however, and there was
no improvement in the fit to the intensities.

When the improved nuclear polarization corrections are
applied, the skin-thickness of the three parameter distribution is
increased to about 2.2 F., a value similar to that obtained for
neighbouring spherical nuclei.   At the same time the quadrupole
moment agrees with the Coulomb excitation result and a better fit
is obtained to the relative intenstities of the X-rays.   The
hyperfine results for deformed nuclei are summarized in Table I.
In section a) only the nuclear polarization in the "model space"
is calculated resulting in the large negative values for β'.   The
average value of t ($=4\log3\ a$) is 1.6 F.   In section a') the
improved nuclear polarization corrections are applied,   the values of t are increased to about 2.2 F. and the β' parameter is no
longer needed.   In sections b) and c) the CERN and Carnegie-
Mellon-Argonne-Binghampton results are listed (4,5).   Their
analyses did not include nuclear polarization corrections and
the skin-thicknesses which they obtained are seen to be unusually
small.

Our analysis illustrates the extreme importance of applying
precise nuclear polarization corrections in interpreting muonic
X-ray data.   It appears that the three-parameter deformed Fermi
distribution is adequate to describe the charge densities at
present.

References

(1) D. Hitlin, S. Bernow, S. Devons, I- Duerdoth, J.W. Kast,
    E.R. Macagno, J. Rainwater, K. Runge, C.S. WU and R.C. Barrett,
    Proceedings of the International Conference on Electromagnetic
    Sizes of Nuclei, Ottawa, 1967, and D. Hitlin, Thesis, Columbia
    University, 1968.
(2) L. Wilets, Dan.Mat.Fys.Medd.29,#3(1954)
    B.A. Jacobsohn, Phys.Rev. 96,1637(1954).
(3) M.Y. Chen, contribution to this conference.
(4) S.A. DeWitt, G. Backenstoss, C. Daum, J.C. Sens and H.L. Acker,
    Nucl.Phys. 87, 657(1967).
(5) R.A. Carrigan,Jr.,P.D. Gupta, R.B. Sutton, M.N. Suzuki, A.C.
    Thomson, R.E. Cole, W.V. Prestwich, A.K. Gaigalas, and S. Raboy,
    Bull.Am.Phys.Soc. 13, 65(1968).

ISOTOPIC VARIATIONS OF THE NUCLEAR CHARGE DISTRIBUTION IN Nd*

E.R. Macagno, S. Bernow, S.C. Cheng, S. Devons,
I. Duerdoth[†], D. Hitlin[††], J.W. Kast, W.Y. Lee
J. Rainwater and C.S. Wu - Columbia University
and
R.C. Barrett - University of Surrey

It is well established that the optical isotope shifts for pairs of isotopes with 88 and 90 neutrons in $_{60}$Nd, $_{62}$Sm and $_{63}$Eu are anomalously large.[1] This is directly related to a change in the neutron-shell configuration. The sudden onset of permanent deformation occurs at a neutron number of 90, as predicted by the nuclear model of Nilsson. It is the increase in nuclear deformation, coupled with the usual increase in nuclear volume, that gives origin to such anomalous isotope shifts.

To explore the variation of the nuclear charge distribution in this interesting transition region, we have made precise measurements of the energies and isotope shifts of the muonic K and L lines in the Nd isotopes (A = 142, 143, 144, 145, 146, 148, 150). The results have been partially reported elsewhere.[2] We wish in this note to present a recent determination of the isotopic variation of parameters of the nuclear charge distribution from our measurements of the even isotopes. A complete discussion will appear in the near future.[3]

We have analyzed the muonic (2p - 1s) and (3d - 2p) transition energies in terms of the parameters c and t of the Fermi distribution

$$\rho(r) = \rho_0 \left[1 + \exp\left(\frac{r-c}{0.228t}\right)\right]^{-1} \tag{1}$$

and in terms of the parameters c, t and β (β fixed to values derived from B(E2) measurements[4]) of the "deformed" Fermi distribution

$$\rho(r) = \rho_0 \left\{1 + \exp\left(\frac{r-c[1 + \beta Y_{20}(\theta,\phi)]}{0.228t}\right)\right\}^{-1} . \tag{2}$$

The analysis included dynamic hyperfine effects [5], which are very pronounced in Nd^{150}, as well as nuclear polarization[6] and Lamb shift[7] corrections. The measured energies and the values of the parameters of the charge distributions are presented in Table 1. The errors quoted for the parameters include uncertainties in the various corrections mentioned above. Plots of the parameters versus mass number A are shown in Figures 1 and 2.

Isotope	Line	Measured Energy	Fermi Distribution		Deformed Fermi Distribution		
		(keV)	c (fm)	t (fm)	c (fm)	t (fm)	β
^{142}Nd	$K_{\alpha 1}$	4352.06 ± 0.40	5.75 ± 0.03	2.38 ± 0.08	5.80 ± 0.03	2.32 ± 0.08	0.104
	$K_{\alpha 1}$	4270.44 ± 0.50					
	$L_{\alpha 2}$	1402.97 ± 0.25					
	$L_{\alpha 1}$ $\alpha 2$	1472.32 ± 0.25					
^{144}Nd	K	4336.38 ± 0.45	5.80 ± 0.03	2.37 ± 0.08	5.85 ± 0.03	2.27 ± 0.08	0.123
	$K_{\alpha 1}$	4254.88 ± 0.55					
	$L_{\alpha 2}$	1402.59 ± 0.25					
	$L_{\alpha 1}$ $\alpha 2$	1471.82 ± 0.25					
^{146}Nd	K	4321.18 ± 0.40	5.77 ± 0.03	2.45 ± 0.08	5.82 ± 0.03	2.42 ± 0.08	0.151
	$K_{\alpha 1}$	4240.50 ± 0.50					
	$L_{\alpha 2}$	1402.58 ± 0.20					
	$L_{\alpha 1}$ $\alpha 2$	1470.99 ± 0.25					
^{148}Nd	$K_{\alpha 1}$	4303.46 ± 0.45	5.80 ± 0.03	2.46 ± 0.08	5.84 ± 0.03	2.40 ± 0.08	0.197
	$K_{\alpha 1}$	4223.76 ± 0.50					
	$L_{\alpha 2}$	1403.54 ± 0.25					
	$L_{\alpha 1}$ $\alpha 2$	1470.97 ± 0.25					
^{150}Nd	K	4266.37 ± 0.40	5.90 ± 0.03	2.42 ± 0.08	5.86 ± 0.03	2.35 ± 0.08	0.279
	$K_{\alpha 1}$	4198.30 ± 0.50					
	$L_{\alpha 2}$	1413.08 ± 0.25					
	$L_{\alpha 1}$ $\alpha 2$	1474.88 ± 0.25					

Table 1. Energy of the K and L lines in Nd and values of the parameters of the charge distributions (eq. (1) and eq. (2)).

It is of great interest to note that even for these isotopes in the transition region, where the shape of nuclei changes from spherical to spheroidal, the skin thickness t does not vary a great deal. As a matter of fact, our results indicate that within the quoted uncertainties, we can consider the skin thickness as constant for the even Nd isotopes. For the deformed model, the skin thickness is generally smaller by a few percent than the corresponding values for the spherical model, as would be expected.

Since the skin thickness t is reasonably constant for the even Nd isotopes, it is possible, as shown in reference 8, to compare the muonic results with corresponding optical measurements, and to thus extract from the measured optical isotope shifts the specific mass shifts. Such a comparison has been made for various optical lines, and the specific mass effects determined; these results appear in reference 3.

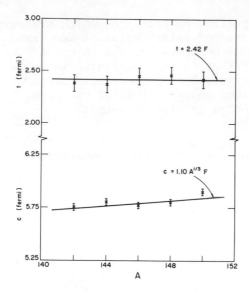

Figure 1. Values of c and t
for the Fermi Distribution,
equation (1), plotted versus
mass number A of the Nd isotopes.

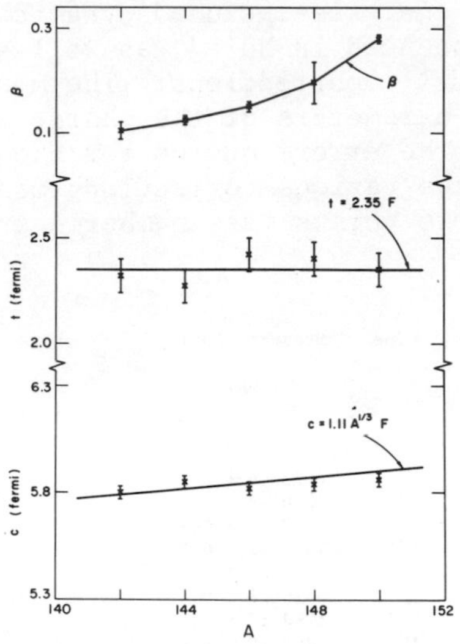

Figure 2. Values of c,t
and β for the Modified
Fermi Distribution, eq. (2),
plotted versus mass number
A of the Nd isotopes.
Values of β are from reference 4.

References

1. D.N. Stacey, Rep. Progr. Phys. 29, 171 (1966).
2. E.R. Macagno et al., Proceedings of the International Con-
    ference on Electromagnetic Sizes of Nuclei, Ottawa (1967);
    C.S. Wu, Proceedings of the International Symposium on the
    Physics of One and Two Electron Atoms, 1968, North Holland
    Publishing Co.
3. E.R. Macagno, S. Bernow, M.Y. Chen, S.C. Cheng, S. Devons,
    I. Duerdoth, D. Hitlin, J.W. Kast, W.Y. Lee, J. Rainwater,
    C.S. Wu and R.C. Barrett, to be published.
4. P.H. Stelson and L. Grodzins, Nuclear Data 1A, 21 (1965);
    P.A. Crowley et al., Bull. Am. Phys. Soc. 13, 79 (1968).
5. D. Hitlin et al., contributed paper to this conference.
6. M.Y. Chen, contributed paper to this conference.
7. R.C. Barrett, Phys. Letters 28 B, 93 (1968).
8. C.S. Wu and L. Wilets, Annual Review of Nuclear Science, 1969,
    in press.

FINITE DISTRIBUTION OF NUCLEAR M1 AND E2 MOMENTS IN MUONIC [115]In, [133]Cs, and [141]Pr†

B. Budick*, W.Y. Lee, S. Bernow, M.Y. Chen, S.C. Cheng,
D. Hitlin**, J.W. Kast, E.R. Macagno, A.M. Rushton and
C.S. Wu

*New York University, NYC; Columbia University, NYC

The distribution of nuclear M1 and E2 moments in the nuclei
[115]In, [133]Cs, and [141]Pr have been investigated by observing the
hfs of their muonic X-rays spectra. As pointed out by Johnson
and Sorenson (1), the nuclear magnetism arising from the orbital
motions is concentrated closer to the center of the nucleus, while
the spin magnetism is more highly localized at the positions of
the individual nucleons and tends to be more uniform. Our results
are listed in Table I. The reductions of the values of $A_1(1s_{1/2})$
from their point-dipole values for all three nuclei studied are
apparent.

We note that in all cases the configuration mixing model
represents a substantial improvement over the point-dipole and
the single particle models. This results mostly from the admixture of configurations whose magnetism is concentrated closer to
the surface of the nucleus, i.e., spin magnetism. The theoretical values in Table I are quoted from Johnson and Sorenson.(2)

Only in the $2p_{3/2}$ state of indium is the E2 hfs constant
large enough to be measured with precision. The experimental
value $A_2(2p_{3/2}) = 2.55$ KeV is in fair agreement with the theoretical value $A_2 = 2.79$ KeV computed using a surface quadrupole distribution, provided the spectroscopic quadrupole moment is taken
as $Q_{spec} = 0.834$ barns. We chose this value in preference to the
value 1.16 barns found in the Table of Isotopes (4), since the
measured atomic beam value(5) is subject to corrections due to
mixing of electronic configurations.(3),(6) Such polarization

**Present address: SLAC, Stanford, California.
†Work supported by U.S.A.E.C. and N.S.F.

corrections, of which Sternheimer(7) corrections are an example, considerably decrease the confidence and precision with which nuclear quadrupole moments are known. The muon, in this case the $2p_{3/2}$ muon, is sensitive to the details of the distribution of the E2 moment, and its interaction with this distribution can be accurately calculated. This suggests that muonic atoms may provide a means of testing Sternheimer as well as other types of electronic polarization corrections for nuclear quadrupole moments.

Table 1

Isotope	State	hfs Constant	(Nuclear Moment Value)	Experimental Value of hfs Constant (keV)	Point-dipole or Surface Quadrupole Model (keV)	Single Particle Model (keV)	Configuration Mixing Model (keV)	BCS Model (keV)
^{115}In	$1s_{1/2}$	A_1	(μ=5.535 nm)	1.65 ±0.07	2.53	1.88	1.56	1.56
	$2p_{1/2}$	A_1		0.55 ±0.13	0.64	0.56	0.46	0.46
	$2p_{3/2}$	A_1	(Q=0.834 b)	2.55 ±0.10	2.79			
^{133}Cs	$1s_{1/2}$	A_1	(μ=2.579 nm)	1.11 ±0.18	1.34	0.91	1.21	1.15
	$2p_{1/2}$	A_1		0.50 ±0.20	0.41	0.28	0.38	0.36
	$2p_{3/2}$	A_1	(Q=0.003 b)		0.01			
^{141}Pr	$1s_{1/2}$	A_1	(μ=4.28 nm)	1.53 ±0.15 (1.52 ±0.07)[a]	2.38	1.62	1.47	1.47
	$2p_{1/2}$	A_1		0.62 ±0.20	0.80	0.60	0.54	0.54
	$2p_{3/2}$	A_1	(Q=-0.06 b)		-0.45			

[a] See reference 2

REFERENCES

1. J. Johnson and R.A. Sorenson, Phys. Letters 26B, 700 (1968).
2. J. Johnson and R.A. Sorenson, Phys. Letters 29B, 420 (1969), and Bull. Am. Phys. Soc. 14, 539 (1969).
3. G. Koster, Phys. Rev. 86, 148 (1952).
4. C.M. Lederer, J.M. Hollander, and I. Perlman, Table of Isotopes, Wiley, 1967.
5. A.K. Mann and P. Kusch, Phys. Rev. 78, 615 (1950).
6. H.L. Acker et. al., Nucl. Phys. 87, 1 (1966).
7. R.M. Sternheimer, Phys. Rev. 105, 158 (1957) and earlier papers.

THE RESONANCE PROCESS AND THE INTENSITY ANOMALY

IN MUONIC ^{127}I

W.Y. Lee, S. Bernow, M.Y. Chen, S.C. Cheng, D. Hitlin,*
J.W. Kast, E.R. Macagno, A.M. Rushton, and C.S. Wu

Columbia University, New York, New York

† Work supported by U.S.A.E.C. and N.S.F.
* Present address: SLAC, Stanford, California

It has been known for some time that the intensity ratio
between the $(2p_{3/2}-1s_{1/2})$ and $(2p_{1/2}-1s_{1/2})$ muonic X-rays for ^{127}I
is anomalously smaller than the theoretical value 1.95. Several
theorists[1],[2],[3] have suggested that this may be explained by
nuclear resonance excitation. We have observed the muonic N, M,
L and K X-rays from ^{127}I, and have calculated a theoretical spectrum within the framework of a resonance process. The resonance
is between the 2p fine structure splitting and the nuclear ground
and the first excited states. A satisfactory fit to the experimental data has been obtained. From this fit we have determined
the M1 and E2 hfs constants for the 1s and 2p states, and have
obtained a quantitative interpretation of the observed intensity
ratios.

The theoretical spectrum was calculated for an initial
choice of the values of the parameters: the 3d and 2p fine structure splittings, the E2 and M1 transition moments, and the hfs
constants A_1 and A_2 of the 2p and 3d states. These parameters were
varied until a best fit (minimum χ^2) to the data was obtained. We
have assumed that the 3d states were statistically populated, and
have not taken into account the effect of excitation to nuclear
states higher than the first. The analysis of our data including
the "nuclear polarization" effect is in progress.

The hfs constants determined are listed in Table I. The
values of $A_1(1s_{1/2})$ and $A_1(2p_{1/2})$ are noticeably smaller than the
values calculated for a point-dipole and a single particle model.
Also the $A_2(2p_{3/2})$ value is smaller than that calculated for a
surface distribution of the quadrupole moment. The predictions of

205

various nuclear models are quoted from Johnson and Sorenson(4).

The intensity ratios are given in Table II. The theoretical values (with the resonance) were calculated from the set of parameters which gave the best fit to our data. We see that the intensity anomaly can be quantitatively interpreted.

TABLE I

hfs Constants	Experimental Columbia 1969	Point Dipole	Surface Quadrupole	Arima Horie	BCS + δ	BCS + δ + phonon (GR = 0)	BCS + δ + phonon, GR=Z/A	Single particle
$A_1(S_{1/2})$	0.87 ± .09	1.40	-	0.95	0.94	0.928	1.11	1.50
$A_1(P_{1/2})$	0.33 ± .08	0.40	-	0.298	0.296	0.293	0.354	0.48
$A_1(P_{3/2})$	0.18 ± .05	0.22	-					
$A_1(D_{3/2})$	0.045 ± .005	0.045						
$A_1(D_{5/2})$	0.030 ± .005	0.031	-					
$A_2(P_{3/2})$	-2.87 ± .01	-	-3.15					
$A_2(D_{3/2})$	-0.251 ± .008	-	-0.255					
$A_2(D_{5/2})$	-0.377 ± .010	-	-0.383					

TABLE II. Intensity Ratios

Ratio	Experimental		Theoretical	
	Columbia '69	CERN '66	With resonance	Without Resonance
$\dfrac{(2p_{3/2}-1s_{1/2})}{(2p_{1/2}-1s_{1/2})}$	1.10±0.08	1.06±0.08	1.18±0.10	1.95
$\dfrac{(3d_{5/2}-2p_{3/2})+(3d_{3/2}-2p_{3/2})}{(3d_{3/2}-2p_{1/2})}$	1.93±.09	3.0±0.50	1.83±0.10	1.95

REFERENCES

1. H.L. Acker, G. Backenstoss, C. Daum, J.C. Sens and S.A. De Wit, Nucl. Phys. 87, 1 (1966).
2. W.B. Rolnick, Phys. Rev. 132, 110 (1963).
3. J. Hufner, Phys. Letters 25B, 189 (1967).
4. J. Johnson and R.A. Sorenson, Phys. Letters 29B, 420 (1969), and Bull. Am. Phys. Soc. 14, 538 (1969).

THEORY OF HIGH ENERGY HADRON-NUCLEUS COLLISIONS*

R. J. Glauber

Lyman Laboratory of Physics, Harvard University

Cambridge, Mass. 02138

I.   INTRODUCTION

The last two or three years have seen a considerable revival
of the use of real nuclei, ones having $A > 1$, as targets in high
energy collision experiments. That has been, I believe, because
of the growing realization that the problem of dealing with nuclei
theoretically is not as complicated as we might have imagined.
Nuclear physics becomes, at billion volt energies, a very much
simpler subject than it is at low energies. A high energy particle
passing through a nucleus tends hardly to be deflected at all in its
passage and is gone before any nuclear rearrangement it induces can
take place. It is so very much more energetic than any nucleon in
the nucleus that, for all practical purposes, when it collides with
individual nucleons the others are simply stationary spectators.

The incident particle tends on the average to lose only a tiny
fraction of its momentum and energy in each individual collision.
That's not to say that there are no catastrophic events which ever
take place, but only that large momentum transfer collisions tend
statistically not to be of great weight. The result is that the
overwhelming proportion of what does take place can be described
in terms rather like those used to describe diffraction phenomena
in physical optics.

The theory we shall discuss here[1,2] is in fact an outgrowth of
optical diffraction theory. It has been given a variety of names
in the nuclear context (eikonal approximation, high-energy approximation, etc.) but the best name for it, I believe, remains diffraction theory. It is simply Fraunhofer diffraction theory generalized
in the ways that are obviously necessary to deal with multiple

scattering processes, with energy losses and with quantum mechanical
variables such as spin and isospin.

Before we begin discussing the theory itself it will be helpful to distinguish between two senses in which collision processes
can be inelastic. One is the explicitly nuclear sense in which a
target nucleus becomes excited and perhaps breaks up while the
particles present do not change in any essential way. The other is
the sense characteristic of elementary particle theory in which the
incident or target particle becomes changed in nature or the number
of particles present is somehow altered. We shall confine most of
our discussion in the present paper to elastic hadron-nucleus collisions and to collisions which are inelastic in the first of the
two senses, i.e. the purely nuclear one. Much of what we shall say
about these relatively simple processes applies as well, at sufficiently high energies, to particle production processes in nuclei.
We shall leave most of the description of such processes to later
speakers but we will have a few words to say about them in concluding.

The angular distribution of elastic scattering by nuclei has
some rather special and easily identified properties. Let us
imagine for the moment that the A nucleons of a nucleus are weak
scatterers, so that an incident particle passing through the nucleus is scattered at most only once. Because it is impossible, as
a matter of principle, to tell which of the A nucleons has been
struck in an elastic collision the amplitudes for scattering by
all of them must be added coherently to form the nuclear scattering
amplitude. The fact that the entire volume of the nucleus contributes coherently to the scattering leads to an intense forward peak
in the angular distribution of scattering. A large part of the
elastic scattering at high energies is collimated within a narrow
cone whose angle is in order of magnitude the ratio of the wavelength λ of the incident particle to the radius R of the nucleus.

Another property of the elastic scattering is indicated by the
fact that in our model the forward scattering amplitude is proportional to A. The magnitude of the forward intensity peak should
therefore increase as A^2. Both of these properties, the forward
collimation and the A^2 dependence, are consequences of the coherent
character of elastic scattering. Indeed elastic scattering is
being widely referred to these days simply as coherent scattering
(and production processes which are elastic in the nuclear sense
are likewise being called coherent production processes.)

Let us write the density function of the nucleus as $\rho(\vec{r})$ with
normalization $\int \rho(\vec{r})d\vec{r} = 1$. Then we can define the nuclear form
factor as

$$S(\vec{q}) = \int e^{i\vec{q}\cdot\vec{r}} \, \rho(\vec{r})d\vec{r} \qquad\qquad (1.1)$$

If we let the amplitude for scattering with momentum transfer ℏ q by a single nucleon be $f(\vec{q})$ then, in the single scattering approximation, the intensity scattered elastically by the nucleus is simply

$$\frac{d\sigma}{d\Omega} = A^2 |f(\vec{q})|^2 \, S^2(\vec{q}) \quad . \qquad\qquad (1.2)$$

An indication of the shape of this distribution for a medium-weight nucleus is given in Fig. 1. It is the form factor $S(\vec{q})$ which gives the distribution its rapid decrease and diffractive shape.

The increase of the forward scattered intensity with A^2 results of course from our assumption that nucleons are weak scatterers as well as from the coherence of their amplitudes. In practice, since nucleons are strong scatterers of incident hadrons multiple collision effects are quantitatively important in nuclei. These processes lead to an A-dependence of the forward scattered intensity which increases rather less rapidly than A^2, but they do not greatly alter the tendency of the distribution to be collimated in the forward direction. It is this strong forward peaking which most clearly characterizes coherent scattering (and an important class of coherent production processes as well).

The fact that the first-excited-state energies of all nuclei are so small (or the order of a few meV at most) compared with multi-GeV incident particle energies means that high-energy elastic scattering by nuclei is exceedingly difficult to isolate, at least by following the familiar path of using accurate energy resolution. That method has been made to work, to be sure, for 1 GeV protons incident on several light nuclei, and probably somewhat better resolution can eventually be achieved. But at appreciably higher energies elastic scattering measurements can only be carried out in practice by detecting the recoil nuclei in some way which ascertains more directly that they are still in their ground states. That is what happens automatically when recoiling deuterons or α-particles are detected, for example, and all of the elastic scattering measurements we have to date at energies above 1 GeV have been made on these nuclei.

The elastic collisions which take place between hadrons and nuclei can lead to a vast variety of final states, both bound and

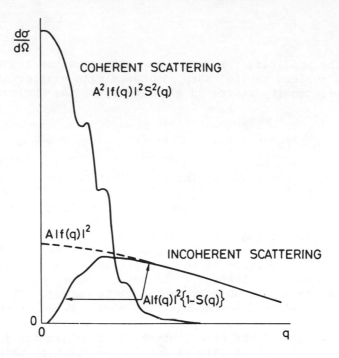

Fig. 1. Differential cross sections for coherent and incoherent single-scattering processes in a medium-weight nucleus. Coulomb scattering is omitted.

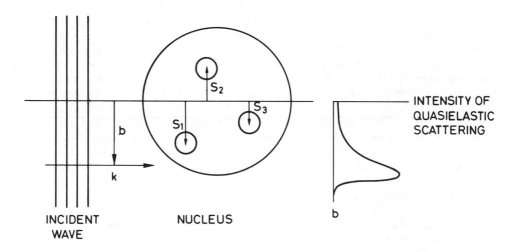

Fig. 2. Schematic picture of nucleus showing definitions of impact vector $\vec{b}$ and transverse components $\vec{s}_j$ of nucleon position vectors. Graph at right shows that quasielastic scattering arises mainly from collisions near the limb of the nucleus.

unbound, for the nuclear system. Since the momenta transferred
in individual collisions are relatively small, however, the energies
transferred are also quite small. The final state of the nuclear
system for small-angle collision processes, in other words, typically has an excitation energy of only a few meV, or perhaps some tens
of meV, values which are quite tiny compared to billion volt incident particle energies. Very little accuracy is lost therefore in
altogether neglecting the energies transferred and using the closure
approximation to sum the differential cross sections corresponding
to all possible nuclear final states.

When we evaluate the summed scattering by means of the closure
approximation and subtract the known elastic cross section we find
a fairly simple expression for the sum of the inelastic cross sections. In the single scattering approximation we have used earlier
it is easily shown to be

$$\left.\frac{d\sigma}{d\Omega}\right)_{inel.} = A\left|f(\vec{q})\right|^2 \left\{ 1 - S(\vec{q}) \right\} , \qquad\qquad (1.3)$$

if we assume that the nucleons are not significantly correlated in
their positions within the nucleus. A cross section of this form
is illustrated in Fig. 1.

One thing we note immediately is that the inelastic cross
section is proportional to the number of nucleons A. That is because it is possible in principle to determine, by observing the
final state, which of the nucleons was struck (or more accurately,
which were the initial and final single particle states between
which the transition took place). The individual nucleons contribute incoherently, in other words, to the summed inelastic scattering
which is therefore called incoherent scattering in current terminology.

For values of the momentum transfer $\hbar\,\vec{q}$ lying outside the
coherence cone of elastic scattering, the form factor $S(\vec{q})$ is
small compared to unity. The dominant inelastic process is the
quasielastic collision, one in which a single nucleon is struck
and in all likelihood knocked out of the nucleus. The differential
cross section for this process, which is just $A\left|f(\vec{q})\right|^2$ according
to Eq. (1.3), has essentially the shape characteristic of the individual nucleon cross sections.

At small momentum transfers, however, the angular distribution
of inelastic scattering must be altogether different in shape. By
restricting our attention to inelastic scattering we are demanding
that the nucleus be left in a state different from the ground state

and, in fact, one orthogonal to it. It is obviously not possible
for a collision with zero momentum transfer to induce such a transition; the scattering matrix element vanishes because of the orthogonality of the initial and final states. The inelastic scattering
near the forward direction is accordingly quite small, at least in
the single scattering approximation for which it vanishes at q = 0.

    If position correlations of the nucleons are to be taken into
account, a further term containing the nucleon pair density must be
added to the cross section (1.3). The exclusion principle leads to
a familiar sort of position correlation, a tendency for identical
nucleons to avoid one another. Its influence on the inelastic
scattering cross section can be seen fairly directly from the fact
that it rules out transitions of the nucleus to a number of the
states of small single-particle excitation which have been included
in the sum (1.3). The effect of the exclusion principle, therefore, at least in the single scattering approximation, is to reduce
the inelastic or incoherent cross section to still smaller values
in the range of small momentum transfers.

    Although it is difficult to perform elastic scattering experi-
ments at high energies there is an alternative way of measuring
nuclear scattering which can be based on the use of relatively
poor energy resolution. If we let the counting equipment accept
particles over a fairly broad range of energies, say 100 meV or so
in the neighborhood of the incident energy, then what we are observing is the summed cross section, the sum of coherent and incoherent
scattering. Nearly all of the scattered intensity observed at
small momentum transfers is elastic, while at large momentum transfers the scattering is almost entirely inelastic. The two distributions are in fact separable to some degree because of their different
shapes. The information one loses by being unable to see the coherent scattering at the larger momentum transfers is more than
made up for by the rather different and in some ways complementary
information which is furnished by the incoherent scattering.

    The single scattering approximation, we must emphasize again,
is hardly more than a qualitative guide to the behavior of nuclear
cross sections. Hadronic interactions are so intense that attenuation effects and multiple collisions have quite a strong influence
on the cross sections we observe. The intensity of incoherent
scattering, for example, is not accurately proportional to A but
rather to a considerably smaller number of nucleons lying within
a fairly small distance of the nuclear surface. The simplest way
to treat such effects is to develop a diffraction theory of multiple
scattering.

II. NUCLEAR DIFFRACTION THEORY

Let us review very briefly some of the elements of diffractive scattering theory.[1,2] We assume, first of all, that the momentum $\hbar \vec{k}$ of the incident particle is high enough so that the wavelength k^{-1} is much smaller than the range of interaction between the particle and a nucleon. It is only when the wavelength is that small in general that the individual scattering amplitudes can be described in diffraction theoretical terms. The forward-peaked character of angular distributions of diffractive scattering implies that the partial waves corresponding to a great many values of the orbital angular momentum contribute coherently to the scattering.

So many partial waves are involved that it becomes an accurate approximation to integrate over them rather than summing. We do just that, in effect, by defining an impact parameter variable b by the relation $kb = \ell + \frac{1}{2}$ and then integrating over b. It is convenient then to replace the complex phase shift parameters δ_ℓ by the function $\chi(b) = 2\delta_\ell$. The function of the phase shifts which occurs in the partial wave expansion of a scattering amplitude is

$$1 - e^{2i\delta_\ell} = 1 - e^{i\chi(b)} \equiv \Gamma(b) . \qquad (2.1)$$

We shall call the function $\Gamma(b)$ the profile function. If the interaction responsible for the scattering process happens not to be azimuthally symmetric about the direction of the incident momentum $\hbar \vec{k}$ it becomes convenient to speak of an impact vector $\vec{b}$ having magnitude b and lying in a plane perpendicular to $\vec{k}$. The phase shift function and the profile function may then be written as $\chi(\vec{b})$ and $\Gamma(\vec{b})$ respectively, i.e. as functions of vector arguments.

The partial wave expansion of the amplitude for scattering by a fixed scatterer is given in diffraction theory by the two-dimensional Fourier integral of the profile function,

$$f(\vec{q}) = \frac{ik}{2\pi} \int e^{i\vec{q}\cdot\vec{b}} \, \Gamma(\vec{b}) \, d^2b \qquad (2.2)$$

This expression represents the scattering amplitude accurately as long as the momentum transfer $\hbar \vec{q} = h(\vec{k} - \vec{k}')$ remains small compared with the incident momentum. Since we assume that it is only in this region that the scattering amplitude is large, we can solve for the profile function in terms of the scattering amplitude by inverting the Fourier transform, i.e. by writing

$$\Gamma(\vec{b}) = \frac{1}{2\pi i k} \int e^{-i\vec{q}\cdot\vec{b}} \, f(\vec{q}) \, d^2q \quad , \tag{2.3}$$

where d^2q is a two-dimensional integration element in a plane perpendicular to $\vec{k}$.

The collision of a high-energy incident particle with a nucleus takes place so quickly that we can think of the nucleons as standing still during it. Let us write the nucleon positions as $\vec{r}_1 \dots \vec{r}_A$ and the projections of these vectors on the plane perpendicular to $\vec{k}$ as $\vec{s}_1 \dots \vec{s}_A$ (see Fig. 2). The profile function for the nucleus as a whole will clearly depend on the coordinates $\vec{s}_1 \dots \vec{s}_A$ as well as on $\vec{b}$. If we assume that the energy transferred to the nucleus by the incident particle is negligibly small at all stages of the collision process then it is not difficult to generalize Eq. (2.2) to deal with inelastic nuclear collisions. The diffraction amplitude for a scattering process in which the nucleus goes from an initial state $|i\rangle$ to a final state $|f\rangle$ while the incident particle undergoes a momentum transfer $\hbar \vec{q}$ is easily shown[1] to be

$$F_{fi}(\vec{q}) = \frac{ik}{2\pi} \int e^{i\vec{q}\cdot\vec{b}} \langle f | \Gamma(\vec{b}, \vec{s}_1, \dots \vec{s}_A) | i \rangle \, d^2b \quad . \tag{2.4}$$

To make use of this expression we must of course have some way of evaluating the profile function for the nucleus. The simplest way to evaluate the function Γ is to think in terms of an optical analogy; we think of the nucleus as a collection of refracting (and absorbing) objects whose refractive indices are not very different from unity. Then the incident wave is not greatly distorted on passing through the nucleus and the phase shifts χ_j produced by the individual nucleons combine additively. The total nucleon phase shift function is

$$\chi(\vec{b}, \vec{s}_1, \dots \vec{s}_A) = \sum_j \chi_j(\vec{b} - \vec{s}_j) \quad ,$$

and the nuclear profile function is consequently

$$\Gamma(\vec{b}, \vec{s}_1, \ldots \vec{s}_A) = 1 - e^{i\chi(\vec{b}, \vec{s}_1, \ldots \vec{s}_A)}$$

$$= 1 - \prod_j e^{i\chi_j(\vec{b} - \vec{s}_j)} \tag{2.5}$$

$$= 1 - \prod_{j=1}^{A} \left[1 - \Gamma_j(\vec{b} - \vec{s}_j) \right]. \tag{2.6}$$

When we expand the latter product we find the combination law for the profile functions

$$\Gamma(\vec{b}, \vec{s}_1, \ldots \vec{s}_A) = \sum_j \Gamma_j(\vec{b} - \vec{s}_j) - \sum_{j<m} \Gamma_j(\vec{b} - \vec{s}_j) \Gamma_m(\vec{b} - \vec{s}_m) + \ldots .$$

$$\tag{2.7}$$

A first approximation to the nuclear profile function is just the sum of the individual nucleon profile functions, but that approximation in a sense includes too much. Because the nucleons cast shadows upon one another or, more generally, because double scattering processes must not be counted twice, we must subtract the sum of products of profile functions taken two at a time. The higher order corrections alternately add and subtract the sums of products of increasing numbers of profile functions until the sum terminates with the A-fold product.

By substituting the expressions (2.6) or (2.7) for the nuclear profile function into Eq. (2.4) we are able to construct the over-all nuclear scattering amplitudes from a knowledge of the profile functions Γ_j of the individual nucleons. Without any fundamental understanding of the strong interactions we do not, of course, have any way of predicting what the profile functions of the nucleons are. The most practical course is therefore to use the inverse Fourier transform Eq. (2.3) to express the profile functions in terms of the scattering amplitudes $f_j(\vec{q})$, many of the properties of which are known directly through measurement. When the profile functions Γ_j are written in this way we see that the combination of Eqs. (2.4) and (2.7) leads to an expansion of the nuclear scattering amplitude in ascending powers of the amplitudes f_j; it is a multiple scattering expansion which terminates with the A-th term.

The explicit expressions for the inelastic nuclear scattering amplitude in terms of the individual nucleon scattering amplitudes are a bit lengthy. Since they have already been written out and reviewed on several occasions,[2,3] we shall forego repeating them here. Their most important property is that they are completely explicit. The nuclear scattering amplitude is given either by a single multi-dimensional integral or, somewhat more conveniently as a rule, by a sum of multiple integrals of lower dimensionality, all of which are of a form that can be computed fairly readily. Of course in evaluating the integrals we need to use some knowledge of the nuclear wave functions, but in practice there has been very little we have really needed to know about them. Both the summed scattering, which is evaluated by closure, and the elastic scattering depend only on the ground state wave function of the nucleus. It has been sufficient for most purposes to date to use the most primitive imaginable picture of the nuclear ground state, to treat the nucleus as a perfect gas of particles all having the same density function, one which has the same shape and more or less the same dimensions as the known nuclear charge density.

We have of course simplified the theory a bit in presenting its elements here. The additivity of the phase shifts which we have assumed, for example, is obviously no longer a correct statement if we are dealing with particles that have internal degrees of freedom. If we include spin or isospin for example in the description of the particles, the profile functions Γ_j become operators which don't in general commute[4-6] and the combination rule (2.6) for the profile functions has to be written in terms of an ordered product. The ordering which must be used is in fact obvious. It is just the order in which the incident particle suffers its successive encounters with the nucleons; i.e. the function Γ for the nucleus depends upon the ordering of the longitudinal coordinates of the nucleons, as well as on the transverse components $\vec{s}_1 \ldots \vec{s}_A$.

While the optical analogy we have used is by far the simplest approach to the theory, other approaches are possible. One of the earliest ones[7-9] to be used in fact is a good deal more systematic than the optical analogy, an advantage gained at the cost of some complication, however. In it one begins by writing out in formal terms the most general expression for the multiple scattering amplitude in a many-particle system. That expression can be simplified considerably because of our assumption that the nucleons in the nucleus effectively stand still during the collision process. The scattering matrix for the nucleus can thereby be expressed just in terms of the propagation function for the incident particle and the scattering matrices for the individual nucleons.

We know of course that the time-independent propagator for an incident particle of total energy E is the familiar function

$$G(\vec{r}, \vec{r}') = - \frac{E}{2\pi\ \hbar^2 c^2}\ \frac{e^{ik|\vec{r} - \vec{r}'|}}{|\vec{r} - \vec{r}'|} \quad . \tag{2.8}$$

On the other hand, if we use this form in the expression for the multiple scattering amplitude we find that the integrals required cannot in general be carried out explicitly. It is a good idea instead to simplify the propagator by approximating it. The way we do that is to return to its Fourier integral representation

$$G(\vec{r},\ \vec{r}') = \frac{-2E}{(2\pi)^3 \hbar^2 c^2} \int \frac{e^{i\vec{p}'\cdot(\vec{r} - \vec{r}')}}{p^2 - k^2 - i\epsilon}\ d\vec{p} \tag{2.9}$$

and to say that because nearly all scattering processes lead to small momentum transfer, the momentum of the incident particle in its passage through the nucleus is never likely to be very different from its incident momentum $\hbar \vec{k}$.

If we write the variable of integration in Eq. (2.9) as $\vec{p} = \vec{k} + \vec{\eta}$ and neglect η^2 in the denominator of the integrand then we may approximate the propagator by the integral

$$G(\vec{r},\vec{r}') \approx - \frac{2E}{(2\pi)^3 \hbar^2 c^2}\ e^{i\vec{k}\cdot(\vec{r}-\vec{r}')} \int \frac{e^{i\vec{\eta}\cdot(\vec{r}-\vec{r}')}}{2\vec{k}\cdot\vec{\eta} - i\epsilon}\ d\vec{\eta} \quad .$$

$$\tag{2.10}$$

If we let z and z' be the components of $\vec{r}$ and $\vec{r}'$ parallel to $\vec{k}$ this integral, which is easily evaluated, may be written as

$$G(\vec{r},\vec{r}') \approx - \frac{i\ E}{\hbar^2 c^2 k}\ e^{i\vec{k}\cdot(\vec{r}-\vec{r}')}\ \delta^{(2)}(\vec{s}-\vec{s}')\ \theta(z-z') \tag{2.11}$$

where $\delta^{(2)}$ is a two-dimensional delta-function and θ is the step-function

$$\theta(z - z') = \begin{cases} 1 & , & z > z' \\ 0 & , & z < z' \end{cases} . \qquad (2.12)$$

The effective propagation function (2.11) of course looks very little like the exact propagator (2.8) unless $\vec{r} - \vec{r}'$ is parallel to $\vec{k}$. In fact it only allows a particle entering the nucleus to propagate in the forward direction. It is a strange looking function but a compact and useful statement of the asymptotic approximation we are making.

Part of the interest in discussing the effective propagation function lies in the value it has recently been shown to have in treating field-theoretical scattering problems. At least half a dozen papers completed in the last several weeks[10-15] have observed that by making use of the four-dimensional form of the effective propagator (2.11) it becomes possible, in the high energy limit, to sum in closed form the series of Feynman amplitudes corresponding to a broad class of ladder diagrams, including ones with intricately tangled "rungs".

The use of the effective propagation function (2.11) leads to vast simplification of the integrals required in the multiple scattering expansion. Not only do the integrals become tractable but a considerable amount of cancellation takes place among the expressions which result. In particular, if the scattering interactions are potentials of any sort, the multiple scattering sum for an A-particle nucleus can be reduced to a form which terminates with the A-th term. Furthermore all of the effects of off-energy-shell propagation of the incident particle in the intermediate states of the collision process cancel out of the expression, leaving only the on-shell scattering amplitudes of the nucleons. It is this elaborate cancellation which explains the simplicity and correctness of the results derived from the optical analogy. Its mechanism has recently been discussed in some detail by Harrington.[16]

Approaching the nuclear collision problem by starting out from the most general form of multiple collision theory may have the advantage of furnishing corrections to the more direct approach via the optical analogy. The elaborate cancellations noted earlier, however, introduce subtle questions into the evaluation of such corrections. Several authors who have examined the double collision terms of the general theory[17,19] have recently suggested that the diffraction theoretical amplitude for double scattering should be supplemented by the off-energy-shell propagation effects which appear to be absent from it. The value of such a correction is

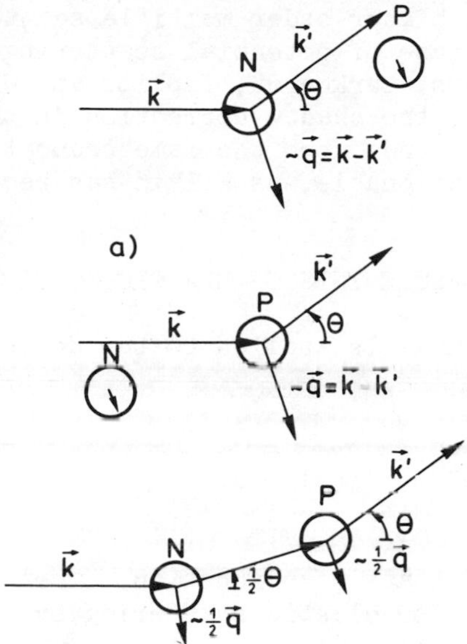

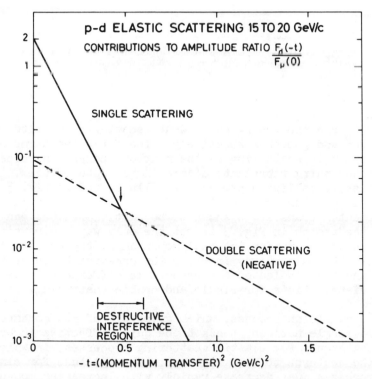

Fig. 3. Three predominant varieties of collisions in which a given momentum $\hbar \vec{q} = \hbar(\vec{k} - \vec{k}')$ is transferred to the deuteron.

Fig. 4. Contributions of single and double scattering to the amplitude for elastic scattering of protons by deuterons.

questionable when the higher order multiple scattering terms which cancel it out in the case of potential scattering remain unexamined. The suggestion by Abers, Burkhardt, Teplitz and Wilkin[19] that, in the high-energy limit, the shadow correction in the deuteron cross section must vanish depends upon the same truncation effect, and is therefore equally questionable, as Wilkin has recently acknowledged.[20]

III. MEASUREMENTS OF ELASTIC SCATTERING

To see how our analysis applies to the deuteron let us define the form factor

$$S(\vec{q}) = \int e^{i\vec{q}\cdot\vec{r}} \, |\varphi(\vec{r})|^2 \, d\vec{r} \quad , \qquad (3.1)$$

where φ is the ground state wave function. Then the diffraction theoretical amplitude for elastic scattering by the deuteron is given[1,4] by

$$F_{ii}(\vec{q}) = f_n(\vec{q}) \, S(\tfrac{1}{2}\vec{q}) + f_p(\vec{q}) \, S(\tfrac{1}{2}\vec{q})$$

$$+ \frac{i}{2\pi k} \int S(\vec{q}\,') \, f_n(\tfrac{1}{2}\vec{q} + \vec{q}\,') \, f_p(\tfrac{1}{2}\vec{q} - \vec{q}\,') \, d^2 q'$$

$$(3.2)$$

where f_n and f_p are the (spin-independent) scattering amplitudes of the neutron and proton respectively. The first two terms of F_{ii} represent single scattering by the neutron and proton respectively while the third represents scattering by both of them. The three varieties of collision process are illustrated in Fig. 3 for the same total momentum transfer $\hbar \, \vec{q} = \hbar(\vec{k} - \vec{k}\,')$. The way in which the momentum transfer is shown to be partitioned between the neutron and the proton in Fig. 3 is not the only way the partition can occur (due to the random internal momentum of the deuteron). It is however the dominant way in which the momentum transfer is partitioned and that indicates a considerable difference between the angular distributions for single and double scattering.

A single collision process, which gives all of its momentum transfer to a single nucleon, tends to be quite effective in breaking up the deuteron. For elastic scattering processes, in which we require the deuteron to hold together, the amplitude for single collision processes must decrease rapidly with increasing momentum transfer. That rapid decrease is provided by the form factors $S(\frac{1}{2}\vec{q})$ in the first two terms of Eq. (3.2).

The double scattering term, on the other hand, describes collisions in which the neutron and proton receive impulses of about the same magnitude. The deuteron has no difficulty holding together when that happens. The double scattering amplitude, then, is relatively insensitive to the structure of the deuteron and looks rather like the amplitude for two successive scatterings by free nucleons.

If we assume $f_n = f_p$ and take these amplitudes, for the moment, to be Gaussian in q and purely imaginary, then single and double scattering amplitudes take the form shown in Fig. 4. The double scattering amplitude is a good deal smaller in magnitude than the single scattering amplitude for q = 0, but it decreases much less rapidly with increasing q and becomes the dominant contribution at the larger momentum transfers shown.

It is most important to note that the double scattering amplitude is opposite in sign to the single scattering amplitude, a property which can immediately be traced back to the alternation of signs in Eq. (2.7). If the nucleon scattering amplitudes were indeed purely imaginary, single and double scattering would cancel precisely at $-t = (\hbar q)^2 \sim 0.5$ (GeV/c)^2, leaving a zero in the differential cross section. Since the neutron and proton scattering amplitudes both have small real parts the cross section should have in fact not a zero but a well defined interference minimum.

The first experiment to examine p-d scattering at angles away from the forward direction was performed by Coleman et al.[21] at 2.0 GeV. Its measurements, made primarily on large-angle collisions, were not extended to small enough angles to fully include the interference region. The measurements were shown to be in excellent agreement with the theory[22] in the angular range dominated by double collisions, but the question of the existence of the interference dip was left open.

That question was presently settled by the analysis of the 1 GeV Brookhaven p-d scattering experiment.[23] The experiment showed an angular distribution which agreed quite well with the theory in both the small and large momentum transfer ranges, but it showed a rather flat shoulder between them instead of the anticipated minimum. An experiment on π^--d scattering at 895 meV/c by the CERN-Trieste group[24] revealed a fairly similar picture, a shoulder in the angular distribution but no evident interference minimum. The comparison of the experimental results with the theory as worked out by Alberi and Bertocchi,[24,25] taking spin dependence of the scattering amplitudes fully into account, is shown in Fig. 5. If

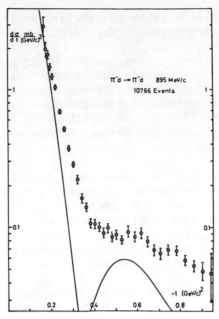

Fig. 5. Initial test of the theory (solid curve) for π^- - d elastic scattering by Alberi and Bertocchi[24,25]. Experimental points are from Bradamante et al.[24]

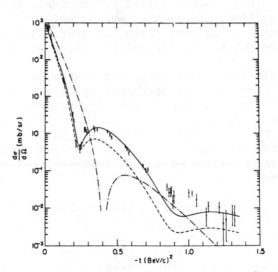

Fig. 6. Analysis by Bassel and Wilkin[29] of p - ^{4}He elastic scattering measurements at 1 GeV by Palevsky et al.[27] Dashed curve: differential cross section predicted with Gaussian form factor for ^{4}He density distribution. Dash-dot curve: improved form factor used to fit recent electron scattering measurements at larger momentum transfers. Solid curve: differential cross section predicted with improved form factor.

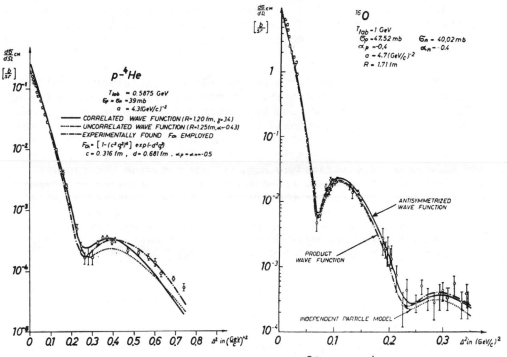

Fig. 7. Analysis by Lešniak and Woźek[30] of p -[4]He elastic
scattering measurements at 587 meV by Boschitz et al.[31]

Fig. 8. Analysis by Lešniak and Woźek[30] of elastic proton
scattering measurements on [16]O at 1 GeV by Palevsky et al.[27]

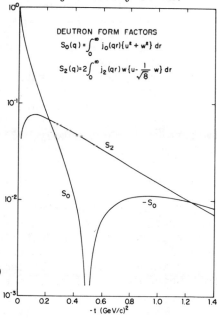

Fig. 9. The spherical form factor
of the deuteron, S_0, and the
quadrupole form factor, S_2, as
functions of the momentum transfer.[40]

any further evidence of disagreement was needed it was provided by the p-d measurements of the Virginia group[26] at 582 meV. These agreed with the previous ones in showing no indication at all of an interference minimum.

Now let us digress for a moment and look at elastic scattering measurements made on some heavier elements. Elastic scattering of 1 GeV protons by ^{4}He had also been measured in the Brookhaven experiment,[27] and it was quickly observed by Czyż and Leśniak[28] and by Bassel and Wilkin[29] that the data were rather well explained by the simple multiple scattering theory. The calculations of Bassel and Wilkin are illustrated in Fig. 6. By choosing the density function for ^{4}He so as to fit the most recent Stanford measurements of the charge form factor they found that the predicted elastic differential cross section is brought into close correspondence with the measured values. Essentially the same conclusion is reached by Leśniak and Wołek[30] (Fig. 7) in treating the p - ^{4}He measurements made at 587 meV by the Virginia group.[31]

The Brookhaven measurements of proton scattering by ^{16}O have had several independent analyses[30,32-34] based on the multiple scattering theory. The calculated cross sections were in all cases found to be in excellent agreement with the measured data. The predicted cross section is found to be rather insensitive to details of the nuclear ground state such as its antisymmetry properties. The calculations of Lesniak and Wołek[30] are shown together with the Brookhaven data in Fig. 8.

While the theory worked well for ^{4}He and ^{16}O, it was clearly missing something for the case of deuterium. There was no lack of suggestions concerning what might be missing. They ranged from guesses about the unknown momentum-transfer dependence of the phase of the nucleon scattering amplitudes to corrections which were rather strong indictments of the basic approximation method. Several of these ideas may turn out to have some physical importance. It is worth keeping in mind, for example, the suggestion that three-body forces might contribute noticeably,[35] or that intermediate states containing inelastically-produced particles may be important.[2,36,37] Fortunately, all of these suggestions had in common the property that invoking them to improve the agreement for deuterium would destroy the agreement already established for ^{4}He and ^{16}O. They could thus be set aside and the search continued.

To simplify the problem a bit let us recall that the interference dip is absent from the deuterium cross section both for incident protons and incident pions. Hence the unknown dip-filling mechanism is not likely to involve the detailed properties of the scattering amplitudes. It looks more like a nuclear effect of some sort than one dependent on the incident particle. Since the theory works quite well, furthermore, for the spin-zero nuclei ^{4}He and ^{16}O, it is reasonable to associate the trouble with the unit spin of the deuteron.

The unique property of a spin-one nucleus, of course, is the fact that it may possess a quadrupole moment. The ground state of the deuteron indeed possesses a small d-state admixture, and its consequences had been largely overlooked in the scattering analysis on the assumption that they would not lead to dramatic changes in the differential cross section. The first observations that the assumption is incorrect came from Coleman and Rhoades[38] and from Harrington[39], whose voices were presently joined by a small chorus of others.[40-42] A considerable number of detailed calculations of deuteron cross sections have been carried out in the last few months as if in expiation for the oversight.

The reason the effect of the d-state on the scattering is not very small is that its primary contribution to the scattering matrix element comes (like the quadrupole moment) from interference of the s-state and d-state wave functions. While the amount of d-state present, i.e. the integral of its squared amplitude, amounts to only 6 or 7 %, the effect of the d-state on the differential cross section is proportional to its amplitude, which is perhaps 25 % as large as the s-state amplitude. We should not be too surprised, therefore, by the possibility that when the s-state contributions form an interference minimum through cancellation, the d-state contributions remain prominently visible.

That is in fact precisely what happens. To understand why it happens let us note that because of the ellipsoidal shape of the deuteron its form factor must depend upon the direction in which its nuclear spin vector J is pointing. This polarization-dependence of the form factor can be represented most conveniently by writing the form factor as the tensor operator[40]

$$\mathcal{S}(\vec{q}) = S_0(q) - \frac{1}{\sqrt{2}}\, S_2(q)\, \left\{ 3(\vec{J}\cdot\hat{q})^2 - 2 \right\}$$

$$(3.3)$$

in which $\hat{q}$ is a unit vector in the direction of the momentum transfer $\hbar\,\vec{q}$. The functions S_0 and S_2 are a pair of scalar form factors given by

$$S_0(q) = \int_0^\infty j_0(qr)(u^2 + w^2)\, dr \qquad (3.4)$$

$$S_2(q) = 2 \int_0^\infty j_2(qr) \, w \left\{ u - \frac{1}{\sqrt{8}} \, w \right\} dr$$

$$(3.5)$$

in which j_0 and j_2 are spherical Bessel functions and u and w are the radial wave functions of the s- and d- states respectively. The behavior of these form factors, whose usefulness has also been noted in connection with electron-deuteron scattering,[43] is shown in Fig. 9 for a model of the deuteron[44] in which the d- state admixture is 6.7 %. The form factor S_0 for the spherically symmetric component of the deuteron density is of course the dominant one in the forward direction where the quadrupole form factor S_2 vanishes. Away from the forward direction, however, S_2 rises rapidly while S_0 falls. The quadrupole form factor is in fact larger in magnitude than the spherical form factor S_0 throughout the range $0.2 \leq -t \leq 0.5 \, (GeV/c)^2$, which contains the interference region of single and double scattering.

Let us assume that the deuteron spins are fully polarized in an arbitrary direction $\hat{p}$ and neglect for the moment any spin dependence of the basic nucleon scattering amplitudes. For each polarization direction $\hat{p}$ we must distinguish between collisions in which the deuteron spin changes direction and those in which it does not; both types are equally elastic. For the case of p-d scattering at 1 GeV, for example, and polarization $\hat{p}$ in the direction of the incident momentum, the angular distribution of non-spin-flip scattering[40] is shown in the "dash-dot" curve of Fig. 10. It has an interference minimum due to the cancellation of single and double scattering which has very much the form we discussed earlier. The contribution of the spin-flip transitions is shown in the dashed curve. The sum of the two angular distributions, which is the differential cross section one would observe, is given by the solid curve. We see that in it the interference minimum furnished by the non-spin-flip processes has been almost entirely filled by the incoherent addition of the spin-flip transitions.

Because the quadrupole form factor S_2 is not small the angular distributions for both spin-flip and non-spin-flip collisions depend rather sensitively on the polarization direction $\hat{p}$. If $\hat{p}$ is chosen to lie in the direction $\hat{q}$ of the momentum transfer, for example, it is easy to show that the spin-flip transitions are much weaker in intensity.[40] A selection rule which is characteristic of the impulse approximation implies that the spin-flip transition cannot then occur in single scattering processes. The angular distribution of spin-flip scattering shown in the dashed

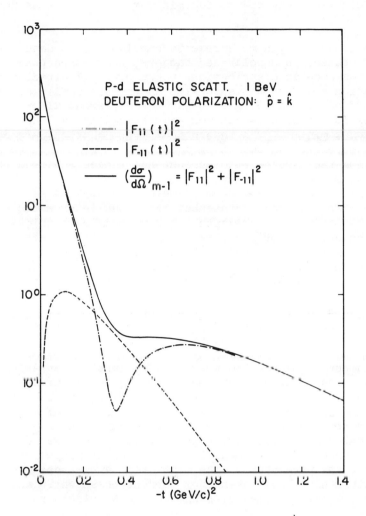

Fig. 10. Contributions to elastic p-d scattering[40] at 1 GeV for
deuteron spins fully polarized in the direction of the incident
proton momentum. Dashed curve: contribution of deuteron spin-
flip transitions. Dash-dot curve: contribution of non-spin-
flip transitions. Solid curve: elastic differential cross section.

curve of Fig. 11 is due to double scattering alone. It is not sufficiently intense to fill the minimum shown in the non-spin-flip
angular distribution, and so the minimum is by no means unobservable.

The cross section measurements carried out to date have been
made, of course, on unpolarized targets; they correspond to averages of the angular distributions taken over all directions of the
polarization $\hat{p}$. When the spin-flip and non-spin-flip distributions
are summed and averaged in this way the differential cross sections
which result for proton-deuteron scattering lose all trace of their
interference minima. The theoretical cross sections are compared
with the experimental data at 1 and 2 GeV in Fig. 12. We see there
that taking the d-state into account has indeed restored the agreement of theory and experiment.

The effect of the d-state has been equally dramatic in dealing
with pion-deuteron measurements. Alberi and Bertocchi,[42] on returning to the analysis of the CERN-Trieste π^--d cross section at 895
meV/c and expanding the calculation to include quadrupole effects,
found the remarkable agreement shown in Fig. 13. The d-state admixture leads to a distribution quite different from that due to
the s-state alone. The latter, of course, is essentially the curve
that was shown in Fig. 5. Since the spin-dependence of the pion
scattering amplitudes is fairly well known it is of course taken
into account in the calculations.

A number of measurements of π^- - d elastic scattering differential cross sections have been carried out at higher incident
momenta by the Minnesota-Iowa group at the Argonne Laboratory[45] and
by the CERN-Trieste group.[46]   The calculations of Michael and Wilkin[41] are compared with five of these measurements ranging in incident momentum up to 15.2 GeV/c in Fig. 14. It is worth emphasizing
that no attempt has been made to fit the experimental cross sections
by varying any parameters. The scattering amplitudes used in the
calculations have been taken from current phase shift and dispersion theoretical analyses of π-nucleon scattering data.

The CERN-Trieste π^--d measurements at 9.0 GeV/c have recently
been extended to considerably larger momentum transfers than are
shown in Fig. 14. The new data[47] are indicated together with those
of two earlier experiments in Fig. 15. The solid curve which passes through many of the 9.0 GeV/c points represents a simplified
form of the theoretical cross section which takes the d-state into
account but neglects the spin dependence of the scattering amplitudes.
It is only intended as a preliminary comparison with the new data.

The CERN-Trieste group has also measured the elastic p-d differential cross section[46,47] at 12.8 GeV/c. The angular distribution

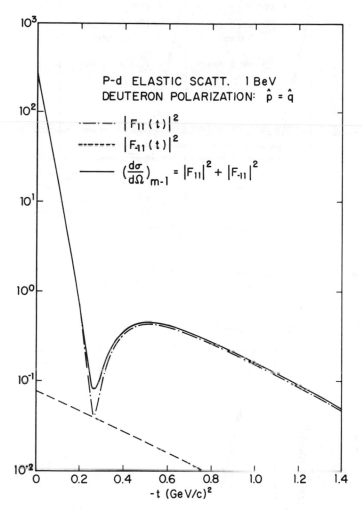

Fig. 11. Contributions to elastic p-d scattering[40] at 1 GeV for
deuteron spins fully polarized in the direction of the momentum
transfer $\vec{q}$. Dashed curve: contribution of deuteron spin-flip
transitions (weak since only double collisions lead to spin-flip).
Dash-dot curve: contribution of non-spin-flip transitions. Solid
curve: elastic differential cross section.

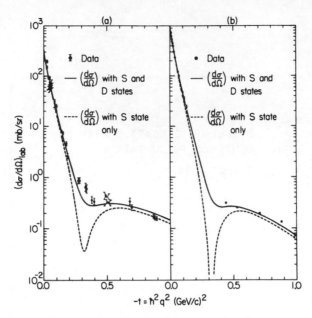

Fig. 12. Comparison of p-d elastic scattering measurements
a) at 1 GeV by Bennett et al.[23] b) at 2 GeV by Coleman et al[21]
with theory[40] including d-state admixture.

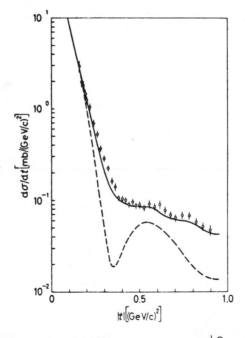

Fig. 13. Analysis of Alberi and Bertocchi[42] of π^--d elastic
scattering measurements at 895 meV/c by Bradamante et al.[24]
The dashed curve takes only the s-state of the deuteron into
account. The solid curve accounts for the d-state admixture as
well as the s-state.

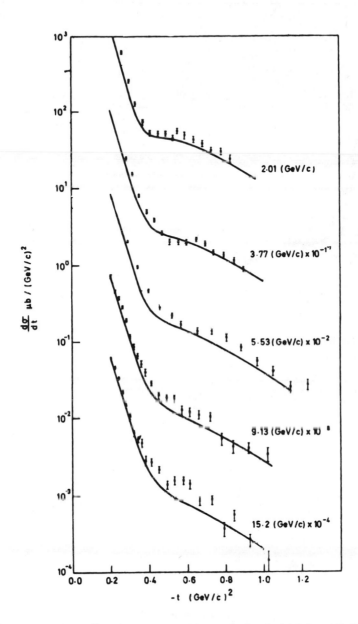

Fig. 14. Elastic π^--d cross sections calculated by Michael and Wilkin[41] (solid curves) compared with experimental data of the Minnesota-Iowa Group[45] (upper three curves) and the CERN-Trieste Group[46] (lower two curves).

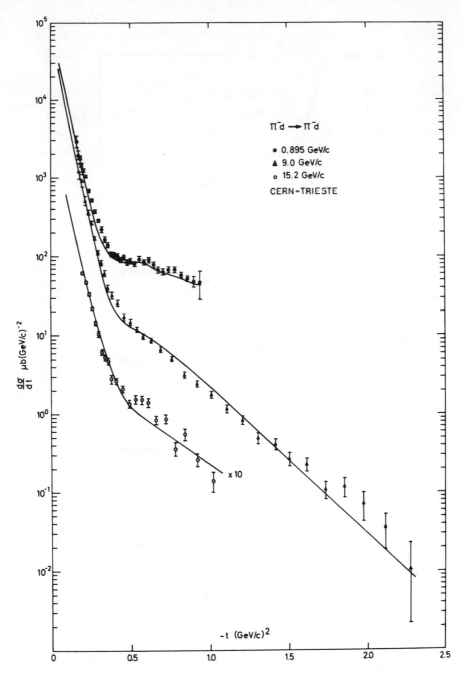

Fig. 15.   Three CERN-Trieste π^--d elastic differential cross sections[24,46,47] and theoretical predictions (solid curve). The 9.0 GeV/c measurements have been carried out over an extended range of momentum transfers.

found is compared with the simplified, spin-independent form of the theory in Fig. 16, and the two are seen to agree reasonably well. The nucleon scattering amplitudes in the calculation have been represented rather crudely as simple exponential functions of t. The slight divergence of theory and experiment seen for large momentum transfers, may be due to the natural breakdown of that representation.

Broadly speaking the distribution of elastic scattering by deuterons can be divided into three angular regions, the small-angle or single-scattering region, the interference region and the large-angle or double-scattering region. From measurements in the small-angle region we learn mainly about the form factor of the deuteron. Since the form factor is already fairly well known in that region, measurements made there serve mainly the practical purpose of determining or checking experimental normalizations. Measurements made in the double scattering region, on the other hand, do tell us something new about the deuteron. Analyzed carefully, they tell us the value of a parameter which is closely related to the mean inverse squared neutron-proton separation in the deuteron. The values of that parameter furnished by rather crude analysis of the CERN-Trieste data seem not to disagree, at least in magnitude, with the values determined by analyzing the shadow corrections in deuteron total cross section measurements.

It is in the interference region that the angular distribution may be expected to furnish the most information. The behavior of the distribution there is obviously sensitive to the phase of the nucleon scattering amplitudes. It is also quite sensitive to the amount of d-state admixture present in the deuteron. While there is no way of separating these dependences in analyzing the experiments which have been done to date, that should not be too difficult to do as soon as the scattering experiments can be carried out on polarized deuteron targets. With a polarized target, as we have seen, the interference dip can be made to appear or disappear by rotating the polarization axis.

Essentially the same information may be gleaned from an alternative experiment proposed by Harrington[48] and Bazin,[49] one which does not rely on a polarized target. The same calculations which explain the angular distribution of scattering by polarized deuterons also show that when the target deuterons are initially unpolarized, the recoil deuterons will be strongly aligned in spin. Their recoil energy is too small for them to be of much use, however. If, on the other hand, we let a high energy deuteron beam fall on an ordinary (unpolarized) hydrogen target, i.e. we just perform a Galilean transformation, the scattered deuterons will have high energy and will still be strongly aligned in spin. The alignment can then be analyzed by means of a second scattering process.

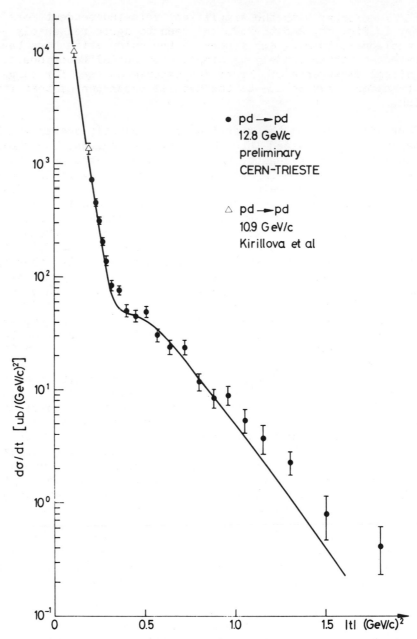

Fig. 16.   CERN-Trieste measurement[46,47] of elastic p-d scattering at 12.8 GeV/c. The theoretical prediction (solid curve) is based on the assumption that the p-p scattering amplitude is Gaussian in shape.

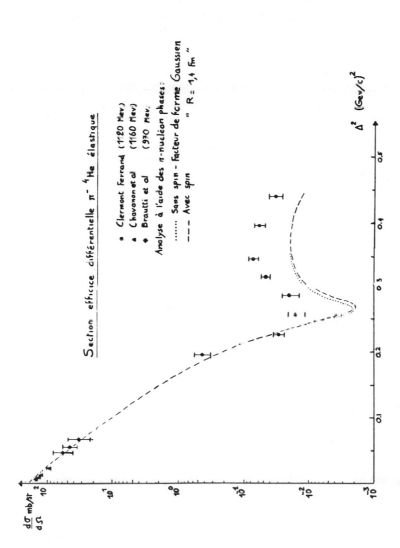

Fig. 17. Preliminary results of measurements the elastic differential cross section for π^- - ^{4}He scattering at 1.25 GeV/c by the CERN-Clermont-Ferrard Group[50]. The theoretical curves assume the density form factor for ^{4}He to be Gaussian and thus omit corrections recently found necessary for large momentum transfers (see Fig. 6).

Still another elastic scattering measurement is being completed presently. It is the first π-- ^{4}He measurement in the GeV region, at 1.25 GeV/c to be precise, carried out by a group at Clermont-Ferrand and CERN. The preliminary results of the measurement,[50] shown in Fig. 17, exhibit the characteristic interference minimum quite clearly. The theoretical curve shown in Fig. 17 is even somewhat more preliminary than the experimental points, however; it is based on the Gaussian expression for the ^{4}He form factor which represents the Stanford measurements only at small momentum transfers. Inclusion of the more recent Stanford form factor measurements should raise the theoretical cross section in the double scattering region, just as it did for p - ^{4}He scattering (Fig. 6). The agreement should then be a good deal closer.

IV. ELASTIC SCATTERING BY THE HEAVIER NUCLEI

Elastic scattering, as we have noted earlier, accounts for almost all of the intense diffraction pattern observed at small angles in the summed scattering cross section. Thus, even where no experimental effort is made to separate the elastic from the inelastic scattering, we do indeed observe it fairly clearly. It is important, therefore, to be able to calculate the elastic scattering accurately.

The most straightforward way of calculating the elastic scattering is by explicitly summing the multiple scattering amplitudes. Indeed, if we were to use an accurate many-particle wave function for the nuclear ground state there would probably be no other way available. The sum terminates with its A-th term so there is no convergence problem in evaluating it, but there is often the practical problem that a good many of those A terms must be taken into account explicitly to retain numerical accuracy. An alternative procedure is to use a simplified representation of the nuclear ground state which permits summing the multiple scattering expansion.

If we neglect position correlations between nucleons and write the many-particle density function for the nuclear ground state in the factorized form $\Pi\,\rho(\vec{r}_j)$, then it is not difficult to show that the elastic scattering amplitude F_{ii} of the nucleus is given by Eq. (2.2) with

$$\vec{\Gamma(b)} \equiv 1 - e^{i\chi_N(\vec{b})} \tag{4.1}$$

$$= 1 - \left\{1 - \frac{1}{2\pi i k}\int e^{-i\vec{q}'\cdot\vec{b}}\,f(\vec{q}')\,S(\vec{q}')\,d^2q'\right\}^A \tag{4.2}$$

where $\chi_N(\vec{b})$ is the nuclear phase shift function and $S(\vec{q})$ is the form factor of the single-particle density ρ. For $A \gg 1$ the phase shift function is closely approximated by

$$\chi_N(\vec{b}) = \frac{A}{2\pi k} \int e^{-i\vec{q}\cdot\vec{b}} f(\vec{q})\, S(\vec{q})\, d^2q \qquad (4.3)$$

and if the range of the high energy interaction is much smaller than the nuclear radius, the latter expression may in turn be approximated as

$$\chi_N(\vec{b}) \approx \frac{1}{2\pi k}\, f(0)\, T(\vec{b}) \qquad (4.4)$$

where $T(\vec{b})$ is the thickness function

$$T(\vec{b}) = A \int_{-\infty}^{\infty} \rho(\vec{b} + \hat{k}z)\, dz \quad , \qquad (4.5)$$

i.e. the integral of the density along a straight-line path at impact vector $\vec{b}$.

The elastic scattering amplitude which results from using the approximation (4.4) in connection with Eqs. (4.1) and (2.2) is the one usually identified with the "optical model". Since the high energy interaction range is not really a great deal smaller than the nuclear radius, it is noticeably more accurate in practice to use the expression (4.3) for the nuclear phase shift. The correction, which amounts to only a few percent for the heavy nuclei, has been shown by Matthiae[51] and Kofoed-Hansen[52] to be fairly important for the lighter elements. Of course, for the lightest elements the "optical model" is not an adequate approximation at all, and the multiple scattering approach, on the other hand, becomes fairly easy to use. There is unfortunately no simple way of saying which value of A separates the small values from the large ones. The multiple scattering analysis is clearly necessary[53] for a nucleus as light as ^{4}He, but a number of careful calculations[30,33,54] carried out for scattering of 1 GeV protons by ^{16}O show agreement of the "optical" and multiple-scattering approaches to within about 10 %.

If we want to take nucleon position correlations into account, there are two sorts of correlations which must be considered. The first is an effect which must be present in any nuclear model; it arises simply from the fact that in the rest frame of the nucleus, the nuclear center of mass remains fixed and so the nucleons cannot move about arbitrarily. That correlation is easily taken into account in the multiple scattering formulas.[2] It is only numerically important, however, for the lightest nuclei. The other sort of correlation is the kind which is specified by the ground state wave function itself. It contains the effect of the exclusion principle and the dynamical effects of nuclear forces. When such correlations are taken into account the expressions (4.3) and (4.4) for the nuclear phase shift function are found to be only the first terms of a series expansion[1] the further terms of which contain higher powers of the nucleon scattering amplitude and integrals of the nuclear correlation functions.

For extremely small momentum transfers the elastic scattering amplitude has the singularity characteristic of Coulomb scattering. That effect is due to small deflections of the incident particle which take place at a considerable distance from the nucleus and so its calculation is not, strictly speaking, a nuclear problem. The amplitudes for Coulomb and nuclear scattering do interfere appreciably, however, and, for the heavier nuclei the two effects become inextricably mixed in the angular distributions observed for charged particle scattering. It is worth pausing for a moment, therefore, to discuss Coulomb scattering, and to show that its effects are easily incorporated into the approximations we have been using.

The simplest way to take Coulomb scattering into account is to think of χ_N as the phase shift produced by the strong interactions alone and to add to it a Coulomb phase shift χ_C, so that the total phase shift becomes

$$\chi(\vec{b}) = \chi_N(\vec{b}) + \chi_C(\vec{b}) \quad . \tag{4.7}$$

It is not too obvious, however, how one should go about calculating χ_C. If the Coulomb potential is $V(r)$ and the incident particle moves with velocity v, then the Coulomb phase shift should be given by the path integral[1]

$$\chi_C(\vec{b}) = -\frac{1}{\hbar v} \int_{-\infty}^{\infty} V(\vec{b} + \hat{k}z) \, dz \, , \tag{4.8}$$

but for a Coulomb potential this integral diverges at both its

limits. A corresponding problem must arise in any treatment of
Coulomb scattering based on the assumption that the wave function
is asymptotically the sum of a plane wave and an outgoing spherical
wave. The long-range character of the Coulomb field means that,
strictly speaking, the asymptotic wave function must take a somewhat different form. To avoid the need for reformulating the
theory, however, we can use the simple and physically meaningful
device of screening the Coulomb field. The nuclei in actual targets
are surrounded, after all, by electron shells which neutralize their
charge completely, so there is nothing unrealistic about the screened
fields. The only subtlety of the argument is the demonstration that
the nuclear scattering measurements in fact tell us nothing about
the screening.

Let us write the screened Coulomb potential as

$$V(\vec{r}) = Ze^2 \int \frac{\rho_C(\vec{r'})}{|\vec{r} - \vec{r'}|} \, dr' \, C(r) \qquad (4.9)$$

where $\rho_C(r)$ is the nuclear charge density normalized to unity,
and $C(r)$ is a function which describes the screening. We assume
that $C(0) = 1$, that the function remains of order unity for radii
smaller than some screening radius a and that for $r > a$ it goes
fairly rapidly to zero. A simple choice, which is adequate as an
illustration, is

$$C(r) = \begin{cases} 1, & r < a \\ 0, & r > a \end{cases}. \qquad (4.10)$$

The form of the screening function $C(r)$ does have a certain
influence on the scattered intensity, but only at scattering angles
of order $\theta_c = (k\,a)^{-1}$ and smaller. Since the screening radius a
is of atomic dimensions the angle θ_c for multi-GeV incident particles
is $\sim 10^{-6}$ to 10^{-7} radians, an angle much smaller than any at which
measurements are made. By measuring the scattering only at angles
$\theta >> \theta_c$ we are in effect only observing particles whose impact
parameters are very much smaller than the screening radius a. To
predict the scattering at such angles it is therefore sufficient,
when we calculate the phase shift by means of Eqs. (4.8) and (4.9),
to expand $\chi_c(b)$ in powers of b/a and to retain only the lowest order
terms. If we do that for example by using the screening function
(4.10) we find, for $a \rightarrow \infty$

$$\chi_c(\vec{b}) = - \frac{2Ze^2}{\hbar v} \int \rho_c(\vec{r}') \ \log \frac{2a}{|\vec{b} - \vec{b}'|} \ d\vec{r}' \qquad (4.11)$$

$$= - \frac{2Ze^2}{\hbar v} \ \log 2a + \frac{Ze^2}{\hbar v} \int \rho_c(\vec{r}') \ \log |\vec{b} - \vec{b}'| \ d\vec{r}' \quad .$$

$$(4.12)$$

We note that the effect of screening is only felt through a constant term

$$\chi_s = - \frac{2Ze^2}{\hbar v} \ \log 2a \qquad (4.13)$$

in the phase shift function. The value of this constant depends upon the choice of the form (4.10) for $C(r)$; different screening functions would lead to different additive constants.

If we introduce the thickness function for the nuclear charge distribution

$$T_c(\vec{b}) = A \int \rho_c(\vec{b} + \hat{k} \ z) \ dz \qquad (4.14)$$

into Eq. (4.12) and carry out the angular integration which remains we find that the Coulomb phase shift function can be written as

$$\chi_c(b) = \chi_s + \chi_\rho(b) \qquad (4.15)$$

where

$$\chi_\rho(b) = \frac{4\pi Ze^2}{\hbar v} \left\{ \log b \int_0^b T_c(b')b' \, db' \right.$$

$$\left. + \int_b^\infty T_c(b') \log b' \, b' db' \right\} \quad . \qquad (4.16)$$

For the particular case of a point charge the phase shift χ_ρ is simply

$$\chi_\rho(b) = \chi_{pt.}(b) \equiv \frac{2Ze^2}{\hbar v} \log b \; , \qquad (4.17)$$

and the scattering amplitude which results has been shown[1] to be

$$f(q) \equiv f_{pt.}(q) \exp(i \chi_s)$$

$$= -\frac{Ze^2}{\hbar v} \frac{2k}{q^2} \exp \left\{ -2i \frac{Ze^2}{\hbar v} \log \left(\frac{q}{2} \right) + 2 i \eta + i \chi_s \right\}$$

$$(4.18)$$

where

$$\eta = \arg \Gamma(1 + i \frac{Ze^2}{\hbar v}) . \qquad (4.19)$$

The amplitude which represents combined nuclear and Coulomb scattering by a finite nucleus is given by an impact parameter integral which converges quite slowly due to the large size of the screening radius. If we subtract from this integral the one which represents Coulomb scattering by a point charge, the difference of the two integrands vanishes outside the nuclear radius and so the convergence of the difference is quite rapid. It is most convenient therefore to construct the nuclear scattering amplitude by adding to this difference the point-charge scattering amplitude given by Eq. (4.18). When we do that we find that the elastic nuclear scattering amplitude may be written as

$$F_{ii}(q) = e^{i\chi_s} \Big\{ f_{pt.}(q)$$

$$-ik \int_0^\infty J_0(qb) \Big[e^{i\chi_N(b) + i\chi_\rho(b)} - e^{i\chi_{pt.}(b)} \Big] b\,db \Big\} \ .$$

$$(4.20)$$

In this expression, which holds for angles $\theta >> \theta_c$, we see that the only residual effect of screening is a constant phase factor in the overall scattering amplitude. To measure the phase factor it would be necessary to observe the interference of scattering by different nuclei, and that is scarcely possible at billion volt energies. It is only the scattered intensity which is measured in practice, and Eq. (4.20) provides a compact way of calculating it.

V. SUMMED SCATTERING

We have already discussed the elastic or coherent component of the summed scattering in some detail, and so we must now turn our attention to the inelastic or incoherent component. For momentum transfers which lie outside the nuclear diffraction cone, i.e. for $q > R^{-1}$ where R is the nuclear radius, it is not too difficult to construct a closed expression for the incoherent cross section.[55] That expression is a fairly complicated integral, but it has the rapidly convergent series expansion[1,2]

$$\sum_{f\neq i} | F_{fi}(\vec{q})|^2 = N_1 |f(\vec{q})|^2$$

$$+ N_2 \frac{1}{\sigma} \int |f(\vec{q}')|^2 |f(\vec{q} - \vec{q}')|^2 \frac{d^2 q'}{k^2} + \ldots$$

$$(5.1)$$

in which the coefficients N_j are given by

$$N_j = \frac{1}{j! \sigma} \int e^{-\sigma T(\vec{b})} \left[\sigma T(\vec{b}) \right]^j d^2 b \quad , \qquad (5.2)$$

σ is the total cross section of a nucleon and $T(\vec{b})$ is the thickness function (4.5).

The first term of the expansion (5.1) represents the effect of single quasielastic collisions; it has the same angular distribution for $q > R^{-1}$ as the single scattering approximation (1.3) since the form factor S is vanishingly small there. The coefficient N_1 however is a good deal smaller than A in general since it takes into account the attenuation of the beam as it passes through the nucleus. In practice the attenuation is so great that the integrand of N_1 is quite small except for a region of impact parameters in the immediate neighborhood of the nuclear surfaces. For multi-GeV protons incident on ^{208}Pb, for example, the number N_1 is only about 10 and all of the 10 nucleons which are effective targets for quasielastic collisions lie in a narrow belt around the limb of the nucleus. (See Fig. 2.)

The second term of the expansion (5.1) corresponds to double quasielastic collisions and the remaining terms to higher order multiple collisions. The entire series is, of course, quite different from the multiple scattering expansions we have considered earlier since each term takes attenuation effects into account. We shall refer to it, therefore, as the shadowed multiple scattering expansion.

While for small momentum transfers the successive terms of the expansion decrease fairly rapidly in magnitude, they have rather different angular dependences. If the cross section $|f(q)|^2$ depends on q as $\exp(-bq^2)$, for example, the double scattering term decreases as $\exp(-\frac{1}{2} bq^2)$, the triple scattering term as $\exp(-\frac{1}{3} bq^2)$, and so on. The higher order multiple scattering terms decrease more slowly, in other words, and there should thus be a succession of angular ranges in which the incoherent scattering is dominated by single, double, triple, and progressively higher order collision processes.[2]

Most of the data available on summed scattering to date come from the experiment of Bellettini et al.[56] which used protons of 19 and 20 GeV/c and eight different nuclear targets. The momentum transfers covered included the nuclear diffraction cone but only a fairly limited range outside it. It was a simple matter to show at the Rehovoth Meeting[2] that the intensities detected at the

larger momentum transfers in this experiment are indeed quite
well represented by the shadowed multiple scattering expansion.
The dominant contributions came from single collisions and their
agreement with experiment represented some confirmation for the
Stanford measurements of the nuclear surface thickness. In the
two years since the Rehovoth meeting a good deal of work has gone
into the theoretical evaluation of the incoherent scattering at
angles lying inside the nuclear diffraction cone, and the experimental measurement of incoherent scattering at angles large enough
for multiple scattering to become significant.

   The easiest way to extend the incoherent cross section to
small momentum transfers is to deal with the successive terms
of the shadowed multiple scattering expansion one at a time. If
we neglect position correlations of the nucleons we find that the
shadowed single scattering term, which should contribute nearly
all of the intensity at small angles is

$$
\left(\frac{d\sigma}{d\Omega}\right)_{inc} = \sum_{f \neq i} \left| F_{fi}(\vec{q}) \right|^2
$$

$$
= \left| f(\vec{q}) \right|^2 \left\{ N_1 - \frac{1}{A} \left| \int e^{i\vec{q}\cdot\vec{b} + \frac{2\pi i}{k} f(0)T(\vec{b})} T(\vec{b})d^2b \right|^2 \right\}.
$$

$$
(5.3)
$$

The first term of this cross section is simply the first term of
the expansion (5.1). The second term goes rapidly to zero outside
the nuclear diffraction cone. For weak scatterers the cross section
(5.3) reduces to the familiar form (1.3). The effects of attenuation and refraction which are accounted for in the cross section
(5.3), however, cause it to behave somewhat differently from the
simple single scattering approximation. The incoherent cross
section (5.3), for example, does not in general vanish for forward
scattering as the cross section (1.3) does. The cross section (1.3)
must vanish for q = 0, as we have noted, since in a pure single
collision process there is no way of transferring the momentum
necessary to raise a nucleon to an orthogonal state. In the "shadowed" collision processes on the other hand both attenuation and
refraction are capable of transferring momentum to the incident
particle wave. Either of these processes may easily combine with
a single quasielastic collision so that the compound process adds
up to incoherent forward scattering.

To take the effect of the exclusion principle into account in evaluating the incoherent scattering, let us write the nuclear ground state wave function as an antisymmetrical combination of the single particle states $\varphi_p(\vec{r})$. Then the appropriate form of the shadowed single scattering approximation is

$$
\left(\frac{d\sigma}{d\Omega}\right)_{inc} = |f(\vec{q})|^2 \Big\{ \int e^{-\sigma T(\vec{b})} \sum_{occ.p} |\varphi_p(\vec{r})|^2 \, d\vec{r}
$$

$$
- \sum_{occ.p,p'} \Big| \int \varphi_{p'}^*(\vec{r}) \, e^{\, i\vec{q}\cdot\vec{b} + \frac{2\pi i}{k} f(o)T(\vec{b})} \varphi_p(\vec{r}) d\vec{r} \Big|^2 \Big\},
$$

(5.4)

where the sums are carried out only over occupied states. It is interesting to note that this cross section still takes on values different from zero in the forward direction. The effects of attenuation and refraction partially defeat the tendency of the exclusion principle to reduce the forward cross section much as they do the effect of orthogonality noted earlier.

Because the proton scattering experiments of Bellettini et al. were so limited in angular range the summed scattering intensities they observed for the heavy nuclei Cu, Pb and U consisted largely of elastic scattering. Goldhaber and Joachain[57] have calculated the elastic scattering for those elements using the optical model. By adding to the elastic scattering rough estimates of the inelastic background they showed that the theoretical distributions of summed scattering for the three heavy elements are at least in fair agreement with the measurements.

Further analyses of the experimental data for 20 GeV/c protons based on detailed calculations of the inelastic as well as the elastic scattering have been carried out by Kofoed-Hansen[52] and by Matthiae.[51] Both Kofoed-Hansen and Matthiae have used the expression (4.3) for the nuclear phase shift rather than the less accurate expression (4.4) in calculating the elastic scattering. They have used Eq. (5.3) to evaluate the shadowed single collision contribution to the inelastic scattering and the higher order terms of Eq. (5.1) to evaluate the contribution of multiple quasielastic collisions. They did use somewhat different nuclear models in their calculations, however.

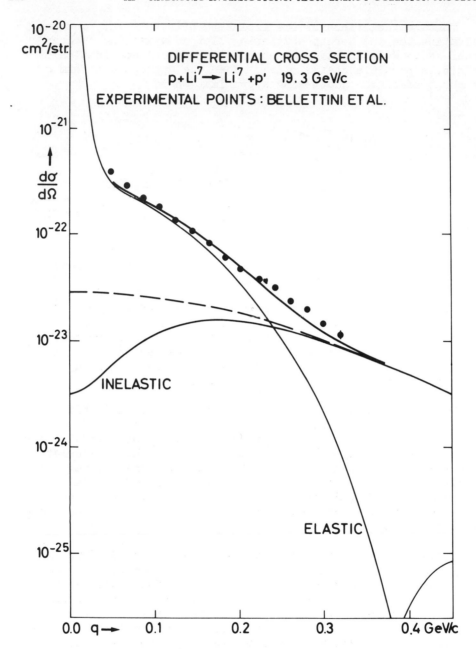

Fig. 18. Elastic, inelastic and summed scattering calculated by Kofoed-Hansen[52] for 19.3 GeV/c protons incident on ^{7}Li. The dashed curve shows the smooth extrapolation toward the forward direction of the large-angle inelastic cross section.

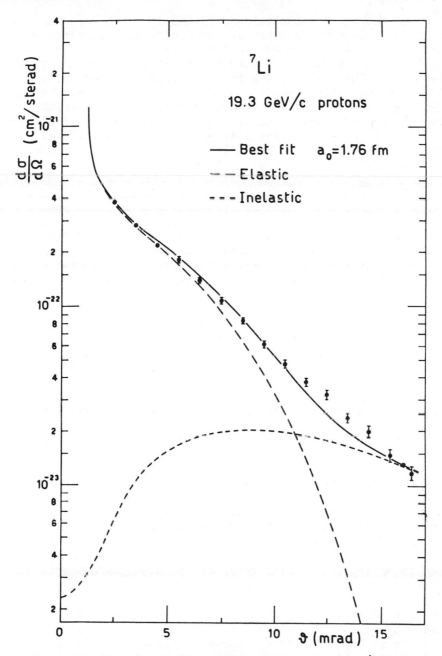

Fig. 19.   Elastic, inelastic and summed scattering (solid curve) calculated by Matthiae for 19.3 GeV/c protons incident on ^{7}Li. Experimental points from Bellettini et al.[36]

Fig. 18 shows the calculations of Kofoed-Hansen for ^{7}Li. They were based on a density function of the Woods-Saxon form chosen to provide the best fit to the Stanford electron scattering experiments. All of the parameters entering the calculation were thus predetermined. The dashed curve in Fig. 18 is the extrapolation toward the forward direction of the incoherent scattering observed at large angles, based on the assumption that the cross section is exponential in t or Gaussian in q. We see that the actual incoherent cross section is some ten times smaller than its extrapolated value near the forward direction. That dip has a perceptible influence on the shape of the summed cross section.

Matthiae's calculation for ^{7}Li is shown in Fig. 19. It is based on a shell-model density distribution constructed from harmonic oscillator wave functions. The radius of the distribution has been used as a variable parameter by Matthiae in seeking the best "least squares" fit to the data. The best fit is seen in Fig. 19 not to differ greatly from Kofoed-Hansen's calculation; the radius it leads to does not differ significantly from the electron-scattering radius.

The other calculations of Matthiae and Kofoed-Hansen also exhibit reassuring similarities. Two further examples are given by Fig. 20, which shows Kofoed-Hansen's calculations for Cu and Fig. 21, which shows Matthiae's calculations for Pb. Both sets of calculations were based on Woods-Saxon density functions in these cases. Kofoed-Hansen again made direct use of the model derived from electron scattering while Matthiae sought an optimal pair of parameters for the distribution by lease squares fitting and was led to values not materially different from those derived from electron scattering.

We have noted earlier that when the summed scattering is observed at larger momentum transfers the dominant contributions should come from multiple quasielastic collisions. A set of measurements has recently been made at CERN which permits us to test this conclusion. These measurements, carried out by Allaby et al. using 19.1 GeV/c protons and four species of nuclei, have been designed to extend the angular range of the older experiments of Bellettini et al.[56]

The preliminary results for the new measurements of summed scattering by ^{12}C are shown together with the older small-angle measurements in Fig. 22. (The apparent discontinuity in the measured cross section at the momentum transfer .33 GeV/c is due to the slightly higher energy at which the small-angle experiment was performed.) Also shown on the same graph are the elastic scattering calculated by means of the optical model, the shadowed

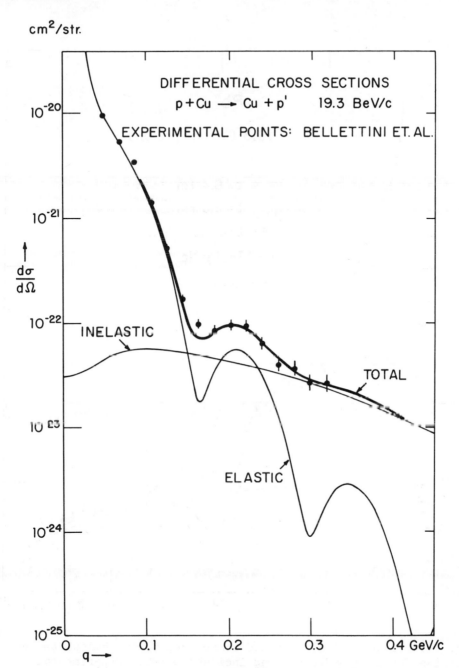

cm^2/str.

DIFFERENTIAL CROSS SECTIONS
p + Cu ⟶ Cu + p'   19.3 BeV/c
EXPERIMENTAL POINTS: BELLETTINI ET. AL.

$\frac{d\sigma}{d\Omega}$

INELASTIC

TOTAL

ELASTIC

Fig. 20. Elastic, inelastic and summed scattering calculated by
Kofoed-Hansen[52] for 19.3 GeV/c protons incident on Cu.

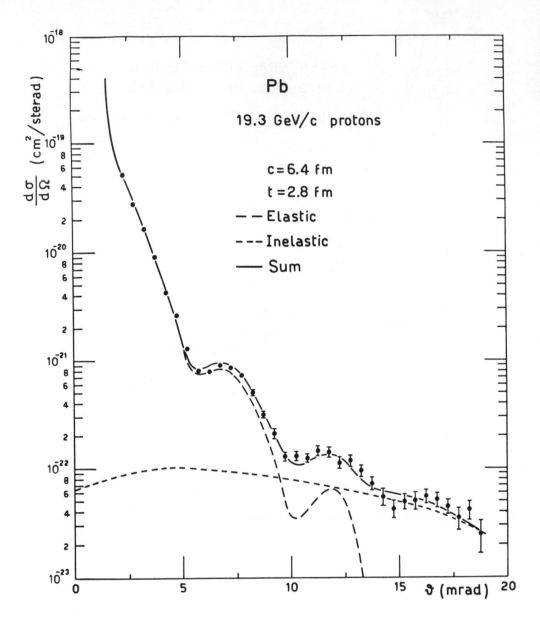

Fig. 21. Elastic, inelastic and summed scattering calculated by Matthiae for 19.3 GeV/c protons incident on Pb. Experimental points from Bellettini et al.[56]

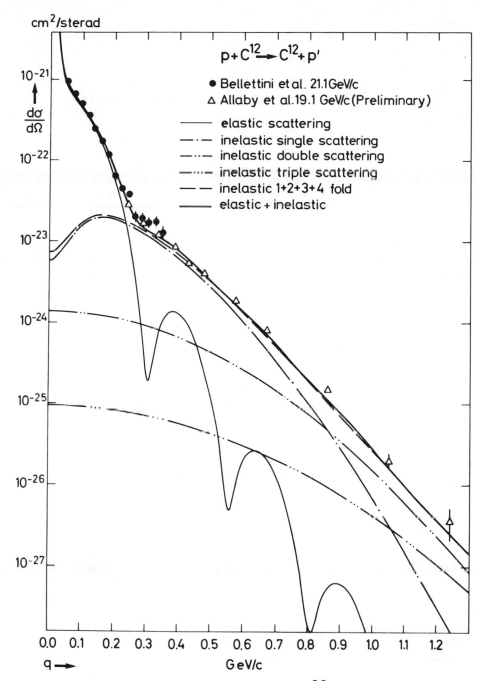

Fig. 22. Summed scattering of protons by ^{12}C; experimental
points and theoretical prediction (heavy, solid curve). Also
shown are calculated contributions of elastic scattering (light,
solid curve) and various orders of multiple quasielastic scattering.

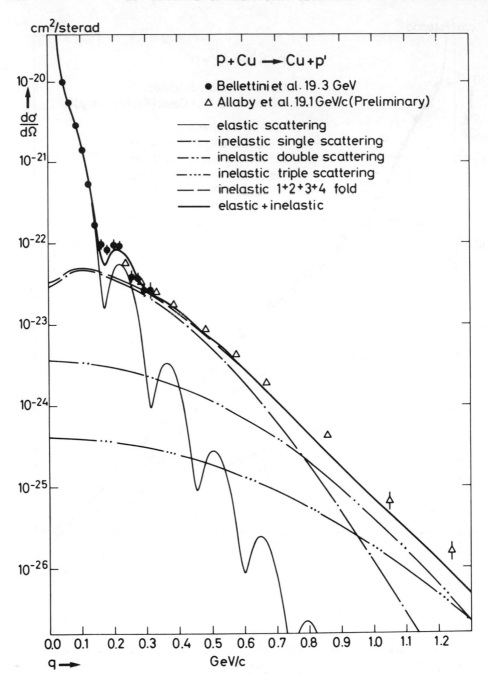

Fig. 23. Summed scattering of protons by Cu; experimental points and theoretical prediction (heavy, solid curve).

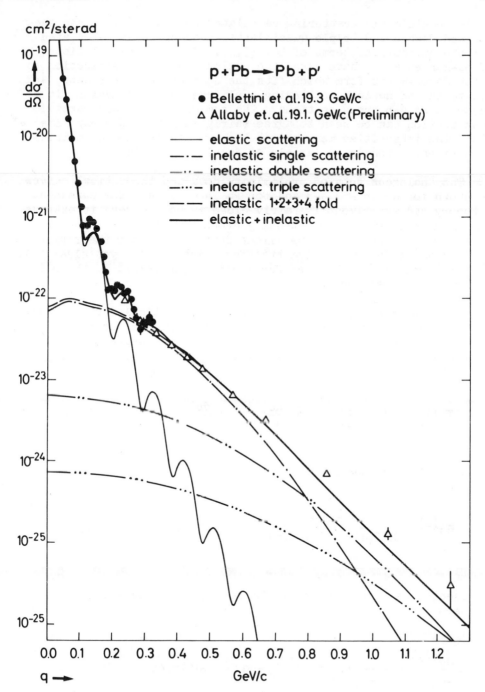

cm²/sterad

$\frac{d\sigma}{d\Omega}$

p + Pb ⟶ Pb + p′

● Bellettini et al. 19.3 GeV/c
△ Allaby et. al. 19.1. GeV/c (Preliminary)

——— elastic scattering
—·—· inelastic single scattering
—··— inelastic double scattering
—···— inelastic triple scattering
——— inelastic 1+2+3+4 fold
——— elastic + inelastic

q ⟶   GeV/c

Fig. 24. Summed scattering of protons by Pb; experimental points
and theoretical prediction (heavy, solid curve).

single quasielastic scattering calculated from Eq. (5.3), and the shadowed double and triple quasielastic scattering as calculated from the appropriate terms of Eq. (5.1). In these calculations Kofoed-Hansen and I have taken the nuclear density distribution to be the Saxon-Woods form which fits the Stanford electron data, and have taken the nucleon scattering amplitude from a Gaussian fit to the measured p-p cross sections. The sum of the various partial cross section can be seen to agree fairly well with the experimental data. The intensities measured at the two largest momentum transfers evidently consisted largely of double scattering.

The analogous sets of preliminary data and theoretical curves are shown for Cu in Fig. 23 and for Pb in Fig. 24. The agreement of theory and experiment is fairly close in both cases; it extends over a range in which the intensity drops by more than a factor of a million. The source of the slight deviation of the theoretical curve from the large-momentum transfer points is not yet clear. It may represent simply the need for a more accurate fit to the p-p differential cross section.

The measurements of Allaby et al[58] are particularly interesting for the case of proton-deuteron scattering. Let us recall that the summed scattering cross section for the deuteron can be written as the sum of the three terms[4]

$$\sum_f |F_{fi}(\vec{q})|^2 = \frac{d\sigma^{(1)}}{d\Omega} + \frac{d\sigma^{(1,2)}}{d\Omega} + \frac{d\sigma^{(2)}}{d\Omega} . \qquad (5.5)$$

The partial cross section

$$\frac{d\sigma^{(1)}}{d\Omega} = 2|f(\vec{q})|^2 \left\{1 + s(\vec{q})\right\} \qquad (5.6)$$

represents the contribution of single collision processes while the term

$$\frac{d\sigma^{(1,2)}}{d\Omega} = -\frac{2}{\pi k} \, \mathrm{Im}\left\{f^*(\vec{q}) \int S(\vec{q} - \tfrac{1}{2}\vec{q}) \, f(\tfrac{1}{2}\vec{q}+\vec{q}') f(\tfrac{1}{2}\vec{q}-\vec{q}') d^2q\right\}$$

$$(5.7)$$

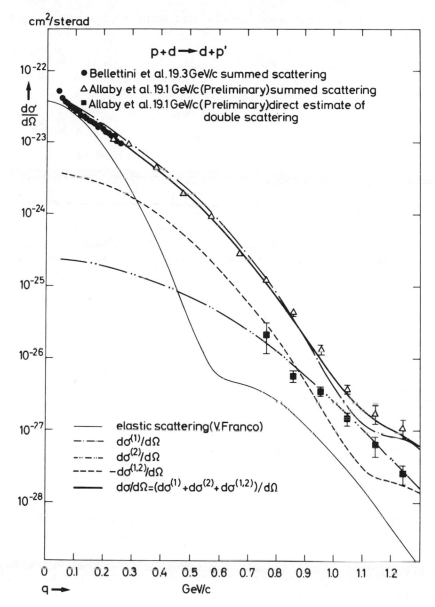

Fig. 25. Summed scattering of 19 GeV/c protons by deuterium; experimental data (circular and triangular points) and theoretical prediction (heavy, solid curve). Also shown are the estimates of the cross section for double scattering (black, square points) found by integrating the area under the double-scattering peak in the momentum spectrum of scattered protons. The calculated curves include the partial cross sections for single scattering, double scattering and interference given by Eqs. (5.6-8).

comes from the interference of single and double collision processes and

$$\frac{d\sigma^{(2)}}{d\Omega} = \frac{1}{(2\pi k)^2} \int d\vec{r} |\varphi(\vec{r})|^2 \left| \int d^2q' e^{i\vec{q}' \cdot \vec{s}} f(\tfrac{1}{2}\vec{q}+\vec{q}') f(\tfrac{1}{2}\vec{q}-\vec{q}') \right|^2$$

(5.8)

represents the contribution of pure double-collision processes.

Fig. 25 shows the recent summed scattering measurements of Allaby et al. on deuterium together with the small-angle data of the older experiment. The angular range covered by the data is now large enough to include the kink in the p-p differential cross section at momentum transfer ~1.1 GeV/c. Kofoed-Hansen and I, in analyzing the experiment, have therefore tried to avoid the need for any analytic fit to the p-p cross section. What we have done instead is to use the saddle-point method to evaluate the integrals (5.7) and (5.8) and that can be done quite simply by using smoothed versions of the experimentally measured p-p differential cross section. The three partial cross sections (5.6-8) are shown in Fig. 25 together with their sum. The excellent agreement of the sum with the measured distribution of summed scattering over the entire range of momentum transfers is a strong confirmation of the assumption that p-n scattering has essentially the same angular distribution as p-p scattering.

The experiment of Allaby et al. detects the protons scattered in any direction by measuring their momentum spectrum. The momentum spectra of protons scattered by deuterons through the larger angles used in the experiment have a two-peaked structure. The peak corresponding to the smaller loss contains those which have been doubly scattered. A partial cross section for double scattering, evaluated by estimating the area under the latter peak, is shown in the solid-square points in Fig. 25. These points agree remarkably well with the predicted cross section $d\sigma^{(2)}/d\Omega$. It is evidently quite possible, by combining accurate p-d and p-p measurements, to solve explicitly for the p-n scattering cross section at large momentum transfers.

VI.   CHARGE EXCHANGE AND PARTICLE PRODUCTION PROCESSES

The approximation methods we have discussed in the earlier sections can all be carried over, more or less intact, to the discussion of particle production processes which are diffractive

in nature. The simplest such cases are the ones in which the
incident particle is converted to one of higher mass with no change
of its intrinsic quantum numbers, e.g. photoproduction of ρ^{0}. The
fact that a mass change takes place means that there is a minimum
value for the momentum transfer in such a process, but at sufficiently
high energies the minimum momentum transfer becomes exceedingly
small. Such production processes then become, in effect, just a
second variety of elastic scattering by nucleons.

The simplest examples of processes in which a quantum number
of the incident particle does change during its passage through the
nucleus are charge-exchange processes. If the target nucleus
initially has Z protons, its final state has proton number Z + 1,
and represents an altogether different nucleus if it happens to
remain bound. There is therefore, strictly speaking, no elastic
component present at all in charge-exchange scattering.

The absence of elastic charge-exchange scattering should make
it simpler experimentally to measure the incoherent component of
the scattering. The incoherent scattering, as we've already noted,
tends to probe the surface region of the nucleus. That would be
quite an interesting thing to do by means of charge-exchange
processes since the incoherent collisions in which $Z \rightarrow Z + 1$ probe
the distribution of protons near the surface.[59] If there is indeed
any appreciable difference in the neutron and proton distributions,
that should be one of the most direct ways of observing it.

We have seen that there is no elastic component present in
charge-exchange scattering, but does that mean there is no coherent
component? The two terms have been used interchangeably in discussing ordinary scattering, and so the question is really: should we
keep their meanings identical when we are dealing with charge-
exchange processes, or with particle production processes more
generally? The answer, I'd suggest, is that we should begin to
make a careful distinction between the meanings of the two terms.
There is a component present in charge-exchange scattering which,
we have already argued, represents the principal feature of coherent
scattering. The novel feature of charge-exchange processes is that
they allow the nuclear wave function to remain essentially unchanged
in the course of an inelastic transition.

Let us suppose that the ground state of the target nucleus has
isospin quantum numbers T and T_3. Then if the charge-exchange
process leaves the nucleus within the same isospin multiplet, i.e.
in either of the states T, $T_3 \pm 1$, all of the position and spin
distributions implicit in its wave function remain unchanged.
Scattering, in other words, which simply rotates the nuclear
isospin, or in the terminology of low-energy nuclear physics leads
to "analogue states", may be regarded as being coherent. A great

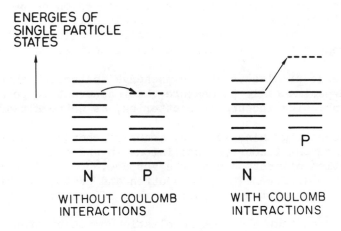

Fig. 26. Schematic picture of energies of occupied single-particle
states in the ground state of a nucleus a) in the absence of
Coulomb interactions b) in the presence of Coulomb interactions.
In coherent charge-exchange scattering an excess neutron is turned
into a proton in an unoccupied state as shown.

many observations of this sort of scattering have in fact been made at low energies in recent years, for the heavier elements as well as the lighter ones.

A little further insight into the structure of the charge-sxchange process can be gained by thinking of the nucleus as a Fermi gas with its lowest lying single-particle states filled by Z protons and N neutrons. Let us imagine that the Coulomb repulsion of the protons can be turned off. Then the energy levels of the filled single-particle states are as shown on the left side of Fig. 26. Each of the excess neutrons present will have a vacant proton state to which it can go with no change of energy, and no change of the space or spin dependence of its wave function. Charge-exchange scattering processes which induce such transitions are clearly coherent, according to our extended definitions.

Of course, when the Coulomb interactions of the protons are turned back on again, the energies of the single-particle states are changed. The energies of the proton states are raised as shown on the right side of Fig. 26, but charge-exchange processes in which an excess neutron becomes a proton in a state having essentially unaltered space and spin dependence are still possible. It may cost a few million volts of energy to induce such transitions, but that sort of energy loss is altogether negligible when the incident particles have billion volt energies. One thing we can see just from the diagram is that the cross section for coherent charge-exchange processes in which $Z \to Z + 1$ increases with the neutron excess. We can also see that, because of the exclusion principle, the processes in which $Z \to Z - 1$ can never have a coherent component unless the target nucleus has a proton excess, i.e. unless it happens to be ^{3}He.

The cross section for coherent charge-exchange scattering can be found by means of the same sort of approximation as we used to evaluate the shadowed single scattering. For a process in which $Z \to Z + 1$ it is given by

$$\frac{d\upsilon}{d\Omega}\bigg|_{\substack{\text{coh.ch.}\\\text{exch.}}} = \frac{d\upsilon}{d\Omega}\bigg|_{\substack{\text{ch.exch.}\\\text{nucleon}}} (T + T_3)(T - T_3 + 1) \times$$

$$\frac{1}{A^2}\left| \int e^{i\vec{q}\cdot\vec{b} + \frac{2\pi i}{k} f(o)T(\vec{b})} T(\vec{b})d^2\vec{b} \right|^2$$

$$(6.1)$$

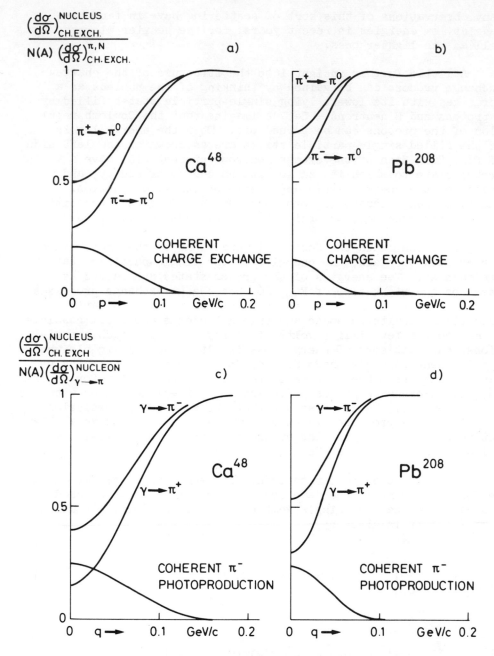

Fig. 27. Coherent and summed charge-exchange cross sections for 6 GeV pions incident on a) ^{48}Ca, b) ^{208}Pb. Coherent and summed $\pi^{\pm}$ photoproduction cross sections for 6 GeV γ-rays incident on c) ^{48}Ca, d) ^{208}Pb.

where the differential cross section on the right is the charge-exchange cross section of a free nucleon, and we have used the convention which defines T_3 for the neutron as $+\frac{1}{2}$. Since for nearly all nuclear ground states T_3 equals T, we have

$$(T + T_3)(T - T_3 + 1) = 2T_3 = N - Z \qquad (6.2)$$

and so the cross section is simply proportional to the neutron excess. The coherent cross section would clearly be a great deal larger if we had available as targets nuclei in other isospin states than $T_3 = \pm T$, but such states are unstable in general.

Two examples of the distribution which should be expected for coherent and incoherent charge exchange collisions of pions are illustrated in the top half of Fig. 27. The pion energy is 6 GeV and the two nuclei are ^{48}Ca and ^{208}Pb. The angular distributions of the coherent charge exchange processes are not too large in magnitude since the neutron excesses are much smaller than N or Z, but they are indeed peaked in the forward direction. The $\pi^- \rightarrow \pi^0$ cross sections are entirely incoherent (since they decrease the nuclear charge) and they show the characteristic dips of incoherent cross sections near the forward direction. The distribution for the process $\pi^+ \rightarrow \pi^0$ is found by adding the coherent cross section to that for $\pi^- \rightarrow \pi^0$. The coherent contribution partially fills the dip in the $\pi^- \rightarrow \pi^0$ cross section and leads to quite a perceptible difference between the forward values of the charge-exchange cross sections for π^+ and π^-.

Photoproduction of charged pions is also a species of charge-exchange process, although a somewhat more complicated one since a change of spin also takes place. If we neglect spin effects for simplicity, then we can calculate the cross sections for coherent and incoherent photoproduction processes just as we calculated the analogous charge-exchange cross sections. The distributions of $\pi^{\pm}$ produced by 6 GeV γ-rays are shown in the bottom half of Fig. 27, again for ^{48}Ca and ^{208}Pb. The existence of the coherent process, as we can see, leads to a noticeable difference between the cross sections for forward production of π^+ and π^-. The magnitudes of these two cross sections are rather sensitive to differences in the surface distributions of protons and neutrons and so their measurement would hold considerable interest.

REFERENCES

1. R. J. Glauber, in Lectures in Theoretical Physics, edited by W. E. Brittin et al. (Interscience Publishers, Inc., New York, 1959) Vol. I p. 315.

2. R. J. Glauber, in High Energy Physics and Nuclear Structure, edited by G. Alexander (North-Holland Publishing Co., Amsterdam 1967) p. 311.

3. C. Wilkin, in Nuclear and Particle Physics, edited by B. Margolis and C. Lam (W. A. Benjamin, Inc. New York, 1968) p. 439.

4. V. Franco and R. J. Glauber, Phys. Rev. $\underline{142}$, 1195 (1966).

5. C. Wilkin, Phys. Rev. Letters $\underline{17}$, 561 (1966).

6. R. J. Glauber and V. Franco, Phys. Rev. $\underline{156}$, 1685 (1967).

7. R. J. Glauber, Phys. Rev. $\underline{91}$, 459 (1953).

8. B. J. Malenka, Phys. Rev. $\underline{95}$, 522 (1954).

9. Ref. 1, p. 339.

10. F. Englert, P. Nicoletopoulos, R. Brout and C. Truffin (to be published).

11. S.-J. Chang and S.-K. Ma, Phys. Rev. Letters $\underline{22}$, 1334 (1969).

12. H. D. I. Abarbanel and C. Itzykson, Phys. Rev. Letters $\underline{23}$, 53 (1969).

13. H. Cheng and T. T. Wu (private communication).

14. E. A. Remler (private communication).

15. M. Lévy and J. S. Sucher, Phys. Rev. (to be published).

16. D. Harrington, Phys. Rev. (to be published).

17. J. Pumplin, Phys. Rev. $\underline{173}$, 1651 (1968).

   E. A. Remler, Phys. Rev. $\underline{176}$, 2108 (1968).

18. V. S. Bhasin and V. S. Varma, Phys. Rev. (to be published).

19. E. S. Abers, H. Burkhardt, V. L. Teplitz and C. Wilkin, Nuovo Cimento $\underline{42}$, 365 (1966).

20. C. Wilkin, Lecture Notes of the Summer School on Diffractive Processes, McGill University, July 1969 (to be published).

21. E. Coleman, R. M. Heinz, O. E. Overseth and D. E. Pellet, Phys. Rev. Letters $\underline{16}$, 761 (1966).

22. V. Franco and E. Coleman, Phys. Rev. Letters $\underline{17}$, 827 (1966).

23. G. W. Bennett et al., Phys. Rev. Letters $\underline{19}$, 387 (1967).

24. F. Bradamante et al., Physics Letters $\underline{28B}$, 191 (1968).

25. G. Alberi and L. Bertocchi, Physics Letters $\underline{28B}$, 186 (1968).

26. E. T. Boschitz, these proceedings.

27. H. Palevsky et al., Phys. Rev. Letters $\underline{18}$, 1200 (1967).

28. W. Czyż and L. Leśniak, Physics Letters $\underline{24B}$, 227 (1967).

29. R. H. Bassel and C. Wilkin, Phys. Rev. Letters $\underline{18}$, 871 (1967).

30. L. Leśniak and H. Wołek, Nuclear Physics $\underline{125A}$, 665 (1969).

31. E. T. Boschitz et al., Phys. Rev. Lett. 20, 116 (1968).

32. J. Formánek and J. S. Trefil, Nucl. Phys. $\underline{B4}$, 165 (1967).

33. H. K. Lee and H. McManus, Phys. Rev. Letters $\underline{20}$, 337 (1968).

34. R. H. Bassel and C. Wilkin, Phys. Rev. $\underline{174}$, 1179 (1968).

35. D. R. Harrington, Phys. Rev. $\underline{176}$, 1982 (1968).

36. J. Pumplin and M. Ross, Phys. Rev. Lett. $\underline{21}$, 1778 (1968).

37. G. Alberi and L. Bertocchi, Nuovo Cimento $\underline{61A}$, 203 (1969).

38. E. Coleman and T. G. Rhoades (private communication).

39. D. R. Harrington, Phys. Rev. Lett. $\underline{21}$, 1496 (1968).

40. V. Franco and R. J. Glauber, Phys. Rev. Lett. $\underline{22}$, 370 (1968).

41. C. Michael and C. Wilkin, Nucl. Phys. $\underline{B11}$, 99 (1969).

42.   G. Alberi and L. Bertocchi, Nuovo Cimento <u>63A</u>, 285 (1969).

43.   N. K. Glendenning and G. Kramer, Phys. Rev. <u>126</u>, 2159 (1962).

44.   M. J. Moravcsik, Nucl. Phys. <u>7</u>, 113 (1958).

45.   M. Fellinger et al. Phys. Rev. Lett. <u>22</u>, 1265 (1969).

46.   F. Bradamante et al., Proceedings of the Lund Conference 1969 (to be published).

47.   M. Fidecaro (private communication).

48.   D. Harrington. Phys. Letters <u>29B</u>, 188 (1969).

49.   M. Bazin (private communication).

50.   M. Querrou et al. (private communication).

51.   G. Matthiae (private communication).

52.   O. Kofoed-Hansen (private communication).

53.   W. Czyz and L. Maximon. Ann. of Phys. 52, 99 (1969).

54.   O. Kofoed-Hansen. Lecture Notes of the Summer School on Diffractive Processes, McGill University, July 1969 (to be published).

55.   Ref. 1, p. 412.

56.   G. Bellettini et al. Nucl. Phys. <u>79</u>, 609 (1966).

57.   A. S. Goldhaber and C. J. Joachain. Phys. Rev. <u>171</u>, 1566 (1968).

58.   J. V. Allaby et al. (private communication).

59.   O. Kofoed-Hansen and B. Margolis. Nucl. Phys. <u>B11</u>, 455 (1969).

MEDIUM ENERGY NUCLEON INTERACTIONS WITH NUCLEI

P. C. Gugelot

University of Virginia

I. Introduction. At this moment it is perhaps appropriate to ask the question what new information about the nucleus is medium energy physics going to give us ? And, if it gives new knowledge will we have to repeat all nuclear reactions which we have been studying at lower energy ? The use of pions as projectiles adds even a new dimension to the wealth of papers and Ph.D. theses which could be produced. As far as protons are concerned I will try to deal with this question by describing to you what I believe to be a few representative cases.

Since this is supposed to be a review talk I would like to present it on the basis of Maurice Goldhaber's concluding remarks at the Rehovoth Conference[1] and consequently, you should not expect new data. These will be presented by the subsequent talks in this session.

After Serber's work in 1947[2] it has often been repeated that a 500 MeV or higher energy nucleon ($k \geq 5.5$ fm^{-1}) would probe nuclear matter much better than a 20 MeV nucleon ($k = 1$ fermi^{-1}). This argument has been used to justify the application of the impulse approximation which reduces a complicated many body problem to a two-body system in which the non-interacting nucleons remain spectators[3]. However, one should keep in mind that the nucleon-nucleon interaction is not a point interaction. The charge radius of a proton is less than 1 fm[4] and the nucleon-nucleon force is about 1.5 fm (if one takes the total cross section for the nucleon-nucleon interaction as a measure of the range of the force). The latter distance is about the average spacing of the nucleons inside nuclear matter. An exception to this is the deuteron with a radius of about 2 fm. The simplest form is the plane wave impulse approximation

which is perhaps applicable only for the deuteron break-up reaction
at not too high momentum transfers. Analogous to the corrections
to the Born approximation applied to low energy physics one can
introduce a distorted wave impulse approximation which takes account
of the presence of the spectator nucleons. Distortions of the in-
going and outgoing waves are certainly as important as in low energy
physics[5]. This is apparent from the magnitude of the parameters
of the optical model describing the nucleon-nucleus interaction.
One knows[6] that the real part of the complex potential is smaller
than at low energy but the imaginary part may be as much as 100 MeV
at 1 GeV incident energy[7]. Such a potential will cause a large
reduction of the cross section obtained from a plane wave approximation (W = 100 MeV at 1 GeV incident energy corresponds to a mean
free path inside nuclear matter of 0.9 fm).

Optical models have served as a semi-empirical description of
nucleon-nucleus interactions. They are presenting average characteristics of nuclei. These are potentials and radii which are
somehow related to nuclear densities[3]. At medium energy the use of
the impulse approximation in which individual scatterings are considered is expected to yield more detailed properties of nuclear
structure. One step in this direction which has appealed strongly
to many experimentalists has been the work on the multiple scattering approximation by Glauber[8] and the subsequent use of this semi-
classical approximation by Bassel and Wilkin[9], and Czyż and Lesniak[10]. Their results showed that one is able to calculate elastic
scattering cross sections from known nucleon-nucleon scattering
amplitudes and nuclear wave functions to obtain a good agreement
with the famous experiments of the Brookhaven cooperation[7,11].

In the multiple-scattering approximation treatment it is
important to know that the nuclear form factor S(q) decreases
rapidly with increasing momentum transfer q, faster than the
nucleon-nucleon scattering amplitude f(q). Therefore at some q,
the double scattering represented by the integral will take over.
At the value of q for which single and double scattering amplitudes
are equally large we see an interference minimum in the cross section because the amplitudes have different signs. As mentioned
before the agreement with the results of the Brookhaven collaboration has been good.

The approximations[12] which were made in the derivation are:
1°. The so called geometric approximations[13], 2°. the nucleons are
assumed to be nailed down and, 3°. the nucleon-nucleon scattering
amplitudes are supposed to be equal to those for free nucleon-
nucleon scattering. The small angle scattering amplitudes are
determined consequently by the nuclear density[14]. This is borne
out by Victor Franco's calculation[15] of the total cross section
which is related to the forward scattering amplitude through the
optical theorem. Fig. 1   shows one of Franco's several results.

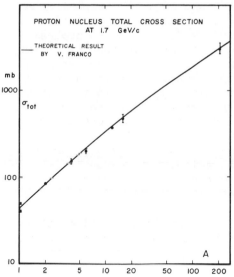

Fig. 1 – V. Franco's result
for the total cross section of
1.7 GeV protons.

The proton energy is 1.7 GeV.
The agreement with the experimental
data[16] is better than can be shown.
No adjustable parameters have
been used. The nuclear radii are
those given by the parametrization
of Elton[17].

If the forward scattering
presents the nuclear density then
the angular distribution of the
elastic scattering should contain
additional information. Fig. 2
shows one representative case,
the result of p-^{4}He by the NASA-
Virginia group[18].

At the diffraction minimum
where single and double scattering
interfere one may expect that
nuclear structure may affect the
depth of the dip. Unfortunately,
the dip is also sensitive to the
approximations in the theory and
to the nucleon-nucleon scattering
amplitudes which were employed.
In particular the inelasticity determines the depth of the minimum.
As an example, I show the experimental points for proton deuteron
scattering at 1 GeV and at 2 GeV[11][19].

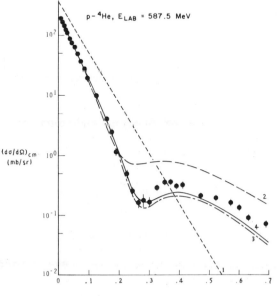

Fig. 2 – The elastic
scattering cross section
of protons on ^{4}He at 600 MeV.
Curve 1,2,3 and 4 are the
results of a multiple scattering calculation for single,
up to double, up to triple
and up to quadruple scattering respectively

The theoretical curves are due to V. Franco[20]. The experimental points show no diffraction dip. Similar results were obtained by the NASA-Virginia cooperation[21] at 600 MeV. Since the nuclear forces are spin dependent which was not taken into account in the earlier analysis one may expect that the six nucleon-nucleon amplitudes will not cancel sufficiently to produce a dip in the cross section. However, the spin dependence is not sufficient to wash out the minimum[22]. Franco and Glauber showed that the inclusion of 6.7% D-state changed the position of the dip for each polarization direction, in addition, the spin flip fills in the minimum. The scattering of protons from **aligned deuterons would be an** interesting experiment. Experimental polarization data[21] is available but no calculation on the basis of the Glauber approximation has been published for 600 MeV. I like to show you without further comment $d\sigma/dt$ versus t for protons and for pions with about the same incident momentum. Even though the fundamental scattering amplitudes are different the cross sections behave very similarly.

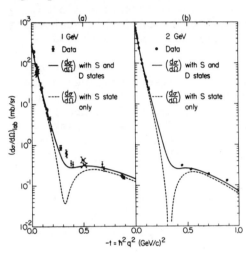

Fig. 3 - The elastic scattering cross section of 1 GeV and 2 GeV protons on deuterium. The the theoretical curves by V. Franco show the effect of the D-wave admixture in the deuteron ground state.

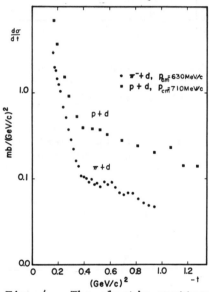

Fig. 4 - The elastic scattering of protons and negative pions by deuterium. $d\sigma/dt$ is plotted vs. t. The proton momentum is $p^p = 0.71$ GeV/c and the pion momentum is $p_{cm}^{\pi} = 0.63$ GeV/c in the center of mass.[23]

One may see from the multiple scattering formalism that the interference between single and double scattering amplitudes is related to the relative location of the two scatterers. Bassel and Wilkin[9]* used this fact to fit the depth of the minimum for $p^{-4}He$

scattering. Unfortunately, the present calculations are still too crude to make possible the untangling of the effects due to what we may call nucleon correlations. Correlations are expected on the basis of two effects: 1°. The Pauli principle, and 2°. the hard-core repulsion. In ^{4}He one has only the correlations due to the repulsive cores. So far no experiment has been able to measure correlations without other obscuring effects. Experimentally, one could measure either the location of the nucleons relative to each other, or the Fourier transform of the wave function in coordinate space which is the momentum distribution. Double scattering will be important whenever two nucleons are close together, however the use of the free nucleon-nucleon scattering amplitudes may be a poor approximation just for this case. We do not know the nucleon-nucleon interaction off the energy shell well enough to tackle this problem.

If we measure the momentum distribution, we have to look for high momentum Fourier components (for instance in neutron pick up). In this case, final state interactions will obscure the interpretations[24]. I will return to this point **later.**

Let us discuss the elastic scattering further. Elastic scattering of nucleons provided us with an **optical potential; electron** scattering resulted in giving us a charge distribution which may be proportional to the density distribution. Kerman, McManus and Thaler[3] were able to relate the nuclear density with the potential using Watson's multiple scattering methods. However, they neglected the virtual excitations in the double scattering. Kujawski[25] considered these; with the help of closure for the virtual states, he derived a second order potential which depends on a correlation function,

$$\tilde{C}(q,q')= \rho_2(q,q')-\rho(q)\rho(q').$$

$\rho(q)$ is the nucleon density as a function of the momentum transfer. $\rho_2(q)$ is the correlated density,

$$\rho_2(q,q')= \langle 0|\ \exp(i\vec{q}\cdot\vec{z}_\alpha)\cdot\ \exp(i\vec{q}'\cdot\vec{z}_\beta)|0\rangle.$$

For ^{4}He Kujawski constructed a correlation function $C(\vec{x},\vec{x}')$ which is the Fourier transform of $\tilde{C}$. This function depends on one parameter λ called the correlation length. A value $\lambda^2= 0.3$ fm^2 was necessary to fit the 1 GeV p-^{4}He data[7]. I like to show you Kujawski's results.

Figure 5 presents the central - and the spin orbit potentials for ^{4}He as seen by 1 GeV nucleons. The effect of the correlations is shown by the full line for which $\lambda^2= 0.3$ fm^2 as compared to the dashed curve. **Figure 6 compares the theoretical curve full line** with the 1 GeV p-^{4}He scattering data[1].

Figure 7 shows the calculated cross section for p - ^{4}He and the polarization calculated at 630 MeV. Three different sets of nucleon

nucleon phase shifts exist at this energy; they are designated as
A2, B3 and C5 by MacGregor. Unfortunately these phase shifts
change fairly rapidly here so that a comparison of the theoretical
curves with the experimental data seems questionable. However,
there is one disturbing fact. At small angles the second order
potentials do not contribute appreciably and at these angles a
simple Glauber type of calculation would reproduce the asymmetry;
it is essentially due to the nucleon-nucleon polarization with
suppressed spin flip probability. It seems obvious that the
experiments as well as the theory have to be improved before we
are able to extract much information about nucleon correlations.
I like to mention that the C5 phase shifts which fit the experimental p-^{4}He best are the phases from the energy dependent solution.
The proton-proton cross section calculated from this solution at
580 MeV agreed very well with a recent p,p cross section measurement
of the NASA-Virginia cooperation (as yet unpublished data).

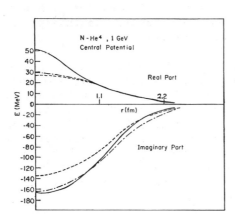 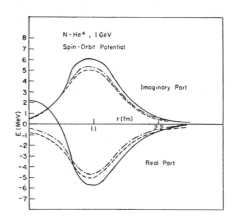

Fig. 5   The nucleon -^{4}He potential as calculated by E. Kujawski.
- - - - -   first order potential
-.-.-.-.-   first and second order potentials with the correlation
              length $\lambda^2 = 0.1$ fm^2
———————   first and second order potentials with the correlation
              length $\lambda^2 = 0.3$ fm^2

     This type of calculations is applicable only at small momentum
transfers. Large angle scattering corresponds to large momentum
transfers. In terms of the multiple scattering theory one might
expect many small angle scatterings to produce the backward scattering. However, if the target is a light nucleus not too many
small angle scatterings are possible. On the other hand in a heavy
element, many small angle scatterings rapidly increase the probability for the nucleon to make an inelastic collision. Backscattering

in ^{3}He and ^{4}He have been measured at 665 MeV. Komarov et al. [26] find lab. cross sections of $4.3\mathrm{x}^{-30}$ cm^2/sr and $4.6\mathrm{x}10^{-31}$ cm^2/sr respectively. We still seem to be very far from interpreting these data.

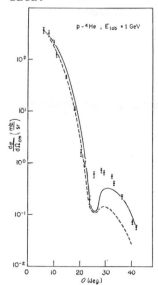

Fig. 6 - Elastic scattering cross section for 1 GeV protons on ^{4}He. ---central potential with correlations, λ^2=0.3 fm^2 _____ central potential and spin-orbit potential with correlations, λ^2=0.3 fm^2.

Proton deuteron scattering shows a peak of 100 µb/sr in the backwards direction at 600 MeV[21]. An increase at very large angles can be understood in terms of a pick-up process; in this case a neutron is exchanged. The pick-up process needs only a momentum transfer q=p/2 as compared to q=2p for ordinary back scattering. However, a simple S-state deuteron does not provide for enough momentum in its ground state to give a sizeable value to the cross section. Kerman and Kisslinger[27] showed that it needs in addition to a 7% D-state the exchange of a D* particle, which is the excited nucleon with T=1/2, J=5/2 at 1688 MeV. An admixture of 2.5% D* has been assumed, but no distortions have been considered. Fig. 8 shows the result of the calculation of Kerman and Kisslinger as compared to the experimental cross section for large angles at 600 MeV.

This result shows most clearly how much information is contained in the scattering from light elements in this energy range. Many facets of nucleon-nucleon scattering and nuclear structure seem to be interwoven.

The theory has been able to predict cross sections at small angles and in one particular case at large angles. Chahound et al[28] made use of the dispersion theory in impulse approximation in calculating p-D scattering. These authors obtain a cross section for the whole angular range. Since they consider only the single scattering and exchange diagrams their cross sections at 90° is almost an order of magnitude too large.

III. Inelastic Scattering. There are too many aspects in inelastic scattering and reactions to be covered in this talk. Tibell and James will discuss some beautiful results shortly afterwards. I will pick out one contribution which I hope Tibell will not discuss himself. It demonstrates the simplicity which medium energy reactions may assume. These are the analyses of the inelastic scattering of 150-200 MeV protons by Tibell[29].

Fig. 9 shows the ^{24}Mg(p,p') reaction exciting the 2$^+$ 1.37 MeV

state and the 4^+ state at 6.01 MeV. These are only 2 examples out
of a sample of 10. Plotting $\lg(q^{-2\ell} \cdot d\sigma/d\Omega)$ vs q^2 he obtains
close to straight lines. In addition, for 4 transitions in ^{27}Al
the ratios of these quantities extrapolated to q=0 are equal to
the ratios of the B(E2). These results can be understood if one
assumes that the distortion is constant for varying q and that the
form factor

$$F(q)= \int \psi_f^* e^{-\vec{q}\cdot\vec{r}} \psi_i dV$$

is represented by a Gaussian over a range of three decades in a
momentum interval 0<q<250 MeV/c One should notice the fact that
this momentum transfer at 200 MeV corresponds only to a scattering
angle of less than 20°. At 20 MeV this q would correspond to
scattering through an angle of 75°. It is likely that distortions
vary little over this angular range. Small momentum transfer reactions or inelastic scattering may be one of the important contributions at medium energies for the investigation of inelastic form
factors.

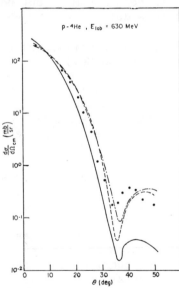 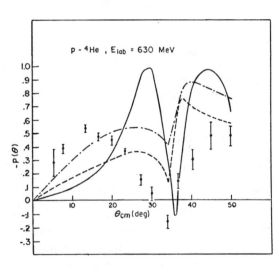

Fig. 7 - Elastic scattering and polarization of protons by ^{4}He.
Kujawski calculated these results for Ep= 630 MeV. The experimen-
tal points are for Ep= 590 MeV for the elastic scattering, and
Ep= 540 MeV for the polarization.
---------- A2 phase shifts
————————— B3 phase shifts
-.-.-.-.-.- C5 phase shifts.

Returning now to one of the major questions of nuclear physics,
we would like to have information about correlations of nucleons in
nuclei. Such investigations need large momentum transfers for which
the probability to scatter from a single nucleon is small. The

so called quasi free scattering of 1 GeV protons from C and Ca as observed by the Brookhaven collaboration and discussed by Corley[30] will provide the necessary high momentum transfers. However, there are too many uncertain factors in the analysis which was carried out by a plane wave impulse approximation that one can extract any pertinent information. Also the (p,pd) reaction, once believed to show correlations is not sufficiently well understood.[31]

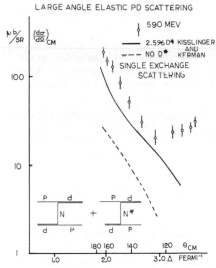

LARGE ANGLE ELASTIC PD SCATTERING

Fig. 8 - Backwards scattering cross section for 600 McV protons on deuterium. Kerman and Kisslinger calculated the cross section for the exchange of D and D* particle.

In the NASA-Virginia collaboration[32] we looked at (p,2p) reactions on deuterium and helium. I would like to discuss the deuterium results for a moment because I may be able to give you an idea of the complications involved in extracting a momentum distribution in what I believe to be the simplest example we could pick.

One observes the 2 protons with equal energy in a geometry coplanar with the incident proton at symmetric angles to the beam. In impulse approximation, the process is considered to be a quasi-free collision between two protons with the neutron as bystander. It is left behind with a small momentum while almost all the energy is shared between the two protons. Since $|\vec{p}_1| = |\vec{p}_2|$ and $\Theta_1 = \Theta_2$ kinematics prescribes that the incoming proton makes a head-on collision with the proton of the deuteron and they scatter in their c.m. system at an angle of 90°. In the plane wave approximation one extracts the momentum density $\rho(\vec{p})$ by making use of the free proton-proton cross section $(d\sigma(pp)/d\Omega)$ at 90°. This density is the square of the Fourier transform of the wave function because the final state consists of free nucleons in the matrix element. This enables one to obtain information about small admixtures to the ground state wave function.

Fig. 10 - shows this distribution which should be symmetric around p= 0 which corresponds to $\Theta_1 = \Theta_2 = 41°$ which is the kinematic angle for a proton being scattered from a free stationary proton. The curves present the density distribution for a S-wave nucleon in various models. A D-wave contribution is also shown. At large momentum the experimental points do not drop off. Does this mean that we find a contribution to very high momentum components (400 MeV/c) At such large momentum transfers the proton is going

far off the energy shell and perhaps we may not use the free nucleon-
nucleon cross section. The neutron is not a bystander anymore. It
takes part in the reaction which one could take into account by a
second scattering. Unfortunately, such a calculation has not been
made for the conditions which apply here. The p,2p results and the
p-deuteron back scattering are clearly correlated. A good understanding of these reactions will give much insight in the deuteron
problem. Similar reactions on heavier nuclei will be more difficult
to treat because the plane wave impulse approximation cannot be used.

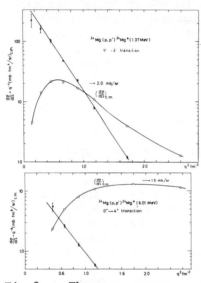

IV. Conclusions. These were the
examples I wanted to present to you.
A first observation one makes looking
at this material is that the theoreticians have had an orgy; of the more
than 30 references I have quoted, only
8 were experimental papers.

There has been considerable progress.
The problems which have been with us
in nuclear physics seem to present a
better handle in this energy range.
The projectile singles itself out
between the target nucleons because
of its high momentum. This resulted
in our ability to calculate nucleon-
nucleus scattering cross sections from
the nucleon-nucleon scattering amplitudes.
The average nuclear properties, density
and nuclear radii can be related, at
least in some light nuclei, to the
optical potential and the radius of the
potential. The degree of accuracy is
not high, but this is to be expected
without further information about nuclear
structure. Let us be satisfied with
this result because if we had already
complete agreement the detailed structure
would still be hidden.

Fig.9 - The c.m. cross
section (open circles)
and the c.m. cross section
dividend by $q^2 \ell$ (filled
circles) as analyzed by
G. Tibell for: a. The
E2 transition to the 1.37
MeV state in ^{24}Mg.
b. The E4 transition to the
6.01 MeV state in ^{24}Mg.

The next step will be difficult; 1°. experiments and theory
have to be improved, 2°. too many different "details" are involved
and mixed up. As a consequence there is a danger of stepping back
in the tracks of low energy physics which is the construction of
nuclear models.

Experiments which do not seem too fruitful at this moment are
high momentum transfer elastic and inelastic scattering from heavy
nuclei. What is left of the cross section is due to complicated

process. I would estimate the limiting momentum transfer to be less than q_{max}^2 = 2mB in which B is the binding energy of a nucleon.

Inelastic scattering in which one particle is kicked out is an interesting tool, especially in light nuclei. In plane wave impulse approximation one measures the Fourier transform of the wave function. It seems profitable now to use much accelerator time for the study of the lightest nuclei with the help of elastic scattering and reactions with nucleons and also with pions.

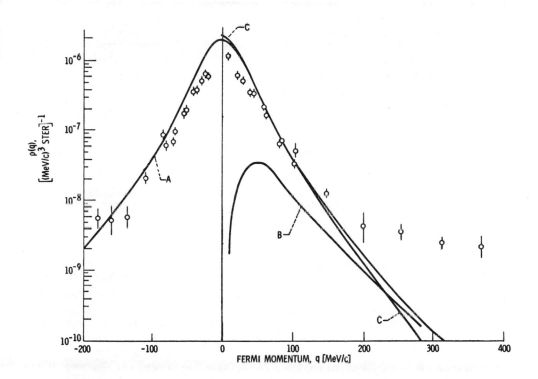

Fig. 10 - The "momentum distribution" for the proton in deuterium in plane wave impulse approximation.
>   Curve A is the square of the Fourier transform of the S-state
>   Hulthén wave function.
>   Curve B is for a pure D-state proton normalized to .07. The
>   same radial dependence as the S-state wave function is
>   assumed.
>   Curve C is for the S-state Hamada-Johnston wave function.

References:

1. M. Goldhaber, High Energy Physics and Nuclear Structure, edited
    by G. Alexander, North Holland Publ. Co., Amsterdam, 1967,
    p. 486.
2. R. Serber, Phys. Rev., 72, 1114 (1947).
3. A. K. Kerman, H. McManus, and R. M. Thaler, Ann. Phys., 8,
    551, (1959).
4. F. Bumiller, M. Croissianx, E. Dally, and R. Hofstadter,
    Phys. Rev., 124, 1623, (1961).
5. H. K. Lee and H. McManus, Phys. Rev. 161, 1087 (1967);
    F. R. Kroll, Ph.D. Thesis, "Quasi free Scattering of protons
    from ^{12}C at 160 MeV and 1014 MeV," University of Maryland (1968).
6. H. Feshbach, Ann. Rev. Nucl. Science, Vol. 8, p. 49, Annual
    Reviews, Stanford 1958; W. E. Frahn and G. Wiechers, Phys. Rev.
    Letters, 16, 810 (1966).
7. H. Palevsky et al., Phys. Rev. Letters, 18, 1200 (1967).
8. R. J. Glauber, "Lectures in Theoretical Physics," edited by
    W. E. Brittin and L. G. Dunham, Interscience Publishers, Inc.,
    New York 1959, Vol. I, p. 315.
    "High Energy Physics and Nuclear Structure," edited by
    G. Alexander, North-Holland Publ. Co., Amsterdam 1967, p. 311;
    V. Franco and R. J. Glauber, Phys. Rev., 142, 1195 (1966).
9. R. H. Bassel and C. Wilkin, Phys. Rev., Letters, 18, 871 (1967);
    R. H. Bassel and C. Wilkin, Phys. Rev., 174, 1179 (1968).
10. W. Czyż and L. Lesniak, Phys. Letters, 24B, 227 (1967).
11. G. W. Bennett et al., Phys. Rev. Letters, 19, 387 (1967).
12. H. Feshbach and J. Huffner, MIT report CTP #87.
    H. Feshbach, "International School of Physics, Enrico Fermi
    38th Course," Academic Press, N. Y. 1967, p.183.
13. D. K. Ross, Phys- Rev. 173, 1695 (1968).
    L. I. Schiff, Phys. Rev., 176, 1390 (1968).
14. W. Czyż and L. C. Maximon, Ann. Phys., 52, 59 (1969).
15. V. Franco - I thank Dr. Franco for letting me use his results
    prior to publication.
16. G. J. Igo et al., Nucl. Phys. B3, 181, 1967. The p,p and p,n
    scattering amplitudes have been averaged. Values measured by
    Bugg et al. (Phys. Rev., 146, 980 1966) have been used.
17. L.R.B. Elton, "Nuclear Sizes," Oxford Univ. Press, Oxford 1961.
18. E. T. Boschitz et al. Phys. Rev. Letters, 20, 1116 (1968).
19. E. Coleman et al. Phys. Rev. 164, 1665 (1967).
20. V. Franco and R. J. Glauber, Phys. Rev. Letters, 22, 370 (1969);
    C. Michael and C. Wilkin, Rutherford Lab. Preprint, RPP/A54, 1968.
    The D-state influence was also proposed by D. R. Harrington,
    Phys. Rev. Letters, 21, 1496 (1968).
21. E. T. Boschitz, these proceedings.
22. E. Kujawski, D. Sachs and J. S. Trevil, Phys. Rev. Letters.
    21, 581 (1968). V. Franco, Phys. Rev. Letters 21, 1360 (1968).
23. F. Bradamante, et al. Phys. Letters, 28B, 193 (1968).

24. G. E. Brown, Comments on Nucl. Part. Phys., 3, 78 (1969).
    W. A. Friedman, Phys. Rev. Letters, 21, 1696 (1968).
    S. T. Taun, L. E. Wright and M. G. Huber, (Phys. Rev. Letters, 23, 174, 1969) discuss the effect of short range correlations on electron elastic scattering.

25. E. Kujawski, Ph.D. Thesis MIT, 1969. I thank Dr. Kujawski for permitting me to use his results before publication.
26. V.I. Komarov et al. J.I.N.R. - P1 - 4373
    V.I. Komarov et al. J.I.N.R. - P1 - 3720
27. A. K. Kerman and L. S. Kisslinger, these proceedings. I thank Dr. Kisslinger for making his results available before publication.
28. J. N. Chahoud et al., Nuovo Cimento, 56, 838 (1968).
29. G. Tibell, Physics Letters, 28B, 638 (1969).
30. D. M. Corley, Ph.D. Thesis, "Quasi-free Scattering of 1 GeV Protons from ^{12}C and ^{40}Ca," University of Maryland, 1968.
31. H. G. Pugh, Phys. Rev. Letters, 20, 601 (1968).
32. C. Perdrisat et al., these proceedings.

DISCUSSION

J. Hufner: In the multiple scattering theory of Kerman, McManus and Thaler (Ref. 3), as in the Glauber theory, one has to consider centre-of-mass corrections. This accounts for the discrepancy between experiments and the calculation of Kujawski (ref. 25) in the case of p-^{4}He.

P.C. Gugelot (in reply to a question): Coulomb corrections, which are quite small in these diffraction experiments on these very light nuclei present no problem.

SPECTRA OF LIGHTEST NUCLEI KNOCKED OUT FROM LIGHT NUCLEI WITH 670 MEV PROTONS AND THE CLUSTERING PHENOMENON

V. P. Dzhelepov

Laboratory of Nuclear Problems, Joint Institute for

Nuclear Research, Dubna, USSR

In recent years the investigation of the interaction of hundreds MeV protons with complex nuclei accompanied with knocking out of lightest nuclei such as deuterons, tritium, helium and heavier ones has become one of the sources of new valuable information on the structure of complex nuclei, and mainly, on the structure peculiarities which are determined by intranuclear nucleon associations-clusters. Indeed, if such knocking out is considered as a result of direct interaction of incident protons with some nucleons of the nucleus, the reaction differential cross sections should be dependent on the parameters describing cluster behaviour and hence, it can be expected that such experiments can allow clarification of the certain features of the physical nature of cluster phenomena in nuclei.

However, until recently all experimental information on the high energy part of the spectra of lightest nuclei knocked out with high energy protons has been confined to the results of experiments run by using the Dubna synchrocyclotron[1] and those performed at Brookhaven[2]. The highest energy part of deuteron spectra corresponding kinematically to quasielastic proton scattering on deuteron clusters in various nuclei has been measured in these experiments.

The experiments made it possible to observe the fast increase of deuteron yeild even with very slight decrease of their momentum with respect to the value corresponding to quasi-elastic pd-scattering. However, the problem of the mechanism forcing this increase remained open. There was no information whatever on the high energy part of the spectra of heavier fragments (e.g. ^{3}He and ^{4}He) knocked out from nuclei. The cross sections of fast proton scattering on free He nuclei at large angles were not known too. It is evident that the acquisition of such data is of great interest from various points of view. One can think, e.g., that proton scattering on 3- and 4-nucleon clusters will result in fast ^{3}He and ^{4}He nuclei knocked out according to the kinematics of quasielastic pHe^3 and pHe^4 scattering, and the comparison of the cross sections of these processes with the appropriate cross sections of elastic scattering by He nuclei will allow evaluation of the probability of the existence of 3- and 4-nucleon cluster states in nuclei and the comparison of the results with calculations made by using theoretical models.

Here the results of two experiments performed recently with the Dubna synchrocyclotron are presented.

1. In the first of these studies carried out by Petrukhin et al. the wide spectrum region of deuterons with the momentum $P_d >$ 700 MeV/C emitted at an angle of 6.5° (lab. s.) was investigated when 670 MeV protons interacted with H,D,Li,C,Al,Cu,Rh and Pb nuclei. Measurements were performed by using the external proton beam of the Dubna synchrocyclotron. Angle beam divergence was $\pm 0.6°$ in the experimental conditions. Secondary charged particles emitted from the target at the above angle were separated according to the momentum by means of the magnetic spectrometer, its resolution being $\Delta P/P = \pm 1.8\%$, and were analysed by the time-of-flight with a 5.6 m path length. The t.o.f. spectrometer had a resolution of $2\tau = 0.9$ nsec. A typical t.o.f. spectrum is shown in Fig. 1 (here the left peak corresponds to deuterons, the right one to protons). When treating deuteron momentum spectrum the following corrections were made: 1) for the dead time of the scintillation counter telescope; 2) for secondary particle loss in the target; 3) for multiple scattering in the scintillation counter, 4) for particle energy loss in the target. The detection threshold of the t.o.f. spectrometer determined by the deuteron range in matter was 600 MeV/c.

The absolute values of the differential cross sections of the deuteron yield were found by using the known cross section of the reaction $P+P \rightarrow d+\pi^+$ [1]measured in the same conditions, and using the CH$_2$-C difference.

The momentum spectrum of deuterons from reaction 1 is shown in Fig. 2. The peaks at 1380 MeV/c and 860 MeV/c momenta correspond to deuteron emission (c.m.s.) in the "forward" direction and "backward" direction. The FWHM of about 45 MeV/c of these peaks describes the momentum resolution of the magnetic spectrometer. The momentum spectrum of deuterons from p+C^{12}- collisions is shown in Fig. 3. Here the peak at Pd∿1600 MeV/c corresponds to the earlier /1,2/ observed proton quasielastic scattering on two-nucleon clusters inside the nucleus. The differential cross section of this reaction obtained in the present experiment is $(3.5+0.6) \times 10^{-27}$ cm^2/sr and agrees well with the previous result /1/.

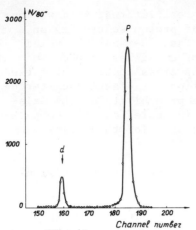

Fig. 1
Time-of-flight spectrum.

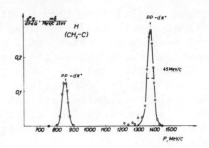

Fig. 2
Momentum spectrum of deuterons
from the p + p → d +π⁺ reaction.

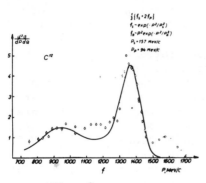

Fig. 3
Momentum spectrum of deu-
terons form p+C^{12} collisions. Solid curve has
been calculated in the
impulse approximation.

In the spectrum region
(Pd<1500 MeV/c) measured for the
first time in these experiments there
is a pronounced peak with
Pd∿1370 MeV/c, the FWHM being about
130 MeV/c, and a wide plateau in
the spectrum area where deuterons
should be observed emitted in reaction
(1) in the backward direction (c.m.s.).
This area seems to be due to deuterons
from reactions: p +<p> → d +π+
p + <n> → d + π° (2), occuring on
bound nucleons of the nucleus. The
difference in the shape of spectra
from reactions (1) and (2) corresponding to free and bound necleons appears
to be due to intranuclear motion of
nucleons. To check this assumption
the momentum spectrum was calculated
in the impulse approximation with
the account of the given momentum
distribution of nucleons in the
nucleus /3/ and with the help of the known angular and energy
dependence of the differential cross section of reaction (1) /4/.
The results of calculations shown in Fig. 3 by a solid line agree
qualitatively with experimenta data. The upper estimate of the
(p,dπ) reaction cross section obtained in the present experiment
is: (32+4) x10^{-27} cm^2/sr.

In order to study the behaviour of deuteron spectra with
respect to the mass number of the nucleus-target the measurements
were performed also for D, Li, Al, Cu, Rh and Pb nuclei. The obtained
sepctra are shown in Fig. 4. The maxima of the spectra with
Pd ∿1600 MeV/c correspond to proton quasielastic scattering on

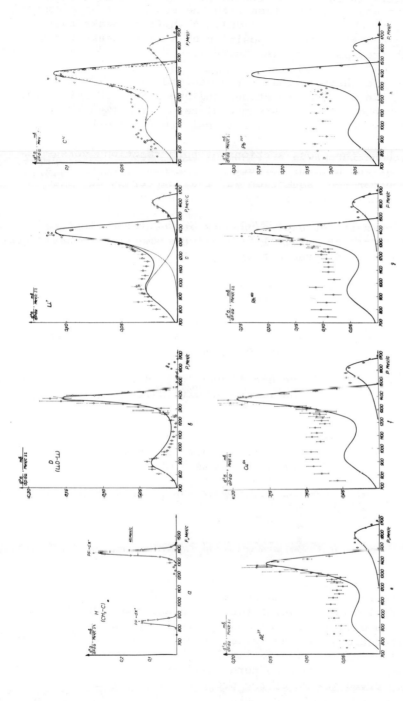

Fig.4. Experimental spectra of deuterons and the results of calculations. ϱ is experimental points. — is calculation ignoring the formfactor for (p,π^+d) and (p,pd) reactions."Forward" and "backward" peaks for the (p,π^+d) reaction are shown separately in Fig.4c. --- (Fig.4d) is calculation with the formfactor for (p,π^+d). --- (Fig.4d) is the same calculation for the (p,π^+d) reaction.

deuteron clusters in the (p,dN) reaction. The peaks at 1370 MeV/c
are a result of proton interaction with the bound nucleon accompanied
by pion emission (the (p,dπ) reaction). A shift of peaks to lower
energies in comparison with the position of similar peaks for
reactions with free particles is observed.

For all nuclei heavier than D the deuteron peak widths are
nearly the same in both the reactions and are about 130 MeV. This
considerably exceeds in the instrumental resolution ($\sim$45 MeV/c).
For deuterium the peak width (p,dπ) is half that for other nuclei.

Fig. 5a shows the cross sections of both reactions versus the
mass number A of the nucleus-target. The agreement with the $A^{1/3}$
law is evidence for the assumption /2/ that deuterons detected in
both the reactions are emitted from the nuclear periphery.

Fig. 5b shows the effective numbers of "deuteron clusters"
compared to the data of refs. /1,2/. All the amount of experimental
information is in good agreement with the $A^{1/3}$ law.

Deuteron spectra for all nuclei were calculated by using the
pole diagram of the dispersion theory /5/.

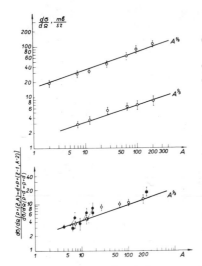

Fig. 5A

Cross sections versus the mass number A.

The upper curve (p,π d).
The lower curve (p,Nd).

Fig. 5b

Effective number of deuteron clusters
versus the mass number A

o – results of the present experiment.

● – Dubna /1/ for 675 MeV.
■ – Brookhaven /2/ for 1 GeV.

For the (p,dπ) reaction the differential cross section of
virtual particle interaction was replaced by the value for the
corresponding real process, while for the (p,dN) reaction the cross
section was assumed to be angular and energy independent. The
results of calculations for the latter reaction (ignoring the form-
factor) are shown in Fig.4 (solid line). The contributions from
(p,dN) and (p,dπ) reactions were normalized irrespective for experimentally observed maxima of the proper deuteron peaks. In the
region of the observed deuteron peaks theoretical curves well describe

experimental data. At low energies where the application of the
pole mechanism is doubtful theoretical curves for all nuclei lie
lower than experimental points. In this case the divergence is
systematically increased in passing over to heavier nuclei. The
introduction of the formfactor into calculations does not give any
essential improvement. As an example, Fig. 4 shows the results
of calculations for the $(p, d\pi^+)$ reaction on carbon with the form-
factor corresponding to $\ell = 1$ (it was assumed that the nucleus remains in the ground state).

Deuteron spectra for D and C nuclei were calculated also in
the impulse approximations. The obtained curves differ slightly
from those predicted with the pole model.

2. The second investigation made using our accelerator by
V.I. Komarov, G.E. Kosarev, O.V. Savchenko is devoted to measuring
the yield of ^{3}He and ^{4}He knocked out from light nuclei at 500 MeV
corresponding to the kinematics of quasielastic pHe^3 and pHe^4
scattering. The general view of the experimental arrangement is
shown in Fig. 6. The proton beam was focused on the target of Li,
Be, C, O. ^{3}He and ^{4}He nuclei emitted from it at an angle of $5.4°$
to the proton direction and (after being bent in the magnetic field)
were separated from the whole amount of emitted particles and were
detected with a system of scintillation counters by time of flight,
energy loss and the range. The effective solid angle of detection
was increased by means of the focusing channel consisting of quadrupole lenses. Fig. 7 shows the measured high energy parts of
^{3}He and ^{4}He spectra. Along with the statistical errors the absolute
values of the measured quantities in Fig. 7 contain some uncertainty
arising in calibration and introducing a correction which equals
15% for ^{3}He and 18% for ^{4}He. Fig. 7 shows a deuteron spectrum in

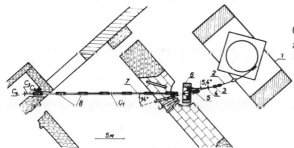

Fig. 6
General view of experimental
arrangement for investigating
^{3}He and ^{4}He knocking out.

neighbourhood of the appropriate momentum, produced by protons from
the CH_2 target. The deuteron peak from the $pp \to d\pi^+$ reaction
shows the instrumental line shape of the $(\Delta p/p \sim 6\%)$ magnetic spectrometer with which the above spectra of He nuclei were measured. The
cross sections of the processes giving rise to such He nuclei are
$10^{-30} - 10^{-31}$ cm^2 sr^{-1}. The observed parts of the spectrum are
characterized by the peak in the He yield from ^{6}Li. It is less
pronounced in the yield from ^{9}Be. In these actual experimental

conditions, no peaks are observable in the ^{12}C and ^{16}O spectra, but only a 'shoulder' in the yeild of He^3 with increasing momentum. The peak in the high energy part of the ^{4}He spectrum can be explained by the contribution of such channels for proton interaction with clusters whose kinematics is close to that of two-particle reactions resulting in He nucleus production in the final state. As a source of fast ^{4}He nuclei production apart from proton scattering on alpha-particles p + <^{4}He> → p + ^{4}He   (3) one can   consider the reaction p + <^{4}H> → n + ^{4}He   (4). However, the probability of a large momentum transfer to such a compact group of nucleons as a cluster of ^{4}He type, should be considerably greater than that of the transfer of the same momentum to the nucleon group of ^{4}H type.   Therefore, one can consider that the basic mechanism of the observed yield of 470 MeV ^{4}He consists in quasielastic proton scattering by alpha-clusters of nuclei.

For ^{3}He knocked out, along with scattering
        p + <^{3}He> → p + ^{3}He (5), there is a possibility of the reaction p + <^{3}H> →   n + ^{3}He (6).   Besides, a small contribution of the reaction P + <^{4}He> → d + ^{3}He (7) on alpha-clusters producing ^{3}He (on free ^{4}He nuclei) with the 1860 MeV/c momentum cannot be excluded experimentally.   The coupling energy of knocked out clusters in the initial nuclei is manifested weakly in experimental conditions.   Nevertheless, the washing-out of the quasielastic scattering peak on increasing the mass number A of the target-nucleus can be explained by the contribution of states with great excitation of the residual nucleus.   In order to evaluate the cross section of quasielastic scattering on clusters in ^{6}Li and ^{9}Be the observed part of the spectra in the region of $P_{He}^3 \approx 1800$ MeV/c and $P_{He}^4 \approx 1920$ MeV/c was approximated by a Gaussian curve.   The normalization and dispersion were found by the least squares method using the experimental points with $P_{He}^3 > 1780$ MeV/c and $P_{He}^4 > 1900$ MeV/c.

In order to compare the obtained cross sections of quasielastic proton scattering with those of elastic scattering on free nuclei the cross sections of elastic pHe^3 and pHe^4 scatterings were measured at the same angle of helium detection as in experiments on helium knocked out from more complex nuclei.   In this case the momentum transferred to He nuclei was $(q, He^3) = 8.8$ fm^{-1}, $(q, He^4) = 9.5$ fm^{-1}.

The comparison of all the obtained data as well as data on deuteron knocking out measured in similar conditions /1/ shows (Fig. 8) that the experimental values of the differential cross section of quasielastic scattering are equal (within a factor 1.5-2.) to the corresponding differential cross sections of elastic scattering multiplied by A/m, where m is the number of nucleons in a cluster.   This result seems to be evident if the knocking out of fast d, ^{3}He, ^{4}He fragments from light nuclei in the energy region of the fragment under discussion occurs by direct interaction of

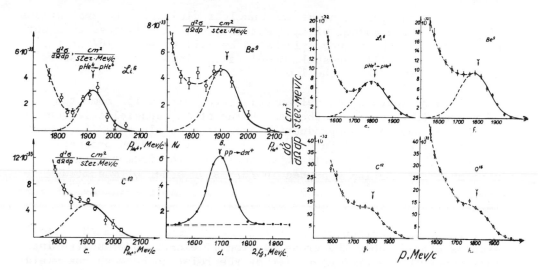

Fig. 7 (a,b,c) - yield of fast ^{4}He nuclei from ^{6}Li, ^{9}Be, ^{12}C; (d) - deuteron yeild (relative units) from the CH$_2$ target in the region close to the effective momentum of He; (e,f,g,h) - yields of fast ^{3}He nuclei from ^{6}Li, ^{9}Be, ^{12}C, ^{16}O.

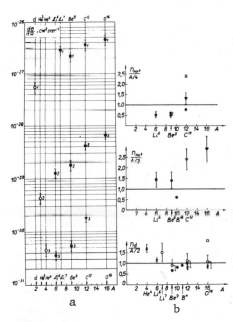

Fig. 8 (a) - differential cross section (lab.s.) of quasielastic scattering (⦶) of 665-675 MeV protons in the A(p,NX) reactions on light nuclei and elastic pX ⟩ pX scattering (⦶) with large momentum transfers to nucleon associations. The points with the "1" index refer to reactions where X = d, θ_d = 7.6° /1/, points with the "2" index refer to reactions X = ^{3}He; "3" - X = ^{4}He, θ_{He} = 5.4° according to the data of the present study. ⊻ - the upper boundary of the cross section of quasielastic scattering. (b) - Ratios (n_x) of the cross sections of quasielastic scattering versus those of free pX → pX scattering. The points ⦶ and ⊻ are the present study; Φ - /1/ and Ⅱ - /2/. The points without errors correspond to the calculation of the effective number of clusters on the basis of the shell model: ● - /6/ with the account of the P-shell only, and □ - /7/ with the account of P- and S-shells.

the incident proton with the intranuclear cluster. The evaluations
of the effective number of clusters obtained in this case turn out
to be close to the values calculated by V.V. Balashov /6/ on the
basis of the oscillator shell model taking into account clusters
knocked out from the P-shell only.

A sharp increase of helium yeild even with a 10% momentum
decrease with respect to the mean value of the quasielastic scattering can be due to mesonproduction processes on clusters (to a certain
extent it recalls a picture observed in fast deuteron production).
The fact that the ^{4}He yield at an angle of 5.8° from the reaction
p + ^{3}He → π^+ + ^{4}He (8) measured in /8/ is an order of magnitude
higher than that of ^{4}He at an angle of 5.4° in elastic pHe^4 scattering is evidence for the above conclusion.

The cross section of the p + d → $\pi°$ + He^3 reaction /8/ also
exceeds considerably the cross section of elastic pHe^3 scattering.
This shows the general character of the relation between the yield
of fast lightest nuclei d, ^{3}He, ^{4}He in elastic scattering on clusters
and in mesonproduction.

The theoretical evaluation of the differential cross section
of backward elastic proton scattering on ^{4}He nuclei made by Prof.
D.I. Blokhintsev /9/ is based on the presentation of direct proton
interaction with density fluctuations. The considerable predomi-
nance of the ^{3}He yield over the ^{4}He one observed experimentally in
the momentum region of 1800-1900 MeV/c is also explained by calculations based on the fluctuation approach.

3. Thus, a conclusion can be drawn that all the observed
results are evidence for the peripheral character of reactions
under study. The cross sections of deuteron knocking-out are
proportional to the $A^{1/3}$ law, the experimental estimates of the
effective number of clusters is noticeably smaller than the theoretical values if the calculation takes into account the cluster
yield not only from the P-shell but also from the S-shell of the
appropriate nucleus (V.G. Neudachin's calculations /7/). Such a
character of cluster interaction reactions can be a consequence of
two reasons: either cluster states occur with a great probability
only in the peripheral region of the nucleus or they occur also
in the central region of the nucleus (which is quite possible),
but the probability of knocking out such clusters due to the superposition of the secondary effects of cluster rescattering is greatly
reduced with increasing nuclear mass number. The dependence of
the $A^{1/3}$ type for the cross sections of the (p,dπ) reactions
occurring on separate nucleons of the nucleus seems to show the
considerable effect of cluster overgrowing even at hundreds MeV
energies of incident protons and knocked out clusters.
The problems touched upon here have been discussed in greater detail
and from the theoretical standpoint by V.V. Balashov in his review

talk at the International Conference on Clustering Phenomena in
Nuclei, Bochum, 1969[10]. It can be expected that further essential
progress in solving the problem in question will be achieved on the
one hand, by running experiments in which the coincidence technique
will be used permitting precise identification of the states of
residual nuclei, and on the other hand, as a result of performing
stricter calculations taking into account cluster re-scattering and
other effects.

REFERENCES

[1] L.S. Azhgirey et al. J. Expt. Theor. Phys. 33, (1957) 1185.
[2] R.J. Sutter et al. Phys. Rev. Lett. 19, (1967) 1189
[3] U. Amaldi et al. Phys. Lett. 25B, (1967) 24.
[4] M.G. Mescheryakov, B.S. Neganov. DAN SSR 100 (1955) 677.
[5] I.S. Shapiro. "Theory of Direct Nuclear Reactions"
 Gosatomizdat, SSSR, 1963.
[6] V.V. Balashov et al. Nucl. Phys. 59, (1964) 417
[7] P. Beregi et al. Nucl. Phys. 66, (1965) 513.
[8] Yu. K. Akimov, O.V. Savchenko, L.M. Soroko. Sov. J. Expt.
 Theor. Phys. 41, (1961) 708.
[9] D.I. Blokhintsev. Sov. J. Expt. Theor. Phys. 33, 1957) 1295.
 D.I. Blokhintev, K.A. Toktarov. Prepring P4-4018, Dubna (1968).
[10] V.V. Balashov. Int. Conf. on Clustering Phenomena in Nuclei,
 Bochum, 1969.

DISCUSSION

V.P.D., in reply to **J. Hufner**: The observation of knock-out D, ^{3}He,
^{4}He, etc., from nuclei, certainly does not, of itself, prove the
existence of these well defined clusters in the target nucleus.

NOTE

The above invited talk of V.P. Dzhelepov includes the substance of
the following communications from the Joint Institute for Nuclear
Research, Dubna, USSR:

  Quasielastic Scattering of 665 MeV Protons on Three and Four-
     Nucleon Clusters in Light Nuclei -- V.I. Komarov, G.E. Kosarev,
     and O.V. Savchenko

  Measurement of the Fast Deuteron Spectrum at 6.5° from the Bom-
     bardment of Carbon by 670 MeV Protons -- L.S. Azhgirey, Z.V.
     Krumstein, Ngo Quang Zui, V.I. Petrukhin, D.M. Khazins, and
     Z. Cisek

  Spectra of Fast Deuterons Produced by 670 MeV Protons on Nuclei --
     L.S. Azhgirey, Z.V. Krumstein, Yu.P. Merekov, Z. Moroz, Ngo
     Quang Zui, V.I. Petrukhin, A.I. Ronzhin, D.M. Khazins, and
     Z. Cisek

ELASTIC SCATTERING OF 600 MeV PROTONS FROM LIGHT NUCLEI

Edmund T. Boschitz
National Aeronautics and Space Administration
Lewis Research Center
Cleveland, Ohio

In 1966 a program was initiated at the NASA-Lewis Research
Center to study the scattering of 600 MeV protons from light nuclei.
Deuterium, Helium (3) and Helium (4) were chosen partly for experimental reasons, and partly because a systematic study of these
nuclei might be most suitable for testing multiple scattering
theories. Such theories require a good knowledge of the elementary
interaction between the incident particle and the target nucleons,
and the nuclear wavefunction (or equivalently Fermi momentum distribution). Therefore, we have supplemented existing nucleon-nucleon
scattering data by measuring the p-p differential cross section and
polarization, and we have determined the momentum distribution for
the proton in the deuteron and in Helium (4) from (p,2p) measurements. In this contribution I will report our results of differential cross section and polarization in elastic p-p, p-d, p-^{3}He,
p-^{4}He scatterings. The (p,2p) results will be presented by Perdrisat
in a separate contribution to this conference. The experiments were
performed at the proton synchrocyclotron of the NASA-Space Radiation
Effects Laboratory in Virginia. Our team, in addition to W. K.
Roberts, J. S. Vincent and myself consisted of M. Blecher, K. Gotow,
P. C. Gugelot, C. F. Perdrisat, and L. W. Swenson of the Virginia
Universities and J. R. Priest from Miami University.

Because the object was the study of the details of the elastic
cross section and the polarization, it was particularly important to
exclude nonelastic events. Figure 1 shows the experimental arrangement. The external proton beam was brought to a focus on the target
by the beam transport system. The intensity of the incident beam was

monitored by an Argon-filled ion chamber (AIC) and by scattering
the particles from aluminum into a pair of range telescopes (M_1
and M_2). The absolute number of incident protons was obtained
by carbon activation using the well known $^{12}C(p,pn)^{11}C$ cross section. The beam direction was monitored with two split ion chambers
(SIC 1 and SIC 2). The targets consisted of CH_2, CD_2, high-pressure
gaseous ^3He and liquid ^4He. Protons scattered from ^4He were detected by multiple counter range telescopes which were mounted on a
precession scattering table. Counter 1 and 2 defined the scattering
volume of the helium targets and the solid angle subtended by the
range telescopes. The angular resolution was $\Delta\theta = 0.75°$ FWHM at
small angles and $\Delta\theta - 2.0°$ at large angles. The overall energy
resolution of the range telescope was about 12 MeV which is sufficient for elastic p-^4He scattering. The efficiency for detecting
elastic events in the range telescope was determined experimentally.
It agreed well with a Monte Carlo calculation for the penetration of
protons through a copper slab. In the case of p-p, p-d, and p-^3He
scattering the recoiling associate particle was detected at the conjugate angle in coincidence with the proton to exclude nonelastic
events. For the CH_2 and CD_2-targets the elastic scattering
from carbon was subtracted for each angle. Background event rates
due to the breakup reactions in CD_2 or ^3He were determined by placing
a sufficient amount of absorber in front of the associate counter
telescope to destroy the true p-d or p-^3He coincidences. As an additional check, at several conjugate angles the proton or the deuteron telescopes were varied with angle while the other one was kept
in a fixed position. For the cross section measurements only one set
of associate telescopes was used.

The polarization was measured in a conventional double-scattering
experiment. The polarized proton beam was produced by an internal
carbon target of the synchrocyclotron. This target was placed so that
protons scattered at $(9±1)$ degrees would enter the external beam
transport system. The polarized proton beam was focused into a 1/2"
× 2" spot at the second target with an intensity of $3×10^7$ protons per

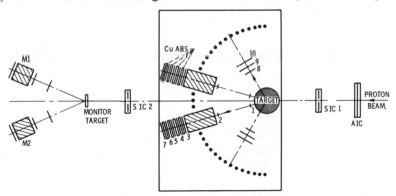

Fig. 1 Schematic diagram of the experimental arrangement.

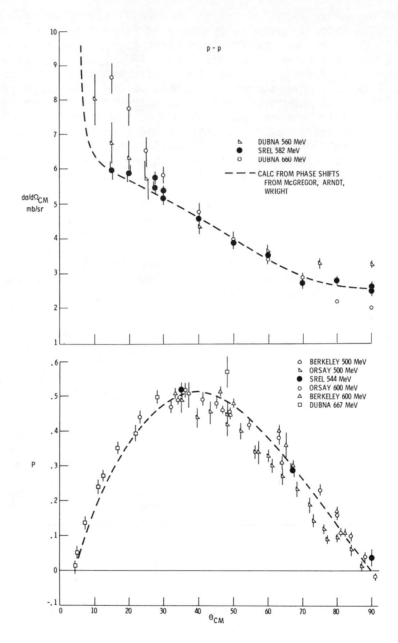

Fig. 2   Differential cross section and polarization in p-p scattering near 600 MeV (Ref. 2).   The dashed curves are the predictions from the phase shift solution at 570 MeV (Ref. 3).

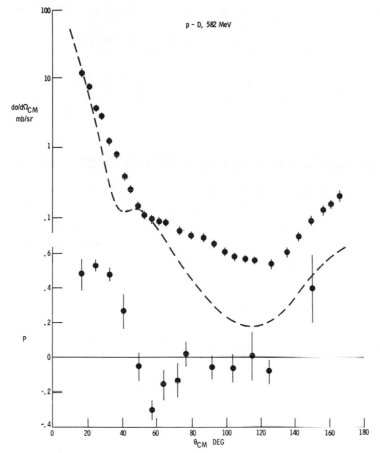

Fig. 3 p-d differential cross section and polarization data ob-
        tained in the present experiment. The dashed curve rep-
        resents the Brookhaven data at 1 GeV (Ref. 4).

second. The beam energy determined by range measurements was 544 MeV
with a spread of 36 MeV FWHM. Two matched telescope systems were positioned at equal angles left and right to record the asymmetries.
The telescopes were interchanged several times during a run, to provide a check on some of the possible instrumental asymmetries. The
overall accuracy of the beam centroid line was estimated to be $\pm 0.03°$,
corresponding to a maximum error of ± 0.010 in the asymmetry. The
beam polarization was determined by measuring the asymmetry in the
scattering from carbon at laboratory angles of $6°$, $8°$, and $10°$ and
from hydrogen at $15°$, using for the analyzing powers the p-^{12}C and
p-p polarization data of Cheng[1]. The beam polarization was found to
be 0.37 ± 0.03.

There is much information in the literature[2] on the polarization
in p-p scattering at our energy but few accurate data exist on differential cross sections. Therefore, we have measured the cross section

between 15° and 90° CM, but taken only a few polarization data, mainly to serve as check points for our beam polarization. The results are shown in Fig. 2 in comparison with the predictions from the phase shift solution for 570 MeV by McGregor, Arndt and Wright.[3]

In Fig. 3 the p-D differential cross section and polarization are shown. The dashed curve represents the Brookhaven p-D data at 1 GeV.[4] The main features, the break in the cross section at forward angles and the peaking at large angles are the same at both energies.

Remler calculated the differential cross section and the polarization in p-D scattering at our energy.[5] Since he has submitted a contribution to this conference, I will say only a few words about his calculation and then show, with his permission, the results. The calculation is based on Watson's multiple scattering equation in the impulse approximation and takes single and "direct" double scattering terms into account. The transition operators describing the free nucleon-nucleon scattering were obtained from the Livermore phase shifts.[3] A typical agreement between the observables predicted by these phase shifts and the experimental data was shown in Fig. 2. For the deuteron wave function, Remler used a fit to the Hamada-Johnston S-state wave function as the sum of three Gaussians. The quadrupole form factor was taken from Glendenning and Kramer with a D-state probability altered to about 7%. The results of these calculations are presented in Fig. 4. The three heavy lines correspond to calculations using the S-state wave function just mentioned. The dotted curve is single scattering only, the dashed one single and double scattering but without the D-state contribution. The solid line is the result when the D-state is included. Although the shape of the cross section and polarization angular distribution is reproduced, the magnitude is not. Since in the beginning there were uncertainties about the experimental normalization of the small angle cross section, a fit to the data beyond the break region was attempted by allowing the computer to vary the shape parameters of the deuteron wave function. An excellent fit to the data at larger angles and, after checking the normalization, also at the smaller angles was obtained with a wave function that is peaked at a smaller radius than the Hamada-Johnston wave function. This may indicate the need for more high momentum components in the deuteron wave function. Since Remler plans to improve upon his calculation this is a tentative result only. Another interesting result of Remler's calculation is the large effect of the D-state component on the polarization.

The comparison of our large angle p-D data with the predictions of the Kerman-Kisslinger model[6] will be presented at this conference in Kisslinger's invited paper.(Session 9)

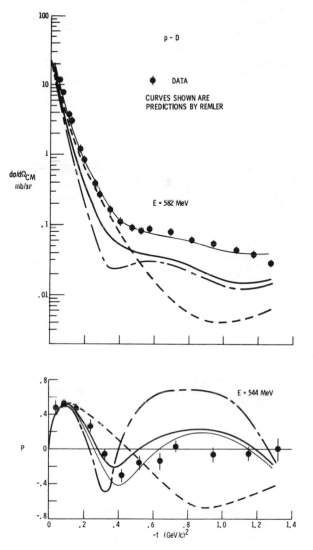

Fig. 4 Comparison of the p-d data with the predictions by Remler
        (Ref. 5). The three heavy lines (dotted, dashed, and
        solid) were obtained with the Hamada-Johnston wave func-
        tion (including single scattering only, single and double
        scattering, and single and double scattering with D-state).
        The light solid line is the full calculation with a mod-
        ified radial distribution in the wave function.

    In Fig. 5 the proton-helium (4) differential cross section and
polarization are shown. The interesting feature is the diffraction
type pattern in both data with the minima coinciding. The curves
are the results of preliminary calculations by Pentz and Ford.[11]

Watson's multiple scattering expansion is used to second order with an approximation in the double scattering term. The spin and iso-spin dependent nucleon-nucleon scattering amplitudes are expressed in Gaussian functions chosen to fit the p-p and n-p cross sections and polarizations. For the α particle wave function one of the forms suggested by Bassel and Wilkin[12] was used which would yield

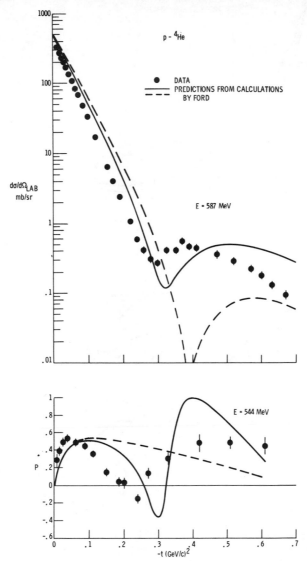

Fig. 5   Comparison of the p-^{4}He data with the predictions by Ford (Ref. 14). The dashed curve corresponds to single scattering only, the solid curve to single and double scattering.

a good fit to the alpha particle form factor recently measured in Stanford.[13] Since higher order scatterings are omitted and the nucleon-nucleon amplitudes are over-simplified there is no good agreement with the data, however, the qualitative features are reproduced. Ford and Pentz also found that the polarization calculated from the Glauber approximation produces too deep a minimum in respect to the data. They take this as a confirmation of the results of a recent paper[14] in which they showed that the Glauber approximation gives good values for the magnitude of the double scattering term but that there is an error in the phase for larger angles. Since, in the interference region the phase relations between the multiple scattering terms are very important one can expect that polarization measurements in this angular range will provide an additional criterion for the phase relations in the various multiple scattering theories.

Figure 6 shows the p-He^3 differential cross section. In the absence of more sophisticated calculations we have followed Czyż and Lesniak's application of the Glauber approximation. The **nuclear density** of ^{3}He was chosen to be the product of Gaussian single-particle densities. The spin and isospin independent scattering amplitudes were chosen to be a parameterized form which is listed in Figure 6

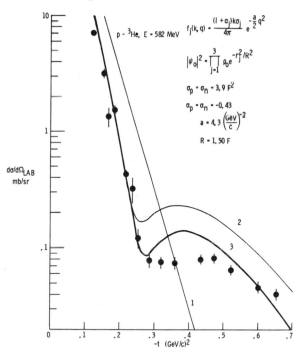

Fig. 6   Comparison of the p-^{3}He data with calculations based on the Glauber approximation. The curves marked 1, 2, 3 indicate that only single (1), single and double (2), and finally also triple scattering is included (3).

k is the momentum of the incident proton, q the momentum transfer, σ the total cross section, α the phase of the nucleon-nucleon amplitude and a the slope of the elastic cross section. σ = 3.9 F^2 is taken from literature[2], a = 4.3 $(GeV/c)^{-2}$ is the average obtained from fitting p-p and p-n data near 600 MeV. The value $|α|$ = 0.43 is the same as was chosen in an earlier calculation to fit the minimum in the p-^{4}He cross section. The curves labeled 1, 2, and 3 are the calculated differential cross sections, if only single (1), single and double (2), and finally also triple scattering (3) is included. Considering the very simple forms used for the elementary scattering amplitude and the ^{3}He-wave function the theoretical prediction is quite good. It remains to be seen, if the lack of a pronounced minimum in the observed cross section can be explained by the spin dependence of the two-body force or if states of higher angular momentum are as important as has been shown for p-d and π-d scattering[7-10].

REFERENCES

1. D. Cheng, UCRL 11926 (1965).
2. M. H. MacGregor, R. A. Arndt, and R. M. Wright, UCRL 50426, (1968).
3. M. H. MacGregor, R. A. Arndt, and R. M. Wright, Phys. Rev. 169, 1161 (1968), Table III.
4. G. W. Bennett, et al., Phys. Rev. Letters 19, 387 (1967).
5. E. A. Remler, William and Mary Report WM-T6, (1966).
6. A. K. Kerman and L. S. Kisslinger, Phys. Review (to be published).
7. D. R. Harrington, Phys. Letters 29B, 188 (1969).
8. E. Coleman and T. G. Rhoades, Effects of D-state in deuterium Scattering, COO-1764-19, (1968).
9. V. Franco and R. J. Glauber, Phys. Rev. Letters 22, 370 (1969).
10. C. Alberi and L. Bertocchi, Phys. Letters 28B, 186 (1968) and Nuovo Cimento 61A, 203 (1969).
11. N. E. Pentz and W. F. Ford, BAPS 14 544 (1969).
12. R. H. Bassel and C. Wilkin, Phys. Rev. Letters 18, 871 (1967) and Phys. Rev. 174, 1179 (1968).
13. R. F. Frosch, et al., Phys. Rev. 160, 874 (1967).
14. W. F. Ford and N. E. Pentz, NASA TM X-52594, (1969).

MEASUREMENTS OF (p,2p) REACTIONS AT 385 MeV

A.N. James

University of Liverpool, U.K.

INTRODUCTION

In reactions where a high energy proton interacts with a nucleus to produce a final state which contains two high energy protons it is normal (1 - 4) to interpret the reaction as a quasi-elastic scattering from a single proton moving within the nucleus. The results of the experiments we have performed at Liverpool are presented in terms of a distorted wave impulse approximation (DWIA) for this knock-out type of reaction.

The cross section for the (p,2p) reaction in DWIA has been given by Jacob and Maris as

$$\frac{d^6\sigma}{dk_1\,dk_2} = \frac{4\hbar^2 c^2 \overline{E_o}^2}{k_o E_1 E_2 E_3} \frac{d\sigma}{d\Omega} P(k)\, \delta(Energy)$$

where the incident proton has momentum $\hbar k_o$ and the two high energy protons in the final state have momenta $\hbar k_1$ and $\hbar k_2$, the momentum of the recoiling core of (A-1) nucleons is $-\hbar k$. The centre of mass proton energy $\overline{E_o}$ and the proton-proton cross section $d\sigma/d\overline{\Omega}$ have been taken to be those corresponding to a system of two protons in the final states k_1 and k_2 (We have used $d\sigma/d\overline{\Omega} = 3.7$ mb sr^{-1}). The momentum and kinetic energy ballancing equations are

$$k = k_1 + k_2 - k_o$$
$$E_B = T_o - T_1 - T_2 - T_{A-1}$$

With a naive single particle model of the target nucleus the energy E_B corresponds to the single particle binding energy and the momentum $\hbar k$ is the momentum of the single particle before it was struck. The function $P(k)$ in a plane wave approximation would be the probability of the struck particle having momentum $\hbar k$ before

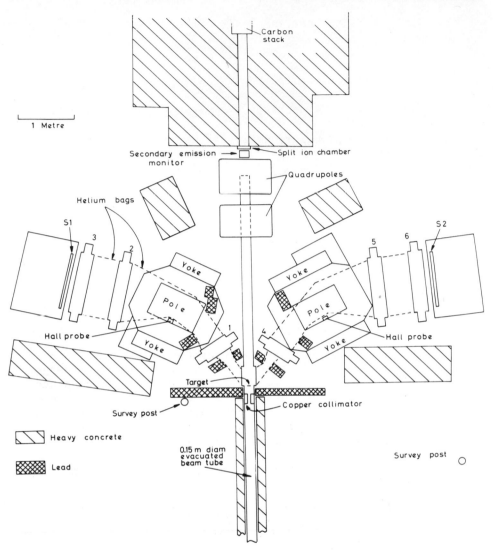

Fig. 1.   Layout of the apparatus used at the Liverpool synchro-cyclotron to study (p,2p) reactions.

the interaction. In the DWIA P($\underset{\sim}{k}$) is called the distorted momentum distribution and it will be reduced in magnitude from the plane wave value by absorption of the proton waves.   Distortion of the waves can also change the effective value of $\underset{\sim}{k}$ by refraction effects, this effect is less important for an experiment at 385 MeV than it has been at lower energies (5).

Experimental measurements of (p,2p) reactions are measurements of $P(\underset{\sim}{k})$ and this is a function of six experimental variables; three components of $\underset{\sim}{k}$, E_B, the ratio $|k_1/k_2|$, and an azimuthal angle. Although this should make it possible to test many of the approximations made in deriving the theoretical expression for the cross section, the practical situation is that the experimental counts are spread so thinly that some further approximation and averaging procedures are necessary. I will discuss this problem after a brief description of our technique.   Before I do this I wish to acknowledge my main co-workers who were Dr. P.T. Andrews and Dr. B.G. Lowe and record the fact that this was the last experiment done using the Liverpool Synchrocyclotron.

EXPERIMENTAL MEASUREMENT

The experimental layout is shown in Fig. 1.   The proton beam was well collimated and its intensity was monitored with a secondary emission chamber, since the targets were thin foils it was not difficult to achieve an accuracy of better than 10% for the cross section scales.   A high energy proton emerging from the target on the left passed through an acoustic spark chamber, the gap (about 10cm) of a simple magnet and then two more acoustic spark chambers. The proton was detected in a 2.5 cm thick plastic scintillation counter.   When a coincidence between the scintillation counters on each side of the beam was detected the spark chambers were triggered and the acoustic delay times and the scintillator pulse amplitudes were recorded on punched paper tape.   From the coordinate data and the magnetic fields it is possible to work out the vectors $\underset{\sim}{k}_1$ and $\underset{\sim}{k}_2$ for the event and then use the pulse heights to identify both particles as protons.

The resolution of this double spectrometer is demonstrated in Fig. 2.   The target used for this data was a deuterated polyethylene foil which contained some hydrogen.   The data displayed is just the simple rate at which events were recorded

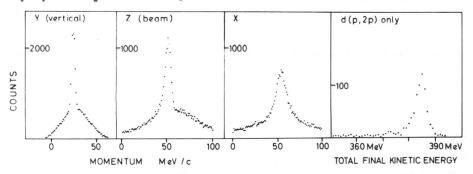

Fig. 2. Experimental resolution in momentum and energy. The momentum resolution determined by free proton proton scattering is better than 15 MeV/c FWHM. The energy resolution is just over 4 MeV FWHM.

within a given recoil momentum ($\hbar$ $\underset{\sim}{k}$) interval.   The sharp peaks
are from free proton proton scattering and show that the momentum
resolution is better than 15 MeV/c FWHM.   The energy spectrum was
obtained by removing events which had a recoil momentum compatible
with free proton proton scattering, it is therefore mainly due to
the d(p,2p)n reaction, the FWHM is just over 4 MeV.

<h2 style="text-align:center">p - STATE IN ^{12}C</h2>

The third figure displays the results we obtained for the
p-state protons in ^{12}C. What has been done here is that the data
has been integrated over the binding energy E_B so as to remove
resolution effects (the p-state is well separated from other
contributions to the reaction) and then the average value of $P(\underset{\sim}{k})$
within cubes 0.1 fm^{-1} to a side in $\underset{\sim}{k}$ space have been evaluated
assuming $P(\underset{\sim}{k})$ to be independent of $|k_1|$ / $|k_2|$.

What one notices first is that there is a minimum in the
distorted momentum distribution exactly at $\underset{\sim}{k}$ = 0.0 and that its
depth,which is greater than earlier measurements,is not in
conflict with theory (6). Gottschalk has already noted (5) at
150 MeV that $P(\underset{\sim}{k})$ depends mainly on the size of $\underset{\sim}{k}$ rather than on
its direction and this is an even better approximation at 385 MeV.
The approximation that $P(\underset{\sim}{k})$ depends only on the size of $\underset{\sim}{k}$ and not
on its direction is consistent with all our data and greatly
simplifies its analysis. We have generally made this assumption.

We treated the region around the peak in $P(\underset{\sim}{k})$ at 0.45 fm^{-1} in
greater detail to attempt to see if $P(\underset{\sim}{k})$ depended on the ratio
$|k_1/k_2|$.   The range of the variables is not large enough to
make a critical test of the impulse approximation but, since we
found a variation less than 10% (consistent with being statistical)
the test does justify our normal procedure of assuming $P(\underset{\sim}{k})$
independent of $|k_1/k_2|$.   We have now reduced the six variables on
which $P(\underset{\sim}{k})$ might depend down to two,the binding energy E_B and the
size of $\underset{\sim}{k}$, the data at this level becomes managable.

One final remark about Fig. 3.   The solid line drawn through
the k_z = 0.0 fm^{-1} data represents the shape of the momentum
distribution for a 1p harmonic oscillator wave function :
$P(\underset{\sim}{k}) \propto k^2$ exp $(-k^2/q^2)$ where q = 0.46 fm^{-1}.   There is no good
reason for doing this except that the harmonic oscillator is a good
zero approximation for a wave function and that the fit to the data
is good. The same shape curve is an equally good fit to the Frascati
^{12}C (e,e'p) data (7) and this can be used according to personal taste
to say that there is little distortion of the shape of the momentum
distribution and/or that there must be distortion in ^{12}C(e,e'p).
You will see that we have used harmonic oscillator wave functions
as a guide to interpreting our data; the fits obtained are better

than we expected.

IDENTIFICATION OF SINGLE PARTICLE STATES

We have taken data for targets of ^{12}C, ^{40}Ca, ^{45}Sc, ^{59}Co, ^{58}Ni, ^{120}Sn, ^{208}Pb and ^{209}Bi and we have reduced the data down to graphs which show how $P(\underset{\sim}{k})$ varies with $\underset{\sim}{k}$ for an average over 5 MeV of binding energy at points separated by 2.5 MeV in E_B.   In Fig. 4 there are samples of such data for ^{40}Ca. $P(\underset{\sim}{k})$ has a sharp maximum at $\underset{\sim}{k}$ = 0 for low binding energies which has been seen before and is characteristic of s-states. At 38.5 MeV there is a very definite

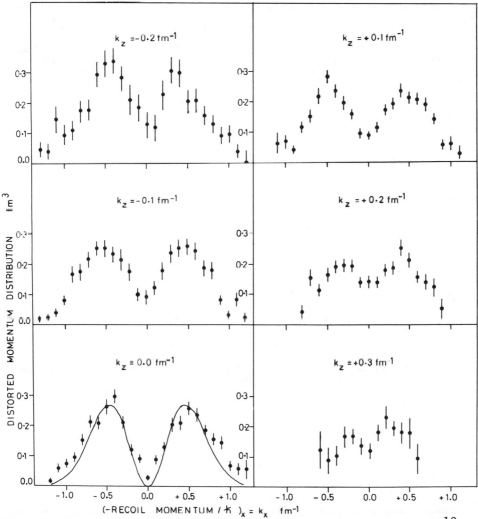

Fig. 3.   Distorted momentum distributions for the p-state of ^{12}C.

minimum in $P(\underset{\sim}{k})$ at $\underset{\sim}{k} = 0$ which indicates a state with $1 \neq 0$.

   The curves shown in fig. 4 are harmonic oscillator momentum distribution shapes $(P_{nl}(\underset{\sim}{k}))$ where the parameter q is chosen to fit the position of the peak in $P(\underset{\sim}{k})$ and the intensity is chosen to look correct on the graph.   The four shapes shown on these graphs are sufficient to fit all the $P(\underset{\sim}{k})$ curves apart from a contribution at high binding energies (60 to 100 MeV) which is roughly constant as a function of $\underset{\sim}{k}$ and E_B. We believe that these four contributions to $P(\underset{\sim}{k})$ are from the 2s,1d,1p and 1s states in ^{40}Ca.   By choosing coefficients A_{nl} in an expansion

$$P(\underset{\sim}{k}) = \sum_{n\ell} A_{n\ell}\, P_{n\ell}(\underset{\sim}{k})$$

for each value of E_B the qualitative behaviour of the strength of these levels as a function of binding energy is found.   The results of an analysis of this sort are shown in Fig. 5, the separation into components is very easy for ^{40}Ca and ^{45}Sc.   For ^{59}Co and ^{58}Ni all the states apart from the 1s state show up distinctively, the addition of the 1s contribution improves the fit to the data but is not essential to fit the data.   The data for ^{120}Sn still shows structure recognisable by harmonic oscillator wave functions at the lower binding energies; both the 1p and 1s contributions

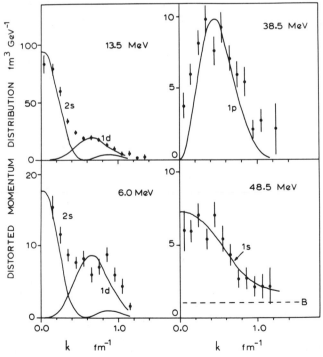

Fig. 4. Some samples of the distorted momentum distributions found for ^{40}Ca. The harmonic oscillator functions shown have $q_{1s} = 0.75$ fm^{-1} and $q_{1p} = q_{1d} = q_{2s} = 0.45$ fm^{-1}

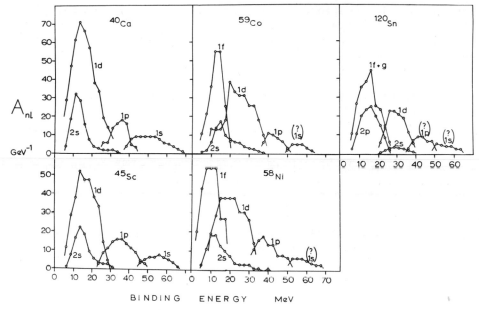

Fig. 5.   The observed variation of the single particle strengths
with binding energy.

improve the fit to the data but again they are not essential to
fit the data.

        When the coefficients A_{nl} are integrated over binding energy
the result using plane wave theory should be the number of active
nucleons belonging to the appropriate single particle state

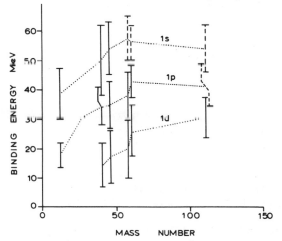

Fig. 6.   The binding energy of single particle levels as a function
of nuclear mass.

Assuming our designations of nl are correct we can therefore estimate the reduction in the cross section caused by distortion. The reduction factors estimated from the data are shown in the table and although the accuracy is poor they do behave qualitatively as expected which supports the single particle assignments.

The result of this experiment is shown in Fig. 6.

Nucleus	REDUCTION FACTORS						$0{\rightarrow}60$ MeV $^{-1}$ $0{\rightarrow}1.2$ fm Errors$\pm$50%)
	1s	1p	1d	2s	1f	2p	
Carbon 12	0.14 $+$.02	0.20 $+$.02					0.22
Calcium 40	0.10	0.034	0.11	0.15			0.10
Scandium 45	0.056	0.038	0.090	0.11			0.09
Cobalt 59	0.029 $+$50%	0.019	0.068	0.10	0.065		0.067
Nickel 58	0.043 $+$50%	0.040	0.076	0.12	0.086		0.080
Tin 120	0.028 $+$70%	0.017	0.030	0.020	0.024*	0.070	0.026

Unless otherwise stated the error in calculating the reduction factors from the data using HO shapes is of the order of 30%. This does not include a 10% error in the cross section measurement or the error introduced by the HO hypothesis. The final column is the reduction factor obtained by integrating the data points $P(\underset{\sim}{k})$ over the given range of $\underset{\sim}{k}$ and E_B.
*This is the average value for the 1f and $1g_{9/2}$ levels.

DISCUSSION follows text of paper entitled "(p,2p) Reactions at 600 MeV" presented by G. Tibell.

REFERENCES

1.   G. Jacob and Th. A.J. Maris, Revs. Mod. Phys. 38 (1966) 121
2.   T. Berggren and H. Tyren, Annual Review of Nuclear Science 16 (1966) 153
3.   M. Riou, Revs. Mod. Phys. 37 (1965) 375
4.   H. Tyren et al. Nuclear Physics 79 (1966) 321
5.   B. Gottschalk, K.H. Wang and K. Strauch, Nuclear Physics 90 (1967) 83
6.   L.R.B. Elton and D.F. Jackson, Phys. Rev. 155 (1967) 1070
7.   Proceedings of Second International Conference on High Energy Physics and Nuclear Structure,Editor G. Alexander, North Holland (1967) page 138.

(p,2p) REACTIONS AT 600 MeV

A.Cordaillat[++], A.Johansson[*], S.Kullander[++], G.Landaud[+],
F.Lemeilleur[+], P.U.Renberg[++], G.Tibell[*], J.Yonnet[+]

+) Université de Caen, ++) CERN, *) University of Uppsala

(Presented by G. Tibell)

INTRODUCTION

In the (p,2p) reaction a nuclear proton is knocked out by an incident proton. Measurements of the momentum vectors of the two outgoing particles, detected in coincidence, make it possible to calculate the separation energy, E_s, of the knocked-out proton, and the recoil momentum, $\vec{q_R}$, of the residual nucleus. This latter quantity can be shown to be closely related to the momentum, $\vec{q}$, of the struck proton before the collision. In the impulse approximation one has $\vec{q} = -\vec{q_R}$. However, various distortion effects disturb this simple model of the reaction, and these effects vary with the energies of incoming and outgoing protons.

If T_o, T_1, T_2 and T_R denote the kinetic energies of incoming and outgoing protons and of the residual nucleus, E_s is defined from

$$E_s = T_o - T_1 - T_2 - T_R \tag{1}$$

The recoil momentum is obtained from the following equation

$$\vec{q_R} = \vec{q_o} - \vec{q_1} - \vec{q_2} \tag{2}$$

In the impulse approximation, the cross section for coincident proton pairs, corresponding to a specific separation energy, can be shown[1] to be given essentially by

$$\frac{d^3\sigma}{d\Omega_1 d\Omega_2 dT_1} = J \times \left(\frac{d\sigma}{d\Omega}\right)^{free} \times \rho(q) \tag{3}$$

where J is a phase space factor and $(\frac{d\sigma}{d\Omega})^{free}$ is the free proton-proton cross section in the center-of-mass system. In the derivation of this simple expression one assumes a completely symmetric experiment, i.e. one in which both energies and scattering angles of outgoing protons are equal. The quantity $\rho(q)$ can be regarded as the momentum distribution of struck protons, more or less distorted depending on the experimental conditions. In certain cases, however, $\rho(q)$ gives rather direct information on the angular momentum, l, of knocked-out protons with a specific value of E_s. A maximum in $\rho(q)$ for $q = 0$ implies a (p,2p) reaction involving a nuclear proton with $l = 0$.

(p,2p) reactions were first studied in Uppsala[2] at 185 MeV and later, in other laboratories[3], at energies ranging from about 50 to 460 MeV. For a few nuclei, experiments have been carried out also at 600 and 1000 MeV. Strongly bound protons or, in terms of the shell model, inner shells, have until very recently been observed only up to ^{16}O.

REASON FOR (p,2p) EXPERIMENTS AT 600 MeV

The main reason for initiating an experiment on (p,2p) reactions at 600 MeV[4] was the hope to obtain information on strongly bound protons for nuclei heavier than ^{16}O. Some results had been obtained from a study of the analogous (e,e´p) reaction[5] but the data suffered from large experimental uncertainties. There are indications, however, that for ls and lp protons E_s grows with atomic number even for the heaviest nuclei studied. The inherent limitations in the electron experiments make it important to investigate strongly bound protons also with (p,2p) experiments.

At 600 MeV the conditions for studying inner shells should be quite favourable, mostly due to the fact that the distortion of outgoing protons is rather small. Inasmuch as the distortion implies an absorption of emitted particles this fact becomes very important since the hole states produced by knocking out strongly bound protons have a short life time and therefore are very broad.

In addition to having access to a favourable proton energy it was possible at the CERN synchrocyclotron to set up an experiment with large acceptance in energies and angles for the outgoing protons. The high quality of the external proton beam as well as the good resolution of the spark chambers used for tracing the magnetically analysed particles made it possible to obtain a high event rate without loosing resolution in energy and angle.

EXPERIMENTAL ARRANGEMENTS

The experimental arrangement is shown schematically in fig. 1.

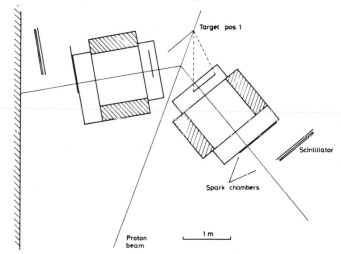

Fig. 1 - Schematic
view of the experi-
mental arrangements.

Coincident protons are detected in scintillators, and the coincidence signal triggers the wire spark chambers which are equipped with
magnetostrictive readout. For each event the spark coordinates, the
pulse heights in four scintillators and the difference in time-of-
flight between the two particles are recorded on magnetic tape. In
the computer analysis of recorded events it is possible to sort out
events which are not due to (p,2p) reactions in the target. Accidental coincidences are sampled in the usual way by delaying one
branch by a time corresponding to the extraction frequency of the
accelerator. Such events are also recorded and can be analysed separately to obtain an idea of their distribution in the energy spectra. Compared with previous (p,2p) experiments the accidental rates
are very low in this experiment, mostly due to the large solid angles used.

PRELIMINARY RESULTS

So far (p,2p) reactions have been studied in ^{12}C, ^{27}Al, ^{28}Si, ^{40}Ca
and ^{90}Zr. For testing the method of analysis we have also investigated free proton-proton collisions. About 300.000 events have been
recorded for each of the target nuclei mentioned. Separation energy
spectra have been plotted, and they show a total energy resolution
of about 7 MeV (FWHM). In the region of E_s where comparisons can be
made our results agree well with those obtained at lower energies.
In addition, they show quite clearly that it is possible, at 600
MeV, to observe also inner shell protons for the heavier nuclei.

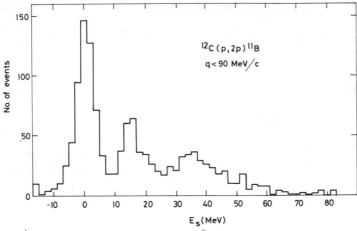

Fig. 2 - Separation energy spectra for the $^{12}C(p,2p)^{11}B$ reactions at 600 MeV.

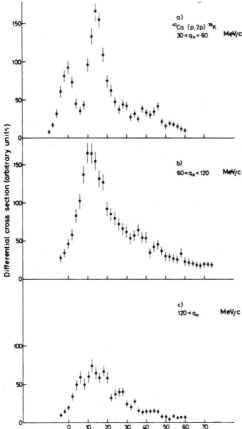

Fig. 3 - Separation energy spectra for the $^{40}Ca(p,2p)^{39}K$ reaction at 600 MeV.

Examples of the energy spectra obtained are presented in Figs. 2 and 3. Fig. 2 shows the E_s spectrum for the $^{12}C(p,2p)^{11}B$ reaction. Only such events in which q is less than 90 MeV/c have been included. The three peaks correspond to free proton-proton scattering from a polyethylene foil mounted on the target and to the knock-out of p and s protons from ^{12}C, respectively. The latter peaks have been observed in previous experiments at lower energies, and the corresponding E_s values agree very well with earlier results.

The data of this experiment in progress will eventually be presented in the form of separation energy spectra for certain values of q_R. By following the dependence on q one hopes to obtain information on the angular momentum, l, of protons belonging to various subshells, characterized by a certain value of the binding energy, E_s. An example of such a family of curves is given in fig. 3 which shows the preliminary and uncorrected energy spectra obtained for (p,2p) reactions in ^{40}Ca for four regions of q_R. The angular acceptance in this case was $37°$ to $43°$ on either side and the energy band went from 200 to 400 MeV for outgoing particles, assuming they were protons. When more data have been analysed one can, of course, decrease the size of the q channels in order to bring out better the differences in the shapes of the spectra. In fig. 3 the peak at E_s = 0 comes from free proton-proton scattering in the polythene foil mounted for energy calibration purposes. The next peak, at E_s equal to about 10 MeV, is mostly due to the knocking out of 2s protons. This is well known from work at lower energies[6] just as is the fact that for large values of q_R (see fig. 3c) the 2s peak dissappears and gives way to contributions from the 1d states at about 8 and 15 MeV, respectively. Above a value of E_s of about 25 MeV the spectra of fig. 3 contain new information on (p,2p) reactions in ^{40}Ca. It is, however, too early now to draw quantitative conclusions from the behaviour of the spectra.

[1]Th.A.J. Maris, Nucl. Phys. 9 (1958/59) 577.
[2]H. Tyrén, P. Hillman and Th.A.J. Maris, Nucl.Phys. 7 (1958) 10.
[3]For an extensive review, see G. Jacob and Th.A.J. Maris, Rev. Mod. Phys. 38 (1966) 121.
[4]A. Johansson and G. Tibell, PH III-67/7, CERN, letter of intention. A. Johansson, G. Tibell, M. Schérer, M. Landaud, A. Cordaillat and S. Kullander, PH III-67/29, CERN proposal.
[5]For references to the (e, e´p) work, see U. Amaldi, Jr., Rendiconti della Scuola Internazionale di Fisica "E. Fermi" - XXXVIII Corso (1967) p. 284.
[6]G. Tibell, O. Sundberg and H. Miklavzic, Phys. Lett. 2 (1962) 100 H. Tyrén, S. Kullander, O. Sundberg, R. Ramachandran, P. Isacsson and T. Berggren, Nucl. Phys. 79 (1966) 321.

DISCUSSION

A.N. James (in reply to questions by H. Feshbach): It is not possible at this stage, to discuss the A-dependence of the binding energy in the neighbourhood of ^{40}Ca. Unfortunately, there are no measurements between ^{12}C and ^{40}Ca. The observation of the 1s-states in Ni, Co and heavier elements is not completely convincing. A.N.J. (in reply to P. Radvanyi): The errors in the energies given for the inner shells are not more than 2 or 3 MeV.

A.N.J. (in reply to a question from N.S. Wall): The results of these experiments for the 1s-states, seem to be in general agreement with the corresponding (e,e' p) experiments. However, a feature of these measurements is the marked spread in momentum, which extends up to the highest binding energies. This is probably due to the nature of the (p,2p) reaction itself -- the result of some multiple interaction process -- and is not some sort of background.

EFFECT OF THE DEUTERON'S D-WAVE COMPONENT IN HIGH-ENERGY SCATTERING

David R. Harrington

Rutgers, the State University

New Brunswick, New Jersey

The electric quadrupole moment of the deuteron was discovered here at Columbia University about 30 years ago.[1] The existence of this moment showed that the deuteron wave function has a D-wave component due to a tensor force in the neutron-proton interaction.[2] In the standard notation a deuteron with angular momentum component $J_z = M$ has the wave function

$$\psi_M(\underline{r}) = \frac{1}{\sqrt{4\pi}\, r}\left[u(r) + \frac{S_{12}(\hat{\underline{r}})}{\sqrt{8}}\, w(r)\right]\chi_{1,M}$$

where

$$S_{12}(\hat{\underline{r}}) = 3(\underline{\sigma}_1\cdot\hat{\underline{r}})(\underline{\sigma}_2\cdot\hat{\underline{r}}) - \underline{\sigma}_1\cdot\underline{\sigma}_2$$

is the tensor operator and u and w are the S- and D-wave radial wave functions respectively. The deuteron's form factors are defined by the relation

$$\int d^3r\, \psi_{M'}^{\dagger}(\underline{r})\, e^{-i\underline{q}\cdot\underline{s}}\, \psi_M(\underline{r})$$

$$= \chi_{1,M'}^{\dagger}\left\{ S_0(\delta) - \left[3(\underline{J}\cdot\hat{\underline{\delta}})^2 - 2\right]\frac{S_2(\delta)}{\sqrt{2}}\right\}\chi_{1,M} ,$$

where

$$S_0(\delta) = \int_0^{\infty} dr\, j_0(\delta r)\left[u^2(r) + w^2(r)\right]$$

is the spherical or monopole form factor and

$$S_2(\delta) = \int_0^{\infty} dr\, j_2(\delta r)\, 2\, w(r)\left[u(r) - \frac{1}{\sqrt{8}}\, w(r)\right]$$

is the quadrupole form factor. These two form factors
have been studied extensively in connection with elastic electron deuteron scattering.[3] As the momentum
transfer δ increases from zero S_0 decreases rapidly
from one, while S_2 increases rapidly from zero with the
result that $S_2 > S_0$ for $\delta > 3$ F^{-1} ($\delta^2 \gtrsim 0.36$ $(GeV/c)^{-2}$).
It should be noted that, given the deuterons quadrupole
moment and the one-pion-exchange potential for large r,
S_2 is relatively model independent up to this momentum
transfer.

When the first calculations of high-energy pion-
deuteron and proton-deuteron scattering were made
using the Glauber approximation, the D-wave component
of the deuteron was completely ignored, which was
quite reasonable at the small momentum transfers then
considered. As experiments and calculations were
carried to higher momentum transfers, however, this
neglect was continued, with the result that for a time
there was an apparent discrepancy between Glauber
theory and experiment. Last year at least three sets
of people finally remembered the D-wave and found that
it eliminated the spurious dip which had been appearing
in the calculations.[4]

By now fairly detailed calculations including the
D-wave have been carried out for pion-deuteron[5,6] and
proton-deuteron[7] scattering. The fits to experiment
are quite good, especially considering the rather large
angles involved and the simplifying assumptions made
concerning the pion-nucleon and nucleon-nucleon amplitudes. Similiar calculations have also been found
to give good fits to the results of deuteron-deuteron
scattering experiments, and here again the D-wave is
important in the dip region.[8]

A very interesting feature of these calculations
is that in the break region ($\Delta^2 = 0.3$-0.4 $(GeV/c)^2$)
the scattering amplitude is dominated by the quadrupole term and is thus strongly dependent on the
deuteron's spin. We can see this by examining the
form of the complete scattering amplitude for momentum
transfer $\underset{\sim}{\Delta}$ in the Glauber approximation:

$$F(\underset{\sim}{\Delta}) = F_{10}(\underset{\sim}{\Delta}) + F_{20}(\underset{\sim}{\Delta}) + F_{12}(\underset{\sim}{\Delta}) + F_{22}(\underset{\sim}{\Delta}),$$

where, denoting the nucleon scattering amplitudes by
$f(\Delta)$ and the deuteron spin operator by $\underset{\sim}{J}$,

$$F_{10} = 2 f(\Delta) S_0(\tfrac{1}{2}\Delta) \, ,$$

$$F_{20} = \frac{i}{2\pi} \int d^2\delta \; f(\tfrac{1}{2}\Delta + \underset{\sim}{\delta}) f(\tfrac{1}{2}\Delta - \underset{\sim}{\delta}) S_0(\delta) ,$$

$$F_{12} = -\left[3(\underset{\sim}{J}\cdot\hat{\underset{\sim}{\Delta}})^2 - 2\right] 2 f(\Delta) \frac{1}{\sqrt{2}} S_2(\delta) ,$$

and

$$F_{22} = \left[3(\underset{\sim}{J}\cdot\hat{\underset{\sim}{k}})^2 - 2\right] \frac{i}{2\pi} \int d^2\delta$$
$$\times f(\tfrac{1}{2}\Delta + \underset{\sim}{\delta}) f(\tfrac{1}{2}\Delta - \underset{\sim}{\delta}) \frac{1}{\sqrt{8}} S_2(\delta) .$$

In the break or dip region F_{10} and F_{20} nearly cancel, F_{12} is of roughly the same magnitude, while F_{22} is considerably smaller. In this region, therefore, the scattering amplitude F is almost proportional to $3(J \cdot \hat{\Delta})^2$ - 2 and thus strongly spin-dependent. It should be emphasized that this spin dependence is present even if the nucleon scattering amplitudes are completely spin-independent. It arises because the orientation of the anisotropic deuteron spatial wave function is determined by the orientation of the deuteron's spin.

Franco and Glauber[7] have illustrated one conse- quence of this spin dependence by calculating the differential cross section for the scattering of pro- tons from polarized deuteron targets. The cross sec- tions in the break region are very sensitive to the state of polarization of the target, and dips can be made to come and go and move about.

I have made closely related calculations for the production of aligned deuterons from randomly oriented initial state deuterons.[9] As shown in Fig. 1, the alignments $A_i = \langle 3J_i^2 - 2 \rangle$ for the momentum transfer (Δ), beam (k), and normal (n) directions become quite large in the break region. In particular, the final deuteron is produced in what is very close to a pure $J_\Delta = 0$ state for $\Delta^2 \approx 0.27$ (GeV/c)^2. This prediction could be checked by analyzing the recoil deuteron's orientation[10] in an experiment with a deuteron target. Perhaps more interesting, however, would be a double scattering experiment with a high energy deuteron beam

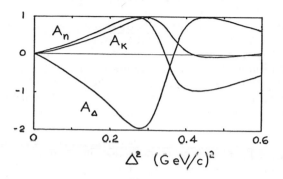

Figure 1
The alignments of A_j of final state deuterons produced
in high-energy proton-deuteron scattering at momentum
transfer Δ from randomly oriented initial state
deuterons. The subscripts indicate the alignment axes:
Δ is the momentum transfer, K the incident momentum,
and n the normal to the scattering plane.

and proton targets. If the first scattering were at
about $\Delta^2 = 0.27$ $(GeV/c)^2$ then the second scattering
would have a strong azimuthal dependence near the same
momentum transfer.

In addition to providing a detailed check of the
Glauber method, these D-wave effects might provide a
useful source of high-energy aligned deuterons. Because there is only alignment, but no polarization,
however, it will take some subtlety to invent interesting experiments which can be done with these beams.

The author would like to thank Prof. M. Bazin for
stimulating conversations.

[1]J.M.B. Kellogg, I.I. Rabi, N.F. Ramsay, Jr., and J.R. Zacharias, Phys. Rev. 55, 318 (1939).
[2]J.M. Blatt and V.F. Weisskopf, Theoretical Nuclear Physics (J. Wiley and Sons, New York, 1952), p.94.
[3]See, for example, J.A. McIntyre and S. Dhar, Phys. Rev. 106, 1074 (1957); N.K. Glendenning and G. Kramer, Phys. Rev. 126, 2159 (1962); and J.E. Elias et al, Phys. Rev. 177, 2075 (1969).
[4]D.R. Harrington, Phys. Rev. Letters 21, 1496 (1968); E. Coleman and T.G. Rhoades (unpublished); and C. Michaeland C. Wilkin (unpublished).
[5]C. Michael and C. Wilkin, Nuc. Phys. B11, 99 (1969).
[6]M. Fellinger et al, Phys. Rev. Letters 22, 1265 (1969).
[7]V. Franco and R.J. Glauber, Phys. Rev. Letters 22, 370 (1969).
[8]A.T. Goshaw, P.J. Oddone, M.J. Bazine, and C.R. Sun, contribution to this conference.
[9]D.R. Harrington, Phys. Letters 29B, 188 (1969).
[10]J.Button and R. Mermod, Phys. Rev. 118, 1333 (1960).

Dr. Claude Bloch: I would like to end the session by perhaps making a few remarks about models.

I was a little disturbed about Gugelot's remark that one should not fall into the pitfall of making models as in low energy nuclear physics. That would put all theoreticians out of business, of course. But certainly, nowadays, there might be some more strigent requirements. Perhaps one should set for all models two conditions.

The first one is that a model should take into account all the experimental data for which it is relevant; and one should also ask what the same model does for other kinds of experiments.

The second condition is perhaps more difficult to satisfy, and even to formulate. An example is the problem discussed by Isabelle and others. To explain some experimental feature, one introduces a particular kind of ingredient into a model. In this particular case it is two-body correlations. Now one must ask whether the experimental data is really directly connected or not with this particular ingredient. In the particular case of electron scattering, for instance, it is not. In other words, it is true that introducing a new interaction into a model modifies every calculation and with sufficient, even quite reasonable, parameters one can get good agreement with experiments. But is the agreement so obtained really significant? Introducing many other different ingredients into the model might not one also obtain agreement? It is very difficult in practice to answer this kind of question; one would really have to make and test many other models - which requires a lot of imagination, and work!

SMALL ANGLE NUCLEON-DEUTERON AND NUCLEON-NUCLEON SCATTERING

L.M.C. Dutton and H.B. van der Raay

Department of Physics, University of Birmingham

England

A series of experiments have been carried out to study pp and pd scattering in the angular range 20 - 70 mr in the laboratory system, at incident momenta of 1.29, 1.39, 1.54 and 1.69 GeV/c.(1-3). These data were recorded with a magnetic spectrometer using sonic spark chambers to define the trajectories of the scattered particles. A system of fast scintillation counters was used to detect the incident particles in coincidence with the scattered particles thus providing a trigger for the spark chambers and enabling absolute cross sections to be determined. The data from the sonic spark chambers was recorded onto magnetic tape by an automatic data handling system in a format suitable for analysis by a computer.

The scattered particle trajectories were determined from the known geometry of the system and the basic spark chamber data. This gave an angular resolution of $\pm$ 1 mr and an overall momentum resolution, including the incident beam spread, of $\pm$ 1%. Good rejection of inelastic events was obtained although elastic and 'quasi-elastic' events could not be resolved when considering pd interactions.

The pp data were analysed to yield α_p, the ratio of the real to imaginary part of the spin independent forward scattering amplitude, by considering the interference between the Coulomb scattering amplitude and the nuclear amplitude. The imaginary part of the nuclear amplitude at zero degrees was obtained from the total cross section using the optical theorem and the angular dependence was expressed as a simple exponential in terms of the momentum transfer squared. A factor, β_p, was also included to

approximate the effects of spin dependent terms. This
parameterization was used to analyse published data on small angle
pp scattering and good agreement was found with the analyses of
previous authors and where applicable with the results of phase
shift analyses, for both α_p and β_p.

The present experimental data were then analysed and compared
with the predictions of dispersion relation calculations (4).
(Fig. I). Although the value of α_p agrees well with these
calculations at 1.69 GeV/c, at lower momenta there is a marked
disagreement which is supported by some earlier Russian data. It
is interesting to note that the apparent discrepancy occurs over
the momentum region where the $N*\frac{3}{2}$ (1236) dominates the pp
interaction. When considering the np system, no departure from
the predictions of dispersion relations was found.

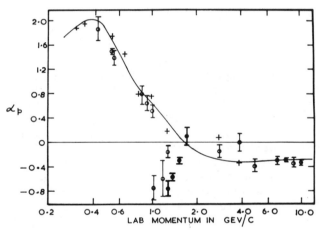

Fig. I. Dependence of α_p on lab momentum. Curve is dispersion
relation calculation + phase shift results, ⏀ previous experiments
⏀ present experiment.

The pd data were analysed according to the Glauber formalism,
by writing the differential cross section in terms of the nucleon-
nucleon amplitudes in a form similar to that used in the pp
analysis taking into account interference and double scattering
terms (5). The values of the parameter α_n were determined from
these analyses and compared with dispersion relation calculations.(6)
(Fig. II). In the present momentum range an ambiguity exists in
these calculations due to a discrepancy between the 0.82 and the
1.25 GeV/c phase shift solutions. The results are in good agreement with the dispersion relation calculations based on the 0.82
GeV/c phase shift analysis and clearly rule out the other.

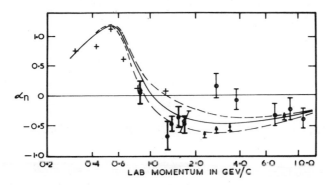

Fig. II.   Dependence of α_n on lab momentum. Curve is dispersion relation prediction, ┼ phase-shift results, ▲ lower limits from charge exchange data, ⓧ previous experiments, ⬤ present experiment.

Finally, data were also obtained on elastic dp scattering at 1.69 GeV/c which is equivalent to pd scattering at 0.845 GeV/c (7). These data, when analysed, gave a value for α_n in good agreement with the phase shift solution at 0.82 GeV/c and consequently with the predictions of dispersion relations.

References

1.   L.M.C. Dutton, R.J.W. Howells, J.D. Jafar and H.B. van der Raay, Phys. Letters, 25B, 245 (1967).
2.   L.M.C. Dutton and H.B. van der Raay, Phys. Letters 26B, 679 (1968).
3.   L.M.C. Dutton and H.B. van der Raay, Phys. Rev. Letters, 21, 1416 (1968).
4.   P. Söding, Phys. Letters 8, 285 (1964).
5.   D.R. Harrington, Phys. Rev. 135, B358 (1964).
6.   D.V. Bugg and A.A. Carter, Phys. Letters 20, 203 (1966).
7.   L.M.C. Dutton, R.J. Howells, J.D. Jafar and H.B. van der Raay, Nucl. Phys. B9, 594 (1969).

QUASI-ELASTIC (p, px) REACTIONS ON LITHIUM AND CARBON AT 156 MeV

D. Bachelier, M.K. Brusel[+], P. Radvanyi, M. Roy,

M. Sowinski[++], M. Bernas and I. Brissaud

Institut de Physique Nucléaire, 91-Orsay, France

We have studied the ^{6}Li (p, p ^{3}He) ^{3}H, ^{7}Li (p, pt) ^{4}He and ^{12}C (p, pd) ^{10}B reactions at 156 MeV with kinematical conditions chosen so as to favour quasi-elastic scattering. The proton and the charged particle x were detected in coincidence; the angle and energy of the proton were fixed, and measurements were performed on particle x at several angles around the angle corresponding to q(recoil) = 0. The protons were analysed with a magnetic spectrometer followed by a plastic scintillator telescope; the particles x were analysed in energy and identified by $\Delta E/\Delta x$ - E silicon solid state detector telescopes; identification was performed with an analogue Goulding identifier or with an on-line computer system.

The ^{6}Li (p, p ^{3}He) ^{3}H angular correlation (figure 1 a) shows a characteristic S shape[1] and is as narrow as the ^{6}Li (p, pd) ^{4}He angular correlation measured by Ruhla et al.[2] The ^{7}Li (p, pt) ^{4}He correlation (figure 1 b) exhibits a neat characteristic L $\neq$ 0 (most likely L = 1) shape; the right wing appears however to be higher than the left one; a detailed analysis is under way. The free p-x scattering differential cross sections are comparable, in the angular range of interest, for these three reactions[3], and a simple comparison can be made in plane wave impulse approximation. The ^{6}Li (p, p ^{3}He) and ^{7}Li (p, pt) reactions have about the same maximum differential cross sections for the kinematical conditions chosen and the ^{6}Li (p, pd) reaction is about 15 times stronger (for a somewhat larger proton scattering angle); but, because of its P shape, the mean q^2 (recoil) value in ^{7}Li (p, pt) is much larger than in the other two reactions, so that - in this approximation - the clustering of ^{7}Li into α + t should be smaller but not very different from the clustering of ^{6}Li into α + d.

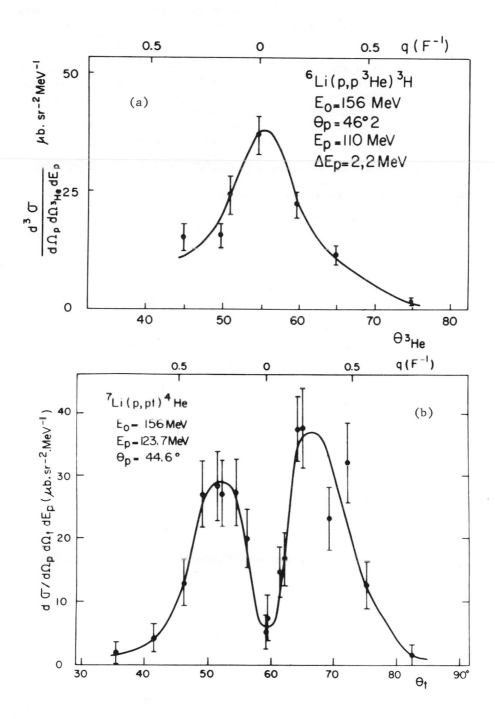

A comparison can also be made with experiments at lower energy.[3] At 55 MeV for $\Theta p = 75^{\circ}$ the maximum cross-section for ^7Li (p, pt) is only a little larger than at 156 MeV.

Energy spectra for ^{12}C(p, pd) ^{10}B at 156 MeV have been measured at q (recoil) = 0 and q (recoil) = 0.38 F^{-1}. The cross-sections are very low. It appears that the ground state of ^{10}B and perhaps some of the first excited low lying states of ^{10}B are the most strongly excited; some excitation appears also around 13-15 MeV in ^{10}B, confirming the predictions of Balashov, Boyarkina and Rotter[4].

- - - - - - - - - -

+ On leave from the University of Illinois, Urbana.
++ On leave from the Institute of Nuclear Science, Swierk (Poland).

[1]D. Bachelier, M. Bernas, C. Detraz, P. Radvanyi and M. Roy, Physics Letters, <u>26B</u> (1968) 283.
[2]C. Ruhla, M. Riou, J. P. Garron, J.C. Jacmart and L. Massonnet, Physics Letters <u>2</u> (1962) 44; C. Ruhla, M. Riou, M. Gusakow, J.C. Jacmart, M. Liu and L. Valentin, Phys. Letts. <u>6</u> (1963) 282.
[3]K. Kuroda, A. Michalowicz and M. Poulet, Nucl. Phys. <u>88</u> (1966) 33; H. Postma and R. Wilson, Phys. Rev. <u>121</u> (1961) 1229; P. Narboni and H. Langevin, private communication.
[4]D.L. Hendrie, M. Chabre and H.G. Pugh, U.C.R.L. Report No. 16.580, (1966) p. 146 (unpublished); P.G. Roos, H. Him, M. Jain and H.D. Holmgren, Phys. Rev. Lett. <u>22</u> (1969) 242.
[5]V.V. Balashov, A.N. Boyarkina & I. Rotter, Nucl.Phys. <u>59</u>(1964) 417.

EFFECT OF CORRELATIONS ON HIGH-ENERGY PROCESSES USING NUCLEAR TARGETS[*]

E. J. Moniz,[†] G. D. Nixon,[†] and J. D. Walecka

Institute of Theoretical Physics, Department of Physics

Stanford University, Stanford, California 94305

Recently a great deal of effort has gone into the problem of using nuclei as targets with known properties in high-energy scattering processes in order to deduce information on hadron-nucleon amplitudes. All of this work essentially follows from the Glauber scattering theory[1] together with a complex optical potential describing the nuclear target. The simplest form of the optical potential is $(2m_p/\hbar^2)V_{opt}(x) = -4\pi A\rho(x)f(0)$ where ρ is the single nucleon density and $f(0)$ is the elementary scattering amplitude. This result assumes that the ground-state density of the target factors into a product of single-particle densities and neglects correlations. The first correction to this result depends on two-particle correlations in the ground-state density, $\Delta(\underset{\sim}{x},\underset{\sim}{y}) = \rho^{(2)}(\underset{\sim}{x},\underset{\sim}{y})-\rho(\underset{\sim}{x})\rho(\underset{\sim}{y}) \equiv \rho(\underset{\sim}{x})\rho(\underset{\sim}{y})(g(|\underset{\sim}{x}-\underset{\sim}{y}|)-1)$ and takes the approximate form[1,2]

$$\frac{2m_p}{\hbar^2} V_{opt}(x) = -4\pi A\rho(x)f(0)[1+(2\pi i/k)A\rho(x)f(0)\int_0^\infty (g(r)-1)dr]$$

In addition, by summing the cross section for excitation of the nucleus over all final nuclear states, the incoherent cross section in many cases takes the form

$$\sum_{f\neq i} [d\sigma/dt]_{f_i} = A_{eff}(d\sigma/dt)_{1-body} + (d\sigma/dt)_{corr.}$$

where the correlation cross section depends on this same function $\Delta(\underset{\sim}{x},\underset{\sim}{y})$.

We have calculated the two-body density by writing

$$\rho^{(2)}(x,y) = (1/A^2)\sum_{ij < k_F} |\psi_{ij}^{B.G.}(\underset{\sim}{x},\underset{\sim}{y})|^2$$

where $\psi^{B.G.}$ is the solution to the Bethe-Goldstone equation for two interacting nucleons in the presence of an infinite nuclear medium. This independent pair approximation forms the basis of theoretical work on the properties of nuclear matter. The two-body correlation function used is sketched in Fig. 1. It was

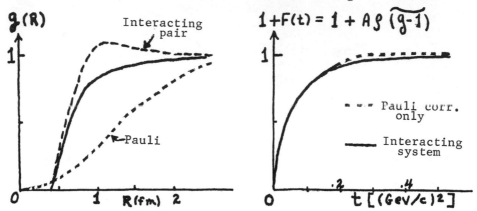

Fig. 1.   Correlation Function and Its Fourier Transform.

computed from a potential containing a hard core repulsion and an exponential attraction fit to low energy scattering.

The qualitative features of our results can be understood if we assume that $f(0)$ is purely imaginary and use the parameters appropriate to nuclear matter, for then the optical potential takes the form

$$(2m_p/\hbar^2)V_{opt}(x) = -ikA\rho(x)\sigma_{eff}$$

$$\sigma_{eff} = \sigma(1+.063f^{-2}\sigma)$$

The anti-correlations increase the effective cross section presented to the incoming projectile.

We have considered the following processes:
1. We have analyzed high-energy N-nucleus (19.7 GeV [3]) and π-nucleus (3 GeV [4]) scattering and have extracted the elementary cross section following the procedure of Drell and Trefil.[5] The results are indicated in Table I. We have used both a Woods-Saxon[6] and uniform single-particle density with $R = 1.3A^{1/3}f$ in arriving at these results.
2. We have re-analyzed the SLAC data on photoproduction of ρ_0 at 8.8 GeV[7] again using the Drell-Trefil procedure. These results are also shown in Table I. As a by-product of this analysis, the γ-ρ coupling constant can be determined from the relation

$$\gamma_\rho^2/4\pi = (\alpha/64\pi)\sigma_{\rho N}^2/[d\sigma(t=t_{min})/dt]_{1-body}.$$

	Woods-Saxon		Constant Density		
	corr.	no corr.	corr.	no corr.	Exp.
N-N	37 ± 5	42 ± 7	38 ± 6	43 ± 7	39.5
π-N	19 ± 9	20 ± 10	20 ± 8	22 ± 10	28.5
ρ^0-N	$26 {+8 \atop -3}$	$30 {+9 \atop -5}$	$26 {+9 \atop -4}$	$30 {+10 \atop -6}$	

Table I. Extracted cross sections in mb. These numbers give the best curves for the A-dependence and pass through all the data points.[8] The quoted errors indicate the range of values for which the curves pass through the experimental error bars.

If the hydrogen data is used for $[d\sigma/dt]_{\text{1-body}}$, any change in $\sigma_\rho N$ is magnified by a factor of 2. In fact, however, this may not be the most model independent way of using all the nuclear data to extract γ_ρ^2.

3. We have analyzed the incoherent photoproduction of ρ_0 and our result takes the form

$$(d\sigma/d\Omega)_A/A_{\text{eff}}(d\sigma/d\Omega)_{\text{1-body}} = 1 + F(q)$$

where the effect of correlations enters both in the calculation of A_{eff} and in the function $F(q)$. This latter function, computed for nuclear matter, is sketched in Fig. 1 for both a non-interacting Fermi gas and for the interacting system. It is clear that the incoherent cross section is not sensitive to the dynamical correlations except through A_{eff}.

*Research sponsored under AFOSR Contract No. F44620-68-C-0075.
†National Science Foundation Predoctoral Fellow.
[1] R. J. Glauber, Lectures in Theoretical Physics, Vol 1 (Interscience, New York, 1959).
[2] L. L. Foldy and J. D. Walecka, Annals of Phys. (to be published).
[3] S. D. Drell and J. Trefil, Phys. Rev. Letters 16, 552 (1966).
[4] G. Bellettini et al. Nucl. Phys. 79, 609 (1966).
[5] M. J. Longo and B. J. Moyer, Phys. Rev. 125, 701 (1962).
[6] R. J. Glauber and G. Matthiae, Instituto Superiore di Sanita, Laboratori di Fisica, Report No. ISS 67/16, (unpublished).
[7] F. Bulos et al., Phys. Rev. Letters 22, 490 (1969).
[8] The Ag point in Ref. 7 has been ignored. The result quoted in Ref.

7 is $\sigma_{\rho N} = 30 {+6 \atop -4}$ mb.

INVESTIGATION OF THE NUCLEON DEUTERON SYSTEM

Edward A. Remler

College of William and Mary, Department of Physics

Williamsburg, Virginia 23185

To the extent that off shell effects may be neglected, a reasonably good description of Nucleon-Deuteron elastic scattering in the forward hemisphere should be possible when one knows the deuteron wave and the two nucleon amplitudes. The upper limit of knowledge of the N-N amplitudes[1] lies at present slightly above the onset of inelasticity ($\sim$0.7 Gev). For high enough bombarding energy it seems physically reasonable that to a good approximation no target constituent will be hit more than once during an elastic scattering process. This idea is already contained in the eikonal approximation so that the success of such calculations[2] lends support to it. In terms of the multiple scattering expansion this implies neglect of triple and higher order contributions.[3] On the other hand, at low enough energies this approximation is clearly invalid.[4] Thus there exists a transition energy above which things simplify considerably in that the multiple scattering series may be truncated and one computes only four types of terms; (1) single scattering, (2) (direct) double scattering, (3) double charge exchange, (4) pickup. This transition energy may be reasonably supposed to be in the neighborhood $\leq$ 0.2 Gev.

Computations of reasonably good accuracy in the aforementioned range, 0.2 Gev $\leq$ lab kinetic energy $\leq$ 0.6 Gev, are therefore now feasible while at the same time an increasingly large body of experimental cross section and polarization data exists with which comparison may be made. One may hope to expect in this manner to obtain new, detailed, and independent information about the hadronic structure of the deuteron as well as any of the other elements in the theoretical mix used to produce these predictions. This procedure would also appear to be a necessary basis for any more than qualitative use of intermediate and high energy scattering of hadrons on nuclei to investigate nuclear structure; to a somewhat lesser extent this applies to its use as a tool in elementary particle physics.

About one year ago I made a first step in the direction outlined above by attempting to theoretically reproduce the forward hemisphere of the p-d data near 0.570 Gev then being obtained at VARC.[5]   The results were compared with the preliminary data of the experimentalists in an unpublished report[6] which contains a more complete description of the calculation than can be given here.   No acceptable agreement could be obtained.   In the process the computer was asked to seek a best fit to a subset of the data excluding the very forward direction by varying shape parameters of the deuteron wave functions. This was done because of the possibility that the very forward direction had normalization difficulties.   Subsequent re-measurements verified this and the fit now runs through the data points as well as can be expected considering the experimental uncertainty and the computational approximations made.   This can be seen in the report to this conference by E. Boschitz.

An independent test of the calculations was made by comparing the predictions in the forward hemisphere, using exactly the same wave function and computer program, to experimental cross-section and polarization data obtained some time ago at 0.146 Gev.[7]   The agreement was about as good as at 0.570 Gev and therefore encourages belief in the procedures being used.

Since these calculations will now be redone with improvements which may possibly change predictions up to $\sim 50\%$ and will be compared with many more experiments in the energy band, it does not seem worthwhile to reproduce the 0.146 Gev curves here.

Keeping in mind the order of magnitude of theoretical error just mentioned, the results do seem very tentatively to indicate that the s-state implied by the data will show greater probability of small nucleon separation than is exhibited by the Hamada-Johnson wave function.

In conclusion, I feel that an encouraging start has been made along this line of investigation and hope that this will in turn stimulate further experiments.

[1]M.H. MacGregor, R.A. Arndt, R.M. Wright, Phys. Rev. 173, 1272 (1968).
[2]R.J. Glauber, this conference.
[3]E.A. Remler, Phys. Rev. 176, 2108 (1968).
[4]R. Aaron, R.D. Amado, Y.Y. Yam, Phys. Rev. 140, B1291 (1965).
[5]E. Boschitz, this conference.
[6]William & Mary report WM-16, (unpublished) Williamsburg, Virginia, (1969).
[7]H. Postma, R. Wilson, Phys. Rev. 121, 1229 (1961).

QUASI-ELASTIC REACTIONS WITH 600 MeV PROTONS

C. F. Perdrisat,[†] L. W. Swenson,[††] P. C. Gugelot,[†††]

E. T. Boschitz,[*] W. K. Roberts,[*] J. S. Vincent,[*]

and J. R. Priest[**]

A measurement of the (p,2p) differential cross section $d\sigma/dEd\Omega_1 d\Omega_2$ at 600 MeV on Deuterium and Helium-4 at the NASA Space Radiation Effects Laboratory is reported [1], [2]. Two coincident protons were detected with identical range telescopes. The energy sum of the outgoing protons was constrained to be equal to the incident energy minus the S-state proton separation energy, 2.2 and 20.4 MeV for D and He-4, respectively. By varying the detection angles symmetrically to the beam, events were selected with nuclear recoil momentum $\vec{q}_r = \vec{p}_o - (\vec{p}_1 + \vec{p}_2)$ along the beam line. Values of q_r up to 370 MeV/c in D and 300 MeV/c in He-4 were observed.

If one assumes the Impulse Approximation, the recoil momentum $\vec{q}_r$ is equal to minus the Fermi momentum $\vec{q}$ of the struck proton. The Plane Wave Impulse Approximation (PWIA) gives the cross section $d\sigma/dEd\Omega^2$ in terms of the free pp cross section $(d\sigma/d\Omega)_{pp}$ and the probability $\rho(q)$ for a target proton to move with momentum q [3]

$$d\sigma dEd\Omega^2 = \frac{4p^2}{p_o} \frac{m^2 + p^2 \sin\theta}{(m^2 + q^2)^{1/2}} (d\sigma/d\Omega)_{pp}^{CM} \frac{N_\ell}{2\ell + 1} \rho(q) \qquad (1)$$

where p_o is the incident, p the outgoing proton momentum and N_ℓ the number of protons in the ℓ-orbital. $\rho(q)$ is the square of the Fourier transform of the spacial wave function.

Each telescope consisted of 4 range channels centered around the mean range. The channel width was 10.4 MeV at 290 MeV. Coincident events were classified according to the stopping channel on each side. Unequal sharing of the energy by the 2 outgoing protons results in the component of the recoil momentum in the scattering

plane and perpendicular to the incident direction (transverse) being
non-zero. No difference between the data with transverse recoil momentum <10 MeV/c and within 10 and 30 MeV/c could be seen in either
reaction. The energy resolution on the outgoing energy sum was 16
MeV. The poor resolution was of no disadvantage for the D-reaction;
however it made it impossible to insure that the recoiling Triton
was in its ground state in the He-4 reaction. The resolution on the
Fermi momentum is due to the horizontal and vertical angular accept-
ances and to the energy resolution. The momentum resolution was
14.5 MeV/c at $q \sim 0$ and 31 MeV/c at $q \sim 300$ MeV/c.

   Fig. 1 and Fig. 2 show the Fermi momentum distributions $\rho(q)$ for
D and He-4 calculated from our data using (1). Values from the literature [4] were used for $d\sigma/d\Omega)^{CM}$. Because of the symmetric kinematics, 90° CM scattering angle ^{pp} was assumed. For the D-data, a
Carbon background from the CD_2 target had to be subtracted. It has
been measured with matched graphite targets. The systematic error on
$\rho(q)$ is smaller than ±25%. No finite resolution correction has been
made.

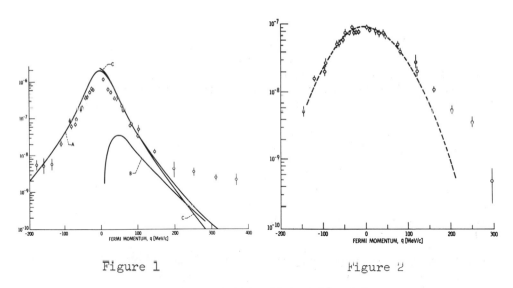

Figure 1   Figure 2

Fermi momentum distribution $\rho(q)$ in $[(MeV/c)^3 sr]^{-1}$ units obtained
from (1) for Deuterium and Helium-4, respectively.

   A change of slope in the D. momentum distribution is seen around
200 MeV/c. This behavior is not expected with the PWIA: the Fourier
transform squared for Hulthén [5] or Hamada-Johnston [6] S-state
wave functions decreases smoothly with q (curves A and C in Fig. 1).
Curve B is for 7% D-state in the Hulthén wave function. It is to be
expected that double scattering contributes to the reaction [7]. As

pointed out by Gottfried [8], the Impulse Approximation is not expected to be valid when Fermi momenta larger than about 250 MeV/c
are involved. Large Fermi momenta occur when two nucleons come
close together. It is then no longer justifiable to assume that
the incident proton interacts with one bound proton only.

A good fit is obtained for He-4 for -120<q<+120 MeV/c with a
Gaussian wave function. The half-width at 1/e is 90 MeV/c (.45 fm^{-1}),
in agreement with other experiments [9], [10]. Above 150 MeV/c a
discrepancy between our data and the PWIA with Gaussian wave function
is observed. Although the discrepancy might seem smaller than in D,
it should be noted that at 250 MeV/c the D and He-4 momentum distributions are equal.

We note finally that the change in slope observed both in D and
He-4 might be comparable to the background of constant $\rho(q)$ values
seen in the Liverpool experiment [9]. Such a background has been
interpreted previously [11] as due to multiple scattering. In D and
He-4 the multiple scattering contribution should be calculable. A
very similar effect has been seen recently at 200 MeV [12].

---

†   College of William and Mary, Williamsburg, Virginia
††   Oregon State University, Corvallis, Oregon
††† University of Virginia, Charlottesville, Virginia
*   Lewis Research Center, N.A.S.A., Cleveland, Ohio
**   Miami University, Oxford, Ohio

[1] A detailed paper will be published in Phys. Rev.
[2] The D-reaction has been studied at 460 MeV by Tyrén et al.
     N.P. 79, 321 (66). Data at 1000 MeV are reported by Simpson
     et al. (Preprint 1969). In both cases the range of recoil
     momenta is about 100 MeV/c.
[3] T. Berggren and G. Jacobs, N.P. 47, 481 (63).
[4] See the recent review by Alexander et al. N.P. B5, 1 (68).
[5] M. J. Moravcsik, N. P. 7, 113 (58).
[6] T. Hamada and I. D. Johnston, N.P. 34, 382 (62).
[7] L. Bertocchi, N.C. 50A, 1015 (67).
[8] K. Gottfried, Annals Phys. 21, 29 (63).
[9] A.N. James, P.T. Andrews, P. Kirkby and B.G. Lowe (Preprint 1969).
[10] M. Riou, Rev. Mod. Phys. 37, 375 (65).
[11] T. Berggren and H. Tyrén, An. Rev. of Nucl. Sc. 16, 153 (66).
[12] C.N. Brown and E.H. Thorndike, Phys. Rev. 177, 2067 (69).

SEPARABLE APPROXIMATION FOR N-D SCATTERING*

E. Harms,[≠] L. Laroze,[†] and J. S. Levinger

Rensselaer Polytechnic Institute

Troy, New York

We present two improvements to Amado's calculation[1] of neutron-deuteron scattering by a spin-dependent central potential with Yamaguchi form factor. i) We use the unitary pole approximation[2] form factor (UPA) for a realistic potential with soft core. ii) We include tensor forces.

We study the accuracy of the UPA for the local Reid singlet S potential by comparing the exact t-matrix $t(p,k;s)$ (from numerical solution of the Lippmann-Schwinger equation) with the UPA result, $t_u(p,k;s)$. Momenta p and k are in F^{-1}, and energy s in units of 41.5 Mev. The Table shows that t and t_u agree surprisingly well over a large range of momenta and energy. This agreement is much better for the Reid potential than it is for a purely attractive potential[3]: e.g., Hulthén local and its UPA (Yamaguchi). The success of the UPA in the Reid case can be understood in terms of the rapid convergence and oscillating character of the unitary pole expansion.[3]

Table 1. UPA for Reid Potential

	s = 0.0	-0.1	-0.5	-3.0
$t(0.57, 0.57; s)$	-7.2	-1.5	-1.0	-0.76
$t_u(0.57, 0.57; s)$	-7.3	-1.5	-1.0	-0.77
$t(0.57, 3.1; s)$	3.2	0.70	0.48	0.40
$t_u(0.57, 3.1; s)$	3.0	0.63	0.43	0.32
$t(3.1, 3.1; s)$	-1.1	-0.02	0.09	0.19
$t_u(3.1, 3.1; s)$	-1.3	-0.26	-0.18	-0.13

Table 2.   k cot(Re δ)

Neutron energy in Mev

	2.45	3.27	5.5	9.0	14.1	49.
Amado[1]	-0.74	-0.67	-0.60	-0.54	-0.32	----
Present Work	-0.41	-0.40	-0.36	-0.26	-0.08	1.05
Experiment[5]	-0.44	-0.37	-0.26	-0.19	-0.09	----

We calculate the phase parameters for $1/2^+$ states of the neutron-deuteron system, up to neutron lab energy 100 Mev. In our preliminary work we use several separable potentials: e.g., the UPA for Schrenk-Mitra's singlet, and the UPA for a modified Hulthén triplet, with 4 percent D state in the deuteron.[4] Both form factors include effects of short range repulsion; calculations are in progress with Reid's soft core potentials, for other scattering states, and at higher energies.

Table 2 shows k cot(Re δ) vs. neutron energy, for the $^2S_{1/2}$ phase shifts. The results presented show that including effects of short range repulsion and of the tensor force in the separable potential improve the agreement between calculation and the phase shifts Van Oers[5] finds from experiments.

We conclude that the UPA can be a good approximation to a local potential, and can give good agreement with experiment on the triton energy,[4] and on some n-d phase shifts.

Supported in part by the National Science Foundation.

≠NSF Trainee. Part of this work is contained in Harms' Ph.D. dissertation, Rensselaer Polytechnic Institute, August 1969. Present address, Fairfield University, Fairfield, Conn.

†AID Fellow, on leave from Universidad Tecnica F. Santa Maria, Chile.

REFERENCES

1. R. Aaron, R. D. Amado, and W. Yam, Phys. Rev. 140, B1291 (1965).
2. M. G. Fuda, Nucl. Phys. A116, 82 (1968).

3. E. Harms, Bull. Amer. Phys. Soc. $\underline{14}$, 493 (1969), and
    Rensselaer dissertation, August 1969. The latter gives many
    more numbers and examples.

4. T. Brady, M. G. Fuda, E. Harms, J. S. Levinger and R. Stagat,
    submitted to Phys. Rev. This potential gives a triton energy
    of -8.45 Mev, and a doublet scattering length of 0.84F.

5. W. T. H. van Oers and J. D. Seagrave, Phys. Lett. $\underline{24B}$, 562
    (1967).

EFFECT OF LONG RANGE CORRELATIONS IN HIGH ENERGY SCATTERING

W. Bassichis, H. Feshbach and J. F. Reading

Laboratory for Nuclear Science and Physics Department

Massachusetts Institute of Technology, Camb., Mass.

An improved Glauber approximation is derived and tested numerically on a potential model. Using this approximation the effect of long range correlations present in spherical nuclei because of the collective motion of the nucleus has been evaluated. The method can be empolyed as long as the Hamiltonian describing the interaction of the target nucleus with the incident particle can be expressed as the sum of terms which are linear and bilinear in the boson operators describing the collective oscillations. An optical potential including these effects has been derived. Inelastic excitations can be readily evaluated. The method is applied to the vibrations of a spherical nucleus where the incident particle-target nuclear interaction is described by a vibrating square well. It is shown that the high energy approximation is valid for the square well in spite of the sudden change in the potential.

ON SCATTERING BY NUCLEI AT HIGH ENERGIES

Herman Feshbach and Jörg Hüfner

Laboratory for Nuclear Science and Physics Department

Massachusetts Institute of Technology, Camb., Mass.

We study how nucleon-nucleon correlations influence high energy scattering by nuclei. Using the multiple scattering theory of Kerman, McManus, and Thaler we formulate the scattering problem in terms of an infinite system of coupled equations. For the calculation of elastic scattering we propose replacing the infinite system by a pair of coupled equations, containing the elastic channel and another effective one which carries all inelastic strength. The coupling potential is proportional to the nuclear pair correlation function and a first approximation to this potential is obtained. These equations are accurate to the extent that the pair correlations and not higher order correlations are important. There are no restrictions to forward scattering nor are any "on the energy shell" approximations made. In order to obtain insight into the structure of the many-channel S matrix for high energies, the infinite system of coupled equations is solved by semi-classical methods and explicit formulae for the S-matrix elements as a function of various potentials and the nuclear pair correlation function are obtained. We show how properties of excited target states may complicate a reliable extraction of correlations in high energy scattering by nuclei. A close relation between our semi-classical solution and Glauber's multiple scattering is established. A numerical study of high-energy nucleon-nucleus scattering using the methods developed in this paper is under way.

REFERENCE

A. K. Kerman, H. McManus, R. Thaler, Annals of Physics **8**, 475 (1959).

OVERLAPPING POTENTIALS AND MULTIPLE SCATTERING

Herman Feshbach

Laboratory for Nuclear Science and Physics Department

Massachusetts Institute of Technology, Camb., Mass.

This paper contains an impact parameter representation of the multiple scattering series in which the total phase shifts produced by the nucleons of a target nucleus is related to the phase shift resulting from scattering by a single nucleon, by two nucleons and so on. The first term of this series yields the additivity assumption of Glauber, Czyz and Maximon. The second term depends directly on the "off energy shell effects". Using an improved version of the Glauber approximation for potential scattering, the second term has been evaluated for the case in which the interaction with each individual particle can be described by a potential. The overlapping of these potentials is shown to have an effect identical with the presence of correlations in target nucleus. When the incident particle is a nucleon, the consequence, for, elastic scattering of the overlapping potentials is estimated to be equal to that of the correlations in the target nucleus in the 500 to 1000 Mev range and to decrease with increasing energy like $(1/k)$ where k is the wave number of the incident nucleon.

REFERENCES

W. Czyz and L. C. Maximon, Annals of Physics $\underline{52}$, 59 (1969).

R. J. Glauber, High Energy Collision Theory in "Lectures in Theoretical Physics", Vol. I, Interscience Publishers Inc., New York, 1959.

DEUTRON-DEUTERON ELASTIC SCATTERING AT SIX LABORATORY MOMENTA

FROM 680 TO 2100 MeV/c

A. T. Goshaw and P. Oddone
Princeton University, Princeton, New Jersey
M. Bazin
Rutgers University, New Brunswick, New Jersey
C. R. Sun
State University of New York at Albany

Glauber multiple scattering theory has been applied recently with success to the π-d, p-d, and $\bar{p}$-d scattering (1). In an experiment of d-d scattering, such as the one described below, one can explore certain features of the multiple scattering theory which exist only when both colliding particles are composite. The study of such composite nuclear systems of familiar particles may help in formulating theoretical models which treat hadron-hadron scattering as collisions between composite systems of unknown subparticles (2).

We have taken 500,000 pictures in the PPA 15" bubble chamber exposed to a separated secondary deuteron beam. The film was scanned for all events of which 70% were rejected as clearly inelastic on the scanning tables including a three dimensional projector. The selected 20,000 events were measured. The resulting dd $\rightarrow$ dd fits were then corrected for scanning efficiency and inelastic background (dd $\rightarrow$ dpn).

The fully corrected differential cross sections are displayed in Fig. 1. Qualitatively the features are quite similar at all six momenta. The d-d elastic scattering amplitude may be written

$$F_{dd}(\vec{\Delta}) = 4f(\vec{\Delta})S^2\left(\tfrac{\vec{\Delta}}{2}\right) + \frac{2i\hbar}{\pi^{3/2}}S\left(\tfrac{\vec{\Delta}}{2}\right)\int f\left(\tfrac{\vec{\Delta}}{2}+\vec{q}\right)S(\vec{q})f\left(\tfrac{\vec{\Delta}}{2}-\vec{q}\right)d\vec{q}$$

$$+ \frac{i\hbar}{\pi^{3/2}}\int S^2(\vec{q})f\left(\tfrac{\vec{\Delta}}{2}+\vec{q}\right)f\left(\tfrac{\vec{\Delta}}{2}-\vec{q}\right)d\vec{q} + higher\ order\ terms$$

where $t \equiv \vec{\Delta}^2$ is the four momentum transfer squared. The very sharp forward peak is dominated by single scattering (first term in the formula); it falls off essentially as S^4 where S is the form

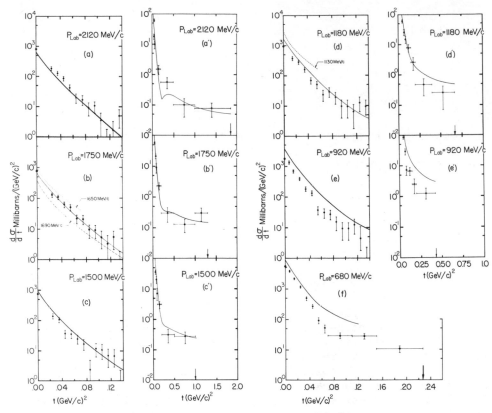

Fig. 1 -- Deuteron-deuteron elastic differential cross sections. The points at t = 0 are optical points calculated from d -d total cross sections measured separately in this experiment.   The arrows indicate the kinematic limits of 90° in the center of mass.

factor of the deuteron. The sharpening of the peaks towards lower incident momenta could be contributed to by the slope of n-p differential cross sections which also gets sharper at lower momenta.

     At large momentum transfer, the dominant term is simultaneous double scattering (last term in the formula). Here the t dependence comes in only through nucleon-nucleon amplitudes (f's) which vary slowly with t. The observation of the flat tail suggests that the simultaneous double scattering process is important in d-d collisions.

     The solid curves are the theoretical predictions of our calculation based essentially on the formalism of Franco (3).   In addition we have put in the D-wave contributions suggested by Harrington. Our calculations include single, sequential double,

and simultaneous double scattering terms. The p-p and n-p total
cross sections and the slopes of their differential cross sections
which we used were extrapolated from data compiled by Wilson (4)
and Hess (5). The ratios of real to imaginary parts in the p-p and
n-p scattering amplitudes were obtained from Dutton et al (6),
neglecting spin dependent terms. The theoretical predictions of
Tubis and Churn (7) are plotted as dotted lines and those of Queen
(8) as dashed curves.

Thus, our experimental data agree qualitatively with the Glauber
theory, although our data extend into the large angular region
where the basic theoretical assumptions may not hold. Further
detailed study awaits our nearly completed experiments with high
statistics at 2.2 and 7.9 GeV/c incident momenta.

REFERENCES

1. V. Franco and R. Glauber, Phys. Rev. Let. 22, 370 (1969);
    D. Harrington, Phys. Rev. Let. 21, 1496 (1968); M. Fellinger
    et al, Phys. Rev. Let. 22, 1265 (1969).
2. D. Harrington and A. Pagmenta, Phys. Rev. 173, 1599 (1968);
    S. Frautschi and B. Margolis, Nuo. Cim. 56 A, 1155 (1968);
    E. Schrauner et al, Phys. Rev. 177, 2590 (1969).
3. V. Franco, Phys. Rev. 175, 1376 (1968).
4. R. Wilson, The Nucleon-Nucleon Interactions, Interscience
    Publishers, N. Y. (1963).
5. W. N. Hess, Rev. Mod. Phys. 30, 368 (1958).
6. L. M. C. Dutton et al Phys. Rev. Let. 25B, 245 (1967).
7. Tubis and Chern, Phys. Rev. 128, 1352 (1962).
8. N. M. Queen, Nucl, Phys. 80, 593 (1962).

NUCLEAR EFFECTS IN MULTIPLE SCATTERING THEORY[*]

Toshitake Kohmura[+]

Department of Physics and Astronomy

University of Rochester, Rochester, New York   14627

Nuclear effects on multiple scatterings should be taken into account when we attempt to obtain some nuclear structure, say nuclear correlations, from scatterings at high energies, whereas the Glauber model neglects essentially the effects. In the Watson multiple scattering theory, the T-matrix for scatterings from nuclei

$$T = \sum_i T_i + \sum_{i \neq j} T_i G T_j + \sum_{i \neq j \neq k} T_i G T_j G T_k + \dots , \tag{1}$$

where

$$T_i = V_i + V_i G T_i, \tag{2}$$

$$G = (E - K - H_n + i\eta)^{-1}, \tag{3}$$

with the interaction V_i between the incident particle and the ith target nucleon, the kinetic energy K of the incident particle, and the nuclear internal Hamiltonian H_n. The nuclear effects on T_i and on G are discussed in the following sections.

[*]Work supported by the U. S. Atomic Energy Commission and the Nishina Memorial Foundation.
[+]On leave of absence from Department of Physics, Tokyo University of Education.

Nuclear effect on transition operators[1] T_i. The transition operator T_i in eq. (3) is rewritten as

$$T_i = t_i + t_i(G-g)T_i,\qquad\qquad\qquad (4)$$

where

$$t_i = V_i + V_i g t_i,\qquad\qquad\qquad\qquad (5)$$

$$g = (E - K - K_n + i\eta)^{-1},\qquad\qquad\qquad (6)$$

with the kinetic energy operator K_n for the nuclear internal system. The first order correction to t_i is $t_i(G-g)t_i$, and

$$G-g = g(V_n + V_n G V_n)g,\qquad\qquad\qquad (7)$$

with $V_n = H_n - K_n$. By using the pole approximation[2] for $V_n + V_n G V_n$ in eq. (7) we have obtained the following results: The nuclear internal potentials reduce the absolute value of the incident particle-nucleon scattering amplitude by a few percent and shifts the phase by about $+10^\circ$. Therefore, the nuclear effect on the absolute value is negligible but the ratio of the real part to the imaginary of the scattering amplitude is enlarged remarkably. The ratio increases with the momentum transfer q. Our result is consistent with the experimental results indicating that the phase of the incident particle-nucleon scattering amplitudes for scatterings from nuclei are deviated from that for scatterings from nucleons.

Nuclear effect on Green Functions[1] G. If we neglect the nuclear effects and take into account the lowest order of q in G, we obtain a formulation equivalent to the Glauber model[3]. The terms due to H_n in G are, however, comparable with the linear terms with respect to q just at small angles where the Glauber model is believed to be valid. Therefore, we should take into account the nuclear effects. Replacing the Hamiltonian H_n-E_o for nuclear excitation by an expectation value and taking into account the lowest order of q, we obtain a correction to the Glauber model. The correction results from the principal parts of G and has a form similar to the formulation of scatterings due to overlapping potentials between the incident particle and each of the target nucleons which is derived in the impact-parameter approximation by Feshbach[4]. "Effective solid angle for scattering", which is defined to be the total cross section divided by the forward differential cross section, is not affected by the principal parts, and is sensitive to the nuclear correlations. Therefore, by analyzing in the Glauber model the data on effective solid angles for scatterings from nuclei, we can obtain the nuclear correlations regardless of the principal parts.

References

1. Part of the first section is based on T. Kohmura, Phys. Rev., (in press), and part of the second section on T. Kohmura, University of Rochester preprint, 1969.
2. C. Lovelace, Phys. Rev., 135, B1225(1964).
3. E. A. Remler, Phys. Rev., 176, 2108(1968)
4. H. Feshbach, MIT preprint CTP 44(1968).

NEW NEUTRON RICH ISOTOPES PRODUCED IN 3 BeV PROTON REACTIONS[*]

G.M. Raisbeck, P. Boerstling, P.W. Riesenfeldt, R. Klapisch,[†]
and T.D. Thomas, Dept. of Chemistry and Princeton-Pennsylvania Accelerator and G.T. Garvey,[‡] Dept. of Physics

Princeton University, Princeton, New Jersey

The study of neutron rich, light isotopes ($7 \leq A \leq 25$) produced
at relatively low kinetic energy (3-10 MeV/Amu) from the reaction
of 3 BeV protons on ^{197}Au targets has continued [1]. The immediate
object of these experiments is to establish the limits of particle
stability for neutron rich isotopes with $Z \leq 8$. The limit of heavy
particle stability represents a sensitive test for the various
procedures used for predicting nuclear masses far removed from the
valley of beta stability [2],[3].

The present experimental setup uses the external 3 BeV proton
beam from the Princeton-Pennsylvania Accelerator in a "parasitic"
mode to bombard a 20 mg/cm^2 Au target. The experiment is capable
of being run whenever the external beam is in use. The heavy fragments resulting from the reaction are identified by measuring their
rate of energy loss in a solid state counter telescope; a method
first employed by the Berkeley group [4]. In addition the flight
time of the fragment is measured to serve as an independent
measure of its mass. The transit time from the target through the
counter telescope is the order of 20 nsec so that the identification of a particular neutron rich species establishes it as stable
against heavy particle decay.

The first two silicon surface barrier solid state detectors in
the counter telescope are thin transmission type detectors which
are followed by a thicker stopping detector and lastly a veto
counter. The detector characteristics are given in Table 1. Only
events producing a coincidence between the 2 transmission detectors
and the stopping counter in anticoincidence with the veto counter
are analyzed. The first transmission counter is 25 cm closer to the

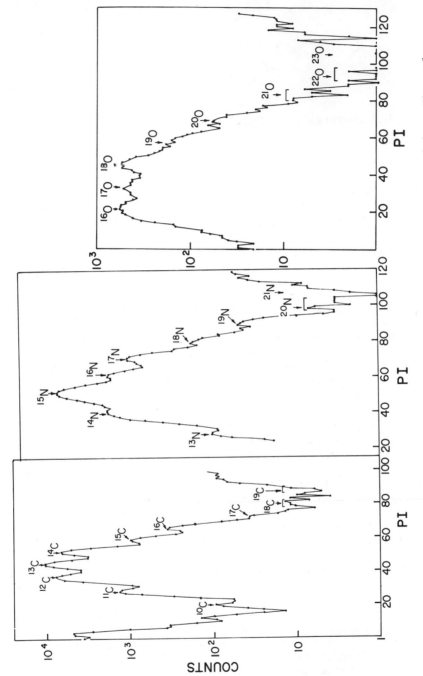

Figure 1 – Particle identifier spectrum for the isotopes of C, N, and O. The values of A, B and C of eq. 2 are chosen in each of the three cases to optimize the resolution within that group of isotopes.

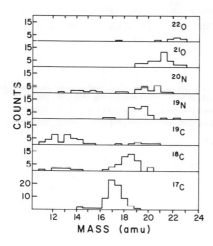

Fig. 2 - Mass spectrum associated with events appearing in the particle identifier spectrum at the location appropriate to the isotope indicated in the insert.   The mass is obtained from $M = K(E_2+E_3)t_{12}^2$ where K is a constant and t_{12} is the flight time between detectors 1 and 2.

Table 1. Counter telescope Parameters.

Detector number	Thickness (microns)	Collimated area (cm²)	Dist. from target (cm)	Discr. level (MeV)	Count. rate (sec⁻¹) Target in	Target out
1	28.0	0.125	30	2	112.0	14.6
2	25.5	0.20	55	2	14.0	4.6
3	181.	0.30	56	3	95.0	39.0
4	300.	0.50	57	1.0	592.0	400.0
1+2+3+4̄					4.4	$< 10^{-3}$

target than the second and provides the flight path referred to above. The data are handled on line by a P.D.P.–9 computer which writes on magnetic tape the four pieces of information from each event; the three energy signals from detectors 1, 2, and 3 and the flight time between detectors 1 and 2. In approximately 1000 hours of running time 7×10^6 acceptable events were examined and 2×10^6 events associated with $Z \geq 4$ were written on tape for subsequent analysis.

In order to eliminate spurious events pile up rejection was used on detectors 1, 2, and 3 and each event was analyzed using the energy loss in detectors 1 and 2 separately to test if they produced a common identification. Our particular mode of particle identification is based on the Bohr–Bethe equation for energy loss from which it can be seen that a product formed from the energy lost in a thin detector times the energy of the particle traversing the detector yields a result proportional to MZ^2 where M is the mass of the particle traversing the detector and Z its charge. Hence each event was initially examined by forming the ratio

$$R = \frac{E_1(DE_1 + E_2 + E_3)}{E_2(FE_2 + E_3)} \tag{1}$$

where E_i is the energy deposited in the ith detector and D and F are adjustable constants chosen to give the smallest spread in R for each element. For those events satisfying the limits set on this ratio condition the particle identification was made by forming the product

$$P = (E_1 + E_2 + C)\ ((E_1 + E_2)A + E_3 + B) \tag{2}$$

where A, B, and C are chosen so as to make P as independent of energy as possible for each element.

The identifier resolution is the order of 4% with most of this attributable to energy straggling and non uniform thickness in the transmission counters, and to residual energy dependence in eq. (2). The timing resolution between detector 1 and 2 is 0.5×10^{-9} sec and as the characteristic transit time is 10 nsec our mass resolution is $\sim 10\%$. Figure 1a, b, and c shows the particle identifier spectrum for the C, N, and O isotopes obtained from analysis of half the data accumulated from a 1000 hour run. The previously unknown isotopes C^{18} and N^{20} are clearly evident. The isotopes O^{21} and N^{19} were previously [1] observed by this group and C^{17} was detected by the Berkeley group [5]. Figure 2 shows the mass spectra associated with particle identifier events corresponding to the listed isotopes. Thus events identifying as C^{17} have energies and flight times consistent with A=17 (some of the 10 fold more abundant C^{16} shows up in this mass spectrum). Particles identifying as C^{18} peak at A=18 with some small contributions from N^{12} and N^{13}. Those events having identifier signals which would correspond to C^{19} and the light N isotopes yield mass peaks at A=19, 12, and 13. Thus C^{19}

appears to exist as a particle stable isotope. It should [2] be the heaviest odd A isotope of C. There is direct evidence in the mass spectrum for N^{20} and O^{22}. The lighter mass objects in the N^{20} spectrum are due to O^{13} and O^{14}.

From this body of data it appears that the cross section for production of the very neutron rich isotopes falls off by a factor of ~10 when an odd neutron is added and by a factor of ~3 where an even neutron is added. Thus producing measurable amounts of O^{28} by this technique do not look too hopeful. Bombardments of heavy targets with high energy (>1 BeV) α-particles coupled with the use of magnetic analyzers and thinner solid state detectors could make the observation of this isotope possible [2].

Further experiments studying the production cross sections, energy spectra and angular distribution of the more abundant isotopes resulting from the bombardment of Au, Mo^{100}, and Ni^{58} with 3 BeV proton have been completed and will be the subject of a future publication [6].

*Supported in part by the U.S. Atomic Energy Commission.
†Permanent Address: Institute de Physique Nucleaire d'Orsay, Centre de Spectrometrie Nucleaire et de Spectrometrie de Mass du C,N.R.S.
‡A.P. Sloan Fellow 1967-1969.

References:
(1) T.D. Thomas, G.M. Raisbeck, P. Boerstling, G.T. Garvey and R.P. Lynch, Phys. Letts. 27B, 507 (1968).
(2) G.T. Garvey and I. Kelson, Phys. Rev. Letts. 16, 197 (1966). G.T. Garvey, W.J. Gerace, R.L. Jaffe, I. Talmi and I. Kelson, (to be published in Rev. Mod. Phys.).
(3) J. Cerny, Ann. Rev. Nucl. Sci. 18, 27 (1968).
(4) A.M. Poskanzer, S.W. Cosper, E.K. Hyde and J. Cerny, Phys. Rev. Letts. 17, 1271 (1966).
(5) A.M. Poskanzer, G.W. Butler, E.K. Hyde, J. Cerny, D.A. Landis and F.S. Goulding, Phys. Letts. 27B, 414 (1968).
(6) G.M. Raisbeck, P. Boerstling, R. Klapisch and T.D. Thomas, (to be published).

PIONS AND NUCLEI

N. W. Tanner

Nuclear Physics Laboratory, Oxford University

England

It is convenient to divide this review of the study of nuclear physics using pions into several sections. Three of the sections are indicated in Fig. 1. which contrasts the pion-nucleon interaction with the proton-nucleon interaction; the fourth section is pion

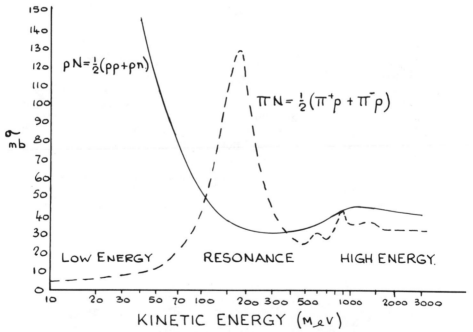

Fig. 1 -- Total cross-section of nucleons for pions and protons.

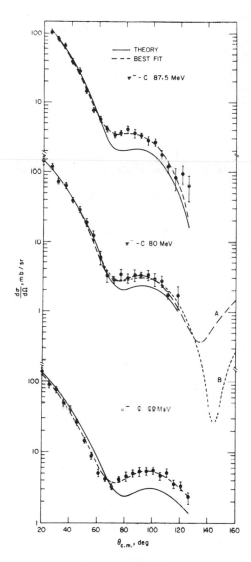

Fig. 2 -- Comparison of π^--C^{12} scattering with the predictions from the Kisslinger potential.

production and absorption. At low energies the pions are weakly interacting and nuclear processes should be well described by the Kisslinger optical potential. In the region of the 3-3 resonance near 180 MeV there is no secure theoretical framework, but much of the experimental data seems to be consistent with the single scattering approximation. At high energies the Glauber multiple scattering formalism offers the prospect of accurate analysis for nuclear reactions at high momentum transfer, i.e. high by the nuclear standard of the Fermi momentum. Pion production and absorption is complicated by the necessity of two nucleons, but may yet yield nuclear physics of interest.

It should be mentioned that virtually all of the experimental data, and most of theory, has been concerned with light nuclei for the good technical reasons of absorption and energy resolution.

1.   Low energy nuclear physics

In terms of momentum transfer pions of low energy, say $\lesssim$ 50 MeV, are equivalent to Van de Graaff protons of 10 or 20 MeV and investigate nuclear momentum components up to about the Fermi momentum ($\sim$170 MeV/c for light nuclei). The difference is that the low energy pion-nucleon interaction is negligible other than in the s- and p-states and both these phase shifts are small. For this reason it is possible to relate the π-nucleus interaction directly to the empirical πN and $2N \rightleftarrows 2N\pi$ data. The relation may be expressed through an optical potential in a form due to Kisslinger[1]

$$V(r) = q(r) - (\nabla \propto (r) \nabla) \qquad (1)$$

which has been developed in considerable detail by Ericson and
Ericson.[2]   Recently Krell and Ericson[3] have determined the parameters (five in their version) of the potential from the level
shifts and widths deduced from pionic X-rays.   The comparison with
the πN scattering lengths is highly satisfactory but there is a
factor of two discrepancy with the parameters related to the absorption processes $\pi^-2p \to np$ and $\pi^+d \to 2p$.

Auerbach, Fleming and Sternheim[4] have compared the predictions
of the Kisslinger potential with experimental π-nucleus elastic
scattering, both with phenomenological parameters and with the theoretical parameters deduced from πN data (π absorption was neglected).
Fig. 2 shows their results for π-C^{12} scattering where the full
curves are for the theoretical parameters, and the dashed curves
are phenomenological fits.   In other cases, notably the 24 MeV
π-He^4 scattering data of Nordberg and Kinsey,[5] the theoretical parameters make a very poor prediction but it seems that it is always
possible to find a phenomenological fit.   Whether the failure of
the theoretical parameters reflects a basic inadequacy in the scattering formalism or poor approximations in the application is not
clear.   Block and Koethe[6] obtain a very satisfactory agreement with
the same 24 MeV π-He^4 scattering data (Fig. 3) by calculating
directly with the πN phase shifts ignoring the Fermi motion of the
nucleons.   This is equivalent to dropping the velocity dependance
of equation 1, a form which according to Krell and Ericson[3] is
essential for the pion X-ray analysis.

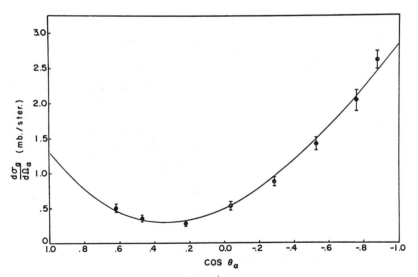

Fig. 3 -- Comparison of π-He^4 scattering with the calculation of Block and Koethe.[6]

There is now quite a lot of low energy π-He^4 scattering data[5,7] aimed at a determination of the pion radius[8] A consistent analysis of the π-He^4 data and the X-ray measurements would go a long way toward establishing the Kisslinger potential experimentally. There remains a need for accurate low energy πd scattering data[3], but this is technically difficult because of the small binding energy of the deuteron.

Given a securely based optical potential it should be possible to analyze inelastic pion-nucleus interactions to establish relations between nuclear states with a precision approaching that provided by β- and γ-transitions near zero momentum transfer. The inelastic processes of particular interest are (π^-,γ), (π,π') and (π,π°), i.e. radiative capture of stopped pions, and inelastic and charge exchange scattering, at a momentum transfer of $\sim$100 MeV/c.

A great deal has been written about the relationship between muon capture, (μ,ν) and (π^-,γ) and the approximate axial vector character of the interaction responsible for pion radiative capture.[9] About 1% of all stopped π^- undergo radiative capture (the remainder are absorbed by (π,2N)) and most give a high energy γ-ray of 100-120 MeV[10], which is consistent with the theory. Deutsch et al[11], using Li^6 and Soergel et al[12] using C^{12} have sought to establish the expected relationship between (π^-,γ) and (μ^-,ν) by counting the residual radio-activity, He^6 or B^{12}, consequent on either reaction. To an accuracy of 20 or 30% the relation has been verified. The Panofsky ratio calculated for He^3, ($\pi^\circ He^3 \to \pi^\circ t$)/($\pi He^3 \to \gamma t$), agrees with the experimental data within the uncertainty of 10%.[13]

From the point of view of nuclear physics (π^-,γ) is much more interesting than (μ^-,ν) because of the difficulties of neutrino spectroscopy. The selection rules for pion capture largely follow the general arguments concerning muon capture.[14] Specifically, Anderson and Eisenberg[9] have given an expression for the radiative pion capture rate from an initial nuclear state i to a final state f as

$$W_\gamma \sim \left| \left\langle f \left| \Sigma_j \tau_j^+ \underline{\sigma}_j \cdot \underline{\varepsilon} \, e^{-i\underline{k}\cdot\underline{r}_j} \, \psi_\pi(\underline{r}_j) \right| i \right\rangle \right|^2 \qquad (2)$$

where $\underline{k}$ and $\underline{\varepsilon}$ are the γ-ray momentum and polarization, $\phi\pi$ is the pion wave function, and $\tau_j^+ \underline{\sigma}_j$ is the Goldhaber-Teller operator for the jth nucleon. The allowed nuclear transitions are $\Delta T = \Delta S = 1$, with a preference for $\Delta L=1$ in case of pion capture from the 1s state and $\Delta L=2$ for the 2p state. Table I gives the calculated transition rates for capture of 1s and 2p pions by O^{16} obtained by Murphy et al[15] using a simple model for the collective states (spin-isospin resonances) of N^{16}. The states are expected to lie 20 or 30 MeV above the O^{16} ground state and to be neutron unstable. Little is known of S=1, T=1 resonances but if they follow the pattern of the familiar S=0, T=1, L=1 resonance of photodisintegration we can expect mixing of S=0 and S=1 resonances and a distribution of strength over several MeV with some fine structure. It should be possible to predict with some confidence the γ-ray spectrum from π^- capture exciting odd parity states of N^{16} using the method of Buck and Hill;[16] even parity states will be more difficult. Experimentally the unravelling of the spin-isospin resonances of various angular momenta will require good γ-ray resolution and γ-neutron correlation measurements. At present there is no high resolution experimental data at all so it may be too much to ask for X-ray-γ-neutron coincidence measurements.

Table I

Transition rates for $O^{16}(\pi^-,\gamma)N^{16}$.

ΔL	J^π	$\lambda(1s)$ $10^{16}sec^{-1}$	$\lambda(2p)$ $10^{12}sec^{-1}$	γ's per 10^4 π [*]		
				1s	2p	(1s+2p)
0	1^+	6.0	6.5	2.5	5.0	7.5
1	0^-	0	5.3	0	4.0	4.0
	1^-	13.6	8.3	5.6	6.3	11.9
	2^-	17.5	23.2	7.3	17.6	24.9
2	1^+	1.0	7.3	0.4	5.5	5.9
	2^+	5.3	23.9	2.2	18.2	20.4
	3^+	7.4	34.5	3.1	26.2	29.3

[*] Calculated from the data in refs. 3 and 35

Pion capture is not restricted to high excited states as Deutsch et al[11] have demonstrated with their observation that 30% of the reaction $Li^6(\pi^-,\gamma)He^6$ leads to the ground state of He^6. The He^6 β-decay to Li^6 is a highly favoured Gamow-Teller transition (log ft = 3) and Deutsch et al have shown that it remains highly favoured at a momentum transfer of ∿130 MeV/c. Not a very surprising result and known in any case from muon capture. There exists, however, the possibility of a more general investigation of inverse β-decay at various momentum transfers through pion charge exchange (not, of course, with stopped π^-). For $(\pi,\pi^°)$ the Fermi transition, $\Delta S=0$ is favoured by about a factor of five over the Gamow-Teller, $\Delta S=1$, transition and the angular distributions are considerably different.   The classic case is the $C^{14}/N^{14}/O^{14}$ problem[18] of anomalously weak β-decays.   A measurement of either $N^{14}(\pi^-,\gamma)C^{14}$ or $N^{14}(\pi^{\pm},\pi^{\circ})O^{14}$ or C^{14} should determine the second forbidden, $\Delta L=2$, matrix element directly as the allowed, $\Delta L=0$, matrix element is negligible for the pion reactions.   Spectroscopy of neutral pions is not easy and it may be expedient to study the equivalent reaction without charge exchange, viz. $N^{14}(\pi,\pi')N^{14}$ (2.3 MeV).

The existing low energy data on charge exchange and inelastic scattering is limited to some total cross-sections measured by Soergel et al[12] for reactions of C^{12} with 70 MeV pions (Table II).

Table II

Cross-sections for πC^{12} reactions

Reaction	(π,π')	(π,π')	(π,π°)	(π,π°)
Final state	4.4MeV,2^+	15 MeV,1^+	ground,1^+	all
π^+	14±1.8mb	<0.7mb	<0.12mb	27±6mb
π^-	17.9±2.9	<0.8mb	0.8±1.8mb	21±6mb

It should be stressed that the extraction of accurate nuclear matrix elements from (π^-,γ), (π,π°) and (π,π') depend on the establishment of a satisfactory optical potential.   According to Guy and Eisenberg[9] the rescattering corrections to (π^-,γ) effect a 40% change.

2.   Resonance Pions

In the neighbourhood of the 3-3 resonance, Fig. 1, the πN interaction is strong and the mean free path of pions in light nuclei is about 1 fermi.   Multiple scattering must be important, but there is no practical scattering theory that can be applied in this energy region.   Lacking anything better, most of the analyses of experimental data have been made in terms of some kind of single scattering approximation.   The qualitative and sometimes quantitative success of this empirical approach is a little surprising.

One of the difficulties impeding the π-nucleus investigations is the lack of accurate and systematic pion-deuteron scattering data.

The deuteron binding energy, 2.2 MeV, is just a little too small to allow the existing magnetic spectrometers[19,20] to resolve the elastic and inelastic scattering clearly.   Bubble chamber measurements cannot easily provide good statistics, but the measurement of Norem[23] at 180 MeV shown in Fig. 4 demonstrates the inadequacy of the impulse approximation calculations.[24]

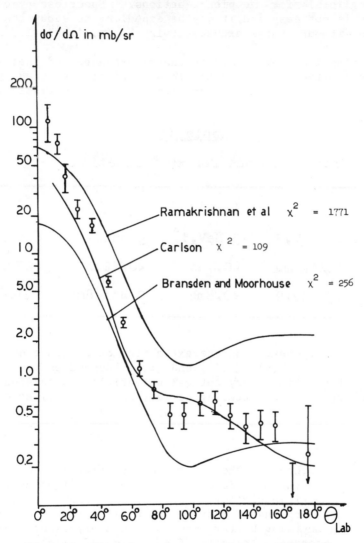

Fig. 4 - Elastic π-d scattering at 180 MeV.

Extensive measurements have been made by Binon et al[19,25] on πC^{12} elastic and inelastic scattering and the total cros--section. Their elastic scattering data near 180 MeV is reproduced in Fig. 5.

The most notable feature is the interference minimum near 50° which
is much deeper at 180 MeV than at the energies immediately either
side (Fig. 5) or elsewhere.   A deep minimum at 180 MeV is consistent
with the prediction from dispersion relations that the real part of
the forward scattering amplitude should go to zero at the resonance
energy.   Elsewhere the scattering cross-section is the sum of the
squares of the real and imaginary scattering amplitudes which have
separate minima which do not coincide and so tend to cancel each
other out.[21,26]

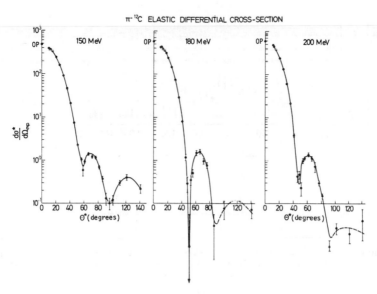

Fig. 5 - Elastic π-C^{12} scattering near 180 MeV.

Attempts to fit the πC^{12} elastic scattering data with a phenomenological optical potential are reported to be unsatisfactory.[25]

The πC^{12} inelastic scattering angular distributions of Binon
et al[25] for 300 MeV/c pions have been compared by Koltun[27] with the
same measurements with protons[28] of 300 MeV/c (45 MeV).   Pions and
protons give much the same form of angular distribution, both for
excitation of the 4.4 MeV, 2$^+$ state of C^{12} and the 9.6 MeV, 3$^-$ state,
with the proton cross-sections a factor of ten larger.

At high momentum transfer the spectra of pions inelastically
scattered by O$^{16(29)}$ closely resemble the form expected for quasi-
free πN scattering, i.e., single scattering.   Two examples of such
spectra are shown in Fig. 6; 45° corresponds to a momentum transfer
q = 300 MeV/c; and 63° to q = 400 MeV/c.   The spectra are centered
about the free nucleon energy transfer of $q^2/2m$ and have a Gaussian
form consistent with scattering by single nucleons in a harmonic

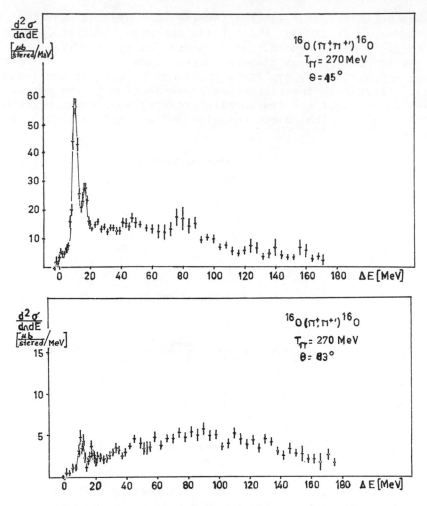

Fig. 6 – Spectra of pions inelastically scattered
by O^{16}

oscillator potential of conventional radius.   At low energy loss
there are binding effects (the ground state and the 6 MeV state of
O^{16} are apparent), and at high energies a non-Gaussian tail which
would be expected for a more realistic wave function than that
obtained from an oscillator potential.   The O^{16} (π,π') inelastic
spectrum at q = 300 MeV/c (45°), Fig. 6, has very much the same form
as the O^{16}(p,p') spectrum[30] at q = 300 MeV/c (180 MeV protons
scattered 25°) if the proton spectrum is spread according to the
pion energy resolution.

The pion charge exchange cross-sections measured by Chivers et
al[31] at 180 MeV by counting residual activity are listed in Table III

Table III

Charge exchange cross-sections

Transition	ΔS	ΔL	$\sigma(\pi,\pi^0)$ Expt.[31]	Theory[33]	Theory[34]	$\sigma(p,n)$ mb Expt[32]
$B^{10} \to C^{10}$	1	0	1.3 ± 0.2			0.65 ± 0.01
$B^{11} \to C^{11}$	0	0	5.3 ± 0.9			3.5 ± 0.2
$C^{13} \to N^{13}$	0	0	3.3 ± 1.0	4.7	2.1	1.9 ± 0.2
$N^{14} \to O^{14}$	1	0	≤0.05	0.06		0.075 ± 0.01
$O^{18} \to F^{18}$	0	0	3.5 ± 0.7			–
$B^{11} \to Be^{11}$	1	0	≤0.5			
$F^{19} \to O^{19}$	0	0	1.3 ± 0.6			–

in comparison with similar (p,n) cross-sections at 155 MeV[32], and with (π^+,π^0) cross-sections calculated by Sakamoto[33] and by Bakke and Reitan[34]; ΔL and ΔS indicate the character of the transition in terms of angular momentum and spin flip. The two calculations for $C^{13}(\pi^+,\pi^0)N^{13}$ use different single scattering approximations but are in better agreement than is suggested by Table III. Fig. 7 shows Sakamoto's cross-section for $C^{13}(\pi^+,\pi^0)N^{13}$ and the experimental $B^{11} \xrightarrow{\pi^+} C^{11}$ cross section as a function of pion energy. Both experiment and theory have a maximum at ∿160 MeV. Bakke and Reitan predict the peak at 110 MeV with a cross-section of 6.5 mb. The differences of the two calculations are thought to be due to the nuclear wave functions. Sakamoto's ratio for the C^{13} and N^{14} cross-sections, ∿10^2, is not in agreement with the estimate ≤ 20 given by Chivers et al[31], but this may also be due to dissimilar nuclear wave functions.

Double charge exchange was not considered for low energy pions for the reasons shown in Fig. 8 - there is no appreciable cross-section.[36] Even at the resonance the process is too weak to allow the identification of specific nuclear states, at least for the next few years. Dr. Becker has reviewed double charge exchange in detail and there is no point in repeating his words here.

For resonance, and presumably high energy pions, the most important nuclear reaction is $(\pi,\pi N)$, analogous to (p,2p). At 180 MeV

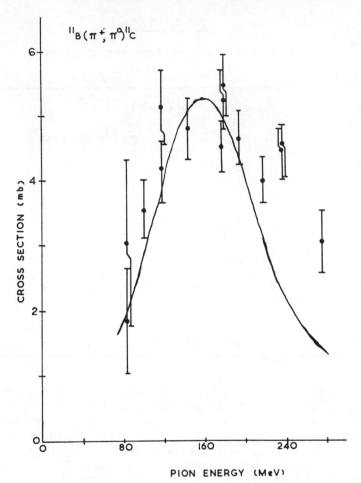

Fig. 7 - Total cross-sections for (π^+, π°). The experimental points are for $B^{11}(\pi^+, \pi^\circ)C^{11}$ and the calculated curve is for $C^{13}(\pi^+, \pi^\circ)N^{13}$.

the total cross-section[3] for $C^{12} \xrightarrow{\pi^+} C^{11}$ is 75 ± 4 mb, $N^{14} \xrightarrow{\pi^+} N^{13}$ 56 ± 6 mb, and $O^{16} \xrightarrow{\pi^+} O^{15}$ 41 ± 4 mb. The energy dependence[31,38] of $C^{12} \xrightarrow{\pi} C^{11}$ is shown in Fig. 9; the curve is from the calculation by Kolybasov[39] for $C^{12} \xrightarrow{\pi^-} C^{11}$. Better fits to the experimental data have been obtained by Selleri[40], Dalkarov[41], and Bertini[42]. The difficulty is that these calculations require the production of C^{11} by π^+ to be about 3 times smaller than the π^- production as a direct consequence of the dominance of the 3-3 resonance. Experimentally the cross-sections for π^+ and π^- are observed to be equal (Fig. 9). A specific measurement[31] of the ratio of π^+ to π^- cross-sections for $C^{12} \to C^{11}$, $N^{14} \to N^{13}$ and $O^{16} \to O^{15}$ yielded 1.03 ± 0.09, 1.05 ± 0.09 and 0.98 ± 0.09 respectively. The predicted ratio is 3

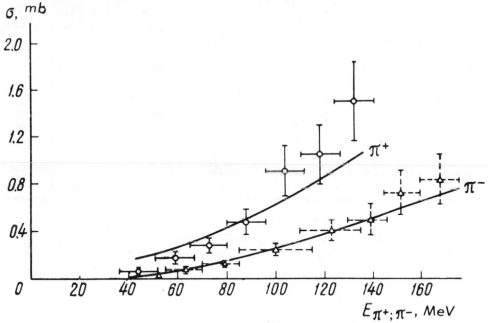

Fig. 8 - Total cross-sections of emulsion nuclei for double charge exchange induced by π^+ and by π^-.

for a quasi-free scattering process, and it will be increased a little because of pion absorption and decreased by charge exchange interactions of the pion or nucleon with the remainder of the nucleus. Hewson[43] has considered the interaction of the scattered nucleon an optical potential and in this way reduced the predicted π^+/π^- ratio to 2. Various other ways of reducing the predicted ratio have been considered[31,43,44] but none are really attractive, and none give a ratio of unity naturally. Measurements[22,45] of π-proton coincidences from the reaction $(\pi,\pi p)$ appear to be similar to $(p,2p)$ measurements, and have not thrown any light on the π^+/π^- ratio.

Apart from this ratio, quasi-free scattering is a reasonable description of π nucleus interactions in the region of the resonance, at least as a first approximation. It would seem wise to check the π^+/π^- ratio, perhaps by measuring the ratio of production of mirror nuclei, e.g. $(\pi^+C^{12} \rightarrow C^{11})/(\pi^+C^{12} \rightarrow B^{11})$ identifying the nuclei by detecting a recognizable γ-ray. Doppler shifts make this a difficult measurement, but probably not impossible for the particular case chosen.

3. <u>High Energy Pions</u>

At high energies the Glauber multiple scattering formalism allows

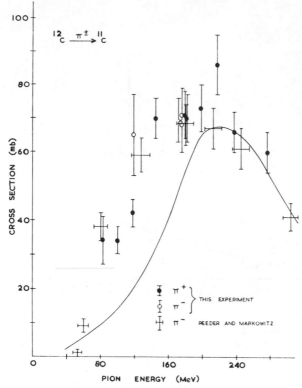

Fig. 9 - Total cross-section for the production of C^{11} from C^{12} by π^+ and by π^-. The curve is calculated for π^-.

the possibility of accurate and relatively simple analysis of small angle scattering. For nuclear physics purposes small angle scattering of, say 2 or 3 GeV/c particles can involve quite high momentum transfers, several times the Fermi momentum.

The approximations of the scattering theory require the incident particle to have an energy and momentum large compared with the energy transfers and momentum transfers of interest, and to have a free nucleon scattering cross-section which is energy independent and which decreases rapidly with scattering angle. Both protons and pions at several GeV/c satisfy these requirements adequately. Bassel and Wilkin[46] have applied the Glauber formalism to proton-nucleus scattering at 1.7 GeV/c with impressive results, but the accuracy of their analysis was limited by uncertainties in the pN scattering amplitudes. The advantage of pions in this connection has been widely discussed and may be summarized:

(a) Pions have spin zero so that πN scattering is much simpler than pN.

(b) Much of the sensitivity to the free nucleon scattering parameters can be avoided by experimentally measuring the ratio (π-Nucleus scattering)/1/2(πp+πn scattering), e.g., $\pi^{\pm}N/1/2(\pi^+p+\pi^-p)$. The equivalent measurement for proton scattering is not practical.

There have been several measurements and analyses of πd scattering. One of these, taken from the paper of Michael and Wilkin[47] is shown in Fig. 10. The two main curves are for a deuteron wave function with and without a d-state, and it is clear that πd scattering

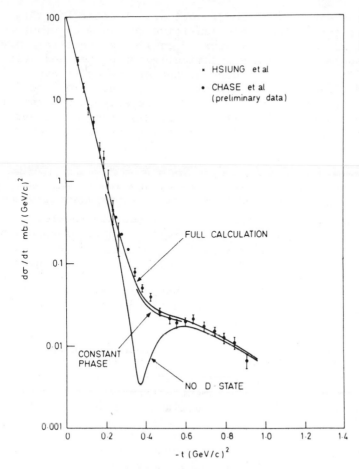

Fig. 10 - π-d scattering at 3.5 GeV/c.

has determined the d-wave component. The short line in Fig. 10 indicates the sensitivity of the calculation to the πN scattering phase.

Other than for πd scattering (from which the recoil can be detected), the experimental prospects are limited by the energy resolution obtainable. With conventional magnets and quadrupoles it should be possible to separate elastic and inelastic scattering of pions from several light nuclei. Undoubtedly much more could be achieved with specially designed equipment giving say $1:10^4$ resolution, but this would be expensive.

The two pion scattering measurements of greatest interest and practicality are the elastic scattering and the sum over all nuclear inelastic scattering excluding meson production and absorption. The inelastic spectra of Fig. 6 for 270 MeV pions indicate the kind of measurement that will be necessary for the inelastic sum. Both elastic and summed inelastic depend only on the ground state wave function of the nucleus (closure is applied to the inelastic sum) and provide, together with electron scattering, three functions of momentum transfer which can be accurately related to the nuclear wave function. Pion measurements should be possible out to a momentum transfer of 500 MeV/c which corresponds to a distance of ∿0.4f, which is about the size of the hard core radius.

Measurement of the (inelastic) sum for pion charge exchange, $\Sigma(\pi,\pi°)$ has been suggested by Eisenberg[48] and by Ericson[21] as a means

of determining the nuclear correlation function.   In fact $\Sigma(\pi, \pi^\circ)$
is likely to be about three times less sensitive to dynamical (but
not Pauli) correlations than $\Sigma(\pi, \pi')$; for (π^+, π°) which concerns only
neutrons, half the nucleon pairs are dynamically correlated and half
are Pauli correlated*; for $(\pi^+, \pi^{+'})$ one quarter of the pairs are
Pauli correlated and three quarters dynamically.   There are also
daunting experimental problems with (π, π°).   An energy resolution of
about 10 MeV will be required to satisfactorily disentangle nuclear
inelastic from meson production at high momentum transfer.   The
prospects of determining the short range behaviour of the nuclear
wave function look much better for $\Sigma(\pi, \pi')$.

The $\Sigma(\pi^\pm, \pi^\circ)$ may be more useful for the study of differences
between the distributions of protons and neutrons in nuclei, as
proposed by Koefed-Hansen and Margolis.[49]   The momentum transfer
required is only about 250 MeV/c so the problems of energy resolution
will be much less severe, particularly if the form of the pion inelastic scattering spectrum is known from (π, π') without charge exchange.

At energies, > 770 MeV, strangeness exchange is possible.   The
cross-section of $\pi^- p \rightarrow K^\circ \Lambda$ rises to a broad maximum near 1 GeV with
a cross-section of more than 100 $\mu b/sr$,[50] and the mirror reaction
$\pi^+ n \rightarrow K^+ \Lambda$ must have the same properties.   There is the possibility,
then, of studying the states of Λ-hypernuclei through reactions such
as $\pi^+ C^{12} \rightarrow K^+ {}_\Lambda C^{12}$.   Assuming a p-shell neutron bound by 18 MeV is
converted into an s-shell Lambda bound by 8 MeV, the sticking proba-
bility is estimated to about 3% for a momentum transfer $p \sim 350$ MeV/c,
i.e., small angle production of K's.   There are 6 neutrons in C^{12}
and allowing a factor of 2 for absorption of pions gives a cross-
section $\sim$10$\mu b/sr$. for the reaction $C^{12}(\pi^+, K^+) {}_\Lambda C^{12}$.   A good (and
expensive) spectrometer, say $1:10^4$, could resolve the energy levels
and determine their positions to 10 or 20 keV accuracy.

4. Pion Production and Absorption

Pion absorption (meaning literally the disappearance of a pion
and not the attenuation of a plane wave in a nucleus) is dominated by
the reaction $(\pi, 2N)$.   Single nucleon absorption is not possible for
free nucleons, and is improbable for bound nucleons as the process
involves a momentum transfer of 500 MeV/c.   Pion production is just
the inverse of absorption by detailed balance.

The two nucleon absorption of pions has been reviewed in detail
by Koltun[27] recently.   Theoretically, the process is complicated by
the final state interactions of the nucleons, and experimentally it
has not been possible to resolve individual states of the residual
nucleus (apart from the ground state in $Li^6(\pi^+, 2p)He^4$).   Qualitatively

* A similar argument can be made that the electron inelastic sum
  will also be insensitive to dynamical correlations.

the observed nuclear excitation spectra are consistent with the calculated 2-hole states.[51] The cross-section for (π^+,2p) is reported[52]
to be about a factor of two larger than a prediction based on the
empirical data for pion absorption by deuterium.   Presumably this
is the same factor of two discrepancy that appears in the πX-ray
absorption widths.[3] There is no doubt that the pion absorption rate
or cross-section can be modified by correlating the nucleus,[53] but
it seems quite unlikely that it will be possible to determine anything unambiguous about correlations from pion absorption.[27,54,55]

Absorption other than by 2-nucleons is quite weak, but not as
weak as predicted.   Single nucleon absorption seems to be the most
important process.   Minehart et al[56] observe a branching ratio of
only $\sim 4 \times 10^{-4}$ for $Li^6\pi^- \to 2t$, whereas Deutsch et al[11] find 1.8% for
$Li^6(\pi^-,n)He^6$(g.s.).   The latter is very much larger than the $\sim 10^{-4}$
branch into the reaction $O^{16}(\pi^-,n)N^{15}$ (g.s.) predicted by
Le Tourneux.[57] Similarly, the cross-section[58] for $C^{12}(\pi^+,p)C^{11}$
(unresolved bound states) for 70 MeV is 0.64 mb which is an order of
magnitude greater than the calculated value.[59]

Measurements of single nucleon absorption through pion production
is much more favourable in terms of intensity and therefore resolution.
The π^+ production spectra[29] from 600 MeV protons shown in Fig. 11,
were measured with a magnetic spectrometer[20] at 0°.   There is no
doubt from the shapes of the spectra and the kinematics of the reaction
that the reaction (p,π^+) to the ground state of the final nucleus is
being observed.   Ground state cross-sections are $pC^{12} \to \pi^+C^{13}$ 0.67μb/sr,
$pC^{13} \to \pi^+C^{14}$ 0.30 μb/sr and $pN^{14} \to \pi^+N^{15}$ 0.24 μb/sr with an uncertainty
of about ±0.2 μb/sr.   Again these cross-sections are a factor of ten
larger than the predictions from the model used by Le Tourneux and
Eisenberg[57,59] for single nucleon absorption.

The factor of ten discrepancy with the model is remarkable for
being so small.   The model assumes a non-relativistic πN interaction
and single scattering so that the cross-section is just proportional
to the momentum probability.   With an uncorrelated harmonic oscillator
the momentum components at the 550 MeV/c required are very weak indeed.
Putting in correlations can easily produce a factor of 10 at 550 MeV/c.
Momentum distributions from the Hulthén wave function or from a hard
sphere Fermi gas give cross-sections about 10^3 larger.

The cross-section for pd $\to$ tπ^+ was also measured by Domingo et al
as 40 μb/sr at 0° (lab).   This appears to be consistent with the old
measurements of this reaction at 12° of 9 μb/sr and 25° of 5 μb/sr
made by detecting the recoiling triton[60,62] and qualitatively consistent with the model of Ruderman and of Bludman[61].   This model is
based on the measured cross-section for pp $\to$ dπ^+ and seems rather more
realistic.

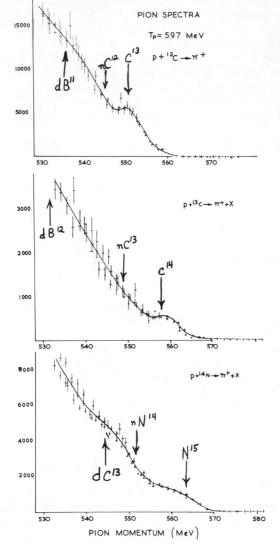

Whether the theory of single nucleon pion absorption can be developed to a level at which nuclear momentum components can be determined from the cross-sections is not yet clear. In view of the history of $(\pi, 2N)$ and correlations, it would be wise to be cautious. It is clear, however, that measurements of (p, π^+) are much easier than $(\pi, 2N)$.

Following the model of ref. 61, one should expect the reaction $C^{13}(p, \pi^-)O^{14}$ to have a cross-section one tenth of the mirror reaction $C^{13}(p, \pi^+)C^{14}$, assuming that the 3-3 resonance dominates the pion production. However, there is no guarantee that the 1/10 will not be 1/3 in view of the ratio of unity obtained for π^+ to π^- production of C^{11} from C^{12}, etc. (see Section 2). At the 1/10 level the $C^{13}(p, \pi^-)$ cross-section should be 0.03 μb/sr at 0° which is probably measurable. It may be that the double charge exchange reaction (p, π^-) will be more profitable than $(\pi^\pm, \pi^\pm)$ has been.

Fig. 11 - Pion production spectra. The arrows indicate the maximum pion momentum for various nuclear states.

REFERENCES

[1] L.S. Kisslinger, Phys. Rev. 98, 761, 1955
[2] M. Ericson and T. Ericson, Ann. of Phys. (NY) 36, 323, 1966
[3] M. Krell and T. Ericson, Nucl. Phys. B11, 521, 1969

[4]E.H. Auerbach, D.M. Fleming and M.M. Sternheim, Phys. Rev. 162, 1683, 1967

[5]M.E. Nordberg and K.F. Kinsey, Phys. Lett. 20, 692, 1966

[6]M.M. Block and D. Koethe, Nucl. Phys. B5, 451, 1968

[7]Block et al. Phys. Rev. 169, 1074, 1968
A. Fainberg, UCRL report 19208, 1969

[8]R.A. Christianson, UCRL report 18881, 1969
Auerbach et al, Phys. Rev. Lett. 21, 162, 1968

[9]A. Fujii and D.J. Hall, Nucl. Phys. 32, 102, 1962
J. Delorme and T. Ericson, Phys. Lett. 21, 98, 1966
D.K. Anderson and J.M.Eisenberg, Phys. Lett. 22, 164, 1966
H. Pietschmann, L.P. Fulcher and J.M. Eisenberg, Phys. Rev. Lett. 19, 1259, 1967
M. Ericson and A. Figureau, Nucl. Phys. B11, 621, 1969, 33, 609, 1967
R. Guy and J.M. Eisenberg, Nucl. Phys. B11, 601, 1969
M. Rho and A. Sherwood, Phys. Lett. 28B, 102, 1968

[10]V.I. Petrukhin and Yu.D. Probaskin, Nucl. Phys. 66, 669, 1965
Davies et al. Nucl. Phys. 78, 673, 1966

[11]Deutsch et al. Phys. Lett. 26B, 315, 1968

[12]V. Soergel, private communication

[13]Zaimidarogu et al. Sov. Phys. JETP 21, 848, 1965 and Ericson and Figureau ref. 9

[14]L.L. Foldy and J.D. Walecka, Nuovo Cimento 34, 1026, 1964

[15]Murphy et al. Phys. Rev. Lett. 19, 714, 1967
(See also M. Kawaguchi, H. Ohtyubo and Y. Sumi, Phys. Lett 27B, 358, 1968 and Guy and Eisenberg ref. 9)

[16]B. Buck and A. Hill, Nucl. Phys. 95, 271, 1967

[17]T.I. Kopalershvili, Phys. Lett. 28B, 163, 1968
Sov. Phys. JETP 8, 351, 1968

[18]H.J. Rose, O. Hausser and E.K. Warburton, Rev. Mod. Phys. 40, 591 (1968)

[19]J.P. Stroot, Proc. Symposium on the use of Nimrod for Nuclear Structure Physics, RHEL/R166, March 1968, p.81

[20]N.W. Tanner, ibid, p.91

[21]T.E.O. Ericson, ibid, p.103

[22]C. Zupancic, ibid, p.67

[23]J.H. Norem, Rutgers Thesis, Dec. 1968

[24]Ramakrishnan et al., Nucl. Phys. 29, 680, 1962
B. Branden & R.G. Moorehouse, Nucl. Phys. 6, 310, 1958
C. Carlson, unpublished.

[25]Binon et al, Proc. Int. Conf. on Nuclear Structure, Tokyo 1967; CERN report 1969, CERN Annual Report, 1968, p.42

[26]T.E.O. Ericson, J. Formanck & M.P. Locher, Phys. Lett. 26B, 91,1967

[27]D.S. Koltun, preprint (Jan. 1969) of a review. The Interactions of Pions with Nuclei for publication in Advances in Nuclear Physics, Vol. 3 (Eds. M. Baranger and E. Vogt).

[28]Peterson et al, Nucl. Phys. A102, 145, 1967

[29]Domingo et al, CERN report, 1969

[30]O. Sundberg and G. Tibell, Upsala preprint
[31]Chivers et al. Nucl. Phys. A126, 129, 1969
(quoted at length by D.H. Wilkinson in Proc. Int. Conf. on
Nuclear Structure, Tokyo, 1967)
[32]L. Valentine, Nucl. Phys. 62, 81, 1965
[33]Y. Sakamoto, Phys. Lett. 29B, 88, 1969; Nucl. Phys. B10, 299, 1969
[34]F.H. Bakke and A. Reitan, Nucl. Phys. B10, 43, 1969
[35]Koch et al. Phys. Lett. 28B, 279, 1969
[36]Batuson et al. Sov. J. Nucl. Phys. 6, 727, 1968
[37]F. Becker - to be published
[38]P.L. Reeder and S.S. Markowitz, Phys. Rev. 133B, 639, 1964
[39]V.M. Kolybasov, Sov. J. Nucl. Phys. 2, 101, 1966
[40]F. Selleri, Phys. Rev. 164, 1475, 1967
[41]O.D. Dalkarov, Phys. Lett. 26B, 610, 1968
[42]A.W. Bertini, quoted by Kottan ref. 27
[43]P.W. Hewson, Nucl. Phys. to be published
[44]V.M. Kolybasov, Phys. Lett. 27B, 3, 1968
[45]Aganyanto et al. Nucl. Phys. B11, 79, 1969
[46]R. Bassel and C. Wilkin, Phys. Rev. 174, 1179, 1968
[47]C. Michael and C. Wilkin, Nucl. Phys. B11, 99, 1969
[48]J.M. Eisenberg, Nucl. Phys. B3, 387, 1967
[49]O. Koefed-Hansen and B. Margolis, Nucl. Phys. B11, 455, 1969
[50]J.E. Rush and W.G. Holladay, Phys. Rev. 148, 144, 1966
[51]Kopaleischmili et al. Sov. J. Nucl. Phys. 7, 198, 1968
[52]Private communication from C. Zupancic concerning the calculations
of Lazard, Ballot and Fourier
[53]Chung M. Danos and M. Huber, Phys. Lett. 29B, 265, 1969
[54]G.E. Brown, comments on particle and nuclear physics 3, 78, 1969
[55]Bressari et al. Nucl. Phys. B9, 427, 1969
[56]Minehart et al, Phys. Rev. 177, 1455, 1969
[57]J. Le Tourneux, Nucl. Phys. 81, 665, 1966
[58]T.R. Witten, M. Blecher and K. Gotow, Phys. Rev. 174, 1166, 1968
[59]J. Le Tourneux and J.M. Eisenberg, Nucl. Phys. 87, 331, 1966
[60]N.E. Booth, Phys. Rev. 132, 2305, 1963
[61]M. Ruderman, Phys. Rev. 87, 383, 1952
S.A. Bludman, Phys. Rev. 94, 1722, 1954
[62]Akimov et al. Sov. Phys. JETP 14, 512, 1962,
Nucl. Phys. 30, 258, 1962

DISCUSSION

J.S. Levinger: What are the arguments that the $(\pi, 2N)$ reaction does
not tell us about correlations?   N.W.T.: The arguments which has
been made (not by me! - for example by Koltun and Brown) by many
people, is essentially that final state interactions (or conditions)
are more important than initial state correlations.   M.G. Huber:
The rather strong statements about the relative importance of final-
state interactions and short-range correlations are based on using
nucleon-nucleon phase-shifts for the description of the outgoing

nucleons.   But these phase-shifts are not appropriate.   Making
shell-model calculations, using optical-model wavefunctions for the
two outgoing nucleons, Chung et al. have shown that both the total
pion absorption rate, and the spectrum of the missing energy,
depend sensitively on, and therefore provide important information
about the details of the short-range correlations.   N.W.T.: This
result (of Chung, Huber and Danos) -- the existence sensitivity to
correlations -- seems very surprising.   M.G.H.: The correlation
function is nothing but the simulation of momentum exchange between
two nucleons, which is just what is happening.   F. Becker: I
should like to mention that calculations of the absorption of pions
by deuterons seem to indicate great sensitivity to the shape of the
wavefunction near the origin.

M.M. Sternheim: Two comments: First, the work of Auerbach and
myself on the optical model analysis of low energy pion nucleus
scatteirng shows that the model works quite well. We were able to
fit nearly all the existing elastic data through 150 MeV with the
Kisslinger model, $V \sim a+bk \cdot k'$, using a and b parameters close to
those derived from pion-nucleon phase shifts, except for the real
part of a. However, Re (a) is proportional to the nearly cancell-
ing sum of the real parts of the π-p and π-n s-wave amplitudes,
so that small corrections to the simplest formulas can produce
large effects. Moyer and Koltun (Phys. Rev. $\underline{182}$, 999, 1969), have
for the case of π-mesic atoms, explained this approximately. For
higher energies, such as the recent 120-280 MeV π-carbon experiment of Mennier et al at CERN, the model seems to work surprisingly
well; and with no parameter adjustment, just the free pion nucleon
phase shifts. Second: Your measurement of the rates for C^{12}
$(\pi^+,\pi^+n)C^{11}$ + $C^{12}(\pi^+,\pi^0p)C^{11}$ and $C^{12}(\pi^-,\pi^-n)C^{11}$ give the same rate
for both at 180 MeV, instead of a 1/3 ratio as suggested by the
free $\pi\pm$-n cross sections. J. Weneser of Brookhaven National Laboratory and I have examined, in a preliminary rough way, an effect
which changes the 1/3 ratio to about 1/2. The effect is this. A
pion enters C^{12} and near the surface strikes a nucleon (the mean
free path is short), which can charge-exchange before it leaves
the nucleus. It has a large cross section for this at low energies.
Faster nucleons tend to eject additional nucleons, so that A<11
nuclei result. N.W.T.: A calculation of this nuclear charge-exchange
effect has been made by P. Hewson (Ref. 43). Using reasonable parameters he obtains a 2:1 ratio but $\underline{not}$ the 1:1 observed experimentally!

A.I. Yavin: It would be interesting to compare the reaction
(π^+,π^0) with (p,n) and (^{3}He,^{3}H) about which much is known at low
energies; in these the analog states are strongly excited. If,
under favorable conditions, these were also excited by (π^+,π^0) one
might study these reactions without observing the π^0, but instead
through the products of decay of the excited analog states.
J. Deutsch: Concerning radiative π-capture: For mass 16 we have

observed remarkably high probability of producing N^{16}. Experimentally one might, through concident -γ detection, detect radiative Π-capture to individual states. N.W.T.: Likewise in the
radiative Π-capture by ^{6}Li, there is about 30% probability of producing ^{6}He (Ref. 11). This is not too surprising: a large G.T.
matrix-element makes radiative capture a very favored process.
T.E.O. Ericson: A brief remark on the variation of the depth of
the minimum, in the elastic scattering by ^{12}C, as a function of
energy. The statement that one can understand this as due to the
scattering-amplitude being purely imaginary at the (3.3) resonance -
as follows from dispersion relations, needs qualification. What
follows from dispersion-relations is that the real part of the
scattering amplitude in the forward direction is zero at resonance.
There is no corresponding statement, a priori, for large angles.
The occurence of the minima suggests, in a way one does not understand, that the entire amplitude is probably imaginary.

π^- ABSORPTION ON SOME LIGHT NUCLEI

F. Calligaris, C. Cernigoi, I. Gabrielli & F. Pellegrini

Istituto di Fisica dell'Universita' - Trieste; Istituto
Nazionale di Fisica Nucleare - Sottosezione di Trieste;
Istituto di Fisica dell'Universita' - Padova; Istituto
Nazionale di Fisica Nucleare - Sezione di Padova

(Presented by C. Cernigoi)

1. INTRODUCTION

We present some unpublished experimental results concerning
the absorption of negative pions on ^{4}He, ^{6}Li, ^{9}Be and ^{12}C; reactions of the type (π^-,np) and (π^-,nd) are investigated for ^{6}Li,
^{9}Be and ^{12}C while on ^{4}He the three reactions, $\pi + {}^4$He = t + n,
$\pi^- + {}^4$He = p 3n are analyzed by the study of the energy spectra of
the neutrons, protons and deutrons. The energy of the correlated
and uncorrelated neutrons was determined by a time-of-flight technique; the energy and the discrimination of protons and deuterons
was done by a combined time-of-flight and range telescope method.

2. EXPERIMENTAL TECHNIQUE

The experimental apparatus was essentially the same as that
reported in ref.[1] where the results on (π^-,nn) reactions are
given. The experiment was performed at the CERN Synchrocyclotron;
an incident beam of 120 MeV negative pions was brought to rest in
targets of about 0.3-0.4 gr/cm^2 in thickness. Since the (π^-,nn)
experiment ran in parallel to the (π^-,np) one, it was possible to
get the experimental value of the ratio $w(\pi^-,nn)/w(\pi^-,np)$ of the
relative absorption rates. As previously mentioned, the discrimination between protons and deuterons was achieved (though with a
certain degree of uncertainty) by measuring simultaneously the
range of the charged particle in a stack of thirty plastic scintillators and its time-of-flight from the target to the first
scintillator of the range telescope. The two sets of information
were memorized and further analyzed by a computer program. The
method has been proved satisfactory, especially in the case of ^{4}He
experiment because of the increased distance of the telescope from
the target, i.e. better time-of-flight resolution. The lower ener-

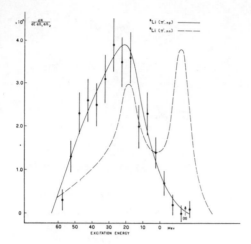

Fig. 1(a) - Excitation energy spectrum of ^{6}Li; (π^-,np) reaction.

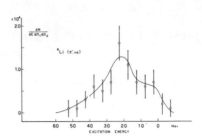

Fig. 1(b) - Excitation energy spectrum of ^{6}Li; (π^-,nd) reaction.

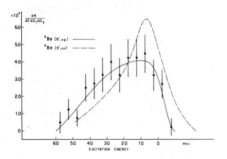

Fig. 2(a) - Excitation energy spectrum of Be; (π^-,np) reaction.

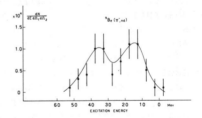

Fig. 2(b) - Excitation energy spectrum of ^{9}Be; (π^-,nd) reaction.

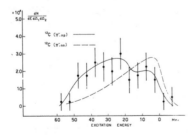

Fig. 3 - Excitation energy spectrum of ^{12}C; (π^-,nd) reaction.

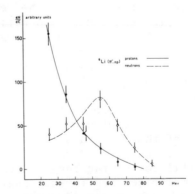

Fig.4(a)- Energy spectra of protons & neutrons in ^{6}Li.

gy limit for the discrimination was determined by the thickness of the first scintillator of the range telescope, and resulted in about 20 MeV for protons and 30 MeV for deuterons.

Following a pion capture, (π^-,np) and (π^-,nd) events were recognized by the detection of coincidences of a charged particle in the range telescope and a nuetron in one of the three larger neutron counters at the opposite side with respect to the target position. The signal corresponding to the stopped pion in a very thin counter near the target was used as the start signal (zero time) of the times-of-flight of the neutron and the correlated charged particle. An electronic system determined how many counters of the range telescope and which of them were involved in the event; so it was possible to count the number of neutrons converted in the range telescope. In the determination of the energy of the charged particles, corrections for the energy losses in the target were calculated.

3. ENERGY SPECTRA OF ^{6}Li, ^{9}Be and ^{12}C

The excitation energy spectra of ^{6}Li for the (π^-,np) and (π^-,nd) reactions are displayed in Figs. 1a and 1b. In Fig. 1a the shape of the (π^-,nn) reaction is reported (dotted line) to show the striking difference between them. Instead of two peaks there is only a broad curve with a maximum corresponding to the value of the second peak in the (n,n) reaction. This fact supports the idea that (p,p) pairs in the nucleus, involved in the capture process, belong to the p-shell of ^{6}Li. In the alpha-deuteron cluster model of this nucleus that means an absorption process inside the alpha core (see ref.[2]). Nevertheless the shape of the two spectra are different in the higher excitation region. No experimental evidence of an eventual ^{4}H nucleus is found; there should be a peak near the zero of our energy scale. This confirms indirectly that the correlation between p- and s-nucleons is very low in ^{6}Li.

More difficult is the interpretation, in the two nucleon absorption process, of the (π^-,nd) spectrum; the zero of the energy scale was fixed by the Q-value of the reaction $\pi^- + {}^6\text{Li} = d+n+{}^3\text{H}$.

Our experiment gives, for the ratio $(\pi^-,nd)/(\pi^-,np)$ a value of 0.30 ± 0.15, which is found practically constant, within our uncertainties, also for the ^{9}Be and ^{12}C nuclei.

When the single correlated particle energy spectra are considered, the situation seems puzzling (see Figs. 4a, 4b, 5a, 5b and 6); the charged particle spectra are different from the neutron spectra. For the charged particles in fact, there is always an excess in the low energy region. This peculiar situation appears also when the absorption process in ^{9}Be and ^{12}C is con-

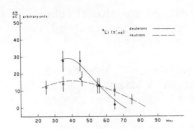

Fig. 4(b)- Energy spectra of
deuterons & neutrons in ⁶Li.

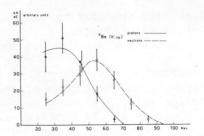

Fig. 5(a) - Energy spectra of
protons and neutrons in ⁹Be.

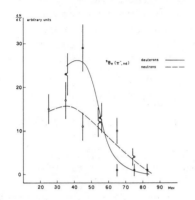

Fig. 5(b) - Energy spectra of
deuterons & neutrons in ⁹Be.

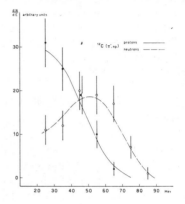

Fig. 6 - Energy spectra of protons and neutrons in ¹²C.

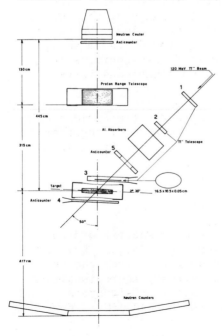

Fig. 7 - Experimental set-up of ⁴He
experiment.

sidered. Probably this is due to the
fact that the final-state interaction
plays an important role when (p,p)
pairs are involved in the process because of their lower correlation with
respect to the (p,n) pairs in the
nucleus.

In Be, the (π⁻,np) excitation
energy spectrum (see Fig. 2a) differs from the corresponding (π⁻,nn)
mainly in the fact that the maximum
is near the zero energy point. This
could be due to the fact that a residual ⁷He is left after the absorption

process whereas in the (π,nn) reaction a ^{7}Li nucleus is left
mainly in high excited states. Although the statistics is not
high, the (π^-,nd) spectrum (see Fig. 2b) suggests the possibility
that a ^{6}He is left in the 15 MeV excited state.

In the case of ^{12}C, only the (π^-,np) spectrum is reported because of low statistics due partly to the fact, already pointed
out in ref.[1], that the relative rate of absorption in this nucleus
is 50% of the corresponding value for Li. In this case the
spectrum is similar to the corresponding (π^-,nn), suggesting the
formation of a ^{10}Be residual nucleus. This is reasonable because
the density of the energy levels of the two isobars, ^{10}B and ^{10}Be,
is comparable; this is certainly not true for ^{7}Li and ^{7}He. Again
the proton spectrum of ^{12}C presents the behaviour above mentioned.
This is in complete disagreement with the result found by the
authors of ref.[3].

This disagreement is even more serious as far as the ratio
w(π^-,nn)/w(π^-,np) is considered. In our results this ratio is
10.0 ± 2.0, which corresponds to a theoretical value of 5.0 ± 1.0
because of the left-right symmetry. Instead the authors of ref.[3]
found an experimental value of 4.6 ± 1.1. An extensive list of
references, related to this important arguement, is reported in
the paper of ref.[4].

4. ABSORPTION ON ^{4}He (*)

Our experimental apparatus is shown in Fig. 7; the only difference with respect to that in ref.[1] is in the increased distance between the counters and the target. Because of the loss of
neutrons converted in the proton range telescope, we preferred to
do the investigation of the neutron spectra separately from the
charged particles. A vacuum pipe (not shown in the figure) was
inserted between the target and the range telescope to avoid
energy losses in air. The construction of the liquid helium target posed difficult technical problems because of its particular
shape (a cylinder of 20.0 cm in diameter and 2.0 cm in height);
this was necessary to obtain the lower part of the energy spectrum.
The helium consumption was of the order of 230 litres/hour of gas.
The electronics enabled us to detect, in the first part of the
experiment, the (π^-,p), (π^-,d), (π^-,np) and (π^-,nd) events and
their relative energy spectra. in the second part of the experiment the (π^-,n) and (π^-,nn) events were detected and analyzed.

The results for the single particle energy spectra of protons
and deuterons are displayed in Fig. 8.

(*) F. Pellegrini did not participate in this experiment.

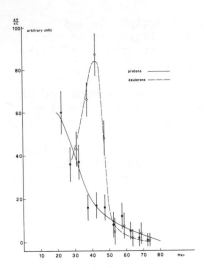

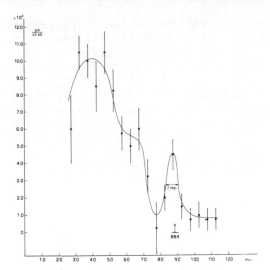

Fig. 8 - Energy spectra of
uncorrelated protons and
deuterons in ^{4}He.

Fig. 9 - Energy spectrum of uncorrelated neutrons in ^{4}He.

From this figure the first thing to be noted is the rather
sharp peak of the deuteron spectrum, centered at about 45 MeV.
The shape and the position of this peak is in quite good agreement
with the theoretical calculations in the paper of ref.[5]. The
proton spectrum is very different from the dueteron one; also here
we have an agreement with the prediction of ref.[5], although not
so good.   In fact, in the energy region ≥ 30 MeV, our ratio, pro-
tons/deuterons, is equal to 0.45 ± 0.10 instead of the calculated
0.35.   Further we find that the relative ratios of the reactions,
$\pi^- + {}^4He = d + 2n$ and $\pi^- + {}^4He = p + 3n$, are respectively $(58 \pm 7)\%$
and $(26 \pm 6)\%$; from this we deduce that about 15% is left to the
reaction, $\pi^- + {}^4He = t + n$. The relative probability of this mode
was a debated matter sometime ago. There were both experimental
and theoretical conclusions attributing to this mode a value going
from 0 to 30%. (see refs.[5,6,7,8,9,and10]). We tried to measure it
directly from the neutron spectrum, hoping that a peak would appear
above the background in the energy region near 90 MeV. Looking at
the Fig. 9 one can see that we succeded. From this result it was
also possible to deduce that the area of the remaining part of the
neutron spectrum is compatible with the normalized total area of
the proton and deuteron spectrum, with an average value of 6% of
the efficiency of our neutron counter. This is in a quite good
agreement with the experimental results and the theoretical predictions for the plastic scintillator efficiency. Due to the expected
decrease of the efficiency at higher energies, we attribute, to the
$\pi^- + {}^4He = t + n$ mode a percentage of $12. \pm 3.$, with respect to the

total absorption process.

Finally comes the argument about the correlated outgoing nucleon pairs. Unfortunately, loss in machine time due to target
troubles, did not allow us to measure statistically significant
spectra. Only tentative values for the experimental ratios,
$w(\pi^-,nn)/w(\pi^-,np)$ and $w(\pi^-,nn)/w(\pi^-,nd)$, are given; they are respectively 9. ± 3. and 7. ± 2.. The first of these values is in
agreement with that reported in ref.[4].

REFERENCES

[1] F. Calligaris, C. Cernigoi, I. Gabrielli and F. Pellegrini, Nucl.
Phys. A 126, 209 (1969).

[2] G. Charpak, J. Fawier, L. Massonet and C. Zupancic, Int. Conf.
on Nuclear Physics, Gutlinburg - Tennessee, Sept. (1966);

C. Zupancic, Proc. Sec. Int. Conf. on High Energy Physics and
Nuclear Structure, Rehovoth (1967);

T. Bressani, G. Charpak, J. Fawier, L. Massonet, W.E. Meyerhof
and C. Zupancic, Phys. Lett. 25B, 409 (1967).

[3] M.E. Nordberg, K.F. Kinsey and R.L. Burman, University of
Rochester, UR-875-210.

[4] D.S. Koltun and A. Reitan, Nucl. Phys. B4, 629 (1968).

[5] S.G. Eckstein, Phys. Rev. 129, 413 (1963).

[6] P. Ammiraju and L.D. Lederman, Nuovo Cimento 4, 281 (1956).

[7] M. Schiff, R.M. Hildebrand and C. Giese, Phys. Rev. 122, 265
(1961).

[8] A.G. Petscher, Phys. Rev. 90, 959 (1953).

[9] M.V. Bortolani, L. Lendinara and L. Monari, Nuovo Cimento 25,
603 (1962).

[10] P. Ammiraju and S.N. Biswas, Nuovo Cimento 17, 726 (1960);

S.G. Eckstein, Phys. Rev. 129, 413 (1963).

'RARE MODE' NUCLEAR ABSORPTION OF PIONS[*]

Kazuo Gotow

Physics Department, Virginia Polytechnic Institute

Since the 1967 meeting of this Conference, a considerable amount of work has been done on the low energy pion nucleus interaction; more refined and detailed work on well-known processes such as $(\pi,2N)$ [1], π-nucleus elastic and inealstic scattering [2] have been performed. Seeing the availability in the near future of low energy pion beams with their intensities a few orders of magnitudes higher than those presently available, it is also important to explore the pion-nucleus processes that have not been studied well mainly because of their low yields. We have made some efforts towards this direction by studying pion absorptions which are relatively rare but may be amenable to theoretical interpretations. This report presents the results of the following two experiments; single proton emission following absorption of π^+ by ^{12}C at 73 MeV, and neutrons following radiative pion capture by ^{12}C, ^{16}O and Ca. The first experiment was done by T. R. Witten, M. Blecher and K. Gotow, the second by W-C. Lam, B. Macdonald, W. P. Trower and K. Gotow.

(1) SINGLE PROTON EMISSION FOLLOWING π^+ ABSORPTION BY ^{12}C AT 73 MeV.

In the reaction ^{12}C$(\pi^+,p)^{11}$C a single neutron in the nucleus absorbs a positive pion when the neutron momentum of the order of 500 MeV/c is available. Letourneaux and Eisenberg [3] made a perturbation calculation using the shell model and their result gives forward cross sections of 80 µb/sr for the $1p_{3/2}$ shell and 20 µb/sr for the $1s_{1/2}$ shell [4] at our pion energy. The only previously studied instance [5] of this process is the reaction $He^4(\pi^+,p)He^3$ in a bubble chamber. Protons from carbon with almost all of the

[*]Supported in part by NASA Grant NGL-004-033

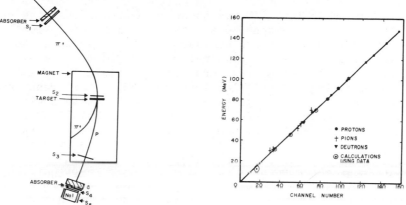

Fig. 1:   Experimental Setup.   Fig. 2:   Calibration of NaI Counter.

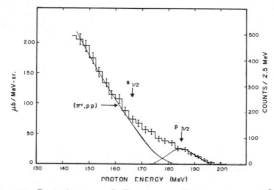

Fig. 3:   Energy Spectrum of Protons Emitted at 0°.

total energy of a positive pion have been observed in a cloud chamber [6]; however, no detailed study was made.

The experimental method is essentially same as that described in the earlier report [7] and Fig. 1 shows the setup. It was designed to detect protons emitted along the incident momentum with energies greater than 130 MeV. The angular resolution was 5° (FWHM). Elimination of backgrounds due to lighter particles was made with a rough magnetic analysis, a lucite Cerenkov anti-counter and π-μ rejection electronics for the NaI total absorption counter. The magnet set a lower limit to the accepted momentum at 340 MeV/c for the particles defined by the counter logic. Fig. 2 shows the calibration of the NaI pulse height vs. the energy deposit. Our typical result is shown in Fig. 3 in terms of the proton energy at the center of the target. The background subtracted is approximately 3%, most of which fall in the lower half of the energy spectrum. The arrows indicate the expected positions for the protons from $1p_{3/2}$ and $1s_{1/2}$ shells. The neutron binding energies used are 19.0 and 37.5 MeV respectively. The curve labeled (π^+,pp) is a calculated distribution of protons resulting from the reaction $^{12}C(\pi^+,pp)^{10}B^*$, normalized at one point. It was obtained assuming that the protons follow three-body phase space distributions with total kinetic energies corresponding to various nuclear levels of ^{10}B up to 40 MeV in excitation. The calculation by Kopaleishvili [8] and the experimental result of Favier et al [9] were used to obtain the relative probability of exciting a particular level in ^{10}B in this reaction. The other curve in Fig. 3 is a gaussian distribution with a width corresponding to our overall energy resolution of 12 MeV (FWHM). This represents what we expect if the single-nucleon emission takes place, leading to the ground state of ^{11}C, with a cross section of 430µb/sr. If we subtract from the data the (π^+,pp) curve and the 430 µb/sr contribution at 185 MeV, we obtain a distribution of the excess counts centered at 174 MeV with an area of 400 µb/sr and a width of 11 MeV. Its origin cannot be determined uniquely due to our poor energy resolution. Also there is no evidence with our statistical accuracy for the single-protons from the $1s_{1/2}$ shell. Similar data were obtained with a slightly different amount of the absorber in front of the NaI counter. The shift of the pulse height distribution thus introduced confirmed the fact that all particles detected were protons; on Fig. 2 the points labeled "calculations from data" were obtained from the pulse height shifts of various portions of the spectra.

In summary, we have obtained a cross section of (430 ± 110)µb/sr for the forward emission of the single-protons from the $1p_{3/2}$ shell. This cross section is five times of the theoretical expectation previously mentioned. Considering a recent observation of an inverse similar reaction, $^{13}C(p,\pi^+)^{14}C$, by Domingo et al [10], a further study of the π^+ absorption by single-nucleons should reveal interesting information on single-hole nuclear states and high momentum components of nucleons in nuclei.

(2) NEUTRONS FOLLOWING RADIATIVE PION CAPTURE BY ^{12}C, ^{16}O and Ca.

In the previous meeting of this Conference Zupancic [11] summarized interesting features of the radiative pion capture by nuclei along with a review of experimental studies on this subject. In order to study the nuclear capture mechanism and the nuclear states involved in this capture mode, we have measured time of flight spectra of neutrons [12] following radiative pion capture in ^{12}C, ^{16}O and Ca. We would like to present results of our preliminary analysis of the data. In this experiment neutron energy distributions have been determined as a function of the angle between neutrons and associated high energy photons. In addition, for the first two nuclei, we have obtained the correlation between neutron time of flight and associated photon pulse height. Fig. 4 shows a typical experimental setup. The photon detectors were lead-glass Cerenkov counters and neutrons were detected with a liquid scintillation counter with a threshold of 100 keV in electron energy.

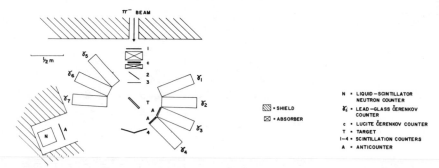

Fig. 4. Experimental Setup for Radiative Pion Capture.

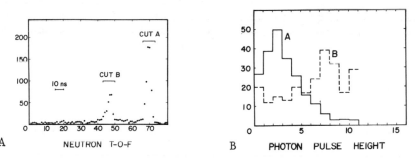

A   NEUTRON T-O-F   B   PHOTON PULSE HEIGHT

Fig. 5,A: Time of flight spectrum from CH_2 in coincidence with γ_1
          counter.
Fig. 5,B: Photon pulse height distributions associated with cut A
          and Cut B.

Fig. 5A shows a time of flight spectrum in coincidence with γ_1 for
a CH_2 target. Peak B is due to 8.9 MeV neutrons from $\pi^- p \rightarrow n\gamma$ and
peak A is mostly due to photons from decaying π^0 produced in $\pi^- p \rightarrow n\pi^0$.
This fact is reflected upon the photon pulse height spectrum associ-
ated with each peak as shown in Fig. 5B. Such data were used for
checking our electronics system and calibrating the time of flight
scale.

    Time of flight spectra obtained with carbon, Fig. 6, demonstrate
qualitative features of the neutron energy distribution. These data
are classified according to the n-γ correlation angle (θ) and the
energy ($E\gamma$) of the associated photons; θ = 180° to 125°; θ = 70° to
25°, $E\gamma$ = 70 to 100 MeV, and $E\gamma \geq$ 100 MeV, where $E\gamma$ is defined as
the most probable photon energy corresponding to the pulse height
cut-offs. At "large correlation angles,"the number of neutrons with
energies greater than 7 MeV associated with "low energy photons" is
substantially higher than that with "high energy photons." (Fig.6A
and B). This fact has been observed previously by Petrukhin et al
[12]. At "smaller correlation angles," however, the neutron energy
distribution does not depend on our photon energy cuts (Fig.6C and D).

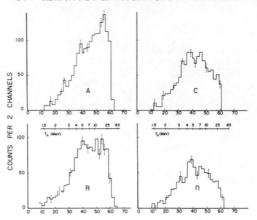

Fig. 6: Neutron time of flight spectra for ^{12}C. A and C are associated with "low energy photons," B and D with "high energy photons." A and B are the spectra at "large correlation angles, C and D are those at "small correlation angles". No correction for the neutron counter efficiency has been applied. The bin width is 3.2 ns.

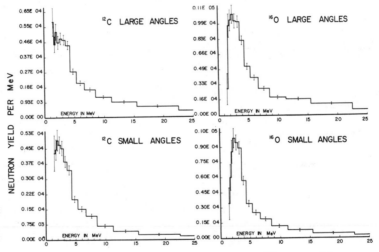

Fig. 7: Neutron energy distributions for ^{12}C and ^{16}O. The data shown here are for all photon energies. The energy bin width corresponds to our time of flight resolution of 4.8 ns. "Small" and "large" angles are as specified for Fig. 6 in the text.

Similar results were obtained also with oxygen nuclei. This immediately suggests that the quasi-free $\pi^-p \to n\gamma$ reaction takes place in the capture process emitting photons according to quasi-free kinematics, and that there exists another mode of the capture which involves nuclear excitations accompanied with photons of lower energies and de-excitation neutrons. For the latter mechanism theoretical studies [13,14] have indicated that the nuclear excitation

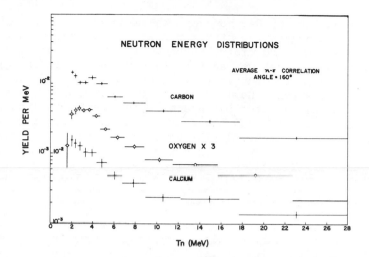

Fig. 8: Neutron energy distributions at "large angles" for ^{12}C and ^{16}O and natural Ca. The data are for all photon energies. The inside scale of the yield is for Ca and Carbon.

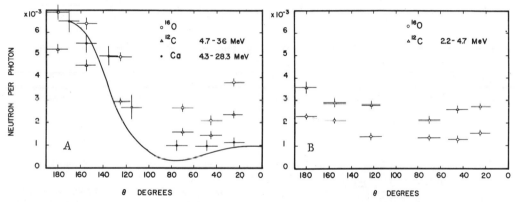

Fig. 9--A: Angular distributions of high energy neutrons. The theoretical curve is normalized to the data at one point.

B: Angular distributions of neutrons in 2.2 - 4.7 MeV. The data are for all proton energies.

should result in the analogues of the giant resonance states of the absorbing nucleus. Fig. 7 shows the neutron energy distributions following radiative pion capture by ^{12}C and ^{16}O. The corrections for the accidental coincidences and for the neutron counter efficiency are included. All distributions show structures around 2 to 4 MeV in energy. This is in contrast with a smooth exponential rise down to 1.5 MeV of the neutron spectrum resulting from non-radiative captures. These structures are observed consistently at all angles. They are at the energy region in which the decay neutrons from the giant resonance states are expected to appear [14]; for ^{12}C most prominently 2.1, 4.6, and 6.7 MeV neutrons from the decaying 23.5 MeV (1$^-$) state of ^{11}B; for ^{16}O 2.7 and 9.0 MeV neutrons from 22.0 MeV (1$^-$) state of ^{16}N, and 4.1 MeV neutrons from 17.5 MeV (2$^-$)state.

In radiative capture by Ca, however, the measured neutron energy spectra shows no suggestive structures at all as one can see on Fig. 8 along with ^{12}C and ^{16}O distributions. It is quite probable with our time resolution that we can not see any details in energy distributions above 7 MeV even if they exist. A theory of the **quasi-free** absorption, $\pi^-p \to n\gamma$, has been given by Dakhno and Prokoshkin [15] using a pole-model and also by Anderson [16] who takes into account nuclear shell-structures and the final state interaction of neutrons in an optical model potential. The angular correlation of neutrons from the quasi-free absorption is expected to peak at 180°. Experimental angular distributions of neutrons with energies greater than approximately 4 MeV are shown on Fig. 8A. The solid curve is the theoretical curve [16] corresponding to the Ca^{40} data. They are in a fair agreement. On the other hand, the angular distributions of neutrons in a 2.2-4.7 MeV energy range are nearly isotropic indicating the majority of neutrons in this energy range come from decays of the excited nucleus. This is seen on Fig. 8B.

Summarizing the results, the radiative pion capture involves two major processes: One is the quasi-free absorption which describes the properties of high energy neutrons adequately. The other is the excitation of the analogue giant resonance states, some evidence of which is presented here. For the second mechanism, however, our results can not positively identify the excited states involved. Finally, I would like to thank for their support the staff of the Space Radiation Effects Laboratory where both experiments were performed.

References

[1]   For example: F. Calligaris, C. Cernigoi and I. Gabrielli;
        Nucl. Phys. A126, (1969) 209, also reference [9].
[2]   For example: D. T. Chivers, E. M. Rimmer, B. W. Allardyce,
        R. C. Witcomb, J. J. Domingo and N. W. Tanner; Nucl. Phys. A126
        (1969) 129, J. P. Stroot; Proceedings of the Symposium on the
        Use of Nimrod for Nuclear Structure Physics, Rutherford High
        Energy Laboratory, edited by C. J. Batty (1968), p. 81.
[3]   J. Letourneaux and J. M. Eisenberg: Nucl. Phys. 87 (1966)331.
[4]   A harmonic oscillator parameter, α, of 125 MeV/c is used in
        the radial wave function, $R_\ell = N_\ell \, r^\ell \, \exp[-\alpha^2 r^2/2]$.
[5]   M. M. Block; Bull.Amer.Phys.Soc. 14 (1969) 52.
[6]   H. Byfield, J. Kessler, and L.M. Lederman; Phys.Rev.86 (1952)17.
[7]   T.R. Witten, M.Blecher and K. Gotow; Phys.Rev. 174 (1968)1166.
[8]   T. I. Kopaleishvili, I.Z. Machabeli, G. Sh.Giorksadze and N.B.
        Krupennikova; Phys.Letters 22 (1966) 181.
[9]   J. Favier, T. Bressani, G. Charpak, L. Massonet, W.E. Meyerhof,
        and C. Zupancic; Phys. Letters (1967) 409.

[10]   N.W. Tanner: Proceedings of the Symposium on the Use of Nimrod
       for Nuclear Structure Physics, Rutherford High Energy Laboratory,
       edited by C.J. Batty (1968) p. 91 and private communication
       from N. W. Tanner.
[11]   C. Zupancic: High Energy Physics and Nuclear Structure,
       edited by G. Alexander (1967), North-Holland Pub.Co.,p.185.
[12]   Petrukhin et al made a similar measurement with a Li target;
       V.I. Petrukhin, V.S. Pogosov, and Yu.D. Prokoshkin; Soviet
       Jour. of Nucl. Phys. 5 (1967) 745.
[13]   L.L. Foldy and T.D. Walecka: Nuovo Cimento 34 (1964) 1026,
       H. Überall, Nuovo Cimento (Suppl.) 4 (1966) 781, and Phys.
       Rev. 139, (1965) B1239.
       J. Barlow, J.C. Sens, P.J. Duke and M.A.R. Kemp; Phys. Letters
       9 (1964) 84.
       V.V. Balashov, V. B. Beliaev, N.M. Kabachnik and R.A.Eramzhian
       Phys. Letters 9 (1964) 168.
[14]   R. Raphael, H. Überall and C. Wernts; Phys. Letters 24B
       (1967) 15.
       F. J. Kelly and Überall: Nucl. Phys. A118 (1968) 302.
       H. Überall: Proceedings of the Williamsburg Conference on
       Intermediate Energy Physics, edited by H. O. Funsten (1966)
       p. 327.
[15]   L. G. Dakhno and Yu D. Prokoshkin; Soviet Jour. of Nucl. Phys.
       7 (1968) 351.
[16]   D. K. Anderson; Private Communication.

INELASTIC INTERACTIONS OF PIONS AND NUCLEONS WITH NUCLEI AT HIGH AND SUPERHIGH ENERGIES

V.S.Barashenkov, K.K.Gudima, S.M.Eliseev
A.S.Iljinov, V.D.Toneev

Joint Institute for Nuclear Research,
D u b n a, USSR

(Presented by V.S. Barashenkov)

1. Detailed calculations showed that the inelastic pion- and nucleon–nucleus interactions are well accounted for by the mechanism of intranuclear cascades in the whole range of the primary particle energies T from a few dozens of MeV to a few GeV.

Good agreement of the cascade calculations with experiment is obtained not only for interactions of elementary particles with nuclei but also for interactions of deuterons with nuclei.

Discrepancies with experiment at $T \lessapprox 1$ GeV indicated in some papers are due to shortcomings of particular versions of the cascade models rather than to the violation of the cascade mechanism itself[1].

However, attempts to extend directly the intranuclear cascade model to the energy range $T > 10$ GeV leads immediately to an essential disagreement with experiment which increases with increasing T.

2. One of the most serious difficulties encountered in calculating the intranuclear cascades is the necessity to introduce detailed information on inelastic particle interactions at different energies into the computer. This imposes high requirements to the computer area storage and in many cases turns out to be completely impossible owing to the absence of necessary experimental data.

The calculation becomes essentially simpler when, as the initial information for the simulation of the picture of an elementary act, one uses instead of the

experimentaly distributions $\omega(\theta)$ and $\omega(p)$ the polynomial approximations of the corresponding integral
distributions $\omega(\geqslant\theta)$ and $\omega(\geqslant p)$. To find such approximations a laborious numerical analysis of a large
number of experimental data is needed. However, the
approximations obtained in such a way can be used in
calculating the cascades in various nuclei and at various energies[2].

It should be noted that the energy dependence of
the characteristics of elementary interactions turns
out to be in this case continuous, which considerably
improves the accuracy of calculations.

Another fact which essentially influences the
accuracy of cascade calculations is the necessity for
the energy and momentum conservation laws to be satisfied when each act of inelastic $\pi\text{-}N$ or $N\text{-}N$
interaction is simulated by the Monte-Carlo method.
In order to calculate the integral average quantities,
like the average multiplicity and the average energy
of secondaries, it is sufficient to take into account
these laws only statistically, i.e. on the average
over a large number of interactions [3]. In so doing
one obtained quite good results also for the total angular and energy distributions.

However the detail calculations shows that the
dispersion of the difference of the total energy before and after interaction near their middle $(\Delta E \simeq 0)$
turns out to be surprisingly large. This may lead to
noticeable errors in characteristics such as the number of particles in a fixed energy interval, the particle
spectrum at a fixed angle and so on. The excitation
energy of the residual nucleus and, consequently, the
number of black prongs in the star are found to be
especially vulnerable.

A rather effective method of simulating the inelastic elementary particle interactions taking exact
account of the energy and momentum conservation laws
has recently been developed in our laboratory.

3. The theoretical average multiplicities $\overline{n}_g$
and $\overline{n}_h$ in proton- and pion-nucleus collisions
which are in good agreement with experiment at T<5GeV
do not reflect the experimentally observed "saturation" at higher energies. The calculated $\overline{n}_s$ values
are very close to the experimental ones up to T$\simeq$20GeV
where there appear noticeable disagreements.

The difference between theoretical and experimental characteristics is revealed more clearly in considering the particle correlations. For example, at

T<5 GeV the dependence of $\bar{n}_s$ on the number of h-prongs in the star is in good agreement with experiment while at higher energies the calculated histograms differ noticeably for the measured ones.

As to the dependence of the average number of grey tracks on the number of s-particles, at T⩾5 GeV it is impossible to speak of even qualitative agreement with experiment.

It whould be noted however that the discrepancy of the energy and especially of the angular characteristics with the experimentally observed ones is noticeably smaller than for the multiplicity. For example, the difference of the experimental and theoretical $\bar{n}_g$ by more than a factor of two in proton-nuclear interactions at T=22.5 GeV is accompanied by only a 5% discrepancy in the average g-particle energies. The difference in the angular particle distributions is found to be even smaller.

At energies T⩾100 GeV the disagreement between the cascade calculations and experiment becomes even more pronounced. For example, at T≃10^3 GeV the theoretical s-particle multiplicity is larger than the experimental one by about a factor of three; some disagreement is observed in the angular distribution too.

4. Thus, the deviations from the ordinary cascade theory start to show first of all in the characteristics of the low-energy component of produced particles at T⩾5 GeV. The estimates show that there are several reasons for such deviations. First, in all cascade calculations yet performed one ignores completely the fact that, as the cascade develops, a still larger number of intranuclear nucleons is involved in it, owing to which the low-energy component of cascade particles meets, on its way, smaller nuclear density. The number of "evaporation" particles decrease, as well.

Another important fact which is disregarded in cascade calculations consists in that at energies higher than a few GeV 'resonons' are produced in π-N and N-N collisions which are then involved in the intranuclear cascade. From the kinetical point of view this is to some extent equivalent to the simultaneous interaction of several "stuck together" particles with an intranuclear nucleon.

With further increasing energy T, owing to relativistic contraction, the angles of emission of particles produced in π-N and N-N collisions become so small that any discrimination of the times of interactions of these particles with an intranuclear nucleon is

meaningless. In other words, there occurs simultaneous
scattering and absorption of several particles on one
nucleon (the absorption of a 'resonon' by a nucleon may
be considered as a particular case of such multiple
particle interactions).

Since at present we know nothing about the properties of multiple particle interactions (MPI) it is ad-
visable to consider the inverse problem: let us attempt
to obtain some information on these interactions from
the analysis of the experimental data of cosmic-ray experiments. It was assumed that all the particles, the
free paths of which end near the center of the intranuclear nucleon at distances shorter than its radius ,
interact simultaneously with this nucleon. The properties of such MPI were supposed to be dependent on only
the value of the "free" energy $\mathcal{E} = \sqrt{(\sum E_i)^2 - (P_i)^2} - \sum M_i$ in c.m.s.
which can be spent for the production of new particles
(E_i , P_i , M_i are the total energy, momentum
and mass of the i-th particle absorbed by a nucleon).

It is clear that such a model gives a rather simplified description of the physical process. However
even in this case one can draw a number of quite definite and rather general conclusions.

The calculations have shown that even at $T \simeq 10$ GeV
the number of MPI in the average emulsion nucleus is
about 20% and at $T \simeq 10^3$ it reaches 40%. In this case
the fraction of particles involved in MPI increases
from 30 to 70%.

It should be however noted that the large contribution of MPI at $T \simeq 10$ GeV appears to be due to the
fact that in calculations the decrease of the nuclear
density was disregarded and all the deviations from
the usual cascade model were ascribed to the MPI effect.
The account of the decrease in the number of intranuclear nucleons is a very complicated problem which we
have just began to solve.

With the aid of computers we have performed several
series of calculations which differ in the assumptions
for the MPI properties. One succeeds in obtaining agreement with experimental data over the range $T \simeq 30-10^3$ GeV
only if the angular and energy distributions of particles
produced in MPI have the asymmetry angular distribution
of created particles (as in inelastic $\pi-N$ interaction at high energies) and besides the existence of a
leading particle carrying away 50-70% of the total
energy is assumed. Presence of the leading particle and
the asymmetry character of the angular distributions of
the remaining particles may be considered to be rather
reliable assumptions.

The table shows good agreement between the calculation results taking into account MPI and experiment.

5. In conclusion we would like to stress once more that the study of the particle—nucleus interaction mechanism at high and superhigh energies depends essentially on the "transition" energy region T=2—30GeV. It is interesting to study not so much integral, ave — rage characteristics as the differential distributions and correlations between various quantities. Particular attention should be given to the low-energy component of produced particles.

TABLE

Comparison of the results of cascade calculations taking into account multiple particle interactions with experiment

T,GeV	Interaction	Characteristics	Theory	Experiment[3]
100	p+LEm	$\bar{n}_s$	7.9+0.4	7.4+0.5
		$\bar{T}_s$,Gev	3.1∓0.2	2.9∓0.3
	p+Em	$\bar{n}_s$	10.3∓0.5	8.0∓0.5
		$\bar{n}_g$	3.6∓0.2	5.0∓1.6
		$\bar{T}_s$,Gev	2.8∓0.2	2.4∓0.9
200	π^-+LEm	$\bar{n}_{s_0}$	9.7∓0.4	8.0∓0.9
		$\theta_{1/2\,s}$	6.5∓0.3	6.2∓0.4
	π^-+Em	$\bar{n}_{s_0}$	11.2∓0.6	10.8∓0.9
		$\theta_{1/2\,s}$	9.0∓0.5	8.3∓0.6
	π^-+HEm	$\bar{n}_{s_0}$	15.4∓0.7	14.7∓2.0
		$\theta_{1/2\,s}$	12.0∓0.6	11.0∓1.1
500	p+Em	$\bar{n}_s$	18.0∓0.9	18.8∓4.2
		$\bar{n}_g$	3.7∓0.2	4.0∓0.8
10^3	p+LEm	$\bar{n}_s$	12.1∓0.6	9.9∓1.4
	p+Em	$\bar{n}_s$	20.5∓1.1	22.5∓3.0
		$\bar{n}_g$	3.6∓0.2	4 ∓1.6

LEm, Em, HEm -- are medium-light, medium, and medium-heavy emulsion nuclei respectively; $\bar{T}_s$ is the mean energy of secondaries (except the leading one); $\theta_{1/2s}^\circ$ is the angle within which half of the s-particles are emitted (lab. system).

R e f e r e n c e s

1. V.S.Barashenkov, K.K.Gudima, V.D.Toneev, Preprints JINR: P2-4313, P2-4346, P2-4402, P2-4302 (1969).
2. V.S.Barashenkov, K.K.Gudima, V.D.Toneev, Preprints JINR: P2-4065, P2-4066 (1968).
3. I.Z.Artykov, V.S.Barashenkov, S.M.Eliseev, Nucl. Phys. B6, 11(1968); B6, 628 (1968).

TOTAL PHOTOABSORPTION CROSS SECTIONS ON HYDROGEN AND COMPLEX

NUCLEI AT ENERGIES UP TO 18 GEV[*]

D. O. Caldwell, V. B. Elings, W. P. Hesse, G. E. Jahn,
R. J. Morrison, and F. V. Murphy

Physics Department, University of California
Santa Barbara, California

D. E. Yount[+]

Stanford Linear Accelerator Center, Stanford University
California

(Presented by R. J. Morrison)

Total cross sections for the photoproduction of hadrons from
hydrogen, deuterium, carbon, copper, and lead have been measured
at SLAC at energies up to 18.3 GeV. These measurements yield
fundamental information about the nature of photon interactions.

On the basis of the hydrogen photoabsorption cross section,
which implies a photon mean free path in nuclear matter of about
800 Fermis, one would expect that the total cross section on
nucleus A, $\sigma_{\gamma A}$, would be just

$$\sigma_{\gamma A} = A \, \sigma_{\gamma p}, \tag{1}$$

where $\sigma_{\gamma p}$ is the total photoabsorption cross section on the proton.
In this picture the nucleus is a thin target which presents A nucleons to the photon beam. On the other hand the vector meson dominance picture gives a qualitatively different behavior. In this
case the photon interacts as a superposition of vector mesons,
dominated by the rho, and these are strongly absorbed. Since the
mean free path of the vector mesons is expected to be short, ~ 3
fermis, the absorption takes place on the surface and the A dependence at high energies should be closer to proportionality to the
nuclear area ~ $A^{2/3}$.[1]

Quantitatively, in the picture of rho dominance[2-6] the total
cross section on nucleus A is closely related to forward rho production from the same nucleus as can be seen from the diagrams:

387

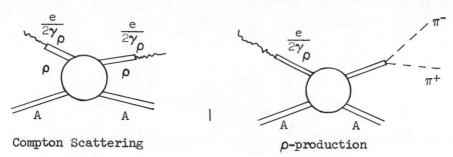

Compton Scattering   ρ-production

At high energies the Compton amplitude at t=0 is $\frac{e}{2\gamma_\rho}$ times the t=0
rho production amplitude where e is the electron charge and γ_ρ is
the rho-photon coupling constant. Using the optical theorem and
the assumption that the amplitude is mostly imaginary one obtains
the relationship

$$\sigma_{\gamma A} = \sqrt{4\pi\alpha \frac{4\pi}{\gamma_\rho^2} \frac{d\sigma}{dt}(\gamma A \to \rho A)\Big|_{t=0}} \, . \tag{2}$$

In this expression $\frac{d\sigma}{dt}(\gamma A \to \rho A)\Big|_{t=0}$ is the rho production differen-
tial cross-section on nucleus A extrapolated to t=0. The essential
point is the proportionality of $\sigma_{\gamma A}$ to $\sqrt{\frac{d\sigma}{dt}(\gamma A \to \rho A)_{t=0}}$.

The experimental setup shown in Fig. 1 was designed to overcome the two main problems encountered in photon total cross
section measurements. The first problem, obtaining beam photons
of known energy, was solved by the method of tagging.[7] A positron
beam of energy $E_0 \pm .005 \, E_0$ and less than a cm in diameter was
incident upon a radiator of thickness .002 radiation lengths. The
beam was focussed to a spot of about 3 mm diameter at the position
of the target and also had the important property that the lower
energy charged particle contamination was less than one particle
per million. A positron which radiated a photon of energy E_γ such
that $.74 \, E_0 < E_\gamma < .94 \, E_\gamma$ was deflected by the tagging magnet into
one of the four tagging counter telescopes. From the curvature in
the magnet the final positron energy E' was determined. A twofold
tagging coincidence indicated that a beam photon of energy $E_\gamma = E_0 - E'$ was incident upon the target.

The second major problem was the huge background from electron-
positron pairs. The number of pairs produced varies with Z and is
200 to 2000 times the true signal in hydrogen and lead respectively.
The total absorption shower counter S1, used as a veto, eliminated
this background. Each tagged photon which was not absorbed in the
target produced a large pulse in S1. Pair production in the target
also resulted in a large veto pulse since the members of the pair

were produced at small angles $\sim \frac{m}{E^{\pm}}$ where m is the electron mass.

SIDE VIEW – NOT TO SCALE

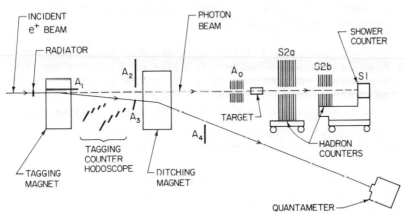

Fig. 1   Schematic Representation of the Experimental Arrangement.

In principle the experiment could have been carried out by counting the number of tagged photons which were absorbed in the target. In practice, however, even with no target 5% of the tagging signals were not accompanied by a large shower in S1. These "false tags" were reduced to 0.3% by the introduction of the anti-counters A_1 and A_2. Tridents and pairs produced in the radiator, which produced a tagging signal but had no accompanying photon, were vetoed by this arrangement.

Even the 0.3% false tag rate was large compared with the rate of true photoabsorption, which was 0.1% for the most favorable case, deuterium. Positive identification of the hadronic final states was provided by the 4-layer, lead-scintillator sandwich detectors S2a and S2b. The requirement of either a 4-fold coincidence or a large shower pulse in at least one of the detectors significantly reduced the accidentals due to low energy electrons but provided an efficiency of better than 99% for the detection of charged pions of energy greater than 2 GeV. These detectors were mounted on separate carts, which could be moved in the beam direction, and which enabled us to change geometry with energy and also to check for lost events and electron-positron pair contamination. The pulse heights from the relevant counters as well as the time over-lap of the main coincidence were recorded on tape for each event.

The effect of pile-up in S1, which viewed the entire bremsstrahlung spectrum, was reduced by the anti-counters A_3 and A_4.
Positrons in these counters were coincident with medium energy
photons which would normally have contributed to accidental vetos
of good events. With A_3 and A_4 in operation this effect was
typically 1%. The shower anti-counter A_O eliminated the few wide
angle photons which were produced in the radiator. The tagging
system with associated veto counters provided a monitor of
absolute accuracy better than 0.3%.

The total cross sections for hydrogen and deuterium are
shown in Fig. 2. The data below 8 GeV are preliminary and are
subject to possible corrections of a few percent. In order to
extract the neutron cross section we have made a rough Glauber
correction of 5 μbarns;[2] a more careful analysis will be performed.

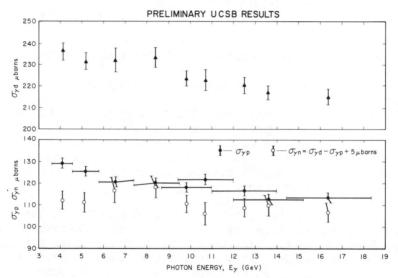

Fig. 2   Total γp and γD cross sections. The neutron cross sections
have been evaluated assuming a Glauber correction of 5
μbarns.

From this data one can see that two rho dominance predictions
are borne out. In the first place the γ-p cross section has an
energy dependence which is typical of the cross sections of strongly interacting particles on the proton, e.g. π-p cross sections.
Secondly, the neutron and proton cross sections are nearly the same.
Fig. 3 shows a quantitative comparison with the predictions of

vector dominance. The measured values of $\sigma_{\gamma p}$ are compared with the predictions of ρ dominance using Eq. 2 and the ρ production measurements of Anderson, et al.[8] The colliding beam value[9] $\frac{\gamma_\rho^2}{4\pi}$ = 0.52 has been assumed and an 18 μbarn contribution from the ω and φ has been included. The comparison with our measurements is good, but better agreement is obtained with $\frac{\gamma_\rho^2}{4\pi}$ = 0.42.[10]

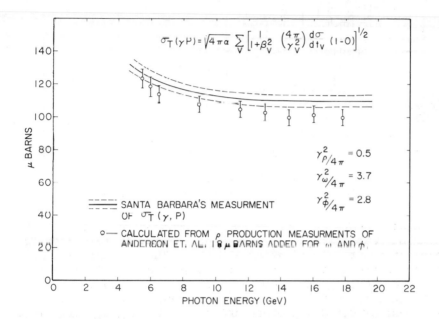

$$\sigma_T(\gamma P) = \sqrt{4\pi\alpha} \sum_V \left[\frac{1}{1+\beta_V^2} \left(\frac{4\pi}{\gamma_V^2} \right) \frac{d\sigma}{dt_V}(1\text{-}0) \right]^{1/2}$$

$\gamma_\rho^2/_{4\pi}$ = 0.5

$\gamma_\omega^2/_{4\pi}$ = 3.7

$\gamma_\phi^2/_{4\pi}$ = 2.8

SANTA BARBARA'S MEASURMENT OF $\sigma_T(\gamma, P)$

○— CALCULATED FROM ρ PRODUCTION MEASURMENTS OF ANDERSON ET. AL. 18 μ BARNS ADDED FOR ω AND ϕ.

μ BARNS

PHOTON ENERGY (GeV)

Fig. 3 Comparison of measured values for $\sigma_{\gamma p}$ and vector dominance predictions using the data of Anderson, et al and colliding beam values of the ρ, ω, and φ coupling constants. The contribution from the $\omega+\varphi$ is assumed to be 18 μbarns.

In concluding the discussion on photoproduction on the nucleon we see in Fig. 4 the compilation of the existing measurements which are in good agreement. The DESY and SLAC/HBC measurements are given in References 11 and 12 respectively. The measurements of SLAC/Group A[13] were made by the technique of inelastic electron scattering with the cross sections evaluated by extrapolating to zero mass photons. When the systematic errors in these measurements are included, good agreement is obtained.

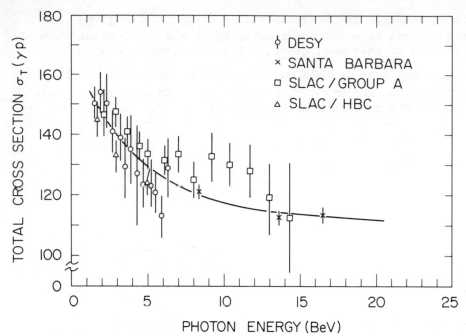

Fig. 4 – Comparison with other measurements.

The measurements on complex nuclei become more difficult with increasing nuclear charge. We have kept the target thickness for all nuclei at about 0.1 radiation length. The rate of pair production background is then a constant while the signal decreases by a factor of 14 in going from hydrogen to lead. Corrections for pair contamination, which were checked experimentally, were 5% for the worst case of 8.4 GeV photons and the lead target. Similarly the empty target subtraction on lead was about 35% and the accidentals were effectively about 14%.

The A dependence of the measurements is shown in Fig. 5.[17] One can see, by comparison with the line $\sigma_{\gamma A} = A \, \sigma_{\gamma p}$, that photons experience strong absorption and therefore do not behave as "simple photons." For comparison the A dependence predicted by rho dominance (using Eq. 2) from the Cornell[14] and SLAC-LRL[15] rho production data on complex nuclei is shown. In this comparison the details of the nuclear structure are not involved. One sees a pronounced disagreement leading to the conclusion that rho dominance is not valid. Rho-production measurements presented at this conference by the DESY-MIT group contradict those of Cornell and SLAC-LRL and appear to be in better agreement with our results and the prediction of rho dominance.

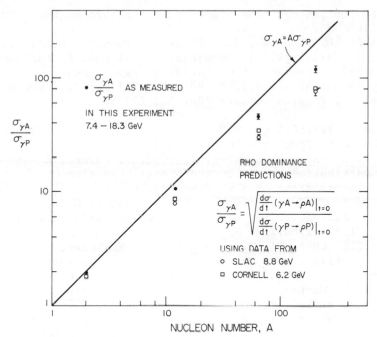

Fig. 5   The ratio of the cross section of nucleus A to that of H
as observed (black points) compared with the result ex-
ected from a purely electromagnetic photon (line) or from
a ρ-dominant photon using Cornell and SLAC-LRL data (open
points).

FOOTNOTES AND REFERENCES

*Supported in part by the U.S. Atomic Energy Commission.

+Now at the Department of Physics and Astronomy, University of
Hawaii, Honolulu, Hawaii.

[1]L. Stodolsky, Phys. Rev. Letters 18, 135 (1967).
[2]S.J. Brodsky and J. Pumplin, Phys. Rev. 182, 1794 (1969).
[3]M. Nauenberg, Phys. Rev. Letters 22, 556 (1969).
[4]K. Gottfried and D.R. Yennie, Phys. Rev. 182, 1595 (1969).
[5]B. Margolis and C.L. Tang, Nuc. Phys. B10, 329 (1969).
[6]Talks given at this conference by D.W.G.S. Lieth, K. Gottfried,
and B. Margolis.
[7]D.O. Caldwell, J.P. Dowd, K. Heinloth, and M.D. Rousseau
Rev. Sci. Instr. 36, 283 (1965) provides a detailed description
of this type of beam.
[8]R. Anderson, D. Gustavson, J. Johnson, D. Ritson, B.H. Wiik.
W.G. Jones, D. Kreinick, F. Murphy, and R. Weinstein, SLAC-PUB-644

(1969, to be published). The Figure is given through the courtesy
of B.H. Wiik.

[9]J.E. Augustin, D. Benaksas, J.C. Bizot, J. Buon, B. Delcoart,
U. Gracco, J. Haissinski, J. Jeanjean, D. Lalanne, F. LaPlanche,
J. LeFrancois, P. Lehmann, P.C. Marin, H. Nguyen Ngoc,
J. Perez-y-Jorba, R. Richard, F. Rumpf, E. Silva, S. Travernier,
and D. Treille, Physics Letters 28B, 503 (1969).

[10]D.O. Caldwell, V.B. Elings, W.P. Hesse, G.E. Jahn, R.J. Morrison,
F.V. Murphy, D.E. Yount (to be published).

[11]H. Meyer, B. Naroska, J.H. Weber, M. Wong, V. Heynen, E. Mandelkow,
D. Notz, Contribution to the International Symposium on Electron
and Proton Interactions at High Energies, Liverpool (Sept. 1969).

[12]J. Ballam, G.B. Chadwick, Z.G.T. Guiragossian, P. Hein, A. Levy,
M. Menke, E. Pickup, P. Seyboth, T.H. Tau, G. Wolf, Phys. Rev.
Letters 21, 1544 (1968).

[13]E.D. Bloom R.L. Cottrell, D.H. Coward, H. DeStaebler, Jr.,
J. Drees, G. Miller, L.W. Mo, R.E. Taylor, J.I. Friedman,
G.C. Hartman, H.W. Kendall, SLAC-PUB-653 (1969, to be published).

[14]G. McCellan, N. Mistry, P. Mostek, H. Ogren, A. Silverman,
J. Swartz, and R. Talman, Phys. Rev. Letters 22, 377 (1969).

[15]Data on C, Cu, and Pb from Ref. 16. The cross section on
hydrogen of 122 ± 12 μbarns/GeV^2 was reported at the Boulder
Conference August, 1969 by the same group. This data was presented
at this conference in a talk by D.W.G.S. Leith.

[16]F. Bulos, W. Busza, R. Giese, R.R. Larsen, D.W.G.S. Leith,
B. Richter, V. Perez-Mendez, A. Stetz, S.H. Williams, M. Beniston,
and J. Rettberg, Phys. Rev. Letters 22, 490 (1969).

[17]Total cross section measurements on complex nuclei at energies
up to 6.3 GeV have been reported at Liverpool. A. Meyer,
B. Naroska, J.H. Weber, M. Wong, V. Heynen, E. Mandelkow, D. Notz,
Contribution to the International Symposium on Electron and Photon
Interactions at High Energies, Liverpool (Sept. 1969). These
measurements appear to be in good agreement although the energies
do not quite overlap. The comparison is shown in the talk by
D.W.G.S. Leith.

DISCUSSION

P. Nemethy: In Fig. 2 it appears that $\sigma_{\gamma n}$ is consistently lower than
$\sigma_{\gamma p}$. Is this significant, or is it within experimental error?
R.J.M.: It may be significant, but the results are "preliminary";
also, better estimates of the Glauber correction need to be made.

RHO MESON PHOTOPRODUCTION FROM COMPLEX NUCLEI*

D. W. G. S. Leith

Stanford Linear Accelerator Center

Stanford University, Stanford, California 94305

I. Introduction

Although the title refers only to photoproduction of rho mesons on complex nuclei , I wish to talk on a rather broader footing. I do this with the justification that in the spirit of the vector dominance model, (VDM), all photo-processes involve the coupling of the photon to the vector mesons. To this end I would like to review the situation on ρ^0 photoproduction on complex nuclei, the total photoabsorption cross section determinations, and the implications of these results for VDM. Finally, I will summarize the experimental results of the searches for heavy vector mesons.

Before going on to talk of photons, let us take a few minutes to review why we use nuclear targets. There are two phenomena — familiar to all of you — which lead us to use complex nuclear targets in high energy physics — diffraction and absorption. The diffraction phenomena,[1] or coherent production from each of the individual nucleons within the nucleus provides a filter for the isolation of reactions involving no change in quantum numbers (i. e., isospin, G-parity and spin-parity series remain unchanged). Such processes may then be studied preferentially to reactions involving meson exchange, spin flip, I-spin exchange, etc. A process may be coherent, and also be inelastic, (in the sense that the incoming and outgoing particles do not have the same mass), if the longitudinal momentum transfer required to make up the mass difference is small compared to the momentum required to break up the nucleus. A list of reactions which could proceed coherently on nuclear targets is given below, and includes examples involving mass changes and also excitation of the spin-parity series:

(a) Elastic scattering, in the forward direction.

(b) $p \rightarrow P_{11}$, where P_{11} is the N*(1400), a $J^P = 1/2^+$ nucleon isobar.

*Work supported by the U. S. Atomic Energy Commission.

(c) $\pi \to A_1$, where A_1 is a 3π mesonic state of mass $\sim$1080 MeV,
          with $J^P = 1^+$.

(d) $K \to K^*_{1300}$, where K^*_{1300} is a $K\pi\pi$ meson of mass $\sim$1300 MeV,
          and with supposed $J^P = 1^+$.

(e) $\gamma \to \rho$, where ρ is the 750 MeV dipion resonance with $J^P = 1^-$.

The second phenomena — the absorption of the outgoing particles — ,
allows us the possibility of determining their attenuation in nuclear matter.
If the outgoing particles live long enough to traverse the nucleus, then by
varying the path length in nuclear matter (which we do by varying the size
of the nucleus) and observing the relative yield, we can deduce the total
cross section. Since we have a whole menagerie of unstable particles in
high energy physics and no other way to measure many of their properties
such as the total cross section, this is a useful technique bridging high
energy physics and nuclear physics. The idea was originally used by
Drell and Trefil[2] to show that the total cross section for protons on protons, derived from an experiment measuring p-A scattering,[3] was in good
agreement with observed pp data. The technique is currently being used
to determine the A_1-p [4] and K^*_{1300}-p [5] total cross sections, in addition
to the experiments to be discussed below on the absorption of rho mesons
by nucleons.

These attenuation investigations were assumed to be independent of
the details of nuclear models, or indeed the details of interaction within
the nucleus. It was hoped to treat the nucleus as a black box which was
our variable thickness absorber in a classical experiment to measure the
total cross section by attenuation. The nuclear physicists present will not
be surprised to find that our naive assumptions were indeed naive! I will
discuss briefly how serious these assumptions are as we come to specific
issues later.

The question of studying photon interactions essentially boils down to
the testing of vector dominance theory — this model has been described
fully before.[6] Basically, the electromagnetic interaction of hadrons is
described by the coupling of the electromagnetic field to the hadronic electromagnetic current —

$$j_\mu^{em} (x) = j_\mu^I (x) + \frac{1}{2} j_\mu^Y (x) \quad , \tag{1}$$

where $j_\mu^I (x)$ and $j_\mu^Y (x)$ are the zero components of the isospin current and
the hypercharge current respectively. The smallness of the coupling constant, $\alpha = e^2/4\pi$, allows one in most cases to treat photoproduction in
lowest order of the electromagnetic interactions.

The vector dominance model then connects the hadronic electromag-
netic current with the fields of the vector mesons ρ^0, ω, and ϕ which
have the same quantum numbers as the electromagnetic current, namely
$J = 1$, $P = -1$, $C = -1$. This connection can be made via the current field

identity —

$$j_\mu^{em}(x) = - \left[\frac{m_\rho^2}{2\gamma_\rho} \cdot \rho_\mu^o(x) + \frac{m_\omega^2}{2\gamma_\omega} \cdot \omega_\mu(x) + \frac{m_\phi^2}{2\gamma_\phi} \cdot \phi_\mu(x) \right] \equiv -\sum_V \frac{m_V^2}{2\gamma_V} \cdot V_\mu(x),$$

(2)

where γ_ρ, γ_ω, γ_ϕ are coupling constants, and m_ρ, m_ω, m_ϕ are the masses of the vector mesons.

The assumption which is being made here is that the vector mesons ρ, ω, ϕ completely satisfy this summation — or that the contribution from the three known vector mesons completely saturates the electromagnetic current. This is a very strong statement and I wish to return to it later in the talk.

According to our field-current identity, any amplitude involving real or virtual photons is a linear combination of vector meson amplitudes, each multiplied by a vector meson propagator. The assumption is then made that the invariant vector meson amplitudes are slowly varying functions of the vector meson mass m_V, — i.e., any energy dependence comes from the propagators and not the coupling constants.

Historically, the vector dominance model (VDM) had its birth with the explanation of the nucleon form factor in terms of vector meson clouds, [7] but has since been generalized to explain the hadronic interactions of the photon. [6] This model may be used to relate many processes involving photons to similar processes involving the vector mesons. It is the hope that all of these relationships will be satisfied by a single value for the vector meson-photon coupling constants. We shall discuss below several experiments attempting to measure this quantity.

II.   Photoproduction of Rho Mesons from Complex Nuclei

The cross section for the photoproduction of rho mesons from nuclear targets, has been measured at DESY[8,9] (2.7-4.5 BeV), at Cornell[10] (6.2 BeV), and at SLAC[11] (at 8.8 BeV). The published results show some disagreement. I wish to describe the SLAC experiment in some detail, present some new but preliminary data, and to discuss the differences between the three experiments.

The scope of the SLAC experiment is shown in Table I. The data up to 10 BeV has been obtained using a monochromatic photon beam which has been described previously. [12] The beam is obtained from two-photon annihilation in flight of 12 BeV positrons on the orbital electrons in a liquid hydrogen target. A typical energy spectrum of the beam is shown in Fig. 1(a). The data above 10 BeV are obtained using a conventional thin target bremsstrahlung beam at 10, 13, and 16 BeV. The energy spectrum at 16 GeV is shown in Fig. 1(b). The exposures were made in such a way that one is able to test for inelastic contributions as one moves away from the end point of the bremsstrahlung spectrum by comparison of the cross sections for several energy cuts (see Fig. 1(c)). This is an especially important point in the bremsstrahlung experiment.

TABLE I

SYSTEMATIC STUDY OF RHO MESON PHOTOPRODUCTION
USING A WIRE SPARK CHAMBER SPECTROMETER AT SLAC

TARGET \ ENERGY (GeV)	MONOCHROMATIC γ's			BREMSSTRAHLUNG		
	5	7	9	10	13	16
H_2	×		×	×	×	×
D_2			×	×		×
B_e	×	×	×	×		×
C	×	×	×			
Aℓ	×	×	×			
Cu	×	×	×			
Ag			×			
Pb	×	×	×			×

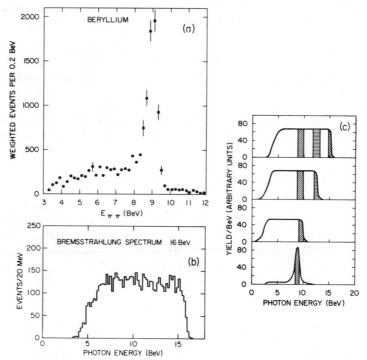

Fig. 1(a) --The energy spectrum from the monochromatic beam at 9 BeV, (b)the thin target bremsstrahlung spectrum for 16 BeV, (c) schematic representation of the photon beam energy spectra from various runs. The shaded areas show energy cuts at corresponding points on the spectra, allowing checks to be made on the possible contribution from inelastic processes. Note that the low energy cutoff of the above spectra is due to the energy acceptance of the spectrometer.

The experimental apparatus used in this experiment is shown schematically in Fig. 2. The wire spark chamber spectrometer and the on-
line IBM 1800 computer system have been described in detail elsewhere.[13]
The photon flux is measured, pulse-by-pulse, by a simple e^+e^- pair spectrometer installed in the last sweeping magnet. To calibrate the absolute
photon flux, the spark chamber system was periodically used as a pair
spectrometer. The properties of the system are: (a) a large mass acceptance of ~ 1000 MeV per setting, with a maximum detectable mass of
~ 3000 MeV, (b) good mass resolution, $\sim \pm 8$ MeV, (c) large decay angular
acceptance, (d) momentum transfer acceptance from 0 to 0.25 $(GeV/c)^2$,
with a resolution of 0.0005 $(GeV/c)^2$ for small t, increasing to 0.002
$(GeV/c)^2$ for large t.

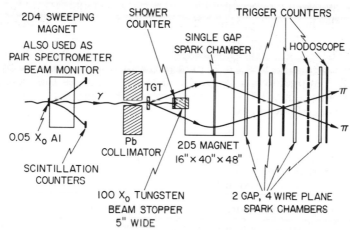

Fig. 2--The spectrometer system, showing the arrangement of the counters,
the magneto-strictive read-out wire spark chambers and the two photon
monitors; a pair spectrometer and a shower counter inside the tungsten
beam stopper. Periodically, for calibrating the 2D4 pair spectrometer, the
beam stopper was removed and the spark chamber system converted into an
electron-positron pair spectrometer. For full description see Ref. 11.

    The large decay angular acceptance allows us to verify that the rho
mesons are indeed transversely polarized, normally an assumption in the
other experiments measuring rho photoproduction. The decay distribution,
evaluated in the helicity system, for rho mesons produced at 9 BeV from
a Be target is shown in Fig. 3. The solid curve is the result of a fit to the
distribution, evaluating the spin density matrix elements using:

$$W(\cos\theta, \phi) = \frac{3}{4\pi} \left[0.5(1-\cos^2\theta) + \rho_{00}(3/2\cos^2\theta - 1/2) \right.$$
$$\left. - \rho_{1-1}\sin^2\theta\cos 2\phi - \sqrt{2}\,\mathrm{Re}\,\rho_{10}\sin 2\theta\cos\phi \right]$$

(3)

assuming

$$\rho_{00} + 2\rho_{11} = 1$$

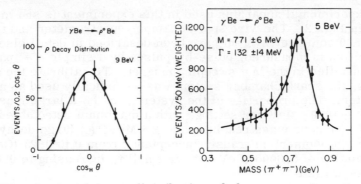

Fig. 3--The observed decay distribution of rho mesons from Be, at 9 BeV, evaluated in the helicity system. The data includes the forward coherent peak (i.e., $t \lesssim .05(\text{GeV/c})^2$).

Fig. 4--The mass distribution of pion pairs produced from Be by photons from the 5 BeV monochromatic peak. The solid line is the best fit to the data using a coherent mixture of resonant and diffractive background amplitudes, as described in text.

    The fit which takes into account the geometrical acceptance of the system, resulted in

$$\rho_{00} = 0.0 \pm 0.1$$
$$\rho_{1-1} = -0.03 \pm 0.05$$
$$\text{Re}\rho_{10} = 0.03 \pm 0.03$$

Clearly the rho mesons are produced with an essentially complete transverse alignment.

    The effective mass distribution of the dipion pairs is measured for each target at each energy. Figure 4 shows the spectrum measured at 5 BeV for Be. The solid line is the result of a fit to the following model:

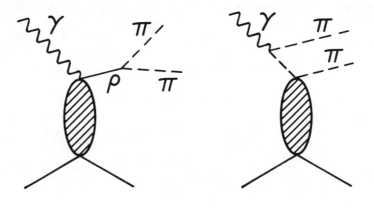

$$\frac{dN}{dM_{\pi\pi}} = C_0 M_\rho \left[\frac{M_{\pi\pi}\Gamma}{(M_\rho^2 - M_{\pi\pi}^2)^2 + M_\rho^2\Gamma^2} + C_1 \frac{M_\rho^2 - M_{\pi\pi}^2}{(M_\rho^2 - M_{\pi\pi}^2)^2 + M_\rho^2\Gamma^2} + C_2 \right] \quad (4)$$

where $\Gamma = \Gamma_\rho \dfrac{M_\rho}{M_{\pi\pi}} \left(\dfrac{M_{\pi\pi}^2 - 4\mu^2}{M_\rho^2 - 4\mu^2} \right)^{3/2}$

and $C_0, C_1, C_2, M_\rho, \Gamma_\rho$ are free parameters, the last two being the mass and width of the rho. This model assumes that the rho meson production amplitude is given by the Breit–Wigner form, and that it interferes coherently with an imaginary amplitude describing the diffractive $\pi\pi$ scattering. This formalism is due to Söding.[14] We find the measured mass and width of the rho meson do not vary as a function of the photon energy, k, or the atomic number, A, and the mean values are found to be;

$$M_\rho \sim 765 \pm 10 \text{ MeV}$$
$$\Gamma_\rho \sim 145 \pm 10 \text{ MeV}$$

The differential cross section is then found by integrating overall decay angles, and all masses around the rho. A study of the differential cross section may then be used to find the total rho–nucleon cross section, $\sigma(\rho N)$, and the vector dominance coupling constant, $\gamma_\rho^2/4\pi$.[15] The equations used in this analysis[16] are given below.

$$\left. \frac{d\sigma}{dt} \right|_{t_{min}} (\gamma A \to \rho^0 A) = \frac{d\sigma}{dt} (\gamma N \to \rho^0 N)\, f\left(\sigma_{\rho N}, \rho(r), t_{min} \right) \quad (5)$$

$$\underbrace{\hspace{4cm}}_{\substack{\text{integral over the} \\ \text{volume of the} \\ \text{nucleus}}}$$

Where

$$t_{min} = \text{minimum 4-momentum transfer} \approx - M_\rho^4/4k^2$$

and

$$f\left(\sigma_{\rho N}, \rho(r), t \right)$$
$$= \left| 2\pi \int_0^\infty \int_{-\infty}^\infty b\, db\, dz\, \underbrace{J_0(q_\perp b)}_{\substack{\text{Nuclear} \\ \text{Shape}}} \underbrace{\rho(b,z)}_{\substack{\text{Mass} \\ \text{Dependence}}} e^{\displaystyle i\sqrt{-t_{min}}\, z - \frac{\sigma_{\rho N}}{2} \int_z^\infty \rho(b,z')dz'} \right|^2 \quad (6)$$
$$\underbrace{\hspace{4cm}}_{\substack{\text{Attenuation of} \\ \text{Rho}}}$$

Now the A-dependence of $(d\sigma/dt)|_{t_{min}}$ $(\gamma A \to \rho^0 A)$ can be used to determine $\sigma_{\rho N}$.

Further, assuming vector dominance and an imaginary forward amplitude, we can write

$$\frac{d\sigma}{dt}\Big|_{t_{min}} (\gamma A \to \rho^0 A) = \frac{\alpha}{4} \frac{4\pi}{\gamma_\rho^2} \frac{\sigma_{\rho N}^2}{16\pi} \ f\left(\sigma_{\rho N}, \rho(r), t_{min}\right) \tag{7}$$

Thus, a measurement of the forward rho production cross section, and of $\sigma(\rho N)$, together with knowledge of the nuclear density distribution, $\rho(r)$, allows the determination of the photon-rho coupling constant, $\gamma_\rho^2/4\pi$ (i.e., the relative A-dependence may be used to determine $\sigma(\rho N)$, and then the absolute value of the differential cross sections together with $\sigma(\rho N)$ may be used to determine the coupling constant).

The nuclear density distribution used for these calculations was a Wood-Saxon distribution!

$$\rho(r) = \rho_0 \left\{1 + \exp\left(\frac{r - C}{a}\right)\right\}^{-1} \tag{8}$$

$$C = C_0 A^{1/3} \ \text{fermi}$$

$$a = .535 \ \text{fermi}$$

where C_0 was taken from electron-nucleus scattering[17] as $= 1.08$, and from nucleon-nucleus scattering[3] as $= 1.18$.

The published 9 BeV results of the SLAC experiment[11] were obtained using the above formulation, taking C_0 to be A-independent, and equal to the two values given above. The results are shown in Fig. 5. However, if the nuclear radii suggested by Glauber and Matthiae[18] are used (i.e., ~15% variation in C_0 as a function of A) then the total cross section determined from the SLAC BeV data is $\sigma(\rho N) = (34 \pm 5)$mb.

At this stage we did not include our hydrogen data for two good reasons — (1) it was not ready, (2) the hydrogen amplitude may include, in principle, contributions other than a purely diffractive amplitude — these contributions come from spin flip amplitudes or from exchange of iso-spin, and have been filtered out in the coherent production using nuclear targets. We have now completed a study of our hydrogen and deuterium data at 9 BeV and see no evidence of substantial spin or iso-spin exchange. We therefore feel free to use this data in our overall fit to the A-dependence.

The mass spectra for the hydrogen and deuterium data is shown in Fig. 6. The solid line represents the best fit using the model described above. The mass and width of the rho are found to be

$M = 760 \pm 10$ MeV $\left.\right\}$ Hydrogen, 9 BeV. $M = 765 \pm 10$ $\left.\right\}$ Deuterium, 9 BeV.

$\Gamma = 135 \pm 10$ MeV $\left.\right\}$ $\Gamma = 152 \pm 10$ $\left.\right\}$

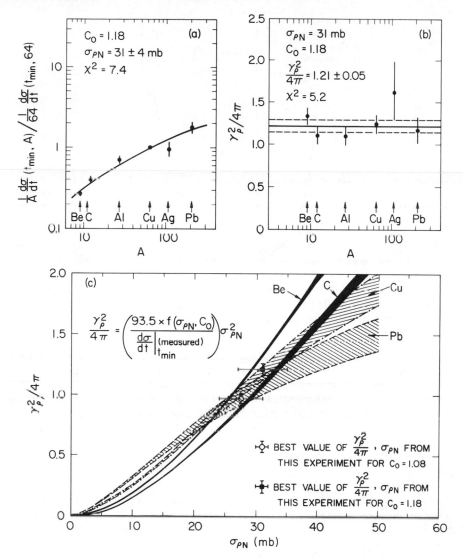

Fig. 5 (a)--The forward cross section as a function of A, relative to Cu. The solid line is the best fit using the optical model described in the text with $C_0 = 1.18$ fermi; the $\sigma_{\rho N}$ deduced from this fit is (31 ± 4) mb. (Statistical error.) (b) The photon-rho coupling constant as determined from our forward cross sections and $\sigma_{\rho N}$. The solid line is the best A-independent fit to the data and gives $\gamma_\rho^2/4\pi = 1.21 \pm 0.05$. (Statistical error.) (c) The dependence of $\gamma_\rho^2/4\pi$ on $\sigma_{\rho N}$ for our measured forward cross sections for several nuclei. In the relation $\sigma_{\rho N}$ is in barns and $d\sigma/dt$ in mb (BeV/c)^2. The best estimate of $\delta(\rho n)$ and $\gamma_\rho^2/4\pi$ was taken as the mean of the values obtained with $C_0 = 1.18$ f and 1.08 f.

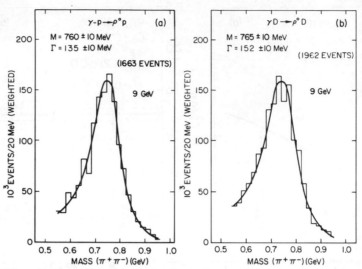

Fig. 6--The dipion mass distributions from 9 BeV photon interactions on (a) hydrogen and (b) deuterium. The solid line is the best fit to the data using a coherent mixture of resonant and diffractive background amplitudes.

The decay distribution of rho mesons produced on hydrogen at 9 BeV, is shown in Fig. 7. The solid line is the best fit to the data, using Eq.(3). The density matrix elements are shown in Fig. 8(a) for the helicity system, [19] and Fig. 8(b) evaluated in the Jackson system. [20] We see that the dynamics of the production are such that the rho meson is transversely polarized in the helicity frame, and not in the Jackson system. Density matrix elements from HBC experiments[21] at 4.3 and 5 BeV are shown for comparison. The corresponding data for deuterium is shown in Fig. 9(a) and Fig. 9(b)

The differential cross section for rho production for both hydrogen and deuterium[22] is shown in Fig. 10. The forward cross sections are found to be:

$$\frac{\gamma\sigma}{\gamma t} \ (\gamma p \to \rho^0 p) = (122 \pm 12) \ \mu b/GeV/c^2$$

$$\frac{\gamma\sigma}{\gamma t} \ (\gamma d \to \rho^0 d) = (430 \pm 30) \ \mu b/GeV/c^2$$

$$\text{and } R = \frac{d\sigma/dt \ (d)}{d\sigma/dt \ (p)} \Bigg|_{t=0} = 3.5 \pm 0.3$$

The value of R expected, after taking into account the Glauber correction, is 3.65, while the Cornell group have measured R, at 6.2 BeV, to be (3.2 ± .2).[23] Our measurement is in agreement with the expected value of R for the case of pure diffraction, but also agrees, within errors, with the Cornell determination. In addition, we see no differences in the spin-density matrix elements in the forward direction for the hydrogen

and deuterium experiments. We therefore conclude that the rho production on hydrogen is dominantly diffractive and use the data in a reanalysis
of the A-dependence. When the hydrogen and deuterium data are included
in the A-dependence study, the observed effective total rho-nucleon cross
section in nuclear matter increases to ~43 mb (see Fig. 11).

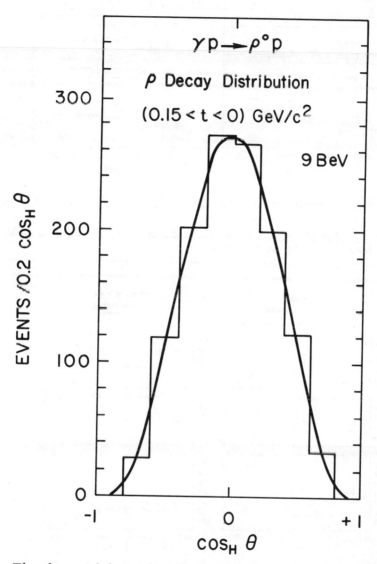

Fig. 7--The observed decay distribution for forward produced rho mesons
(i. e., < 0. 15 GeV/c^2) on hydrogen at 9 BeV. The solid line is a best fit
to the data.

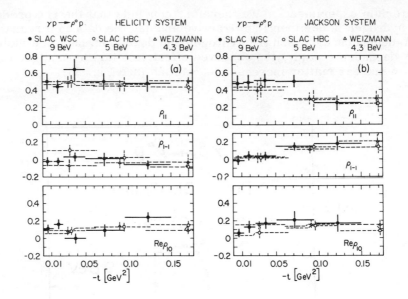

Fig. 8--The spin density matrix elements for rho decay evaluated in (a) helicity frame, and (b) the Jackson frame. The data is from the SLAC wire spark chamber experiment at 9 BeV, the SLAC HBC group at 5 BeV and the Weizmann HBC group at 4.3 BeV.

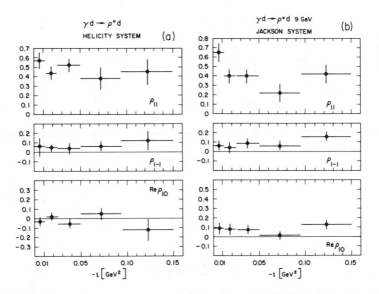

Fig. 9--The spin density matrix elements for rho mesons produced on deuterium at 9 BeV, evaluated in (a) the helicity frame, and (b) the Jackson frame.

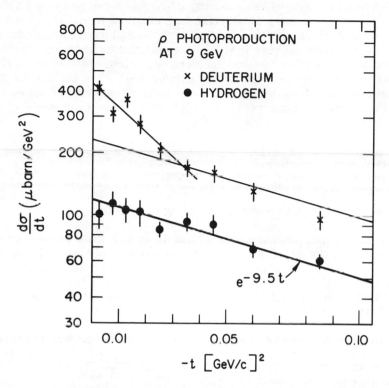

Fig. 10--The differential cross section for rho production on hydrogen and deuterium at 9 BeV.

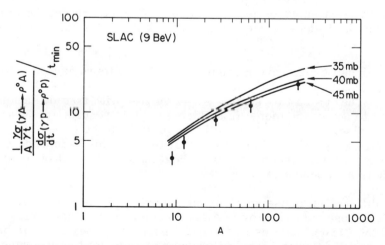

Fig. 11--The A-dependence of the SLAC 9 BeV forward rho cross section.

An alternate method of determining $\sigma(\rho N)$ has been proposed by Silverman,[24] in which the ρ absorption is compared to the observed total cross sections of π's, K's, N's. In Fig. 12 I have plotted the square of the total cross section on nuclear targets, normalized to hydrogen, as a function of atomic number for several incident particles. New data on K_2^0-A total cross sections, presented to this Conference,[25] is included. The solid lines are optical model calculations from Silverman,[24] based on the measured attenuation cross sections of $K^{\pm}$, $\pi^{\pm}$, $\bar{p}$, and p on nucleii, at 10.8 and 19 BeV/c, from Galbraith et al.[26]

Also plotted are the ratio, $\partial\sigma/\partial t(\gamma A \rightarrow \rho^0 A)\big/\partial\sigma/\partial t(\gamma p \rightarrow \rho^0 p)$, for the SLAC and Cornell experiments.[9,10] This quantity is related, through VDM, to the ratio $\left[\partial\sigma/\partial t(\rho^0 A \rightarrow \rho^0 A)\big/\partial\sigma/\partial t(\rho^0 p \rightarrow \rho^0 p)\right]$, which in turn is related to $\left[\sigma_T^2(\rho^0 A)/\sigma_T^2(\rho^0 p)\right]$ by the optical theorem. It is clear from the plot that $\sigma(\rho N)$ is more nearly 40 mb than 25 mb.

There is now preliminary data available at 5, 7, and 16 BeV on complex nuclei, from the SLAC group. The forward, and extrapolated differential cross sections are given in Table II. An overall fit to the A-dependence of all the SLAC complex nuclei data yield a total cross section, $\sigma(\rho N)$, ~ 37 mb. If the hydrogen cross section is included in the fit the best estimate of the cross section increases to ~ 42 mb.

Preliminary values of the forward cross section for several targets, as a function of the photon energy, k, is shown in Fig. 13. The solid curve is the prediction of the Drell-Trefil formalism, normalized to the highest energy data points. The measured energy dependence shows good agreement with the model. The extrapolated t=0 differential cross section is shown in Fig. 14 for several targets. The data is certainly in agreement with an energy independent cross section, as would be expected for a diffractive process. However, perhaps more important, the SLAC experiments at different energies (5, 7, 9 and 16 BeV) agree rather well on the k-dependence and A-dependence and seem to represent a self-consistent set of data.

The Cornell group have measured the rho photoproduction cross sections from H_2, D_2, Be, C, Mg, Cu, Ag, Au, and Pb targets at 6.2 BeV photon energy.[10] They use a two-magnet spectrometer with a scintillation counter hodoscope to detect the pions; this setup and their results are described more fully elsewhere.[10,23] Cornell derive $\sigma(\rho N)$ to be (38 ± 4) mb, and $\gamma_\rho^2/4\pi = (1.2 \pm .2)$. It should be noted that they use an A-dependent radius in their analysis (as discussed above), and include the hydrogen cross section in their A-dependence study.

The differential cross sections measured by Cornell are somewhat larger than those published by SLAC. To compare the experiments, I have analyzed the Cornell data using the same phenomenology that has been followed for the SLAC data. The differences are in: (a) the resonance line shape — the SLAC group use a p-wave, energy dependent width for the rho meson, rather than the s-wave, constant width shape of Cornell, (a 7% effect);

(b) the treatment of the background contribution to the dipion effective mass distribution — SLAC fit using a coherent diffractive background while Cornell ignore any background (10-15% effect); (c) the energy dependence of the forward cross section — the forward rho cross section on nuclear targets should show the same energy dependence as the cross section on protons, shown in Fig. 15 (an 11% effect between 6 and 9 BeV). In Table III I have listed the measured Cornell data at 6.2 BeV, the data adjusted (as described above) to show what should be measured at 9 BeV if the Cornell measurements are correct, and finally, the SLAC measured data at 9 BeV. There is clearly good agreement between these experiments.

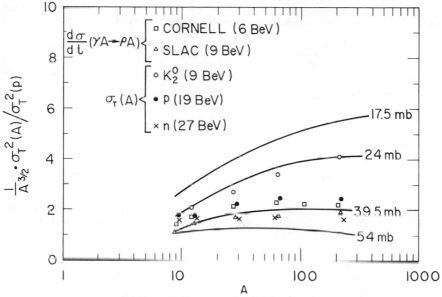

Fig. 12--The A-dependence of the total cross section for K_2^0, p and n compared to rho mesons. The curves are optical model calculations.

TABLE II
SLAC PRELIMINARY DATA ON THE
FORWARD AND EXTRAPOLATED DIFFERENTIAL CROSS SECTIONS FOR $\gamma A \rightarrow \rho^0 A$ (in mb/GeV/c^2)

Photon Energy k (BeV)	4.83		6.88		8.83		15	
Atomic Number A	$\theta = 0^0$	t = 0	$\theta = 0^0$	t = 0	$\theta = 0^0$	t = 0	$\theta = 0$	t = 0
9	$3.49 \pm .66 \pm .20$	$4.40 \pm .83 \pm .25$	$4.21 \pm .94 \pm .38$	$4.72 \pm 1.05 \pm .43$	$3.48 \pm .27$	$3.73 \pm .29$	$4.72 \pm .47$	$4.76 \pm .47$
12	$6.45 \pm 1.23 \pm .40$	$8.31 \pm 1.59 \pm .52$	$8.30 \pm 1.83 \pm .70$	$9.45 \pm 2.08 \pm .80$	$6.85 \pm .74$	$7.39 \pm .80$	-	-
27	$20.0 \pm 5.1 \pm 1.7$	$28.1 \pm 7.2 \pm 2.4$	$33.5 \pm 8.2 \pm 3.7$	$39.7 \pm 9.3 \pm 4.4$	26.7 ± 2.7	29.6 ± 3.0	-	-
64	$72.5 \pm 15.2 \pm 5.9$	$119 \pm 25 \pm 10$	$91 \pm 24 \pm 13$	$116 \pm 31 \pm 17$	90.4 ± 9.2	104.7 ± 10.2	-	-
108	-	-	-	-	147 ± 29	178 ± 35	-	-
208	$268 \pm 67 \pm 45$	$635 \pm 159 \pm 107$	-	-	525 ± 69	677 ± 89	530 ± 67	586 ± 74

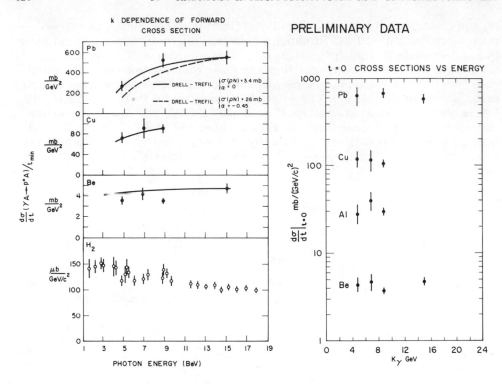

Fig. 13--The energy dependence of the forward rho cross section for Be, Cu and Pb as measured by the SLAC wire chamber group. The forward hydrogen cross section is shown for comparison.

Fig. 14--The energy dependence of the extrapolated, t=0, differential cross section for rho production.

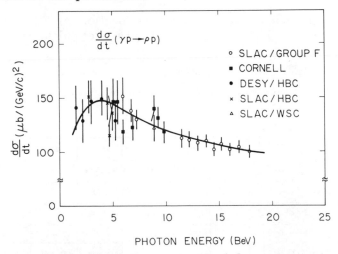

Fig. 15 --The energy dependence of the forward rho cross section on hydrogen.

TABLE III
CORNELL-SLAC COMPARISON

$$\left[\frac{d\sigma}{dt} (\gamma A \rightarrow \rho^0 A) \Big|_{t=0} \right]$$

Atomic Number A	Cornell Measured Cross Section (mb/GeV/c^2)	Cornell Adjusted[†] Cross Section (mb/GeV/c^2)	SLAC Measured Cross Section (mb/GeV/c^2)
Be	5.4 ± .16	4.01 ± .15	3.7 ± .3
C	9 ± .27	6.7 ± .20	7.4 ± .8
Aℓ	33.3 + 1.0	24.8 + 0.7	29.6 + 3.0
Cu	154 ± 4.5	114.5 ± 3.6	104.7 ± 10.2
Pb	890 ± 26	662 ± 20	677 ± 89

[†]This correction takes into account the different treatment in the mass fitting:
(a) A different Breit-Wigner line shape; (b) Background subtraction. This
amounts to 17%. In addition, the k dependence of the $\gamma p \rightarrow \rho^0 p$ reaction has
been used to compare 6 BeV with 9 BeV. This is an additional correction of 11%.

The third group performing these measurements (DESY/MIT) found results[8] in contradiction to the above experiments. They have now new results from a very beautiful and very systematic experiment which became available only yesterday.[27] They wish to present their data themselves and so Dr. Knasel will present these results himself at the end of this talk. The old experiment did not have data conveniently available for comparison — so very little could be understood of this discrepancy.

We have discussed the new experiment and the data quite extensively and find that the same phenomenology has been followed by both the DESY and SLAC groups, in treating the data (with the exception of evaluating the geometrical efficiency of the respective apparatus), the same features are observed for the background in the dipion mass spectra, the same measured mass and width of the rho and the same measured rho production cross sections on hydrogen. We do, however, seem to differ in our quoted cross sections for nuclear targets.

The results quoted for this new experiment show $\sigma(\rho N) \sim 26$ mb, and $\gamma_\rho^2/4\pi \sim 0.5$ in good agreement with their own previous experiment,[28] but in disagreement with the SLAC and Cornell experiments. The origin of this discrepancy must be found soon, and will only be found by detailed discussion among the three groups involved.

Before leaving this topic, I'd like to discuss briefly two effects which could change the evaluation of the above experiments; — (a) two-body correlations in nuclear matter, and (b) the presence of a large real part in ρ-N scattering.

The existence of two-body correlations in nuclear matter has long been recognized, but the effects have been assumed to be small and the independent particle model of the nucleus has been taken to be adequate.

However, groups at Stanford[29] and McGill[30] have submitted papers to this Conference giving detailed evaluation of this effect on calculations of particle absorption in nuclear matter. The two-body correlation function was evaluated using a "hard-core" repulsive potential together with a short range attractive force determined from the fitting to the low energy nucleon-nucleon scattering data. The two calculations differ in their quantitative conclusions, but both imply $\sim 10\%$ correction for the case of the rho absorption experiments discussed above. This implies that the $\sigma(\rho N)$, deduced from an experiment observing the absorption of rho mesons in nuclear matter should be reduced by $\sim 10\%$ to give the real "free nucleon" cross section.

The effects of a large real part in rho-nucleon scattering have been considered by Talman and Schwartz.[31] They fixed the value of the coupling constant, $\gamma_\rho^2/4\pi$, to the upper limit of the storage ring determination,[32] (i.e.,~ 0.65), and then adjusted the rho-nucleon cross section and α, the ratio of the real to imaginary parts of the ρ-N scattering amplitudes, until they were able to fit the A-dependence of the forward cross section in $\gamma A \rightarrow \rho^0 A$. They were able to fit the Cornell data at 6.2 BeV with $\sigma(\rho N)$ ~ 27 mb and $\alpha = -.45$.

I would like to make some observations on the likelihood of such a large real part. The total photon cross section shows an energy dependence very similar to that of the average of $\pi^\pm p$ cross sections. Figure 16 shows the data[33] on $\sigma^T(\gamma p)$, while the solid line is $1/200$ of the average of the $\pi^+ p$ and $\pi^- p$ total cross sections. The best fit to the data above 2 BeV is found to be $\sigma^T(\gamma p) = (96.5 + 70/\sqrt{k})\mu b$. Even with this rather arbitrary normalization, the agreement on both magnitude and k-dependence between the photon and hadron cross sections is extraordinarily good. Such an observation would lead one to expect that the real parts involved in the two processes should be comparable; (i.e., $\alpha_{\gamma N} \sim \alpha_{\rho N} \sim \alpha_{\pi^\pm p} \sim -0.2$ at 5 BeV).

Secondly, the total cross section data (extrapolated smoothly to high energies) may be used, with the optical theorem and dispersion relations to evaluate the real part in Compton scattering. This calculation of the dispersion integral has been performed by Damashek and Gilman,[34] and the calculated real part (minus the Thomson term, e^2/m) is shown in Fig. 17. Again this supports a real part in γp (and hence from the vector dominance model in ρN scattering) of $\alpha \leq -0.2$ at 5 BeV.

Finally, the inclusion of this large real part only becomes valid if experiments at many energies can be satisfied by approximately the same values of $\sigma(\rho N)$ and α (when allowance is made for their expected smooth, slow energy dependence). The energy dependence of the preliminary SLAC lead and copper data (see Fig. 13) shows disagreement with the predictions of the Cornell parameters. More striking, however, is the A-dependence at high energy. The quantity, $\Delta\sigma(\rho N)/\Delta\alpha$, varies about a factor of ~ 3 between 5 BeV and 16 BeV. The SLAC 16 BeV data could not be made compatible with a $\sigma(\rho N)$ of ~ 26 mb without $|\alpha|$ being >1.

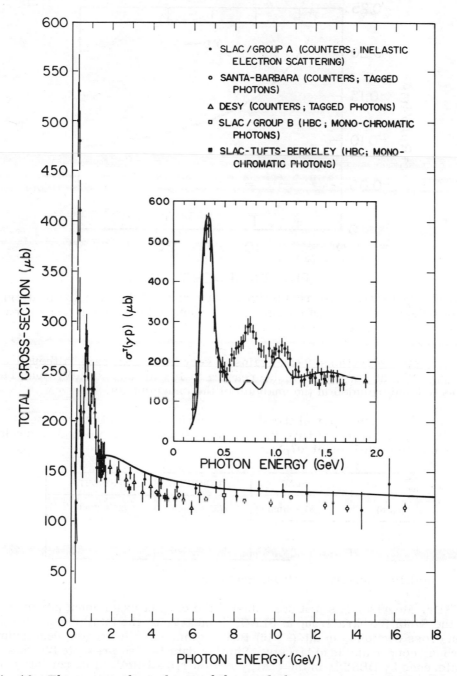

Fig. 16--The energy dependence of the total photon cross section.   The
solid line is 1/200 of the mean of the π^+p and π^-p total cross sections.

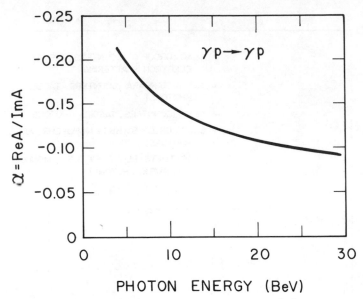

Fig. 17--The ratio of the real to imaginary parts of the Compton scattering amplitude as a function of energy. (The Thomson term, e^2/m , has been subtracted from the calculated value.)

These observations speak against a very large real part, although values ~ 20% at low energies, decreasing to 10% around 20 BeV, must be expected and included in the analysis of the experiments.

Finally, a summary of the situation in rho photoproduction experiments is given below. I have included the effects of the two-body correlations on the evaluation of $\sigma(\rho N)$.

	SLAC	CORNELL	DESY/MIT
$\sigma(\rho N)$	(35 ± 4) mb	(35 ± 5) mb	(26 ± 2) mb
$\gamma_\rho^2/4\pi$	~1.2	~1	~0.5

III. Total Photoabsorption Cross Sections

Dr. Morrison has just described[35] the elegant experiment performed by the Santa Barbara group at SLAC,[36] measuring the total photon absorption cross sections, in the (8 - 16) BeV region, from hydrogen, deuterium, carbon, copper and lead targets. Similar data is also presented to this conference by DESY[37] in the energy region 1.5 - 6 BeV. I do not intend to describe these experiments, but wish to discuss the results within the framework of vector dominance.

The detailed theoretical calculations for the absorption of photons have been described elsewhere, [38] but may simply be expressed in terms of the following two processes:

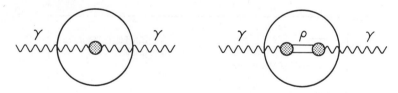

In process (a), the Compton scattering proceeds through the single step of direct interaction with one of the nucleons, whereas in (b) there is an intermediate state of the rho meson — a two-step process. At low energies, the phase difference between these two diagrams is rather large, being given by $\left\{\exp\left[(m^2_{\pi\pi}/2k) \cdot R\right]\right\}$, and therefore only the left-hand process contributes. This means that the photon is very weakly absorbed and the A-dependence will go as A. At high energies, the phase difference becomes negligible, but the diagrams are 180° out of phase, and so there is complete cancellation. This results in an A-dependence characteristic of the absorption cross section of the strongly interacting particle in the intermediate state of process (b). The A-dependence of the cross section would then be expected to go as the surface area asymptotically, and $\sim A^{0.8}$ at the energies of these experiments. The transition between the 1-step and 2-step domain, or between the cross section varying as A and $A^{0.8}$, is predicted to be in the region 4-8 BeV. The DESY experiment gives the A-dependence as $A^{0.95+0.02}$, while UC SB say it is of order $A^{0.9}$. Both experiments yield a value which is neither in one domain nor the other. There is also no observed change in slope as a function of energy. In Fig. 18, we show the energy dependence of the total cross section for several nucleii normalized to hydrogen; the black dots refer to the Santa Barbara experiment while the open circles represent the data from DESY. The consistency between the two experiments is very evident.

The curves show the energy dependence as calculated in the model of Brodsky and Pumplin[38] for various values of the rho-nucleon cross section. The data clearly agree with a cross section of 17 mb or less, and disagree with the 35 mb cross section that would be expected from the SLAC-Cornell determination of $\sigma(\rho N)$. In addition it should be noted that the data do not agree with the 26 mb that is predicted by DESY/MIT experiment.

One comment should be made at this stage on the spectral functions used in the above detailed calculation. [38] The evaluations of the A-dependence of the total photon cross sections have been made using a symmetric, 110 MeV wide mass distribution for the rho, as seen by the storage rings. [32] Now clearly the dipion mass spectrum, as seen in photoproduction, is a very different shape, showing a large shoulder at low $\pi\pi$ masses. These 300-400 MeV dipions will be coherently produced and thereby cause appreciable shadowing, at very low photon momenta.

Therefore, the transition region calculated by these models will in fact be much more gradual than has been presented to date, although at high energies the results will be unchanged.

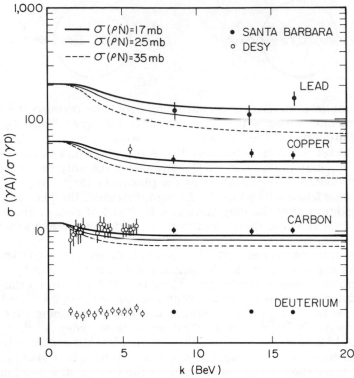

Fig. 18--The energy dependence of the total photon cross section for several nucleii. The curves show the calculated k-dependence for different values of the rho-nucleon total cross section, $\sigma(\rho N)$, using the model of Brodsky and Pumplin with $r_0 = 1.13$ f.

In Fig. 19, the A-dependence of the total cross sections, normalized to hydrogen, is plotted for energies between 5 and 18 BeV. The upper and lower lines reference the case of zero absorption and complete absorption respectively. The total cross section data agree very well between experiments, and for different energies, on a slope just a little less than the zero absorption limit. The other data points on the plot are the square root of the rho photoproduction cross sections normalized to hydrogen, from the SLAC and Cornell experiments. The straight line through these points is the expected A-dependence for a total rho-nucleon cross section of 35 mb.

From the optical theorem and rho dominance[39] it can be shown:

$$\frac{\sigma_T(\gamma A)}{\sigma_T(\gamma p)} = \sqrt{\frac{\frac{d\sigma}{dt}(\gamma A \to \rho^{\circ} A)}{\frac{d\sigma}{dt}(\gamma p \to \rho^{\circ} p)}} \Bigg|_{t=0} \tag{9}$$

This comparison of ratios is independent of nuclear physics and the absolute values of the vector dominance parameters — both sides of the equation can be experimentally measured. The equality of Eq. (9) is badly violated, as clearly shown in Fig. 19, where for lead, the left-hand side is measured as ~140, while the right-hand side gives ~70. This discrepancy has serious implications for vector dominance.

Let us consider, at this stage, a simple model which violates rho dominance, but not the spirit of vector dominance, and which allows a simple description of the above phenomena. First we return to the current field identity:

$$j_\mu^{em}(x) \equiv -\sum \frac{M_v^2}{2\gamma_v} \tag{10}$$

and rather than making the assumption that the rho meson saturates the electromagnetic current, we postulate a series of additional vector mesons of higher and higher masses (or a continuum of p-wave pion pairs) which also couple to the photon. We parameterize this additional contribution as an "equivalent meson," such that:

$$\rho' = \sum V$$
$$M_{\rho'} = \overline{M}_v$$
$$\gamma_{\rho'}^2 = \sum \gamma_v^2$$

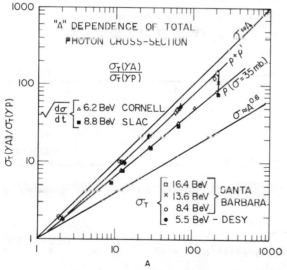

Fig. 19--The A-dependence of the total photon cross section. Also included is the square root of the t = 0 rho cross sections. These should be equal within the framework of VDM. The upper and lower lines represent the A-dependence expected from zero absorption and asymptotic strong absorption of the photon. The intermediate lines are the predictions of a simple model of photon interactions described in the text.

In this picture the hadronic interaction of the photon is mediated by either the ρ or the ρ' — the relative amounts being given by the coupling constants $\gamma_\rho^2/4\pi$ and $\gamma_{\rho'}^2/4\pi$. The absorption of the photon will depend on the masses of the ρ and ρ' (i.e., only have "strong" absorption when the phase difference $\exp\left[\left(M_V^2/2k\right) \cdot R\right]$ is small), and their total cross sections, $\sigma(\rho N)$ and $\sigma(\rho'N)$.

If the mass of the ρ' is greater than 2 BeV, then at present energies it will not give rise to a coherent amplitude in photoabsorption. This means that the photon absorption will have a contribution which has essentially zero absorption (the ρ' amplitude) and a contribution which has strong absorption (the ρ amplitude). The A-dependence, and the k-dependence of the total photon cross section may be used to determine the relative amounts of the ρ and ρ' amplitudes, and also the minimum mass of the ρ'. Measurement of the coherent rho photoproduction cross section may be used to fix the parameters of the ρ amplitude, since even if $\rho' \rightarrow \rho$ coupling is substantial, the ρ' amplitude is not coherent at these energies and does not contribute.

Quantitatively, the coherent rho experiment gives $\sigma(\rho N) \sim 35$ mb, and $\gamma_\rho^2/4\pi \sim 1.2$. The fit to the total cross section data implies the ρ and ρ' amplitudes are equal and that the effective mass of the ρ' be greater than 3000 MeV. Figures 19 and 20 show the A-dependence and k-dependence respectively, as calculated from this model. They are in good agreement with the data.

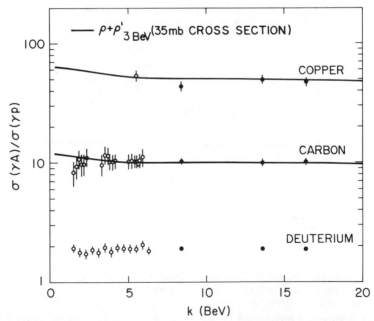

Fig. 20--The A-dependence of the total cross section, with the predictions of the simple model of photon interactions described in the text.

We return now to the discrepancy between Eq. (9) and the measured data, discussed above. The equation should be rewritten as

$$\frac{\sigma^T(\gamma A)}{\sigma^T(\gamma p)} = \frac{\sqrt{\dfrac{1}{\gamma_\rho^2} \dfrac{d\sigma}{dt}(\gamma A \to \rho^0 A)} + \sqrt{\dfrac{1}{\gamma_{\rho'}^2} \dfrac{d\sigma}{dt}(\gamma A \to \rho' A)}}{\sqrt{\dfrac{1}{\gamma_\rho^2} \dfrac{d\sigma}{dt}(\gamma p \to \rho^0 p)} + \sqrt{\dfrac{1}{\gamma_{\rho'}^2} \dfrac{d\sigma}{dt}(\gamma p \to \rho' p)}} \tag{11}$$

The RHS of this equation is $70 + 208/1 + 1 = 139$, for the lead case. The denominator has equal contributions from the ρ and ρ' amplitudes as required by the model fits discussed above, while in the numerator the 70 comes from the measured ρ cross section on lead, [10,11] and the 208 is the ρ' amplitude contribution with no absorption or shadowing (i.e., heavy mass ρ' has essentially zero absorption and consequently $\sigma(A) \propto A$). We see, then, that the new form of the equation is satisfied.

To show that this model also works for hydrogen data, consider the relationship (omitting the ρ' amplitude, for the moment),

$$\sigma^T(\gamma p) \propto \sqrt{\frac{1}{\gamma_\rho^2} \frac{d\sigma}{dt}(\gamma p \to \rho^0 p)} + \sqrt{(\text{term for } \omega \text{ and } \phi)} \tag{12}$$

Here the total photon cross section on hydrogen is related to the forward Compton amplitude, by the optical theorem, which in turn is related to the forward vector meson cross sections by VDM. This relationship has been shown to work well for $\gamma_\rho^2/4\pi = 0.1$. Within our simple model, we now rewrite this equation as

$$\sigma^T(\gamma p) \propto \sqrt{\frac{1}{\gamma_\rho^2} \frac{d\sigma}{dt}(\gamma p \to \rho p)} + \sqrt{(\text{term for } \omega \text{ and } \phi)} + \sqrt{\frac{1}{\gamma_{\rho'}^2} \frac{d\sigma}{dt}(\gamma p \to \rho' p)} \tag{13}$$

This relationship is also well satisfied for $\gamma_\rho^2/4\pi \sim 1.2$ and equal ρ and ρ' amplitudes.

We have shown that with a simple model which assumes there are contributions to the hadronic interaction of the photon in addition to the ρ meson, that the new data on total photon cross sections can be explained and made compatible with the coherent rho production data. In addition, we have shown that this model is consistent with the hydrogen photoproduction data.

IV. Search for High Mass Vector Mesons

Several groups have searched for evidence of the photoproduction of high mass vector mesons. These experiments fall into two classes — those measuring directly the 2π decay channel with some spectrometer arrangement, [40, 41, 42] and those measuring the effect averaged over all decay modes by searching for peaks in a missing-mass experiment. [43]

In performing these searches, it is important to use the highest energy photons possible to avoid problems from phase space inhibition or kinematic effects (like minimum momentum transfer). In Fig. 21, the 2π effective mass spectrum from the SLAC experiment[42] is shown. This is preliminary data from an experiment at 16 BeV, on a Be target, and using the same spectrometer, described in Sect. II above. No strong structure is seen above the rho meson, up to masses of about 2 BeV. The differential cross section for three dipion effective mass regions are shown in Fig. 22 where the diffractive contribution is clearly seen to decrease as the effective mass increases.

The above experiments set limits of the photoproduction cross sections relative to the ρ meson, for possible vector mesons of width ~ 100 MeV and masses ≤ 1800 MeV of $\sim 2.10^{-3}$ in the 2π mode, and $\sim 4.10^{-2}$ for all decays.

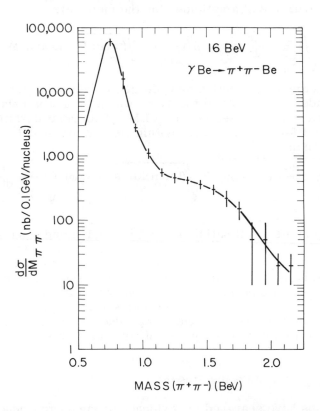

Fig. 21--The effective mass distribution for pion pairs produced by 16 BeV photons on a Be target.

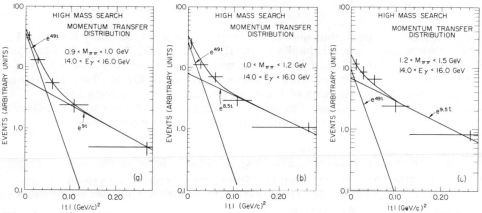

Fig. 22--The differential cross section of dipions, produced by 16 BeV photons on a Be target, and with masses of (a) 0.9-1.0 BeV, (b) 1.0-1.2 BeV and (c) 1.2-1.5 BeV.

V.   Conclusion

We have reviewed the experimental situation on ρ^0 photoproduction and photon absorption in complex nucleii, and shown that there are some disagreements with the VDM. A simple model which extends rho dominance to a generalized vector meson dominance was introduced and was able to take these disagreements into account. We have also shown that for masses up to 1800 MeV, there is no evidence of other vector mesons which couple strongly to 2π and therefore, if this model is correct, the additional contributions to the photon interaction must come from even higher mass states, or from the general continuum of p-wave dipion states.

This leads us to the disagreement between the three experiments measuring coherent ρ^0 production. The question remains whether $\sigma(\rho N)$ = 25 mb and $\gamma_\rho^2/4\pi \sim 0.5$, or $\sigma(\rho N)$ = 35 mb and $\gamma_\rho^2/4\pi \sim 1.2$ are the correct set of parameters to be drawn from these experiments. The three groups involved have a very real and pressing responsibility to sort out this problem. The situation is upsetting to everybody and can only be sorted out by detailed discussion of the data by the three groups involved. Hopefully this will be done very soon.

In conclusion, I think these experiments using nuclear targets have proven to be very useful tools in understanding high energy physics phenomena. The problems of nuclear models and nuclear physics theory, although present, are indeed much less than have been imagined — for experiments trying to determine elementary particle properties to an accuracy of 10 percent, the understanding of nuclear physics, I am sure, is quite sufficient. Finally, I'm sorry that I can't paint a pretty picture of how everything goes together and how vector dominance works beautifully, but it just doesn't at this stage.

VI. Acknowledgements

The SLAC experiment described here is the work of many people. I would like to give special thanks to R. R. Larsen, W. Busza, E. Kluge, S. Williams and R. Giese for all their work in reducing the new data presented above. I would also like to thank S. Brodsky for his help in running his programs for evaluation of the total photon cross sections and for several useful discussions. Finally, I wish to thank Professor F. J. Gilman and Dr. R. R. Larsen for their helpful comments and discussion.

VII. References and Footnotes

1. M. L. Good and W. D. Walker, Phys. Rev. 120, 1857 (1960);
   A. S. Goldhaber and M. Goldhaber, Preludes in Theoretical Physics
   (North Holland, Amsterdam, 1966).

2. S. D. Drell and J. S. Trefil, Phys. Rev. Letters 16, 552 (1966).

3. G. Bellettini et al., Nuclear Physics 79, 609 (1966).

4. A. S. Goldhaber et al., Phys. Rev. Letters 15, 802 (1969).

5. Berkeley-BNL-Orsay-Milan Collaboration, private communication
   (H. H. Bingham).

6. J. J. Sakurai, Ann. Phys. (N.Y.) 11, 1 (1960); N. M. Kroll,
   T. D. Lee and B. Zumino, Phys. Rev. 157, 1376 (1967); H. Joos,
   Schladming Lectures (1967), Acta Physica Austriaca, Suppl. IV (1967);
   D. Schildknecht, DESY Reports 69/10 (1969).

7. Y. Nambu, Phys. Rev. 106, 1366 (1957); W. R. Frazer and J. R. Fulco,
   Phys. Rev. 117, 1603 (1960).

8. J. G. Asbury et al., Phys. Rev. Letters 19, 865 (1967); Phys. Rev.
   Letters 20, 227 (1968).

9. H. Blechschmidt et al., Nuovo Cimento 52A, 1348 (1967).

10. G. McClellan et al., Phys. Rev. Letters 22, 377 (1969).

11. F. Bulos et al., Phys. Rev. Letters 22, 490 (1969).

12. J. Ballam et al., Stanford Linear Accelerator Center Report No.
    SLAC-PUB-530 (1968), submitted to Nucl. Instr. and Methods.

13. F. Bulos et al., "Photoproduction of Rho Mesons at 9 BeV," in Proc.
    of the Fourteenth Int. Conf. on High Energy Physics, Vienna, Austria
    (1968), unpublished; M. Beniston, "On Line Analysis of Wire Spark
    Chamber Data," in Fifteenth IEEE Nuclear Science Symposium,

Montreal, Canada (1968), to be published; R. Russell, "On-Line Wire Spark Chamber Data Acquisition System," in Fifteenth IEEE Nuclear Science Symposium, Montreal, Canada (1968), to be published; F. Bulos, et al., in Proceedings of the Int. Symposium on Nuclear Electronics and Institute for High Energy Physics, Versailles, France (1968), to be published.

14. P. Söding, Phys. Rev. Letters 19, 702 (1966).

15. We define the photon-rho coupling constant to be $em_\rho^2/2\gamma_\rho$ but, by convention, will deal with the quantity $\gamma_\rho^2/4\pi$.

16. The optical model analysis has been discussed in detail by: R. Glauber, Lectures in Theoretical Physics (Interscience, N. Y.), Vol. I (1959); S. D. Drell and J. S. Trefil, Phys. Rev. Letters 16, 552 (1966); K. S. Kölbig and Margolis, Nuclear Physics B6, 85 (1968).

17. R. Hofstadter, Ann. Rev. Nucl. Sci. 4, 231 (1957); H. R. Collard, L. R. B. Elton, R. Hoffstadter, and H. Schoper, in Landolt-Börnstein, Numerical Data and Functional Relationships in Science and Technology, edited by K.-H. Hellwege (Springer-Verlag, Berlin, 1967), New Series Group I, Vol. 2.

18. R. G. Glauber and G. Matthiae, Istituto Superiore di Sanità, Rome, Italy, Report No. ISS 67/16, 1967 (unpublished).

19. The coordinate system is defined with z-axis as the line-of-flight of the rho in the rho center-of-mass.

20. The coordinate system is defined with the z-axis as the line-of-flight of beam particle in the rho center-of-mass.

21. J. Ballam et al., Stanford Linear Accelerator Center Report, to be published; Y. Eisenberg et al., Phys. Rev. Letters 22, 669 (1969).

22. F. Bulos et al., Paper submitted to the APS Boulder Conference on Particles and Fields (1969).

23. G. McClellan et al., Phys. Rev. Letters 22, 374 (1969).

24. A. Silverman, Cornell University Report No. CLNS-73 (1969).

25. W. L. Lakin et al., Stanford University Report No. HEPL-614 (1969).

26. W. Galbraith et al., Brookhaven Report No. BNL-11598.

27. H. Alvenslaken et al., Paper presented at the Third Int. Conf. on High Energy Physics and Nuclear Structure, Columbia, New York (1969).

28. It should be noted that the recent DESY analysis includes the effects of a 20% real part in the ρ-N amplitude, and also the effects of two-body correlations on the evaluation of $\sigma(\rho N)$.

29. E. J. Moniz, G. D. Nixon and J. D. Walecka, Paper presented to the Third Int. Conf. on High Energy Physics and Nuclear Structure, Columbia, New York (1969).

30. G. Van Bochmann,   B. Margolis and C. L. Tang, Paper presented to the Third Int. Conf. on High Energy Physics and Nuclear Structure, Columbia, New York (1969).

31. J. Swartz and R. Talman, Cornell University Report No. CLNS-79 (1969).

32. J. E. Augustin et al., Phys. Letters 28B, 508 and 513 (1969).

33. E. D. Bloom et al., Stanford Linear Accelerator Center Report No. SLAC-PUB-653 (1969); D. O. Caldwell et al., Paper submitted to the 1969 Int. Symposium on Electron and Photon Interactions at High Energies, Liverpool, England (1969); M. Meyer et al., Paper submitted to the 1969 Int. Symposium on Electron and Photon Interactions at High Energies, Liverpool, England (1969); J. Ballam et al., Paper submitted to the 1969 Int. Symposium on Electron and Photon Interactions at High Energies, Liverpool, England (1969); J. Ballam et al., Phys. Rev. Letters 21, 1541 (1968).

34. M. Damashek and F. J. Gilman, Stanford Linear Accelerator Center Report to be published.

35. R. Morrison, Contribution to the Third Int. Conf. on High Energy Physics and Nuclear Structure, Columbia, New York (1969).

36. D. O. Caldwell et al., Paper submitted to the Third Int. Conf. on High Energy Physics and Nuclear Structure, Columbia, New York (1969).

37. M. Meyer et al., Paper submitted to 1969 Int. Symposium on Electron Photon Physics at High Energies, Liverpool, England.

38. K. Gottfried and D. R. Yennie, Cornell University Report No. CLNS-51; S. J. Brodsky and J. Pumplin, Stanford Linear Accelerator Center Report No. SLAC-PUB-554 (1969); B. Margolis and C. L. Tang, Nucl. Phys. B10, 329 (1969); M. Nauenberg, Phys.Rev. Letters 22, 556 (1969).

39. At these energies ($\sim$10 BeV) the inclusion of the expected amount of real part would have less than a 10% effect on these numbers.

40. N. Hicks et al., Phys. Letters 9, 602 (1969).

41. G. McClellan et al., Cornell University Report No. CLNS-59(1969).

42. F. Bulos et al., Preliminary results were presented at APS Boulder Conference on Particles and Fields (1969).

43. R. Anderson et al., Stanford Linear Accelerator Center Report No. SLAC-PUB-644 (1969).

DISCUSSION

H. Pilkuhn: Does not the $A^{2/3}$ dependence of the total photon-interaction cross-section result from any strong-interaction theory, whatever the particular type of meson(s) dominance, as long as these have finite mass and the energy is sufficiently high? D.W.G.S.L.: However the A dependence is not 2/3 but 0.92 (± few per-cent), and it is this one has to struggle to explain.   A. Dar (interpreted by K. Gottfried): Do you see, at 0°, an E dependence of the cross-section (for some particular heavy nucleus) which is characteristic of the variation of the minimum-momentum transfer (i.e., agreement with Drell-Trefil formula)?   D.W.G.S.L.: Essentially yes; over the whole range of energy and A.   L.M. Lederman: Apropos the possibilities of high mass 1⁻ di-pion states, our experimental search for di-muons indicate that the yield of such 1⁻ states, at say 3 GeV, is some 3 orders-of-magnitude less than for ρ-production.

K. Gottfried: Does your experiment agree with Cornell?   D.W.G.S.L.: Yes.   The experimental results themselves are in agreement.

PHOTOPRODUCTION OF RHO MESONS FROM COMPLEX NUCLEI

T. M. Knasel

Laboratory for Nuclear Science

Massachusetts Institute of Technology

I would like to report a recent experiment on ρ meson photo production performed at the DESY 7.5 GeV electron-synchrotron by the DESY-MIT collaboration.[1] We made a study of one million $\pi\pi$ events in the ρ mass region from 14 elements, Hydrogen through Uranium. These events are organized into a four dimensional data matrix where the dependencies of the cross-section on mass, on momentum, and on momentum transfer are all explicitly separated. Our resolution is $\Delta m = \pm 15$ MeV, $\Delta p = \pm 200$ MeV/c and $\Delta t = \pm 0.001$ GeV^2.

Because we have made measurements over a wide range of the kinematical quantities, we can present cross-sections free of theoretical assumptions.

Figure 1 shows a 3-dimensional projection of our data matrix, the $\pi\pi$ invariant mass spectra as a function of A for p and t fixed. This represents about 5% of our data. Note that the $\pi\pi$ spectra are dominated by ρ production, but that there exists also non-resonate background that is a function of A. In order to extract the ρ cross section, and in order not to make a systematic error in the relative yield of ρ's as a function of A, we have fit all our data for a ρ term (leaving the mass and width free) and for the amount of background at each A.

Figure 2 shows the cross-section $d\sigma/d\Omega dm$ for p fixed as a function of m and t for 13 elements. As you know, ρ photo-production is described as scattering off nuclei of virtual ρ mesons; and this gives a diffraction pattern which is indicative of the size of the nucleus seen by the ρ. This is completely

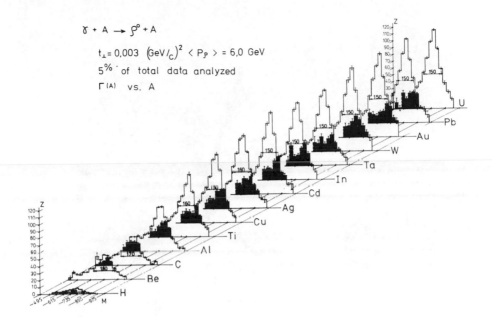

$\gamma + A \longrightarrow \rho^{0} + A$

$t_{\perp} = 0.003 \; (GeV/_c)^2 \quad \langle P_P \rangle = 6.0 \; GeV$

5% of total data analyzed

$\Gamma^{(A)}$ vs. A

Figure 1

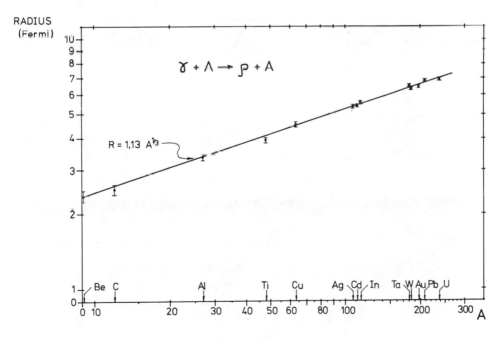

RADIUS (Fermi)

$\gamma + \Lambda \longrightarrow \rho + A$

$R = 1.13 \; A^{1/3}$

Figure 3

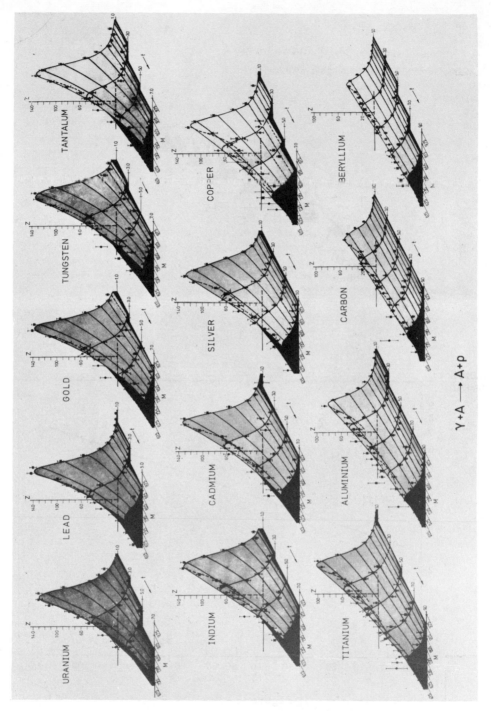

$\gamma + A \longrightarrow A + \rho$

Figure 2

analogous to the case of   nucleus scattering.   Note that the cross-
section falls with increasing t much more rapidly as A increases.

To obtain the radius of the nuclear target, and the rate of
re-absorption of the ρ's in the nucleus, we used an optical model
calculation of Kolbig and Margolis[2].   We included a ratio of real
to imaginary scattering amplitude of -0.2 as given by γp total cross-
section measurements at the same energy, and anti-correlation
effects in the nuclear density function.

Figure 3 shows our results for the Woods Saxon radii of the
elements Be through U.   Our best fit is R = $1.13A^{1/3}$ to 2% accuracy
for the radius.   This is the first accurate measurement of the
nuclear density seen by the scattering of the ρ resonance.

We also obtain the ρ-nucleon total cross section $\sigma(\rho n)$ to be
26 ± 2 mbarns, and the coupling constant in the Vector Dominance
Model to be $\gamma \rho^2/4$ =   0.54 ± 0.1.   These values agree with the
original DESY experiment.   Our determination of $\sigma(\rho n)$ is in good
agreement with the quark model, and our determination of $\gamma \rho^2/4\pi$ is
in agreement with values obtained in leptonic decay of the ρ, and
from storage ring determinations where the virtual photon is on the
mass shell of the ρ.

REFERENCES

[1]The members of the DESY-MIT collaboration are: H. Alvensleben,
U. Becker, William K. Bertram, M. Chen, K.J. Cohen, T.M. Knasel,
R. Marshall, D.J. Quinn, R. Rohde, G.H. Sanders, H. Schubel and
Samuel C. C. Ting.

[2]K. S. Kolbig and B. Margolis, Nucl. Phys. B6 (1968) 85.

DISCUSSION

J. Trefil: How do these new ρ-production results affect the relationship of the total photon cross-sections and vector dominance?
T.M.K.: A detailed analysis of these very recent results has not yet
been made.   A preliminary check indicates that our values of $\partial\sigma/\partial t$
for ρ-photoproduction are in very good agreement with those of the
Santa-Barbara group.   K. Gottfried: This new $\sigma(\rho n)$ cross-section
(25 m.b), may be all right as far as the total photo-cross-section
is concerned, but the incoherent cross-section (π-production in particular) may require a still small cross-section, for compatibility
with vector dominance.   T.M.K.: Agreed.   (c.f. also Session 9).

R. Glauber: Was an optical model used for all nuclei, including the
lightest? T.M.K.: In fitting $\sigma(\rho$,Nucleon) and $\gamma \rho^2$ we excluded first
H, and then successively Be, C, and Al; but there was little change
- showing that we are rather insensitive to the fit with these very
light nuclei.

COHERENT AND INCOHERENT PARTICLE PRODUCTION IN NUCLEI

B. Margolis

McGill University, Montreal 110, Canada

We are interested in the study of high energy reactions in nuclei for the following reasons.

1. There is much information that we can gain for particle physics. This includes

   a) The determination of unstable particle cross sections and scattering amplitudes.

   b) Information on coupling constants e.g. the vector dominance coupling constants γ_γ^2 ; also possibly couplings between unstable particles.

   c) Information on hadron-hadron diffraction mechanisms e.g. does double pomeron exchange have meaning ?

   d) An answer to the question : Is vector dominance a good approximation over large ranges of "photon mass" ?

   e) Is the A_1 a particle or a kinematical enhancement of an uncorrelated π-ρ system ? [14]

2. There is considerable scope for the study of nuclear properties. One can get answers to the questions :

   a) Are neutron and proton distributions the same?

   b) What is the nature of correlations in the nucleus ?

3. There are analogies to be drawn in hadron interaction physics. The question of the importance of multiple scattering or absorptive corrections to Regge pole exchange has drawn heavily on the theory of high energy nuclear reactions.

4. A big nucleus may be useful for producing beams of particles with one or more units of strangeness through

cascade processes.

Category 3 is outside the scope of this conference and
so I will content myself on matters concerning this subject by giving references in the written proceedings. [1]
Category 4 will perhaps be   discussed by Dr. Pilkuhn. [2]
I will then discuss mainly category 1 and to some extent
category 2.

The Theory of Coherent and Incoherent Reactions.

The Glauber multiple scattering theory [3] has been
generalized [4] to include one step particle production
at energies high enough so that longitudinal momentum
transfer effects due to mass differences are unimportant.
Using the simplest form for the ground state nuclear wave function one obtains simple expressions for coherent
processes (no nuclear excitation) and for processes summed over all nuclear final states. There is an equivalent
optical model for these processes which in fact allows for
the calculation of coherent and incoherent (nuclear excitation) processes at finite energies where mass differences are important. This also allows for multi-step
processes. This optical model, for diffractive processes
(no spin or iso-spin dependence of amplitudes) is described by the following coupled equations :

$$(\nabla^2 + k^2 - m_\alpha^2) \psi_\alpha = \sum_\beta U_{\alpha\beta} \psi_\beta \qquad (1)$$

$$U_{\alpha\beta} = -4\pi f_{\alpha\beta}(o) A \rho(\vec{r}) \qquad (2)$$

Here $f_{\alpha\beta}(o)$ is the two body amplitude for producing α
on a nucleon if β is incident. The quantity $\rho(\vec{r})$ is the
single particle density corresponding to a nuclear wave
function squared

$$|u_I (r_1 \ldots r_A)|^2 = \prod_{\alpha=1}^{A} \rho(r_\alpha) \qquad (3)$$

in Glauber multiple scattering theory, neglecting C.M.
motion.

The scattered or produced waves are strongly peaked
forward. We therefore first solve the equations (1) as
one dimensional wave equations in the incident direction.
Neglect of the weak backward reflected wave in the integral representation

$$\psi_\alpha(b,z) = \frac{e^{ik_\alpha z}}{2ik_\alpha} \int_{-\infty}^{z} e^{-ik_\alpha z} \sum U_{\alpha\beta}\psi_\beta dz + \frac{e^{-ik_\alpha z}}{2ik_\alpha} \int_{z}^{\infty} e^{ik_\alpha z} \sum U_{\alpha\beta}\psi_\beta dz$$

(4)

leads to the first order differential equation for
$\varphi_\alpha(b,z) = \exp(-ik_\alpha z)\psi_\alpha$

$$\frac{d}{dz} \varphi_\alpha (b,z) = \frac{1}{2ik_\alpha} \sum U_{\alpha\beta}(b,z) \, e^{i(k_\beta-k_\alpha)z} \varphi_\beta(b,z)$$

(5)

and an elastic scattering or coherent production amplitude (the matrix element provides the angular deflection)

$$F_{\alpha 1} = \frac{1}{4\pi} \int e^{-ik_\alpha \cdot r} \sum U_{\alpha\beta}\psi_\beta d^3 r \equiv \frac{k_\alpha}{2\pi i} \int e^{iq\cdot b} d^2 b [\varphi_\alpha(b,\infty)-\delta_{\alpha 1}]$$

(6)

$$\frac{d\sigma_\alpha^{(c)}}{d\Omega} = |F_{\alpha 1}|^2$$

(7)

One has included here any number of back and forth transitions among particles that can be connected coherently.

Incoherent production, corresponding in multiple scattering theory to closure, that is to summing over all final nuclear states, can be treated by dealing with intensities of waves rather than amplitudes. The answer for the intensity $I_\alpha(b,z)$ in any channel α is

$$I_\alpha(b,z) = A\rho(b,z) \Big| \sum_{\alpha'\alpha''} \psi_{\alpha'}^{(1)}(b,z) f_{\alpha'\alpha''}(q^2) \psi_{\alpha''}^{(\alpha)}(b,-z) \Big|^2$$

(8)

where $\psi_\alpha^{(\gamma)}(b,z)$ is the wave amplitude in channel α for an incident beam in channel γ. This allows for any number of coherent steps (no nuclear excitation) but only one incoherent step, and hence is not valid for $\exp(-aq^2) \ll 1$ where $\sqrt{a}$ is the typical range of the two body amplitudes $f_{\alpha'\alpha''}$. In fact some processes may be dominated at all momentum transfers by multi-step incoherence. Again as we shall see, this expression must be corrected for correlations at small momentum transfers. The incoherent cross section with the above restrictions satisfied is then

$$\frac{d\sigma}{d\Omega}^{(I)} = \int I_\alpha(b,z) \ d^2b \ dz \qquad (9)$$

As examples of the use of these coupled equations we consider here the calculation of (1) Coherent and incoherent production of particles A_1 and A_3 assuming a coupled π, A_1, A_3 system and neglecting all other couplings.[5] (2) Elastic scattering of photons assuming vector dominance.[6,7] (3) The calculation of coherent and incoherent production of ρ^0 mesons assuming vector dominance.[7,8] Calculations for these processes are shown in figures 1 to 4.

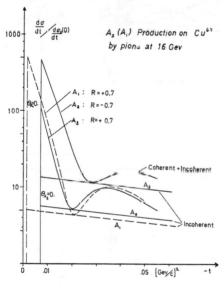

Fig. 1: Coherent and incoherent production of A_1 and A_3 using the model described. Correlations and 1/A coherent corrections will affect the small t incoherent behaviour.

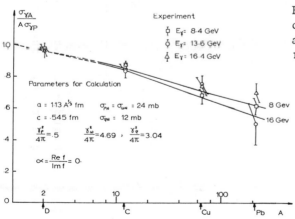

Fig. 2: Total photon cross section calculations and data from reference (11).

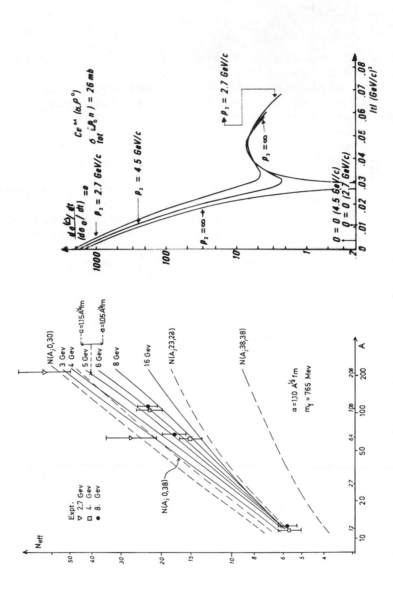

Figure 3. Incoherent Production of ρ^0 mesons. Data from reference (12).

Figure 4. Coherent production of ρ^0 mesons from reference (4).

It is of interest to examine processes (1) and (2) at infinite energy. For A_1 production we get to good accuracy the forward amplitude

$$F_{\pi A_1}(o) = f_{\pi A_1}(o)[N_1(A,\sigma/2) - \frac{f_{\pi A_3}(o)f_{A_3 A_1}(o)}{f_{\pi A_1}(o)f_{A_3 A_3}(o)} N_2(A,\sigma/2)]$$

(10)

where we have assumed the total cross sections of π, A_1 and A_3 on nucleons all to be σ.

$$N_m(A, \tfrac{1}{2}\sigma) = \frac{1}{m!}\, \frac{2}{\sigma} \int [\, \frac{\sigma}{2}\, T\,(b)\,]^m\, e^{-\sigma/2\, T(b)} d^2 b$$

(11)

Experiments of Morrison et al[9] yield $f_{\pi A_3}(o)/f_{\pi A_1}(o)=.35$, $f_{\pi A_1}(o)/f_{\pi\pi}(o)=.27$ at 8 Gev/c.
Since these processes are diffractive, we can expect little energy dependence of these ratios. In order to calculate A_1 production in this model we also need $\frac{f_{A_3 A_1}(o)}{f_{A_3 A_3}(o)}$. Calling the coefficient of N_2 in (10) R_{A_1}, this is then a parameter in the calculation. One can expect that R_{A_1} is not bigger than 0.1. A_1 production is expected to go predominantly as a one-step process. The more general one step formula which allows for a different cross section for the π-meson and A_1 is[4]

$$F_{\pi A_1}(q^2) = f_{\pi A_1}(o)\frac{2}{\sigma_2-\sigma_1} \int [e^{-\frac{\sigma_1}{2} T(b)} - e^{-\frac{\sigma_2}{2} T(b)}]e^{iq\cdot b} d^2 b$$

(12)

An expression similar to (10) with A_1 and A_3 interchanged holds for A_3 production. Here the quantity R_{A_3} is expected to be much larger. As can be seen in figure 1, the sign of $R = R_{A_3}$ drastically affects the production rate. This sign could be negative if multi-pomeron exchange occurs. There is then, interesting dynamical information possibly available from diffractive production in nuclei. By measuring production rates across the periodic table one can expect to learn not only about unstable particle cross sections but coupling constants as well. The actual situation for strong interactions may

well be more complicated in that other resonances may
contribute.

A similar discussion can be made for photo-induced
reactions. Coherent photo-production of ρ^o mesons is a
one step process, given at infinite energy by

$$F_{\gamma\rho}(q^2) = f_{\gamma\rho} \frac{2}{\sigma} \int e^{iq\cdot b}(1 - e^{-\sigma/2\ T(b)})\ d^2b \qquad (13)$$

Figure 4 shows the energy dependence that one gets due to
longitudinal momentum transfer. <u>The accurate measurement
of diffractive production of ρ-mesons is a crucial experiment for vector dominance.</u> One can determine from measurements across the periodic table
1) the nuclear radius from the diffractive slope; (in fact
one can get two relationships between the surface thickness and radius parameters for any two parameter family
of radial functions, e.g. Wood-Saxon, by looking at the
diffractive slope at fixed energy and the energy dependence in the forward direction, say.)
2) The cross section σ_{on} from the relative A dependence in medium to heavy nuclei.
3) The coupling constant γ_ρ^2 from the absolute photo-production cross section.

One can calculate the total photon cross section from
elastic scattering of photons and the optical theorem. E-
lastic photon scattering is expected to be dominantly a
mixture of one and two step processes. The one amplitude
is in the forward direction

$$F_{\gamma\gamma}^{(1)}(o) = A\ f_{\gamma\gamma}(o) \qquad (14)$$

The two step amplitude is complicated at finite energies
but at infinite energy, assuming vector dominance, it is

$$F_{\gamma\gamma}^{(2)}(o) = -\ f_{\gamma\gamma}(o)[A - \frac{\sum\limits_{v} N(o,\sigma_{v/2})\ \sigma_{v}\ (\gamma_{v}^2/4\pi)^{-1}}{\sum\limits_{v} \sigma_{v}\ (\gamma_{v}^2/4\pi)^{-1}}\ ]$$

$$(15)$$

The sum is over the vector mesons ρ, ω and φ . If we neglect the ω and φ contributions we have

$$F_{\gamma\gamma}(o) \equiv f_{\gamma\gamma}(o) \; N(o, \sigma_\rho/2) \qquad (16)$$

The photon scatters like a vector meson at very high energy. We have determined the ρ-nucleon total cross section from available total photon cross section data[10,11] from 6 to 16 Gev. It comes around 24 to 25 mb and appears to support the vector dominance hypothesis.

Incoherent production of ρ-mesons can also be calculated on the vector dominance hypothesis as a combination of one and two step processes. At relatively small energies the one step dominates and we find

$$|F_{\gamma\rho}^{(I)}(t)|^2 = |f_{\gamma\rho}(t)|^2 \; N(o, \sigma_\rho) \qquad (17)$$

at infinite energy

$$|F_{\gamma\rho}^{(I)}(t)|^2 = |f_{\gamma\rho}(t)|^2 \; N(\sigma_\rho, \sigma_\rho) \qquad (18)$$

i.e. the photon again acts like a vector meson. Calculations at intermediate energies are shown in figure 3. The data[12] is not inconsistent with vector dominance if one assumes vector dominance on the nucleon in the formula for $f_{\gamma\rho}(t)$. i.e. if one uses

$$\frac{d\sigma_{\gamma\rho}}{dt}(o) = \frac{1}{16} \frac{\alpha}{4\pi} (\gamma_\rho^2/4\pi)^{-1} \sigma_\rho^2 \qquad (19)$$

to determine σ_ρ as a function of energy. In this case, neglecting real parts, one finds that σ_ρ falls from about 30 to 23 or 24 mb in going from 3 to 20 Gev/c photon energy, taking $(\gamma_\rho^2/4\pi)^{-1} = .5$.

There are many other applications of the above described formalism. I will close with brief remarks on the reliability of the theory. The optical model presented here follows from Glauber multiple scattering theory using a product wave function. We have estimated the effect of correlations in the nuclear wave function.[13]

Define the correlation function $g(r_1, r_2)$ through

$$\int d\vec{r}_3 d\vec{r}_4 \ldots d\vec{r}_A \; |u_I(r_1 \ldots r_A)|^2 = \rho(r_1) \cdot \rho(r_2)[1 + g(r_1, r_2)] \qquad (20)$$

$$\int d\vec{r}_2 d\vec{r}_3 \ldots d\vec{r}_A |u_I(\vec{r}_1 \ldots \vec{r}_A)|^2 = \rho(\vec{r}_1) \qquad (21)$$

Then the effect is to change σ in formula (13) for coherent production to

$$\sigma_E = \sigma [1 - \xi \eta(\sigma/2)\sigma] \qquad (22)$$

$$\xi = \frac{1}{16\pi a} \int e^{-b^2/4a} g(b,z) dz \, d^2b \qquad (23)$$

$$\eta(\sigma/2) = \frac{\int e^{-\sigma/2 \, T(b)} Q(b) \, d^2b}{\int e^{-\sigma/2 \, T(b)} T(b) \, d^2b} \; ; \quad \begin{array}{l} Q(b) = A^2 \int_{-\infty}^{\infty} \rho^2(b,z) dz \\[2ex] T(b) = A \int_{-\infty}^{\infty} \rho(b,z) dz \end{array} \qquad (24)$$

where $a = 8 \, (\text{Gev}/c)^{-2}$ is the two body production range. One must also change the two body amplitude $f(o)$ to

$$f^E(o) = f(o) \frac{\sigma_E}{\sigma} \qquad (25)$$

There are also corrections to incoherent production through

$$\frac{d\sigma^{(I)}}{d\Omega} \lesssim |f(t)|^2 [\; 1 + \eta(\sigma) \, G(t) \;] \, N_{eff} \qquad (26)$$

where $G(t)$ is the Fourier transform of $g(\vec{r})$. The approximation $g(r_1, r_2) \simeq g(r_1 - r_2)$ has been made. For coherent production σ_E can be expected to differ from σ by a millibarn or two. The appearance of $G(t)$ in the incoherent term means that we can get correlation information in principle from measurements of incoherent processes. The Pauli principle will play a strong role here.

References
(1) S. Frautschi and B. Margolis, Nuovo Cimento 56A, 1155 (1968), S. Frautschi, O. Kofoed-Hansen and B. Margolis, Nuovo Cimento 61A, 41 (1969); M. Jacob and S. Pokorski, Nuovo Cimento 61A (1969) 233. R.C. Arnold and M. L. Blackmon, Phys.Rev. 176,2082 (1968)

(2)   See "Production of Fast Pions and Hyperons in Nuclei"
      B. Margolis and H. Pilkuhn, Karlsruhe Report 3/68 - 7.
(3)   R.J. Glauber, Boulder Lectures in Theoretical Physics,
      Vol. 1 (1958) (Inter Science Publ. Inc.,New York,1959)
(4)   K.S. Kölbig and B. Margolis, Nuclear Physics B6 (1968)
      85.
(5)   G. von Bochmann and B. Margolis, to be published.
(6)   B. Margolis and C.L. Tang, Nuc.Phys. B10 (1969) 329.
      M. Nauenberg Phys. Rev. Lett. 22, 556 (1969), S.J.
      Brodsky and J. Pumplin SLAC-Pub.-554, Stanford U.
      (1969).
(7)   K. Gottfried and D.R. Yennie CLNS-51 (1969) Cornell U.
(8)   G. von Bochmann and B. Margolis, unpublished.
(9)   D.R.O. Morrison et al, Nucl. Phys. B8 (1968) 45.
(10) H.Meyer, B.Naroska, J.H.Weber and M.Wong, DESY pre-
      print.
(11) D.O. Caldwell, V.B. Elings, W.P. Hesse, G.E. Jahn,
      R.J. Morrison, F.V. Murphy, D.E. Yount SLAC preprint.
(12) S.C.C. Ting et al, Phys.Rev. Lett. 19, (1967) 865.
      A. Silverman et al Phys.Rev. Lett. 23, 554 (1969).
(13) G. von Bochmann, B. Margolis and C.L. Tang, DESY
      preprint to appear in Physics Letters.
(14) A.S. Goldhaber, C.J. Joachain, H.J. Lubatti and
      J.J. Veillet Phys. Rev. Lett. 22, 15 (1969) 802.

K_L-K_S REGENERATION AT HIGH ENERGIES

H. Foeth, M. Holder, E. Radermacher and A. Staude
P.I.T.H. Aachen, Germany

P. Darriulat, J. Deutsch, K. Kleinknecht, C. Rubbia and
K. Tittel
CERN, Geneva, Switzerland

M.I. Ferrero and C. Grosso
Istituto di Fisica dell'Università and INFN, Torino, Italy

(Presented by P. Darriulat)

The phenomenon of regeneration consists of a transformation of an incident long-lived neutral K_L wave into a linear superposition of short- and long-lived states, K_S and K_L. It is expected to occur for any process in which K_0 and $\overline{K_0}$ interact differently. In particular, the amplitude for K_S regeneration from nuclei at momentum transfer q, is the half difference between the K_0 and $\overline{K_0}$ nuclear scattering amplitudes $\frac{1}{2}\{f(q^2) - \overline{f}(q^2)\}$. This process is coherent in the very forward direction (transmission regeneration). Both modulus and phase of $f(0) - \overline{f}(0)$ can be deduced from the observation of the interference between the regenerated K_S and the transmitted K_L waves.

We have measured the K_S intensity observing their decays into $\pi^+\pi^-$ mode after lead regenerators of various thicknesses exposed in a neutral beam derived from the high-energy extracted proton beam of the CERN Proton Synchrotron. The detector [1] and the experimental method [2] have been described in previous publications to which we refer for details. Transmission regeneration is identified by the characteristic sharp peak observed in the forward direction. We have also measured the q^2 dependence of the differential nuclear regeneration cross-section up to momentum transfers of 0.16 GeV/c, far beyond the first diffraction minimum. Results are presented in Fig. 1. They are compared to the predictions of calculations made in the framework of an optical model. The proton and neutron nuclear densities are described with Fermi distributions of the form $\{1 + \exp[(r-R)/d]\}^{-1}$. Imaginary parts of the kaon-nucleon amplitudes are determined from measured total cross-sections of charged kaons on nucleons (optical theorem and isospin symmetry) and real parts are deduced from the same total cross-sections with forward dispersion relations.

The q^2 dependence of the differential regeneration cross-section is modified by multiple scattering and by the presence of coherent regeneration along the path of the kaon before and after scattering. The effect results in a widening of the diffraction peak and in a filling of the minimum. It has been taken into account exactly.

The shape of the differential regeneration cross-section is not well reproduced when both proton and neutron distributions are taken equal to the nuclear charge distribution from electron scattering and muonic X-ray experiments [3]. It is relevant to remark that, since neutrons have a substantially stronger regeneration power than protons (by a factor 5 at 4 GeV/c) K_L-K_S regeneration is particularly sensitive to the neutron distribution inside the nucleus. Agreement can be achieved by increasing artificially the neutron radius R_n from 6.66 fm to 7.85 fm or the neutron surface thickness d_n from 0.50 fm to 1.02 fm. However, in both cases the predicted forward regeneration amplitude turns out somewhat larger than experimentally observed, but still within the uncertainties of the kaon-nucleon scattering amplitudes. The q^2 dependence of the differential regeneration cross-section is very sensitive to nuclear dimensions and practically unaffected by changes in the kaon-nucleon amplitudes. This sensitivity is enhanced by the opacity of the nucleus which causes regeneration to occur preferentially on its periphery.

Table 1

Parameters of the neutron distribution

Conditions	$d_n = d_p = 0.50$ fm $R_p = 6.66$ fm	$R_n = R_p = 6.66$ fm $d_p = 0.50$ fm
Fit to $\lvert f(0) - \bar{f}(0) \rvert$	$R_n = (6.6 \pm 0.6)$ fm	$d_n = (0.5 \pm 0.3)$ fm
Fit to $(d\sigma/dq^2)(q^2)$	$R_n = (7.85 \pm 0.12)$ fm	$d_n = (1.02 \pm 0.10)$ fm
Mean value	$R_n = (7.65 \pm 0.12)$ fm	$d_n = (0.86 \pm 0.10)$ fm

DISCUSSION follows paper presented by H. Pilkuhn entitled "Two-Step Reactions in Nuclei".

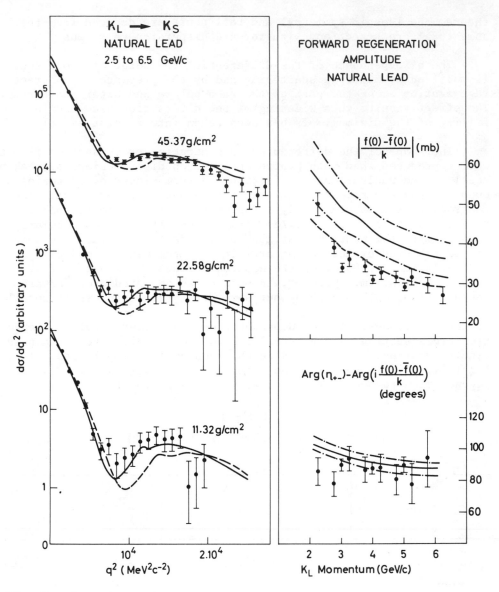

Fig. 1 - Experimental results compared to optical model predictions: for equal neutron & proton distribution (dashed line), or taken from Table 1 (full line).

REFERENCES

[1] H. Faissner et al., Colloque International sur l'electronique nucleaire, Versailles, 10-13 September, 1968.
[2] A. Boehm et al., Phys. Letts. 27B, 594 (1968) and references therein.
[3] H.L. Anderson et al., E.F.I.N.S. Report, (Chicago), 1969.

TWO-STEP REACTIONS IN NUCLEI

H. Pilkuhn

University of Karlsruhe

Germany

1. INTRODUCTION

In the application of Glauber theory for particle production on nuclei, one normally includes only elastic scattering in the initial and final states as modifications of the elementary reaction.[1] There are, however, reactions for which this does not work. One example is double charge exchange of the type π^+ ^{10}Be$\to$ π^- ^{10}C, which proceeds in two steps, via transformation of the pion into a neutral meson. Another example is Ξ-production in p-nucleus collisions, which could in principle go via e.g. pp$\to$ KKΞp, but in which nevertheless the incident baryon is frequently transformed into a hyperon of zero hypercharge ($\Lambda, \Sigma, \Sigma^*$ etc.) before going to the final state. Finally, there are reactions such as π^+N$\to\eta$N' or η'N', where intermediate states such as the ρ-meson yield important corrections.

2. DOUBLE CHARGE OR HYPERCHARGE EXCHANGE

Let us consider the coherent reactions π^+ ^{10}Be$\to\pi^-$ ^{10}C, π^+ ^{14}C$\to$ π^- ^{14}O, π^+ ^{18}O$\to$ π^- ^{18}Ne, where the initial and final nuclear states belong to the same isotopic multiplet. For ^{18}O, only the two protons outside the closed shells participate. The particle physicist may compare the phases of different reactions. Let

$$-\sqrt{2}f^{(-)} = f(\pi^- p \to \pi^\circ n) \tag{2.1}$$

denote the non-spin-flip part of the πN charge exchange amplitude, and $f_{\pi\eta}$ that of the reaction $\pi^- p \to \eta n$. Then, in the forward direction, the amplitude contains the combination

$$2f^{(-)}(\bar{q})f^{(-)}(-\bar{q}) - f_{\pi\eta}(\bar{q})f_{\pi\eta}(-\bar{q}) \tag{2.2}$$

which depends upon the phase between $f^{(-)}$ and $f_{\pi\eta}$. In practice, the situation will be slightly more complicated, since also $\rho°$ and $\omega°$ intermediate states will contribute (the latter again with a minus sign). Also, the energy cannot be taken too high, since the cross section decreases with increasing energy as p_{lab}^{-2} or faster. Finally, there could be nuclear effects other than pair correlations. For example, there could be a small component of the Δ-isobar among the nucleons in the nuclear wave function. For that component, double charge exchange could go in one step, namely $\pi^+ n \rightarrow \pi^- \Delta^{++}$ or $\pi^+ \Delta^- \rightarrow \pi^- p$. The reaction has been observed to have a small and narrow forward peak.[5] This is rather exceptional and could be due to a final state interaction which need not occur for a virtual Δ.

Next, two comments about hypercharge exchange. One may look for neutron-rich hyperfragments in reactions such as $K^- \, {}^6Li \rightarrow \pi^+ \, {}^6_\Lambda H$, $K^- \, {}^7Li \rightarrow \pi^+ \, {}^7_\Lambda H$, or $K^- \, {}^9Be \rightarrow \pi^+ \, {}^9_\Lambda He$. Similarly, one may search for double hypernuclei in reactions such as $K^- \, {}^{10}B \rightarrow K^+ \, {}^{10}_{\Lambda\Lambda}Li$, $K^- \, {}^6Li \rightarrow K^+ \, {}^6_{\Lambda\Lambda}H$. Of course, the cross sections for these reactions will be small, 0.1 μb/sr or 0.01 μb/sr in the forward direction for double-hypernucleus production in a 3-4 GeV/c K^--beam. Moreover, the $K^- - K^+$ momentum difference must be measured to a high precision. On the other hand, this is probably the only systematic way of searching for such hypernuclei.

3. HYPERON PRODUCTION IN PROTON-NUCLEUS COLLISIONS

The reactions p+nucleus$\rightarrow$fast Ξ^-+ anything are presumably of little theoretical interest, but are considered for the construction of Ξ^- beams from a PS. The cross section for the reaction $pp \rightarrow \Xi^-$+ anything is probably less than 15 μb at high energies. Thus, if one just multiplies this cross section by an effective nucleon number N_1 for the nucleus, one gets very little Ξ^--production. In this situation, production in two steps will dominate.[2] The cross section for fast Σ or Λ production in pp-collisions is about 2.5 mb. By SU_3-arguments, the cross section for Ξ production in Σ or Λ collisions will be of the same order of magnitude. This then yields about 56 μb, to be multiplied by the effective nucleon number N_2 of double collisions:

$$\sigma(\Xi^-) \approx (15N_1 + 56N_2)\mu b$$

Thus, to produce a Ξ^- beam, one should take a heavy target nucleus, e.g. ^{238}U.

The situation is similar but less pronounced for Σ^- production. Here the important intermediate states are neutrons and neutral hyperons.

As yet no experiments have been performed on these reactions. Instead, another peculiar reaction has been investigated at Serpukhov, namely the production of antideuterons in p-Al collisions.[3]   There it has been found that the ratio $\bar{d}/\pi^-$ is about 0.002 $(\bar{p}/\pi^-)^2$, independent of the absolute values of these ratios. This suggests that $\bar{d}$ production is a two-step process, the two antinucleons coming from collisions with different target nuclei. In that case again, one would expect larger $\bar{d}$ yields from heavier targets.

4.   η AND η' PRODUCTION

In the reactions discussed so far, the suppression of the direct (= one-step) production was rather obvious.   We now look at reactions in which the direct production dominates, but where nevertheless two-step production does modify the details.[4]   Candidates for such cases are $\pi^+N\rightarrow\eta N'$ (or $\pi^+N\rightarrow\eta$ 'N'). Here the direct reaction is $\pi^+n\rightarrow\eta p$ ($\pi^+n\rightarrow\eta$ 'p').   Around 5 GeV/c, the reaction $\pi^+n\rightarrow\rho^\circ p$, which has a large pion-exchange contribution, is almost 10 times larger than $\pi^+n\rightarrow\eta p$.   The second step would then be $\rho^\circ n\rightarrow\eta n$ ($\rho^\circ n\rightarrow\eta'n$), which is of course unknown experimentally.   However, its cross section can be estimated using the ρ-dominance model of photoproduction, $\sigma\sim0.1$ mb.   The two-step production is then about 20% of the double-collision term containing one elastic pion scattering.   For example, this may fill the dips in coherent production.

References

[1]See e.g. K. S. Kolbig and B. Margolis, Nucl. Phys. B6, 85 (1968)

[2]B. Margolis and H. Pilkuhn, Kernforschungszentrum Karlsruhe, ext. report 3/68-7 (1968)

[3]F. Binon et al., IHEP-CERN Collab., pres. at the Int. Conf. on Elem. Particles, Lund, June 1969

[4]G. Ebel and H. Pilkuhn, Nucl. Phys. B7, 147 (1968)

[5]P. M. Dauber et al., Phys. Letters 29B 609 (1969)

DISCUSSION (after P. Darriulat):

R. Glauber: This is a very subtle probe.   The regeneration amplitude results from the almost complete cancellation of two nearly equal terms: it is very much smaller than the scattering amplitude.   It does, therefore, appear necessary to use optical-model approximations with great care and circumspection.   P.D.: Yes.   The regeneration amplitude is only 1/10 of the scattering amplitude.

A MODEL FOR INELASTIC EFFECTS IN HIGH-ENERGY PION-DEUTERON SCATTERING

David R. Harrington

Rutgers, the State University

New Brunswick, New Jersey

If useful nuclear structure information is to be obtained by using the Glauber method to analyze high-energy hadron-nucleus scattering experiments, it is essential that the corrections to this approximation be well understood. One class of corrections which should be present even in the high energy limit is that due to inelastic intermediate state contributions to the double scattering amplitude.[1] We have constructed a highly idealized, but simple and internally consistent, model for these inelastic state corrections and applied it to pion-deuteron scattering.[2] In our model the pion is treated as the ground state of a system of two constituents (quark and anti-quark, say) having non-relativistic internal motions. High-energy pion-deuteron scattering is then treated as a slightly generalized version of deuteron-deuteron scattering using the Glauber method.[3]

The propogation of the quark and anti-quark between the neutron and proton can be described in two equivalent ways: as sums over complete sets of plane waves or over eigenstates of the quark - anti-quark Hamiltonian. Approximations can be generated by keeping only a finite number of terms in the latter series. In particular, the elastic approximation is obtained by keeping only the ground state, i.e. the pion itself.

We have made illustrative numerical calculations for a simplified case. We ignore spin, isospin, and the D-wave component of the deuteron, and assume the quark - anti-quark system can be treated as a three-dimensional

446

harmonic oscillator. Using units of $(GeV/c)^2$, we take

$$f(\Delta) = \bar{f}(\Delta) = 3i \exp(-2.5 \, \Delta^2),$$
$$S_d(\Delta) = \exp(-33.3 \, \Delta^2)$$
$$S_\pi(\Delta) = \exp(-4.0 \, \Delta^2)$$

for the quark-nucleon scattering amplitudes and for the
deuteron and pion elastic form factors, respectively.
The pion inelastic form factors are determined by the
same parameter as the elastic form factor, which is
essentially fixed by requiring that the break in the
pion-nucleon differential cross section come at the
right point.[4]

    The results are shown in Fig. 1. The most striking
features are: (1) Near the forward direction inelastic
corrections are of the order of 0.001% of the uncorrected
results, and thus completely negligible for practical
purposes. (2) Inelastic contributions are significant
for larger momentum transfers, and tend to wipe out the
structure in the elastic approximation which reflects
the structure in pion-nucleon scattering. (3) Adding
the correction from the n = 2 oscillator states (n = 1
states do not contribute if the quark and anti-quark have
the same interactions) considerably extends the region
in which the Glauber approximation is accurate.

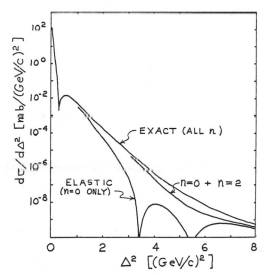

Fig. 1. Pion-deuteron differential cross sections
calculated from the model exactly, in the elastic
approximation, and with the inelastic corrections
from the n = 2 excited states of the pion.

The model presented here might, at best, approximate
reality at very high energies. At lower energies different sorts of corrections will probably be more important. Inelastic effects, however, certainly should be
present once production thresholds are reached, and the
above results may give some indication of their more
general features.

Part of this work was done while the author was a
summer visitor at Brookhaven National Laboratory. A more
detailed report is in preparation.

References:

[1] Previous estimates of inelastic effects have been made
by J. Pumplin and M. Ross, Phys. Rev. Letters 21, 1778
(1968) and by G. Alberi and L. Bertocchi, Nuovo Cimento
61A, 201 (1969).

[2] For a recent and quite complete discussion of pion-
deuteron scattering in the Glauber approximation see
C. Michael and C. Wilkin, Nucl. Phys. B11, 99 (1969).

[3] V. Franco, Phys. Rev. 175, 1376 (1968).

[4] D. R. Harrington and A. Pagnamenta, Phys. Rev. 173,
1599 (1968).

EXPERIMENTAL STUDY OF MANY-BODY REACTIONS WITH ^{8}Li PRODUCTION

IN SLOW NEGATIVE PION CAPTURE BY ^{12}C, ^{14}N, ^{16}O NUCLEI

N.M. Agababian, Yu.A. Batusov, S.A. Bunyatov, Kh.M. Chernev,[*]
P. Cuer,[**] J.-P. Massue,[***] V.M. Sidorov, V.A. Yarba

[*]   Institute of Physics of the Academy of Sciences, Sofia,
      Bulgaria
[**] Laboratory of Corpuscular Physics, C.N.R.
      Strasbourg-Cronenbourg, France
[***] On leave of absence from the Laboratory of Corpuscular
      Physics, C.N.R. Strasbourg-Cronenbourg, France

      Laboratory of Nuclear Problems
      Joint Institute for Nuclear Research, Dubna, USSR

Experiments on negative pion capture by light nuclei are a
source of information on the role of nucleon clusters in light
nuclei. In order to study some reactions with ^{8}Li production in
negative pion capture by light nuclei (C,N,O) the authors of the
present paper have made an experiment using photoemulsion chambers
at Dubna synchrocyclotron. The experimental arrangement has been
described in ref. /1/. The reactions resulting in one- or two-
prong σ-stars have been studied previously /1-3/. Here the results
of the kinematic analysis of 3-prong σ-stars in photoemulsion are
given in Table 1. As many as 1000 3-prong events have been measured
and calculated according to the kinematic programme. The number of
events corresponding to various reactions is given in the Table.
For the events which do not satisfy the kinematics of the reaction
with one neutron or without it the missing mass was calculated under
the assumption that they should correspond to reactions 4,9,15.
The result of the analysis shows that the number of events corresponding to various reactions is equal to the total number of the
analysed events with statistical accuracy. This confirms the conclusion that practically all σ-stars having hammer tracks belong to
reactions on light nuclei in photoemulsion. The obtained relative
probability values are given in the Table.

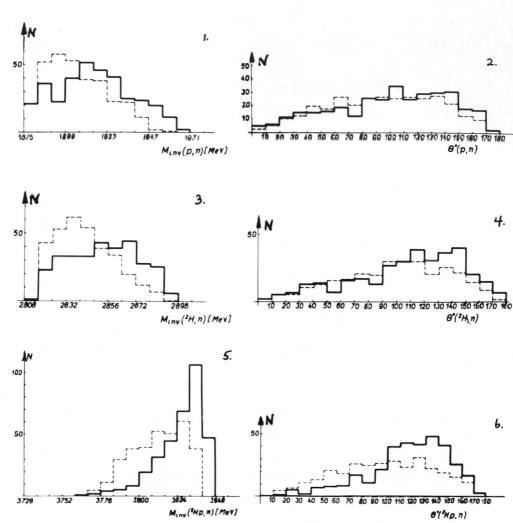

Figs. 1-3-5. Distribution according to effective masses for reaction (2). (Dashed curve is phased space.)

Figs. 2-4-6. Distribution according to angular correlations for different combinations of final particles for reaction (2). (Dashed curve is phased space.)

TABLE I

Reactions	Number of events	Relative probabilities	Channel
$\pi^- + {}^{12}C \rightarrow {}^{8}Li + {}^{2}H + {}^{2}H$	21	1.4×10^{-4}	1
$\rightarrow {}^{8}Li + {}^{2}H + {}^{1}H + n$	324	2.1×10^{-3}	2
$\rightarrow {}^{8}Li + {}^{3}H + {}^{1}H$	24	1.5×10^{-4}	3
$\rightarrow {}^{8}Li + {}^{1}H + {}^{1}H + 2n$	271	1.7×10^{-3}	4
$\pi^- + {}^{14}N \rightarrow {}^{8}Li + {}^{3}He + {}^{3}H$	5	1.1×10^{-4}	5
$\rightarrow {}^{8}Li + {}^{4}He + {}^{2}H$	7	1.5×10^{-4}	6
$\rightarrow {}^{8}Li + {}^{3}He + {}^{2}H + n$	54	1.2×10^{-3}	7
$\rightarrow {}^{8}Li + {}^{4}He + {}^{1}H + n$	66	1.5×10^{-3}	8
$\rightarrow {}^{8}Li + {}^{3}He + {}^{1}H + 2n$	155	3.4×10^{-3}	9
$\pi^- + {}^{O16} \rightarrow {}^{8}Li + {}^{4}He + {}^{4}He$	0	2.3×10^{-5}	10
$\rightarrow {}^{8}Li + {}^{6}Li + {}^{2}H$	7	6.2×10^{-5}	11
$\rightarrow {}^{8}Li + {}^{7}Li + {}^{1}H$	2	3.5×10^{-5}	12
$\rightarrow {}^{8}Li + {}^{6}Li + {}^{1}H + n$	12	1.2×10^{-4}	13
$\rightarrow {}^{8}Li + {}^{4}He + {}^{3}He + n$	16	1.4×10^{-4}	14
$\rightarrow {}^{8}Li + {}^{3}He + {}^{3}He + 2n$	15	1.3×10^{-4}	15

Since for some reactions there is sufficiently rich statistics one can obtain some information on the mechanism of these reactions from the characteristics of secondary particles.

Figs. 1-3 show the distribution of the number of events according to the effective masses and angular correlations of various combinations of final particles for reaction 2. Figs. 2 and 3 show that there is a noticeable difference of experimental distribution from the curves corresponding to phase space. A large number of events having the large values of effective masses and obvious angular correlations show that the important contribution to reaction 2 is made by negative pion absorption in ${}^{4}Li$ and ${}^{3}He$ clusters in the ${}^{12}C$ nucleus. The role of 4-nucleon ${}^{4}Li$ cluster in the ${}^{12}C$ nucleus for reaction 2 is most essential.

Thus from the results obtained and from refs. /1,2/ it is seen that the negative pion capture on ${}^{4}Li$ cluster is of importance for the reactions with ${}^{8}Li$ production on carbon.

In order to give the quantitative evaluation of the significance of various nucleon clusters in negative pion capture by light nuclei a more detailed analysis of the experimental results obtained is being carried out.

REFERENCES
1. Yu.A. Batusov, S.A. Bunyatov, V.M. Sidorov, V.A. Yarba. Yad. Phys. 6,1151,1967; Yad.Phys. 6,1149,1967.
2. Yu.A. Batusov, S.A. Bunyatov, V.M. Sidorov, Yu.S. Chaiks, Kh.M. Chernev, V.A. Yarba. Preprint p1-4309, Dubna, 1969.
3. J.-P. Massue. Thesis, Strasbourg, 1964.

ABSORPTION OF POSITIVE PIONS BY DEUTERON

C. LAZARD[+], J. L. BALLOT[+] and F. BECKER[++]

+   Institut de Physique Nucléaire , Division de
Physique Théorique . 91-Orsay- France

++   Departement de Physique . Centre de recherches
Nucléaires . 67 Strasbourg -France

      The purpose of the present paper is to show to
what extend a non phenomenological model can account for the
experimental results over a large energy range . Total cross sec
tion, angular distribution and polarization will be estimated from
threshold to 300 Mev for the laboratory pion energy .

      Our starting point will be the non relativistic limit
of the PS(PV) pion-nucleon interaction and we take into account,
the Born term, the forward and backward rescattering pion terms,
as well as the nucleon-nucleon rescattering in the final state; all
these corresponding graphs are given in figure 1 .

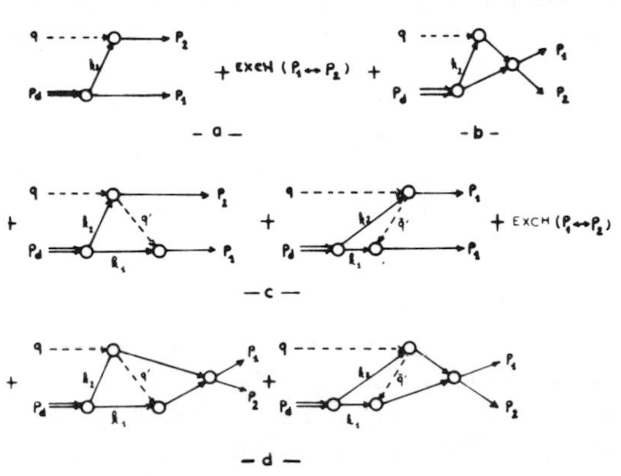

Figure 1.

452

In a first step, we restrict ourselves to the usual approximation "S" and" P" waves for the pion that we denote " Approximation I ". In this case, as it is well known, the angular dependence has a $(A + \cos^2 \Theta)$ shape for the differential cross section and a $\sin \Theta / (A + \cos^2 \Theta)$ shape for the polarization, Θ being the pion nucleon angle in the center of mass system . Then in a second step, we consider all the pion partial waves; however the final state interaction is treated only for "S" and"P " pion waves (Approximation II) . Here , the angular distribution can be expanded in a power of $\cos^2 \Theta$ and the polarization angular dependence will be more complex than in the previous case . We use also different wave functions for the Deuteron and for the two proton final state . These functions differ only at short distances and it is interesting to see how these differences affect the theoretical predictions . Note that in the two proton final wave function we have introduced the cut-off function

$$F_{\ell}(r) = (1 - \exp(- Z r))^{\ell+1}$$

depending on a cut-off parameter Z which allows to modify the asymptotic scattering wave functions near the origin;   two reasonable values $Z = \mu$ or $Z = 2\mu$ simulate roughly a well with a range of 1. 4 or 0. 7 Fermi respectively.

The results for the total cross section are given on figures (2 and 3). In figure 2 , we show the cross section calculated using the set 6 of the deuteron Gourdin's parametrization .

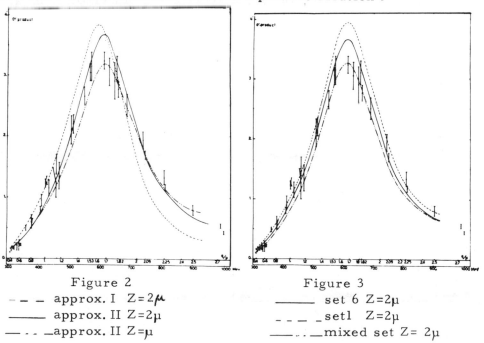

Figure 2
- - - approx. I Z=2μ
_____ approx. II Z=2μ
— .. _approx. II Z=μ

Figure 3
_____ set 6 Z=2μ
_ _ _ _set1 Z=2μ
—. .. _mixed set Z= 2μ

The difference between the two approximations appears, even at low energy; the approximation II gives a better position of the peak at the resonance. The result is sensitive to the value of z.

In figure 3, we give the total cross-section obtained in approximation II, for three parametrizations of the deuteron wave-function, the cut-off parameter being Z= 2 μ. The so-called mixed set is made of the Mac Gee S-wave function and of the D part of set 1 of Gourdin. Since both sets differ essentially near the origine,it appears that the total cross-section is sensitive to the behaviour of the deuteron wave-function near the origine. This fact is also true for the differential cross-section,as it is seen on figure 4.

At very low energy, in the case of approximation I, we have compared our results to Rose's fit and to the values obtained by Koltun and Reitan. For the α coefficient, our result is in good agreememt with both the Rose's fit and the Reitan's value; our value for the β coefficient is notably different from the Reitan's result and is a little higher than the Rose's fit.

The angular distributions given by approximation II is no more linear in $\cos^2\theta$ contrary to approximation I which contains only the S and P waves of the pion. On figure 4, we give the differential cross-section found in approximation II, for the three previous deuteron parametrizations at 750 Mev proton energy.

For the polarization, we have compared the angular dependence of the coefficient $\Lambda(\theta)$ with the Akimov and al experimemts. Contrary to the total and differential cross-sections, the polarization is not very sensitive to the different nuclear wave functions. Our results are qualitatively in agreememt with the experimental data.

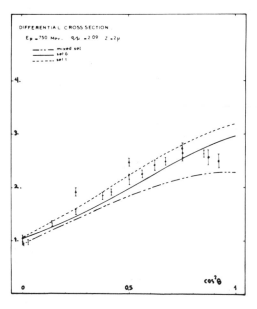

Figure 4

$\pi^{\pm}$ ABSORPTION CROSS SECTIONS AND NEUTRON DENSITY PARAMETERS

Morton M. Sternheim

University of Massachusetts

Amherst, Massachusetts 01002

A recent reanalysis[1] of the old 700 MeV $\pi^{\pm}$ - Pb inelastic scattering experiment of Abashian, Cool, and Cronin[2] led to a clear contradiction with current suggestions of a neutron-rich surface or "halo" unless one introduced physically unreasonable neutron distributions. This report will present estimates of the sensitivity of similar experiments to nuclear density parameters for representative nuclei as a function of energy.

The basic idea of such experiments is the following: at energies for which $\lambda \equiv \sigma\,(\pi^{+}\,n)/\sigma\,(\pi^{+}\,p) = \sigma\,(\pi^{-}\,p)/\sigma\,(\pi^{-}\,n) \geq 2$, both π^{-} are strongly absorbed in the interior of a large nucleus, while in the surface region the π^{+} are mainly absorbed by neutrons and the π^{-} by protons. The quantity $q \equiv \sigma(\pi^{-}\text{-A})/\sigma(\pi^{+}\text{-A})-1$, where the σ's are absorption cross sections, is consequently sensitive to the properties of the surface region. (A similar situation occurs for $\lambda \leq 1/2$.) For example, changing R_n, the neutron half density radius, by ten percent changes the calculated value of q for Pb by about .06 at the most favorable energies. Fig. 1 suggests $75 < T < 350$ MeV and $500 < T < 900$ MeV as promising regions for such experiments.

The calculations were performed using the ABACUS-M optical model code[3] with a simple (nonderivative) optical potential

$$U \equiv 2\,E_{\pi}\,V_{optical} = -\,Z\,b_{p}\,\rho_{p} - N\,b_{n}\,\rho_{n} \qquad (1)$$

Here $b_{p} = 4\pi f(0)$, where $f(0)$ is the forward π-p scattering amplitude, etc. We used the CERN phase shifts[7] and averaged over the motion of nucleons within the nucleus using a Fermi gas model.[8]

This model cannot give good quantitative fits to pion nucleus

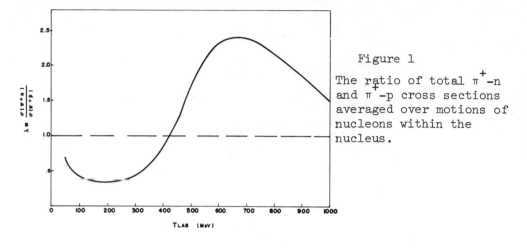

Figure 1

The ratio of total π^+-n and π^+-p cross sections averaged over motions of nucleons within the nucleus.

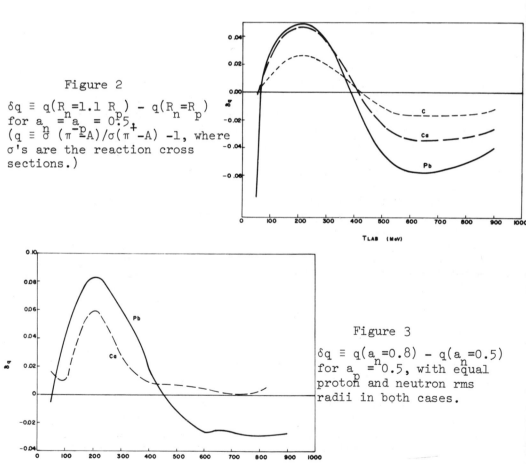

Figure 2

$\delta q \equiv q(R_n = 1.1\ R_p) - q(R_n = R_p)$ for $a_n = a_p = 0.5$, ($q \equiv \sigma\ (\pi^-A)/\sigma(\pi^+-A) - 1$, where σ's are the reaction cross sections.)

Figure 3

$\delta q \equiv q(a_n = 0.8) - q(a_n = 0.5)$ for $a_p = 0.5$, with equal proton and neutron rms radii in both cases.

elastic differential cross section data for $T \leq 200$ MeV because of
the dominant role of the P_{33} pi-nucleon phase shift. The Kisslinger
model[4,5] does take this into account very well for $T \leq 100$ MeV by
including a derivative potential $b' \nabla \cdot (\rho \nabla \Psi)$. Our unpublished a-
nalysis of the recent experiments[6] on C^{12} shows that this model
works surprisingly well even at the resonance itself. For our
present purposes, however, the simple optical model of (1) should
be sufficient to estimate the general sensitivity of pion scattering to variations in ρ_p and ρ_n.

The change δq resulting from increasing the half density radius
R_n from 1.0 R_p to 1.1 R_p with skin thickness parameters $a_n = a_p$
constant is shown in Figure 2, and increases from a maximum of 2.5%
for C to 6% for Pb. The δq resulting from varying a_n from .5 to
.8f - with the R_n adjusted to keep the rms neutron and proton
radii equal - is shown in Fig. 3; it is as much as 8% for Pb.

Thus the ratio of π^+/π^- absorption cross sections is sensitive
to details of the neutron density, assuming the proton density is
known already, e.g., from electron scattering. An accuracy of $\leq 1\%$
in q is required for interesting results. The validity of the optical model used can be verified by studying the energy dependence
of the cross sections[8] and of the ratios[7] and is itself an interesting question worth further study, especially near the 33 resonance.
Note that the computed ratios are quite insensitive to the optical
parameters used.[1]

Differential elastic $\pi^\pm$ scattering data is also sensitive to
density details. This will be explored in a later paper.

*Research supported by the National Science Foundation. Part of
this work was done while the author was a Visiting Staff Member
at Los Alamos Scientific Laboratory.

1. E.H. Auerbach, H.M. Qureshi, and M.M. Sternheim, Phys. Rev.
    Letters 21, 162 (1968).
2. A. Abashian, R. Cool, and J.W. Cronin, Phys. Rev. 104, 855
    (1956); L.R.B. Elton, Nuclear Sizes (Oxford University Press),
    1961, p. 93.
3. E.H. Auerbach and M.M. Sternheim, Brookhaven National Laboratory
    Report 12696, July 1968.
4. E.H. Auerbach, D.M. Fleming, and M.M. Sternheim, Phys. Rev. 162,
    1683 (1967), and Phys. Rev. 171, 1781 (1968).
5. L.S. Kisslinger, Phys. Rev. 98, 761 (1955).
6. Binon et al., Proceedings of the International Conference on
    Nuclear Structure, Tokyo, 1967.
7. Lovelace et al., CERN preprint.
8. M. Crozon et al., Nucl. Phys. 64, 567 (1965).

PROMPT GAMMA RAYS FROM π-BOMBARDMENT OF C^{12} NEAR THE 3-3 RESONANCE

J. Lieb and H. Funsten

William and Mary College

Williamsburg, Virginia

Preliminary data on prompt gamma rays from discrete nuclear levels from pion bombardment of C^{12} has been obtained. Measurement of such gamma rays give a method of indirectly studying the scattering of pions on nucleii in a manner similiar to that employed by Clegg et. al[1], in studying proton reactions at the Harwell synchrocyclotron. The use of pions differs from protons in having 0 spin, unit isospin and displaying an isospin dependant nucleon resonance. Gamma ray measurements obviate necessity of using a good resolution pion beam but entail some abiguities in assignment of nuclear levels.

The data was taken using several negative pion beam configurations at the NASA-W&M Space Radiation Effects Laboratory synchrocyclotron, giving energies of 100, 250, and 315 MeV. The experimental set-up, Fig. 1 consisted of a three counter beam telescope ahead of a 4" x 4" C^{12} target. Gamma rays were detected by both a 5" x 5" NaI and a 30 c.c. Ge (Li) detector. Anti-coincidence shielding was used with both gamma detectors. Outgoing charged particles were detected by an array of four pairs of 12"x 12" plastic scintillators surrounding the target with one pair in each quadrant. Additionally, an absorber, usually 1/4" aluminum, was placed between the scintillators of a pair. By demanding for a valid event a scattered charged particle to traverse both scintillators of a pair, background due to neutron and proton knockouts was reduced. Hence a valid event was a gamma ray coincident with a pion entering and leaving the target. Muon contamination was less than 10%.

Straight-forward electronics was used; discriminator threshold on all charged particle counters were adjusted by moving
the counters into the direct pion beam and adjusting triggering
just below the resulting voltage band of pulses produced by the
pions. An eliminator circuit vetoes events in which two pions
entered the system within the rise time of the NaI gamma counter.
Good events were routed to an analyzer to record a gamma spectrum
for each of the four quadrant directions. For the two higher
energies, the target thickness was chosen to place the mid-target
pion energy at the pion-nucleon resonance, -200 MeV. For the 100
MeV, run, the target thickness was chosen to give a pion energy of
70 MeV. at target exit, a compromise between increased gamma yield
due to thickness and background due to pions stopping in the target.

The NaI and Ge (Li) spectra were smoothed with a three point
parabolic fitting function to facilitate choices of initial parameters for guassian + background least-square fits to the original
spectra. Fig. 2 shows the gamma ray peaks observed with possible
assignments to levels in residual nucleii, the energies listed
being averages over the four quadrant directions. Some ambiguity
due to the poor resolution of the NaI detector was removed by comparison with the Ge (Li) detector spectra. The latter however,
exhibited poorer statistics due to the relatively small germanium
detector efficiency.

The most pronounced peak occurs at 4.43 MeV; its assignment
to the B^{11} II-0 transition can possibly be ruled out due to the
lack of a companion line at 4.32 MeV. for C^{11} II-0 which would be
expected to be stronger than the B^{11} transition. Also measured
peak energy shift with quadrant direction indicates two body
kinematics. Ratioes of quadrant directions for backward/forward
π scattering taken at 100 and 310 MeV indicate a decrease in the
backward/forward scattering closer to resonance in agreement with
the inelastic C^{12} cross-section measurements of Stroot[2] at CERN.

Two levels are assigned to single nucleon knock-out reactions
leaving mass-11 nucleii. The level at 2 MeV. is assigned to the
B^{11} I-0 transition. This state is clearly seen on the Ge (Li)
spectra with a 2.12 MeV. C^{11} I-0 component of less than one third
that of the B^{11} AT E_π=250 MeV. The III-0 transitions at 5 MeV
indicate a factor of 2 for C^{11} over B^{11}. Although the above germanium results with π^- at the 3-3 resonance indicate a greater
yield for neutron over proton knockout by at least a factor of 3/1,
the expected ratio for these two levels is 9/1 since single charge
exchange would not be observed with the apparatus. Hence the
above results would not contradict those of Tanner[3] who observed
C^{11} production ratio of 1/1 for π-vs. π^+ reactions instead of an
expected ratio of 3/1.

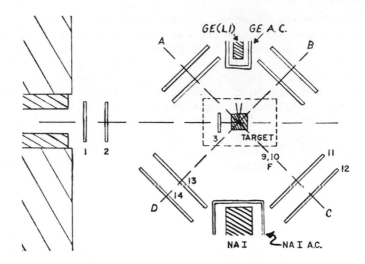

Fig. 1 Experimental set up

ENERGIES OF GAMMA RAYS

FROM 100 MeV PIONS ON C^{12}

E_{gamma}	Int.(rel.)	ASSIGNMENT
1.00*#	400	B^{10} II-I, 1.23; Li^8 I-0, .975
1.11*#	100	
1.32#	110	
1.49*	70	
2.02*#	130	C^{11} I-0, 1.995
2.21*(av.)	120	Li^8 II-0, 2.26
2.65#	70	Be^{10} II-I, 2.59
3.33# 3.84	60	Be^{10} I-0, 3.37;
4.44*#	600	C^{12} I-0, 4.439; B^{11} II-0, 4.444
4.75*#	50	C^{11} III-0, 4.79
5.04*	15	B^{11} III-0, 5.02
7.6#		
8.8		
9.5		
11.6		

* denotes level also seen at E_{pi} at 310 MeV

\# denotes level also seen at E_{pi} at 250 MeV

Fig. 2 Gamma ray energies observed.

References:

1. A. B. Clegg et.al. Proc. Phys. Soc. 78, 681 (1961)
2. T. P. Stroot Proc. Nimrod Conf. (1968)
3. N. W. Tanner et. al. Phys. Lett. 26B 573 (1968)

PRELIMINARY RESULTS ON INTERACTIONS OF LOW ENERGY PIONS ON CARBON NUCLEI

E. Bellotti, S. Bonetti, E. Fiorini, A. Pullia and

M. Rollier

Istituto di Fisica and I.N.F.N. - Milan

This report is a preliminary step of a study on nuclear structure of carbon by means of pion interactions at ten different momenta ranging from 0.25 to 2.0 GeV/c, in the Ecole Polytechnique bubble chamber BP3 (1), filled with propane. We have measured about 2,000 events at the lowest energy (156 to 126 MeV, the spread being due to energy loss in the fiducial region). They can be divided by means of kinematical analysis into four channels: a) elastic scattering on free proton; b) coherent scattering on carbon nucleus; c) incoherent interactions on carbon; and d) pion absorptions by carbon.

Events of type a), which constitute the background of our experiment, have been selected by fitting interactions with one secondary pion and proton. Their angle and momentum distributions check very well with spectra obtained by interpolation of results in hydrogen at nearby energies (2). We have also evaluated the number of those events where the proton is not visible due to its low momentum; these events would appear in channel b).

Coherent events have been selected by fitting all single pion events without blobs or other evidence of nuclear breakup. Their cross section is (228 ± 30) mb per carbon nucleus, when the background due to pion decays and to scattering on hydrogen with low momentum transfer is subtracted. The distribution of the square of four momentum transfer, shown in Fig. 1 presents the well known exponential shape also found at higher energies (3). The slope is $(78 \pm 4 \text{ GeV}^{-2}$, which corresponds to a radius of $(3.01 \pm .08)$ fermis. Note that the smearing of the exponential peak due to experimental resolution is much lower than in corresponding experiments at higher energies (3).

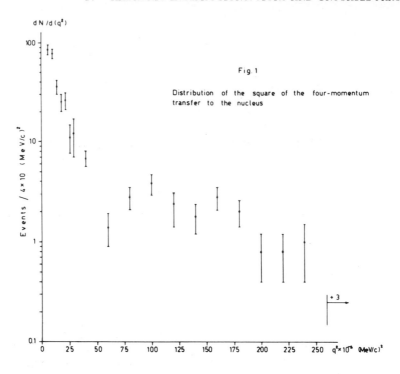

Fig. 1

Distribution of the square of the four-momentum transfer to the nucleus

Incoherent events could be divided into channels: c_1 ($\pi^+p \rightarrow \pi^+p$); c_2 ($\pi^+n \rightarrow \pi^+n$); and c_3 ($\pi^+n \rightarrow \pi^\circ p$); their cross sections are 157 ± 10; 24 ± 4 and 31 ± 5 mb, respectively. We have included in channel c_3 the number of those events where neither γ-ray materialize in the chamber. The percentages of events in these three channels have been compared with the results of a rough Montecarlo calculation, where a single pion in interaction in the nucleus and Fermi distributed nucleon momenta have been assumed. The results are in reasonable agreement with our experimental data.

All the remaining events show no evidence for the emission of a pion, and are therefore candidates for nuclear absorption. The distributions of the difference between the kinetic incoming energy and the visible kinetic energy of the secondaries are reported in Fig. 2 for events with different numbers of emitted protons. Note that one should subtract to this difference at least the average binding energy multiplied by the number of emitted protons. The forward distribution of the one proton events reflects the fact that most of these interactions are incoherent events with one π° where neither γ-ray materialize in the chamber. The events with more than one proton are mostly concentrated in the negative region of the spectrum, with a minimum corresponding approximately to the pion mass, as expected. Measurements and improved calculation are in progress to clear up if the bump in the negative region for three proton events can be due to pion

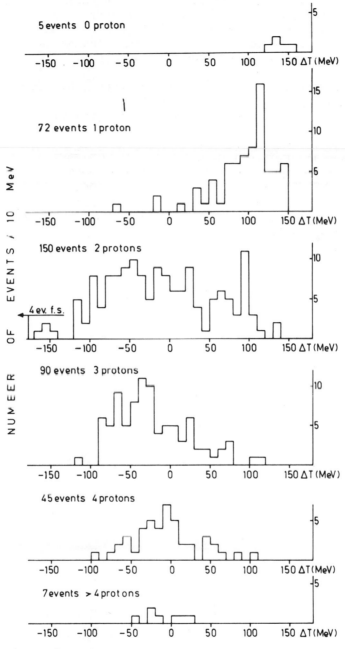

Fig. 2   Distribution of the difference $T_{inc} - T_{out}$ for pion absorptions

absorption by an alpha particle inside the nucleus. After
correcting for events with no visible electron-positron pair we
obtain an absorption cross section of (216 ± 12) mb; corresponding
to about 30% of the total cross section. The average proton
multiplicity in pion absorption is 2.4.

It is a pleasure to thank our colleagues of Ecole Polytechnique and Orsay HLBC groups for the exposure of the chamber
BP3 at Saclay, which made this experiment possible.

1) M. Bloch, A Lagarrigue, P. Rancon and A. Rousset: Rev. Sci.
    Instr., 32, 1302 (1961).
2) See for instance: G. Kallen - Elementary Particle Physics -
    Addison-Wesley publ. Co. - Palo Alto (1964).
3) Berkeley-Milan-Orsay-Saclay Collaboration: Topical Con-
    ference on High Energy Collisions of Hadron - CERN - January
    1968, pag. 537.

MEASUREMENTS OF THE K_L^o-NUCLEUS TOTAL CROSS

SECTION AT GeV ENERGIES[†]

W. L. Lakin[×‡], L. H. O'Neill[×§], J. N. Otis[×§],

E. B. Hughes[*], and L. Madansky[**‖]

*High Energy Physics Lab, Stanford Univ., Stanford, Calif.

**Stanford Linear Accelerator Ctr., Stanford Univ.,

Stanford, California

Experimental results on hadron-nucleus scattering at high energy can be used to investigate nuclear structure or, conversely, if the nucleus is sufficiently well understood, to determine hadron-nucleon amplitudes. In the analysis of such experiments the nucleus is described by a complex optical potential and the collision treated according to Glauber scattering theory. The data available at high energy, involving the stable or relatively long-lived hadrons, are not plentiful. It was with the aim of increasing the variety of the available data, against which the basic theoretical approach can be tested, that the experiments to be described were carried out.

The goal of the experiment, in its initial phase, was to measure the total cross section for the interaction of K_L^o mesons with selected nuclei at GeV energies. The experiment was done at the Stanford Linear Accelerator Center using a neutral beam significantly richer in K_L^o mesons, relative to neutrons, than is presently available at other accelerators. A 3° neutral beam, taken from a beryllium target struck by 16 GeV electrons, was chopped into 0.3 nS pulses, spaced by 50 nS, permitting a time of flight measurement on each K_L^o detected. The total cross sections were measured by a good geometry technique. The K_L^o detector consisted of a sandwich of steel absorbers and plastic scintillators and the cross sections were measured as a function of time of flight. The maximum resolvable K_L^o momentum was ~ 8 GeV/c, corresponding to a timing resolution of 0.75 nS (FWHM) over a flight path of 139 meters.

Preliminary values of the observed cross sections are listed in Table I. These data correspond to times of flight smaller than

TABLE I

Nucleus	Total Cross Section		Woods-Saxon Parameters	
	Observed	Calculated	c (Fermi's)	a (Fermi's)
C^{12}	190 ± 6	190.4	2.36	0.44
Al^{27}	385 ± 12	393.9	3.09	0.50
$Cu^{63.5}$	840 ± 16	845.2	4.30	0.57
$Pb^{207.2}$	2307 ± 55	2242.0	6.68	0.52

The observed and calculated K_L^0-nucleus total cross sections at 9.0 GeV/c. The elementary K-nucleon total cross sections and ratios of real to imaginary forward scattering amplitudes used in the optical model calculation are as follows: K^+-n (17.55 mb, -0.11), K^+-p (17.24 mb, -0.23), K^--n (20.45 mb, 0.0), K^--p (22.86 mb, 0.0).

1.2 nS, or to K_L^0 momenta larger than 7.1 GeV/c. The relative neutron intensity in the beam in this timing window (neutron momentum larger than 13.4 GeV/c) is estimated to be 4.0 %, and the tabulated K_L^0 cross sections have been correspondingly adjusted using the values of the neutron-nucleon total cross sections at 10 GeV/c given in Engler et al.[2] The neutron admixture is determined from the known neutron momentum spectrum[3] and from considerations of the apparent cross section versus time of flight.

According to the optical model, the K_L^0-nucleus total cross section is determined by the real and imaginary forward scattering amplitudes of the elementary K^0- and $\overline{K^0}$-nucleon interaction and by the spatial extent of the nuclear potential. The cross sections predicted by the optical model, in the Glauber or eikonal approximation, are included in Table I. The elementary K^0 and $\overline{K^0}$ cross sections were obtained, using charge independence, from published K^+- and K^--nucleon cross sections[4] and ratios of real to imaginary parts of the forward scattering amplitudes taken from dispersion relation calculations.[5] The nuclear potential was assumed to follow the Woods-Saxon distribution

$$\rho(r) = \rho(0) \left[1 + \exp\left(\frac{r-c}{a}\right) \right]^{-1}$$

with half-density radius c and skin thickness t = 4.4 a typical of electron scattering results.[6]

TABLE II

Nucleus	Total Cross Section			Woods-Saxon Parameters	
	Observed	Engler et al	Calculated	c (Fermi's)	a (Fermi's)
C^{12}	362 ± 9	340 ± 3	349.6	2.36	0.44
Al^{27}	666 ± 16	683 ± 3	676.6	3.09	0.50
$Cu^{63.5}$	1382 ± 27	1364 + 14	1354.7	4.30	0.57
$Pb^{207.2}$	3071 ± 79	3146 ± 50	3125.9	6.68	0.52

The observed and calculated neutron-nucleus total cross sections at 3.8 GeV/c and the results of Engler et al at 10 GeV/c. The elementary neutron-nucleon total cross sections and ratios of real to imaginary forward scattering amplitudes used in the optical model calculation are as follows: n-p (42.55 mb, -0.51), n-n (42.20 mb, -0.33).

From the experimental data at relatively long times of flight (8 - 16 nS), it was also possible to measure the neutron-nucleus total cross section at GeV energies. The admixture of detected K_L^0's in this timing window is apparently quite small. Preliminary values of the neutron-nucleus cross sections at an average neutron momentum of 3.8 GeV/c are given in Table II together with cross sections computed from the optical model. In this calculation the elementary neutron-nucleon cross sections were taken from the measurements of Bugg et al[7] and ratios of real to imaginary parts of the forward scattering amplitudes from dispersion relation calculations.[8] Table II also includes, for comparison, the neutron-nucleus cross sections reported by Engler et al[2] at 10 GeV/c.

ACKNOWLEDGEMENT

We are indebted to Professor O. Kofoed-Hansen of CERN for discussions on the optical model calculation.

† Work supported in part by the U. S. Office of Naval Research, Contract [Nonr 225(67)] and the National Science Foundation Grant GP-9498.

‡   Now at the Stanford Linear Accelerator Center.

§   National Science Foundation Predoctoral Fellow.

‖   National Science Foundation Senior Postdoctoral Fellow on leave
    from Johns Hopkins University.

REFERENCES

1. R. J. Glauber, Lectures in Theoretical Physics (Interscience
    Publishers, Inc., New York, 1959) Vol. I, p. 315.

2. J. Engler et al, Phys. Letters 28B, 64 (1968).

3. A. D. Brody et al, Phys. Rev. Letters 22, 966 (1969).

4. W. Galbraith et al, Phys. Rev. 138, B913 (1965).

5. M. Lusignoli et al, Nuovo Cimento XLV, A792 (1966) and IL,
    A705 (1967).

6. H. R. Collard, L. R. B. Elton, R. Hofstadter, and H. Schoper
    in Landolt-Börnstein, Numerical Data and Functional Relation-
    ships in Science and Technology, edited by K.-H. Hellwege
    (Springer-Verlag, Berlin, 1967), New Series, Group I, Vol. 2.

7. D. V. Bugg et al, Phys. Rev. 146, 980 (1966).

8. P. Söding, Phys. Letters 8, 285 (1964), and A. A. Carter and
    D. V. Bugg, Phys. Letters 20, 203 (1966).

RECENT EXPERIMENTAL WORK ON PIONIC X-RAYS

G. Backenstoss

CERN, Geneva, Switzerland, and

University of Karlsruhe, Germany

An atomic nucleus in the field of which a π-meson moves is called a pionic atom. This implies the fact that the properties of such a system are closely related to the properties of an ordinary electronic atom. This is the case because in both systems the electromagnetic interaction plays a dominant role. Since with available meson intensities it is very improbable to have more than one meson Coulomb captured by one nucleus, only the relatively simple situation as it is known from hydrogen-like atoms must be considered. After Coulomb capture the mesonic atom is automatically left in a highly excited state so that its de-excitation can be observed. The aforesaid is true, of course, also for muonic atoms where one has found qualitatively all the effects known from the optical spectroscopy of hydrogen-like atoms such as fine structure, hyperfine structure, isotope shift, finite size effect, etc. But frequently the quantitative order of magnitude of these effects is different due to the heavy mass of the muon. This is the reason for the success of the studies on muonic atoms.

The pion is even heavier than the muon; therefore, the pionic atom should exhibit the same effects even more markedly except for the ones connected with the spin-½ of the fermions. However, these electromagnetic effects are better studied with muonic atoms since the pion, unlike the muon and the electron, is also subjected to the strong interactions which may mask these effects. At first sight it might even be astonishing that the electromagnetic interaction plays an important role. The reason is the much longer range of the electromagnetic field, and the fact that the mesons are captured predominantly into high angular momentum states with a small probability to be inside the range of the strong nuclear field.

Only after many transitions to lower atomic states does the pion
come into the range of the strong interacting forces, which then
rapidly change the picture. Therefore, the main interest has to
be concentrated on such atomic levels that are still present due
to the electromagnetic interaction but which are already influenced
by the strong interaction. It is rather obvious but nevertheless
it should be remarked that, in order to investigate the deviations
of pionic atoms from the purely electromagnetic behaviour as exhibited by the muonic atom, a comparison with the latter is the
most straightforward method, and a good understanding of the muonic
atom is often vital.

In this way the study of pionic atoms provides a powerful
tool to investigate the strong pion-nucleus interaction, which in
its turn has significant bearing on the structure of the nucleus,
as well as relations to the elementary pion-nucleon interaction.
Apart from this problem, the pionic atom also provides information
on the pion itself, providing, as is generally known, the best
measurement of the pion mass to date.

The deviations from the purely electromagnetic level scheme
result mainly in two effects. Firstly, the energies of the pionic
levels are shifted. Secondly, the pions can be absorbed by the
nucleus. This causes a broadening of the levels out of which
absorption takes place which may be measured in many cases from
the natural Lorentzian line width of a pionic X-ray transition.
Since the range of the strong interaction is short, there will be
only one level strongly affected in a given atom, the effect in
the next higher level $(n + 1, \ell + 1)$ being about three orders of
magnitude smaller. This implies that with present techniques
neither energy shifts nor line shapes can be measured for more than
one X-ray transition in one particular pionic atom, and that the
energy shifts and line widths of an X-ray transition are almost
wholly caused by the lower level of the transition. However
the width of the upper level, although three orders of magnitude
smaller, may be determined by a measurement of the intensity of a
transition which depends on the ratio of the probability for pion
absorption, $W_a = \Gamma_a/\hbar$, to the transition probability, W_x, for
electric dipole radiation. Therefore, the main emphasis of the
experimental work is to measure exact energies, line shapes and
intensities.

The earliest studies on pionic atoms were performed with NaI-
detectors by Camac et al.[1] and Stearns et al.[2] who measured X-ray
energies and yields, and by West and Bradley[3] with proportional
counters who obtained also one line width. During recent years
new measurements have been performed with Ge(Li)-detectors at
various laboratories and have provided more accurate and new
results[4-17]. Although the experimental techniques are conven-

tional I would like to say a few words concerning the experimental difficulties one faces which are typical for π-mesic X-ray measurements. Compared with muonic X-ray work the background in the interesting energy region of ∿ 100 keV is usually higher and many disturbing lines are present making evaluation of the spectra more difficult. Stopping pions undergo nuclear reactions; γ-rays produced after pion capture are prompt and hence cannot be rejected. Muons are present in the π-beam producing high yield muonic X-rays, whereas the interesting pionic X-rays have yields below 10%. Furthermore, broadened lines are more sensitive to the shape of the underlying background.

In Tables 1 and 3 the data of the 2p-1s and the 3d-2p transition energies obtained by the various groups are presented. I have omitted quoting values for the strong interaction shift as these are given frequently. These numbers are only obtained indirectly by calculating the point nucleus energies with the Klein-Gordon equation, applying corrections for the finite size nucleus and the vacuum polarization. The difference between an energy calculated in this way and the measured energy is then defined as strong interaction shift. However, this method is not clean, since the mentioned effects are not independent. The strong interaction deforms the pion wave function leading to different finite size and vacuum polarization corrections. The comparison between experiment and theory, therefore, should be made with the calculated energy values directly. In Figs. 1 and 2 the shifts ΔE are shown only in order to demonstrate the size of the effects. The s-level shifts are all negative and decrease the binding energy, whereas the p-level shifts and all higher level shifts are positive, and correspond to an increased binding energy.

The agreement between the different measurements is good. The only disagreement occurs for Mg where we concluded that the line in question is not a broadened pionic line (Fig. 3). The isospin effect proportional to 1/A (A = mass number) is clearly established for the 2p-1s transitions. It can be shown that all T = 1/2 nuclei are shifted towards smaller energies compared with the T = 0 nuclei. This effect can be directly seen for the pair ^{10}B and ^{11}B and most clearly for ΔT = 1 for ^{16}O and ^{18}O where an isospin shift of about 5 keV occurs. Although a few measurements on heavier isotopes exist, for example, ^{40}Ca and ^{44}Ca as well as ^{58}Ni and ^{60}Ni for the 2p level and seven Sn isotopes for the 3d level, this effect could not be seen so far for those levels, except perhaps for the Ca isotopes.

The measurements of line shapes require gamma detectors of high resolution. The natural line shape is of Lorentzian type and is folded with the instrumental line width, which is essentially of Gaussian shape. At the moment, the smallest Lorentzian line widths accessible for measurement are about 10-20% of the instru-

Table 1

Energies of pionic 2p-1s transitions in keV

Target (1)	Berkeley[a] (2)	CERN[b] (3)	W & M[c] (4)	Theory (5)
^{4}He			10.69 ± 0.06	
^{6}Li	23.9 ± 0.2		24.18 ± 0.06	24.05
^{7}Li	23.8 ± 0.2		24.06 ± 0.06	23.91
^{9}Be	42.1 ± 0.2	42.38 ± 0.20	42.32 ± 0.05	42.09
^{10}B	64.9 ± 0.2	65.94 ± 0.18	65.79 ± 0.11	65.61
^{11}B	64.5 ± 0.2	64.98 ± 0.18	65.00 ± 0.11	64.91
^{12}C	93.3 ± 0.5	92.94 ± 0.15	93.19 ± 0.12	93.13
^{14}N	123.9 ± 0.5	124.74 ± 0.15		124.61
^{16}O	160.6 ± 0.7	159.95 ± 0.25		159.37
^{18}O		155.01 ± 0.25		155.13
^{19}F	196.5 ± 0.5	195.9 ± 0.5		194.44
^{23}Na	277.2 ± 1.0	276.2 ± 1.0		277.57
Mg	330.3 ± 1.0	no X-ray line		

a) Refs [4,5]; b) Ref. [6]; c) Refs [7-9].

Table 2

Widths of pionic levels

Target (1)	Γ_{1s} (keV)				Γ_{2p} (eV)	
	Berkeley[a] (2)	CERN[b] (3)	W & M[c] (4)	Theory (5)	CERN[d] (6)	Theory (7)
^{4}He						
^{6}Li	0.39 ± 0.36		0.15 ± 0.05	0.13		
^{7}Li	0.57 ± 0.30		0.19 ± 0.05	0.17		
^{9}Be	0.85 ± 0.28	1.07 ± 0.30	0.58 ± 0.05	0.54	0.16 ± 0.03	0.07
^{10}B	1.4 ± 0.5	1.25 ± 0.25	1.68 ± 0.12	1.30	0.32 ± 0.06	0.26
^{11}B	2.3 ± 0.5	1.87 ± 0.25	1.72 ± 0.15	1.36	0.27 ± 0.04	0.33
^{12}C	2.6 ± 0.5	2.96 ± 0.25	3.25 ± 0.15	2.88	1.02 ± 0.29	1.0
^{14}N	4.1 ± 0.4	4.48 ± 0.30		5.02	2.1 ± 0.3	2.8
^{16}O	9.0 ± 2.0	7.56 ± 0.50		6.75	4.7 ± 0.8	5.3
^{18}O		8.67 ± 0.70		6.00	3.8 ± 0.7	6.1
^{19}F	4.6 ± 2.0	9.4 ± 1.5		8.22	11.2 ± 1.9	10
^{23}Na	4.6 ± 3.0	10.3 ± 4.0	6.2 ± 1.2	16.0	34.6 ± 7.6	37

a) Refs [4,5]; b) Ref. [6]; c) Refs [7-9]; d) Ref. [15,16].

mental line width, provided sufficient statistics and a good knowledge of the instrumental line shape exist. Since for small widths
even small deviations from the Gaussian shape are important, a good
calibration is vital, for which purpose muonic X-ray lines with
practically zero natural line width produced simultaneously in the
pionic spectra are nearly ideal. In cases where they are not
available at sufficiently similar energies, suitable radioactive
γ-rays may be fed in simultaneously by accidental coincidences with
the beam telescope in the X-ray spectrum. The difficulties in
measuring the broad lines are due to two reasons. The Lorentz
curve has long tails and, therefore, an error in the background
subtraction enters sensitively in the determination of the Lorentz
width. Furthermore, the broad lines caused by strong pion absorption in the lower level are subjected also to a significant pion
absorption out of the upper level, implying low yields.

Results obtained by the different groups are shown in Table 2
for Γ_{1s}, in Table 4 for Γ_{2p}, in Table 6 for Γ_{3d} and in Table 8 for
Γ_{4f}. Tables 6 and 8 contain nuclei with magnetic dipole and electric quadrupole moments which cause hyperfine structure splittings
of the lower level which are not negligible compared with the widths

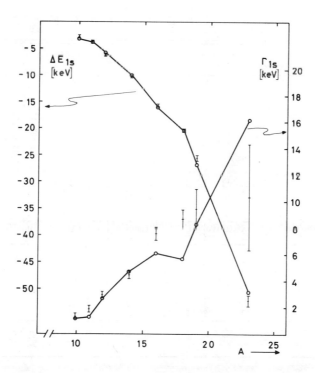

Fig. 1—Energy shifts and widths of the pionic 1s level versus mass number A. The full line is calculated with constants of column 1, Table 10.

Table 3

Energies of pionic 3d-2p transitions in keV

Target (1)	Berkeley[b] (2)	CERN[c] (3)	W & M[d] (4)	Theory (5)
$^{27}_{13}$Al	87.53 ± 0.07[a]	87.40 ± 0.10		87.37
^{14}Si		101.58 ± 0.15		101.48
$^{31}_{15}$P		116.78 ± 0.10		116.69
$^{32}_{16}$S	133.2 ± 0.3	133.06 ± 0.10		132.98
^{17}Cl		150.55 ± 0.15		150.36
^{19}K	188.6 ± 0.3	188.77 ± 0.18		188.48
$^{40}_{20}$Ca	209.3 ± 0.3	209.66 ± 0.18		209.24
^{44}Ca	(208.94 ± 0.10)			
^{22}Ti		253.98 ± 0.20		253.77
$^{51}_{23}$V	278.2 ± 0.4	277.85 ± 0.20		277.76
^{24}Cr	302.5 ± 0.5	302.75 ± 0.25		302.95
$^{55}_{25}$Mn	328.5 ± 0.8	329.12 ± 0.25		329.02
^{26}Fe	356.9 ± 1.0	356.43 ± 0.30		356.57
$^{59}_{27}$Co	384.6 ± 1.0	384.74 ± 0.35		384.65
$^{58}_{28}$Ni		415.23 ± 0.70	414.11 ± 0.48	414.56
$^{60}_{28}$Ni			414.08 ± 0.51	
^{29}Cu		446.1 ± 2.0		444.12
^{30}Zn		478.2 ± 3.0		475.99

a) Crystal diffraction spectrometer, Ref. [18];
b) Refs [4,5,10]; c) Ref. [11]; d) Ref. [12].

Table 4

Widths of pionic levels

Target (1)	Γ_{2p} (keV)				Γ_{3d} (eV)	
	Berkeley[a] (2)	CERN[b] (3)	W & M[c] (4)	Theory (5)	CERN (6)	Theory (7)
$^{27}_{13}$Al		0.11 ± 0.08		0.11	0.02 ± 0.01	0.02
^{14}Si		0.18 ± 0.08		0.17	0.01 ± 0.04	0.03
$^{31}_{15}$P		0.20 ± 0.08		0.25	0.09 ± 0.05	0.06
$^{32}_{16}$S	0.8 ± 0.4	0.79 ± 0.15		0.36	0.07 ± 0.06	0.10
^{17}Cl		0.89 ± 0.25		0.51	0.30 ± 0.13	0.18
^{19}K	1.9 ± 0.15	1.45 ± 0.15		1.03	0.6 ± 0.3	0.49
$^{40}_{20}$Ca	2.29 ± 0.13	2.00 ± 0.25		1.52	0.5 ± 0.3	0.74
^{44}Ca	2.07 ± 0.15					
^{22}Ti		2.89 ± 0.25		2.66	2.5 ± 0.7	2.02
$^{51}_{23}$V		3.66 ± 0.25		3.51	1.7 ± 0.9	3.22
^{24}Cr		4.46 ± 0.35		4.27	4.9 ± 1.1	4.28
$^{55}_{25}$Mn		6.38 ± 0.40		5.01	6.7 ± 1.4	5.97
^{26}Fe	6.0 ± 2.5	8.65 ± 0.60		6.38	9.2 ± 2.2	8.10
$^{59}_{27}$Co		7.37 ± 0.70		7.94	12.9 ± 7.0	11.6
$^{58}_{28}$Ni			7.6 ± 1.4	9.70	12.2 ± 4.3	14.9
$^{60}_{28}$Ni			8.5 ± 1.5			
^{29}Cu		15.9 ± 4.0		11.2	18.4 ± 6.8	21.3
^{30}Zn		16.8 ± 6.0		13.4	29.5 ± 12.4	27.7

a) Refs [4,5,10]; b) Ref. [11] and new data; c) Ref. [12].

Table 5

Energies of pionic 4f-3d transitions in keV

Target (1)	Berkeley[b] (2)	CERN[c] (3)	Theory (4)
$^{40}_{20}$Ca	72.352 ± 0.009[a]		
^{22}Ti	87.651 ± 0.009[a]		
$^{89}_{39}$Y	278.2 ± 0.3		
$^{93}_{41}$Nb	307.6 ± 0.3	307.7 ± 0.2	307.64
^{42}Mo		323.2 ± 0.2	323.06
$^{103}_{45}$Rh	370.9 ± 0.4		
^{49}In	442.1 ± 1.1	442.9 ± 0.5	442.62
$^{116}_{50}$Sn	460.9 ± 0.6		
$^{117}_{50}$Sn	460.4 ± 0.6		
$^{118}_{50}$Sn	460.4 ± 0.6		
$^{119}_{50}$Sn	460.3 ± 0.6		
$^{120}_{50}$Sn	460.5 ± 0.6		
$^{122}_{50}$Sn	460.3 ± 0.6		
$^{124}_{50}$Sn	460.2 ± 0.6		
$^{127}_{53}$I	519.1 ± 1.1	520.8 ± 0.8	520.17
$^{133}_{55}$Cs	560.5 ± 1.1	562.0 ± 1.5	561.55
^{57}La	603.6 ± 0.9	604.9 ± 2.0	604.61
$^{140}_{58}$Ce	626.1 ± 2.0		
$^{141}_{59}$Pr	649.5 ± 2.0	648.1 ± 2.0	649.61

a) crystal diffraction spectrometer, Ref. [19];
b) Ref. [5]; c) Ref. [13].

Table 6

Width of pionic 3d levels in keV

Target (1)	Berkeley[a] Without hfs (2)	CERN[b] Without hfs (3)	CERN[b] With hfs (4)	Theory (5)
$^{89}_{39}$Y	0.3 ± 0.6			
$^{93}_{41}$Nb	0.6 ± 0.4	0.52 ± 0.10		0.41
^{42}Mo		0.56 ± 0.10		0.51
$^{103}_{45}$Rh	1.2 ± 0.6			
^{49}In		2.8 ± 0.6	2.5 ± 0.6	2.0
$^{116}_{50}$Sn	1.9 ± 1.2			
$^{117}_{50}$Sn	2.1 ± 1.2			
$^{118}_{50}$Sn	2.5 ± 1.2			
$^{119}_{50}$Sn	1.9 ± 1.2			
$^{120}_{50}$Sn	2.7 ± 1.2			
$^{122}_{50}$Sn	2.0 ± 1.2			
$^{124}_{50}$Sn	2.3 ± 1.2			
$^{127}_{53}$I	4.2 ± 1.8	4.6 ± 1.5	4.4 ± 1.5	3.67
$^{133}_{55}$Cs		3.3 ± 1.5		4.71
^{57}La	5.8 ± 3.8	6.2 ± 2.0	6.2 ± 2.0	6.24
$^{140}_{58}$Ce				
$^{141}_{59}$Pr	6.7 ± 2.8	5.4 ± 2.5		8.11

a) Ref. [5]; b) Ref. [13].

Γ_{3d} and Γ_{4f} respectively. To what extent this is the case, may be seen by comparing columns 3 and 4 in Tables 6 and 8 [13]. In column 4, the hfs is taken into account for the evaluation of the width, in column 3 it is not. In particular for [181]Ta no single line fit is possible[13,14]; hfs splitting must be taken into account. The spectra have been evaluated by determining one single width Γ_{4f}, which is common to all hfs components. This is of course an assumption which may not be valid. In a deformed nucleus the overlap of the pion with the nucleus depends on the relative orientation between nuclear spin and pion angular momentum. It would be extremely interesting if one could measure different widths for the different hfs states. But at present the experimental techniques do not seem to be sufficiently developed. The agreement between the width measurements is considerably less than for the corresponding energy measurements. Real discrepancies exist for Γ_{1s} in Be, F, Na, for Γ_{2p} in K and for Γ_{4f} in Th. In the case of F and Na, nuclear γ-rays are present which may influence the determination of the widths.

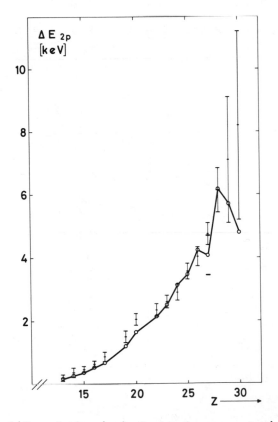

Fig. 2 Energy shifts of the pionic 2p level versus atomic number Z.
Full line calculated with constants of column 1, Table 10.

Table 8

Widths of pionic 4f levels in keV

Target (1)	Berkeley a) Without hfs (2)	CERN b)		Theory (5)
		Without hfs (3)	With hfs (4)	
$^{181}_{73}$Ta			0.5 ± 0.2	0.27
$_{78}$Pt		1.8 ± 1.0		0.59
$^{197}_{79}$Au		1.1 ± 0.3	1.0 ± 0.3	0.70
$_{80}$Hg		1.4 ± 0.5		0.77
$_{81}$Tl		1.0 ± 0.2		0.90
$^{206}_{82}$Pb		1.2 ± 0.4		1.01
$^{nat}_{82}$Pb		1.1 ± 0.3		1.06
$^{209}_{83}$Bi	1.7 ± 1.0	1.7 ± 0.5	1.7 ± 0.5	1.22
$^{232}_{90}$Th	6.0 ± 0.9	4.5 ± 0.8		2.6
$^{238}_{92}$U	6.1 ± 1.0	6.8 ± 0.8		3.2
$^{239}_{94}$Pu	9.1 ± 2.5			4.0

a) Ref. [5]; b) Ref. [13].

Table 7

Energies of pionic 5g-4f transitions in keV

Target (1)	Berkeley a) (2)	CERN b) (3)	W & M c) (4)	Theory (5)
$^{181}_{73}$Ta	453.1 ± 0.4	453.90 ± 0.20	453.4 ± 0.3	453.51
$_{78}$Pt		519.34 ± 0.24		519.?2
$^{197}_{79}$Au	532.5 ± 0.5	533.16 ± 0.20		532.?7
$_{80}$Hg		547.14 ± 0.25		546.?3
$_{81}$Tl		561.67 ± 0.25		560.?8
$^{206}_{82}$Pb		575.62 ± 0.30		575.?8
$^{nat}_{82}$Pb		575.56 ± 0.25		575.?6
$^{209}_{83}$Bi	589.8 ± 0.9	590.06 ± 0.30		589.?6
$^{232}_{90}$Th	698.0 ± 0.5	698.4 ± 0.4		696.7
$^{238}_{92}$U	731.4 ± 1.1	732.0 ± 0.4		729.?
$^{239}_{94}$Pu	766.2 ± 1.5			762.?

a) Ref. [5]; b) Ref. [13]; c) Ref. [14].

In Fig. 4 the spectrum of ^{23}Na is shown. On top of the broad 2p-1s
line appear two γ-rays having instrumental line widths. It is hard
to distinguish these γ-lines. However, in Fig. 3 the spectrum of
Mg is shown. There the same two γ-lines are present in a region
where no pionic line is expected. Since in many cases the same
γ-lines are observed in the spectra of neighbouring elements[20])

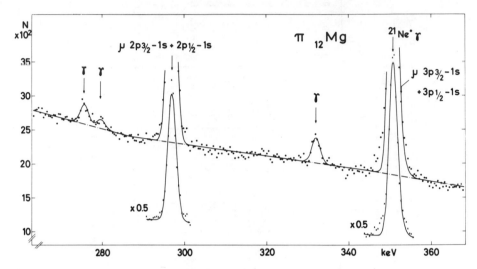

Fig. 3 Mg-spectrum in the region of the pionic 2p-1s line.

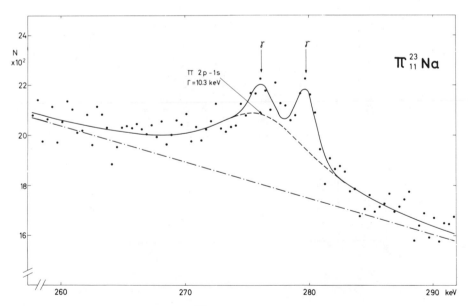

Fig. 4 Pionic 2p-1s line in ^{23}Na.

(e.g. the ^{19}F γ-lines are present in the spectra of ^{27}Al, ^{24}Mg, ^{23}Na) the comparison of different spectra is very helpful.

As we have seen, the region of directly measurable line widths lies between 0.1 and $\sim$ 10 keV. The pion absorption from the upper level causes a width of the order of 0.1-10 eV, which cannot be measured with currently used solid state detectors. However, those widths may be comparable to the radiative widths originating from electric dipole transitions of the pion. Therefore, yield measurements of pionic X-ray transitions contain information about pion absorption. The yield, Y, is given by $Y = (W_X P)/(W_X + W_A + W_a)$ where W_X, W_A and W_a are the transition rates for electric dipole transition, Auger transition and pion absorption, respectively, and P is the population of the upper level. W_X and W_A can be calculated exactly, the latter being negligible in most cases. If one knows the population P and measures Y one obtains $W_a = \Gamma/\hbar = W_X(P/Y-1)-W_A$. This method works best if W_X/W_a is of the order of one. The simplest way is to determine P/Y directly[15,17] by measuring P, which is the sum of the transition intensities populating this level, where usually Auger transitions can be neglected, as well as the intensity Y of the transition in question. This is a relative measurement where absolute errors, such as those due to geometrical quantities, cancel. Unfortunately this direct method works for the 2p-1s transitions only for ^{19}F and ^{23}Na, whereas for lower Z nuclei the energies of the $3 \to 2$, $4 \to 2$, ... transitions (L series) are too small ($\lesssim$ 30 keV) to be measured with sufficient accuracy, because corrections for self absorption in the target and for the detector efficiency are large. However, for higher transitions (3d-2p, etc.), the energy difference of the L and M series becomes smaller and the method is therefore more favoured.

In the other cases an indirect method, particularly suited for the 2p-1s transitions, must be applied[15-17]. In order to avoid an absolute yield determination with all the difficulties such a measurement contains, it is easier to compare the pionic yields with the yield of the always present muonic X-rays. Assuming the K series of muonic X-ray spectra totals up to 100%, it is easy to determine the yield of any particular line of the K series suitable for comparison. Since the muonic yields are usually higher by almost an order of magnitude, only a small muon contamination is needed. If the fraction of muons stopped in the target is known, the yield of the pionic line can be obtained immediately. There are various methods of determining this fraction. From the range curve of momentum-selected particles the contribution of the muons at the place of the pion peak can be estimated. A more direct method is the comparison of muonic X-ray transitions with those pionic transitions for which the yield is known, i.e. where no absorption takes place and the Auger transition rate is known to be

small. Finally, the number of stopped muons can be directly
measured by the μ-e decay if the target is replaced by a scintillator. All three methods have consistently given the fraction of
stopped muons as $(6.8 \pm 0.4)\%$. With this number we obtain the
2p-1s yields listed in Table 9, column 1, where also old measurements taken with NaI crystals are shown for comparison (columns 2
and 3).

The population P has been determined assuming a similar cascade in pionic and in muonic atoms for levels in which no pion
absorption occurs. From the muonic K series of a particular
nucleus the population of the 2p level was determined. An initial
distribution function of the muons over the various ℓ-substates
(ℓ = 0 to 13) for n = 14 was found, which gives a good fit to the
K-series intensities. The assumption was made that the same distribution is realized in the n = 17 state of the corresponding
pionic atom. From this initial distribution the population P was
calculated. For the nuclei of Table 9 these populations are given
in column 4. Since the above assumption is not very well justified —
deviations between muonic and pionic cascades have been observed[21] —
an error of 10% has been attributed to the quantity P.

The electric dipole transition rate W_X, with which the pion
absorption rate W_a is compared, can be calculated according to

$$W_X = C \, E^3 \, |< \phi_f \, |r| \, \phi_i >|^2 \, ,$$

where E is the transition energy and ϕ_i and ϕ_f are the radial parts
of the pion wave function's initial and final states. Neglecting
the effect of the strong interaction one obtains for W_X the numbers
given in column 5 of Table 9. With the strong interaction present
one has to take into account the shift of the energy levels changing E as well as the deformation of ϕ giving a different dipole
matrix element. In the case of the 2p-1s transitions, for example,
E is decreased whereas the matrix element is increased owing to a
repulsive potential in the 1s state and an attractive potential in
the 2p state. The data for W_X using the exact pion wave functions
are given in column 6 of Table 9. With these data one obtains
finally the absorption widths Γ_{2p} shown in column 6 of Table 2,
where a 10% error in the values of P(2p) has been included. The
values for ^{19}F and ^{23}Na are the mean values of the data obtained
with both methods mentioned above and were consistent within the
errors quoted.

A very similar procedure applied to the yields of the 3d-2p
transitions leads to the data for Γ_{3d} as shown in column 6 of
Table 4. In all cases the direct method of comparing the 3d-2p
yields with the $4 \rightarrow 3$, $5 \rightarrow 3$, ... intensities was applicable.

Table 10

Parameters of the nuclear optical potential

	From experiment 1	Predicted by theory 2	
b_0	-0.030	-0.028	$[x_\pi]$
b_1	-0.08	-0.09	$[x_\pi]$
c_0	+0.22	+0.19	$[x_\pi]^3$
c_1	+0.18	+0.18	$[x_\pi]^3$
Im B_0	+0.040	+0.017	$[x_\pi]^4$
Im C_0	+0.08	+0.07	$[x_\pi]^6$

Table 9

Pionic 2p-1s yields Y (per cent), populations of the pionic 2p levels P(2p) (per cent) and 2p-1s X-ray transition rates (sec⁻¹)

	CERN [15,16] 1	Camac [1] 2	Stearns [2] 3	P(2p) 4	$W_X \times 10^{14}$ el.mag. 5	el.mag. +strong 6
Be	10.5 ± 1.4	15.0 ± 1.6	15.5	67	0.44	0.44
^{10}B	11.0 ± 1.6	10.9 ± 1.4	13.9	61	1.07	1.06
nat B	12.5 ± 1.0	7.6 ± 0.9	11.7	62	1.07	1.04
C	7.5 ± 2.0	4.0 ± 0.8	3.4	62	2.22	2.14
N	6.8 ± 0.8	3.1 ± 1.0	3.4	62	4.12	3.95
^{16}O	4.9 ± 0.7			57	7.02	6.70
^{18}O	5.6 ± 0.8			55	7.02	6.48
F	4.1 ± 1.0		4.1	70	11.2	10.6
Na	3.3 ± 0.7			78	25.1	22.0

Although the indirect method, where comparison with muonic transition and the muonic cascade is made, depends on certain assumptions, it was nevertheless applied as a check for ^{19}K, ^{51}V, ^{55}Mn and ^{59}Co and the values given are mean values of the two methods. Also W_X was calculated including the strong interaction. For the 3d-2p transition W_X is increased by 0.8%, 2% and 7.5% for $_{14}$Si, $_{20}$Ca and $_{30}$Zn respectively as compared with the pure electromagnetic interaction.

The experimental situation has now reached a state where one has a reasonably good knowledge about the trends caused by the strong interaction throughout the periodic system. A very success-ful method to explain these data was given by T. and M. Ericson[??]), who introduced an optical potential to account for the pion-nucleus interaction.

Since the pion is a spin zero particle the pionic atom is described by the Klein-Gordon equation

$$\left[\frac{\hbar^2}{2\mu} \Delta + \left(E - V_F(r) \right) \left(1 + \frac{E - V_F(r)}{2\mu c^2} \right) \right] \psi = 0 , \tag{1}$$

where μ is the reduced pion mass, E its energy and V_F the potential produced by the extended Fermi-type charge distribution of the nucleus. The additional effect of the strong pion-nucleus inter-action will be accounted for by a nuclear optical potential V_N and the left side of Eq. (1) is put equal to $V_N\psi$. This nuclear poten-tial cannot be a simple local potential since such a potential will not reproduce repulsive and attractive interactions in different states. Therefore, a non-local, velocity-dependent part must be introduced. Since the potential should also describe absorption, both parts must be complex. In this way one arrives at a potential of the following type

$$V_N(r) = -2\pi \frac{\hbar^2}{2\mu} \left[b_0\rho(r) + b_1 \{\rho^n(r) - \rho^P(r)\} + i \text{ Im } B_0\rho^2(r) + \right.$$

$$\left. + \vec{\nabla}f \{c_0\rho(r) + c_1 \left[\rho^n(r) - \rho^P(r)\right] + i \text{ Im } C_0\rho^2(r)\} \vec{\nabla} \right] , \tag{2}$$

where $f = \left[1 + 4\pi/3\{c_0\rho + c_1(\rho^n - \rho^P) + i \text{ Im } B_0\rho^2\}\right]^{-1}$ is due to short range correlations between the nucleons.

$\rho(r) = \rho^n(r) + \rho^P(r)$ is the nuclear density and $\rho^n(r)$ and $\rho^P(r)$ are the neutron and proton density. The nice feature of this potential is that for the pion 1s state the local part dominates, $\vec{\nabla}\psi_{1s}$ being very small, and that for the pion 2p state the gradient term is leading. Therefore, the 1s level shift is mainly deter-mined by the constant b_0 and the 1s level width by Im B_0. The term with b_1 gives an additional shift for nuclei with isotopic spin $T \neq 0$, as can be seen for ^{16}O and ^{18}O. Similarly c_0 mainly affects the 2p level shift and Im C_0 the p level width. The term

$c_1\left[\rho^n(r) - \rho^p(r)\right]$ produces a shift for $T \neq 0$ nuclei essentially in the higher levels (2p, 3d, 4f and 5g). Attention should be paid to the fact that the absorptive parts are proportional to ρ^2, since the absorption of pions takes place mainly on two nucleons for reasons of energy and momentum conservation.

An exact solution of the Klein-Gordon equation has been obtained numerically for such a potential by M. Krell[23] and energies and widths of most transitions experimentally observed have been calculated[24]. A set of parameters has been chosen which gives good agreement with experimental data. The parameters are given in column 1 of Table 10. The values thus calculated are the ones listed in Tables 1 to 8 in the columns marked "Theory".

As can be seen from a comparison of the experimental with the calculated data (Tables 1-8, Figs. 5, 6) the agreement is remarkably good. All trends are shown correctly. The values of Γ_{2p} are described correctly over five orders of magnitude and the values of Γ_{3d} over six orders of magnitude. In particular, the isospin effect (parameter b_I) is proved to be necessary for a good understanding of the data. It should be noted that the parameters c_0 and Im C_0 are largely responsible in explaining the shifts and widths in the 2p level as well as in the 3d, 4p and 5g levels. Agreement for the heavier neutron-rich nuclei for which the higher levels have been

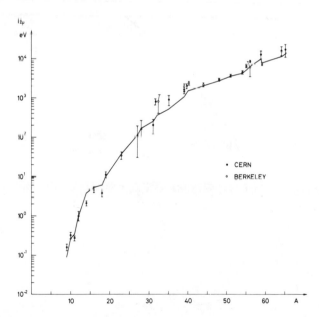

Fig. 5 Γ_{2p} versus mass number A for $A \leq 23$ from 2p-1s intensities and for $A \geq 27$ from 3d-2p line shapes. Full line calculated with constants of column 1, Table 10.

observed can only be reached if in addition an isospin parameter c_1
is introduced. The short range correlations have been taken into
consideration in the experimental determination of the parameters
so far. In column 2 of Table 10, the values of the parameters predicted by the multiple scattering theory are shown. The agreement
looks impressive except for the parameter $Im\ B_0$, which turns out to
be about 2.5 times larger than that theoretically deduced from the
pion absorption in deuterium. Furthermore, the observed Γ_{1s} is considerably less for the heaviest nucleus observed (^{23}Na) as predicted
by an $Im\ B_0$ chosen to fit the other Γ_{1s} values. It is now of great
interest to study details of shifts and most particularly of the
width for special nuclei. As an example the Γ_{1s} values of ^{16}O and
^{18}O are mentioned. Experimentally Γ_{1s} in ^{18}O is larger by 15% as
compared with Γ_{1s} of ^{16}O which just becomes significant in view of
the errors involved. Similar trends are indicated for ^{10}B, ^{11}B and
perhaps ^6Li, ^7Li and ^{19}F. However, any simple description by an
optical potential gives an opposite effect since the additional

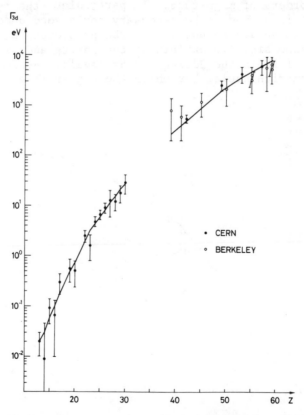

Fig. 6 Γ_{3d} versus atomic number Z for Z $\leq$ 30 from 3d-2p intensities and
for Z $\geq$ 39 from 4f-3d line shapes. Full line calculated with constants
of column 1, Table 10.

energy shift caused by the isotopic spin term b_1 corresponds to a
repulsive effect on the pion wave function reducing the absorption.
Another approach[25] to calculating the pion absorption is a microscopic description based on an extended shell-model treatment where
the short-range nucleon-nucleon interaction is taken into account
by a two-body correlation factor, which is a function of a characteristic momentum exchanged between the two nucleons involved in
the absorption process. In the momentum range between 250 and
350 MeV/c, Γ_{1s} and Γ_{2p} amount to about half of the measured value
for ^{16}O. However, the same calculation repeated for ^{18}O yields an
absorption rate[26] increased by about 20% in the 1s level and decreased by about 10% in the 2p level, in good agreement with the
experiment.

Another feature worthwhile mentioning is the strong dependence
of the absorption widths on the nuclear density. An increase of
the r.m.s. radius of the nucleus by 5% yields a decrease in Γ_{1s} and
Γ_{2p} by $\sim$ 15%. Therefore, the measurement of the widths should provide a powerful tool to determine nuclear radii as compared with
charge radii as soon as the optical potential parameters are sufficiently well understood.

Another aspect, linked to the work on pionic atoms by the experimental technique and physically by its relevance to the pion absorption process, is the observation of nuclear γ-rays assigned to
nuclei which are produced after pion absorption[20]. There have
been nuclei observed which have lost as many as eight nucleons. The
dominant process expected is the one where two nucleons have been
emitted. For example, in ^{23}Na the γ lines of ^{21}Ne are most fre-
quent. In heavier atoms, however, the dominance of this process
seems to be reduced. These measurements do not, of course, yield
directly the distribution of the residual nuclei but show only
those nuclei formed via their excited states. But most likely the
residual nuclei will be excited, as could be proved in the case of
muon absorption[27]. It would be very interesting if the final distribution of the nuclei left after pion absorption took place could
be related to the initial two-hole state produced.

A further field of investigation in pionic atoms is connected
more with atomic physics or the physics of the condensed state and
I will, therefore, mention it only briefly. It is the study of
the pionic cascade, which may depend on the chemical compound in
which the atom under investigation is present or on its physical
modification[10,21]. Since differences have been found[21] between
muonic and pionic atoms, there are certainly effects involved which
cannot be studied in muonic atoms only.

To conclude one can say that the experimental situation in the
field of pionic atoms has reached a state where the important quantities such as energies, line shapes and line intensities have been

measured on a representative sample of nuclei with a sufficient
accuracy that the trends are clearly shown and a detailed analysis
of the data using an optical pion-nucleus potential can be performed.
Whereas from the determination of the optical potential parameters
conclusions about the pion nucleon interactions at zero energy might
be reached, it is conceivable that with the same parameters at hand,
information about the nuclear density distribution, $\rho(r)$ can be obtained particularly where two Γ's have been measured. Moreover,
the measured values of Γ_{1s}, Γ_{2p}, etc., form important direct information which allows theoretical approaches to calculating pion
absorption by various models to be checked. In particular for the
nucleon pair correlations the pion absorption widths should be a
sensitive quantity.

   In future, after the present survey of data available on pionic
atoms, more accurate and more detailed measurements are needed.
These should be extended to separated isotopes. The cascade process
should be studied in more detail in order to be able to reduce the
uncertainties in the widths obtained from intensity measurements.
A measurement of a width performed by the two methods (line shape
and intensity) would be of great interest because it is assumed at
present that a broadening of a line by reasons other than by absorption is excluded, an assumption which could be checked by such
a comparison. However, this would mean measuring directly a line
width of about 30-50 eV with about a 10% accuracy, which seems difficult to reach even with bent crystal spectrometers at present
pion intensities. Finally, a more detailed study of nuclear γ-rays
occurring simultaneously with the pionic X-rays should be of interest.

REFERENCES

1. M. Camac, A.D. McGuire, J.B. Platt and H.J. Schulte, Phys.
       Rev. 99, 897 (1955).
2. M. Stearns and M.B. Stearns, Phys.Rev. 103, 1534 (1956), and
       107, 1709 (1957).
3. D. West and E.F. Bradley, Phil.Mag. 2, 957 (1957).
4. D.A. Jenkins, R. Kunselman, M.K. Simmons and T. Yamazaki,
       Phys.Rev.Letters 17, 1 (1966).
5. D.A. Jenkins and R. Kunselman, Phys.Rev.Letters 17, 1148 (1966).
6. G. Backenstoss, S. Charalambus, H. Daniel, H. Koch, G. Poelz,
       H. Schmitt and L. Tauscher, Phys.Letters 25B, 365 (1967).
7. R.J. Wetmore, D.C. Buckle, J.R. Kane and R.T. Siegel, Phys.
       Rev.Letters 19, 1003 (1967).
8. R.J. Harris, W.B. Shuler, M. Eckhause, R.T. Siegel and
       R.E. Welsh, Phys.Rev.Letters 20, 505 (1968).
9. G.H. Miller, M. Eckhause, W.W. Sapp and R.E. Welsh, Phys.
       Letters 27B, 663 (1968).
10. A.R. Kunselmann, Thesis, Berkely 1969, UCRL Report 18654.
11. G. Poelz, H. Schmitt, L. Tauscher, G. Backenstoss, S. Charalambus,
       H. Daniel and H. Koch, Phys.Letters 26B, 331 (1968).

12. D.A. Jenkins, R.J. Powers and G.H. Miller, preprint.
13. H. Schmitt, L. Tauscher, G. Backenstoss, S. Charalambus,
    H. Daniel, H. Koch and G. Poelz, Phys.Letters 27B, 530 (1968).
14. R.A. Carrigan, W.W. Sapp, W.B. Shuler, R.T. Siegel and
    R.E. Welsh, Phys.Letters 25B, 193 (1967).
15. H. Koch, G. Poelz, H. Schmitt, L. Tauscher, G. Backenstoss,
    S. Charalambus and H. Daniel, Phys.Letters 28B, 279 (1968).

16. H. Koch, M. Krell, C.v.d. Malsburg, G. Poelz, H. Schmitt,
    L. Tauscher, G. Backenstoss, S. Charalambus and H. Daniel,
    Phys.Letters 29B, 140 (1969).
17. H. Koch, Thesis, University of Karlsruhe 1969, unpublished.
18. A. Astbury, J.P. Deutsch, K.M. Grove, R.E. Shafer and R.E. Taylor,
    Comptes Rendus du Congrès Int. de Phys.Nucl., Paris 1964,
    vol. 2, p. 225.
19. R.E. Shafer, Phys.Rev. 163, 1451 (1967).
20. G. Poelz, H. Schmitt, L. Tauscher, G. Backenstoss, S. Charalambus
    and H. Daniel, To be published in Zeitschrift f. Phys, 1969.
21. L. Tauscher, G. Backenstoss, S. Charalambus, H. Daniel, H. Koch,
    G. Poelz and H. Schmitt, Phys.Letters 27A, 581 (1968).
22. M. Ericson and T.E.O. Ericson, Ann.Phys. (N.Y.) 36, 323 (1966).
23. M. Krell and T.E.O. Ericson, Nucl.Phys. 11B, 521 (1969).
24. G. Backenstoss, S. Charalambus, H. Daniel, H. Koch, M. Krell,
    G. Poelz, H. Schmitt and L. Tauscher, Proc.Int.Conf. on the
    Physics of One and Two Electron Atoms (A. Sommerfeld Centen-
    nial Memorial Meeting),Sept. 1968, North Holland Publishing
    Company, Amsterdam.
25. K. Chung, M. Danos and M.G. Huber, Phys.Letters 29B, 265 (1969).
26. K. Chung, Private communication.
27. G. Backenstoss, S. Charalambus, H. Daniel, U. Lynen,
    C.v.d. Malsburg, G. Poelz, H. Povel, H. Schmitt and
    L. Tauscher, Contributed paper to this conference.

DISCUSSION

A. Roberts: Recent measurements of the absorption-width of the 2p-
level in ^{4}He fit nicely with the results presented in heavier
nuclei.   V. Telegdi: Is there any evidence for H.F.S. that might
be attributable to spin dependent π-nuclear interaction (as
distinct from electromagnetic H.F.S.)?   G.B.: Nothing of this
sort has been found.   Some earlier discrepancies in level-width
measurements, which might have suggested such a H.F.S., turned out
to be instrumental.

THEORETICAL DEVELOPMENTS IN THE FIELD OF π-MESIC ATOMS

T.E.O. Ericson

CERN, Geneva, Switzerland

Three connected topics will be discussed in this report:

a)   Is there an anomaly in π-nuclear interaction as a function of nuclear size with a change of its sign at a certain critical nuclear radius? [1]

b)   Why is the π-nuclear s-wave interaction repulsive?

c)   A few results from a recent numerical analysis of π-mesic atoms [2]. Up to now we have been confined to the use of per-turbative methods and mass distributions of the square-box type for purely technical reasons. These are certainly not good enough for quantitative discussions. Anybody working on μ-mesic atoms will recall the tremendous breakthrough in that field when proper analysis became possible, and how that field was held up for nearly two years awaiting those methods. The corresponding stage has just been passed for π-mesic atoms.

The non-absorptive π-nuclear interaction at low energy has two dominant features. (a) The interaction for s-wave pions is repul-sive as exemplified by the negative 1s level shifts of π-mesonic atoms. Since the coherent sum of the s-wave scattering lengths over the nucleons is close to zero for nuclei of zero isospin, this repulsion arises mostly from the fluctuation scattering by the (polarized) nuclear medium, i.e. from the discrete nature of nu-clear matter [3] (the nuclear polarization is very small for μ-mesic atoms but very important in π-mesic atoms). (b) For higher angu-lar momenta the interaction is observed to be attractive, since the measured level shifts of 2p, 3d, and 4f-states of π-mesonic atoms are positive. For a p-wave pion this is due to

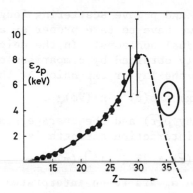

Fig. 1 – Interaction
energy-shift pionic
2p-state.

the elementary p-wave πN interaction: a coherent sum of the p-wave
amplitudes gives attraction. The attractive p-wave interaction dominates contributions from higher πN waves also when the π-nuclear
angular momentum $\ell > 1$. Even in very light elements the coherent
sum of πN amplitudes with $\ell > 1$ is negligible compared with contributions in which the pion is in a p-state with respect to individual nucleons.

The observed interaction shift for the pionic 2p-state is
shown in Fig. 1 as a function of Z. It is a growing function of
Z up to Z = 30, the heaviest nucleus for which observations exist.
An argument will now be given that the proper extrapolation of this
curve is a dramatic downward trend with a change of sign for $Z \approx 36$.
The observed attraction should therefore reverse into repulsion for
a slight further increase of nuclear radius. The origin of this
predicted effect is the following: for a sufficiently small nucleus,
a pion in a relative p-state with respect to the nucleus will also
be very predominantly in a relative p-state with respect to the
individual nucleons. For a larger nuclear radius, the pion can increasingly be in a relative s-state with respect to the nucleons,
and thus feel the repulsive interaction of this state. There can
therefore be a critical nuclear radius for which the attraction
and repulsion exactly balance (with no net interaction). A very
simple but in essence correct estimate of the critical nuclear size
can be made as follows: the π-nuclear interaction strength is given
approximately by the expectation value of πN scattering amplitudes
on the individual nucleons in the nucleus. In practice only
s- and p-wave πN scattering is of importance, as mentioned above.
For a single nucleon and a plane wave

$$\exp \{i\vec{k}\cdot\vec{x}\} = 1 + i\vec{k}\cdot\vec{x} + \ldots \ .$$

The interaction strength is then

$$I_1 \simeq a_s + 3a_p k^2 + \ldots \ ,$$

where a_s and a_p are πN s- and p-wave scattering lengths. For the bound wave function $\phi(\vec{r})$ we have to take proper s- and p-state contributions for the individual nucleons. In the neighbourhood of a point $\vec{r}$ this is immediately obtained by comparison of the plane-wave expansion above with the Taylor expansion of the wave function,

$$\phi(\vec{r} + \vec{x}) \simeq \phi(\vec{r}) + \vec{x} \cdot (\vec{\nabla}\phi)_{\vec{r}} \ .$$

With a nuclear density $\rho(\vec{r})$ and the average scattering lengths $\bar{a}_s$ and $\bar{a}_p$, the A-particle interaction strength is

$$I_A \simeq \int \left[\bar{a}_s \phi^2(\vec{r}) + 3\bar{a}_p (\vec{\nabla}\phi)^2 \right] \rho(\vec{r}) d\vec{r} \ .$$

The parameters $\bar{a}_s$ and $\bar{a}_p$ are to be interpreted as the effective πN scattering lengths in the nucleus. Fluctuation scattering by the discrete nuclear medium produces an effective field at each scatterer, which causes $\bar{a}_s$ and $\bar{a}_p$ to differ from a simple coherent average of πN scattering lengths. It is obvious from the expression for the interaction strength I_A that the net interaction strength may be repulsive or attractive depending on the relative size of the integrals containing ϕ^2 and $(\vec{\nabla}\phi)^2$, since $\bar{a}_s$ and $\bar{a}_p$ have opposite signs. These integrals depend strongly on angular momentum and nuclear radius.

An insight into this dependence is obtained by taking the matter distribution to be uniform out to a radius R and the pion wave function $\phi(r) \propto r^\ell Y_{\ell m}$ over nuclear dimensions. These are gross oversimplifications, of course, of the real situation. An elementary calculation gives the critical radius R_ℓ

$$R_\ell^{\ 2} = -(3\bar{a}_p/\bar{a}_s)\ell(2\ell + 3) \ .$$

It is quite obvious that the higher the angular momentum, the larger the critical radius. This leads to the following predicted critical radii: $R_1 = 6$ f, $R_2 = 11$ f and $R_3 = 15$ f, which correspond to A ~ 100, A ~ 250-350 and A ~ 1000. These indicate that for real nuclei the effect should be observable for $\ell = 1$, possibly for $\ell = 2$, while effects in $\ell \geq 3$ seem impossible to detect at present. A more detailed numerical analysis on a more sophisticated basis indicates that the effect should occur around $Z \approx 36$, i.e. for $A \approx 80$.

Since the experimental interaction shifts do not apparently show any effect at Z = 30, we decided to plot $\epsilon_{2p}/\Gamma_{2p}$, the ratio of shifts to widths versus Z. The rationale for this was that this should eliminate the probability of the presence of the pion in the nucleus, which scales strongly with Z as the Bohr orbit shrinks. The result is seen in Fig. 2, which strongly indicates that our prediction may be true. We did not rely on any parameter-dependent calculation to obtain this curve.

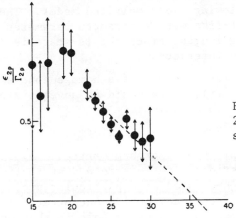

Fig. 2 - Ratio of
2p-level energy-
shifts to widths.

An interaction that changes from attraction to repulsion as a function of nuclear size is unusual in nuclear physics. Since the amplitude changes sign even in the Born approximation, the nature of the effect is completely different from the well-known change of sign of neutron scattering lengths at a nuclear size resonance (anomalous dispersion). A direct measurement, whether this effect exists or not, would therefore be very interesting.

An interesting corollary of this little calculation is worth mentioning. It is clear that the energy shifts in 3d- and 4f-states depend almost entirely on the p-wave πN interaction. Since $a_p^{\pi^-n} \approx 10\, a_p^{\pi^-p}$, when averaged over spin, the interaction is almost entirely with the neutrons and hardly at all with the protons. Any radial effect in these states necessarily measures nuclear neutron radii directly.

We now turn to the second question: why is the s-wave π-nuclear interaction repulsive at low energy? [2,3] This can be seen rather easily from multiple scattering theory.

Low-energy multiple scattering is largely the antithesis of the Glauber picture. The scatterers are small compared to wavelength, so that only one or a few partial waves contribute in each scattering. The scattered waves do not obey ray optics from one scattering to the next, but are spherical outgoing waves. It is then possible to scatter more than once on the same particle. This gives rise to a modification of the local field, due to the nuclear polarization. Such effects, which are associated with the discrete structure of the nucleus (or correlations), can be looked at in terms of virtual nuclear excitations. The local effective field is a very important effect at low energy, in contrast to the high-energy case.

The most striking effect of the granularity of the nuclear medium appears in the 1s-state of the mesic atom. The physics is obtained from the following oversimplified model. The πN scattering is taken to be point-like with a strength parameter a given by the πN s-wave scattering length, since the 1s-state is dominated by the elementary s-wave πN interaction.

If the nucleus were a piece of a homogeneous and uniform medium with a density $\rho(\vec{r})$, there would be an equivalent potential $V(\vec{r})$ for the elastic interaction:

$$2 \mu V(\vec{r}) = -4\pi a \rho(\vec{r}).$$

A nucleus is <u>not</u> a piece of amorphous matter. A pion within the nucleus clearly experiences the discrete distribution of matter within. In the neighbourhood of a particular nucleon, the pion sees that the mass of the nucleon is concentrated into a point with a corresponding absence of medium (a hole in the medium) surrounding this nucleon (see figure). (More exactly, this hole is described by the nuclear pair-correlation function.) Since a structure of this kind can only be seen by virtually exciting the system, this effect is exactly equivalent to calculations of nuclear polarization using excited states of the nucleus.

The important effect of the hole is now this: in the uniform medium there is an average wave function $\Phi(\vec{r})$. The wave incident on a scatterer in the hole is not $\Phi(\vec{r})$, but an effective wave $\Phi^{eff}(\vec{r})$. In the long wavelength limit, these are related by

$$\Phi^{eff} = \Phi - a \left\langle \frac{1}{r} \right\rangle_{hole} \Phi^{eff}.$$

The effective field in the middle of the hole has thus two contributions: i) the average field $\Phi(\vec{r})$ that would result from the background medium itself; ii) the local modification by the hole, which is a removal of uniform matter close to $\vec{r}$ in accordance with our ansatz. At the centre of the hole this contribution is proportional to the exciting field Φ^{eff} over the hole, the strength of scattering a for one particle, and the average of the inverse distance of the centre to the various parts of the hole (from the propagation of a spherical wave). The sign is negative since these contributions have been overcounted in $\Phi(\vec{r})$. Hence

$$\Phi^{eff} = \frac{1}{1 + a \left\langle \frac{1}{r} \right\rangle_{hole}} \Phi.$$

Since the derivation of the potential for a uniform medium supposed that the exciting wave function was Φ and not Φ^{eff}, the potential is modified due to the correlations:

$$2 \mu V = -4\pi \frac{a}{1 + \left\langle \frac{1}{r} \right\rangle_{hole} a} \approx -4\pi \left[a - a \left\langle \frac{1}{r} \right\rangle_{hole} a + \dots \right].$$

This correction is always <u>repulsive</u> in this order since it depends on a^2.

We now generalize this to a nucleus with neutrons and protons for which virtual charge exchange also can occur. We identify $<1/r>_{hole}$ with $<1/r>_{corr}$, the expectation value of $1/r$ over the normally defined pair correlation function. For a $T = 0$ nucleus the leading order terms are

$$2 \mu V = -4\pi \left[\frac{a_n + a_p}{2} - \frac{2(a_n^2 + a_p^2) + (a_n - a_p)^2}{4} \left\langle \frac{1}{r} \right\rangle_{corr} \right] \rho(\vec{r}).$$

The remarkable thing for pions is now that $(a_n + a_p) \approx 0$. In the absence of the higher order term, nuclei would be nearly transparent to s-wave pions. The elastic scattering occurs predominantly (to about 70%) by the second order term. The important correlations are the Pauli correlations, which have the largest range. <u>The repulsive interaction is thus caused by the granular structure of the nucleus.</u> This result has also recently been emphasized by Koltun in somewhat different terms [4].

The third topic   results from a recent analysis of experiments on pionic atoms [2]. Multiple scattering theory leads to an equivalent potential for the π-nuclear interaction:

$$2 \mu V = q + \vec{\nabla} \cdot (1 + \frac{\alpha_0}{3})^{-1} \alpha_0 \vec{\nabla} .$$

The terms q and α_0 depend on the mass densities ρ_n, ρ_p, and $\rho = 1/2(\rho_n + \rho_p)$ and on constants b_0, b_1, etc., which are simply related to s- and p-wave πN scattering lengths and production amplitudes.

The leading terms are

$$q = -4\pi \left[b_0 \rho + b_1(\rho_n - \rho_p) + i \ \text{Im} \ B_0 \rho^2 \right] + \text{corrections}$$

$$\alpha_0 = 4\pi \left[c_0 \rho + c_1(\rho_n - \rho_p) + i \ C_0 \rho^2 \right] + \text{corrections} .$$

Absorption is an essential phenomenon in the π-nuclear interaction. It merits a comment since it is an embarassment to the multiple scattering description. Let us take the most naïve viewpoint possible, that pions are absorbed in nuclei as they are in the deuteron but for scaling factors. If there is a short-range interaction which is the same in nuclei and for the deuteron, the deuteron absorption rate is $\beta_{11} |\Phi_D(0)|^2$. In the nucleus the local absorption rate is proportional to the same constant β_{11} and the local chance encounters of a neutron and a proton: $\beta_{11} \ 3/4 \ \rho_n \rho_p$. It is not necessarily true that two nuclear and the two deuteronic nucleons engaged in π-absorption behave in the same way, but we try to get away with it. A smaller contribution from singlet nucleon states has also to be added. The analysis proceeded as follows: we first took a series of measured 1s and 2p shifts and widths from pure isotopes with

reasonably well determined charge distributions. We then assumed
that neutron and proton distributions coincide: this is an innocent
assumption for 1s- and 2p-states for shifts and for T = 0 nuclei,
but it may not be quite so good for widths of T ≠ 0 nuclei.   The
constants of the potential can then be directly determined, and
they are in fact strongly overdetermined. The results are given
in Table 1.   The unimportant constant c_1 was introduced theoretically. The results of the analysis agree surprisingly well with
previous analysis of less sophistication.

     To obtain a feeling for the behaviour of the non-local poten-
tial, we transformed it into a local representation. The corresponding local potential for ^{40}Ca is seen in Fig. 3. The Coulomb
interaction is included for comparison. The gradient interaction
corresponds to a spread-out dipole layer with attraction far out
in the nuclear surface, repulsion inside. Since higher angular
momentum states are small in the nuclear inside, they reflect this
attraction in the energy shifts. The local repulsive potential q
is of the magnitude of the Coulomb potential over the nuclear region,
but with opposite sign.

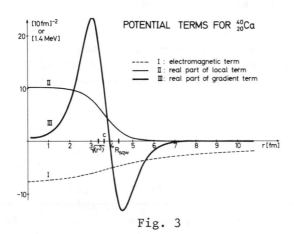

Fig. 3

Table 1

	Exp.	Theor.
b_0	-0.030	$-0.025 + a^{(+)}$
b_1	-0.08	-0.09
c_0	0.22	0.19
c_1	[0.18]	0.18
Im B_0	0.040	0.017
Im C_0	0.08	0.07

     It is interesting for many applications to know how the wave
function is changed by the strong interaction. Figure 4 shows the
probability density of a pion in the 1s-state of ^{16}O for various
assumptions about the interaction. It shows reduction by a factor
of 3 in the nuclear interior, but this reduction is nearly model-
independent. Outside the nucleus the wave function is fully determined by the measured energy, and it is seen to deviate strongly
from its unperturbed value very far outside the nucleus. Inside the
nucleus any interaction giving the correct energy shift gives similar
values for the probability density. It is mainly the repulsion in

the 1s-state that decreases the wave function, while the absorption plays no noticeable role. The gradient interaction enhances the repulsion in the nuclear interior somewhat by decreasing the mean free path.

The parameters are not much correlated. An extreme example is seen for the 1s-states in Fig. 5 where the energy shifts have been fitted using an interaction without absorption and without gradient terms (which is a wild change). The fit is nearly identical with nearly identical values for b_0 and b_1. In the absorption, important changes occur if the real interaction is turned off. This is to be expected, however, since the probability of presence in the nucleus is then quite wrong, as just discussed.

The limitation on the determination of b_0 is mainly the experimental uncertainty in $a^{(+)} = (a_n + a_p)/2$ for πN scattering. The remaining terms of b_0 come mainly from nuclear polarization as we have already discussed. One may try to turn this uncertainty around and use it to predict $a^{(+)}$ from π-mesic atoms. We then find that $-0.010 \ \mu^{-1} < a^{(+)} < 0$.

Most other corrections to b_0 are under control, although the dispersive correction from π-absorption (believed to be small) is not satisfactorily evaluated at present.

Figure 6 shows the extent to which the 1s shifts can be described. They are displayed on a reduced scale. The strong isospin dependence of the interaction is particularly apparent in boron and oxygen. The parameters are in general in good agreement with predictions. The exception is Im B_0, which theoretically is

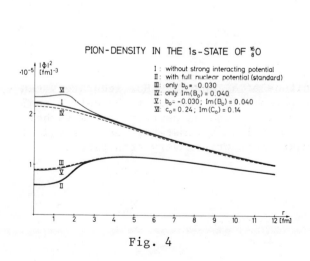

PION-DENSITY IN THE 1s-STATE OF $^{16}_{8}$O

I : without strong interacting potential
II : with full nuclear potential (standard)
III : only $b_0 = -0.030$
IV : only Im(B_0) = 0.040
V : $b_0 = -0.030$; Im(B_0) = 0.040
VI : $c_0 = 0.24$; Im(C_0) = 0.14

Fig. 4

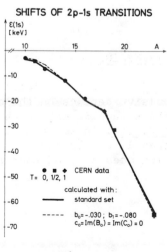

SHIFTS OF 2p-1s TRANSITIONS

ε(1s) [keV]

● ■ ♦ CERN data
T= 0, 1/2, 1

calculated with:
——— standard set

- - - - $b_0 = -.030$; $b_1 = -.080$
$c_0 = $ Im(B_0) = Im(C_0) = 0

Fig. 5

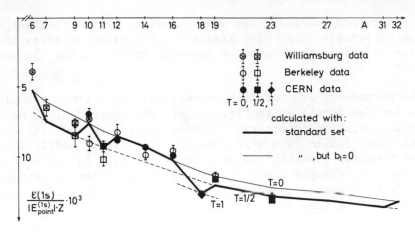

Fig. 6 - 1s level shifts: Experiment & Theory.

about a factor of 2 too weak to describe the 1s widths.   Since the
2p absorption widths are well described for a wide range of nuclei,
one may take the viewpoint that this is extremely good for the
simple two-nucleon absorption model used.   One may, however, also
wonder if it does not indicate an interesting modification of this
interaction in the nucleus, since the effect seems real.

REFERENCES

[1] M. Ericson, T.E.O. Ericson, M. Krell, Phys. Rev. Letts. 22, 1189
     (1969).
[2] M. Krell, T.E.O. Ericson, Nuclear Phys. B11, 521 (1969).
[3] M. Ericson and T.E.O. Ericson, Ann. Phys. (N.Y.) 36, 323 (1966).
[4] D. Koltun, Rochester preprint (1969).

DISCUSSION

M. Rho: Your prediction that the shift goes from attractive to repulsive depends on the magnitude of $\bar{a}_s$ and $\bar{a}_p$. How accurately can
these quantities, which are presumably A dependent, be determined?
T.E.O.E.: These quantities, which depend on neutron excess rather
than A, can be calculated from optical parameters, but this calculation does not, of course, make their origin very transparent.

GAMMA-SPECTRA FROM RADIATIVE PION CAPTURE IN CARBON

James A. Bistirlich, Kenneth M. Crowe, Anthony S.L. Parsons,* Paul Skarek,** and Peter Truoel***

Lawrence Radiation Laboratory, University of California Berkeley, California

(Presented by Peter Truoel)

We have completed an experiment at the Berkeley 184 inch-cyclotron set up to study radiative capture of pions on nuclei, i.e. the process $\pi^- N(A,Z) \to \gamma N^*(A,Z-1)$.

This process is believed to be theoretically well understood, and a number of authors have made very detailed predictions, mainly making use of the close relation of the matrix-elements governing this process with the ones appearing in muon-capture. The analogy with the latter process should manifest itself in the excitation of collective states in the residual nucleus, namely $\Delta T_3 = -1$ analogous states to the giant resonances in the parent nucleus, as observed in inelastic electron scattering.[1] The lack of experimental information, however, did not allow, until now, to check these models in detail. The expected fine structure in the energy spectrum requires a high-resolution and, therefore, low efficiency detector, which in connection with the low pion intensities available make the experiment rather time consuming.

Our experimental set up, as shown in Figure 1, included a large size 180°-pair-spectrometer. It consisted of two 18" x 36" C-magnets with a common poletip of 18" x 86" and a 13" gap, allowing a maximum field of 10 kG, known to an accuracy of .2% throughout the volume. The γ-rays were converted in a .0045" thick gold foil, inserted between two sets of four gap-, thin (.0005") aluminum plate spark chambers to detect the electron positron pairs. The trigger signal was given by a stopped pion, a neutral particle entering the spectrometer and the electron positron pair detected in two of the six trigger counters in front of the magnet. The tracks are recorded optically and the pictures are being measured on semi-automatic scanning machines. We were able to obtain 10,000

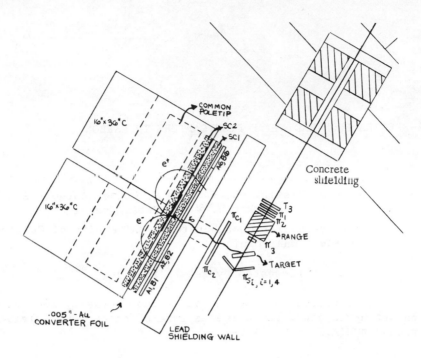

Fig. 1 - Experimental set-up.

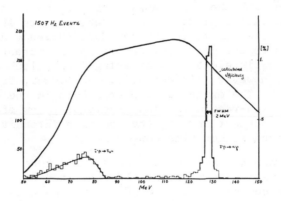

Fig. 2 - γ^- spectrum for pion-capture on hydrogen.
Solid line: calculated spectrometer efficiency.
The theoretical yield for mesonic capture has been
calculated using the experimental resolution given
by the radiative capture, the calculated efficiency
and the Panofsky ratio.

events showing a pair, for each of the elements, ^{4}He, ^{6}Li, ^{12}C, ^{16}O, ^{24}Mg and ^{40}Ca. In addition, the properties of the spectrometer were studied extensively using a liquid hydrogen target. The reaction $\pi^- p \rightarrow n\gamma$ with pions at rest allows a measurement of the resolution and provides, also, an independent calibration of the energy scale. The spectrum from mesonic capture $\pi^- p \rightarrow \pi^0 n$ serves as a check on the low energy cut off in our efficiency curve. The resulting spectrum is shown in Figure 2 together with the energy dependent efficiency of our spectrometer as given from a Monte-Carlo calculation. Our energy resolution is 2 MeV (FWHM). The Panofsky-ratio obtained from these data by use of the calculated relative efficiency agrees with the known value within our statistical errors. The same holds for the total capture rate on hydrogen. Including the efficiency our total solid angle was 3.8 x 10^{-4} sterradians, resulting in an event rate of 50 - 100 per hour. We obtained the following rates for radiative pion capture with the emission of a γ-ray with energy greater than 50 MeV relative to the total capture rate: ^{4}He 1.40%, ^{6}Li 2.07%, ^{12}C 1.68%, ^{24}Mg 2.74%, ^{40}Ca 2.20%. The estimated systematic error is 10%. While our value for carbon agrees with the one given by Davies et. al.[2] of 1.6 ± .1%; we disagree with their value for ^{6}Li of 3.3 ± .2%.

So far, we have been able to scan only a small sample of our data, concentrating on hydrogen and carbon. The carbon spectrum, including 25% of our total data for this element, is shown in Figure 3a. We compare our result to a simple phase space calculation, assuming a ^{11}B + n + γ final state and a prediction by use of a Fermi gas model. Both calculations give no representation of our data, the Fermi gas model has already been shown to give too high absorption rates by Anderson et. al.[3] Finally, we compare our result with the predictions of the giant resonance model by Kelly and Uberall.[1] We have folded the spectrum, which they obtained for 1s-capture using the model of Kamimura et. al.[4] and which includes the widths of the contributing states, with our experimental resolution, as given by the hydrogen line. The result is displayed in Figure 3b. The relative strengths of the states appearing in 2p-capture from the same model is indicated, also. We have omitted their calculations based on an alternative model by Lewis and Walecka.[5] The detailed nature of the experimental spectrum is not described too well by this model, the general features, however, besides the relative strength of the matrix-elements for the different states, seem to be present. The 3 MeV broad structure centered around 122 MeV corresponds to excitation energies between 2.7 and 5.7 MeV in ^{12}B, and, including a shift due to Coulomb energy of 1.7 MeV, to 18 - 21 MeV in ^{12}C. This group probably corresponds to the strong resonance seen in inelastic electron scattering at 19.6 MeV (together with two smaller excitations around 20.4 and 18.1 MeV.)[6] Both the Kamimura and the Lewis model require a strong 2^- spin-isospin resonance in this region. The two peaks at 118 and 116 MeV correspond to about 25 and 23 MeV excitation energy in ^{12}C and can

be associated with the giant dipole resonance at 22.8 MeV seen in
photoabsorption and its inverse reaction ^{11}B (p, γ_0) ^{12}C,[7] and the
peak at 25.5 MeV, also seen in the latter process and in the reaction $^{11}B(p, \gamma_1)$ ^{12}C. [7] Both states are well explained by the
Kaminura model. This model as well as the Lewis model does not
give any account of a dominant structure around 112 MeV (28 MeV in
^{12}C). There is, also, no evidence of this in the reactions mentioned
above, nor from inelastic electron scattering. If we follow Kelly
and Uberall's speculation, we can attribute this to a positive
parity state, arising from spin-isopin-oscillation in a giant quadrupole mode. Positive parity states are expected[8] to contribute
even more strongly to pion-capture than the negative parity states.
With our full data sample available we hope to arrive at more
definite conclusions. We then will also be able to shed light on
the question, whether there is any structure visible around 105 MeV,
where both models place a $J = 1^-$ $(p_{1/2}s_{1/2})-$ state. This energy
region has not yet been investigated by the other reactions. At
present, however, the conclusion is possible that radiative pion
capture like muon capture proceeds mainly through the excitation
of collective states.

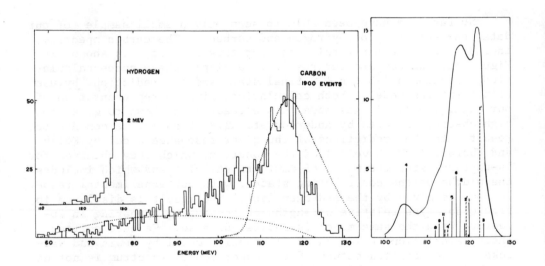

Figure 3a

γ^- spectrum from radiative pion-capture in carbon. The two curves
are a phase-space calculation assuming a $B^{11}n\gamma$ final state and a
prediction from the Fermi gas model. For comparison, the experimental resolution is shown, also.

Figure 3b

Theoretical predictions of Kelly and Uberall[1] folded with the experimental resolution.

REFERENCES

1. F. J. Kelly, H. Uberall, Nucl. Phys. A118 (1968), 310
2. H. Davies, H. Muirhead, T. N. Woulds, Nucl. Phys. 78 (1966), 673
3. D. K. Anderson, T. M. Eisenberg, Phys. Letters 22 (1966), 164
4. M. Kamimura, K. Ikeda, A. Arima,   Nucl. Phys. A95 (1967), 129
5. F. H. Lewis, J. D. Walecka, Phys. Review 133 (1964), B 849
6. T. W. Donnelly, J. D. Walecka, I. Sick, E. B. Hughes, Phys. Rev. Lett. 21 (1968), 1196
7. R. G. Allas, S.S. Hannah, L. Meyer-Schutzmeister, R. G. Segel, Nucl. Phys. 58 (1964), 122
8. J. D. Murphy, R. Raphael, H. Uberall, R. F. Wagner, D. K. Anderson, C. Werutz, Phys. Rev. Lett. 19 (1967), 714

*Present address: Rutherford High Energy Laboratory, Chilton, Berkshire, England
**Permanent address: CERN, Geneva, Switzerland
***On leave from: SIN, Zurich, Switzerland

MEASUREMENTS OF KAONIC X-RAY SPECTRA AND THE CAPTURE OF KAONS ON NUCLEAR SURFACES

Clyde E. Wiegand

Lawrence Radiation Laboratory

University of California, Berkeley, California 94720

K⁻-mesons, when stopped in matter, are captured by nuclei into hydrogen-like systems that we call kaonic atoms. When kaonic atoms undergo de-excitation they emit characteristic x rays in the same manner as do muonic and pionic atoms. We have studied the x-ray emission of 29 elements ranging from $Z = 3$ to $Z = 92$[1] and conclude from the preliminary measurements on the medium and heavy elements that nuclear matter -- very probably neutrons -- extends to larger radii than the radii of the charge distributions given by the classic electron scattering experiments of Hofstadter[2] and the Stanford group.

That neutrons should dominate nuclear surfaces is not a new idea:
Johnson and Teller[3] (1954) and Wilets[3] (1956) suggested that neutrons should populate the nuclear surface. Since then several authors have proposed that the nuclear surface should be rich in neutrons. [4]

Muonic and pionic atoms[5] have also been studied for many years and have presented us with a wealth of information on the internal structure of the nuclei, but as Wilkinson[6] pointed out, kaons should be sensitive probes of the nuclear surface. Kaons are sensitive to the low-density regions of nuclei because they strongly interact with single protons or single neutrons. A cursory analysis of the kaonic x-ray measurements suggests that the neutrons are distributed in a low-density tail. The parameters of the tail are relatively insensitive to the radii at half the central matter density.

502

To study the kaonic x rays of medium and heavy elements we used an experimental arrangement similar to that of Wiegand and Mack. [7] The external proton beam of the Bevatron was used to generate a secondary beam of negative kaons that were brought to rest in our targets. In fixed positions a few centimeters on either side of the targets we placed two lithium-drifted germanium detectors. Goulding, Landis, and Pehl[8] of the Chemistry Nuclear Instrumentation Group of our laboratory supplied us with the solid state detectors and their associated amplifiers. When a kaon stopped in the target, digital numbers proportional to the energies deposited in the detectors were stored on magnetic tape. A computer processed the data into spectra.

Figure 1 shows three examples of the kaonic x-ray spectra. No backgrounds have been subtracted and no corrections for detector efficiency or target absorption have been applied. Consider the lead spectrum. It is the result of stopping about a million kaons. The detector was lithium-drifted germanium in the planar configuration with an effective volume of 13 cm^3. The lines at about 80 keV are the $_{82}$Pb K alpha and K beta x rays from the emptying of the atomic electron K-shell. The next peak at 117 keV represents the kaonic transition from n = 12 to n = 11. Transitions to successively lower n values continue to the last observable peak at 427 keV, which corresponds to the n = 8 to 7 transition.

The kaons must be absorbed from orbits of maximum angular momentum (circular orbits) for our assumption that the kaons are absorbed on the nuclear surface to be valid. [6] Absorption from elliptical orbits would involve a nuclear volume effect and spoil the whole argument. To appreciate the argument for capture from circular orbits, consider the schematic energy level diagram of kaonic lead atoms shown in Fig. 2. Kaons are preferentially captured into atomic states of high angular momentum approximately in proportion to the statistical weight $2\ell+1$. Once a kaon enters a circular orbit it is trapped; there is no escape mechanism before capture. The existence of such a spectrum as the Pb series is strong evidence that the kaons are absorbed from circular orbits, because kaons in a given n level but with $\ell < n - 1$ are much more strongly absorbed than those with $\ell = n - 1$. Thus kaons in elliptical orbits cannot contribute at the end of a series. Therefore, we believe that if a kaon survives down to the eighth or ninth level of a kaonic Pb atom, it is surely in a circular orbit. Also the observed intensities of the lines are in approximate agreement with the calculations of Eisenberg and Kessler, [9] except where the kaons are strongly absorbed.

Figure 3 presents a chart of the principal kaonic lines that we measured in this survey.

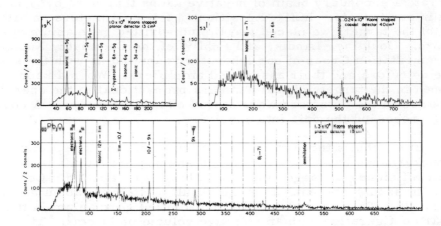

Fig. 1 - Examples of kaonic x-ray spectra.

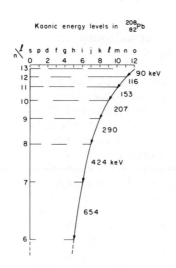

Fig. 2 - Schematic diagram
of the energy levels of
kaonic lead atoms.

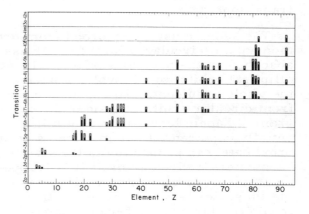

Fig. 3 - Chart of the principal kaonic
x-ray lines measured.   The heights of
the bars are proportional to the
observed intensities with the shaded
areas representing estimates of the
errors.   The distance between the horizontal lines corresponds to 1 x-ray
per stopped kaon.

The main point of the chart is to indicate where the various transitions are cut off. Consider the band that contains 5_g to 4f transitions: we see intense lines from $_{17}$Cl, $_{19}$K, $_{20}$Ca, and $_{22}$Ti, but at $_{28}$Ni the transition is overwhelmed by nuclear capture. In $_{29}$Cu the line is unobservable. In this first experiment we did not find the cutoff of the 6h → 5g lines, but it lies some-where between $_{42}$Mo and $_{53}$I. The 7i → 6h series ends at 64Gd (or $_{65}$Tb, which we did not measure). The transition was unobservable in $_{66}$Dy. We interpret the diminished intensity of the Gd 7i → 6h line to indicate that the nuclear capture rate is equal to or greater than the radiation rate from the state n = 7, l = 6 to n = 6, l = 5.

The radiation rates of the circular orbit transitions serve as built-in calibrations of the capture rates. When the kaon is about to be absorbed, the kaonic atom is very nearly hydrogenic and we have no hesitation in applying the hydrogen atom formulas. In the kaonic Gd atom, the radiation rate from n = 7, l = 6, to n = 6, l = 5 amounts to 1.2×10^{16} sec^{-1}. Therefore, the capture rate must be about 10^{16} sec^{-1}.

To relate the kaon absorption rate to nuclear size, we choose as a model of nuclei the Saxon-Woods distribution of nuclear matter

$$\rho_p(r) = \frac{\rho_p(0)}{1 + \exp[(r-C)/z]} \, ,$$

where the radius at half the central density is $C = 1.07 \, A^{1/3} F$ and the skin thickness parameter is $z = 0.55$ F. According to the Stanford data these parameters of the charge distribution apply with good accuracy over the range of medium and heavy elements.

The capture rate can be approximated by the overlap integral

$$P_{cap} = \frac{2W}{\hbar} \int \frac{\rho(0)}{1 + \exp[(r-C)/z]} R^2_{n, n-1} \, r^2 dr,$$

where $\rho(0)$ is the nuclear matter density at the center of the nucleus in units of particles per F^3, $R_{n, n-1}$ is the normalized hydrogenic radial eigenfunction for K$^-$, and W is the imaginary part of the kaon-nucleus potential. By invoking such an overlap integral we have made W independent of r and have reduced the kaon to a point with no interaction radius. However, these simplifications are probably not as serious as the assumption we must make for the value of W. To calculate our overlap integrals we rather arbitrarily set 2W equal to 100 MeV. Using the standard central density of 0.14 nucleons per F^3, 2W = 100 MeV, $C = 1.07 A^{1/3} F$, z = 0.55 F, R^2_7 (n = 7, $_{64}$Gd), the above overlap integral gives a capture rate of 2.5×10^{14} sec^{-1}. This is a factor

of 50 less than the experimentally calibrated capture rate of
1.2×10^{16} sec^{-1}. It is unlikely that W can account for the
entire discrepancy. Because kaons react about equally on neutrons and protons through the reactions K$^-$ + N → hyperon + pion,
we have strong evidence that the kaons are absorbed by nucleons
above the conventional nuclear surface.

We can bring the capture rate into agreement with the experiment by introducing a distribution of neutrons with different parameters than the proton parameters. Figure 4 shows one such
distribution for $^{157}_{64}$Gd. We show the conventional proton distribution ρ_p and a neutron distribution ρ_n with the same radius but
with a neutron skin thickness twice that of the proton skin thickness. The proton distribution requires $\int \rho_p dV = Z$ and the neutron
distribution $\int \rho_n dV = A - Z$. The curve on the right-hand side of
the illustration is the kaon distribution for n = 7, ℓ = 6. The
heavy solid curve in the center portion of the picture is the capture rate. The integral of the capture rate expression equals that
of the radiation rate: 1.2×10^{16} sec^{-1}. In this nucleon distribution practically all the contribution to the overlap integral comes
from the neutrons.

Figure 5 shows neutron and proton distributions made with conventional parameters and some modified distributions. The kaon-
nucleus interaction energy is 2W = 100 MeV for all the curves.
The curves with the skin thickness parameters increased to 2.3
times the conventional value give the experimentally determined
capture rate. Bethe[4] has suggested that the neutrons outside a
nucleus (r $\gtrsim$ C) should be distributed by

$$\rho_n(r) = (1/2) \rho_n(0) \exp[-(8M\epsilon)^{1/2} (r - C)/\hbar] ,$$

where M is the neutron mass and ϵ is the neutron binding energy. Bethe's proposed exponential tail appears to give too small
a contribution to the capture-rate overlap integral , if we keep
2 W = 100 MeV. However, there is apparently no reason to believe
that W cannot be as high as 500 MeV. An increase in 2 W by a
factor of 10 would bring Bethe's theory into agreement with the
experiment at Z $\simeq$ 64. Bethe's exponential falloff has a solid
theoretical foundation based on the neutron binding energy whereas
our Saxon-Woods distribution is only a convenient model that is
similar to the measured charge distribution.

In these experiments the capture rate is probably larger
than the radiation rate for the last transition observed in a
series. In the next phase of the experiment we will try to determine how the intensities of the lines diminish as the series approach
cutoff. Another part of the next experiment will be to stop kaons
in targets composed of single pure isotopes. Any observable

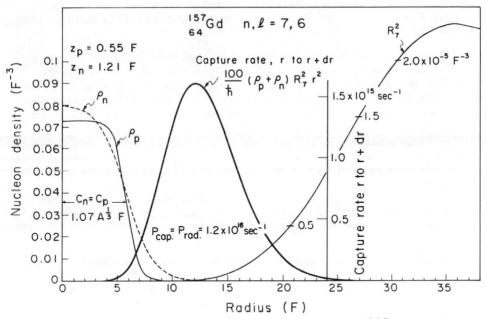

Fig. 4 – Example of a distribution of nucleons in $^{157}_{64}$Gd that agrees
with the experimental kaon capture rate.   The proton distribution
is that of Saxon-Woods with Stanford parameters.   The neutron distribution has the same radius but double the skin-thickness parameter.   Also shown are the kaon distribution for n = 7, ℓ = 6,
and the overlap function.

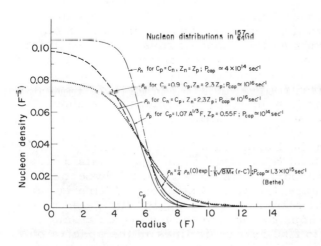

Fig. 5 – Several examples
of nucleon distributions
for $^{157}_{64}$Gd that are based
on the Saxon-Woods model
and fit the experimental
kaon capture rate.   Also
shown is an exponential
distribution suggested
by Bethe which gives the
required capture rate if
W is increased to about
500 MeV.   After the
illustration was prepared,
we learned that Bethe
prefers the exponential
falloff to be attached
at $\rho_n = \frac{1}{2} \rho_n(0)$.

change in the intensities of the lines, as we add neutrons, will be valuable information. We also expect to try to observe differences in the isotopes of Eu and Gd, where nuclear deformation might be a contributing factor.

The energies of the lines were measured by comparing them with the calibrated emissions of selected radioactive isotopes. We found the energies of the x-ray lines to be the energies given by the Klein-Gordon equation plus corrections for the vacuum polarization. The present accuracy of the experimental calibration is insufficient to establish differences due to strong interactions or other effects.

An interesting byproduct of the energy measurements of the kaonic x rays is an independent check on the mass of the kaons. Our present energy calibration establishes the kaonic mass at 493.97 ± 0.22 MeV,[10] only 0.15 MeV higher and well within the accepted value, 493.82 ± 0.11 MeV.[11]

Now let us focus our attention on the potassium spectrum in Fig. 1. An interesting feature is the series of lines 6h → 5g, 7h → 5g, and 8h → 5g. These transitions of $\Delta n = 1$, 2, and 3 have relative intensities in fair agreement with the predictions of Eisenberg and Kessler.[9]

At 186 keV there is a peak that corresponds to the pionic transition 3d → 2p. Several pionic lines have been seen among the kaonic lines. This is not surprising, because hyperons are generated when kaons react with nucleons. The subsequent decay of the hyperons is a source of pions that can lead to pionic x rays.

Some of the Σ^--hyperons that were generated in the targets through the reactions $K^- + n \rightarrow \Sigma^- + \pi^0$ and $K^- + p \rightarrow \Sigma^- + \pi^+$, were captured by target nuclei into hyperonic atoms. As the Σ^- particles cascaded toward the nuclei, x rays should have been emitted in the same manner as the kaonic x rays. We looked for the hyperonic transitions and found one line in the spectrum taken with a $_{19}$K target. In the spectrum resulting from one million kaons stopped in $_{19}$K there is a peak at 136 keV that corresponds to the Σ^--hyperonic transition 6h → 5g. Its intensity is about 0.015 x rays per stopped kaon whereas the kaonic transition 5g → 4f has an intensity about 40 times greater. This is a reasonable intensity based upon the European K^- Collaboration[12] experiment which reports that 0.08 K^- stopped in nuclear emulsion lead to the generation of Σ^--hyperons. X rays from the hyperonic transition n = 7 to 6 are expected at 82.2 keV, but the line is not observed. The n = 5 to 4 transition is probably eliminated by nuclear absorption. We hope to confirm the hyperonic atoms in the next experiment and to find hyperonic lines in the spectra of heavy elements. Σ^--hyperonic lines of the heavy elements are

especially interesting because they should show a fine structure
due to the magnetic moment of Σ^-. For example, in U the splitting of the n = 11 to 10 hyperonic transition at 467 keV should
amount to 0.64 keV according to SU_3 theory. With more intense
kaon beams and larger germanium detectors of improved resolution, we may be able to measure the Σ^- magnetic moment.

In summation we have a new tool for the investigation of the nuclear surface, and we may be able to make a significant and independent measurement of the kaonic mass. Study of Σ^--hyperonic
atoms should eventually result in the measurement of the Σ^- magnetic moment. The study of antiprotonic atoms is another possibility.

I wish to thank the many persons who made the experiment possible, especially: Frederick Goulding and Richard Pehl for the detector system, Rory Van Tuyl and Jack Walton for the electronics,
and the Bevatron crews for the kaon beam. Raymond Kunselman,
with the help of Donald Brandshaft, made the computer programs
for the overlap integrals. Professors Emilio Segrè and Lincoln
Wolfenstein contributed many valuable ideas to the interpretation
of the data.

REFERENCES

1. Clyde E. Wiegand, Phys. Rev. Letters 22, 1235 (1969).
   Clyde E. Wiegand and Raymond Kunselman, submitted to
   Proceedings of the International Conference on Hypernuclear
   Physics, (Argonne National Laboratory, May 5-7, 1969).

2. Robert Hofstadter, Ann. Rev. Nucl. Sci. 7, 231 (1957), and
   R. Herman and R. Hofstadter, High Energy Electron Scattering Tables (Stanford University Press, Stanford, 1960).

3. M. H. Johnson and E. Teller, Phys. Rev. 93, 357 (1954);
   Lawrence Wilets, Phys. Rev. 101, 1805 (1956).

4. P. B. Jones, Phil. Mag. 3, 33 (1958); R. G. Seyler and C.
   H. Blanchard, Phys. Rev. 131, 355 (1963); E. H. S. Burhop,
   Nucl. Phys. B1, 438 (1967); H. A. Bethe, Phys. Rev. 167,
   879 (1968); G. W. Greenlees, G. J. Pyle, and Y. C. Tang,
   Phys. Rev. 171, 115 (1968); J. P. Schiffer, "Coulomb
   Energies," paper presented at Conference on Isobaric Spin,
   Asilomar, California, March 1969.

5. E. H. S. Burhop, Mesonic Atoms, in High Energy Physics
   E. H. S. Burhop, ed. (Academic Press, New York, 1969),
   Vol. 3, p. 110.

6.   D. H. Wilkinson, Phil. Mag. <u>4</u>, 215 (1959); D. H. Wilkinson,
     <u>Proceedings of the Rutherford Jubilee International Confer-</u>
     <u>ence, 1961</u> (Heywood and Co., Ltd., London 1961), p. 339;
     D. H. Wilkinson, <u>Proceedings of the International Confer-</u>
     <u>ence on Nuclear Structure, Tokyo, 1967</u> (Physical Society of
     Japan), p. 484.

7.   Clyde E. Wiegand and Dick A. Mack, Phys. Rev. Letters <u>18</u>,
     685 (1967).

8.   F. S. Goulding, D. A. Landis, and R. H. Pehl, <u>Semiconduc-</u>
     <u>tor Nuclear-Particle Detectors and Circuits</u> (National Acad-
     emy of Sciences, Washington, 1969) p. 455.

9.   Y. Eisenberg and D. Kessler, Phys. Rev. <u>130</u>, 2352 (1963).

10. Raymond Kunselman and Clyde E. Wiegand, submitted to
     <u>Proceedings of the International Conference on Hypernuclear</u>
     <u>Physics</u> (Argonne National Laboratory, May 5-7, 1969).

11. Particle Properties, Lawrence Radiation Laboratory Report
     UCRL-8030, January 1969.

12. European K^- Collaboration Experiment, Nuovo Cimento <u>14</u>,
     315 (1959).

DISCUSSION

<u>R.E. Peierls</u>: It seems essential for your analysis of the capture
ratio to assume that the Kaonic wavefunction is essentially hydrogenic.   Is this justified for the calculation of the absorption
near the nucleon surface where the wavefunction is modified?
<u>C.E.W.</u>: We have assumed that since the Kaon is far from the nucleon
centre, the effect is small.   Ericson has estimated the effect for
pionic atoms - where it is larger, because capture takes place at
shorter distances.

<u>D.C.J. Lu</u>: What kind of charge distribution did you use in computing
the vacuum-polarization corrections?   <u>C.E.W.</u>: We used a point-
distribution.   Estimates made with more realistic distributions
showed that this effect would be very small.   <u>T.E.O.E.</u>: The corresponding effect has been examined for pionic atoms.   Typically, the
effect of vacuum polarization in that case modifies the strong interaction energy shifts by 1-2%.   These small effects which are due
to the distortion of the wavefunctions outside the nuclear interaction region by the strong interaction, are much smaller than the
ordinary vacuum-polarization effects.

C.E.W. (in reply to question by D. Hamilton): The Kaon beam
started with 500 MeV/c, was degraded in 50 gm/cm^2 of Carbon, and
about 10 K's per second were stopped.   In Lead some 1/4 million,
and in Potassium about 1 million were stopped.

MUONIC, PIONIC, KAONIC AND Σ^- X-RAYS, ATOMIC DE-EXCITATION

CASCADES, AND NUCLEAR ABSORPTION RATES IN He, Li, Be, AND C.

S. Berezin, G. Burleson, D. Eartly*, A. Roberts, T. White

Argonne National Laboratory, Northwestern University

University of Chicago, National Accelerator Laboratory

(Presented by A. Roberts)

EXPERIMENTAL ARRANGEMENT

Large xenon-filled proportional counters were used to obtain
sufficient solid angle for efficient x-ray detection from a 12-in.
long liquid He target. Mesons came from an unseparated 850-MeV/c
beam produced at 0° by the external proton beam. Muons or pions
were selected by a threshold Cerenkov counter, kaons by means of a
pair of Fitch-type Cerenkov counters in a time-of-flight arrangement
that provided a stopping K^- signal of greater than 99.5% purity.[2]
True and delayed (background) coincidence rates were simultaneously
recorded.

RESULTS: HELIUM

With nearly 10^6 kaons stopped in helium, we saw no identifiable
x-ray lines, contrary to Burleson et al.[1] The capability of detecting
kaonic x-rays in He was demonstrated by a wide variety of tests.
Exhaustive background studies, including the use of a dummy
absorptive (styrofoam) target designed to deplete the background
spectrum at low (<8 KeV) x-ray energies, were made. These data
provided a set of consistency relations which established the
presence of weak background effects, and also the absence of x-rays
from helium, within the errors of observation. Assuming corrected
Klein-Gordon energies and widths, the observed (null) yields are
for K_α, 0.06 ± .05 x-ray per kaon, or 0.08 ± .07, assuming the line
to be respectively unshifted and narrow, or shifted and broadened
as suggested by the K^- - He scattering data of M. Block et al.[3]
(41.0 KeV, Γ = 6.7 KeV). Corresponding yields for unshifted K_β,

* Now at National Accelerator Laboratory.

L_α, and L_β are $0.02 \pm .07$, $0.03 \pm .04$.

Pion and Muon Capture Data in He and Other Light Elements

Despite the disadvantages of the ZGS as a source of stopped μ's or π's, we utilized the μ and π content of the beam not merely to check the operation of the apparatus, but also to obtain the previously unmeasured yields of x-rays from He, and to check existing data in other light elements.

We find a total K-series yield in He for muons and pions respectively of $0.79 \pm .03$ and $0.18 \pm .05$. We also find the intensity ratios, K_α to the sum of all K-series lines, $0.58 \pm .05$ and $0.40 \pm .05$ respectively, in agreement with earlier measurements.[4] Yields and intensity ratios can both be used to check the predictions of cascade calculations, and to discover which input data yield the correct predictions.

Cascade Calculations in He

Previous K-series measurements of intensity ratios in He[4] indicated a considerable deviation from the simple circular-orbit theory of mesonic cascades. We have made extensive numerical Monte Carlo computations for the $(He-M)^+$ ion to see how these deviations may arise. It turns out that no reasonable choice of starting level, no ℓ-distribution of initial capture substates, and no variation of (external) Auger rates over two orders of magnitude can duplicate the observed yields and intensity distributions. To achieve agreement with observation, one must add to the cascade mechanism weak collisional Stark mixing, in the form of "sliding" transitions (i.e., $n, \ell \pm 1$) as discussed by Ruderman[5] and by Leon and Bethe.[6] When this is done, a single value of the collisional mixing parameter will give agreement with the experimental data in muonic, pionic and kaonic He atoms.

The cascade parameters so obtained also lead to values of lifetimes (widths) for the observed x-ray transitions in good agreement both with the limited experimental data, and with predictions of the shell model; and in fair agreement with predictions of widths from pion-He[8] and kaon-He[3] scattering. The calculated pion and kaon cascade times do not match the observed moderation times in He.[9,10]

Kaon Capture and X-ray Spectra in Li, Be, and C

Unlike helium, x-rays were readily detected from kaon capture in Li, Be, and C. In Be and C we confirmed Wiegand and Mack's[11] observation of the L_α and M_α lines. In Li we saw at least four lines where they identified only one, the L_α ; and in both Be and

C we see one additional line. In addition, further unresolved
structure is present in both Be and C, perhaps also in Li.

Two of the Li lines are readily identified as the kaonic L_α
and L_β; the other two cannot be identified as originating from
kaonic atomic transitions. By eliminating all other interpretations
so far suggested, we conclude that these lines, at 12.90 ± .30 KeV,
are most likely the M_α and M_β lines, both shifted 1 KeV, of Σ^- -
atoms. Stopped kaons absorbed in light nuclei produce Σ^- - hyperons
in about 15% of absorptions. The slower Σ^-'s can be stopped in the
target, so that we find the Li Σ^- M x-ray yield to be 0.4 ± 0.7.
Similarly the Be line at 21.0 KeV seems best identified as the $\Sigma^- M_\alpha$
On the other hand, the C line at 18.0 KeV, which fits no reasonable
Σ^- transition, seems most likely the pion L_α, which is possible
since kaon absorption produces slow pions as well as hyperons. It
is not clear why in that case pion lines are not also observed in
Li or Be. Additional work with higher kaon flux and better resolution is clearly needed.

The production of Σ^- atoms with reasonable intensity could
facilitate the study of the hyperon-nuclear interaction. The
observed shift of the M-lines in Li, for example, indicates an
attractive Σ^- - Li interaction in the 3d state of 0.95 ± .30 KeV.

Cascades in the Light Elements; Muon and Pion Spectra

Our pion and muon data in Li, Be, and C are all remeasurements
of previously observed quantities, and are in good agreement with
them. We did Monte Carlo calculations for Li, Be, and C too, for
pions, muons and kaons. In these elements the α-member of a series
does not manifest the anomalously low intensity it shows in helium.
Consequently, no major departure from the statistically uniform
distribution of capture substates is required to match the yields
and intensity ratios seen; the Eisenberg-Kessler[5] parameter α is
best fitted as $0.20 \pm {}^{.1}_{.2}$. It is possible to evaluate 2p absorption
rates for pionic atoms, and 3d rates for kaonic atoms, and to
demonstrate their agreement with other experimental data and with
shell-model calculations.

More detailed descriptions of this work are in press and in
preparation.[12]

REFERENCES

[1]G. Burleson, D. Cohen, R. Lamb, D. Michael, R. Schluter, and
 T. White, Phys. Rev. Letters 15, 70 (1965).
[2]D. Eartly and A. Roberts, Nucl. Instr. and Meth. in press.
[3]M. Block et al., (to be published).
[4]R. Wetmore, D. Buckle, J. Kane, R. Siegel, Phys. Rev. Letters, 19,
 1003 (1967).

[5]M. Ruderman, Phys. Rev. 118, 1632 (1960); also Y. Eisenberg
and D. Kessler, Nuovo Cim. 20, 1195 (1960).
[6]M. Leon and H. Bethe, Phys. Rev. 127, 636 (1962).
[7]M. Ericson and T.E.O. Ericson, Ann. Phys. 36, 323 (1966);
M. Ericson, C.R. Acad. Sci. 258, 1471 (1963).
[8]J. Boyd and V. Viers, Bull. Am. Phys. Soc. 14, 525 (1969);
to be published.
[9]M. Block, T. Kikuchi, D. Koetki, J. Kopelman, C. Sun,
R. Walker, G. Culligan, V.L. Telegdi, and R. Winston, Phys.
Rev. Letters 11, 301 (1963); and J. Fetkovich, E. Pewitt,
Phys. Rev. Letters 11, 290 (1963).
[10]M. Block, J. Kopelman, C. Sun, Phys. Rev. 140 B, 143 (1965).
[11]C. Wiegand and D. Mack, Phys. Rev. Letters 18, 685 (1967).
[12]S. Berezin, et al., Phys. Letters in press; ibid, Nuclear
Physics, in press; ibid, to be published.

HYPERNUCLEI

A. Gal

Lab. for Nucl. Sci. and Phys. Dept., M.I.T., Mass. and

Dept. of Theor. Phys., The Hebrew Univ. of Jerusalem [*]

1. INTRODUCTION

The aim of the present exposition is to bring some highlights from the recent International Conference on Hypernuclear Physics held at Argonne National Laboratory on May 5-7 this year. Many exciting experiments and interesting theoretical analyses were discussed at that Conference. However, due to time and space limitations, only a part of the topics will be dealt with here and the more curious reader is urged to look at the Proceedings of the A.N.L. Conference, where a detailed discussion and a rather complete list of references is available.

2. LAMBDA PROTON FINAL STATE INTERACTIONS

In addition to direct Λp low-energy scattering experiments [1,2], the lambda proton system has been extensively studied in the reaction

$$K^- d \rightarrow \pi^- p \Lambda \tag{1}$$

Recently Tan [3] observed a strong enhancement in the partial rate for the reaction (1) at rest for an effective Λp mass just below the ΣN threshold. The sharp peak is followed, on the decaying slope, by a tiny shoulder. Similar observations were made for the reaction (1) with 400 MeV/c kaons (Cline et al [4]) and 910, 1106 MeV/c kaons (Alexander et al. [5]). In these experiments the peak below the ΣN threshold is enhanced for pions in the forward direction, which affirms the $^3S_1$ assignment (This assignment follows from the measured, in ref. 3, polarization properties of the Λ in the rest frame of the Λp system with respect to the outgoing pion.) to the Λp system in this energy region. It is natural to

expect an enhancement at the ΣN threshold region from a consideration of the two-step process

$$K^-d \rightarrow \pi^- \underbrace{\Sigma(N)}_{\Lambda P} \qquad\qquad (2)$$

For on-mass-shell intermediate baryons the second step gives rise
to the v^{-1} type dependence near and above the ΣN threshold. This
apparent singularity is suppressed as one goes below the ΣN threshold since the intermediate ΣN system, in this analysis, then
necessarily becomes more and more distant from its mass-shell.
The resultant curve should resemble a resonance curve, but the
exact position of the peak depends to a high degree on the on-shell
and off-shell assumptions incorporated in the calculation[5,3]. It
is not clear at the moment whether or not the mechanism discussed
above is sufficient to explain all existing data. The question
whether a genuine $^3S_1$ resonance or only a cusp phenomenon plays
the main role in the Λp system just below the ΣN threshold is still
open, and awaits more precise direct Λp scattering experiments at
that energy region (where the present statistics and resolution
are rather poor).[6]

     In the low Λp effective mass region the direct Λ production
competes with the two-step process[2]. The corresponding analysis[3]
gives for the low-energy scattering parameters of the triplet Λp
system

$$a_t = -2.0 \pm 0.5 \text{ fm}, \quad r_t = 3.0 \pm 1.0 \text{ fm} \qquad (3)$$

This is in accordance with the best fits obtained from direct Λp
scattering[1,2].

     On the theoretical side Fast, Helder and de-Swart[7] reproduce,
within a two-channel formalism, available experimental data on Σp
reactions[6] with One Boson Exchange (OBE) potentials. The pseudoscalar and vector meson nonets were taken into account. The fitting
parameters used by these authors were the hard-core radii for triplet
and singlet, $T = 1/2$ and $T = 3/2$, YN states. They get a $^3S_1-^3D_1$
resonance in Λp scattering just below the ΣN threshold. If they
fix it to appear at 3 MeV below the $\Sigma^0 n$ threshold (as the experimental data might suggest) then $a_t(\Lambda p) = -1.5$ fm. Downs and coworkers[8], using coupling constants of a scalar and a vector meson
unitary singlet, as well as the mass of the scalar meson and the
radius of a hard core as fitting parameters, also get a $^3S_1-^3D_1$ Λp
resonance just below the Σ production threshold. Their scattering
lengths are: $a_s = -2.25$ fm and $a_t = -2.12$ fm.

3.   THE HYPERON-NUCLEON FORCE AND BINDING ENERGIES OF Λ IN HYPERNUCLEI

In this section we shall outline results of calculations of

Λ-binding energies (BE_Λ) in hypernuclei. In general there exists
at present a discrepancy between the ΛN interaction parameters
derived from low-energy Λp scattering and between those parameters
that yield, in a model calculation, the binding energies of many
hypernuclear systems. This discrepancy is most evident in cases
where the ΛN triplet interaction is favorably weighted. In these
cases (such as $_\Lambda He^5$, nuclear-matter) calculations with ΛN central
forces that fit the low energy Λp scattering data give too big
binding energies. Whether the reason for such a suppression of
the free ΛN force when it operates in hypernuclei, is due to the
presence of a strong tensor component in the ΛN force, or the
appearance of essentially repulsive three body ΛNN forces, or, to
a lesser extent, isospin suppression[9] has not yet been determined,

3.1. Nuclear Matter
        Binding energies of Λ in heavy hypernuclei, as studied in
emulsions, lead one to conclude that the binding energy D for nuclear
matter is around 30 MeV with an error of 2-3 MeV. It is true that
the exact number depends somewhat on the extrapolation procedure,
but since the kinetic energy of the Λ in heavy hypernuclei is not
too big (of the order of a few MeV) and the same is probably true
for surface effects, the accepted value for D cannot be too far
from reality. Theoretical calculations[10] with two-body central
ΛN forces that fit the low energy scattering data give, however, a
much higher value for D, roughly of 60 MeV. Recently, Bodmer and
Rote[11] performed coupled-channel g-matrix calculations for D in
order to check the effects of (i) a tensor component in the ΛN
force (the two coupled channels being for S and D waves) and (ii)
the ΣN (in addition to the ΛN) channel. For (i) they consider a
central potential (outside a hard core of 0.43 fm.) arising from
an exchange of an isoscalar meson ($m = 3m_\pi$). A short range tensor
potential (as well as the central part) due to K meson exchange is
added. The strengths of the triplet central and tensor potentials
are so related that the triplet ΛN scattering length is -2 fm. The
singlet potential equals the triplet one in the absence of K meson
exchange. Bodmer and Rote then compute D for a whole range of tensor-
strengths compatible with the above requirements. They find that
the value for D is only slightly reduced and the conclusion is that
one expects very little suppression in nuclear matter for the short
range ΛN tensor forces predicted by meson theory. The effect of
such short range tensor forces is probably very similar for a free
scattering and for nuclear matter. For (ii), Bodmer and Rote add
to the central ΛN potential the strong OPE tensor coupling between
the ΛN and the ΣN channels. The same constraints as in (i) are
kept. They find a considerable suppression, as much as 10 MeV for
reasonable values of the $\Lambda \Sigma \pi$ coupling constant. This suppression
originates from the long range of the OPE tensor coupling; the
corresponding moments are to a high degree inaccessible due to the
Pauli exclusion principle. We remark that a consideration of the

ΣN channel amounts to taking into account a big part of the three body ΛNN force in the one channel formalism[12]. There are other means for further reducing the binding D, like p-state suppression and charge symmetry breaking (CSB) effects. The latter cannot be reliably estimated since no Λn scattering data are available.

3.2.   s-Shell Hypernuclei

ΛN central forces that fit both the low energy scattering and binding energies (in a detailed variation-type calculation) of $_\Lambda H^3$, $_\Lambda H^4$ and $_\Lambda He^4$ have been considered by Herndon and Tang[13] and discussed recently by Tang[14]. The above authors use CSB terms in the ΛN force to account for the splitting in the isospin doublet ($_\Lambda H^4$, $_\Lambda He^4$). They allow also for a ΛNN force due to two pion exchange (TPE). However, the part considered by them is probably the weakest one among others. The conclusion of these investigations is that the central ΛN force is only weakly spin-dependent and that this force is of an intrinsic range around 1.8 fm and contains a hard core of a 0.45 fm. The angular distribution of the Λp scattering events suggests that for such forces the strength in odd-parity states is about half of that in even-parity states. The overall data favors to some extent a repulsive ΛNN force but in view of what was stated above this conclusion is rather preliminary. Nevertheless it helps in reducing the Λ-binding of $_\Lambda He^5$, which otherwise would come as high as 6 MeV (experimentally 3 MeV).

The problem of overbinding in $_\Lambda He^5$ is quite similar to that in nuclear matter (the previous subsection). Law, Gunye and Bhaduri[15] found that overbinding of $_\Lambda He^5$ persists even on inclusion of a short range tensor force, a conclusion similar to that of Bodmer and Rote[11] for nuclear matter. On the other hand it had been estimated[16] that with all terms of the TPE-ΛNN force it is possible to bring the calculated binding energy of $_\Lambda He$ in accord with the experimental quantity. A detailed calculation should take into account nuclear correlations in He^4 since the strong parts of the TPE-ΛNN force are rather singular.

Another comment appropriate at this point is that calculations of the type reported by Tang should be more alert to the effects of CSB forces. These forces, as pointed out by Hartt and Sullivan[17], contribute also in the $_\Lambda H^3$ hypernucleus by coupling the T=0 state with the low lying T=1 state. This is analogous to the S' state considered some time ago by Bodmer[9], but seem to be of more importance since the same spatial symmetry components are connected now in the T=0 and T=1 states. When the most general central expression (two parameters) for the CSB is taken into account it might happen[17] that the particular combination contributing in the A=4 doublet is not as important as the other one. This might lead to strong effects on the scattering lengths. Also, with such strong CSB forces one should treat more carefully their range, and both possible spin assignments to the ground state of $_\Lambda He^4$ should be considered.

3.3 p-Shell Hypernuclei

Λ-binding energies (BE_Λ) in the nuclear p-shell are quite well known [18] through $_\Lambda C^{13}$. Earlier attempts [19] to relate these BE_Λ values in a shell-model analysis in terms of central two-body ΛN forces alone proved unsuccessful. The reason for this failure is mainly due to the exceptionally low binding of $_\Lambda Be^9$ relative to its neighbors (BE_Λ ($_\Lambda Be^9$) = 6.63 $\pm$ 0.04 MeV, while $BE_\Lambda(_\Lambda Li^8)$ = 6.80 $\pm$ 0.05 MeV and $BE_\Lambda(_\Lambda Li^9)$ = 8.25 $\pm$ 0.13 MeV). The shell model approach has been recently applied with the most general two-body ΛN forces, including tensor as well as spin-orbit forces, by Dalitz, Gal and Soper [20]. A reasonable fit to the data can be achieved but only with a huge antisymmetric spin orbit ΛN force. Also, the more familiar parameter V which is the matrix element of the central spin-independent ΛN force between a 1s Λ and a 1p nucleon, turns out in these fits to be around 0.7 MeV, while an estimate consistent with the scattering data would give 1.5 - 2.0 MeV. The central spin dependent matrix element Δ is rather big in these fits. The other alternative considered by these authors is to use only a symmetric spin-orbit ΛN force (analogy with the NN case) but no other non-central ΛN forces in addition to the central parameter V (Δ is put equal to zero) and to introduce three body ΛNN force parameters that come out of the TPE mechanism. There are five such parameters and if one really sticks to the TPE model all of them are related. The common strength of the TPE ΛNN force is left as a free parameter. The resultant fits of the BE_Λ values are about as good as in the case of pure two-body forces. The strength of the ΛNN force turns out to be of the same order of magnitude as derived from the TPE diagram and for V one gets values around 1.5 MeV, which is a considerable improvement over the two-body fits. One has to remember that the TPE-ΛNN force possesses some features which are not the most general ones (spin and isospin dependence). The spin-orbit parameter is quite big, about the same magnitude as in the NN problem but the sign is opposite. The validity of these fits can partly be tested when BE_Λ values beyond $_\Lambda C^{13}$ are determined, since in the $p_{1/2}$ shell the spin-orbit ΛN force and the strong non-central components of the ΛNN force combine in a different way than in the $p_{3/2}$ shell [21]. Knowledge of hypernuclear spins and excitation energies is of immense value for discriminating against some fits. Unfortunately, the only experimentally fixed [18,22] spin in the p-shell is J=1 of $_\Lambda Li^8$, where an element of inconclusiveness comes in due to nuclear excitation (about 0.5 MeV). It is hoped that soon, as statistics improve, the spin of $_\Lambda B^{12}$ will be experimentally established by a measurement of the angular correlation of the decaying π^- with respect to the normal to the 3α-plane in a $_\Lambda B^{12}$ decay that goes through a certain level of C^{12} [22]. With the very selective criteria applied by the K^- Collaboration group for established BE_Λ values, the evidence[18] for an isomer $_\Lambda He^7$ is not really positive. At the same time the asymmetry in the Be_Λ distribution for $_\Lambda Li$ might suggest some isomerism[23]. Direct counter experiments for detecting hypernuclear γ decays have not yet been performed. However, the efficiency of such experiments depends to a great extent on production rates.

Very interesting results have been recently reached by the K^- Collaboration group [24] with regard to the following uniquely determined processes

$$K^- C^{12} \rightarrow \pi^- p \, _\Lambda B^{11} \qquad\qquad (4)$$

$$K^- N^{14} \rightarrow \pi^- p \, _\Lambda C^{13} \qquad\qquad (5)$$

The sharp peaking of the outgoing π^- momenta was interpreted as an evidence for ~ 10 MeV excitations (proton unstable) of $_\Lambda C^{12}$ and $_\Lambda N^{14}$ respectively. Dalitz [25] conjectured that these excitations correspond to the trapping of the Λ particle in a p-state. Quantitatively, the support for this conjecture is the rather good overlap between 1p state nucleon and Λ orbitals obtained in a Hartree-Fock calculation [26].

4. MEASUREMENTS IN A HELIUM BUBBLE CHAMBER

Recently, a lot of experimental data [27,28] came from stopping K^- in the ANL-CMU helium bubble chamber. Here we state briefly only the following results:
(i)   The lifetime of $_\Lambda H^3$ was measured [29] to be $(2.64 \, ^{+0.84}_{-0.52})$ 10^{-10} sec. This is in agreement with the theoretical calculation [30] which finds only a slight change from the free Λ lifetime (measured in this experiment to be $(2.58 \pm 0.10)10^{-10}$ sec.). However, previous measurements, notably that of Block [31] in a He bubble chamber, do not converge to the same value.
(ii) Production rates for $_\Lambda H^4$ and $_\Lambda He^4$ were established to be tween 1 and 2 percent [27]. The binding energy of $_\Lambda He^4$ was determined not only from decay, as is usually the case in emulsion studies, but also by the unique production

$$K^- He^4 \rightarrow \, _\Lambda He^4 \, \pi^- \qquad\qquad (6)$$

The results [27], 1.83 ± 0.62 MeV from production and 1.99 ± 0.37 MeV from decay, do not positively suggest that $_\Lambda He^4$ is produced both in its ground state and in an expected low lying spin-flip excited state. However, the errors are still too large to draw more definite conclusions. Such conclusions will provide

valuable information on the relative magnitude of the CS and CSB
spin dependent central ΛN forces.

REFERENCES

*   Present address: Dept. of Theor. Phys., The Hebrew Univer-
        sity of Jerusalem.
1.   G. Alexander et. al. Phys. Rev. 173, 1452 (1968).
2.   B. Sechi-Zorn et. al., Phys. Rev. 175, 1735 (1968).
3.   T. H. Tan, Phys. Rev. Lett. 23, 395 (1969).
4.   D. Cline et. al., Phys. Rev. Lett. 20, 1452 (1968).
5.   G. Alexander et. al. Phys. Rev. Lett. 22, 403 (1969).
6.   G. Alexander, Invited talk at the International Conference
        on Hypernuclear Physics (ICHP), ANL (May, 1969).
7.   G. Fast, J. C. Helder and J. J. de-Swart, Phys. Rev. Lett.
        22, 1453 (1969).
8.   B. W. Downs, Invited talk at ICHP, ANL (May 1969).
9.   A. R. Bodmer, Phys. Rev. 141, 1387 (1966).
10.   For an extensive list of references see ref. 11.
11.   A. R. Bodmer and D. M. Rote, Invited talk ICHP, ANL (May 1969).
12.   Y. Nogami, Invited talk at ICHP, ANL (May 1969).
13.   R. C. Herndon and Y. C. Tang, Phys. Rev. 153, 1091 (1967).
        ibid 159, 853 (1967) and ibid. 165, 1093 (1968).
14.   Y. C. Tang, Invited talk at ICHP, ANL (May 1969).
15.   J. Law, M. R. Gunye and R. K. Bhaduri, Contributed paper
        to ICHP, ANL (May 1969).
16.   R. K. Bhaduri, B. A. Loiseau and Y. Nogami, Ann. Phys. (N.Y.)
        44, 57 (1967).
17.   K. Hartt and E. Sullivan, Contributed paper to this Conference.
18.   D. H. Davis and J. Sacton, Invited talk at ICHP, ANL (May 1969).
19.   R. H. Dalitz, in Proc. Int'l School of Phys. "Enrico Fermi",
        Course 38, Academic Press N.Y. (1967), Edited by T.E.O.
        Ericson.
20.   R. H. Dalitz, A. Gal and J. M. Soper, in preparation.
21.   A. Gal, Invited talk at ICHP, ANL (May 1969).
22.   P. Vilain et. al. to be published in Nucl. Phys.
23.   J. Pniewski et al. Nucl. Phys. B2, 317 (1967).
24.   K$^-$ Collaboration, private communication (Aug. 1969).
25.   R. H. Dalitz, Summary talk at ICHP, ANL (May 1969).
26.   W. H. Bassichis and A. Gal, to be published in Phys. Rev.
27.   J. McKenzie, Invited talk at ICHP, ANL (May 1969).
28.   G. Keyes, Invited talk at ICHP, ANL (May 1969).
29.   G. Keyes et. al., Phys. Rev. Lett. 20, 819 (1968).
30.   M. Rayet and R. H. Dalitz, Nuovo Cimento 46, 786 (1966).
31.   M. M. Block et. al., Proc. Sienna Conf. El. Part. 62 (1963).

DISCUSSION

<u>M. Rho</u>: I understand that the 3-nucleon force in nuclear matter is
a consequence of Pauli-exclusion correlations, and similarly for
the Λ-nuclear interaction.   If so, is there any reason to expect
larger corrections for ΛNN than for NNN?   <u>A.G.</u>:   The two cases
are not closely comparable.   One has to consider not only ΛN, but
also ΣN.   (Subsequent discussion -- the record of which is
undecipherable -- between <u>M.R.</u> and <u>A.G.</u> indicated a divergence of
views as to the different nature of the '3-body force' in the
hypernuclear and ordinary nuclear cases.)

<u>K. Hartt</u> commented on the possibility of strong C.S.B. forces play-
ing a significant role in p-shell hypernuclei, with possible agreement with experiment for binding energies.

<u>J.D. Walecka</u>: I would like to recall a suggestion by de Shalit, at
the first conference in this series (1963).   A Λ in an odd-A very
heavy nucleus can be in a $S\frac{1}{2}$ doublet level.   The doublet spacing
decreases as A increases.   The first excited state of such a
hypernuclei could, then, be at a very low energy and have a correspondingly long life-time.   Experimentally such investigations may
not be presently practicable, but they would be very interesting in
providing information on the spin-dependence.

NUCLEAR INTERACTIONS AND STUDY OF UNSTABLE PARTICLES

A. A. Kuznetsov

Joint Institute for Nuclear Research

Dubna, USSR

I would like to report the results of some work done in the Laboratories of High Energies and of Nuclear Problems of JINR.

I. OBSERVATION OF RESONANCES IN THE Λ°p SYSTEM
by B. A. Shahbazian

Since 1962 a search for resonances in the Λp-system has been in progress.   The 55 cm long propane bubble chamber was exposed to the neutron beam of a 7.5 GeV/c average momentum.   The first result, obtained with rather limited statistics of one-pronged interactions with carbon nuclei containing a Λ-hyperon and only one proton (Fig. 1), revealed a significant peak near the sum of Λ-particle and proton masses.   The existence of this peak was further confirmed with increased statistics, not only in one-pronged events, but in three-pronged events with only one proton and

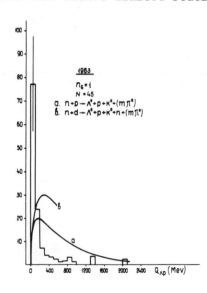

Fig. 1 - The distribution over the total kinetic energy Q $_{\Lambda p}$ in the Λp rest system for one-pronged interactions (1963). Curves a and b are phase space volume distributions.

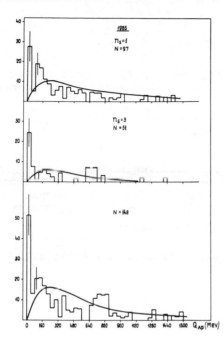

a Λ-hyperon (Fig. 2)[1]. The results obtained by the end of 1967[2] (Fig. 3) permitted this peak to be identified as a resonance on a virtual level of Λp-system at (4.8±1.1) MeV.

The corresponding scattering parameters, supposing independence on total spin states of the Λp-system, are:

Fig. 2 - The same as in Fig. 1 for one- and three-pronged events, and the total distribution, with only one proton and a Λ-hyperon (1965).

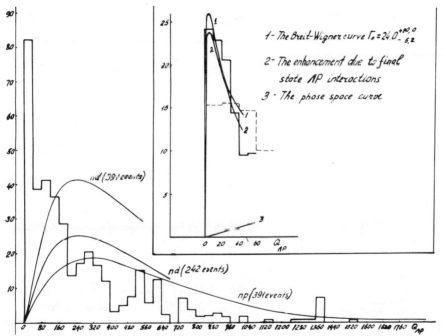

Fig. 3 - The $Q\Lambda_p$ distribution for 1-,2-,3-,5-pronged events with only one proton and a Λ-hyperon (1968). The smooth curves are total phase space volumes for 12 reaction channels including Δ_{33} and Y^*_{1385} formation as well as 1-3 pion criterion.

$$a = (-2.82 \pm 0.70) \cdot 10^{-13} \text{ cm}$$
$$r = (2.50 \pm 0.60) \cdot 10^{-13} \text{ cm.}$$

The angular distribution of Λ-hyperons, with respect to the line of flight of the Λp-system, in the first 10 MeV interval is isotropic and symmetric, the asymmetry parameter being

$$\frac{F-B}{F+B} = 0.055 \pm 0.020.$$

It was of prime interest to search for this resonance under different conditions. The same author continued the search in interactions of 4.0 GeV/c π⁻-mesons with carbon nuclei in the same propane bubble chamber. The results, presented in Fig. 4[3], show the same peak partly smeared out because of fake Λp combinations.

The scattering parameters, obtained from the neutron-carbon and pion-carbon combined data, are

$$a = (-2.63 \pm 0.60) \cdot 10^{-13} \text{ cm}$$
$$r = (2.60 \pm 0.50) \cdot 10^{-13} \text{ cm}$$

The negative sign of scattering length indicates the impossibility of bound states in the Λp-system. Simultaneously it indicates the existence of a new resonance with baryonic number B=2 and hypercharge Y=1.

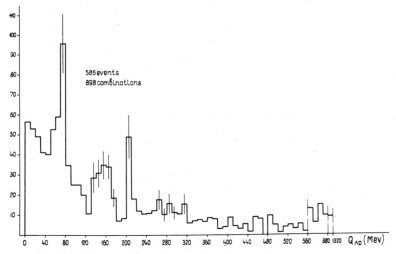

Fig. 4 - The Q $_{\Lambda p}$ distribution for 4.0 GeV/c momentum π⁻-carbon interaction. The total number of combinations (898) is normalized to the total number of Λ-hyperons.

The pronounced peak at 2127 MeV is ascribed to two possible causes, or to their superposition. The first is the formation of a Σ-hyperon strongly interacting with a proton (ΣN resonance like the above Λp resonance) with subsequent $\Sigma N \rightarrow \Lambda p$ conversion. Alternatively, one cannot exclude the existence of a geniune Λp resonance slightly below the ΣN threshold.

The significant enhancement in the $Q_{\Lambda p}$ = (130 - 180) MeV interval was observed earlier by the author in neutron carbon interactions[1] containing a Λ-hyperon and more than one proton. Perhaps, this peak is due to a new Λp-resonance. Another pronounced significant peak is seen at 2257 MeV with $\Gamma \leqslant 15$ MeV.

In neutron-proton interactions the same author observed a peak at 2573 MeV, $\Gamma \leqslant 80$ MeV which cannot be ascribed neither to kinematical effects, nor to the reflections of known resonances[1]. No significant enhancements were observed in the pp effective mass spectrum[1]. This work on search for resonances with B > 1 continues.

II. ELASTIC π^-, ^{4}He SCATTERING AT 3.48 GeV/c
by A.A. Nomofilov, I.M. Sitnik, L.A. Slepets, L.N. Strunov.

Some preliminary data on the measurement of elastic π^-, ^{4}He scattering cross sections in the range of four-momentum transfers of $0.73 \times 10^{-2} < |t| < 0.42 \times 10^{-1}$ (GeV/c)^2 are now presented (Fig. 5). The t-resolution is on an average equal to 1×10^{-3} (GeV/c)^2.

The process $\pi^- {}^4$He $\rightarrow \pi^- {}^4$He is interesting, since it makes possible to study elastic scattering of spinless particles and permits a spin-independent part of πN scattering amplitude in the framework of the Glauber model to be estimated. Using this model for our interval of small t, where single interaction predominates, it was found that πN spin-flip amplitude gives no contribution to the coherent process[4].

In order to detect small angle π^-He scattering coherent events, we have adopted a method using recoil α-particles. This method was used earlier[1,2] for measuring elastic π^-P scattering in the range of small values of t. Recoil α-particles were recorded in a $50 \times 50 \times 15$ cm^3 magnetic cloud chamber filled with helium at 1.12 atm., through which a pion flux of 4×10^4 particles per cycle passed. The chamber was insensitive to individual relativistic particles of the beam. Use was made of the fact that recoil α-particles ionization exceeded that of beam pions 100 times for the above interval of t.

The direction of the beam particles was determined in a special exposure in which the chamber recorded relativistic particles.

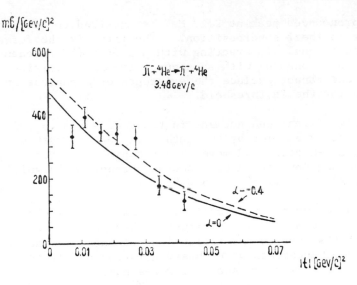

Fig. 5 - The differential cross section of the elastic
$\pi^-{}^4$He scattering. The curves are calculated by W. Czyz
and L. Lesniak[3] by the Glauber method using the value of
nuclear radius ^{4}He $R_{4_{He}}$ = 1.37 fm and the N-scattering
amplitude parametrisation.

$$f_{\pi n} = \frac{(i + \alpha)\sigma p}{4\pi} e^{-\frac{a}{2}|t|}$$

where p=3.48 GeV/c, σ = 28.4 mb, a = 7 $(GeV/c)^2$. The
Coulomb amplitudes effects, which are negligible for
$|t| > 0.01$ $(GeV/c)^2$, are not subtracted from the experimental results.

About 4000 photographs were twice scanned.   The single scanning efficiency was 90%.   Elastic events were identified and separated from background, by momentum-angle kinematic criterion.   The
background level was one-fifteenth of the height of the elastic peak
which contains 870 events satisfying the selection criteria (stop-
ping in the chamber gas; recoil α-particle momentum 75$\leqslant$p$\leqslant$215 MeV/c;
track projection length on the XY plane of the chamber $\geqslant$1 cm).
The momentum of the recoil particle was measured by range.

An absolute measurement of the pion flux was made with the help
of the nuclear photoemulsion, placed behind the chamber and inter-
cepting the entire beam, and also with the help of an integral electronic system to measure the flux for each momentum.   The
(3.48$\pm$0.05) GeV/c pion beam contained (7 1)% μ-mesons and (2.4$\pm$0.3)%
electrons.   A value of $n.n_{He}L$ was determined with an accuracy of
3%(n-pion flux, n_{He}-helium density, L - length of the chamber
active volume).

Further results include additional events in the t-range of Fig. 5, and measurements at He pressure of ~4 atm. which extend the t range to 0.1 GeV^2/c^2. At these higher t values, the α–particle momenta may be measured by magnetic deflection.

III. MEASUREMENT OF THE CROSS SECTION OF NEUTRAL STRANGE PARTICLE PRODUCTION IN 5.1 GeV/c π^-p INTERACTION
 by Yu. A. Budagov, V.B. Vinogradov, A.G. Volodko, V.P.Dzelepov G. Martinska, V.S. Kladnitsky, Yu. F. Lomakin, V.B. Flyagin, P.V. Shlyapnikov.

The experiment was made by using a propane bubble chamber. The results of measuring the cross sections of the reactions in which

Table 1 – Production cross-sections for π^--p interactions (a) 0-prong events, (b) 2-prong events.

Channel	π^- – meson momentum (GeV/c)					
	3.0 [5]	4.0 [6]	4.65 [7]	5.1 present work	6.0 [8]	10.0 [9]
$\Lambda^0 K^0$	31±14	93±14	40	26±8	41±4	26
$\Sigma^0 K^0$	86±25			17±7		
$\Lambda^0 K^0 \pi^0$	141±33	88±33	120	72±16		11
$\Sigma^0 K^0 \pi^0$				38±14		
$(\Lambda/\Sigma)^0 K^0 K^0 \bar{K}^0$	–	–	–	38±26		
$(\Lambda/\Sigma)^0 K^0$ + neutral's	110±28	88±31	90	19±44		72
$\sigma_{tot}(\Lambda/\Sigma + K^0 +$ neutral's)	368±58	269±47	250	214±25		
$K^0 \bar{K}^0 + n$	52±13	176±62	190	71±18		
$K^0 \bar{K}^0 n + \pi^0$	17±7			84±18		
$\sigma_{tot}(K^0 \bar{K}^0 +$ neutral's)	69±14	176±62	190	155±25		

Channel	π^--meson momentum (GeV/c)				
	3.0[5]	4.0[6]	4.65[7]	5.1 present work	10.0[2]
$p\pi^- K^0 \bar{K}^0$	19±6	51±14		21±9	
$p\pi^- K^0 \bar{K}^0$		20±20		7±7	
$\pi^+\pi^- K^0\bar{K}^0 n$		64±37		55±15	
$\pi^+\pi^- K^0\bar{K}^0 n$ + neutral				22±28	
$\mathfrak{G}$ tot.				105±21	
$K^+\pi^- \Lambda^0$	136±21	133±21	30	59±7	
$K^+\pi^- \Sigma^0$				14±7	
$K^+\pi^- \Lambda^0\pi^0$	96±16	93±17	130	70±12	
$K^+\pi^- \Sigma^0\pi^0$				33±15	
$K^+\pi^- \left(\Lambda/\Sigma^0\right)+m\pi^0$		45±23		49±19	
$\mathfrak{G}$ tot.	232±24	271±27	160	225±13	
$\pi^+\pi^- \Lambda^0 K^0$	72±15	195±21	240	64±12	34
$\pi^+\pi^- \Sigma^0 K^0$	86±19			30±17	
$\pi^+\pi^- \Lambda^0 K^0\pi^0$	60±20	63±26		66±12	53
$\pi^+\pi^- \Sigma^0 K^0\pi^0$				27±12	
$\pi^+\pi^- \Lambda^0 K^0\pi^0\pi^0$				25±18	129
$\pi^+\pi^- \left(\Lambda/\Sigma\right) K^0 \, m\pi^0$				31±42	
$\mathfrak{G}$ tot.	218± 33	258±30	240	242±27	

Cross sections in microbarns.

mainly neutral strange particles are produced in π^- proton interactions. The chamber having the working volume of 100x50x40 cm^3
and the 17 kG magnetic field[1] was exposed to the negative pion
beam with the momentum (5.1±0.1)GeV/c from the Dubna synchrotron.
The mean detection efficiency of a single gamma-quantum was 22%.
As many as 230,000 pictures have been treated with an average number
of 6 π mesons per picture. Zero-prong and two-prong events with one
or two V^0-particles were selected for the analysis. These events
were measured and treated according to the programmes of geometric
reconstruction and kinematic identification. Ionization measurements were also made.

The procedure developed for reaction channel identification
allowed the following degree of separation: (30-50)% for zero-
prong events and (70-90)% for two-prong events. The method of
identification allows the safe selection of interactions on quasi-
free carbon nuclei. The analysis shows that these contribute, for
the majority of reactions, about 20% of the total number of events,
which agrees well with the estimates given in refs. [3,4].

As seen from Table 1a and 1b, the use of propane bubble chambers
made it possible to obtain new rich material on strange particle
production and in addition to discriminate in most cases the reaction
channels with 3- and 4-neutral particle production. This could be
done successfully with hydrogen bubble chambers.

The conclusion is that the relative contribution of many-body
states tends to increase in some channels with increasing energy.

REFERENCES

1. V. F. Vishnevsky, V. I. Moroz, B. A. Shahbazian, JETP Pis'ma,
    v.5, 307 (1967), JINR Preprint P1-3169, Dubna, 1967.
2. Proceedings of the XIV Int. Conf. on High Energy Phys. p. 173,
    report of R. D. Tripp, Vienna, 1968, JINR Preprint E1-4022,
    Dubna, 1968.
3. Proceedings of the Int. Conf. on Elem. Particles, Lund, 1969
    (in press), JINR Preprint E1-4584.
4. V. A. Nikitin, A. A. Nomofilov, V. A. Sviridov, L. A. Slepets,
    I. M. Sitnik, L. N. Strunov, Jadernaya Fizika (USSR) 1, 183
    (1965); Phys. Lett. 22, 350 (1966).
5. N. N. Govorun, I. V. Popova, L. A. Smirnova, T. V. Ryltseva,
    V. A. Nikitin, A. A. Nomofilov, V. A. Sviridov, L. A. Slepets,
    I. M. Sitnik, L. N. Strunov, P.T.E. N 4,44 (1966).
6. W. Czyz and L. Lesniak, Private Communication, The method of
    calculation one can see in paper W. Czyz and L. Lesniak, Phys.
    Lett. 248, 227 (1967) and in Ref. (4).
7. J. Formanek and J. S. Trefil, Nucl. Phys. B3, 155 (1967).

8.   Yu. A. Budagov et al. Prib. i Techn. Exper. N 1,1964,61.

9.   V. S. Kladnitsky, V. B. Flyagin, Prib. i Techn. Exper. N 1, 1965, 24.

NUCLEAR STRUCTURE INFORMATION FROM PION ABSORPTION EXPERIMENTS

K. Chung, Michael Danos and Max G. Huber

University of Erlangen, Germany and National Bureau

of Standards, Washington, D.C.

The study of the absorption of π mesons by complex nuclei is particularly suited to reveal interesting and new information on nuclear structure properties. The reason for this is that a large energy, i.e. 140 MeV, and very little momentum are released and have to be absorbed by the nucleus. This is only possible via the high momentum components of the nucleon wave functions, or in r-space, by the short-range correlations between two nucleons. Presumably they arise from the short range part of the nucleon-nucleon interaction which causes a modification of the currently most successful model, the independent particle model. These effects in practice cannot be taken into account by a straight-forward application of shell model techniques. A more practicable way is the use of correlated Jastrow functions[1]; the correlated many-body wave function, $\hat{\Psi}$, is given by

$$\hat{\Psi}(1,2,\ldots,A) = \Psi(1,2,\ldots,A) \, \Pi f(r_{ik})$$

where $\Psi(1,2,\ldots,A)$ is the shell model wave function, e.g. a Slater determinant of independent particle wave functions, and $f(r_{ik})$ is the correlation factor. In order to study the short range corre-lation effects, it is useful to expand $f(r)$ in a Fourier-Bessel-series:

$$f(r) = 1 - \int dq \, w(q) \, j_0(qr)$$

The function $w(q)$ describes the distribution of momentum to be exchanged between otherwise independent particles.

In this formulation the quantity of interest to be determined by the experiments is $w(q)$. It has been shown[2] that the absorp-

533

tion rate of bound 1s and 2p pions in ^{16}O can be explained if the exchanged momentum distribution is centered between 250 MeV/c and 400 MeV/c. More detailed information could be gained from an analysis of the spectrum of the missing energy. Namely, it turns out that the pion is mainly absorbed by a pair of p-shell or s-shell nucleons; the corresponding ratio of the absorption rates varies between 3 and 1 depending upon the detailed form of $w(q)$ as long as it lies within the above range of q-space.

There are further experiments which will reveal detailed information on $w(q)$. For example, in the one-nucleon emission processes, the high momentum components of the nucleon wave functions are tested. These are, in turn, to a large extent determined by $w(q)$. Another such experiment is the quasi-deuteron effect induced by fast π-mesons; for example, the reaction (π^+, pp). As can be seen, the essential quantity here is the relative momentum Δk of the outgoing particles:

$$\Delta k = k_1' - k_2'$$

which because of momentum conservation is related to the momenta of the initial state particles by $\Delta k = k_1 - k_2 + k_\pi + 2\hbar q$. Since $\hbar q$ is much larger than any of the shell model momenta involved, Δk is essentially determined by the value of $\hbar q$. By judiciously varying the kinematic conditions of the experiment, one actually can vary Δk. In this manner one can obtain detailed information on the function $w(q)$.

This way we see that the presently available nuclear physics information makes a detailed use of π mesons as a tool of studying subtle nuclear effects very worthwhile.

[1] R. Jastrow, Phys. Rev., 98, 1479 (1955)

[2] K. Chung, M. Danos, M. G. Huber, Phys. Letters 29B, 265 (1969)

CHARGED PARTICLE EMISSION FOLLOWING PION CAPTURE IN NUCLEI[*]

L. Coulson, R.C. Minehart, and K. Ziock

University of Virginia

Charlottesville, Virginia

The absorption of a negative pion by a nucleus frequently results in the emission of one or several charged particles, an event that in nuclear emulsion gives rise to the occurrence of the well known pion stars. Rabin et al.[1] found that the singly charged particles emitted following pion capture in nuclear emulsions consist of protons, deuterons, and tritons. Protons and deuterons occur with approximately equal probability and account for most of the observed tracks. The measurement of Rabin et al. suffers from the well known shortcomings of the emulsion method, inability to identify the absorbing nucleus other than by classifying it as light (C,N,O) or heavy (Ag,Br), poor mass resolution for the emitted particles, and poor statistics.

We have stopped a negative pion beam from the 600 MeV synchro-cyclotron at the Space Radiation Effects Laboratory (SREL) of the NASA in Newport News, Virginia in thin targets made of ^{6}Li, ^{7}Li, ^{9}Be, ^{12}C, and ^{27}Al. The charged particles emitted following the capture of a pion were identified by radius of curvature in a 20"x20" bending magnet and by range in a 28 gap range chamber.[2] Particle trajectories were determined from the location of sparks in four sonic spark chambers, two on each side of the magnet. The four sonic chambers and the range chamber were triggered by a coincidence between a stopping pion and a pulse from a planar proportional counter, located between the third and fourth sonic chambers.

When the radius of curvature is plotted as a function of range, three distinct branches are resolved. We attribute the three

[*] Work supported by the National Aeronautics and Space Administration

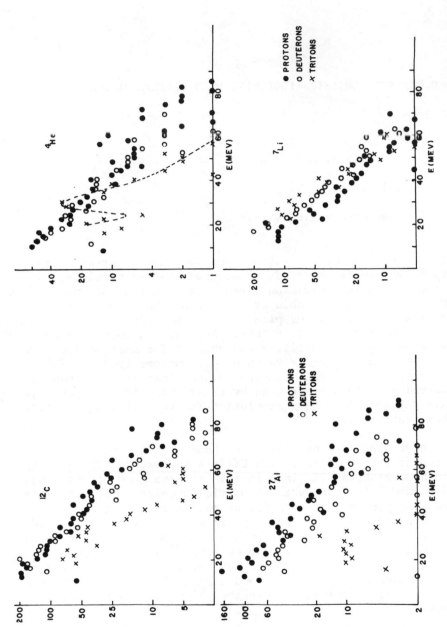

Fig. 1 - Energy spectra of protons, deuterons, and tritons emitted after pion capture.

branches to protons, deuterons, and tritons although with our present detection scheme a ^{3}He branch would be superimposed on the ^{2}H branch and ^{4}He would appear superimposed on ^{3}H. However, at a given point on these branches the helium isotopes would have considerably higher energy than the hydrogen isotopes. Specifically, the ^{3}H would have an energy 8/3 times that of the corresponding deuteron and the ^{4}He would have an energy 3 times that of the corresponding triton. Because of the higher energy of the helium isotopes at any point on the branches we think that the relative contribution from the helium isotopes is small. Furthermore, separate data that we have obtained from an E-ΔE system also indicate that the contribution of the helium isotopes is small.

Figure 1 shows the spectra obtained from ^{7}Li, ^{12}C, ^{27}Al, and from ^{4}He. The ^{4}He events, taken simultaneously with the other data, originated in the helium atmosphere surrounding the targets. The data from the other targets have been corrected for a helium background, with the subtraction being most significant for emission of tritons from the aluminum target. The bulk of the triton emission from ^{4}He appears to be contained in a peak centered around an energy of 30 MeV. This peak can be attributed to the two body final state reaction, $\pi^- + {}^4\text{He} \rightarrow {}^3\text{H} + n$. The triton from this reaction has a unique laboratory energy of 30.6 MeV, thereby establishing an absolute normalization of our energy scale.

The spectra have been corrected for a variation of solid angle with momentum, and for energy losses suffered by the particles travelling through the system. The spectra are shown for energy bins taken to be 2 MeV wide. The line drawn through the triton spectrum from ^{4}He is not a statistical fit but is only intended to guide the eye.

All the spectra, with the exception of the triton spectrum from ^{4}He, seem roughly exponential in shape. The fact that the deuteron spectra closely resemble the proton spectra and extend to energies of more than 40 MeV makes it unlikely that the deuterons are predominantly the result of nuclear boil-off reactions. Our results suggest the possibility that negative pions frequently may be absorbed by clusters of three or more nucleons.

---

[1]N.B. Rabin, A.O. Weissenberg and E.D. Kolganova, Phys. Letts. 2, 110 (1962).
[2]Similar to the one described by L. Coulson, W. Grubb, R.C. Minehart, and K. Ziock, Nucl. Instr. and Meth. 61, 209 (1968).

NEUTRONS FROM RADIATIVE PION ABSORPTION ON O^{16} and Si^{28} NUCLEI[†]

M.M. Holland, R.C. Minehart, and S.E. Sobottka

University of Virginia

Charlottesville, Virginia

Recently, much interest has centered on demonstrating that the excitation of giant resonances plays a dominant role in radiative pion absorption on complex nuclei.[*] After stopping in the target material, the pion is generally captured into highly excited, circular atomic orbits from which it cascades to lower orbits via Auger and radiative dipole transitions. For the light nuclei of interest, absorption occurs predominantly from the 2p atomic orbit accompanied by the emission of a high energy photon. One would expect the residual nucleus to be excited to the isobaric analogs of the giant resonance states of the capturing nucleus, which subsequently decay via neutron emission[1]. Observation of the energy spectra of the emitted particles would provide a means of studying these particular states. In addition, by exploiting the similarities and differences between this process, muon capture, and the electromagnetic excitation (both γ and (e,e'))of giant resonances in complex nuclei, one can hope to learn much about the structure of the giant resonance region.

We have measured the energy spectra of neutrons and gamma-rays emitted at a relative angle of 80° following pion capture on O^{16} and Si^{28} nuclei.[**] The experimental arrangement is depicted in Figure 1. Each of the beam telescope detectors, i.e. C1,2,3,4, and 5 and the charged particle anti-detectors A1 and A2 was plastic scintillator. A pion stop in the target was identified by a $C(123\overline{45})$. The water and silicon targets were $3gm/cm^2$ each. The

[†]Work supported by the National Aeronautics and Space Administration

[*]See additional references contained in reference 1.
[**]Work performed at the Space Radiation Effects Laboratory.

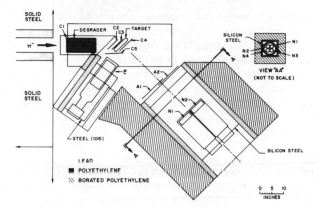

Figure 1: Schematic of experimental apparatus and layout.

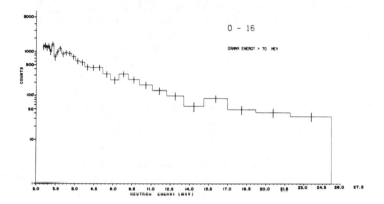

Figure 2: Energy spectrum of neutrons emitted following pion absorption on O^{16}.

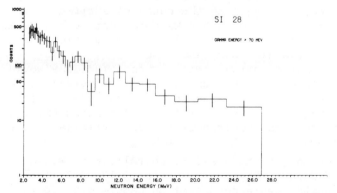

Figure 3: Energy spectrum of neutrons emitted following pion absorption on Si^{28}.

Cerenkov total absorption gamma-ray detector ($\hat{C}$) utilized an optically polished cylinder of Schott lead glass (critical energy=18.4 MeV, radiation length=3.06cm, density=3.60 g/cm^3, refractive index=1.62) which was five inches in diameter and twelve inches long. One end was viewed by a RCA4522 phototube, and the other surfaces were coated with a titanium oxide reflective coating. Energies were determined by pulse height analysis. Calibration of the detector yielded an energy resolution of 35% FWHM for 100 MeV gamma-rays.

The neutron detectors (N1,2,3, and 4) consisted of an array of four NE213 liquid scintillator cells, each 6 inches in diameter and two inches deep, and each viewed by a 58 AVP phototube. A zero-cross pulse shape discrimination system permitted substantial reduction of background due to gamma-rays. Neutron energies were deduced through time-of-flight measurements over a 41 inch flight path. An energy resolution of 1.3 and 1.6 MeV at 10 MeV was achieved for the silicon and oxygen targets, respectively. The neutron energy threshold was approximately 1 MeV. The resulting neutron energy spectra were corrected for detection efficiency[2] as a function of energy, and a random background, typically 12% and 5% for oxygen and silicon, respectively, was subtracted. All of the detectors were magnetically shielded by concentric cylinders of mu-metal and silicon or low carbon content steel, and shielded against background radiation with lead, and borated polyethylene, as indicated in figure 1.

The energy spectra for neutrons associated with gamma-rays of energies in excess of 70 MeV are displayed in figures 2 and 3. The data are plotted in arbitrary units of counts/MeV. The bin widths are one-half the FWHM resolution. The errors indicated represent statistical uncertainties only. The theoretical calculations for oxygen[1], based upon the generalized Goldhaber-Teller model of the giant multipole resonances, predict peaks centered at 4.5 and 8 MeV, arising from the decay of excited states of N^{16} to definite final states of N^{15}. We note that such structure may be present in figure 2. No analogous calculations have been performed for the Si^{28} nucleus. Therefore, little can be said about the expected energy spectra except that one would expect to see structure between 2 and 10 MeV attributable to giant resonance excitation. We see indications of peaks centered at 7.5 and 13 MeV which may result from direct neutron emission.[3] Better statistics would be required before any definite conclusions could be drawn.

[1]   J.D. Murphy, R. Raphael, H. Uberall, R.F. Wagner, D.K. Anderson, and C. Werntz, Phys. Rev. Letters, 19, 714 (1967).
[2]   R.J. Kurtz, Laurence Radiation Laboratory Report UCRL-11339, 1964 (unpublished).
[3]   D.K. Anderson, VPI, private communication.

THEORY OF RADIATIVE np CAPTURE

R.J. Adler,* B.T. Chertok, H.C. Miller

*Virginia Polytechnic Institute, The American University

The experimental np capture cross section for thermal neutrons is now measured to be $\sigma = 334.5 \pm .5$ mb.[1] This is in disagreement with conventional quantum mechanical calculations using the Bethe Longmire approximation with small corrections.[2] These result in $\sigma = 302.5 \pm 3.99$ mb. Several authors[3-5] have made dispersion theoretic calculations in the hope of resolving this 10% discrepancy. Their results are in agreement with each other, but it has been shown [2,3,6] that the dispersion result corresponds to a zero range deuteron wave function and an incorrect singlet effective range in the conventional wave function treatment. As such, the dispersion approach is not yet accurate enough to be compared numerically with the experimental result, and the nonrelativistic Bethe Longmire approach is favored.

Two attempts have been made to treat the meson exchange current shown in Fig. 1.

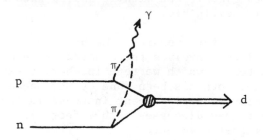

Fig. 1. Meson exchange contribution.

Skolnick[4] has applied a polology approach to a dispersion
calculation and obtains an increase of 2 to 3%. We find that an
attempt by Kaschluhn and Lewin[6] neglects crucial spin factors and
as a consequence gives no reliable quantitative results.

Using a phenomenological approach developed in a previous
paper an electrodisintegration[7] we have made a very simple calculation of the pion exchange effect which we believe is sufficiently
accurate to give meaningful numerical results. We obtain the
main contribution from Fig. 2;

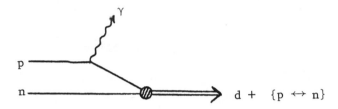

Fig. 2. The main contribution.

in the notation of references[2] the matrix element is, in agreement
with previous results,

$$M = \int_0^\infty U_g(r)U_S(r)dr = 4.00 \ f \tag{1}$$

To this the pion current of Fig. 1 adds a contribution which
agrees <u>numerically</u> with Ref. 4

$$M_{\pi\pi} = \frac{G^2}{48M\pi} \int_0^\infty U_g(r) \frac{e^{-m_\pi r}}{r} \left[1 - \frac{m_\pi r}{2} \right] U_S(r)dr = .05 \ f \tag{2}$$

This gives an increase of 2 or 3% in cross section: $\sigma = 309.5 \pm 5$ mb.
There remains a discrepancy of about 7%.

We have also calculated the next most relevant exchange
diagrams, similar to Fig. 1, but with $\omega\pi$ and $\rho\eta$ exchanged. The
effect is completely negligible.

A discrepancy therefore remains. We believe the most likely
source of more amplitude involves the consideration of excited
nucleons in the deuteron, which we term incoherent excitation
effects. Preliminary work has been done by Stranahan[8] on the 33
resonance, who obtains several percent increase in the cross
section. Kisslinger[9] has discusses such effects in connection
with the deuteron magnetic moment.

We feel that the processes represented by Figures 1 and 2
will yield no further light on the discrepancy, and that the

excited nucleons are now of primary interest.

[1]A. Cox, S. Wynchank, and C. Collie, Nucl. Phys. $\underline{74}$, 481 (1965).
[2]H.P. Noyes, Nucl. Phys. $\underline{74}$, 508 (1965).
[3]B. Sakita and C. Goebel, Phys. Rev., $\underline{127}$, 1787 (1962).
[4]M. Skolnick, Phys. Rev. $\underline{136}$, B1493 (1964).
[5]B. Boscoe, C. Ciocchetti, and A. Molinari, Nuovo Cimento $\underline{28}$, 1427
 (1963).
[6]F. Kaschluhn and K. Lewin, Nucl. Phys. $\underline{87}$, 73 (1966).
[7]R.J. Adler, Phys. Rev. $\underline{169}$, 1192 (1968) and erratum $\underline{174}$, 2169 (1968).
[8]C. Stranahan, Phys. Rev. $\underline{135}$, B953 (1964).
[9]L.S. Kisslinger, preprint.

LOS ALAMOS MESON PHYSICS FACILITY[*]

Robert L. Burman

Los Alamos Scientific Laboratory
University of California
Los Alamos, New Mexico

A medium-energy physics "Mecca on the Mesa" is currently under construction at the Los Alamos Scientific Laboratory in New Mexico. This facility consists of a high-intensity, 800-MeV linear proton accelerator and of the experimental halls and equipment to handle experiments utilizing the primary proton beam and various secondary and tertiary beams. Today, I will concentrate on the experimental facilities of LAMPF.

The experimental area of LAMPF is shown in Fig. 1. Leading into several experimental halls is a half-mile long linear accelerator, consisting of a 750-keV Cockcroft-Walton injector, a 100-MeV drift-tube section, and a final 800-MeV wave-guide section. Provision is made for the future addition of an isotope separator, a biomedical research facility, and a pulsed neutron research facility. Simultaneous use of experimental areas will be possible through the simultaneous accelerator of both H^+ and H^- beams. We expect to accelerate, in alternate phases, 1-mA H^+ and 100-μA H^- beams, and, additionally, to be able at a later date to accelerate a 60-600 nA polarized beam. Other beam characteristics will be: energy, variable from 100 to 800 MeV; energy spread, $\pm$ 0.4%; transverse phase space, $\pi/3$ mrad-cm; duty factor, 6% (500 μsec macropulse, 120 cps repetition rate); rf microstructure, 0.25 nsec pulses 0.5 nsec apart. In normal operation, the H^+ and H^- beams would be separated in the switchyard. The 1-mA H^+ beam would go into Area A, where up to three meson production targets could be used to provide a number of secondary and tertiary beam lines. The 100 μA H^- beam would be directed towards Areas B and C, where by successively stripping portions of the H^- beam, both experimental areas can operate simultaneously.

[*] Work conducted under the auspices of the U.S. Atomic Energy Comm.

The meson production targets in Area A will be covered by a
stacked steel and concrete shielding wall, approximately 15 ft high,
and 20 ft thick from the proton beam line. Secondary beam lines
will extend through the main shielding wall into separate experimental caves surrounded by individual stacked shielding. Radiation levels outside the stacked shielding will be kept below the radiation
worker tolerance level of 2.5 mrem/hr. In order to keep the floor
space relatively uncluttered, utilities for the secondary lines and
experimental equipment will be supported on overhead bridges. A
30-ton crane above the utility bridges will service the entire area.

Since we anticipate high residual radioactivity in the target
cells, on the order of 4 kCi for any individual component, we have
provided for remote handling of equipment in the target cell and the
immediate vicinity via a movable hot cell;[1] its operation is depicted in Fig. 2. After the hot cell is rolled into position above the
afflicted target cell, shielding blocks are rolled aside to provide
access. The hot cell is then lowered into position, allowing the
use of remote handling equipment.

We are anticipating operating several pion and muon secondary
lines simultaneously. The design of a low energy muon line to be
used primarily for stopped muons is close to completion.[2] As shown
in Fig. 3 it consists of a two-bend pion channel, a quadrupole alter-
nating-gradient pion-decay and muon collection section, and a muon
analyzer. The large spatial extent of the muon beam at the exit of
the analyzer will allow creation of two simultaneous beams by means
of septum magnets. The channel has been designed to achieve high intensity, high purity, large polarization, and a momentum range tun-
able from 0 to 250 MeV/c; details of the expected performance are
given in Table I.

Several pion secondary channels are under active consideration.
The performance characteristics of two are shown in Tables II and
III. The low energy pion channel (Table 2) is designed to operate
over pion energies from 0-200 MeV either with high resolution ($\Delta p/p$
= ± 0.1%) or high flux ($\Delta p/p$ = ± 5.0%). This channel is designed as
a vertical, four-bend channel with a 60° pion production angle, in
order to take advantage of the small vertical target dimension and
the reduced background at large production angles.[3] The high-energy
pion channel is designed primarily for elementary-particle experiments; it is a horizontal, two-bend channel with a 10° production
angle. Both of these channels utilize a 50-MeV thick pion production
target. In addition, consideration is being given to a medium-energy
very high intensity ($\sim 10^{11} \pi^+$/sec) channel, and to a high-energy
high-resolution channel and pion spectrometer.

As discussed earlier, we intend to operate two other experimental areas at reduced proton beam intensities, by the sumultaneous acceleration of a H⁻ beam. Area B, intended primarily for the study

TABLE I. MUON CHANNEL

Central Momentum	Forward Decay Momentum	Flux	Muons Pol.	Backward Decay Momentum	Flux	Muons Pol.
90 MeV/c	115 MeV/c	0.7×10^7	90–100%	50 MeV/c	0.6×10^7	50–80%
180 MeV/c				110 MeV/c	7×10^7	50–80%

Note: Flux in number/sec-MeV/c; 50-MeV thick Cu production target.

TABLE II. LOW ENERGY PION CHANNEL

(Ref: R. Burman and M. Jakobson, MP-6/RB,MJ-1)

	$\Delta p/p = \pm 0.1\%$	$\pm 1.0\%$	$\pm 5.0\%$
π^+ Flux:			
T_π = 500 MeV:	1.5×10^7/sec	0.6×10^9	1.5×10^9
100 MeV:	3×10^7	1.2×10^9	3.5×10^9
150 MeV:	2×10^7	0.8×10^9	2×10^9
Solid Angle:	5 msr	20 msr	10 msr
Beam Spot:	1.5 x 4 cm	3 x 4 cm	6 x 8 cm
	30 x 80 mr	60 x 160 mr	80 x 120 mr
Length:	14.4 meters		

TABLE III. HIGH ENERGY PION CHANNEL

(Ref: P. Gram)

	$\Delta p/p = \pm 0.5\%$	$\pm 5.0\%$
π^+ Flux:		
T_π = 550 MeV:	4×10^6	8×10^7
400 MeV:	9×10^7	2×10^9
200 MeV:	2×10^8	3×10^9
Solid Angle:	1.6 msr	3 msr
Beam Spot:	1.6 x 1.5 cm	13 x 7 cm
	40 x 20 mr	75 x 24 mr
Length:	21 meters	

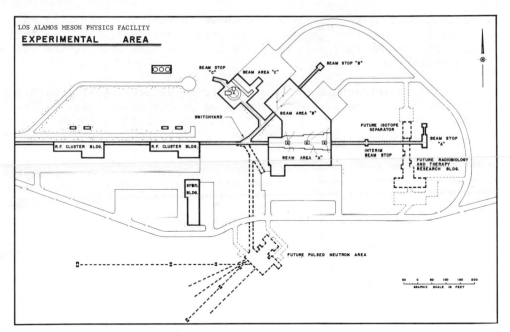

Figure 1.   LAMPF experimental facilities.

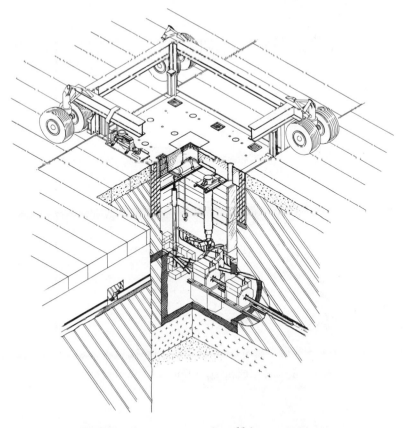

Figure 2.   Remote handling system.

of nucleon-nucleon interactions, is shown in Fig. 4. Provision is
made for the production and use of polarized and unpolarized neutron
beams (produced by charge-exchange on deuterium) as well as the use
of the primary proton beam.

Area C is intended to handle nuclear structure research with
the proton beam as well as other types of proton experiments.[4] The
layout of Area C is shown in Fig. 5. It is designed to house the
High Resolution Spectrometer (HRS) on one side, and to handle a variety of possible experimental setups on the other side. The HRS
would occupy the left half of the building, whose 40-ft x 80-ft
floor space would allow 0-180° coverage. A rail system in the floor
would guide the spectrometer about a pivot. Water and power would
be led to the pivot through a floor trench; signal calling could use

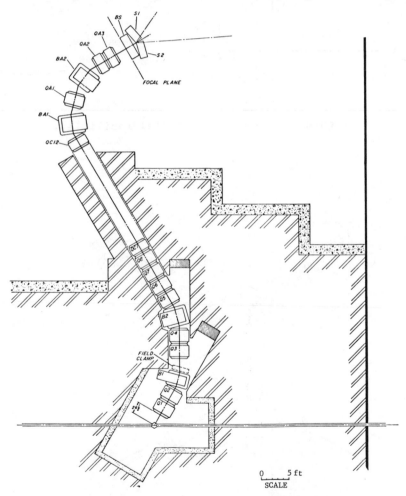

Figure 3. Stopped muon channel.

either the floor trench or an overhead tray. The floor area to the
right of the beam line would be available for other experimental
setups not using the HRS. In order to accommodate a variety of experimental layouts, and still keep the size and cost of the building
within reasonable bounds, a large section of the wall on the right
side is open. This section would be filled in with movable shielding blocks, as indicated, so as to enclose a particular experiment.
The HRS itself will be described in detail later in this session by
H. A. Thiessen.

REFERENCES

[1] M. T. Wilson, IEEE Transactions on Nuclear Science, Vol. NS-16,
P. 588 (1969).

[2] V. S. Hughes, et al., "Stopped Muon Channel, Design Status,"
July 2, 1969, Los Alamos Scientific Laboratory Internal Report.

[3] R. Burman and M. Jakobson, Los Alamos Scientific Laboratory
Internal Report MP-6/RB,MJ-1, June 1969.

[4] "LAMPF High Resolution Proton Spectrometer Status Report,"
July 18, 1969, Los Alamos Scientific Laboratory Internal Report.

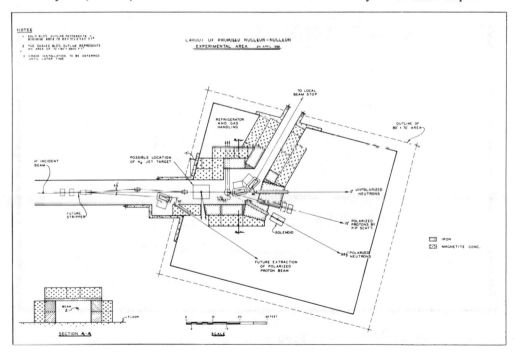

Figure 4.   Area B.

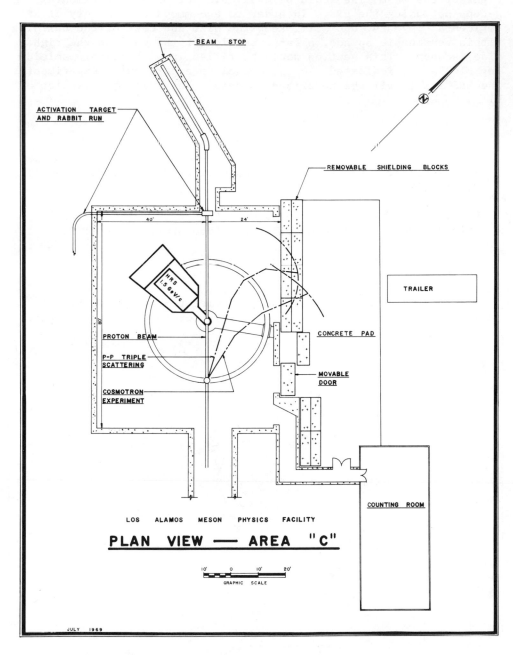

Figure 5.   Area C.

THE 700 MEV HIGH INTENSITY PHASOTRON (Design Status)

V. P. Dzhelepov

Laboratory of Nuclear Problems, Joint Institute
for Nuclear Research, Dubna, USSR

Nobody now doubts that the design and construction of high
intensity proton accelerators for energies below 1 GeV, the so-
called "meson factories" will provide us with a new powerful tool
by using which many acute problems of present day nuclear physics
as well as of some other fields of science can be tackled.

In this connection the scientists from the Laboratory of
Nuclear Problems in collaboration with those from the Research
Institute Of the State Atomic Energy Committee are developing a
conversion design of the 680 MeV synchrocyclotron with an output
current of 2.3 micro-A into a small "meson factory" with a proton
beam average current up to 50 micro A at 700 MeV[1]. The employ-
ment of more effective extraction and focusing systems in new conditions promises an increase of meson beam intensities of 50-100
times, and external proton beam intensities 200 times greater than
the ones available at present.

The main idea of conversion is to transform our synchrocyclotron to the relativistic cyclotron[2] under development. The new
accelerator employs the space variation magnetic field increasing
along the radius. Designs based on this method are also being
developed at some laboratories in the USA: the Columbia University[3], the Carnegie Institute of Technology[4].

The application of the above said field results in dynamic
stability of particle motion and ensures the reduction of r.f.
voltage frequency variation (earlier one of 25.5-14.3 MHz, to the
new one of 18.18-14.41 MHz) and thus to simplify the conditions
for frequency variator operation. The expected intensity increase
for the phasotron with magnetic field variation compared to an

ordinary synchrocyclotron is described according to estimations as
follows:

$$i_{oph} = i_{os} \sqrt{\left(\frac{V_{ph}}{V_s}\right)^3 \cdot \frac{K_{os}}{K_{oph}}} \qquad . \qquad (*)$$

Here V is the amplitude of the accelerating voltage, while K = 1

$$K = 1 - \frac{n}{1+n} \cdot \frac{1}{\beta^2}, \quad \overset{\circ}{n} = \frac{\tau}{H} \cdot \frac{dH}{d\tau} \quad \text{and} \quad i_{oph}, \ i_{os}$$

are currents in the central region of the field variation phasotron and the ordinary synchrocyclotron, respectively. The formula
(*) shows that intensity is increased by increasing frequency modulation (which is proportional to the accelerating voltage amplitude) and by extending the phase stability region which is proportional to $(V/K)^{1/2}$. The current of 50 micro-A can be obtained
with V = 50 kV (instead of V = (12-15) kV), with voltage frequency modulation F = 500-600 Hz instead of available 100-110 Hz
and the parameter $K_{0\ ph}$=0.32 instead of $K_{0\ s}$=1.8.

As has been shown earlier[1] the required magnetic field variation is produced with iron shims cut in an Archimedian spiral
and the law of average magnetic field increase is ensured by the
proper profile of pole tips and the change of the angular width
of spiral shims.

The r.f. system of the new accelerator is a half-wave homogeneous line tuned with the rotating capacitance frequency variator. In order to determine the final parameters of the accelerator its magnetic system is modelled (the magnet scale 1:5.2,
see Fig. 1) as well as the r.f. system (originally approximately,
by using the 1:4 scale and soon by employing a natural size model
together with the models of frequency variators). The results obtained by using the magnet model satisfy the requirements for
variation. However, the parameters will be determined more precisely in order to obtain the required average magnetic fiels. To
reach the law of the average field variation satisfying the conditions of good particle focusing and the variation of the accelerator central region is especially difficult. This difficulty is

aggravated by the fact that the
reduced (comparing to ordinary
synchrocyclotrons) value of the
parameter "K" starts to essentially affect the phase process of
the first phase oscillation and
results in strict tolerance for
the mean value of the field. For
the sake of avoiding some unexpected complications in the accelerator working chamber three pairs
of current coils at radii R≤100 cm
are planned to be included.

Fig.1 Magnetic system model.

The characteristics of the magnetic field H_z, H_r and H_ϕ are measured with precision magnetometers developed at our Laboratory: NMR[5] (their accuracy is 0.3-0.7 Oe with 0.5-24 kOe fields); and the Hall unit[6] (their accuracy is (0.02-0.05)%, whereas for H_z and H_ϕ it is 1.10^{-3}%).

The results of measurements are analyzed by using the computer. The "current" and "field" stabilization of the magnetic field will be made simultaneously. "Current" stabilization is made by using an electronic stabilizer to ±0.03% accuracy. "Field" stabilization is performed by employing the system based on electron paramagnetic resonance[7] to a 10^{-4} accuracy.

In simulating the magnetic field, mechanical forces affecting various parts of the magnetic system were determined. For this purpose magnetic fields separated by 3 cm from the appropriate surfaces of these parts were measured. Two important circumstances were established as a result: 1) magnetic forces press the chamber lids to magnet poles, 2) the forces are so strong that they completely eliminate pressure on the lids produced by the atmosphere in vacuum pumping. This, in turn, eliminates lid downwarping in the working state.

Both internal and external targets are proposed for producing secondary beams. Also about 50% particle extraction efficiency is planned in the new accelerator.

The latter problem is quite difficult and delicate to solve in our conditions. A design of the particle extraction system is developed which is based on the principle of parametric resonant swing of radial oscillations by two local inhomogeneities of the magnetic field (a peeler and regenerator) separated by an angle of about 60° (Fig. 2). The phasotron with an average magnetic field increasing with radius makes it possible to control self-oscillation frequencies at full radii and thus to effectively employ the regenerative method of particle extraction from the chamber at frequencies $Q_r = 1.05 - 1.09$.

In designing the frequency variators, since the phasotron average current is proportional to the ratio of acceleration time to modulation period, a variator was designed where this ratio is large (about 0.75). The number of variators is 2, the number of rotor turns is 3600. The model test showed that voltage across the variation rotor over the whole frequency range is smaller than the accelerating voltage. It is 30 kV at low frequency (Fig. 3). The change of the variator capacitance is close to the calculational one. The capacitance of the collector condensor $C_c >$ $10 . C_{max.var.}$. Voltage distribution along the dee accelerating edge is nearly uniform (reduction near edges is smaller than 10%).

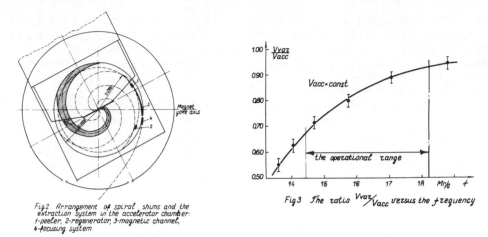

Fig.2 Arrangement of spiral shims and the
extraction system in the accelerator chamber:
1-peeler, 2-regenerator, 3-magnetic channel,
4-focusing system

Fig.3 The ratio V_{vaz}/V_{acc} versus the frequency

The electric field strength between the variator plates with $V_{acc.}= 50$ kV is 8 kV/mm. The voltage across the rotor shaft does not exceed 3 kV. A self-excited oscillator with GU-65A tubes of 500 kW at 30 MHz is used for the r.f. system power supply. The oscillator is coupled with the resonance line by two parallel feeders and two coupling variators.

The system of 20 magnetoprobes at the rotor shaft serves as a time device for controlling the acceleration cycle and maintaining the specific modes of accelerator operation (accumulation, single mode, etc.).

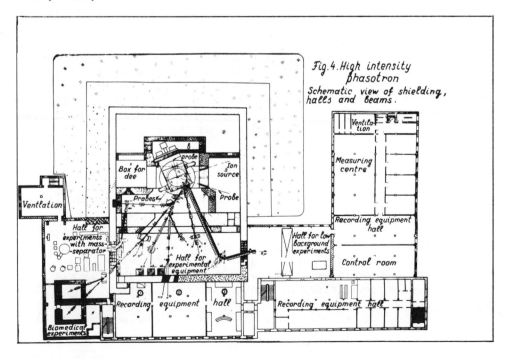

Fig.4. High intensity
phasotron
Schematic view of shielding,
halls and beams.

The schematic view of the accelerator, beam layout, radiation shielding (including the additional one), experimental areas, etc., is shown in Fig. 4. At the present time a considerable amount of working drawings have been prepared and the manufacture of huge installations is supposed to start in the near future.

References

1. A.A. Glazov, Yu.N. Denisov, V.P. Dzhelepov, V.P. Dmitrievsky et al. Proc. of the VI Int. Conf. on High Energy Accelerators. Cambridge, USA, p. 303 (1967); Preprint 9-3951, Dubna, 1968; Atomn. Energ. 27, 1, 16 (1969).

2. M.A. Gashev, A.A. Glazov, Yu.N. Denisov, V.P. Dzhelepov, V.P. Dmitrievsky et al. Proc. IV Int. Conf. on Accelerators, Dubna, USSR (1963), M.(1964), p. 547;
   V.P. Dzhelepov, V.P. Dmitrievsky, B.I. Zamolodchikov, V.V. Kolga. Usp. Fiz. Nauk. 85, Issue 4, 651 (1965).

3. R. Cohen, J. Rainwater. IEEE Trans. on Nucl. Sci. No. 5-13, 4, 552 (1966).

4. M.H. Foss. IEEE Trans. on Nucl. Sci. No. 5-13, 4 (1966).

5. Yu.N. Denisov, S.A. Ivashkevich. Izmer. Techn. 2, 56 (1968).

6. D.P. Vasilevskaya, Yu.N. Denisov, N.I. Dyakov. Prib. i Techn. Exper. 5, 203 (1966).

7. Yu.N. Denisov, S.A. Ivashkevich, V.V. Kalinichenko. Prib. i Techn. Exper. 1, 158 (1966).

DISCUSSION

V. P. Dzelepov (in reply to H. L. Anderson): To deal with the increased level of radioactivity, we shall install remote controlled manipulators for changing chambers, D's, etc. A special channel will be installed for radiospectrometry, and this will provide some shielding. It will be possible to transfer the target to a special shielded vault for chemical processing.
V.P.D.(in reply to J. Rainwater): The shutdown for the present cyclotron is expected to be in mid-1972.

THE SIN RING CYCLOTRON PROJECT

STATUS REPORT

J.P. Blaser, H.A. Willax, H.-J. Gerber, SIN group *)

SIN - Schweizerisches Institut für Nuklearforschung
(Swiss Institute for Nuclear Research)
Neunbrunnenstrasse 85 - 8050 Zürich

Abstract

A two stage isochronous accelerator for proton-beams of at least 100 µA at 585 MeV is under construction at the Swiss Institute for Nuclear Research at Zurich. As previously described it consists of a sector focussed isochronous cyclotron for 70 MeV, acting as injector, and an isochronous ring cyclotron.

A prototype of the C-shaped sector magnets and of the r.f. accelerating cavities have been tested successfully. Accelerator components are being manufactured. Building construction work has started and installation of the machine will begin in 1971.

1. Description of the SIN-Accelerator

As a device for the production of a high intensity beam of protons with an energy well above the π-production threshold we proposed in 1962 a combination of an AVF-cyclotron (70 MeV) with a sector focussed ring cyclotron (500 MeV), both operating isochronously.

*)H. Baumann   W. Hirt   R. Reimann   J. Zichy
  B. Berkes   W. Joho   G. Rudolf
  L. Besse   P. Lanz   Th. Schaub
  H. Braun   H. Oschwald   U. Schryber
  H. Frei   A. Paulin   H. Stüssi
  H.J. Gerber   Ch. Perret   M. Werner

The reference design which was used for development of the
basic machine components has been described formerly [1,2].

In 1967 the specifications of the injector cyclotron were
changed from a fixed energy to a multi-particle variable
energy machine, whereas the ring device remained at fixed
energy. The injector cyclotron is being manufactured by
Philips, Holland.

1.2 The Ring Cyclotron

The 70 MeV proton beam is injected into the ring device
whose basic characteristic features are advantageous for
high intensity beams of intermediate energies:

- separated magnets with small gaps producing a low
  average field and rather strong axial focussing,

- a rather large energy gain per revolution, provided by
  separated r.f. cavities for very high voltages.

In such a device the conditions for a good general control and quantitative extraction of a CW-beam of high
quality are fulfilled.

The SIN ring cyclotron has 8 C-shaped magnets of $\sim 18^{\circ}$
azimuthal width. The pole gap decreases from 9 to 5 cm
over a radial range from 200 to 460 cm. For reasons of
easier and efficient machining, the pole contours were
chosen to be circular arcs, providing a spiral angle of
$\sim 32^{\circ}$. The pole field increases radially from about 15 -
20.6 kG for a final energy of 585 MeV. Sets of low power
pole face windings will provide the necessary field corrections. The stainless steel vacuum chamber is directly
joined to the magnet poles by a flexible welding joint.
There are different sections of the vacuum chamber containing probes, collimators, injection and extraction
devices.

Four cavities of 40 cm width in beam direction, 500 cm
radial length and 330 cm height, with an accelerating
gap of 15 cm provide voltages of about 500 kV each at
50 MHz, which is the 6th harmonic of the particle revolution frequency in this machine. They are excited by
individual 250 kW r.f. power stages jointly driven by a
highly stabilized master oscillator. Voltage and phase
control are provided in the circuits.

The beam is injected in the mid-plane through a 90° bending magnet followed by a magnetic injection channel,
and brought onto equilibrium orbit by a correction

channel. Since there is a complete separation of the orbits at injection radius, beam loss is negligible.

With about 1.5 MeV energy gain per revolution near extraction radius the radial gain is 6 mm, increasing to
~ 8 mm at the extraction point due to the turnover of
the magnetic field. Incoherent radial beam amplitudes
will be on the order of 3 - 4 mm in this region. For an
extraction device consisting of an electrostatic channel
of 50 kV/cm with a 0.1 mm septum (120 cm long), a magnetic focussing element 45° downstream and an extraction
septum magnet 90° downstream, extraction rates above 90%
have been computed.

The vacuum chamber is partly stainless steel, partly
aluminium. The individual sections can be joined either
with metal gaskets or inflatable bellows carrying radiation resistant elastomeres. Since oil contamination of the
surfaces in the r.f. cavities seem to limit the maximum
voltage achievable, a combination of turbo-molecular
pumps with titanium sublimators and ion-getter pumps,
directly connected to the cavities is used.

2. Status of the Project

2.1 Magnet design

After extensive studies on 2 magnet models 1 : 5 scale
in connection with computations on beam stability, a
full scale prototype of one sector was built [3].

After the first machining of the poles, which was done
by Brown, Boveri Company Switzerland, the theoretical
field has been achieved within $\pm$ 3 o/o at nominal excitation. With this field configuration the final energy
was 525 MeV as earlier planned.

Since there was still time to attempt a more ambitious
program it was decided early this year to try a pole gap
machining for 585 MeV final energy. Simultaneously
another set of 1 : 5 scale measurements was started for
this final energy. The first results show that the necessary field profile is achievable. Using the same sector geometry for this higher final energy, the beam has
to pass twice the non-linear x z-coupling resonance
$v_r = 2 v_z$. However, with extensive numerical computation
it has been shown that in our case there should be only
an increase in the z amplitude of 10% in the worst case.

2.2 r.f. Design

The development of the r.f. cavities and the power stages
went through the work on 1 : 5 scale models, a full scale
working model operating in vacuum and, finally the prototype cavity excited by a home-built power stage delivering
60 - 80 kW at 50 MHz [2]. The first r.f. power system for
250 kW is under construction at AEG-Telefunken, West Germany, and will be delivered next spring.

The prototype cavity is a welded construction of 20 mm
aluminium sheet with casted supporting ribs for mechanical rigidity. The inner surfaces are treated by rolling
with polished steel rolls. By this method it was possible
to reduce the r.f. power loss to about 70% of the originally expected value. The Q value of this cavity turned
out to be 32000. External cooling is used.

With a combination of a 2000 m^3/h turbo-molecular pump,
a titanium sublimator and an ion-getter pump of total
10000 l/s effective pumping speed at 10^{-6} torr we can
reach vacua of 1 x 10^{-6} torr after a few hours of pump-
down. After a few weeks pumping the pressure came down
to 6 x 10^{-8} torr.

The cavity has been baked in and was operating at about
400 kV. It was interesting to notice the X-ray level
being down by a factor of 10 compared to the first working model at equivalent voltage.

A careful theoretical investigation of the problem of
beam loading was carried out in cooperation with the
Accelerator Group of Karlsruhe. It has been shown that
there is no stability problem as long as the beam power
per cavity does not exceed about 1/4 of the r.f. power
loss in the cavity. If a fast amplitude and phase regu-
lating system can be applied, the beam power could be
carried beyond this limit.

2.3 Mechanical Design

The most serious problem was considered to be the joint
of the vacuum chamber with the spiralled magnet poles.
The 60 cm thick pole itself is a part of the chamber wall
and has to be tightly connected to the stainless steel
chamber wall. This is done by means of a double-collar
of thin stainless steel sheet, one part welded to the
pole, the other to the chamber wall. A full scale prototype of such a section has been built and successfully
tested.

The design of injection and extraction elements as well
as beam probes and collimators has started.

2.4 Control System

It is planned to have a small digital computer as an aid
for operating it will be mainly used for automatic log-
ging of data in the second phase for parameter settings
and limit control. In the third phase of operation it
possibly could enter as an active element into a part of
the control function.

2.5 Buildings and Main Installations

The laboratory will be built in Villigen, 35 km northwest of Zurich, close to the Swiss Federal Institute for
Reactor Research (EIR), on the river Aare. See figure for
general layout. The main parts will be:

2.5.1 the experimental hall with 85 x 48 m floor space
   and 18 m height. A 60 ton crane spanning the whole
   width will service the area from 12 m height. The
   hall contains the 2 accelerators in vaults at the
   north end. The shielding walls of the accelerator
   will be partly cast concrete, partly movable blocks.
   The roof shielding consists of removable concrete
   beams. The experimental area available outside the
   vaults is 2700 m^2.

2.5.2 the operations building adjoining the northeast part
   of the experimental hall. It contains control room,
   counting rooms, offices and workshops (a total of
   ~ 2000 m^2 useful area).

2.5.3 the service building adjoining the northwest part
   of the experimental hall. It contains general utilities, the cooling system, the main power conversion system and a small workshop for special purposes (a total of 1500 m^2 useful area).

2.5.4 the laboratory building, situated about 70 m northwest of the main hall. It contains a total useful
   area for offices and laboratories of about 1800 m^2.
   Further buildings of this type can be added later.

A bridge over the river Aare will connect the SIN laboratory with the labs of the Swiss Federal Institute for
Reactor Research, allowing many services to be operated
in common. Building construction work has started. In
connection with the design of the building the main power

distributions and the cooling water circuits have been
designed. There is a total of 10 MW electric power and
6 MW cooling power available in the first phase of operation. This capacity can be expanded. Cooling water is
taken from the river Aare.

3. Secondary beams

Secondary beams will be produced essentially on external targets. For low intensity special beams (e.g. test
beams) in internal target in a field free sector may be
interesting. The attached table gives the calculated
characteristics of the expected beams. They are based on
100 μA extracted current. In addition it is planned to
use a polarized ion source on the injector in order to
obtain polarized proton beams at 585 MeV.

The macroscopic duty-cycle of 100% is advantageous whenever the experiment is intensity-limited by the microstructure of the beam, or at intensities, where the microstructure is unimportant. The accelerator gives one
bunch every 20 ns or 5.10^7 bunches/sec. As an example, a
medium resolution π^- -beam ($\Delta E/E = 4\%$) carries about
2.10^7 pions/sec. These pions arrive nicely separated.
($P (> 1$ per bunch$) = 6\%$) The same applies for a typical
polarized neutron beam.

For some types of experiments, the variation of the length
in time of an rf-bunch will be useful, with a slight loss
in intensity we can have a spillout time of the bunch as
big as 2 ns, or - at about 30% of the intensity - as
short as 1/2 ns, while one gets about 1 ns at normal
operation.

4. Layout of the hall

To reserve maximum flexibility, accelerators, external
secondary beam production targets and experiments are
inside a large hall. (See figure.) The primary proton
beam shall not leave the heavy shielding so that biological tolerances with beam on can be maintained in the hall
for most experiments. The shielding around experiments
is movable and serves to reduce background. Mostly no
roofs should be needed. Most of these beams will be derived from 3 external targets; these are: a thick target
for high intensity secondary beams (middle of the hall),
a thin target in the same beam line mainly for pion beams
with high resolution, and a target using a reduced intensity proton beam or for a high resolution spectrometer

NUKLEONENSTRAHLEN

Teilchen	Energie (ΔE) [MeV]	Intensität [sec^{-1}]	Bemerkungen
p	585 ($\pm$ 0.7)	6.10^{14}	100 µA
p↑ 40%	580 ($\pm$ 3)	3.10^{10}	C_{12}, 6 gr/cm^2 $10°$, 3.10^{-3} sr
n↑ 33%	450 ($\pm$ 30)	1.10^7	D_2, 3 gr/cm^2, $27°$ 2.10^{-5} sr, 20 µA
n "Monoenergetisch"	580 ($\pm$ 2)	3.10^8	D_2, 1.7 gr/cm^2, $0°$ 2.10^{-4} sr, 20 µA

MESONENSTRAHLEN

Teilchen	Energie (ΔE) [MeV]	Intensität [sec^{-1}]	Bemerkungen
π^+	150 ($\pm$ 3)	$2 \ .10^8$	100 µA, C_{12}, 9gr/cm^2 6.10^{-3} sr, 10 m
π^-	150 ($\pm$ 3)	$2 \ .10^7$	
μ^-	60 ($\pm$ 10)	$1.5.10^8$	Auf 100cm^2, Solenoid 10 m, 15 cm Ø, 50 kG
μ^-	stop	$2 \ .10^5 \ (gsec)^{-1}$	
μ^+	ca. 4–8 × mehr je nach Energie		

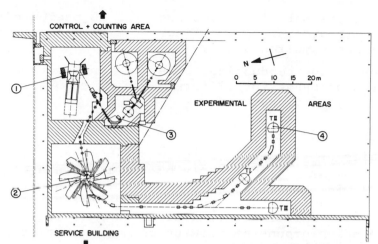

General Layout - 1) Injector cyclotron with variable energy, 2) Ring
cyclotron for 580 MeV protons, 3) Transport system for beams with
variable energy, including 110° analyzing magnet, 4) Production
targets T I, II, III for secondary beams.

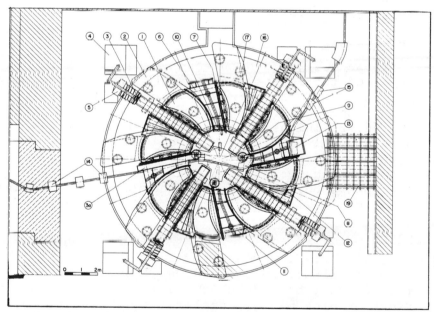

Ring Cyclotron - 1) Sector magnet pole, 2) Sector magnet yoke, 3)
r.f. accelerating cavity, 3a) 3rd harmonic cavity, 4) r.f. amplifier,
5) vacuum pumps, 6) vacuum chamber, 7) inflatable seal, 8) main injection magnet, 9) secondary injection magnet, 10) injection inflec-
tor, 11) extraction deflector, 12) compensating channel, 13) extraction
septum magnet, 14) injection beam optics, 15) extracted beam optics.

facility. The hall is dimensioned for 10 medium sized experiments sitting on the floor with several of them able to run simultaneously.

A probably superconducting, solenoid-type μ-channel and several pion beams for different energy ranges and resolution are being designed.

5. Research program

The choice of a research program out of the many possibilities is determined mainly by the interests of the users and the specific capabilities of the accelerator. The SIN, although being a national laboratory to be used by the Swiss universities, will welcome collaboration with research groups of foreign countries. Up to now the interests of the future users have been focussed on

- work with stopped μ's (μ-atoms, muonium, μ-capture),
- nucleon-nucleon and nucleon-nucleus scattering, especially neutron work,
- pion-nucleon and pion-nucleus interaction,
- biological and medical research with pions.

6. Time-schedule

A regrettable delay has occured in decision-making for the choice of the injector cyclotron. As a consequence construction of the buildings has seriously started only early 1969. though the project was fully funded since spring 1966. Buildings should be ready to allow installation of equipment and accelerators in the middle of 1971. The injector is scheduled to deliver beam in the course of 1973. The ring cyclotron and experimental support installations should be ready by that time so that experimental exploitation should start end of 1973.

References

[1]   H.A. Willax, J.P. Blaser
      Progress Report on the 500 MeV Isochronous Cyclo-
      tron Factory of ETH-Zurich. V International Confe-
      rence on High Energy accelerators, Frascati, Sep-
      tember 1965. Proceedings page 413.

[2]   J.P. Blaser, H.A. Willax
      IEEE, NS 13, Vol. 4, 194, 1966
      Proceedings of the International Conference on
      Isochronous Cyclotrons.

[3]   P.A. Tschopp, H.A. Willax
      Design of the Guiding Magnet for the ETH 500 MeV
      Isochronous Cyclotron.
      International Conference on Magnet Technology,
      Oxford, 1967.
      Proceedings page 262.

DISCUSSION

J. P. Blaser (in reply to H.L. Anderson): (1) The total cost is
$22 $\times$ 10^6; about $3 $\times$ 10^6 for the injector, and $8 $\times$ 10^6 for the
main ring.   The remainder is for buildings and facilities; (2)
The practical limit of this type of "C.W." machine design is the
proton rest-mass ($\sim$1 geV).

V. Telegdi et al: Comments on relative cost, reliability and
flexibility of superconducting and ordinary channels.

TRI-UNIVERSITY MESON FACILITY

J. B. Warren

Director, TRIUMF, Vancouver, Canada

In April 1968 a group of four Universities in Western Canada started to build a meson workshop and facility for intermediate energy physics and chemistry. This centres around the construction of a six sector A.V.F. cyclotron designed to accelerate 100 μA of H^- ions to an energy of 500 MeV or 400 μA to 450 MeV.[1] The final beam may be extracted with essentially 100% efficiency by stripping off the two electrons by passing the H^- ions through a thin foil, so the resulting protons swerve out of the machine. By varying the radial position of this extractor foil, the energy of the extracted protons may be continuously varied from 150 MeV to 500 MeV. There are six valleys in the magnet, so the possibility exists of extracting six (or even more) beams simultaneously, but we propose to start with provision for two. Being a fixed frequency cyclotron, the macroscopic duty cycle is 100%; at the outset the energy resolution of the beam is expected to be ±500 kV and the RF phase acceptance ±30°, giving a micro-duty factor of 7 nsec in 44 nsec.

H^- DISSOCIATION

The binding energy of the second electron in the H^- ion is only 0.755 eV. In moving through a magnetic field, it is subjected to an electric dissociating force $E(MV/cm) = 0.3 \, \gamma\beta B(kG)$ which tends to remove one electron and leave a neutral atom. For a given energy or velocity β, this sets a limit on B_{max} that can be used in the cyclotron, since there is a limit on the number of neutrals as these produce radioactivity in the machine. These neutrals spill out in a fan-shaped beam from the ions circulating in the outer orbits, for essentially all the stripping occurs in the 450–500 MeV range in the hill regions. In addition, there is some stripping from collisions with residual gas atoms.

Our criterion, based on experience at other machines, is that it should be possible to service the machine with a total beam spill of 20 µA at 500 MeV, about 4 µA equivalent coming from gas stripping if the pressure is kept to 10^{-7} Torr, and 16 µA from electric stripping. These neutrals activate a band about 2 cm wide in the 0.25" thick wall of the vacuum vessel at the median plane, and would result in a residual radiation field 24 hours after shutdown, following prolonged running, of 2 Rad/h at a distance of 1 metre from the tank wall and 540 mRad/h at the centre of the cyclotron,[2] which figures have to be compared with a permissible dose rate of 100 mRem/week. After 30 days cool-off the 540 mRad/h has only dropped to 290 mRad/h. Obviously, with remote handling procedures, and carefully conceived servicing, one might learn how to live with higher levels of activation, but it is this residual level which essentially sets the upper limit to the intensity of the available beam. With this loss of 10 kW in beam power we expect to extract a 100 µA beam at 500 MeV.

The theory of the dissociation gives an exponential dependence of lifetime of the ion

$$\tau \simeq \frac{10^{-14}}{E} e^{49/E} \text{ sec}$$

where E(MV/cm) is the electric dissociating field.[3,4] To settle the parameters with the precision required for the design of the magnet, a quite direct measurement of the lifetime has been made[5] at the same $\nabla \wedge \bar{B}$ products it was expected to use at TRIUMF, using 50 MeV H^- ions accelerated in the P.L.A. at the Rutherford Laboratory together with magnetic fields of the order of 20 kilogauss. The maximum field has been set at 5.76 kG, based on these measurements.

THE MAGNET DESIGN

The design has been carried through with a series of 1 : 20 scale models, energized by two single coils linking all the sectors, the field being measured every 1° azimuthally and 1/8" radially with a Hall probe. At the end of March 1969 a shape, which would give a focussed isochronous beam out to 500 MeV on the full-scale machine at a radius of 312" with a 6% stripping loss, had been established, giving a final magnet weight of 4200 tons. It is this very large size which poses most of the design challenge, along with the problems of radiation shielding and activation. Cross-section and plan views are shown in Fig. 1.

It is expected that the final magnet, made mostly of 3" plate, will be within 2% of the performance of the model but a 1 in 10 scale model is now being built to check the scaling and decide the

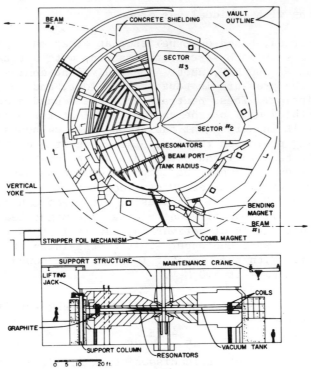

Fig. 1   Cyclotron Cross Sections

shimming procedure. The energizing coils for each sector will be
fabricated of rectangular cross-section aluminum bar 18" x 3/4" and
welded together on site. To achieve the precise field pattern
required, 54 trim coil pairs are being provided, together with 12
harmonic coil pairs at different radii for each of the 6 magnet
sectors.

RF SYSTEM

The RF system adopted is an unusual one; the dees are in the
form of $\lambda/4$ resonant cavities, and to keep down the size the RF is
operated at the 5th harmonic of the ion rotation frequency, i.e.
at 23 MHz so that $\lambda/4$ is about 10 feet.

One dee is to be fed by a single loop joined to the transmitter
by a resonant line $5\ \lambda/2$ long. The coupling between the top and
bottom resonators is very tight because of the filling in of the
sides by the flux guides and this also makes the voltage distribution along the edges very uniform. The coupling between dees is
also so tight as to make a separate loop for each unnecessary. The
ions circulate in the 4" gap between the top and bottom resonators;
the dee-to-dee voltage has to be maintained at 200 kV, the power

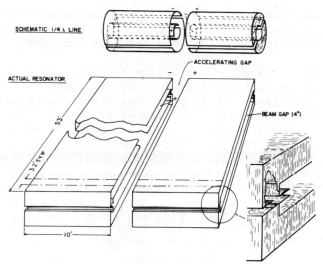

Fig. 2   RF Resonator System

required being in the vicinity of 1.4 MW. The structure involves
no insulators; it is to be fabricated from a silver aluminum roll-
bond sheet in which water channels are formed. For tuning, adjustable flaps on the ground plane at the high voltage end will be used.
The system has been checked for Q and uniformity of voltage in a
series of models, the last being half-scale of one dee.

In order to achieve a wider phase acceptance, it has always
been the dream of cyclotron designers to flat top the RF by introducing higher harmonics. With this kind of resonant structure,
model tests show that it is possible to introduce 3rd harmonic via
a second loop appropriately placed. By introducing 16% amplitude
of 3rd harmonic of the RF, the phase acceptance should increase
from 60° to 90°.

INJECTION AND CENTRAL REGION

Ehlers[b] in 1965 built a hot filament, Penning plasma arc H⁻
ion source with extraction perpendicular to the field, which
delivered 5 mA with a gas flow rate of 30 cc/min. A commercial
version of this is now available which should deliver the 2 mA we
require within the emittance $\pi/4$ $\Delta x \cdot \Delta \theta_x$ of 42 mm mrad at an energy
of 300 keV, the planned injection energy. The ion source will be
placed in a room outside the shielding with a straight horizontal
run to the centre of the machine above the shielding, thence
through a 90° bend to the centre where it will be inflected by an
electrostatic deflector. Electrostatic quadrupoles will be used in
this beam transport system: the residual magnetic field above the

cyclotron roof shielding is low enough to be handled by conventional screening and steering. The 45 ft vertical section to the centre of the machine provides adequate room for beam bunching.

Focussing in this central region is dominated by the electric field. The magnetic focussing is relatively weak ($v_z \simeq 0.2$). Extensive calculations are in progress; these show that a 60° wide phase band can be accepted. In view of the great importance of the central region for achieving a good beam quality, we are constructing a full-size central region model. It will be the largest lowest energy cyclotron ever! 2.5 MeV 60" diameter. It will enable the RF performance at full voltage of 2 resonator sections to be checked as well as the beam optics of the central region. While relatively costly, it is hoped to shorten the schedule of the project by getting the central region tested in this way.

VACUUM SYSTEM

The huge size of the vacuum pill-box introduces mechanical problems both from the 2700 ton force squeezing it, and from the expansion which it undergoes when heated, which makes the 56 ft diameter seal a difficult problem. Sky hooks and ground hooks, together with a central pillar of high-strength stainless steel, hold the shape of the top and bottom plates. Polyurethane or viton O-rings should withstand the irradiation for 6 to 12 months at a time, but a more permanent simple solution to the seal problems is still elusive. For servicing the interior, the whole top assembly of magnet, top lid and upper resonator may be raised 42" by jacks.

To achieve a pressure of less than 10^{-7} Torr to keep the gas dissociation below 4%, it is proposed to use a cryogenic pumping system using helium-cooled pipe at 20°K mounted inside a liquid air-cooled surface. This will be augmented by some 10" diffusion pumps to remove the hydrogen load. The main decision required here is whether the heat load will be small enough to be handled by Stirling engines or whether a turbine is necessary, which only comes in larger sizes. The reliability of such refrigerator plants, particularly of the turbine, is rather unknown.

THE FACILITY

Civil Engineering

At the outset two beam lines are being provided, one feeding a Meson Area and terminating in a target arrangement providing a slow neutron facility, the second feeding current, up to 10 µA, to a

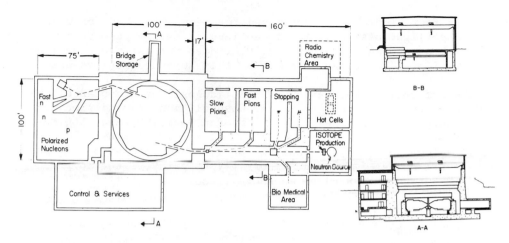

Fig. 3   Typical TRIUMF Beam Distribution

Proton Area; both areas will be used for the construction of the
cyclotron. Practically all shielding will be removable blocks so
it should be possible to change beam lines without much difficulty.
The cyclotron will be on its own pad in a skeletal concrete box;
the 16 ft of concrete shielding on the two sides through which the
beams emerge will be on a separate pad. The experimental floor is
20 ft below grade so as to benefit from earth shielding as much as
possible, the beam height 4'6" above this. The set up area for the
Local Control Rooms is at grade level, above the experimental floor
instead of to one side as in most accelerators.

Two 50-ton cranes each with two 25-ton hooks which can be syn-
chronized will cover the whole area. In addition, each experimental
region may, as part of the experimental equipment, have a small 2-
ton crane for local work, eliminating the need for removing the
roof members for most purposes. The main cranes will be used to
transfer lead flasks containing very active targets or cyclotron
components to the hot cells or storage areas.

Meson Area, Targets and Secondary Beams

The first target is likely to be 4 g/cm^2 of liquid paraffin or
water, providing plenty of hydrogen and hence high flux of π^+, as
well as carbon or oxgyen for a good yield of π^- via (p,n). A Be
target may be an alternate, since the flux of π^- from it is 70%
more for the higher forward momenta pions. The use of a large
volume of circulating liquid reduces the residual activity, and no
heat transfer problems arise. This target will feed two channels:

(i)   a fast pion channel optimized for pions $\sim$150 MeV for which the long flight path required to achieve an energy resolution of ±100 keV at 200 MeV, or $\Delta p/p < 3.10^{-4}$, does not introduce serious attenuation. The design parameters it is hoped to achieve are:

Take off angle <20°, perhaps 0°		Solid Angle of Acceptance 5 msr	
Channel Length	15 m	Channel Pion Transmission 0.4	
Pi Energy	200 MeV	Proton Beam 100 μA at 500 MeV	
π^+ Flux	$4\ 10^7$ sec^{-1} MeV^{-1}	π^- Flux 1/10 π^+ Flux	

(ii) a lower energy pion channel optimized for pions $\sim$50 MeV using a take-off angle of 70° and a short flight path for the $\Delta p/p$ <1.2 10^{-3} needed to achieve an energy resolution of +100 keV at 50 MeV;

Channel Length	5 m	Pi Transmission 0.5
π^+ Flux	$1\ 10^7$ sec^{-1} MeV^{-1}	π^- Flux 1/10 π^+ Flux

The second target, 20 g/cm , most likely will be of dense graphite, but other possibilities are Be or a heavy, dense metal. There is some uncertainty as to how to produce the maximum number of very slow pions; this target would feed 3 channels:

(i)   A stopping pion channel as short as possible to reduce muon contamination with bends to remove neutrons and optimized to give as large a stopping rate as possible within small amounts of isotopically pure material.

(ii) A muon channel optimized to give the largest flux of stopping muons in a small target, as there is less interest in a fast muon channel at TRIUMF. Design efforts at present are centred on a helical channel, i.e. a long quadrupole with the pole tips rotated so as to describe a helix; a rather simple configuration which can be made out of stampings is shown in Fig. 4; its transport qualities are currently being examined.

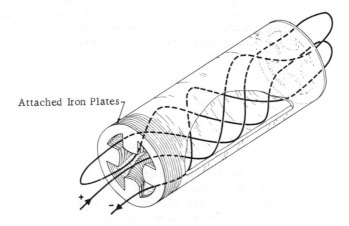

Attached Iron Plates

Fig. 4   Helical Muon Channel

(iii) A π^- channel of large aperture, maximum flux and good purity, energy-variable from 30-110 MeV, feeding a radiobiological and medical therapy area.

The third target is a beam dump, a slow neutron facility and an irradiation facility. The target will be lead bismuth eutectic cooled by light water: the moderator will be heavy water surrounded by graphite.

Proton Area

The configuration in this area is again flexible, and it is expected that some of the targets and set-ups will change quite frequently. The whole floor can take 17 ft of shielding and at the outset it is expected to shield an area 40 ft x 24 ft in which there may be

(i) A scattering chamber for reaction studies with particle identifiers and time-of-flight coincidence, using targets ∿100 keV thickness.

(ii) A target for the production of a beam of polarized protons by scattering at 15° from liquid hydrogen.

(iii) A target for production of fast neutrons by scattering from liquid deuterium at 0° and of polarized fast neutrons by scattering at 27°.

(iv) A proton irradiation chamber before the beam dump.

(v) A chamber for studying, for example, π production and for looking for rare events.

FURTHER DEVELOPMENTS OF THE FACILITY

Beam Lines and Experimental Areas

While it will be possible to provide up to six external beams, since maximum beam can only be provided to one target at one time the use of more of the pions from this thick target may be one of the first developments. The provision of a high resolution proton beam of low intensity is also high on the list of priorities for intermediate energy structure physics, and the qualities of this beam will decide the parameters of any high resolution magnetic spectrometer for particle analysis that may go into the P Area.

Cyclotron Performance

With the present design, if all trim coils are fully powered and the RF frequency raised 3%, it may be possible to run a small beam at higher energy without exceeding the activation limit; this would provide a useful extension of the maximum pion momentum available.

The current maximum, set by the maximum output of 2 mA of the H^- ion source, the matching of the emittance of the source to admittance of the cyclotron (a 15% phase acceptance factor) and the transmission of the quadrupoles (90%), amounts to 270 μA. By bunching, a factor of two increase can reasonably be expected. By adding 3rd harmonic to the RF, the phase factor should increase to 26%. With a pressure 1 10^{-8} in the tank so that gas stripping is negligible, and an electric dissociation of 6% at 500 MeV, for the same cyclotron activation a beam of 330 μA could be accelerated to 500 MeV, while 900 μA would be the limit imposed by the ion source at 450 MeV. These challenging figures, however, ignore the problems of the stripper foil. The two stripped electrons go round back into the foil owing to the magnetic field, and it looks as if the maximum power such a foil can radiate away is in the region of 250 watts.

The energy resolution[7] of the raw beam is set by the amplitude of the radial oscillations, which is governed by the injected admittance, first harmonic field errors and energy gain per turn. Calculation gives for the energy spread of the 500 MeV raw beam using a wide stripper foil, i.e. with a width greater than two turn separations so that the whole beam is intercepted, $\Delta E = \pm 0.60$ MeV. Low energy defining slits placed at 70" radius, 15 MeV energy, reduce the emittance, microduty factor and intensity but also reduce ΔE. These slits will be built in from the start. With such slits set at 0.025" ΔE amounts to ± 0.10 MeV at 500 MeV which seems to be a very useful energy resolution at the expense of a $\pm 2°$ phase acceptance.

Separated turn acceleration is possible with such slits only at the expense of a very narrow phase acceptance and very tight tolerances on average field, RF frequency and voltage. However these tolerances are much alleviated if the RF wave form is flat-topped by admixture of 11.3% of 3rd harmonic as shown below:

	Fundamental Only	Fundamental Mixed with 11.3% 3rd Harmonic
Energy Resolution	± 50 keV	± 25 keV
Average Magnetic Field	1/3 10^6	1/4 10^5
RF Frequency	1/6 10^6	1/10^6
Resonator Voltage	1/10^4	1/2 10^4
Phase Acceptance	$\pm 0.5°$	$\pm 7°$

The tolerances in the righthand column are well within the present state of the art.

Other schemes have also been proposed including extraction of H$^-$ and energy dispersion outside the machine; this possibility exists with the current TRIUMF design. The potentialities both in intensity and energy resolution are thus extremely attractive: we would welcome a simple scheme for achieving a 100% microduty cycle!

REFERENCES

1. E.W. Vogt and J.R. Richardson. "An H$^-$ Meson Facility", IEEE Trans. on Nuclear Science, NS-13, (4), 262, (1966).

2. I. Thorson. "Shielding and Activation in a 500 MeV H$^-$ Cyclotron Facility", TRI-68-4, (1968).

3. J.R. Hiskes, Private Communication.

4. D.L. Judd, Nucl. Instr. & Methods 18, 70 (1962).

5. G.M. Stinson et al. "Electric Dissociation of H$^-$ Ions by Magnetic Fields, TRI-69-1 and Nucl. Instr. & Methods (to be published).

6. K.W. Ehlers, IEEE Trans. Nucl. Science NS-12 (3), 811, (1965)

7. J.R. Richardson, "Energy Resolution in a 500 MeV H$^-$ Cyclotron", TRI-69-6.

DISCUSSION

J.B. Warren (in reply to question by V.P. Dzelepov): Gas stripping of the H$^-$ ions should be no more than 4% of the total, for a pressure of 3-4 $\times$ 10^{-8} Torr. This is based on experimental measurements up to 50 MeV, and theoretical extrapolation to 500 MeV.

INDIANA 200 MeV CYCLOTRON PROJECT

Summary (by Editor) of a Report by:

M. E. Rickey

University of Indiana

The 200 MeV accelerator in course of development is a variable energy machine intended for the acceleration of a variety of particles from protons to "heavy ions". It employs separate sector magnets (like S.I.N.), but in a pure "Thomas" configuration, for which the radial oscillation frequency is independent of amplitude.

The equilibrium orbit is a rounded square, rather than a circle, reflecting the four pole magnet geometry. There is no "middle" to this main accelerating ring. There are large open "valleys" providing space for R.F. (27-35 MHz.), injection, and beam extraction.

The injector is in many respects, a one-third scale model of the main accelerator. It too has no "middle" so that the ion source is external and can be different for different ions, including the possibility of a Tandem Van der Graaff for heavy ions.

The large accelerator magnets weigh some 1800 tons in all, and the main ring is about 40 feet across the diagonal. The injector ring is one-third size linearly.

Expected Performance

Protons	:	15 MeV to 200 MeV (nominal).
Neg. H ions	:	Up to 160 MeV (ion dissociation limit).
Heavy ions	:	Up to Z^2/M MeV, with Z/M from one (proton), down to one-eighth.
Good beam optics:		(1/2 m.m. - millirad. phase-space).
Energy spread	:	Between $1/10^3$ and $1/10^4$.
Orbit separation:		4 mm. 100% beam extraction.
Current	:	Max. 50 microamp. Space-charge limit.

<u>Cost</u>   : About $M5 for the 200 MeV machine, (about $M3 for
          100 MeV).

<u>Status</u> : Building just begun.   Completion expected in 1970.

DISCUSSION

<u>M. E. Rickey</u> (in reply to a question): The energy limit for this
particular geometry is 240 MeV.   This is a dynamical limit set by
a resonance at that energy.

ELECTRON ACCELERATORS

THE SACLAY LINEAR ACCELERATOR (ALS)

A.M.L. MESSIAH

Department of Nuclear Physics, Saclay

France

This is a short report on the electron linear accelerators of the new generation, i.e. medium energy, high intensity and high duty cycle. At present, there are two projects of this sort under way, one at Saclay, the other at MIT. The Saclay linac (ALS) has started operation this year and is in the preliminary stage of utilization for physics. The MIT project is by now under construction and should be ready for work within a couple of years. In this talk, I will concentrate on the Saclay project, with only passing references to MIT ; but one should bear in mind that the two projects are quite comparable in size and capability, and that they will cover essentially the same new field of physics.

The primary purpose of these linacs is to probe the high energy and short distance regions of the nucleus with the electromagnetic interaction, thus to be able to perform precision experiments with real or virtual photons of energy-momentum ranging from 100 to 500 MeV or more. This leads one to energies ranging up to 500 MeV, say, for the primary electron beam, together with order of magnitude improvements in the other beam characteristics, notably in intensity and duty cycle.

But these linacs may serve another quite different purpose. Their electron beams of high intensity turn out to be very good sources of slow pions and muons, in fact much better than those in operation at present, and, hopefully, not too far behind those to be expected from the forthcoming meson factories.

In short, these linacs are fit to probe the nucleus either with high energy photons (virtual and real), or with slow pions and slow muons.

MAIN CHARACTERISTICS OF THE SACLAY (ALS) AND MIT LINACS

They are summarized in the two following tables

ALS

- Traveling Wave -- 3000 MHz
- 30 Accelerating sections, each about 6 m long, powered by 15 klystrons of 60 kW per unit
- Length of beam pulse -- 10 μs
- Operates at 1% or 2% duty cycle
- Expected performance of electron beam[*] :

Duty cycle	Repeti- tion Rate (Hz)	Average Power of Klystrons (kW)	Performance					
			Unloaded E_{max} (MeV)	100kW of Beam		Max. beam power		
				E_{max} (MeV)	$\bar{I}$ (μA)	E (MeV)	$\bar{I}$ (μA)	$\bar{P}$ (kW)
1%	1000	40	640 (550)	590 (500)	170 (200)	420	800	330
2%	2000	40	450 (380)	420 (340)	240 (300)	300	1120	330

- Energy spread of the primary beam : $\Delta E/E \lesssim 10^{-2}$
- Expected performance of positron beam[*] :

Duty cycle	E_{max} (MeV)
1%	550 (475)
2%	390 (325)

Current of about 0.5 μA in an energy band of width $\Delta E/E \simeq 2 \times 10^{-2}$.

[*] The values quoted are taken from the observed performances of the components of the linac, with indication in parenthesis below of the values warranted by CSF, the constructor of the machine.

MIT

. Traveling Wave -- 2856 MHz

. 23 Accelerating sections, five about 4 m long, and 18 about 8 m long, powered by 10 klystrons of 80 kW per unit

. Length of beam pulse -- 5 - 15 μs

. Operates between 2% and 6% duty cycle

. Expected performance of electron beam :

Duty cycle	Repetition Rate (Hz)	Average Power of Klystrons (kW)	Performance		
			Unloaded E_{max} (MeV)	60 kW E_{max} (MeV)	of Beam $\bar{I}$ (μA)
2%	5000	80	430	400	150
6%	5000	60	215	200	300

. Energy spread of the primary beam : $\Delta E/E \lesssim 0.4\%$

Contrary to the MIT project, the ALS has been designed from the start to accelerate positrons, in order to produce beams of monochromatic photons, following the method proposed long ago by Tzara. Apart from this, the two projects are quite similar, although MIT put somewhat more emphasis on duty cycle whereas Saclay aimed at somewhat greater energy and intensity. In any event, the performances of both linacs represent a considerable improvement in duty cycle and beam intensity over all existing equipment, as can be seen from the survey diagram of Fig.1 .

ALS - RESULTS OF FIRST OPERATION

The Saclay linac has started operation this year. A rather comprehensive description of the machine together with an account of its results of first operation can be found in Dr. F. Netter's report at the Yerevan Conference (7th Int. Conf. on High Energy

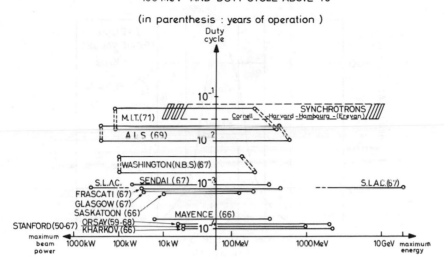

Fig. 1

Accelerator, Aug. 27th - Sept.2nd 1969). Let us here summarize the
most significant results.

Electron Acceleration.   For electron acceleration, the ALS
is found to be a quite good machine.

Various tests have been conducted either at low repetition
rate (6.25 Hz), or at the regular rates of 1,000 Hz and 2,000 Hz ,
i.e. full duty cycle but with a mean beam power voluntarily limited to 100 kW . The observed performances in energy and intensity
are in agreement with the calculated values quoted in the table
above.

The other characteristics of the beam - emittance, definition
in energy, stability in time - are found quite satisfactory, in
fact often beyond expectation, and make this machine well fit for
precision work. Typically, an emittance of 34 mrad.mm has been
found in a 500 MeV run. The definition in energy is found in all
cases significantly better than 1% FWHM . Fig.2 gives a typical
energy spectrum observed during a test run. Stability of the beam
is such that its mean energy stays constant to well within 1% in
one day of continuous operation, as illustrated in the example of
Fig.3 .

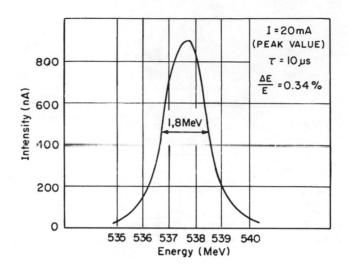

Fig. 2 - Typical energy spectrum of the electron beam

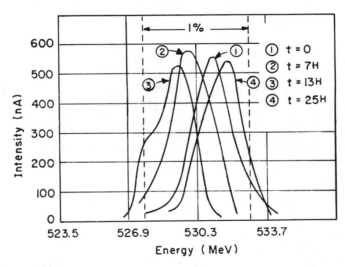

Fig. 3 - Evolution of the energy spectrum during a typical 25 hours run

Positron Acceleration.   The positrons are produced by impact
of the primary electron beam on a tungsten target located right
after the 6th section, and accelerated in the 24 remaining sections.
Adequate magnetic focusing of the positrons along the axis of the
machine is provided by solenoids half of the way, and by triplet
of quadrupoles located between sections in the second half.

Positron operation has proved hard to adjust, and should not
yet be considered completely through with its youth disease. The
maximum average current today detected in the experimental room
at 1% duty cycle is 0.2 µA for 466 MeV positrons with a spectrum
1.4% FWHM . It is obtained with 25 kW average power on the conversion target and corresponds to an efficiency for c^+/e^- conversion
close to 10^{-3} . Operation at lower energy is less efficient. Nevertheless, a current of 0.05 µA has been lately obtained at 217 MeV
within a band width of 1% , and is considered good enough to start
experimental work with monochromatic photons. Improvements are
expected.

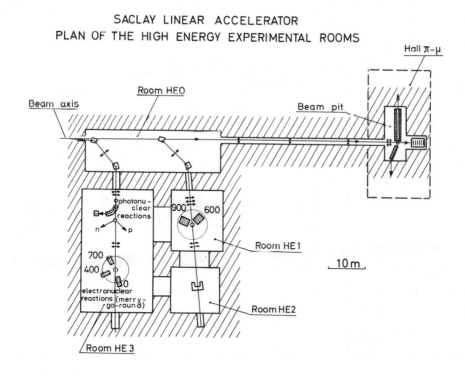

SACLAY LINEAR ACCELERATOR
PLAN OF THE HIGH ENERGY EXPERIMENTAL ROOMS

Fig. 4

ALS - EXPERIMENTAL FACILITIES AND PROGRAM

The accelerator and its experimental facilities are under ground. The accelerating sections are in a vault, 200 m long, beam axis about 6 m below ground level. The beam can be switched after the 12th section into a smaller experimental room for "low energy" work, or continue all the way up to the main experimental area. The plan of the latter is given Fig.4, at first a set of four rooms separated by concrete walls 4.80 m thick (Rooms HE 0,1,2,3), then after a tunnel 40 m long a pit housed by a hall at ground level to be used for pion and muon work.

Experiments with photons.   Either bremsstrahlung photons or, preferably, the monochromatic beam from positron annihilation will be used. The "low energy" room is best suited for photons below 100 MeV . For higher energy, the experiments are to be conducted in room HE 3 where an experimental set up for (γ,p) experiments is accomplishing test runs at present. (γ,np) and $(\gamma,\pi^{\pm})$ experiments are contemplated later on.

Experiments with electrons.   The accelerator is ideally fit for precision scattering experiments with electrons, either without coincidence, or in coincidence with another particle - i.e. (e,e'X) X being $\pi^{\pm}$, p, d or even α .

Two experimental stations are prepared, one in room HE 1 for very high resolution spectroscopy, the other in room HE 3 more flexible and with less resolution.

The station in HE 1 will comprise two very big spectrometers, the so-called "600" and "900" rotating around the same vertical axis. Their characteristics are compared in the Table below with those of other big magnets in use at present for electron spectroscopy.

	Stanford "72"	Orsay "500MeV A"	Orsay "1.3 GeV"	NBS	ALS "900"	ALS "600"
Radius (m)	1.8	1.1	2.17	0.75	1.8	1.4
P_{max} (MeV/c)	1 040	600	1 300	250	900	600
Resolution($10^{-3}\times$)	2.5	8	3.5	0.3	0.2	1.5
Solid angle (msr)	15.6	8.5	4.4	7	5.5	7
Acceptance $\frac{\Delta p}{p}$	10%	7%	5%	2%	8%	40%

Magnet "900" is fit for (e,e') reactions at a resolution of 2×10^{-4}. The ensemble "900"+"600" is fit for (e,e'X) reactions - notably (e,e'p) - at a resolution of 2×10^{-3} . The two spectrometers are now under construction. The station should be operational by mid-71 .

The station in HE 3 is a "merry-go-round" of three different spectrometers, whose characteristics have been chosen to fit a wide range of experiments, the so-called "400", "700" and "0" . "400" is specifically designed for detecting scattered electrons up to 450 MeV/c with resolution 10^{-3} . "700" can detect up to 700 MeV/c with a similar resolution and greater momentum acceptance, in view of heavier particle detection. "0" has been suitably shaped for detection of particles scattered very close to 0° . (e,e') experiments with the "400" should be ready to start this fall. The station completely equipped with its three magnets should be operational next spring, permitting in particular a series of (e,e'p) experiments at good resolution.

Pions and muons。   A target for producing pions can be placed in the beam pit shown on Fig.4, from which pions or/and muons can be collected and transported up to ground level in the π-μ experimental hall. The pit is large enough to house two, even perhaps three, such transport systems. At present, one pion channel has been designed and should be operational by mid-70 .

In the energy range from 30 to 120 MeV, the intensities of pions or muons to be expected are at least one good order of magnitude larger than those obtained on existing machines, which makes the ALS - and the MIT linac as well - eligible members of the meson factory club. As an illustration of this point, one can see on Fig.5 the result of calculations of the pion rates to be expected from proton and electron machines under the same (somewhat arbitrary) channel conditions. Primary beam intensity and energy have been chosen in keeping with existing projects, LAMPF and Nevis (broken line) on the one hand, ALS and MIT linac on the other.

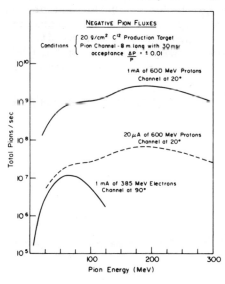

Fig.5 - Typical fluxes of π^{-} to be obtained under the same channel conditions from proton and electron accelerators

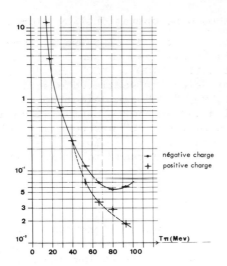

Fig.6 - Relative intensity of e^- to π^- and e^+ to π^+ to be expected in the forthcoming ALS pion channel

Experiments on pion production have been performed at Saclay, preparatory to the pion channel design. They do confirm the calculations on pion rates. Better, they show that it is actually possible, under suitable conditions, to obtain such rates, while keeping the electron contamination at an acceptable level (e.g. 10% at 50 MeV). In the actual design at ALS, the 400 MeV electron beam hitsfirst a bremsstrahlung target of 0.3 radiation length, then a carbon target 2 cm long (i.e. 3.8 g/cm^2) for pion production. The pions are collected at 120° within 16 msr and with $\Delta p/p = 5\%$ maximum momentum acceptance in a channel 8 m long. At 70 MeV, i.e. the maximum of the spectrum, the expected rate of π^- in the 5% range is 2×10^6/sec., assuming 600 µA in the primary beam. Fig.6 gives the e^- (or e^+) contamination to be expected at the end of such a channel as a function of pion energy. Although not specifically designed to this end, the same channel can give muons at respectable rates, notably for experiments with stop muons : in 2 g/cm^2 of stopping material, say, it will be possible to stop muons at a rate of 10^5/sec. .

THE PRINCETON-PENNSYLVANIA 3 GEV PROTON ACCELERATOR AS A NUCLEAR TOOL

Milton G. White

Princeton-Pennsylvania Accelerator

Princeton University

INTRODUCTION

I would like to describe the Princeton-Pennsylvania Accelerator as a tool for nuclear research. Some people tend to think of the PPA as being too high in energy for nuclear research and, therefore, regard it as only an elementary particle research tool. While much of our work has, in fact, been related to the study of mesons, especially K-mesons, and to the study of the invariance principles, such as C, P, T, about 15% of the research has been either purely nuclear, or with strong nuclear overtones. As examples of the purely nuclear, I mention spallation reactions at 3 BeV and the study of the resulting neutron deficient nuclei. On the other side of the valley are the baryon stable, neutron rich, light fragments, ^{19}N and ^{21}O, which have been discovered in confirmation of Garvey's mass formula. Negative pion scattering by twelve different complex nuclei from Be to Au over the energy range 600-1600 MeV has been used to determine optical model potentials. Neutron total cross sections over the momentum range 500-2800 MeV/c are now being studied for a wide variety of target nuclei, including the Physical Review! An especially beautiful experiment, begun as a missing mass search in the reaction $p+d \rightarrow He^3+X^0$ not only studied things of interest to elementary particle physics, the so-called ABC region, but simultaneously, and inadvertently, provided a hadronic probe, the proton, at correlation distances far smaller than ever achieved before. The distances probed ranged down to 10^{-14} cm, yet He^3 fusion still occurs, albeit with a cross section of only 10^{-34} cm^2. I fully expect this trend towards the joining up of nuclear structure and elementary particles to continue and to accelerate. My object in talking here today is to make better known the properties of the PPA which are of value to those whose main interest lies in nuclear structure itself. It is also my belief that the fields of nuclear physics and elementary particles, which have for some years been undergoing fission, are now ripe for fusion. When this happens, I believe that both fields will experience a great enlightment.

This paper will be divided into three sections: (1) PPA Facilities and Properties Now; (2) Major Improvements Now Under Way; (3) A Proposed Heavy Ion Improvement Program.

PRESENT PPA FACILITIES

Basically the PPA is a 3 BeV proton synchrotron which cycles at 20 Hz. At the moment our proton beam current is 5×10^{10} protons per pulse or 10^{12} protons per second, maximum, limited by space charge near injection time. The internal beam spills on a platinum or beryllium target 6x6x25 mm, the spill duration being 8 msec over a rounded sine wave top. The resulting energy variation of 10% is time correlated permitting an energy resolution of 1 MeV.

There are five secondary beams of π^+, π^- and neutrals currently being fed by this target. They come off at 13°, 20°, 34°, 49° and 90°. Because of our nanosecond bunch structure it is easy to use time-of-flight techniques after momentum analysis for separating and identifying various particles, as shown in Figure 1. Data obtained in this manner for various particles, over a range of momenta, are presented in Figure 2 which is taken from the paper of Piroue and Smith [Phys. Rev. 148, 1315 (1966)]. It is worth pointing out that there is still a quite useable π^+ flux up to 2.3 BeV/c and π^- to 2.1 BeV/c. The K^- flux is too low for most purposes, but the K^+ and K^0 fluxes are very useful (Table I). The time-of-flight technique is also especially valuable for determining the momentum of uncharged particles, e.g., neutrons, K^0.

The external proton beam is produced by slow, resonant extraction with about 50-60% of the beam being extracted and focussed to a 6 mm diameter spot 25 m from the synchrotron. While the bunch structure of the protons is retained in the external beam, it is not quite as good as that of the internally targeted beam. Still, it is quite useful for separating particles. The external beam building, measuring 36x36 m^2, is well equipped to accommodate a wide variety of experimental setups. For example, recently there were eight secondary beams being fed from three targets in the external proton beam. All these beams were available simultaneously, though not all eight experiments could be run truly simultaneously due to differing primary beam requirements. Of the fourteen experiments set up and ready to take data about 4-5, on the average, are actually running simultaneously.

In Table I is given a resume of actual beam fluxes which experimenters obtain on a regular basis. Note that these figures are
quoted for the external proton beam and would be somewhat higher,
in most instances, for beams obtained from the internal target.

IMPROVEMENTS NOW UNDER WAY

By suitable changes in the radio-frequency system we can accelerate deuterons and alpha-particles. We have had deuteron beams as
intense as proton beams up to 500 MeV and by February 1970 will
reach 2.4 BeV. Alpha-particle beams of good intensity have been
successfully injected by making He^+ in our 3 MeV Van de Graaff,
stripping to He^{++} at 3 MeV, and injecting into the PPA. By March
1970 we expect to reach 4.8 BeV energy and at currents not far below our proton currents. Of course both particles will open up
numerous fields of interest to nuclear structure, e.g., the deuteron
will produce, by stripping off the neutron, a monochromatic neutron
beam of 1.2 BeV energy, and well collimated. The alpha-particle
beam will be used to look for coherent production of mesons in multi-
particle collisions and its spin zero, isospin zero character should
make it interesting and unique.

Our most far ranging improvement will be the introduction of
magnet flat topping this winter. We are now installing equipment
which will give a 50 msec flat top at 10 cps (or 33 2/3 msec at 12
cps, 16 1/3 msec at 15 cps, 1 msec at 20 cps). We expect to hold
the top flat to ±0.05% initially and ±0.01% eventually. Therefore
the H·ρ of the synchrotron will be constant to the same degree.
During flat top one may either leave on the radio-frequency to main-
tain a bunch structure or one may remove it, adiabatically, to smooth
out the structure. Since frequencies are easy to measure and hold
constant we anticipate final energy spreads in the range of about
±300 KeV. Probably better, say ±150 KeV, can be achieved if there
is a demand for it. Still further reduction in energy spread is
not inconceivable but the ultimate limit will be the spread in injection energy which lies in the low tens of kilovolt range. Of
value to the nuclear scientist is the fact that we are able to run
at any energy lower than the maximum merely by lowering the magnetic
field and changing the radio-frequency program suitably. This can
be done in a matter of a few minutes.

HEAVY ION IMPROVEMENT PROGRAM

Of great interest to nuclear and elementary particle scientists
alike is our Heavy Ion Improvement Program now before the U. S.
Atomic Energy Commission. If this proposal is accepted both areas
of interest will benefit and will, I believe, be drawn into a fruitful symbiotic relationship. Work under this new proposal will be

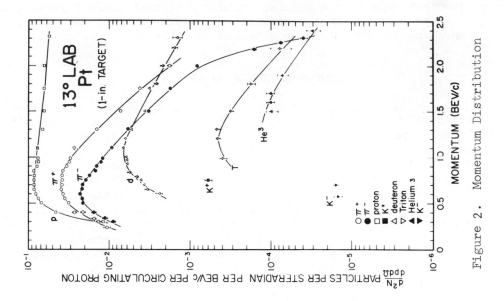

Figure 2.   Momentum Distribution

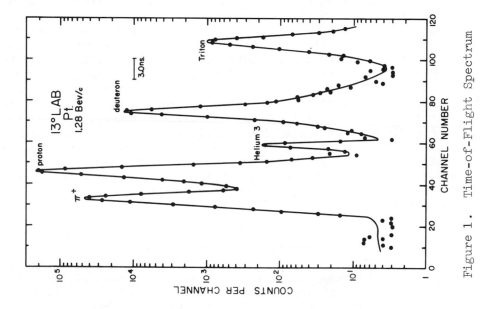

Figure 1.   Time-of-Flight Spectrum

divided into four main phases, each constituting a substantial ad-
vance in PPA scientific utility and each being complete within its
own scope. The emphasis of the first two phases is on searching
for the super heavy transuranic elements and on those nuclear structure problems which can be studied with very heavy projectiles with
energies in the region of 5-50 MeV per nucleon. Even in these first
two phases, light to medium heavy nuclei can be accelerated to much
higher energies, up to 1200 MeV per nucleon. The last two phases of
this program make available the heaviest ions at the maximum PPA
energy, e.g., uranium at 800 MeV per nucleon, which would exceed by
far any other heavy ion accelerator thus far proposed. At these
high nuclear energies one may discover new forms of collective, or
coherent, production of various particles and a new range of nuclear
studies will be opened up. The center-of-mass energy of a 190 GeV
uranium nucleus on uranium is approximately 90 GeV and to the extent, if any, that the entire nucleus behaves as a unit this energy
will be available to create new particles.

   Phase I, by substituting for our present epoxy chamber a ce-
ramic vacuum chamber capable of reaching an average system pressure
of 3×10^{-9} torr, would provide at minimal expense and time delay,
heavy ions with sufficient energy to search for the super heavy
transuranic elements which are believed by some theorists to be
stable against heavy particle decay. For example, Phase I would
yield partially stripped xenon with an energy in the range 7-28 MeV
per nucleon and probably uranium up to 7.9 MeV per nucleon, and at
an initial current of about 10^{10} particles per second. The cost of
Phase I, depending upon options involving speed of achievement, varieties of ions available, and efficiency of synchrotron utilization
varies from $500,000 to $1,050,000. In Phase I some types of light
ions, e.g., fully stripped carbon and nitrogen could equally well
be produced and accelerated and to much higher energies, e.g., 1200
MeV per nucleon. Phase I could be operational in fifteen to eigh-
teen months if funded by November 1969.

   Phase II, by adding a small booster synchrotron (see Figure 3)
would provide a much wider choice of light and heavy ions, at higher currents, in higher charge states and, therefore at higher final
energies than in Phase I. In Phase II medium heavy and even some
types of heavy ions could be accelerated to the region around 10-100
MeV per nucleon and probably much higher depending upon ion source
developments and the use of stripping after acceleration in the
booster. In addition proton, deuteron and alpha-particle currents
would be at least twenty times more intense than now available.
Phase II would cost, depending upon the ultimate current objectives,
between $2,000,000 and $2,675,000.

   The booster synchrotron proposed in Phase II is essentially the
same as that described in our original booster proposal of 1968
(PPAD 646 D). It consists of a 6 sector, rapid cycling (20 Hz),

Beam	P max	$\Delta p/p)_{min}$[a]	Image[b]	Measured Flux [c]
1. 8°Ext	2.5 GeV/c	± 4%	1.9 X 1.5 cm	250,000/sec π^- @ 1440 MeV/c, $\Delta p/p$ = ±5%
2. 25° Ext	1.5 GeV/c	± .6%	3.3X1.5cm	40,000/sec π^- @ 1350 MeV/c, $\Delta p/p$ = ±2%
3. K⁺ Beam	—	—	—	1200/sec K⁺stops[d] 475 MeV/c, $\Delta p/p$ = ±2%
4. 49°Ext	≈500MeV/c	—	—	1 X10⁶ π^+/sec @ 300MeV/c, $\Delta p/p$ = ±6%
5. 90°Ext	300MeV/c	± 3%	—	1.2 X10⁶ π^+/sec[e] @ 150MeV/c, $\Delta p/p$ = ±6%
				3500 μ^+/sec in a 3″X2″X5gm/cm² tgt
				from π decays in the beam line.

a- Half width at half maximum.

b- For $\Delta p/p$ = ±0.01

c- 3 inch Pt external target

d- In a berylia target 1.3 cm diameter X 1.3 cm long.

e- Entering beam line. 3/4 decay before reaching end of line.

f- π^+/π^- flux ≈ 2

Table I.   Beam Fluxes Obtainable at the PPA

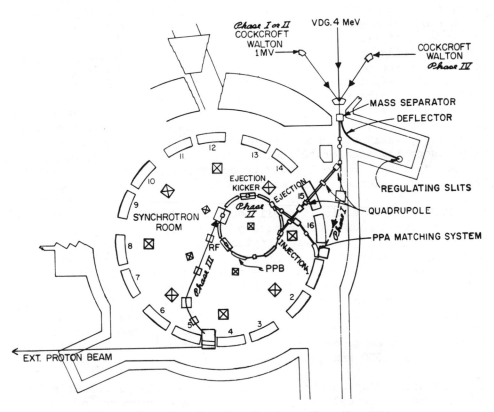

Figure 3.   Booster Synchrotron for the PPA

	NOW Epoxy Vac.Ch. 4MV V.dG. Sept. 1969	FLAT TOP + R.F. Changes Jan. 1970	PHASE I Ceramic Vac.Ch.+R.F changes Jan. 1971	PHASE II Booster Inj. +Cockroft-Walton Sept. 1971	PHASE III Stripping in PPA-Storage in PPB-Accel in PPA Sept. 1972	PHASE IV 2nd C-W +Polarized proton ion source Jan. 1973	
P	10^{12} — 3BeV	10^{12} — 3BeV	10^{12} — 3BeV	2×10^{13} — 3BeV	2×10^{13} — 3BeV	2×10^{13} — 3BeV	Phase III not used.
P(↑)	—	—	—	—	—	10^{8}-10^{9} — 3BeV	Phase IV only.
d	10^{12} — 0.5BeV	10^{12} — 2.4BeV	10^{12} — 2.4BeV	2×10^{13} — 2.4BeV	2×10^{13} — 2.4BeV	2×10^{13} — 2.4BeV	Phase III not used
α	3×10^{11} (10^{11}) — 0.5BeV	3×10^{11} (10^{11}) — 4.8BeV	3×10^{11} (10^{11}) — 4.8BeV	10^{13} — 4.8BeV	10^{13} — 4.8BeV	10^{13} — 4.8BeV	Phase III not used.
C^{6+} / N^{7+}	—	8×10^{10} — 14.4Bev / (2×10^{10}) — 16.8Bev	8×10^{10} — 14.4Bev / (2×10^{10}) — 16.8Bev	3×10^{11} (a) — 14.4Bev / 16.8Bev	3×10^{11} — 14.4Bev / 16.8Bev	3×10^{11} — 14.4Bev / 16.8Bev	(a) PPB accel. & stripping before PPA injection. Phase III not used.
Ne^{3+}	—	—	3×10^{11} (10^{11}) — 34BeV	10^{11} (b) — 24BeV	10^{11} — 24BeV	10^{11} — 24BeV	(b) PPB accel. & full stripping before PPA injection.
Ni^{4+} or Ni^{5+}	—	—	2×10^{11} (c) (2.5×10^{10}) — 3.1Bev	5×10^{10} (d) — 17Bev	2×10^{11} (4×10^{10}) (e) — 67Bev	2×10^{11} (4×10^{10}) — 67Bev	(c) Ni^{5+}, 100μa inj. (d) Ni^{4+} PPB accel. stripping to Ni^{12+} before PPA inj. (e) PPA full strip, then store PPB, PPA accel.
Kr	—	—	2.5×10^{11} (5×10^{10}) (f) — 14Bev	3×10^{11} (g) — 3 Bev / 5×10^{10} (3×10^{10}) (h) — 12 Bev	1.5×10^{11} (3×10^{10}) (e) — 80Bev	1.5×10^{11} (3×10^{10}) — 80Bev	(f) Kr^{4+} 200μa inj. (g) Kr^{6+} PPB, PPA accel. (h) Kr^{6+} PPB accel. & strip to Kr^{12+} before injection PPA & accel.
Xe	—	—	2.5×10^{11} (5×10^{10}) (i) — 0.9Bev	2.5×10^{11} (j) — 3.7Bev / 3×10^{10} (k) — 22 Bev	1.2×10^{11} (2×10^{10}) (e) — 120 Bev	1.2×10^{11} (2×10^{10}) — 120 Bev	(i) Xe^{4+} 250μa inj. PPA (j) Xe^{8+} PPB accel. & PPA (k) Xe^{9+} PPB accel & strip to Xe^{201}, PPA accel.
U	—	—	1.2×10^{11} (2×10^{10}) (l) — 2.1 Bev	1.2×10^{11} (i) — 2.1 Bev / 2.5×10^{10} (10^{10}) (m) — 8.3Bev	1.2×10^{11} (1.2×10^{10}) (n) — 190 Bev	1.2×10^{11} (1.2×10^{10}) — 190 Bev	(l) U^{8+} ~100μa inj. (m) U^{7+} PPB accel & strip to U^{16+} PPA accel. (n) PPA full strip, store PPB, PPA accel.

(1) Energies given are for whole nucleus.

(2) Currents are for 20 Hz pulse rate.

(3) Heavy ion currents are predicted space charge limits (particles/sec). Currents in parenthesis are estimated present day ion-source limited currents. Stripping losses are included.

(4) Assume approximately 10% of PHASE II beam survives full stripping and subsequent PPB storage and final PPA acceleration.

Table II — Particles which will be available at completion of Phases I-IV of PPA Heavy Ion Improvement Program.

alternating gradient magnet, 3.66 m in average radius. We are now testing laminations from the National Accelerator Laboratory booster magnet to verify our belief that they will be suitable for our booster. If so, we will run at 7.5 Kg maximum on the orbit. The magnet aperture will be 5.5x10.0 cm^2. The vacuum chamber will be designed for 10^{-9} torr, or better. Injection of protons and heavier ions into the booster will be made by our present Van de Graaff, upgraded from 3-4 MV, or by a Cockcroft-Walton type of injector, using whatever charge state can be produced in sufficient quantity. After acceleration in the booster we may use some additional stripping before injecting into the PPA.

Phase III, by adding ion stripping and beam transport equipment to Phases I and II, would permit the acceleration of many types of heavy ions, fully stripped, to energies in the 800-1200 MeV range. Currents would be in the range of 10^{10}-10^{11} particles per second. For example, uranium could be accelerated to 800 MeV per nucleon, or 191 BeV for the uranium nucleus as a whole. The cost of Phase III is $800,000 while the time to accomplish is about twelve months after Phase II is operational.

Phase IV, by adding a second Cockcroft-Walton injector for polarized protons, deuterons and heavy ions, and by permitting additional heavy ion source development work, would greatly enhance the capabilities of the PPA and would make the most effective use of the synchrotron. The cost of this phase is $1,200,000 and the work would extend two or more years past the completion of Phase II.

Table II summarizes the PPA capabilities at each phase for a representative selection of light to heavy nuclei. Upon the completion of any of the foregoing phases, the energy of the various particles could be readily varied over a wide range and the present highly successful beam extraction system should perform equally well on all ions at all energies. With the completion of flat topping, this coming winter, it will be possible to have long beam spill times, with particles either bunched or unbunched, a 50% duty cycle, and an energy spread, ultimately, of 0.01%. Thus, with regard to the most important parameters of a high energy nuclear research tool, i.e., energy, energy spread, variability of energy, duty cycle, available particles, quality of external beam, the PPA would compare favorably with any existing or proposed accelerator. For the study of elementary particles, the improved PPA would be as good as any other in its energy range.

The cost of all four phases would be approximately $5,050,000, though it should be reiterated that the construction of any given phase does not imply the necessity, only the desirability, of proceeding with the next phase. The total time to complete all four phases is estimated to be about three and one-half to four years.

HIGH RESOLUTION PION AND PROTON SPECTROSCOPY AT LAMPF*

Henry A. Thiessen

Los Alamos Scientific Laboratory
University of California
Los Alamos, New Mexico

Of the various beam and experimental facilities which are being developed as part of LAMPF, this report deals with two :

i) A high resolution proton beam and spectrometer, (HRS), for the investigation of proton induced reactions;

ii) A beam channel and spectrometer for relatively energetic pions (EPICS).

PROTON SPECTROMETER

The spectrometer is to be capable of measuring a wide range of reactions including (p,p'), (p,d), (p,t), and (p,2p) from a wide range of nuclei. We have found that the optical requirements for the system are dictated almost entirely by the (p,p') reaction. Once we are able to measure this reaction, the others can be measured as well with minor perturbations in the system. In order to study (p,p') reactions, one must in general determine the momenta of both the incoming and outgoing protons as well as the scattering angle θ. From these observables, one can compute the mass of the residual nucleus (M), the momentum transfer ($\sqrt{t}$), the total energy in the c.m. system ($\sqrt{s}$), or any other parameters which may be useful to describe the scattering amplitude. The LAMPF proton beam will have a small transverse phase space (0.075 π cm mrad will contain 96% of the beam) but the energy spread of ± 3.5 MeV is far too large for the 100 keV energy resolution desired. We must, therefore, take account of the energy of the incident proton in our measurements of

*Work conducted under the auspices of the U. S. Atomic Energy Commission.

these reactions.   The most direct approach to this problem, namely
building a spectrometer which is capable of measuring both the
incident proton and scattered proton momenta (an "absolute" spectrometer), has been considered and rejected for economic reasons.

If we consider the (p,p') scattering amplitude to be a function of s, θ, and M, we find that for each value of M, there is a
strong dependence on θ but not on s. In other words, we need good
resolution on θ and M, but would be satisfied with $\pm$ 3.5 MeV resolution on s. In order to see how we might measure M well without
measuring s, let us write the kinematical expression for M as a
function of P_I (incident proton momentum), P_s (scattered proton
momentum), and θ, namely

$$M = f(P_I, P_s, \theta).$$

In order to compute the resolution required, we also have
computed the partial derivatives

$$\frac{\partial M}{\partial P_I}, \frac{\partial M}{\partial P_s}, \text{ and } \frac{\partial M}{\partial \theta}.$$

If we take the case of small scattering angles and/or heavy targets, we find that

$$\frac{\partial M}{\partial P_I} \approx - \frac{\partial M}{\partial P_s}.$$

Thus if we have a spectrometer system which measures $(P_s - P_I)$, we
have a system which also measures M.   This is the basis for the
energy loss spectrometer (or more precisely, the momentum loss
spectrometer).

In order to see how we might construct such a spectrometer
system, let us consider a beam analysis system which provides a
dispersion D_I on the target followed by a spectrometer with magni-
fication -M and the dispersion D_s from the target to the focal
plane. Ignoring the small phase space of the beam, we can write
the position of the particle on the focal plane as follows:

$$x = - Mx_I + D_s \delta_s = - MD_I \delta_I + D_s \delta_s$$

$$\text{where} \quad \delta_I = \frac{P_I - P_{Io}}{P_{Io}} \text{ and } \delta_s = \frac{P_s - P_{so}}{P_{so}}$$

If we adjust D_I such that $MD_I = D_s$

then we have: $x = D_s (\delta_s - \delta_I)$

$$\propto (P_s - P_I).$$

Thus we will have an image in the focal plane in which the displacement is proportional to the excitation of the residual nucleus, and the cross sections for the excitation of a particular state will be an average over the range of momenta in the incident beam.

When we measure at angles other than 0°, we find that

$$\frac{\partial M}{\partial P_I} \neq - \frac{\partial M}{\partial P_s}.$$

The required matching can still be accomplished if the dispersion of the beam is adjustable. The ratio of dispersion of the beam analysis system to dispersion of the spectrometer must be adjustable by $\sim 25\%$ to study light nuclei at large scattering angles. [No more than $\sim 10\%$ adjustment is required to study (p,d) and (p,t) reactions].

At large angles, the measurement of the scattering angle and the momentum difference are independent. This is so because our configuration, V H V (vertical dispersion, horizontal scattering, vertical analysis), separates the scattering angle and momentum analysis into orthogonal planes in the spectrometer. The term in the excitation energy $\partial M/\partial\theta$ is nonzero at large angles and this places a requirement on our ability to measure the scattering angle. Because this term is larger for light nuclei than heavy nuclei, we have chosen ^{24}Mg at 45° and 800 MeV as a compromise case for which we will require our resolution on the excitation of the residual nucleus to be 100 keV or less. (For lighter nuclei or larger angles the resolution will be worse; conversely, for heavy nuclei and small angles the resolution will be better). To satisfy this requirement, the resolution in scattering angle must be 1 mrad or better. In the proposed spectrometer, the angle measurement is accomplished by parallel to point imaging from the target to the focal plane of the spectrometer. Thus a single detector (such as a spark chamber) placed in the focal plane which gives two-dimensional readout of the position of a particle can give a readout of the scattering angle and momentum loss of the particle in the target.

The effects of resolution and quantization introduced by the detectors must be considered. We believe that it is possible to build any one of three types of detectors which can give 0.5 mm resolution for particles passing normal to the plane of the detector, and we are designing the spectrometer to work with any one of the three. These detectors are: wire spark chambers, Charpak-type proportional chambers, and filament scintillation hodoscopes. Proportional chambers and scintillation counter hodoscopes have the problem that their output is quantized in units of the resolution. Several points on the resolution curve are needed to get the line shape for purposes of resolving nearby peaks. For this reason, we are required to have a dispersion large enough that with an 0.5 mm spacing between points, we will obtain three points on a resolution curve. This number can be doubled by placing the two sets of detectors symmetrically about the focal plane rather than placing one detector on the focal plane. Further improvement can be obtained by using different resolution on each of the two detectors. Thus, we expect no difficulty with getting enough points on each resolution curve. In addition, the resolution of the spectrometer will not depend strongly on the resolution of the detectors. The spectrometer design we have chosen is shown in figure 1. The radius of curvature is 4 meters. The solid angle is ~ 2.5 msr. This system will achieve 75 keV resolution for ^{24}Mg at 45° and 50 keV resolution for ^{208}Pb. It should be noted that

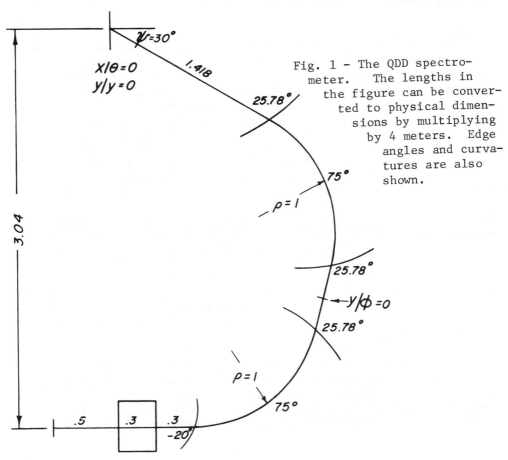

Fig. 1 - The QDD spectrometer. The lengths in the figure can be converted to physical dimensions by multiplying by 4 meters. Edge angles and curvatures are also shown.

the figures for resolution are tentative since final aberration
calculations are not complete.   However, our experience so far
indicates that appropriate corrections (in the form of curved pole
faces) can be made to keep the aberrations small enough.   For
higher resolutions, it will be possible to cut the solid angle of
the spectrometer and momentum spread of the beam to make the effects
of aberrations negligible, but with a loss of counting rate.

PROTON BEAM ANALYSIS SYSTEM

The function of this system is to provide a beam at the target
which is dispersed vertically and "matched" to the spectrometer in
such a way that "momentum loss" is measured directly to the required accuracy.   In addition, the angular spread of the beam at
the target must be small enough to keep the undertainties of the
scattering within ~1 mrad. fwhm. (for ^{24}Mg at 45°).

A possible system which meets most of the requirements for
beam analysis is shown schematically in figure 2.   This schematic
is not to scale but does indicate the system elements, the angle of
bend in the various bending magnets, and the location of the
various vertical and horizontal images.   The elements labeled B_1
through B_5 are bending magnets with angles of bend as indicated.
Bending magnets B_4 and B_5 have non-zero edge angles. Q_1 through Q_3
are quadrupole doublets. Q_2 and Q_3 provide the means for adjusting

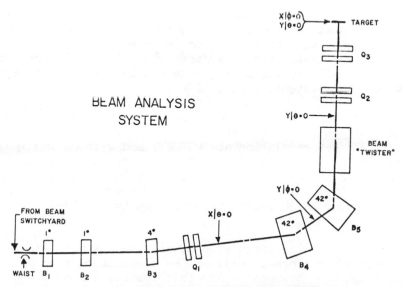

Fig. 2 - Proton Beam Analysis System Preliminary Layout

the dispersion and angular dispersion to "match" the spectrometer.

The "beam twister" is an arrangement of five quadrupoles rotated 45º about the beam. Its purpose is to effect the transformation y → x, x → y, i.e., of interchanging y and x in order to disperse the beam vertically at the target with magnets that actually provide dispersion in the horizontal plane. Such a beam twister has the economic advantage that the horizontal bending magnets which are required by the site layout can also be used to provide the necessary dispersion.

Second order aberrations have been examined and there is no large contribution that cannot be handled. Further details on higher order effects are not warranted at this time because the present system is not the final system. Modifications to provide the capability of an achromatic beam at the target are being considered. In addition, it appears that making a waist instead of a focus at the target should improve the beam divergence considerably. A new improved beam analysis system design incorporating these ideas will be presented at the September design review meeting. It is expected that the basic concept of horizontal dispersion with rotation to the vertical plane will be found to be sound, and that the new system will involve only minor changes in the present design.

PION SYSTEM

Because we have only begun the detailed design work on EPICS, the only information available at this time is of a general nature, namely basic properties of the system and specifications that we hope to meet. As in the proton system, resolution is of utmost importance. In addition, counting rate problems will be severe, even with the high intensity of LAMPF.

The basic requirement on the pion channel is that it provide both π^+ and π^- beams in the energy region of 100 to 500 MeV, with better than 100 keV resolution at 300 MeV, and with maximum possible intensity. A dispersed beam at the scattering target is essential, since it may be possible to use a 1% momentum spread with 0.03% momentum resolution, effectively increasing the beam intensity a factor of 30 over that which can be obtained in a monochromatic system.

The matching of the dispersion of the beam to the dispersion of the spectrometer probably will not be possible because of the large phase space of the beam. For this reason, an "absolute" spectrometer will be a leading candidate for the job. Such a spectrometer works by first providing an image of the target at a detector, then measuring the momentum of the scattered particle.

From the knowledge of the origin of the particle on the target and
the dispersion of the beam, the incident momentum can be determined.   The momentum matching can be accomplished in the data
reduction.   The economic argument against such a system in the
proton case, i.e., the cost of the required building, is not a factor for a pion system.

An absolute spectrometer has the property that its resolution
depends primarily on the resolution of its detectors, and this
resolution cannot be improved by a simple cut in counting rate.   In
order that the entire system be capable of ~ 30 keV resolution in
some cases, the spectrometer must be capable of this resolution in
all cases.   However, it will be possible to achieve comparable
resolution in the beam by cutting both the production target size
and the aberrations, which can be accomplished with a loss in
counting rate.   Note that aberrations are not a problem in an absolute spectrometer since there will be detectors before and after
the dispersive element.   In such a situation, it can be shown that
it is possible to correct for aberrations to all orders during the
data reduction.

In order to perform precise π^+ experiments, it will be necessary to use an electromagnetic separator to remove the large background of protons from the pion beam.   In the LASL pion production
experiment, a proton to pion ratio of ~ 20:1 was observed at
500 MeV/C and 30° production angle from a carbon target.   We require that this be reduced to ~ 1:10 by an appropriate separator.

The yield of pions can be calculated for a reasonable set of
assumptions about the channel, namely: 1) a 3.5 cm carbon target;
2) solid angle of 10 msr;   3) $\Delta p/p$ of ± 0.5%;   4) channel length
of 20 meters; and 5) production angle of 30°.   The pion yields of
such a channel are presented in the table below (cross sections
were taken from preliminary results of the LASL pion production
experiment[2]).

Estimated Pion Yield in Pions/sec

$E\pi$	π^+	π^-
400	1.6×10^8	1.1×10^7
300	5.6×10^8	5.6×10^7
200	4.7×10^8	7.3×10^7
100	1.4×10^8	3.6×10^7

A similar calculation can be performed for the yield of scat-

2.   E. D. Theriot, et al., to be published.

tered pions into the spectrometer assuming: 1) a 100 keV copper
scattering target; 2) solid angle of 10 msr; and 3) a 10 meter path
length spectrometer. The results are presented below:

Scattered Pion Yield
(counts/minute) / (μb/sr)

Eπ	π^+	π^-
400	4.0	0.3
300	14.	1.4
200	9.9	1.5
100	2.2	.56

Most of the above assumptions are reasonable, but whether they can
be achieved in a practical design remains to be seen. One possible
improvement remains, however, and that is the production of π^+ from
the reaction $p+p \rightarrow d+\pi^+$ in a CH_2 or H_2O target. Such a target
would improve the 400 MeV π^+ yield by a factor of ~ 5.

Acknowledgments

     Much credit for the existence of the HRS must go to George Igo
who has gathered interest and organized all of the early activity
on this project.   Most of the credit for the details of the present
design belongs to H. Enge and S. Kowalski of Deuteron Inc. and Karl
Brown of SLAC.   We must also recognize valuable discussions with
R. Burman, R. J. Macek, B. Zeidman, N. Hintz, H. Palevsky, and
J. Friedes.

NEVIS SYNCHROCYCLOTRON CONVERSION PROJECT[†]

R. Cohen, E. Martin, J. Rainwater, R. Schneider,
K. Ziegler - Columbia University, New York City
S. Ohnuma - Yale University, New Haven, Connecticut

The Columbia University Nevis Synchrocyclotron was planned and constructed before 1950. It presently produces an internal beam of protons which are accelerated to ~ 380 MeV at 70 Hz repetition cycle rate and ~ 1.5 µA time average current. Present operation mainly uses a Be vibrating (internal) target to give long duty factor external pion and muon beams via the cyclotron fringe field focusing properties. It is also used for pulsed neutron time of flight spectroscopy.

The revision program planning began in 1965 and completion of the conversion is expected during 1971. The most complete description is given in a series of papers at the 1969 Particle Accelerator Conference held in Washington, D.C.[1] and the International Conference on Cyclotrons held in Oxford.[2] The conversion will change the machine to a 550 MeV machine having a long duty factor (5 to 40 µA time average) external beam facility.

Plans call for retention of the basic 2000 ton steel magnet but with the addition of a 10-in. steel band around the yoke to lower its magnetic reluctance. Pole iron within 30 in. of the median plane will be replaced by new iron which will provide a three-fold symmetrical azimuthally varying field (AVF) for the beam. The system will remain a synchrocyclotron rather than become a fixed frequency (CW) machine. The azimuthal average magnetic field <B> will increase from ~ 17 kG near the center to ~ 20 kG near 80 in. radius. This implies a reduced FM frequency range for the dee from ~ 26.5 MHz at injection to ~ 18.5 MHz at maximum energy. Additional field excitation will be provided by the addition of new auxiliary magnet coils between the main coils and outside the 170-in. pole diameter.

The azimuthal field variation will be produced using spiral ridge

603

sector iron having a three-fold symmetry. We can control the shape
of the <B> vs r curve by varying the azimuthal extent of the spiral
hill iron, and by using a radial variation of the pole face gap which
will change from a 44 in. gap near the center to a 16 in. gap from 80
to 85 in. radius. Since one of our objectives is to achieve strong
magnetic axial focusing starting at r < 2 in., we shall have quite
small median plane gap between top and bottom sector iron pieces. The
gap spacing will be < 1 in. starting at r < 1 in., and increase to a
few inches at larger radii. This does not leave space for an accelerating dee electrode between the sector hill iron. We will, therefore, divide the top and bottom sector iron each into two layers. One
lies within about 7 in. of the median plane (the "floating shims"),
and the other part is further than 10 in. from the median plane. One
of the three sets of floating shims will be mounted on Al_2O_3 insulators and will form part of the rf dee structure.

The design gives strong magnetic axial beam focusing starting at
r < 2 in. from the center, which should raise the internal beam current space charge limit to >> 20 µA time average. We expect to accelerate ≥ 20 µA time average internal beam current of 550 MeV protons. We expect to operate at ≥ 300 Hz FM repetition frequency, with
about 40% rf on-time during the acceleration portion of each FM cycle.
The rf dee voltage will be > 30 kV peak near injection and can be reduced smoothly to about 70% of its peak value at the low frequency
end without loss of the phase bundle. The crucial region is at a
small radius where the sum of the electrical plus magnetic focusing
reaches a minimum. This should occur near r = 1.6 in. for our design.
The large azimuthal field flutter starting at r < 2 in. also raises
(v_r-1) safely above unity. which helps assure radial beam stability.

The planned dee-line rotating capacitor resonator system will
use two rotating capacitors in parallel at the far end of the dee-
line system, situated outside the magnet coils.

For the rotating capacitors, we plan to use short rotors 48 in.
in diameter at the outside of the blade circle having 4 layers of 12
blades. Our rotating capacitor design differs from others in that we
plan to have the stator blade ring at dee-line rf potential rather
than at ground potential. The large area central rotor face will be
≤ 0.010 in. from a ground face to provide a much greater capacitance
from rotor to ground than from rotor to stator. This leads to a structure where the rotor drive seal and bearings are relatively near ground
potential. The connection of the stator ring to the dee line is via
four transmission lines which extend from the end of the main vacuum
chamber to the capacitor housings. The connections are at points 90°
apart around the stator ring. This topology permits the use of a
thick iron case for the capacitor housing. It provides almost complete magnetic shielding of the rotor from the cyclotron fringing magnetic field. This design puts the rotor drive shaft (on the chamber
side) in air and readily accessible.

The capacitor voltage ranges from ~ 90% of the dee voltage at
the high frequency end to about 50% of the dee voltage at the low
frequency end. Tests using a half scale model of the system, including the model rotating capacitors, show that the tuning properly
covers the desired frequency range.

The extraction studies have been underway since 1965 using detailed computer orbit tracing programs. We expect to use a peeler
system near the middle of the dee azimuth (at the sector hill). An
extraction channel (current channel) comes next (outside the dee azimuth) followed by a regenerator. The peeler and regenerator will use
fixed iron structures. A "current bump" coil situated opposite to
the regenerator will repel the bunch from the peeler-regenerator so
the "bunch" can be "parked" at an optimum energy-radius by turning
off the main rf to the dee. The bump then gradually decreases and
the proton bunch slowly moves into the extraction system. The inherent variation of energies and radial oscillation amplitudes present should make this a gradual "peeling off" process for a long duty
factor beam. The final radial turn spacing should be nearly 1 in. at
the extraction channel entrance. Calculations indicate that appreciably better than 50% extraction efficiency should be achieved.

The beam subsequently has a long path within the chamber and
after the channel. The 16 in. vertical aperture for r > 85 in. will
permit iron pieces to be placed to modify the fringe field behavior to
provide considerable radial focusing to yield an external beam focus
where a target for pion, muon and scattered proton production will be
situated. These secondary beams will then be led along curved paths
through the thick iron shield wall, using bending and quadrupole magnets embedded within the iron shield wall. The main bunch will continue into the side wall to an underground beam dump and chemistry
irradiation facility. The scattered, lower intensity, proton beam to
the experimental floor will also contain polarization which increases
its experimental utility.

The building extension was finished in 1968. Auxiliary coils and
the new vacuum chamber have been ordered. The band to surround the yoke
has been delivered.

[1] I.E.E.E. Trans. Nucl. Science, Vol. NS-16, #3, p. 421 (1969).

[2] Proceedings of the International Conference on Cyclotrons, Oxford
(1969), to be published.

[†] Research supported in part by the National Science Foundation.

THE POSSIBILITY OF USING STORAGE RINGS WITH INTERNAL THIN TARGETS

S. T. Belyaev, G. J. Budker, S. G. Popov

Institute of Nuclear Physics

Novosibirsk, U.S.S.R.

Storage rings, apart from the colliding-beam experiments, give some unique possibilities for operation with thin internal targets. When one operates with an extracted beam, the choice of the target thickness is a compromise between two contradictory requirements; namely, increase of the reaction total yield and necessity of a high energy resolution. In the storage ring these requirements reconcile naturally. The target thickness may be chosen as small as necessary for a single passage of the particles through the target, but the effective thickness (real thickness multiplied by number of revolutions) is nearly the same as for commonly used targets, and in many cases even much larger. The particles energy losses in the period between two successive passages of the target are compensated by r.f. cavity, and accumulated angular and energy spreads may be stabilized by damping effects (radiation for electrons, artificial beam-"cooling" for protons.[1]

Let us consider the storage-ring operation regime when the losses of particles are determined by processes of their interaction with the target. The beam life-time τ (it is convenient to consider dimensionless time in terms of the periods of revolution) is determined by the cross-sections of the processes δ_i and by the target thickness n_0 (number of particles per cm^2)

$$\tau^{-1} = n_0 \Sigma \delta_i = \Sigma \tau_i^{-1}$$

The quantity $n_{eff} = n_0 \tau$ does not depend explicitly on n_0 and is determined by the total cross-section of all processes which result in particle losses: $n_{eff} = (\Sigma \delta_i)^{-1}$.

It is useful to distinguish "single" processes from multiple
ones.   The latter's influence, being accumulated for many revolutions, results in time-linear increase of the mean squares of
corresponding parameters (amplitude of oscillation): $< a_i^2 > = \alpha_i t$.
The rising time of $< a_i^2 >$ up to maximum admissible value
$\tau_i = (\max a_i)^2 / \alpha_i$ determines the beam lifetime connected with a
given process.   The influence of the "single" processes is
different, in principle.   They result only in single particle
losses but, on the contrary to "multiple" processes, do not
influence essentially other beam parameters; namely, angular and
energy spreads, transverse beam size.   These important parameters
may be essentially improved in the "super-thin target" regime when
the "multiple" processes are suppressed by damping effects.

Let τ_{di} be the damping time of oscillations a_i.   Decreasing
the target thickness n_o, one may have lifetime τ to be greater than
τ_{di}.   In this case, the damping effects limit the rise of amplitude
a_i at the equilibrium value $a_i^o = < \alpha_i \, \tau_{di} >^{1/2} < \max a_i$.   The effective beam thickness n_{eff} increases by jump in this case because the
corresponding "multiple" process (δ_i) ceases to work.

The "super-thin target" regime was investigated in part on the
electron storage ring VEP-1 at our Institute in the process of preparation for the operation with the colliding beams.   The experimental results (n_{eff} value, stable beam parameters) are in good
agreement with the calculations.[2]

For investigation of the transition from the "thin target"
regime to the "super-thin target" regime special experiments were
performed.[3]   Aluminium foils of various thicknesses (thin target)
as well as a quartz filament of 1μ thickness (nearly a super-thin
target at 135 MeV) were used.   At the given target thickness the
dependence of the effective thickness on radial aperture was
measured.   The quantity n_{eff} was characterized by the bremsstrahlung intensity.   The bremsstrahlung quanta with the energy of
$(0.96 \pm 0.003)E_o$ were selected by coincidence with recoiled electrons
of the corresponding energy (E_o-primary electron energy).   The
electrons were detected by a counter located inside a uniform mag
netic field of the storage ring (Fig. 1) so that the 1800-focusing
took place[4] (the so-called monochromatic γ-quanta).   The calculated
curves as well as experimental results for the aluminium and quartz
targets are shown in Fig. 2.   The straight line I with $n_{eff} \simeq$
0.05 r.1. is for super-thin target.   The parabolic curve II is for
the thin target (corrections were made on uncompensated energy
losses).   The transition to the "super-thin target" regime for the
quartz target is obvious.

In conclusion we summarize the possibilities and the advantages
of the storage rings with the internal target:

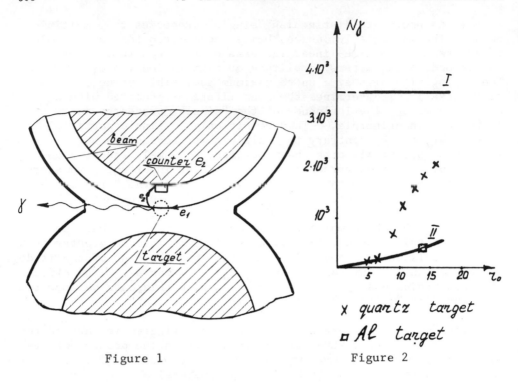

| Figure 1 | Figure 2 |

i)   The target thickness may be decreased without decrease
     of the reaction total yield.

     Let us consider an example (with the synchrotron B3-M[5] as the
injector into the storage ring). The number of injected particles
is 2.10^{11} per puls; repetition rate — 5 cps at the energy of 100-
500 MeV. In this energy range $n_{eff} \simeq 0.2$ r.l. (10^{22} cm^{-2} for Al).
The luminosity is of 10^{34} sec^{-1} cm^{-2} at the stable energy spread of
50-150 KeV (the energy resolution is $3 \div 5.10^{-4}$). The best competitor
to the storage ring seems to be a high-current linac. If the
average current is of 10 µA, and the energy spread is of the order
of 1% (it corresponds to the target thickness of 10^{-2} r.l.) then
the luminosity is equal to 3.10^{34} sec^{-1} cm^{-2}. A higher energy
resolution demands the decrease of the luminosity. At the energy
resolution of 3.10^{-3} the luminosity will be 10^3 times lower.

ii) The advantages for the reactions with particles produc-
     tion near threshold are obvious (for example, pions
     electroproduction).

iii) For several experiments the advantage of the storage
     rings is their continuous operation. The intensity of
     monochromatic quanta for a linac is limited by its

repetition rate (50–500 cps) because the electronic
equipment practically may detect not more than several
tens of recoiled electrons per pulse of 1 μsec duration.
On the storage rings (multi-channel modification of
the experiment shown in Fig. 2) practically all electrons may be converted into polarized monochromatic
quanta.

iv) The storage rings give the possibility, in principle,
for experiments with unique primary particles (polarized
electrons and positrons[6], antiprotons) as well as with
unique targets (polarized gas target, electrons, neutron
beams and so on).   The possible high energy resolution
makes experiments on the proton storage rings with
various nuclear targets to be of interest (for example,
the missing-mass experiments at 2–3 GeV, accompanied by
hyper-nuclei production.

REFERENCES

[1] G. J. Budker, Atomic Energy (USSR) 22, 346, 1967.

[2] V. L. Auslender et al., Proceedings of the International symposium
on electron and positron storage rings (Saclay, 1966 p VIIb).

[3] G. I. Budker, A. P. Onuchin, S. G. Popov, G. M. Tumaykin.
Nuclear Phys. (USSR) 6, 775, 1967.

[4] L. S. Korobeynikov, L. M. Kurdadze, A. P. Onuchin, S. G. Popov,
G. M. Tumaykin. Nucl. Phys. (USSR) 6, 84, 1967.

[5] E. A. Abramjan et al., Proceedings of the International conference
on high energy accelerators (Atomizdat, 1964, 284)

[6] W. N. Bayer, W. M. Katkov.   JETP (USSR) 52, 1422, 1967.

PRELIMINARY REPORT ON A 500-MEV LINAC-POWERED RACE-TRACK

MICROTRON FOR THE UNIVERSITIES OF NORTHWEST ITALY

L. Gonella, G. Manuzio, R. Malvano

Istituto Nazionale Fisica Nucleare and Universities of

Genoa, Pavia, Torino

The electron beams required nowadays for nuclear research must offer high intensity, in a narrow energy bin, at high duty cycle. In order to get these characteristics within cost limits consistent with University budgets we decided for a linac-powered race-track microtron (1-3), consisting of a linac section placed between two constant-field magnets so as to drive the electrons through n cotangential race-track orbits. The system behaves like a transformer multiplying by n the rated energy of the linac and dividing by n its rated current: moreover the recycling magnets endow the linac with a strong phase focusing so that the machine yields the energy of a n-section linac with a much sharper spectrum. This spectrum compaction is not obtained by simply throwing away part of the electrons against energy-defining slits, but by keeping them to oscillate very close to a "synchronous" value of the accelerating voltage, $\Delta V_s = N\lambda Bc/2\pi$, determined by the magnetic field B and independent of the external or internal fluctuations of the linac peak voltage ΔV_p: λ is the rf wavelength, N the differential harmonic number, the harmonic number of the n-th orbit being given by $N_n = N_0 + nN$. In principle, therefore, with a well-centered injection bunch the whole rf power transfered to the beam will be spent on the utilized electrons, which fact, added to the drastic reduction in copper losses with respect to a n-section linac, makes the installation and operation costs much lower than those required by a beam with the same energy, and current within the given narrow energy bin, obtained with a conventional linac and its analyzing magnet. Indeed the efficiency of the rf power transfer to the output beam may be measured by the average input rf power required per unit average beam power within a 0.1% energy bin: this figure is 56 - 65 for the design data of

the MIT linac (4), while a sample computation for our machine
yields a value 16.   Another figure of merit is the beam power lost
on the energy-defining slits per unit beam power within a 0.1%
energy bin, which is a measure of the high energy background at
the entrance of the beam transport system: this is 7 for the
quoted data of the MIT linac and 0.5 for our sample computation.

The parameters of the machine are not yet definitely decided:
we are thinking of an output of about 500 MeV, with average useful
beam power of 10 - 20 kW at 3% duty cycle, to be reached in 25 -
30 orbits with a magnetic field of about 1.5 Wb/m^2, an injection
energy of 12 - 20 MeV, and a gain per turn of 15 - 20 MeV, obtained
with two conventional L-band linac sections excited with two 75-kW
klystrons.   The duty cycle is limited by the cooling and the
available klystron power.   We choose an external high-energy
injection in order to exploit the higher beam-loading tolerance
that seems allowed by working at N=1 (a low-energy injection is
possible only with N>>2), and to make available the injection
linac as an independent tool for extra-nuclear research.

The design of the machine does not call for major physical
or technological breakthroughs: it is not a matter of a new acceleration method as much as of a new optimal arrangement of well-
known devices.   The main design problem consists in determining
the focusing parameters in space and phase, where many questions
require particular attention since the adiabatic conditions no
longer hold, which in conventional cyclic machines allow to
write differential linearized equations.

A number of focusing devices (edges and/or quadrupoles) may
be used on each orbit to achieve spatial focusing.   On the basis
of the similar and simpler synchrotron beam stability problem we
worked out a generalization suitable to our case, obtaining beta-
tron beam oscillations of limited extent.   Requirements on the
one-turn transfer matrices to get a first order separation between
the spatial and the phase focusing problems were also imposed.
First-order computer calculations show that, with a quadrupole
focusing system helped by a reduction of the recycling magnets
fringing field, the spatial focusing is achieved with quadrupole
gradients lower that 150 G/cm and tolerances of the order of 2%
on the field gradient and 2 mm on the positions.

For the phase focusing, the first problem is to find size and
position of the injection buckets, i.e. the range of phase and
energy within which an electron must be injected in order to get,
through the n orbits, a gain per turn close enough to ΔV_s not to
be stopped by the energy-defining slits on the magnet edges. These
buckets are found to be different in size and shape at different
values of the "sychronous" phase ϕ_s = arcsin ($\Delta V_s/\Delta V_p$) of the

accelerating voltage, and to overlap only partially: a behavior quite different from the conventional "adiabatic" machines. As ϕ_s changes during the acceleration, owing to the variation of ΔV_p, i.e. to a given beam load in presence of reasonable external fluctuations. Preliminary results show that with N = 1 injection bunches 5-degree wide, with fairly large energy tolerance, can be accelerated without losses through ΔV_p excursions allowing for a 1% external fluctuation and a 4-7% beam load. On the other hand at N=2 the buckets overlap only for very small variations of ΔV_p so that 2 - 3 degree injection bunches can be accomodated. With the wide energy-defining slits used in this sample computation, the energy spread of the whole accelerated bunch, jittering through the whole injection tolerance, keeps within ± 0.07 - 0.12%. One can reasonably expect still better values with optimized parameters and external compensation of the beam-loading buildup during the machine filling time.

We like to express our deep appreciation to: M. De Mutti, E. Mancini, M. Pilo of Genova; F. Orlandino, R. Rabagliati, S. Rebola of Torino; A. Gigli, T. Pinelli of Pavia; and to all other members of the working group.

REFERENCES

1. L. I. Schiff, Rev. Sci. Instr. __17__ (1946) 6
2. B. H. Wiik and P. B. Wilson, Nucl. Instr. & Methods __56__ (1967) 197
3. Proposal to construct a 30-MeV Superconducting Linac as a Step Toward a 600-MeV Superconducting Microtron, Phys. Dept., University of Illinois, March 1967
4. W. Bertozzi, J. Haimson, C. P. Sargent, W. Turchinetz, IEEE Trans. on Nucl. Science __NS-14__ No. 3 (1967) 191

"PRECETRON" - A PRINCIPLE FOR OBTAINING PION-PION AND MUON-MUON

COLLISIONS†

Robert Macek* and Bogdan Maglic**

University of Pennsylvania, Department of Physics

Philadelphia, Pennsylvania

A device for pion-pion collisions in the mass region of ρ^- meson has been proposed and its parameters calculated. It operates on the following idea: over 10^{13} π^+ and π^- in the momentum range P_π = 310-550 MeV/c are produced by a short burst of 6×10^{15} protons of 800 MeV incident onto a metal target placed in the center of a 400 Kgauss (pulsed) magnetic field, extending over a radius of $R \sim 5$ cm and shaped to contain $\sim$ 100 turns of pions. Pion orbits are circles tangential to the target; π^+ and π^- orbits precess in opposite directions, the locii of their centers being circles of radius $r \simeq \frac{1}{2}R$ whose origin is at the target. (Fig. 1A,B,C) The π^+-π^- collisions of interest take place in the peripheral region of this "magnetic pot", where the crossing angles of the intersecting precessing orbits are smallest. This orbit-precession device, called "precetron", has a large momentum acceptance of pions, $\Delta_p/_p \simeq 50\%$ and large angular acceptance, 0° to 180° in horizontal and $\pm 10°$ in vertical plane, in contrast to a conventional storage ring which would require well-defined momentum and small angular divergence in order to have the pions captured.

Single particle orbits have been calculated to first order; they show the precessing character of the orbits and the stability of the free oscillations. Pion densities in the colliding region -- an outer shell of thickness ηR -- have been calculated as a function of η; number of protons/sec, N; magnetic field, B; proton pulse length, g, in units of pion laboratory life-time, (which is

†Work supported by U.S. Atomic Energy Commission. Further details in Princeton-Pennsylvania Accelerator Report No. PPAR14 , June 1969.
 *Now at LASL.
**Now at Rutgers University.

$\sim 10^{-7}$ sec); number of proton pulses/sec, b; and other parameters, eq. (28). The π-π collision rate is proportional to $N^2 B^3$ and inversely proportional to bg^2, i.e, the poorer the duty cycle of the proton accelerator, the higher the rate. If a proton linac with g = b = 1 can be made with the same average power as the Los Alamos Meson Factory (but with 600,000 times poorer duty cycle), there will be about 20,000 π-π collisions/hour. However, this b = g = 1 regime can be also accomplished without changing the machine parameters, by storing the proton beam in a storage ring for ~ 1 sec and dumping it onto the precetron target in one turn.

All backward scattered pions from $\pi^+\pi^- \to \pi^+\pi^-$, which represent $\sim 50\%$ of the collisions, will leave the magnetic pot because of the change of the sign of the $\vec{v} \times \vec{B}$ force; their momenta will be measured outside and their directions after scattering reconstructed, to obtain the π-π effective mass of the collision (no knowledge of the crossing angles and energies before collision is needed to obtain the effective mass).

After several pion life-times, the muons from the decaying pions will be orbiting in similar precessing orbits and similar collision rates will be obtained.

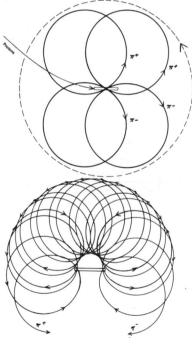

Fig. 1A- First turn for π^+ and π^- of median energy emitted at 30° to proton beam. The dashed line gives the direction of the π^+ orbit precession. B is directed into the plane of the paper; it is centered at the target center.

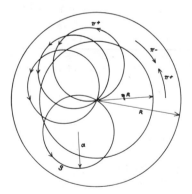

Fig. 1B- Typical π^+ and π^- colliding orbits of <u>one</u> momentum showing preces- sion.

Fig. 1C - Typical π^+ orbits showing the shell from which the interactions are accepted. The shell is the region of radii from ηR to R.

CURRENT DEVELOPMENTS IN THE STUDY OF
ELECTROMAGNETIC PROPERTIES OF MUONS

E. Picasso

CERN, Geneva, Switzerland

INTRODUCTION

A review of the status of this problem has been given by
Farley [1] and a general survey of the muon by Feinberg and
Lederman [2]. In my talk I will weigh up the various subjects with
different emphasis; this is not because of the differing interest
in the subjects but because my knowledge of them is not equally
balanced.

1. ELECTROMAGNETIC PROPERTIES OF MUONS AT LOW MOMENTUM TRANSFER

1.1 The Anomalous Magnetic Moment of the Muon [3]

The muon is, in the present view, simply a heavy electron with
a mass m_μ, some 207 times larger than m_e. A determination of the
g-factor of the muon is of great importance, because it may indicate
that there are some basic differences in the interaction properties
of muons and electrons; in other words it may indicate the existence
of a possible new coupling of the muon to another field which might,
perhaps, explain the mass difference between electrons and muons.

The measurement of the g-factor of the muon provides also one
of the most critical tests of quantum electrodynamics at low momentum transfer and high precision. Due to the muon "energy scale"
which is m_μ/m_e times that of the electron, a determination of the
g-factor of the muon tests the quantum electrodynamics (QED) at
shorter distance.

The theoretical expression of the g-factor can be written in a power series of the fine structure constant α (and hence of e^2). Due to radiative corrections, the g-factor of the muon has a value slightly different from two. The anomaly, $a_\mu \equiv (g_\mu - 2)/2$, measures the deviation of g from the predicted value two of the Dirac theory. It can be written

$$a_\mu^+ = a_\mu^- = \sum_{n=1}^{\infty} A_n \left(\frac{\alpha}{\pi}\right)^n , \qquad (1)$$

where the A_n's are constants depending on masses. Due to the fact that we believe the muon interacts with the electromagnetic field in the same manner as the electron, the second-order radiative correction should be the same as for the electron since it is mass-independent; thus $A_1 = \frac{1}{2}$. The fourth-order radiative corrections must contain a contribution equal to that of the electron $[-0.32848 (\alpha/\pi)^2]$ since this is mass-independent. An additional fourth-order contribution, for the muon, comes from the fact that a virtual photon emitted by a muon may polarize the vacuum via a virtual electron-positron pair. The electron pair contributes to the anomalous magnetic moment of the muon by $+1.09426 (\alpha^2/\pi^2)$; the over-all value of A_2 is $+0.76578$.

We now consider briefly the sixth-order radiative corrections. Leading contributions to the α^3 term in $(a_\mu - a_e)$, excluding the photon-photon scattering diagram, were calculated [4] to be $+2.82 (\alpha/\pi)^3$. Recently Aldins, Brodsky, Dufner and Kinoshita [5] did a remarkable piece of work in calculating the contribution of the photon-photon (electron loop three-photon exchange) diagram, obtaining the surprisingly large value

$$\delta a_{photon-photon} = (18.4 \pm 1.1) (\alpha/\pi)^3 . \qquad (2)$$

Thus the total α^3 term in $(a_\mu - a_e)$ is $+21.2 (\alpha/\pi)^3$. I want to emphasize that all Feynman diagrams from QED, which contribute to $(a_\mu - a_e)$ through sixth order, have been calculated or bounded with an uncertainty of 0.6×10^{-8} [5]. The α^3 term in a_e has been estimated to be $0.18 (\alpha/\pi)^3$, so the over-all value of A_3 is $+21.4 (\alpha/\pi)^3$ (QED only) [5].

A virtual photon emitted by a muon may polarize the vacuum with strongly interacting particles. This contribution to the $(g - 2)/2$ of the muon has been calculated recently by Gourdin and de Rafael [6] using the recent data of the Orsay e^+e^- colliding-beam experiments, and they found

$$\frac{1}{2}(g - 2)_{strong} = 5.02 \left(\frac{\alpha}{\pi}\right)^3 = (6.5 \pm 0.5) \times 10^{-8} .$$

where α [7] is:

$$\alpha^{-1} = 137.03608 \pm 1.5 \text{ ppm} .$$

The uncertainty comes from the experimental error on the width of the ρ, ϕ, and ω.

The possibility of bounding the hadronic vacuum polarization contribution to the muon magnetic moment anomaly has been considered recently by J.S. Bell and E. de Rafael [8]. These authors point out that improved electron-electron scattering experiments will yield a useful empirical bound and it becomes more and more clear that improved $(g_\mu - 2)/2$ experiments should be complemented by improved experiments on ee scattering at large momentum transfer, and on e^+e^- annihilation at all available energies. More recently J. Bailey [9] gave an upper bound to hadronic vacuum-polarization contribution to $(g - 2)/2$ using a field-current identity plus one simple hypothesis; this result tells us that probably such a contribution will be $\sim 10\%$ of the main contribution of the low-energy term.

The experiment on the anomalous magnetic moment of the muon is sensitive to the existence of interactions which may have escaped detection by other experiments. In fact a neutral boson particle coupled to a charged muon current with a coupling constant f, modifies the g-factor by the following amount

$$\delta g = \frac{1}{3\pi} \cdot \frac{f^2}{4\pi} \left(\frac{m_\mu}{M}\right)^2 \tag{3}$$

(in the limit $M >> m_\mu$ and if the neutral boson is a vector boson field) where M is the mass of the particle.

The theoretical value of the anomalous magnetic moment predicted by the QED, excluding the photon-photon scattering diagram, is therefore

$$a_\mu = \frac{\alpha}{2\pi} + 0.76578 \frac{\alpha^2}{\pi^2} + 3.00 \frac{\alpha^3}{\pi^3} + \ldots = 116558.0 \times 10^{-8} . \tag{4}$$

If one adds the contribution of photon-photon scattering one finds

$$a_\mu = 116581.0 \times 10^{-8} \tag{5}$$

with an uncertainty $\pm 2.2 \times 10^{-8}$. This uncertainty includes the uncertainty in the sixth-order correction, in the value of $\alpha/2\pi$ ($\pm 0.2 \times 10^{-8}$) and in the photon-photon calculation ($\pm 1.4 \times 10^{-8}$). If one adds the contributions due to the strong interaction as calculated by Gourdin and de Rafael [6] (only taking into account the contribution to a_μ due to the ρ, ω, ϕ resonances) one gets

$$a_{\mu,\text{theory}} = 116587.5 \times 10^{-8} \tag{6}$$

with an over-all uncertainty estimated at $\pm 2.7 \times 10^{-8}$. The uncertainty quoted does not take into account the uncertainty coming from the vacuum polarization contribution of higher mass hadron. The experiment carried out at CERN gives the following result:

$$a_{\mu,exp} = (116616 \pm 31) \times 10^{-8} \ . \tag{7}$$

The comparison between experiment and theory gives

$$a_{exp} - a_{th} = +(29 \pm 31) \times 10^{-8} \ . \tag{8}$$

The theoretical and experimental value agree within one standard deviation experimental accuracy. The experiment is well known and it has been described elsewhere [3b,c]. In Fig. 1 we show the distribution of decay-electron events as a function of time. The lower curve shows rotation frequency of the muon at early time.

An experiment is now proposed at CERN [10], based on a new method, to determine the (g - 2)/2 quantity with a precision of

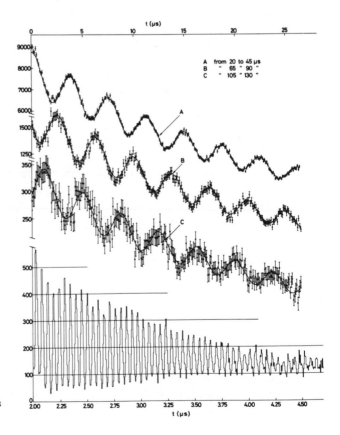

Fig. 1 Distribution of decay-electron events as a position of time.

10-20 ppm in "a". Let me point out that in order to interpret properly any future (g - 2) experiment of the muon, we must consider the
following :

a) Every hadronic state contributes a positive effect to δa_s;
    therefore the total correction will be somewhat greater. New
    information from $e^+ e^-$ colliding-beam experiments up to a few
    GeV centre-of-mass energy will be available; therefore the
    contribution of higher mass hadrons to the anomaly can be
    evaluated. The new experimental error will be around three
    times smaller than the hadronic effect so that it will be
    clearly confirmed. A reduction of the experimental error in
    the determination of the anomalous magnetic moment of the muon
    will be not only a further test of QED but it will also bound
    the cross-section for $e^+ e^-$ annihilation.

b) A modification of the photon propagator

$$\frac{1}{q^2} \rightarrow \frac{1}{q^2} - \frac{1}{q^2 + \Lambda^2} \tag{9}$$

    lowers the theoretical value of a by

$$\frac{\delta a}{a} = \frac{2}{3} \frac{m_\mu^2}{\Lambda^2} , \tag{10}$$

    where Λ is the mass of an unstable particle with a negative
    metric, "heavy photon", as suggested recently by T.D. Lee and
    G. Wick [11], in order to remove divergences and not to violate
    unitarity. The proposed theory [11] predicts a strong reso-
    nance in $e^+ e^-$ collision when the total centre-of-mass energy
    equals Λ. The Frascati colliding-beam experiments should
    begin to see the tails of this resonance if $\Lambda \sim$ 6-8 GeV so they
    will provide an independent check. New colliding-beam experi-
    ments now planned should see the tail of this resonance even
    if $\Lambda \sim$ 13 GeV.

c) The weak interactions are estimated to contribute to

$$\delta a_w \simeq -2 \times 10^{-8} \tag{11}$$

    if the intermediate boson exists, but the result depends
    strongly on the magnetic moment of the boson.

d) A new coupling of the muon can either increase or decrease
    the value of a.

     It must be emphasized that the $e^+ e^-$ colliding-beam experiments
must complement any new determination of the (g - 2) experiment,
but we have to note that, on the other hand, data obtained from

e^+e^- colliding-beam experiments apply only to particles that couple directly or indirectly to the electron. The couplings of the
muon may be different and this can be tested only by the muon
(g - 2) experiment and by the μ-pair production by muons.

To summarize, I assume that colliding-beam experiments will
check the Lee-Wick hypothesis and measure the cross-section for
hadron production; these results will enable their contributions
to the anomalous magnetic moment of the muon to be calculated.
A new result of the determination of the (g - 2) quantity would
therefore test the following possibilities:

a)   a new coupling of the muon;

b)   a contribution of the weak interaction;

c)   a larger contribution from hadrons, implying peaks in hadron
     structure at energies not yet attained.

If the new (g - 2) measurement agrees still with theory, one
would conclude that the cut-off for QED would be higher than
13 GeV. Of course, several of these effects may be operating together, leading to cancellation or to reinforcement.

To conclude this section let us suppose the photon propagator
has been modified according to Eq. (9). With 95% confidence the
following limit was established

$$\Lambda_\gamma > 5 \text{ GeV} .$$

We must note that this value of Λ_γ would lower the experimental
result by 31×10^{-8} in contrast to the actual result which is
slightly higher [12].

It should be pointed out that the muon (g - 2) experiment tests
the high-q^2 behaviour of QED theory only indirectly. In fact the
presence of the high q^2 virtual photon contributes to the static
value; therefore an accurate experiment tests if these virtual
photons are present in the amount predicted by the theory.

1.2   The Muonium Atom [13]

Muonium is the atom consisting of a positive muon and an electron. It is an ideal system for studying the electromagnetic interaction of the muon and electron; in fact the energy levels of muonium should be calculable from the QED [14,15] of the electron, muon,
and photon fields with the Bethe-Salpeter equation. The QED corrections to the hyperfine splitting in the muonium ground state are
∿ 200 ppm, in addition to the radiative corrections due to the free
electron moment; the muonium hfs provides, therefore, a reasonably

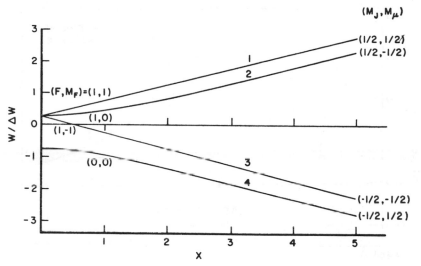

Fig. 2 The diagram of the energy levels of the muonium ground state in a magnetic field. $x = (g_J - g_\mu) \mu_{e_0} H/\Delta W$. At zero field the energy separation is the hfs interval.

good test for QED. One of the most important aspects of muonium physics is based on the fact that it is the simplest system involving the muon and the electron, and therefore a comparison of the theoretical and experimental values of the hfs may indicate whether there are basic differences in the interaction of the electron and muon. Figure 2 shows the Zeeman energy level diagram of muonium in its ground state. The separation at zero magnetic field between $F = 1$ and $F = 0$ states is the hfs interval $\Delta\nu$. Transitions have been observed by a Yale group [16,17] at a relatively high magnetic field of 6 kG between levels 1 and 2 corresponding to $\Delta M_\mu = +1$ (M_μ is the muon magnetic quantum number), and also $\Delta F = \pm 1$, $\Delta M_F = \pm 1$ at a weak magnetic field of 3 G.

The high magnetic field experiment consists of inducing a transition from the substate 1 to the substate 2 of Fig. 2. The accurate experiment done by the Yale group gives as a final result of the high field experiment [16]

$$\Delta\nu(\exp) = (4463.15 \pm 0.06) \text{ MHz} . \qquad (12)$$

The error of 1 standard deviation is primarily due to statistical counting errors and to magnetic field inhomogeneity. One of the main difficulties in these experiments is the shift of $\Delta\nu$ with argon pressure due to collisions between the muonium and argon atoms (~ 0.1 MHz/10 atm). Data are taken as a function of pressure and linearly extrapolated to zero pressure.

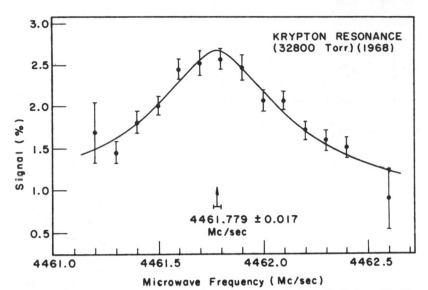

Fig. 3 A typical resonance curve for the transition $(F,M_F) = (1.1) \leftrightarrow$
(0,0) and $(1,-1) \leftrightarrow (0,0)$ in krypton buffer gas at 32,800 Torr. The
error bars (1 st. dev.) are due to counting statistics. The solid line
is a best fit to a theoretical Lorentzian line shape. The fitted centre
is 4463.260 ± 0.013 MHz.

    A low field experiment [17a,b] was used to measure $\Delta \nu$ in the
transition $(F,M_F) = (1,1) \leftrightarrow (0,0)$ and $(1,-1) \leftrightarrow (0,0)$. The experiments were done at a weak magnetic field ($\sim$ 3 G) and at a very
weak magnetic field ($\sim$ 10 mG). The principal difficulties of these
experiments are the relatively small signal and the requirement of
a homogeneous stable and small magnetic field. Figure 3 shows a
typical resonance curve for the unresolved transitions $(F,M_F) =$
$(1,1) \leftrightarrow (0,0)$ and $(1,-1) \leftrightarrow (0,0)$ in krypton buffer gas at 32,800
Torr. The Yale group was able to obtain the following results

$$\Delta \nu = 4463.260 \pm 0.013 \text{ MHz (from Kr data)} \qquad (13)$$

$$\Delta \nu = 4463.220 \pm 0.020 \text{ MHz (from Ar data)} . \qquad (14)$$

All the systematic errors are negligible compared with the random
errors. The authors claim that the difference of two standard
deviations between the Ar and Kr measurements may be evidence of
a non-linearity in the argon pressure shift.

    A weighted average of the high field measurement in Ar and of
the low field measurements (Ar, Kr) gives

$$\Delta \nu = 4463.248 \pm 0.031 \text{ MHz (1 st. dev.)} . \qquad (15)$$

The indicated 1 standard deviation errors in the individual measurements are due to counting statistics, which is the dominant
source of error. The error in the average value for $\Delta\nu$ has been
increased to account for the variation of the two measurements
about their weighted average. The average value $\Delta\nu$ is in very
good agreement with their previously quoted value of

$$\Delta\nu_{\text{exp}} = 4463.26 \pm 0.04 \text{ MHz} . \tag{16}$$

Recently, a Chicago group [18] measured the ground-state hyper-
fine interval $\Delta\nu$ of muonium in Ar by measuring the $(1,1) \leftrightarrow (1,0)$
Zeeman frequency ν_1 at that external field value ($B = 11.353$ kG)
where $\partial\nu_1/\partial B = 0$. The measurements were done at lower argon pressures than before (down to 3.15×10^3 Torr); this fact reduces
uncertainties in extrapolating $\Delta\nu(p)$ to zero pressure. The Chicago
result is

$$\Delta\nu(0) = (4463.317 \pm 0.021) \text{ MHz } (\pm 5 \text{ ppm}) , \tag{17}$$

which disagrees with the value found by the Yale group for Ar, but
is consistent with their result for Kr. Among the technical solutions used by the Chicago group I want to mention the use of that
external magnetic field value that makes the transition frequency
ν_1 field-independent to the first order. As a function of B, the
frequency goes through a maximum at a value $B_0 = 11.353$ kG; therefore inhomogeneities have only a second-order effect: an r.m.s.
$\delta B/B_0$ of 1% induces a shift $\delta\nu_1/\nu_1$ of -7 ppm.

Following the Chicago method we can deduce the value for the
ratio of the magnetic moment of the muon to that of the proton,
μ_μ/μ_p and compare it with the value of f_μ/f_p as measured by
Hutchinson et al.

$$\frac{\Delta\nu(\mu e)}{\Delta\nu(pe)} = \frac{\mu_\mu}{\mu_p} \left(\frac{1 + m_e/m_p}{1 + m_e/m_\mu} \right)^3 (1 + \delta_p - \delta_\mu) \tag{18}$$

where the δ_μ (179 ppm) and δ_p (35 ppm). Using the value of the
hydrogen hyperfine splitting [19]: $\Delta\nu(pe) = 1420.40575$ MHz, they
get:

$$\mu_\mu/\mu_p = 3.183351 \pm 0.000016 \left(\begin{matrix} \pm 5 \text{ ppm} \\ 1 \text{ st.dev.} \end{matrix} \right) . \tag{19}$$

An alternative method for deducing the ratio μ_μ/μ_p is given by
equating $\Delta\nu_{\text{theor}}$ to the $\Delta\nu_{\text{exp}}$ measured by the Yale group, assuming
the a.c. Josephson value of α [7]. The value obtained by the Yale
group is

$$\mu_\mu/\mu_n = 3.183306 \pm 0.000028 \quad \left(\begin{array}{c}\pm 9 \text{ ppm}\\ 1 \text{ st.dev.}\end{array}\right) ,$$

which agrees within 1.6 standard deviation with the value obtained
by the Chicago group and within 2.6 standard deviations with the
value measured in water by Hutchinson [20].

1.3   The 2s-2p Energy Difference of Muonic Hydrogen

I should now like to discuss one experiment that has not yet
been done, but which I believe will be done in the near future.
The CERN-Pisa group is studying the possibility of doing a kind
of Lamb-Retherford experiment in H and He, in which transitions
are induced between the levels 2s and 2p by a submillimetre wave
generator.   The experiment consists of detecting the 2s-2p resonance
frequency by observing a resonant enhancement in the counting rate
of the 2p-1s γ-rays of 1.9 keV.   The results of the experiment
according to the analysis done by Di Giacomo [21], will provide a
measurement to some per cent of accuracy of order α^2 of the electron vacuum polarization.   Therefore a measurement of the 2s-2p
energy difference in the µp atom would provide a good test of
pure quantum electrodynamics, apart from the possibility that a
proton halo could exist (i.e. the charge distribution around the
proton may have a weak positive halo with characteristic range about
8 fermi, and effective charge about 1% of the proton charge [22a,b]).

The CERN-Pisa group is doing preliminary tests to establish
the various effects that can depopulate the 2s level and to detect
γ-rays coming from the 2p-1s transition.

1.4   Electric Dipole Moment

The existence of an edm for an elementary particle would be
inconsistent with invariance under parity and time reversal [23].
The latest direct measurement to establish a limit on the edm has
been made by Charpak et al. [24].   They found

$$\text{edm} \leq e \times (0.6 \pm 1.1) \times 10^{-17} \text{ cm} .$$

It would be desirable to reduce the limit established by Charpak
et al. on the edm of the muon, although we know that the edm of the
electron, if it exists, must be less than [25]

$$(\text{edm})_e \leq e \times (3 \times 10^{-24}) \text{ cm}$$

(at 90% confidence level).

2.   ELECTROMAGNETIC PROPERTIES OF THE MUON AT HIGH MOMENTUM TRANSFER

2.1 Generality

The high-energy experiments examine whether the muon follows
QED at high momentum transfer. The idea is to test QED in a situation where the four-momentum transfer, characteristic of the experiment, is as large as possible.

In a particular model for the breakdown of QED one assumes that
a violation occurs for large absolute values of the momentum transfer associated with the photon propagator or the muon propagator or
muon-photon vertex function. These propagators can be modified by
multiplying them by a single-pole structure factor $(1 - q^2/\Lambda^2)^{-1}$,
where the cut-off constant can be different for muon, photon propagators and vertex function. Since the characteristic constant Λ
can be different in all three cases, it is possible that violation
of QED could be seen only in particular experiments and not in
others. However, it should be emphasized that this is a very simplified model, since the lepton propagator and the vertex function
cannot be modified separately and independently without violating
certain basic theorems in field theory [26]. Such a simple model
of breakdown serves the practical purpose of allowing a comparison
between different QED experiments, although it seems to me that the
different parameters Λ obtained in the study of different reactions
are hard to compare since, even assuming adequacy of the fitting
formula, they usually refer to different explored structures.

2.2 Elastic Muon-Nucleon Scattering

Recently the Columbia-Rochester group [27] have performed a
test on muon-electron universality by comparing their muon-proton
elastic scattering cross-sections with similar electron-proton
results. Any difference in the results must be attributed to the
lepton vertex, because the influence of the nucleon vertex and the
photon propagator on the cross-section is the same. Measurements
of the muon-proton cross-section for μ^+ and μ^- at 6 and 11 GeV/c
and μ^- at 17 GeV/c were used. To facilitate the comparison, the
Columbia-Rochester group extracts a single form factor from the
slope, assuming that $G_E = G_M/\mu \equiv G$. The form factor $G(q^2)$ has a
convenient parametrization as a function of q^2. The analysis is
done comparing the Columbia-Rochester data with the data of
Janssens et al. [28] for the electron. The electron and muon data
were fitted separately with the function

$$G_{e,\mu}(q^2) = F(q^2) \frac{N_{e,\mu}}{1 + (q^2/\Lambda^2_{e,\mu})} . \qquad (20)$$

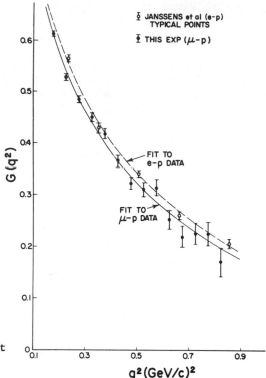

Fig. 4 The measurements of the form factor $G(q^2)$ versus q^2 for the Columbia-Rochester experiment and for the e-p data of Janssens et al.

The factor $F(q^2)$ is common to both fits and any μe structure difference is contained in the expression $N_{e,\mu}/(1 + q^2/\Lambda^2_{e,\mu})$ where $\Lambda_{e,\mu}$ is a cut-off parameter and $N_{e,\mu}$ is an arbitrary normalization. Figure 4 shows the measurements of the form factor $G(q^2)$ versus q^2 for this experiment and for the ep data of Janssens et al. The conclusions are independent of the particular form used for $F(q^2)$. Equation (20) was fitted separately to the electron and muon data. To compare the two fits the following ratio was used:

$$\frac{G_\mu}{G_e} = \frac{N}{1 + q^2/\Lambda^2} \tag{21}$$

where $N = N_\mu/N_e$ and $1/\Lambda^2 = 1/\Lambda^2_\mu - 1/\Lambda^2_e$. The results of the fit are summarized as follows:

i) No relative normalization difference ($N_\mu = N_e = 1$). The following equation was obtained

$$1/\Lambda^2 = 0.148 \pm 0.024 \ (GeV/c)^{-2} \ . \tag{22}$$

ii)   The shapes of the electron and muon data were supposed to be
      the same $(1/\Lambda_e^2 = 1/\Lambda_\mu^2 = 0)$. The relative normalization was
      found to be

$$N = 0.960 \pm 0.006, \tag{23}$$

that gives 6 standard deviations from unity. The systematic
error on the determination of q is less than 0.5% yielding to
an error in the cross-section of 2.5%. The difference in the
cross-sections due to the 4% suppression of the ratio G_μ/G_e is
8%. The authors were unable to find a single systematic error
that can explain such a deviation in the cross-sections, but
a combination of systematic errors could explain, perhaps,
such a difference.

iii) The authors, in view of systematic uncertainties, prefer a
      conservative point of view in fitting the data. They, in fact,
      left both N and $1/\Lambda$ as free parameters. They found

$$\begin{aligned} N &= 0.976 \pm 0.017 \\ 1/\Lambda^2 &= 0.064 \pm 0.056 \ (GeV/c)^{-2} . \end{aligned} \tag{24}$$

In the last case the parameter $1/\Lambda$ is not significantly different from zero. They quoted, with 95% confidence, a limit
of $\Lambda > 2.4$ GeV/c.

The same group tested the validity of one-photon exchange
approximation in the muon-proton elastic scattering experiments
[29]. It has been shown [30] that this approximation is valid
in the electron-proton elastic scattering. The cross-sections
of muon-proton elastic scattering have been determined with
an accuracy of 2% in the range $0.15 < q^2 < 0.85$ $(GeV/c)^2$ with
μ^+ and μ^- beams of 6 and 11 GeV/c and a μ^- beam of 17 GeV/c.

The Columbia-Rochester-Harvard-NAL group have now proposed
studying the muon-proton interaction from $q^2 \simeq 1(GeV/c)^2$ up to
$q^2 \sim 8(GeV/c)^2$, to compare the data with electron scattering data.
The comparison will be done using elastic and inelastic scattering;
in fact a relative mapping of q^2 and $E_{\mu,e}$ for electrons and muons
would be good enough to discover any possible difference.

2.3 Muon Pair Production

A detailed and lucid review on the subject of the muon pair
production has been given by Prof. Weinstein [31] at the International Symposium on Electron and Photon Interactions at High
Energies in Stanford in 1967. Prof. Stein will talk in the morn-
ing session about a new determination of the muon pair production.
Therefore I will leave him to deal with this subject; I only want

to make a few remarks here. Relatively large deviations from the
predictions of QED could be missed in the muon-proton elastic scattering, because either the new elements could be strongly coupled
to the lepton propagator, or the domain of variables should be time-
like rather than space-like. Most of the work on space-like leptons
has been done by the use of pair-photoproduction on nuclei (mainly
carbon), both for wide angle electron-pairs (WAEP) and muon-pairs
(WAMP). Usually symmetric pairs are investigated with one exception
[32]. The advantages of using a symmetric arrangement are based,
mainly, on the following considerations:

a) The charge-conjugation invariance [33] requires that the inter-
    ference between Bethe-Heitler and virtual Compton diagrams
    vanishes.

b) A low nuclear recoil is produced. Such a fact permits one to
    minimize the corrections due to the nuclear form factors. It
    is known that both virtual Compton contributions and inelastic
    nuclear form factors introduce uncertainties of some per cent
    in the interpretation of wide angle pairs as a QED process.

c) When the invariant mass M of a produced muon pair is near the
    ρ^0 mass the photoproduction of ρ^0 and the subsequent decay in
    $\mu^+\mu^-$ can contribute to the measured yield. A separation can
    be made, to some extent, since the dependence on the pair aper-
    ture angle is rather different in the Bethe-Heitler and ρ^0 con-
    tribution.

     A Cornell group [34] has measured the photoproduction of wide-
angle muon pairs from carbon in order to test the validity of quantum electrodynamics at small distances. The data extend to a muon-
pair invariant mass of about 1150 MeV and are consistent with quantum
electrodynamics. If one simulates a possible breakdown of QED with
a term of the form $\lambda M_{\mu\mu}^4 \, \sigma_{BH}$, where $M_{\mu\mu}$ is the invariant mass of
the $\mu^+\mu^-$ final state

$$M_{\mu\mu}^2 = (p_+ + p_-)^2 \, ,$$

one finds

$$\lambda = -0.09 \pm 0.05 \text{ GeV}^{-4} \, .$$

The authors feel that the uncertainty in the theory (for example,
the exact shape of the mass distribution of the rho and other sources
of uncertainty) could account for the non-zero value of λ given by
the least-squares fit.

     I will try to summarize the knowledge that I have on the sub-
ject by saying that new and old data [34,31] of the photoproduction
of the muon pair appear to be in substantial agreement with QED.
In other words the test that we are doing on intermediate space-
like charged muons removed from the mass-shell seems to agree with
the QED prediction.

2.4 Bremsstrahlung of Muons at Large Angle

In the muon bremsstrahlung processes one of the two dominant amplitudes involves a time-like virtual muon. The two amplitudes are comparable, but since only the sum is gauge invariant, their relative importance depends on the gauge that has been chosen for the calculation. Any measurement of the yield at large values of q_t^2 provides a test of the validity of QED.

Recently Liberman et al. [35] have measured muon bremsstrahlung in an experiment to observe the process by which 9 to 13 GeV/c muons lose most of their energy in a carbon target. Several corrections were applied to the yield predicted by QED in order to compare the expected total number of events with the experimental number. A normalization uncertainty of 11% was estimated by the authors. The ratio of measured events to predicted events is R = 0.91 ± 0.12 for both beam polarities, where the error includes statistics. A deviation of R from unity involves the following possibilities:

a) A breakdown of QED could be of the form

$$R = A \ (1.0 \pm Q^4/\Lambda^4) \ . \tag{25}$$

Liberman et al. [35] found

$$R = 0.93 \ \left[(1.0 \pm 0.045) - (0.69 \pm 1.37) \ 10^{-12} \ (\text{MeV}/c)^{-4} \ Q^4\right] \ .$$

This corresponds to a $\Lambda > 0.73$ GeV/c, with 95% confidence limits. The errors included in making this fit are only statistical errors.

b) Existence of a heavy muon with mass $Q/c = M$, which decays into a muon and a photon. The experiment does not show any statistical evidence for a heavy muon with mass less than 600 MeV/c^2. The experiment of Liberman et al. agrees, therefore, with the predictions of QED; the time-like muon propagator seems not to present anything anomalous.

2.5 Muon Tridents

One reaction in which two identical muons are produced in the final state is the direct production of a muon pair by a muon in the field of a nucleus. The muon trident processes are interesting mainly for two reasons: i) they give a good test of the validity of QED at small distances; ii) they give the opportunity of observing the effect of the statistics of the muons. The effect of the statistics is essentially a limitation on phase space in any final state. In regions of phase space in which two identical particles overlap, the states of parallel spins will be excluded by Fermi-Dirac statistics and the cross-section [36] will be depressed.

Studying the muon trident process is equivalent to studying the muon-muon scattering process; therefore any structure in the muon will show up as a modification of the cross-section versus q^2. An advantage of such a configuration is that any hypothetical neutral meson which decays into muon pairs will show up as a bump in the invariant mass distribution of the produced pair. The trident experiments are sensitive to the mass M of such a hypothetical neutral meson, while the (g - 2) experiment is sensitive to the ratio of the mass to the coupling constant. Recently J.J. Russell et al. [37] have studied the direct production of a muon pair by an incident muon in the field of a heavy nucleus. The observed total cross-section for the production of the muon trident is 60.4 ± 6.1 nanobarns (or 89 ± 9.5 events) and it is in good agreement with the predicted value of 55.7 ± 1.3 nanobarns (or 82 ± 2 events). A new experiment has been proposed by a Columbia-Harvard-Rochester-NAL group at Brookhaven.

2.6 Relative Electron, Muon Form Factors for Time-Like Photon

There have been measurements of $B_{\rho\mu} = (\rho \rightarrow 2\mu/\rho \rightarrow all)$ and $B_{\rho e} = (\rho \rightarrow 2e)/(\rho \rightarrow all)$ as by-products of lepton pair production experiments. These measurements test for a difference in eγe vertex and $\mu\gamma\mu$ vertex cut-off momenta in the time-like region, as (e,p) versus (μ,p) scattering does in the space-like region. The average value of the branching ratio is [38]

$$R = \frac{\rho^0 \rightarrow e^+e^-}{\rho^0 \rightarrow \mu^+\mu^-} = 0.97 \pm 0.17 \ . \tag{26}$$

The γ-ray at the lepton vertex has $q^2 = 0.75(\text{GeV})^2$. Thus μ and e form factors are equal at $q^2 = m_\rho^2$ to within 9% (1 standard deviation).

2.7 Colliding-Beam Experiments

Electron-positron scattering experiments at high momentum transfers provide a very good dynamical test of quantum electrodynamics; particularly the reaction

$$e^+ + e^- \rightarrow \mu^+ + \mu^- \tag{27}$$

can be useful also to test the muon structure.

At lowest order the reaction proceeds through the exchange of one photon in the time-like region. This reaction, therefore, provides time-like virtual-photon interaction with muon vertex. I want only to remark here that to determine a breakdown of the theory in $e^+ + e^- \rightarrow \mu^+ + \mu^-$, according to a single pole model, one has to

measure the absolute cross-section; in fact the angular dependence
is independent of a possible breakdown as long as the one-photon
channel dominates. For instance with $F(q^2) = 1/(1 + q^2/\Lambda^2)$ the
cross-section change of $|F(-4E^2)|^2 \simeq 1 + 8E^2/\Lambda^2$ and for $8E^2/\Lambda^2 \simeq 10\%$
it is possible to explore up to $\Lambda \sim 9E$, that means for $E \sim 1$ GeV
to explore distances around 3×10^{-2} fermi. The reaction (27) looks
like one of the most promising experiments to investigate the $\mu\gamma\mu$
structure. The interpretation of this experiment requires an accurate evaluation of the radiative corrections, that have been discussed
by Furlan, Gatto and Longhi [39]. The reaction (27) could show also
a structure in the cross-section in the region of centre-of-mass
energy comparable to the ϕ^0 mass; in fact the presence of a ψ^0 in
the vacuum polarization contribution to two muon annihilation should
give a peculiar wavy structure to the cross-section [38].

T.D. Lee and G.C. Wick [11] in listing some of the experimental
consequences of their theory, point out that if the heavy photon
pole exists with, for example, a mass $m \simeq 10$ GeV, then at $(q^2)^{\frac{1}{2}} \sim$
~ 6 GeV the deviation in transition probability from usual quantum
electrodynamics for the process (27) is $[1/(1 + q^2/m^2)]^2 \simeq 50\%$.
As pointed out earlier, colliding-beam experiments $(e^+e^- \to \mu^+\mu^-)$
should show if such a heavy photon pole exists with a mass around
9 GeV, measuring with an accuracy of a few per cent the cross-
section in the region around $q^2 \sim (3 \text{ GeV})^2$.

3. CONCLUSIONS

The muon looks like a heavy electron. The muon-electron puzzle was established in 1945 in Rome by the Conversi, Pancini,
Piccioni [40] experiments. Soon thereafter its fermion nature [41]
was established. As remarked by L.M. Lederman [42], from this time
on the "puzzle" of a large mass difference between two otherwise
identical particles with no indication of strong interaction has
been sufficient to receive a paragraph in the summarizing epoch of
every international conference.

In the last 20 years great progress has been made on the understanding of weak, strong and electromagnetic interactions, and in
the discovery of deeper symmetries in the organization of strongly
interacting particles. In contrast the muon-electron puzzle still
persists in spite of the experimental efforts made. I have to point
out here that quite a lot of the progress made in understanding the
limit of validity of electrodynamics has been possible because of
the desire to solve the muon-electron puzzle. Being physicists we
cannot accept the fact that such a muon-electron puzzle will persist for ever without being understood. What are the activities I
foresee in the near future as being the most promising to lead to
an understanding of the electromagnetic properties of the muons?
There are mainly two roads of approach to the problem:

the high-energy road, that is followed primarily in the USA, and
the low-energy road, that is the road followed at CERN [measuring
of the muon (g - 2)] and also in the USA (studying the properties
of the muon bound to the electron). The high-energy road approaches
the understanding of the muon-electron puzzle in studying more in-
timate collisions (higher q^2), in the hope that whatever causes the
muon-electron mass difference, some reflection of this will show up
in the electromagnetic interactions. The low-energy road is followed at CERN with the same hope: that whatever will be found to
cause the muon to be different from the electron, it will be detected in studying with high accuracy the static properties of the muon,
and in particular its anomalous magnetic moment.

The new proposal of the Columbia-Rochester-Harvard-NAL group
to study the high-energy collisions of the muon is very impressive.
I believe a deep knowledge of the electromagnetic properties of the
muon at high q^2 will be found by such experiments. In particular I
want to mention the muon trident as one of the more promising roads.

I hope that the colliding-beam experiments will soon give some
results on the time-like photon interaction with the muon vertex.

The new proposal of CERN to measure the anomalous magnetic
moment of the muon with higher accuracy will permit the deeper ex-
ploration of the muon structure and in particular the examination of
a region where the strong, electromagnetic and weak interactions
will be mixed up together.

I want to conclude my talk by calling your attention to the recent calculation of Aldins, Brodsky, Dufner and Kinoshita on the
light by light contribution to the anomalous magnetic moment of the
muon. It is a surprising, but welcome result, partly because the
theory has moved to meet the experiment and partly because it is an
example of how an unexpected result can still be obtained in the
QED framework if we explore deeply the structure of one particle
(the muon).

Acknowledgements

I should like to thank Drs. J. Bailey and F. Combley, and
Professor F.J.M. Farley for their helpful criticism in preparing
this talk.

REFERENCES

1. F.J.M. Farley, Prog.Nucl.Phys. 9, 257 (1964).
2. G. Feinberg and L.M. Lederman, Ann.Rev.Nucl.Sci. 13, 431 (1963).
3. a) G. Charpak, F.J.M. Farley, R.L. Garwin, T. Muller, J.C. Sens
        and A. Zichichi, Nuovo Cimento 37, 1241 (1965).

    b) F.J.M. Farley, J. Bailey, R.C.A. Brown, M. Giesch,
        H. Jöstlein, S. van der Meer, E. Picasso and
        M. Tannenbaum, Nuovo Cimento 45, 281 (1966).
    c) J. Bailey, W. Bartl, G. von Bochmann, R.C.A. Brown,
        F.J.M. Farley, H. Jöstlein, E. Picasso and R.W. Williams,
        Phys.Letters 28 B, 287 (1968).
    d) J. Bailey and E. Picasso, Prog.Nucl.Phys. 12 (in press).
4. For fourth-order radiative correction see, for example:
    a) C. Sommerfield, Phys.Rev. 107, 328 (1957).
    b) A. Peterman, Helv.Phys.Acta 30, 407 (1957), Phys.Rev. 105,
        1931 (1957), Nucl.Phys. 3, 689 (1957).
    c) H.H. Elend, Phys.Letters 20, 1 (1966), 21, 720 (1966).
For sixth-order radiative correction see, for example:
    d) S.D. Drell, Proc. 13th Int.Conf. on High Energy Physics,
        Berkeley, 1966 (Univ. of Calif.Press, Berkeley, 1967),
        p.85.
    e) T. Kinoshita, Nuovo Cimento 51 B, 140 (1967), Cargèse Lec-
        tures in Physics (ed. M. Levy) (Gordon and Breach, New
        York, 1968), vol. 2.
    f) B.E. Lautrop and E. de Rafael, Phys.Rev. 174, 1835 (1968).
5. J. Aldins, S.J. Brodsky, A.J. Dufner and T. Kinoshita, Phys.
    Rev.Letters 23, 441 (1969).
6. M. Gourdin and E. de Rafael, Nucl.Phys. B10, 667 (1969).
7. B.N. Taylor, W.H. Parker, and D.N. Langenberg, to be published
    in Rev.Mod.Phys. 41 (July 1969).
8. J.S. Bell and E. de Rafael, Nucl.Phys. B11, 611 (1969).
9. J. Bailey, private communication.
10. J. Bailey, F.J.M. Farley, H. Jöstlein, G. Petrucci, E. Picasso
    and F. Wickens, CERN Internal Note PH I/COM-69-20, 19 May
    1969.
11. T.D. Lee and G.C. Wick, Nucl.Phys. B9, 209 (1969).
12. a) F.J.M. Farley, "The status of quantum electrodynamics",
        invited paper at the first meeting of the European
        Physical Society, Florence (1969).
    b) S.J. Brodsky, Status of quantum electrodynamics, 1969
        International Conference on Electron and Photon Inter-
        actions at High Energies (preprint).
13. V.W. Hughes, Ann.Rev.Nucl.Sci. 16, 445 (1966).
14. G. Källen, Quantenelektrodynamik, Handbuch der Physik (ed.
    S. Flügge) (Springer-Verlag, Berlin, 1958), vol. 5/1,
    p. 169-364.
15. H.A. Bethe and E.E. Salpeter, Quantum mechanics of one- and
    two-electron systems, Encyclopedia of Physics (ed.
    S. Flügge) (Springer-Verlag, Berlin, 1957), vol. 35/1,
    p. 88-436.
16. W.E. Cleland, J.M. Bailey, M. Eckhause, V.W. Hughes, R. Prepost,
    J.E. Rothberg and R. Mobley, Phys.Rev.Letters 13, 202 (1964).

17. a) P.A. Thompson, J.J. Amato, P. Crane, V.W. Hughes, R.M. Mobley,
         G. zu Putlitz and J.E. Rothberg, Phys.Rev.Letters 22, 163
         (1969).
     b) P. Crane, J.J. Amato, V.W. Hughes, D.M. Lazarus, G. zu Putlitz
         and P.A. Thompson, Recent muonium hyperfine structure
         measurements, submitted to III Int. Conf. on High Energy
         Physics and Nuclear Structure, Columbia Univ., New York
         (Sept. 8-12, 1969).
18. R.D. Ehrlich, H. Hofer, A. Magnon, D. Stowell, R.A. Swanson
         and V.L. Telegdi, EFINS 69-71.
19. S.B. Crompton, D. Kleppner and N.F. Ramsey, Phys. Rev. Letters
         11, 338 (1963).
20. D.P. Hutchinson, J. Menes, G. Shapiro and A.M. Patlach, Phys.
         Rev. 131, 1351 (1963); see also G. McD. Bingham, Nuovo
         Cimento 27, 1352 (1963).
21. A. Di Giacomo, Nucl.Phys. B11, 41 (1969).
22. a) R.C. Barrett, S.J. Brodsky, G.W. Erickson and M.H. Goldhaber,
          Phys.Rev. 166, 1589 (1968).
     b) D.R. Yennie, Proceedings of 1967 International Symposium on
          Electron and Photon Interactions at High Energies
          (Stanford, 1967), p. 32.
23. J.S. Bell, Phys.Rev. 92, 997 (1953).
24. G. Charpak, F.J.M. Farley, R.L. Garwin, T. Muller, J.C. Sens
         and A. Zichichi, Phys. Letters 1, 16 (1962).
25. M.C. Weisskopf, J.P. Carrico, H. Gould, E. Lipworth and
         T.S. Stein, Phys.Rev.Letters 21, 1645 (1968).
26. N.M. Kroll, Nuovo Cimento 45 A, 65 (1966).
27. L. Camilleri, J.H. Christenson, M. Kramer, L.M. Lederman,
         Y. Nagashima and T. Yamanouchi, Phys.Rev.Letters 23, 153
         (1969).
28. T. Janssens, R. Hofstadter, E.B. Hughes and M.R. Yearian,
         Phys.Rev. 142, 922 (1966).
29. L. Camilleri, J.H. Christenson, M. Kramer, L.M. Lederman,
         Y. Nagashima and T. Yamanouchi, Phys.Rev.Letters 23, 149
         (1969).
30. G. Weber, J. Pine and R.E. Taylor in Proceedings of the Third
         International Symposium on Electron and Photon Interactions
         at High Energies, Stanford Linear Accelerator Center, 1967
         (Clearing House of Federal Scientific and Technical Informa-
         tion, Washington D.C. 1968).
31. S. Weinstein in Proceedings of the Third International Symposium
         on Electron and Photon Interactions at High Energies,
         Stanford Linear Accelerator Center, 1967 (Clearing House of
         Federal Scientific and Technical Information, Washington
         D.C. 1968).
32. D.J. Quinn and D.M. Ritson, Phys.Rev.Letters 20, 890 (1968).
33. A.S. Krass, Phys.Rev. 138, B1268 (1965).
34. S. Hayes, R. Imlay, P.M. Joseph, A.S. Keizer, J. Knowles and
         P.C. Stein, Phys.Rev.Letters 22, 1134 (1969).

35. A.D. Liberman, C.M. Hoffman, E. Engels, Jr., D.C. Imrie,
    P.G. Innocenti, R. Wilson, C. Zajde, W.A. Blanpied,
    D.G. Stairs and D.J. Drickey, Phys.Rev.Letters 22, 663 (1969).
36. M.J. Tannenbaum, Phys.Rev. 167, 3108 (1968).
37. J.J. Russell, R. Sah, M.J. Tannenbaum, W.E. Cleland, D.G. Ryan
    and D.G. Stairs, Muon trident production in lead, submitted
    to the Daresbury Conference, 1969.
38. C. Bernardini, High Energy Experiments on QED, LNF/68/42.
39. G.R. Furlan, R. Gatto and G. Longhi, Phys. Letters 12, 262
    (1964).
40. M. Conversi, E. Pancini and O. Piccioni, Phys.Rev. 71, 205 (1947).
41. a) J. Steinberger, Phys.Rev. 75, 1136 (1949).
    b) R. Leighton, C.D. Anderson and A.J. Seriff, Phys.Rev. 75,
       1432 (1949).
42. L.M. Lederman: "Prospects for the muon-electron problem" in
    Old and new problems in elementary particles (ed. G. Puppi)
    (New York, Academic Press, 1968).

DISCUSSION

V.W. Hughes: The new method for measuring g-2 proposed by CERN, using electric field focussing, has been examined as a possibility for a similar experiment at LAMPF. Whilst one cannot use the same high-energy γ-value, one might be able to exploit the method at lower muon energies, where the apparatus would be smaller, but the accuracy attainable, comparable.

THE POSITIVE MUON MAGNETIC MOMENT

D. Hutchinson, A. Kanofsky[*], F. Larsen, N. Schoen,
D. Sober
Princeton-Pennsylvania Accelerator

[*]Lehigh University & Princeton-Pennsylvania Accelerator

The magnetic moment of the positive muon has been measured
at PPA by determining its precession frequency in a magnetic field.

Longitudinally polarized muons are obtained from forward
decays in flight along a 158 MeV/c pion beam. As shown in
Figure 1, the beam enters the precession magnet, passes through
1 1/8 inch copper pion and slow muon filter and stops in a water
target. The vertical magnetic field causes the muon spin to precess with a frequency $\omega_\mu = 2\mu_\mu \cdot H/\hbar$. The muon moment,
$\mu_\mu, = g \frac{e_\mu}{2m_\mu c} \cdot \frac{\hbar}{2}$ where g represents the muon g factor. An event

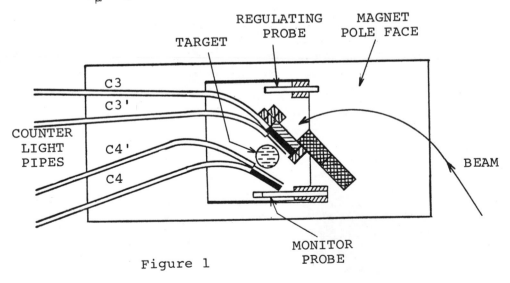

Figure 1

is counted when the muon decays into a positron which passes through counters C4' and C4. There is only one measurement for each **event**, the time interval between the muon stop and the decay positron. The time distribution is given by

(1) $dN = e^{-t/\tau} (1 - a \cos(\omega_\mu t + \alpha))$

where τ is the muon lifetime, 2.2 μsec, and the cosine term arises from the $\vec{S}_\mu \cdot \vec{P}_e$ correlation in muon decay. The parameter a is reduced from its maximum value of 0.3 to about .04 in this experiment due to incomplete polarization of the muon beam, further depolarization upon stopping in the target, and effects of time resolution and solid angle of the C4'-C4 telescope.

The method of measuring the decay time is similar to that of a previous measurement at Nevis Laboratory, Columbia University.[1] The decay positrons are accepted during a 4.0 μsec gate after the muon stop, which is subdivided at 2.0 μsec. Those decays in a .05 to 2.0 μsec interval are called early, those from 2.0 to 4.0 μsec are late. The time for each event is measured modulo a reference frequency, ω_{Ref}, which is almost equal to the muon precession frequency. This has the effect of folding the distribution so that only one cycle of the modulation frequency remains. The initial phase, α, of the cosine function is arbitrary, determined by delays in the logic, but $\Delta\alpha$, $\alpha_{late} - \alpha_{early}$, is equal to $(\omega_{Ref} - \omega_\mu) \cdot T$. T is the length of the early gate, 1.95 μsec. Then, to find ω_μ, the magnetic field, as measured by the frequency f_p of proton NMR in water, is varied in the region of $\omega_{Ref} \sim \omega_\mu$ and $\Delta\alpha$ found for each field setting. A plot of $\Delta\alpha$ versus f_p, Figure 2, has $\Delta\alpha=0$ intercept where $\omega_\mu = \omega_{Ref}$. The value of f_p at $\Delta\alpha=0$ is 51.09950±.00036 Mhz, and $\omega_{Ref}=162.66618±.00005$ Mhz x 2π, giving a value of $\omega_\mu/\omega_p = 3.183322±23$ (statistical error).

This initial result is subject to the following corrections and errors.

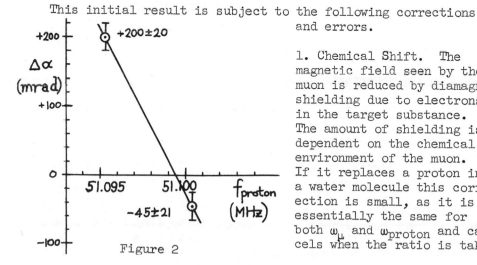

Figure 2

1. Chemical Shift. The magnetic field seen by the muon is reduced by diamagnetic shielding due to electrons in the target substance. The amount of shielding is dependent on the chemical environment of the muon. If it replaces a proton in a water molecule this correction is small, as it is essentially the same for both ω_μ and ω_{proton} and cancels when the ratio is taken.

However, Ruderman[2] has suggested that the muon forms $(H_2O-\mu-H_2O)^+$ and calculated that the muon shielding could be reduced by as much as 20 ppm. As it is difficult to make a reliable calculation of this shift, due to dimer formation, and difficulty in knowing molecular geometry, we change the initial value of ω_μ/ω_p by -10 ± 10 ppm to cover both cases.

2. Systematic Error in the Clock. The muon lifetime limits the time available for frequency measurement to the order of $2\mu sec$, so care is needed in the clock calibration as an error of 2×10^{-12} sec will produce 1 ppm error in the measurement. The muon stopping time is simulated by a pulse generator and the decay time simulated by a second, random time random voltage, pulse generator. Each pulse is linearly mixed with a common RF voltage at the calibration precession frequency and then triggers a discriminator. These two pulses are now time correlated at a known frequency as the RF voltage modulates the probability of a discriminator trigger. The probability of time interval t between the two pulses is

$$(2)\ p(t)dt \sim e^{-t/T_{p2}}\left(1+\sum_{j=1}^{\infty}a_j\cos(j\omega_{RF}t+\theta_j)\right)dt$$

T_{p2} is the average period of the random pulser and is set equal to $2.2\mu sec$. The summation is a Fourier series representation of an arbitrary correlation, but as only the first term is used in the data analysis[2] reduces to $(3)\ P(t)=e^{-t/2.2\mu sec}(1+a_1\cos(\omega_{RF}t+\theta_1))dt$.

This is of the same form as equation (1), except that ω_μ is replaced by the known ω_{RF}. Calibration of the clock at intervals during data taking indicated a constant systematic error in $\Delta\alpha$ of (-6 ± 4) ppm. Therefore ω_μ/ω_p is decreased by 6 ± 4 ppm.

3. Monitor-Target ΔH. The correction due to the field difference between the target average and the monitor position field is $+24\pm1$ ppm.

4. Target Field Variation. An error of 5 ppm is allowed due to field variation over the target volume and uncertainty in the muon stopping distribution.

5. Time Variation of H. An error of 5 ppm is allowed due to time fluctuations of the magnetic field and the absolute resolution of the NMR signal due to shape distortion.

The net correction is +8 ppm with a total error of 15 ppm, yielding $\omega_\mu/\omega_p = 3.18335\pm.00005$.

The Columbia measurement[1] was published prior to Ruderman's model[2] of muon chemistry. If we change it by -10 ± 10 ppm to account for uncertainty in the chemical shift we have $\omega_\mu/\omega_p=3.18335\pm.00005$, in agreement with the present measurement.

There have been two recent measurements of muonium hyperfine splitting[3,4] which is a function of the muon magnetic moment. If the theory for the splitting is assumed to be correct, these measurements can be used to calculate the muon moment. V. Telegdi and coworkers find $\frac{\omega_\mu}{\omega_p}$ =3.183351±.000016 (5 ppm)[3] and V. Hughes and coworkers find ω_μ/ω_p = 3.183314±.000034[4] (11 ppm).

The agreement between these two measurements and the above direct measurements then supports the calculated value of the muonium hfs to 16 ppm, with almost all of the error due to the muon moment measurement.

References

1. D. P. Hutchinson, J. Menes, G. Shapiro and A. M. Patlach, Phys. Rev. 131, 1351 (1963) also Nevis Report 103.

2. M. A. Ruderman, Phys. Rev. Lett. 17, 794 (1966).

3. R. D. Ehrlich, H. Hofer, A. Magnon, D. Stowell, R. A. Swanson, V. L. Telegdi, Phys. Rev. Lett. 23, 513 (1969).

4. P. A. Thompson, J. J. Amato, P. Crane, V. W. Hughes, R. M. Mobley, G. Zu Putlitz, J. E. Rothberg, Phys. Rev. Lett. 22, 163 (1969).

DISCUSSION

V. Hughes: A question and a comment regarding the chemistry. Is it still your belief that for sufficient OH concentration you will form muonium type water (μ^+OH)? Ruderman's model is certainly interesting, but the starting point is a muon bound to something. My guess is that the starting point is muonium at 1 keV or thereabouts, but what happens from this point -- down to some stationary state is obscure. D.P.H.: To understand the physical chemistry of this process (and the corresponding diamagnetic shifts), would be, we are told, a difficult and lengthy business, and the outcome none too reliable! We are presently investigating experimentally the effect of OH$^-$, to see if the production of the positive ion (as proposed by Ruderman) is suppressed, by large concentrations. Preliminary results indicate some possible effect. V. Telegdi: There was found, many years ago, a dependence of residual μ-polarisation on pH.

VACUUM POLARIZATION IN MUONIC ATOMS

Herbert L. Anderson

University of Chicago

Chicago, Illinois

My purpose in presenting this report is simply to point out that the precision in measuring muonic x-ray lines is now becoming high enough to provide a significant test of the vacuum polarization effects in muon physics not only in the lowest but also in higher orders. For this purpose it is best to choose one of the higher transitions which are little affected by the finite nuclear size. In general, finite nuclear size corrections are small enough in muonic atoms for transitions with energies less than 500 keV.

The situation for Pb^{206} is outlined in greater detail in Fig. 1 which lists the various corrections which contribute to the observed energies of three of the more prominent transitions. Here we show the transition energies as calculated by solving the non-relativistic Schrodinger equation, listing separately the effect of taking into account the reduced mass and the effect of spin and relativity, obtained by solving the Dirac equation for a point nucleus. The finite nuclear size is a major effect in the case of the 2p-1s transition, amounting to 10 MeV. For the 4f-3d transition it has fallen to 4 keV but amounts to only 10 eV or so for the 5g-4f transition.

On the other hand, the vacuum polarization is a long range effect, being determined by the electron Compton wavelength. In lowest order it amounts to 34.8 keV for the $2p_{\frac{1}{2}}$ - $1s_{\frac{1}{2}}$ transition. It is already larger than the finite size effect for 4f-3d and is 200 times larger for 5g-4f. Other corrections such as the Lamb shift and the correction for the anomalous magnetic moment are less than 10 eV for this transition.

Except for these there remain the electron shielding and higher order radiative corrections. The higher order radiative corrections according to a recent calculation by Fricke[1] are given in Fig. 2. In the case of the 5g-4f transitions for Pb these amount to 0.04 keV, about ½ as much as the correction due to electron shielding.

Figure 1 indicates the experimental accuracy obtained by the Chicago-Ottawa group.[2] In their measurements on the Pb isotopes made two years ago the accuracy obtained was ±0.15 keV for the 5g-4f transition. This can certainly be improved, perhaps by as much as an order of magnitude using improved detectors and techniques now available in an experiment specifically designed to measure the higher order transitions. In fact, in a paper submitted to this Conference Backenstoss, et al[3] report a measurement of the vacuum polarization effect in muonic Bi and Pb in which the stated accuracy is down to the ±0.040 keV level.

In the Chicago-Ottawa experiment,[2] a comparison of the observed and calculated energies for the transitions $4f_{5/2}-3d_{3/2}$, $4f_{7/2}-3d_{5/2}$, $5g_{7/2}-4f_{5/2}$, and $5g_{9/2}-4f_{7/2}$ was made. The lowest order vacuum polarization corrections were increased by 2.2% to take into account the higher order effects as given by Fricke.[1] It was then found that the vacuum polarization effect was overestimated by 3.9 ± 2.0%. Thus, the higher order corrections were just at the limit of the accuracy of this experiment.

Now, Backenstoss, et al[3] reporting measurements of the same lines in Bi^{209} and in Pb^{206} find that the weighted mean of the relative deviations of the experimental vacuum polarization values from those calculated amounts to 0.2 ± 0.9%. By omitting the higher order terms in the calculated values this value changes to -1.5 ± 0.9%. Hence, the agreement between theory and experiment is improved significantly by including the higher order terms.

The importance of vacuum polarization effects in muonic states was first demonstrated by Koslov, Fitch and Rainwater.[4] They studied the 3d-2p transition in phosphorous where finite size effects are quite small. Although limited to NaI as a detector, by utilizing the K-absorption edge in Pb they were able to measure the transition energy accurately enough to establish the effect.

An extension of this technique yielded a precision measurement of the muon mass. This was made possible by a more precise determination of the energy dependence of the x-ray absorption coefficient in lead by Bearden[5] and the calculation by Petermann and Yamaguchi[6] of various corrections to the 3d-2p transition energies in phosphorous. The experiments by Lathrop et al[7,8] and Devons et al[9] yielded as an average

$$m_\mu = (206.76 \pm 0.02) \, m_e$$

These measurements still stand as the most accurate determination
of the muon mass via the muonic Rydberg and are in good agreement
with the more precise value 206.765 ± 0.003 obtained from the g-2
experiments.[10,11,12] The argument has been turned around[10] to show
that if the g-2 value of the mass is used in interpreting these
pioneer muonic x-ray experiments, the vacuum polarization effect
appears to be correctly estimated to ± 4%. The work reported here
by Backenstoss et al[3] using Li drift Ge detectors has achieved better
accuracy strengthening the view that the opportunity for further
improvements is at hand.

If the experimental accuracy can be brought to the electron
volt range, it will be necessary to calculate the radiative and the
electron screening corrections with greater care. Literally, dozens
of lines are available to check the self-consistency of these corrections as well as the manner of handling the experimental problems
having to do with line shape, background and energy calibration of
the detector. For each element, a whole series of lines are
accessible, and one may choose the Z of the nucleus over a wide
range. Thus, possible level shifts due to the accidental coincidence
of a nuclear with an atomic level, which have their own interest,
should be easy to sort out. The single quantity that emerges from
all this, once the corrections are correctly made, is the muonic
Rydberg, and hence the muon mass. This lends itself to a direct
comparison with the electron mass because of the presence of the
511 keV annihilation radiation which is generally quite prominent
in muonic x-ray spectra.

One of my graduate students, Madhu Dixit, recently calculated
the electron screening correction for a number of the more useful
transitions in a variety of muonic atoms between Z = 40 and Z = 82.
For the electronic charge densities in the atom, he used the ones
obtained by Liberman, Waber and Cromer[13] who solve the self-
consistent field equations using Dirac wave functions, taking
exchange into account using Slaters' $\rho^{1/3}$ approximations. The shift
in energy levels was calculated in first order perturbation theory
by calculating the integral,

$$\Delta E_{n,\ell,j} = \int_{0}^{\infty} V_e(r) \{f^2_{n,\ell,j} + g^2_{n,\ell,j}\} \, dr \quad ,$$

taken for Z-1 electrons about a nucleus with charge Z-1 where $V_e(r)$
is the potential resulting from the electron charge density distribution and $f_{n,\ell,j}$ and $g_{n,\ell,j}$ are the major and minor point nucleus
Dirac wave functions for muons, normalized such that

CORRECTIONS TO THE ENERGY LEVELS IN Pb^{206}

TRANSITION	$2p_{1/2}-1s_{1/2}$ keV	$4f_{5/2}-3d_{3/2}$ keV	$5g_{7/2}-4f_{5/2}$ keV
SCHRÖDINGER	14186.65	919.51	425.60
REDUCED MASS	−7.81	−0.51	−0.23
SPIN AND RELATIVITY	+1427.42	+50.11	+10.28
FINITE SIZE	−9857.20	−4.00	−0.01
VACUUM POLARIZATION	+34.80	+6.71	+2.17
LAMB SHIFT	−1.71	−0.02	0.00
ANOM. MAG. MOMENT	−0.68	+0.05	+0.01
ELECTRON SHIELDING	−	−0.06	−0.09
TOTAL CALCULATED	5781.47	971.79	437.72
EXPERIMENTAL	5788.33 ±0.48	971.74 ±0.20	437.86 ±0.15
$E_{expt.}-E_{calc.}$	+6.9 ± 2.3	−0.05 ± 0.20	+0.14 ± 0.15

Figure 1

HIGHER ORDER CORRECTIONS TO THE VACUUM POLARIZATION[a]

2nd Order:	$2p_{1/2}-1s_{1/2}$	$3d_{3/2}-2p_{1/2}$	$4f_{5/2}-3d_{3/2}$
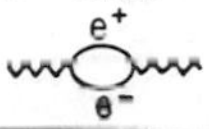 e^+ / e^-	34.80 keV	21.83 keV	6.71 keV
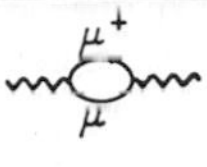 μ^+ / μ^-	0.20	0.05	0.00
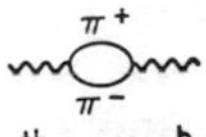 π^+ / π^-	0.03	−	−
4th Order:[b]			
	0.28	0.15	0.03
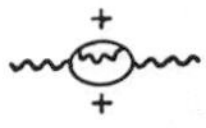	0.41	0.19	0.05
Coulomb Green's function correction[c]	0.24	0.22	0.07
TOTAL HIGHER ORDER	1.16 keV	0.61 keV	0.15 keV

a B. Fricke Z. Physik 218, 495 (1969).
b G. Källen and A. Sabry Dan. Mat. Fys. Medd. 29, No. 17 (1955).
c E. Wichmann and N. M. Kroll, Phys. Rev. 96, 232 (1954).

Figure 2

$$\int_{o}^{\infty} (f^2_{n,\ell,j} + g^2_{n,\ell,j}) dr = 1 \quad .$$

The present calculations do not extend below n = 3, although there would be no special difficulty in calculating these cases as well. For heavy elements the effect of the finite extension of the nuclear charge on the muon wave function should then be taken into account. This is a serious correction mainly for n = 1 in the heavy elements, but the electron shielding correction is quite small here.

In Table 1 we give examples of the electron shielding correction for some of the principal lines in the case of Pb, Z = 82 and Nb, Z = 41. The correction lowers the transition energy by about 4 parts in 10^5 for the 4f-3d transition, by 18 x 10^{-5} for 5g-4f, and by 48 x 10^{-5} for 6h-5g. Use of the higher lines is made increasingly more difficult by the presence of weaker transitions close enough in energy to the principal transition to make a precise energy determination less reliable.

The electron shielding effect may be calculated approximately by considering only the two 1s electrons in the atom. In Pb this usually underestimates the effect by about 10% or so. Adding the 2s electrons already overestimates the effect, and the overestimate worsens as the effects of the remaining electrons are successively added without taking into account their mutual shielding. For the most precise work a self-consistent field calculation is required.

In Table 2 we list the more useful transitions in Pb with an intensity which is greater than 1% per μ-capture and for which the vacuum polarization effect is dominant. We give the vacuum polarization effect calculated to lowest order only and the electron shielding correction is given to the nearest eV. We give the values for Pb as an example because our measurements and calculations for this element are most complete. However, the measurements on elements of lower Z may turn out to be more favorable, and in any case should be included in a test of the higher order corrections and a precision determination of the muon mass.

TABLE 1
Electron Shielding Correction in Mesonic Atoms

	Z = 41			Z = 82		
Transition	E keV	ΔE eV	ΔE/E × 10^5	E keV	ΔE eV	ΔE/E × 10^5
$4f_{7/2}-3d_{5/2}$	258.4	−9.0	−3.5	938.2	−44.6	−4.8
$5g_{9/2}-4f_{7/2}$	107.4	−17.3	−16.2	431.4	−77.9	−18.0
$6h_{11/2}-5g_{9/2}$	58.1	−27.9	−48.0	233.3	−114.2	−49.0

TABLE 2

Electron Screening and Vacuum Polarization Corrections
for the Major Higher Transitions in Muonic Pb

(lines with intensity greater than 0.01 per μ-capture)

Transition	Approx. Energy keV	Fraction per μ-Capture	Vac Polarization (Lowest Order) eV	Electron Screening eV
$5f_{5/2}-3d_{3/2}$	1404.7	0.022	8567	− 162
$5f_{7/2}-3d_{5/2}$	1366.6	0.024	7965	− 164
$4f_{5/2}-3d_{3/2}$	971.9	0.197	6712	− 44
$4f_{7/2}-3d_{5/2}$	938.2	0.282	6165	− 45
$4f_{5/2}-3d_{5/2}$	929.0	0.015	6058	− 43
$6g_{7/2}-4f_{5/2}$	671.1	0.018	2882	− 251
$6g_{9/2}-4f_{7/2}$	663.5	0.023	2788	− 252
$5g_{7/2}-4f_{5/2}$	437.8	0.157	2165	− 78
$5f_{7/2}-4d_{5/2}$	437.1	0.022	2441	− 95
$5g_{9/2}-4f_{7/2}$	431.4	0.176	2082	− 78
$6g_{7/2}-5f_{5/2}$	238.3	0.020	1027	− 133
$6g_{9/2}-5f_{7/2}$	235.2	0.026	988	− 133
$6h_{9/2}-5g_{7/2}$	235.0	0.119	956	− 114
$6h_{11/2}-5g_{9/2}$	233.0	0.146	838	− 114

REFERENCES

1) B. Fricke, Z. Physik 218, 495 (1969).

2) H. L. Anderson, C. K. Hargrove, E. P. Hincks, J. D. McAndrew,
    R. J. McKee, R. D. Barton, and D. Kessler, Phys. Rev., in press.

3) G. Backenstoss, S. Charalambus, H. Daniel, H. Koch, Ch.v.d.
    Malsburg, G. Poelz, H. Schmitt, and L. Tauscher, Contributed
    paper to this conference. I thank Dr. Backenstoss for a
    preprint of this report.

4) S. Koslov, V. Fitch, and J. Rainwater, Phys. Rev. 95, 291 (1954).

5) A. J. Bearden, Phys. Rev. Letters 4, 240 (1960).

6) A. Petermann and Y. Yamaguchi, Phys. Rev. Letters 2, 359 (1959).

7) J. Lathrop, R. A. Lundy, V. L. Telegdi, R. Winston and D. D.
    Yovanovitch, Nuovo Cimento 17, 109 (1960).

8) J. Lathrop, R. A. Lundy, S. Penman, V. L. Telegdi, R. Winston,
    D. D. Yovanovitch, and A. J. Bearden, Nuovo Cimento 17, 114
    (1960).

9) S. Devons, G. Gidal, L. M. Lederman, and G. Shapiro, Phys. Rev.
    Letters 5, 330 (1960).

10) G Charpak, F.J.M. Farley, R. L. Garwin, T. Muller, J. C. Sens,
    and Z. Zichichi, Nuovo Cimento 37, 1241 (1965).

11) F.J.M. Farley, J. Bailey, R.C.A. Brown, M. Geisch, H. Joestlein,
    S. Van der Meer, E. Picasso, and M. Tannenbaum, Nuovo Cimento
    45A, 281 (1966).

12) J. Bailey, W. Bartl, G. van Bochmann, R.C.A. Brown, F.J.M.
    Farley, H. Jostlein, E. Picasso, and R. W. Williams, CERN
    preprint (1968).

13) D. Libermann, J. T. Waber, and D. T. Cromer, Phys. Rev. 137,
    A27 (1965). We thank these authors for making their tables
    available to us.

π^0 DECAY

Stephen L. Adler

Institute for Advanced Study

Princeton, New Jersey

I wish to describe some recent theoretical work on $\pi^0 \to 2\gamma$ decay, which helps to resolve puzzling questions which have arisen over the years, and which may shed light on the nature of possible fundamental constituents of matter, such as quarks. Purely kinematic considerations tell us that the matrix element and the decay rate for this process are

$$\mathcal{M}(\pi^0 \to 2\gamma) = k_1^{\xi} k_2^{\tau} e_1^{*\sigma} e_2^{*\rho} \epsilon_{\xi\tau\sigma\rho} F \ ,$$

$$\tau^{-1} = (\mu^3/64\pi)F^2 \ , \tag{1}$$

with (k_1, e_1), (k_2, e_2) the momentum and polarization four-vectors of the two photons, μ the pion mass, and F an intrinsic coupling constant. The job for the theorist, of course, is to try to calculate F. An important step towards this goal was taken in 1949 by Steinberger,[1] who considered a model in which the π^0 dissociates (via pseudoscalar coupling) into a proton-antiproton pair, which emit the two photons and then annihilate. In lowest order perturbation theory there are only two Feynman diagrams, the triangle diagram in Fig. 1a and the corresponding diagram with the two photons interchanged. Although this diagram appears to be linearly divergent, the presence of the γ_5 in the Fermion trace causes all divergent terms to vanish identically, and a straightforward calculation gives (neglecting small terms of order μ^2/m_N^2)

$$F \approx - \frac{\alpha}{\pi} \frac{g_r}{m_N} \ , \tag{2}$$

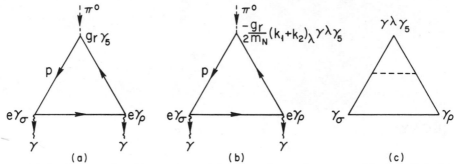

Fig. 1(a) Triangle diagram with pseudoscalar coupling. (b) Triangle diagram with pseudovector coupling. (c) A virtual meson
correction to the triangle diagram.

$$\alpha = \text{fine structure constant} = e^2/4\pi \approx 1/137 \quad ,$$
$$g_r = \text{pion-nucleon coupling constant} \approx 13.6 \quad ,$$
$$m_N = \text{nucleon mass} \;.$$

Substituting Eq. (2) into Eq. (1), one finds a decay rate $\tau^{-1} \approx 14$ eV,
in fairly good agreement with the experimental rate $\tau^{-1}_{\text{expt}} =$
$(1.12 \pm 0.22) \times 10^{16}$ sec^{-1} = (7.37 ± 1.5) eV.   That such a naive calculation should work so well is, in fact, rather puzzling, since we
know that Eq. (1) is just the first term in a power series in the
strong coupling g_r, and there is no obvious reason why one should
be able to get away with the neglect of all of the higher terms.

A second puzzle also emerged from Steinberger's calculation.
In addition to calculating $\pi^0 \rightarrow 2\gamma$ decay using pseudoscalar
coupling, Steinberger also repeated the calculation with pseudovector (axial-vector) coupling, by evaluating the diagram shown in
Fig. 1b.   This diagram is actually linearly divergent, and must be
evaluated by regulator techniques to insure photon gauge invariance.
On the basis of the pseudoscalar-pseudovector equivalence theorem,
one would expect Fig. 1b to give the same rate as Fig. 1a, but actual calculation shows that Fig. 1b gives a $\pi^0 \rightarrow 2\gamma$ amplitude F
smaller than that of Eq. (2) by a factor $\mu^2/6m_N^2$.   In the limit of
zero pion mass, the pseudovector amplitude F actually vanishes!

During the last ten years, extensive and very successful calculations on soft pion emission have been done using the partially-
conserved axial-vector current (PCAC) hypothesis.   This hypothesis states that, apart from certain equal-time commutator terms
(which do not enter into our problem), soft pions behave as if they
were coupled to nucleons by pseudovector, rather than pseudo-

scalar, coupling. When we turn to Steinberger's calculation, PCAC thus leads us to the troublesome conclusion that the answer $F \approx 0$ should be chosen over the numerically reasonable answer for F given by Eq. (2)! This conclusion is independent of the perturbation theory model used by Steinberger, and is easily derived in a completely general fashion.[2] All we need do is to sandwich the PCAC equation

$$\partial_\lambda \mathcal{F}_3^{5\lambda} = (f_\pi / \sqrt{2}) \phi_{\pi^0} ,$$

(3)

$$f_\pi = \text{charged pion decay amplitude} ,$$

between the two photon state $< \gamma(k_1, \epsilon_1) \gamma(k_2, \epsilon_2)|$ and the vacuum $|0>$. The matrix element of the right-hand side of Eq. (3) is proportional to $\mathcal{M}(\pi^0 \to 2\gamma)$, while a purely kinematic analysis of $< \gamma(k_1, \epsilon_1) \gamma(k_2, \epsilon_2)| \mathcal{F}_3^{5\lambda} |0>$ shows that the matrix element of the left-hand side of Eq. (3) is proportional to $k_1^\xi k_2^\tau e_1^{*\sigma} e_2^{*\rho} e_{\xi\tau\sigma\rho} \times (k_1 + k_2)^2$. Thus, in a model-independent way, PCAC predicts $F \propto (k_1 + k_2)^2 = \mu^2$, just as was found from the pseudovector triangle graph in perturbation theory.

To summarize, the theory of π^0 decay presents the following three puzzles:

(1) Naive calculation using the lowest order pseudoscalar coupling triangle diagram, and neglecting possible strong interaction renormalization effects, gives a surprisingly good result.

(2) The pseudoscalar-pseudovector equivalence theorem breaks down. Pseudovector coupling predicts that $\pi^0 \to 2\gamma$ decay is strongly suppressed.

(3) PCAC implies, in a model independent way, that $\pi^0 \to 2\gamma$ decay is strongly suppressed.

Recent theoretical work,[3] which I will now briefly describe, has helped to resolve these puzzles. The key observation is that when very singular diagrams (e.g. triangle diagrams) are present, formal field-theory results such as Ward identities, the pseudoscalar-pseudovector equivalence theorem, and the PCAC equation itself, break down. Consider, for example, the pseudoscalar-pseudovector equivalence theorem, which asserts that the diagrams of Figs. 1a and 1b should give identical results. The theorem is formally derived by taking the vacuum to two photon matrix element of the divergence equation

$$(g_r/2im_N)\partial_\lambda j^{5\lambda} = g_r j^5 \; ;$$

$$j^{5\lambda} = \bar{\psi}\gamma^\lambda\gamma_5\psi \; , \quad j^5 = \bar{\psi}\gamma_5\psi \; . \tag{4}$$

[We can neglect meson terms in Eq. (4) because no virtual mesons appear in Figs. 1a and 1b.] The matrix element of the right-hand side of Eq. (4) corresponds to Fig. 1a, while the matrix element of the left-hand side of Eq. (4) corresponds to Fig. 1b, and Eq. (4) asserts that they should be equal. This formal derivation breaks down because the local product of operators $\bar{\psi}(x)\gamma^\lambda\gamma_5\psi(x)$ is singular, and the naive manipulations of equations of motion which lead to Eq. (4) are incorrect. The correct answer can be obtained by regarding the axial-vector current and the pseudoscalar current as limits of nonlocal, gauge-invariant currents, [4]

$$j^{5\lambda}(x) = \lim_{\epsilon \to 0} \bar{\psi}(x+\epsilon)\gamma^\lambda\gamma_5\psi(x-\epsilon)\exp\left[-ie \int_{x-\epsilon}^{x+\epsilon} d\ell \cdot A(\ell)\right] ,$$

$$j^5(x) = \lim_{\epsilon \to 0} \bar{\psi}(x+\epsilon)\gamma_5\psi(x-\epsilon)\exp\left[-ie \int_{x-\epsilon}^{x+\epsilon} d\ell \cdot A(\ell)\right] , \tag{5}$$

$$A = \text{electromagnetic field,}$$

giving, after a careful calculation

$$\partial_\lambda j^{5\lambda} = 2im_N j^5 + (\alpha/4\pi)F^{\xi\sigma}F^{\tau\rho}e_{\xi\sigma\tau\rho} ,$$

$$F^{\xi\sigma} = \partial^\sigma A^\xi - \partial^\xi A^\sigma = \text{electromagnetic field-strength tensor .} \tag{6}$$

The matrix element of the extra term in Eq. (6) precisely accounts for the difference between Figs. 1a and 1b as calculated by Steinberger! An alternative procedure[3] is to directly calculate the difference of Figs. 1a and 1b in momentum space. If one freely translates the loop integration variables one "deduces" that the difference is zero, but if one pays careful attention to the fact that in linearly divergent integrals the origin of integration cannot be freely shifted, one reproduces the right-hand side of Eq. (6). We see, then, that the pseudoscalar-pseudovector equivalence theorem breaks down for triangle diagrams because of the presence of singular (linearly divergent) integrals, and the breakdown is compactly summarized by Eq. (6).

Clearly, the phenomenon which modifies Eq. (6) will also affect the PCAC equation, Eq. (3). Let us consider a particular field-theoretic model, the σ-model of Gell-Mann and Lévy. This model

consists of a neutron n and a proton p interacting with the pions
(π^+, π^0, π^-) and with a scalar, isoscalar meson σ via $SU_2 \otimes SU_2$
symmetric couplings. In the absence of electromagnetism, Eq. (3)
is satisfied as an operator identity, with ϕ_{π^0} the canonical pion
field. In the presence of electromagnetism, the singularity
triangle diagram changes Eq. (3) to read

$$\partial_\lambda \mathcal{F}_3^{5\lambda} = (f_\pi/\sqrt{2})\phi_{\pi^0} + \tfrac{1}{2}(\alpha/4\pi)F^{\xi\sigma}F^{\tau\rho}e_{\xi\sigma\tau\rho} \, . \tag{7}$$

In other words, the PCAC equation for the neutral pion must be
modified in the presence of electromagnetic interactions. Just as
we did above, let us take the matrix element of Eq. (7) between
the vacuum and the two-photon state. In the soft pion limit, the
matrix element of the left-hand side makes no contribution, but instead of deducing that the $\pi^0 \to 2\gamma$ amplitude F vanishes, we now
find that F is proportional to the matrix element of the extra
term in Eq. (7),

$$F\Big|_{(k_1+k_2)^2 = 0} = -\frac{\alpha}{\pi} \frac{\sqrt{2}\,\mu^2}{f_\pi} \, . \tag{8}$$

Because Eq. (8) has been derived without recourse to perturbation
theory, it is an <u>exact low energy theorem for</u> π^0 decay.[5] Using
the experimental value $f_\pi \approx 0.96\,\mu^3$ and substituting Eq. (8) into
Eq. (1), we find the decay rate $\tau^{-1} = 7.4$ eV, in excellent agreement with experiment. It is interesting to compare Eq. (8) with
Steinberger's lowest order perturbation theory result [Eq. (2)] by
using the Goldberger-Treiman relation,

$$\frac{\sqrt{2}\,\mu^2}{f_\pi} \approx \frac{g_r}{m_N g_A} \, , \tag{9}$$

g_A = nucleon axial-vector coupling constant ≈ 1.22,

to rewrite Eq. (8) in the form

$$F\Big|_{(k_1+k_2)^2 = 0} \approx -\frac{\alpha}{\pi} \frac{g_r}{m_N} \frac{1}{g_A} \, . \tag{10}$$

We see that the effects of higher orders of perturbation theory are
entirely contained in the factor g_A^{-1}, which is numerically close to
unity; as a result Steinberger's calculation, which neglects the factor g_A, is a fairly good first approximation.

We see, then, that the modified PCAC equation resolves the
puzzles noted above. At the same time, however, some new

questions and problems are raised.   Let us now consider these problems, as well as some of the experimental implications of Eq. (7).

1. We have just gone through some rather subtle reasoning to avoid the prediction, following from the unmodified PCAC equation, that F is suppressed by a factor $\sim \mu^2/m_N^2$. However, one can always ask how one knows that π^0 decay is *not* really a suppressed decay. One argument is the theoretical one, that with the extra term PCAC gives a good answer for π^0 decay, which means that without this term, the rate would be much too small to agree with experiment. There is also an interesting experimental test,[6] which strongly suggests that π^0 decay is not suppressed. To see this, let us return to the suppression argument following Eq. (3), in the altered situation in which one of the photons is off-mass-shell, say $k_1^2 \neq 0$. Some simple kinematics shows that the vacuum to two photon matrix element of $\partial_\lambda \mathcal{F}_3^{5\lambda}$ is now proportional to $k_1^\xi k_2^\tau e_1^{*\sigma} e_2^{*\rho} \times e_{\xi\tau\sigma\rho}[\mu^2 + \beta k_1^2]$, with β of order unity. We see that while the on-shell part of the amplitude is suppressed by a factor μ^2, the off-shell dependence is not suppressed. Since the off-shell amplitude is measured in the reaction $\pi^0 \rightarrow e^+ e^- \gamma$, our suppression argument predicts that the k_1^2 dependence of this process will have the form $1 + (\beta/\mu^2)k_1^2$, which has a much larger slope than the form $1 + (\beta/m_\rho^2)k_1^2$ expected in the absence of suppression of the $\pi^0 \rightarrow 2\gamma$ decay. A measurement of this slope has been reported by Devons et al.,[7] who find a matrix element $1 + a k_1^2$, $a = (0.01 \pm 0.11)/\mu^2$. Clearly, this is strong evidence against $\pi^0 \rightarrow 2\gamma$ suppression.

2. The argument that Eq. (8) is an exact low energy theorem is not as simple as we have made it sound. To be sure that Eq. (8) is exact, we must be sure that strong interaction modifications of the triangle diagram, such as illustrated in Fig. 1c, do not renormalize the extra term in Eq. (7). This can in fact be demonstrated, to any finite order of perturbation theory.[5] The reason that virtual meson corrections do not modify Eq. (7) is that they always involve Fermion loops with more than three vertices (Fig. 1c involves a 5-vertex loop), which satisfy normal Ward identities because they are highly convergent. There is still the possibility that Eq. (7) is modified by nonperturbative effects, such as contributions from triangles involving bound states of the fundamental fields. Our neglect of possible nonperturbative modifications is pure assumption.

3. The spectacularly good agreement of Eq. (8) with experiment is

somewhat fortuitous, both because of the large error in the experimental $\pi^0 \to 2\gamma$ rate and because of the usual 10-20 percent extrapolation error involved in PCAC arguments. For example, use
of the Goldberger-Treiman relation to replace Eq. (8) by Eq. (10)
alters the theoretical prediction by 20 percent, to $\tau^{-1} = 9.1$ eV.

4. The constant $\frac{1}{2}$ appearing in front of the term $(\alpha/4\pi)F^{\xi\sigma}F^{\tau\rho} \times e_{\xi\sigma\tau\rho}$ in Eq. (7) arises from our particular choice of fundamental
Fermion fields, and differs in different field-theoretic models.
Quite generally, if $\mathcal{F}_3^{5\lambda}$ is expressed in terms of fundamental
fields by

$$\mathcal{F}_3^{5\lambda} = \sum_j g_j \bar{\psi}_j \gamma^\lambda \gamma_5 \psi_j + \text{meson terms} \ , \tag{11}$$

then the modified PCAC reads

$$\partial_\lambda \mathcal{F}_3^{5\lambda} = (f_\pi/\sqrt{2})\phi_{\pi^0} + S(\alpha/4\pi)F^{\xi\sigma}F^{\tau\rho}e_{\xi\sigma\tau\rho} \ , \tag{12}$$

$$S = \sum_j g_j Q_j^2 \ ,$$

where the charge of the jth fermion is Q_j. All we are doing, of
course, is adding up the contributions of the triangle diagrams involving the various Fermions. The $\pi^0 \to 2\gamma$ low energy theorem
derived from Eq. (12) is

$$F\Big|_{(k_1+k_2)^2 - 0} = -\frac{\alpha}{\pi}\frac{\sqrt{2}\,\mu^2}{f_\pi}\,2S \approx -\frac{\alpha}{\pi}\frac{g_r}{m_N}\frac{1}{g_A}\,2S \ , \tag{13}$$

which reduces to our previous result when $S = \frac{1}{2}$.

The comparison which we have made above with the experi-
mental π^0 decay rate tells us that $|S| \approx 0.5$, but does not de-
termine the <u>sign</u> of S. However, there are a number of different
ways of determining the sign of S, all of which, fortunately, seem
to agree! The first method is by analysis of $\pi^+ \to e^+\nu\gamma$ decay,
the vector part of which is related by CVC to F and the axial-vector part of which can be estimated by hard pion techniques. Using
the experimentally measured vector to axial-vector ratio for this
process, Okubo[8] finds that S is positive. A second method is to
make use of forward π^0 photoproduction, where one can observe
the interference between the Primakoff amplitude (which is proportional to F) and the forward purely strong interaction amplitude.
The sign of the latter can be determined by finite energy sum rules
from the known sign of the photoproduction amplitude in the (3, 3)
resonance region; the analysis has been carried out by Gilman,[9]
who finds S positive. A third method consists of comparing Eq.
(13) with an approximate expression for the $\pi^0 \to 2\gamma$ amplitude de-

rived by Goldberger and Treiman[10] (as corrected by Pagels[10]).
These authors applied a pole dominance argument to proton Compton scattering dispersion relations, obtaining the relation

$$F \approx - 4\pi\alpha \; \frac{\kappa_p}{g_r} \; \frac{1}{m_N} \; ,$$

(14)

κ_p = proton anomalous magnetic moment = 1.79,

which gives a $\pi^0 \to 2\gamma$ rate of 2.0 eV, in fair agreement with experiment. Comparison of Eq. (14) with Eq. (13) again gives S positive. A fourth method which has been proposed[8] is to use Compton
scattering data on protons to try to measure the interference of the
pion exchange piece (proportional to F) with the nucleon and nucleon
isobar exchange pieces. The problem with this proposal[11] is that
one does not know whether to take the pion exchange piece in its
Born approximation form, $tF/(t-\mu^2)$, or in the polology form,
$\mu^2 F/(t-\mu^2)$. Since t is negative in the physical region, this uncertainty leads to a sign ambiguity and renders the method dubious.
In any case, with fair certainty one learns from the first three methods that S is positive.

   Armed with the experimental knowledge that S = + 0.5, we can
now use Eq. (12) to test various models of the hadrons which have
been proposed. One very popular model is the triplet model, consisting of an SU_3-triplet of Fermions (p, n, λ) interacting by meson
exchange. The charges of (p, n, λ) are $(Q, Q-1, Q-1)$ and the corresponding axial-vector couplings are $(g_p, g_n, g_\lambda) = (\frac{1}{2}, -\frac{1}{2}, 0)$. One immediately finds $S = \frac{1}{2} Q^2 - \frac{1}{2}(Q-1)^2 = Q - \frac{1}{2}$, and so $S = \frac{1}{2}$ requires
Q = 1 [i.e., integral triplet charges (1, 0, 0)]. Note that the frac-
tionally charged quark model has Q = 2/3, S = 1/6, and so the
quark hypothesis is strongly excluded. Another integrally charged
triplet model which is allowed is the Han-Nambu-Tavkhelidze[12]
model, which has three triplets, S, U, B, with respective charges
(1, 0, 0), (1, 0, 0), (0, -1, -1) and with axial-vector couplings $(\frac{1}{2}, -\frac{1}{2}, 0)$
for each triplet.

5. The ideas which we have developed can also be applied to the
$\eta \to 2\gamma$ and the $X^0 \to 2\gamma$ decays.[13] Unfortunately, the experimental situation here is worse, and the theoretical situation is also
worse, because the soft η and soft X^0 approximations involve a
much larger extrapolation from the physical region than does the
soft pion approximation. Nonetheless, pursuing this track, Gla-
show et al.[13] find a connection between the $\pi^0 \to 2\gamma$, $\eta \to 2\gamma$ and
$X^0 \to 2\gamma$ decay rates, which predicts

$$\tau^{-1}(X^0 \to 2\gamma) \approx 350 \text{ keV} \quad \text{quark model } (Q = \tfrac{2}{3}) \ ,$$

$$\approx 120 \text{ keV} \quad \text{integrally charged } (Q = \pm 1)$$
$$\text{triplet model } . \tag{15}$$

Present experiments do not distinguish between the two alternatives in Eq. (15).

REFERENCES

1. J. Steinberger, Phys. Rev. 76, 1180 (1949).
2. D. G. Sutherland, Nuclear Phys. B2, 433 (1967).
3. S. L. Adler, Phys. Rev. 177, 2426 (1969); J. S. Bell and R. Jackiw, Nuovo Cimento 60, 47 (1969).
4. J. Schwinger, Phys. Rev. 82, 664 (1951); C. R. Hagen, Phys. Rev. 177, 2622 (1969); B. Zumino, in Proceedings of the Topical Conference on Weak Interactions, CERN, Geneva (1969), p. 361.
5. S. L. Adler, Ref. 3; S. L. Adler and W. A. Bardeen, Phys. Rev. (in press).
6. R. F. Dashen, private communication.
7. S. Devons et al., Phys. Rev. 184, 1356 (1969).
8. S. Okubo, Phys. Rev. 179, 1629 (1969).
9. F. J. Gilman, Phys. Rev. (in press).
10. M. L. Goldberger and S. B. Treiman, Nuovo Cimento 9, 451 (1958); H. Pagels, Phys. Rev. 158, 1566 (1967).
11. A. Hearn, private communication.
12. M. Y. Han and Y. Nambu, Phys. Rev. 139, B1006 (1965); A. Tavkhelidze, in High Energy Physics and Elementary Particles, International Atomic Energy Agency, Vienna (1965), p. 763.
13. S. Okubo (to be published); S.L. Glashow, R. Jackiw and S.S. Shei (to be published).

DISCUSSION

A. Dar: Recent measurements, by the Primakoff effect, of the π^0 decay rate give τ^{-1} = 11.7 eV (+ 10%). S.L.A.: Using $g^2/4\pi$ = 14.6, the theoretical estimate is in the range 7.4 - 9.1 eV, but as indicated there is ~ 20% uncertainty in the PCAC extrapolation. Incidentally, the new width is in still poorer agreement with the fractionally-charged quark model. V. Telegdi: Would one expect to see, in η-decay, evidence for a non E.M. isospin symmetry breaking interaction? S.L.A.: One can invent such an interaction to explain the 3π decay of the η. The extra terms I discussed will not affect this decay mode.

PHOTOPRODUCTION OF MUON PAIRS

S. Hayes, R. Imlay, P.M. Joseph, A.S. Keizer,
J. Knowles, and P.C. Stein

Laboratory of Nuclear Studies, Cornell University
Ithaca, New York

(Presented by P. C. Stein)

A series of experiments measuring the photoproduction of
muon pairs has been performed at the Cornell 10 Gev electron
synchrotron. In an earlier experiment[1], we made measurements
of muon pair photoproduction for muon pair invariant masses of
from 650 to 1150 Mev to test the validity of quantum electrodynamics. No significant deviations from the predictions of QED
were observed. The set of measurements reported on here increased
the invariant mass range measured to 2000 Mev. The object of these
experiments was to measure the branching ratio of $\phi \rightarrow \mu^+ + \mu^-$,
to search for possible production of new vector mesons, and to
look for deviations from the predictions of QED at higher masses.
Preliminary results of these measurements are given below.
Final analysis is in progress.

The apparatus is shown in Fig. 1. It is virtually identical
with the apparatus described in Ref. 1. Muons produced in a
6.55 gm/cm^2 target were detected in pairs by hodoscopes H_L and H_R.
The hodoscopes spanned horizontal production angles of 10.0 to
15.0 degrees in eight equal steps. The range in iron of each
muon was determined by means of eight range counters R_L 1-4 and
R_R 1-4. Four additional counters T_L, T_L', T_R, and T_R' were used
to reduce accidentals. Aluminum was placed in front of the
hodoscopes to attenuate the large flux of incident pion pairs.

Two runs were taken. In the first, each arm of the telescope
was set to accept muons from 2.18 to 3.43 Gev. In the second run ,
the telescopes accepted muons from 3.34 to 4.63 Gev. The
two runs thus reasured muon-pair masses from 850 to 1600 Mev, and
from 1300 to 2000 Mev respectively. In the first run, care was
taken to ensure high efficiency of the counter system and low

random coincidence contamination. This data was used to determine
the branching ratio of $\phi \rightarrow \mu^+ + \mu^-$, and to investigate the
validity of QED. The data from the second run was used to search
for the possible production of a new vector meson.

In the analysis of the data, contributions from five
processes were considered. These were: elastic production (EL BH),
or

$$\gamma + C \rightarrow C + \mu^+ + \mu^- \quad , \tag{1}$$

inelastic production (IN BH),

$$\gamma + C \rightarrow C' + \mu^+ + \mu^- \quad , \tag{2}$$

where C' consists of any excited state of six protons and six
neutrons,
inelastic production with pion production (IN PI),

$$\gamma + C \rightarrow N^{12} + (\text{pions}) + \mu^+ + \mu^- \tag{3}$$

rho production (RHO), which consists of the photoproduction
of rho mesons followed by either the direct or indirect decay
of the rho into two muons

$$\gamma + C \rightarrow C + \{ \rho \rightarrow (\mu^+ + \mu^-) \text{ or } (\pi^+ + \pi^- \rightarrow \mu^+ + \mu^- + \nu + \bar{\nu}) \} \tag{4}$$

and phi production (PHI),

$$\gamma + C \rightarrow C + (\phi \rightarrow \mu^+ + \mu^-) \quad . \tag{5}$$

All contributions were integrated over the apparatus using
Monte-Carlo techniques, taking into account multiple scattering
and range straggling.

The inelastic contributions (2) were evaluated using the
Drell-Schwartz[2] sum rule for inelastic electron scattering
applied to pair production, using the formalism of Drell and
Walejka[3]. The effects of nucleon correlations were introduced
using the calculations of McVoy and Van Hove.[4]

The pion inelastic contributions (3) were calculated
assuming that the twelve nucleons act incoherently. The 1238 Mev
resonance and the higher mass region were treated independently.

The rho contribution (4) was calculated assuming a rho cross
section of the form

$$d\sigma/dt = 8.5(e^{47t} + 0.043e^{10t}) \text{ mb/Gev}^2 \quad , \tag{6}$$

and a $\rho \rightarrow 2\mu$ branching ratio of 5.0×10^{-5}. The mass distribution
of the rho was taken to be that of a relativistic Breit-Wigner
with a width of 120 Mev. The direct decay $\rho \rightarrow 2\mu$ was 2/3 of
the contribution to (4).

Extensive data on the angular distribution of the photoproduction cross section for the phi is not available. We
parameterized the phi cross section using the form

$$d\sigma/dt = A(e^{bt} + Ce^{5t}) \tag{7}$$

with the cross section[5] $(d\sigma/dt)_{\theta=0} = 272 \, \mu\text{b/Gev}^2. \tag{8}$

For a preliminary analysis of the data taken in the first

run we fit to a total theory of the form

$$EL\ BH\ +\ D(IN\ BH\ +\ IN\ PI)\ +\ RHO\ +\ B\times PHI \qquad (9)$$

where B is the phi to two muon branching ratio. Choosing C=0.1 and b = 58 GeV^{-2} for the phi, a preliminary fit to the invariant mass distribution gives D = 1.2 and B = $(2.6 \pm 0.4)\times 10^{-4}$. The error quoted is statistical only. The results of this fit are shown in Fig. 2.

Our results may be compared with the previous measurements of the phi leptonic branching ratios:

$$B(\phi \rightarrow e^+ + e^-) = (2.9 \pm 0.8)\times 10^{-4}$$

from photoproduction of phi mesons[6], and[7]

$$B(\phi \rightarrow e^+ + e^-) = (3.96 \pm 0.62)\times 10^{-4}$$

assuming that $B(\phi \rightarrow K_S^0 + K_L^0) = 0.308$

or

$$B(\phi \rightarrow e^+ + e^-) = (3.08 \pm 0.66)\times 10^{-4}$$

assuming that $B(\phi \rightarrow K_S^0 + K_L^0) = 0.389$

from $e^+ + e^- \rightarrow \phi \rightarrow K_S^0 + K_L^0$. $\qquad (10)$

To investigate a possible breakdown in QED, we multiplied the Bethe-Heitler terms in (9) by a factor of $1 + \lambda M_{\mu\mu}^4$, where $M_{\mu\mu}$ is the mass of the final two muon system. Our best fit to the data gave

$$\lambda = -.01 \pm .03\ GeV^{-4}$$

The excellent agreement with the predictions of QED may be fortuitous, since the calculations of the inelastic terms are somewhat uncertain. The results quoted here are preliminary, and no attempt has been made to estimate the effects of systematic errors.

The data in the second run was analyzed according to the same prescription as given above, except that there was no phi contribution. The results are shown in Fig. 3. There is clearly no evidence for any significant deviation from theory.

We are therefore in the position of putting an upper limit on any possible ρ' production. A Monte-Carlo calculation was performed to calculate rates for a series of assumed masses from 1400 to 2000 MeV. The production cross section was assumed to be $A(e^{47t} + 0.043e^{10t})$. The experiment can only measure the product AB, where B is the branching ratio for $\rho' \rightarrow \mu^+ + \mu^-$. Two standard deviation limits on AB varied from 2.4×10^{-5} at a mass of 1400 MeV to 0.4×10^{-5} mb/GeV^2 at a mass of 2000 MeV. The corresponding value for the ρ is 50×10^{-5} mb/GeV^2.

These results can be translated into lower limits on $\gamma_{\rho'}^2$, where $em_{\rho'}^2/2\gamma_{\rho'}$ is the direct photon-ρ' coupling constant. Assuming that the ρ'-carbon scattering cross section is equal to the ρ-carbon scattering cross section, we obtain

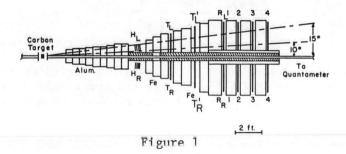

Figure 1

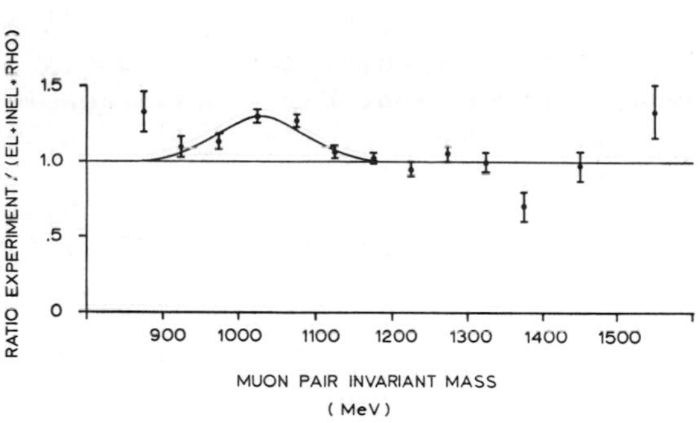

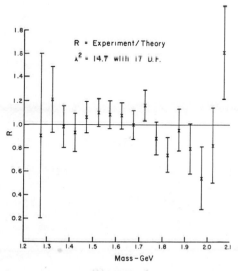

Figure 2

Figure 3

$$\left(\frac{\delta_{\rho'}}{\delta_{\rho}}\right)^4 = \frac{(BA)_\rho}{(BA)_{\rho'}} \frac{m_{\rho'}}{m_\rho} \frac{\Gamma_\rho}{\Gamma_{\rho'}}$$

We assume that there is no direct connection between δ_ρ' and $\Gamma_{\rho'}$.

If we assume that $\Gamma_\rho = \Gamma_{\rho'}$, we obtain a series of lower bounds on the ratio $(\delta_{\rho'}/\delta_\rho)^2$ which, at the two standard deviation level, vary from 6.5 to 18.5 at a mass of 1400 MeV and 2000 MeV respectively.

REFERENCES

1) S. Hayes, et al., Phys. Rev. Letters 22, 1134(1969)
2) S. Drell and C. Schwartz, Phys. Rev. 112, 568(1958)
3) S. Drell and J. Walecka, Ann. Phys.(N.Y.) 28, 18(1964)
4) K. McVoy and L. Van Hove, Phys. Rev. 125, 1034(1962)
5) R. Talman, private communication
6) U. Becker, et al., Phys. Rev. Letters 21, 1504(1968)
7) J. Augustin, et. al., Physics Letters 28B, 517(1969)

Lincoln Wolfenstein

Carnegie-Mellon University, Department of Physics

Pittsburgh, Pennsylvania 15213

1. BASIC ASSUMPTIONS OF WEAK INTERACTION THEORY

Let me begin with a brief summary of the present status of weak interaction theory. The basic interaction is described in terms of charged currents

$$H_W = \frac{G}{\sqrt{2}} \, g_\lambda \, g_\lambda^{\ +} \tag{1}$$

$$g_\lambda = \bar{\nu}_\mu \, \gamma_\lambda (1+\gamma_5) \, \mu + \bar{\nu}_e \, \gamma_\lambda (1+\gamma_5) e + J_\lambda \tag{2}$$

where J is the hadronic current assumed to be of the Cabibbo form

$$J_\lambda = \cos \theta (J_\lambda^{\ 1} + iJ_\lambda^{\ 2}) + \sin \theta (J_\lambda^{\ 4} + iJ_\lambda^{\ 5}) \tag{3}$$

where the superscripts define components of an SU_3 octet. Here we will be interested only in the strangeness-conserving current $\cos \theta (J_\lambda^{\ 1} + iJ_\lambda^{\ 2})$ where 1,2 are isospin indices. Beyond the assumed isovector property we desire to specify the hadronic current as completely as possible. This is much harder than in the case of the leptonic current (the first two terms of Eq.(2)) because we do not know what elementary particles to use to describe the current. A standard assumption is that this current has the basic properties described by the quark representation

$$\cos \theta (J_\lambda^{\ 1} + iJ_\lambda^{\ 2}) \equiv V_\lambda + A_\lambda \propto \bar{p} \, \gamma_\lambda (1 + \gamma_5)n \tag{4}$$

where p,n are the field operators of the quarks. Unfortunately, this leads to very few conclusions without a knowledge of the strong interactions of quarks.

One simple consequence of Eq. (4) is the absence of second-class currents.[1] This means the currents have well-defined transformation properties under G, which is the product of charge conjugation times charge symmetry,

$$G \, V_\lambda \, G^{-1} = V_\lambda \; ; \quad G \, A_\lambda \, G^{-1} = - \, A_\lambda \tag{5}$$

Another conclusion is the conserved vector current hypothesis (CVC); this means that the vector current is an isotopic spin partner of the isovector part of the electromagnetic current. This allows a simple relationship between matrix elements determined by electromagnetic experiments to those involving the weak vector current. Both this relationship and Eq. (4) are difficult to test in nuclear beta-decay because of the low momentum transfer. Another consequence of CVC is the lack of renormalization of the "charge"; this means that at zero momentum transfer in a pure vector interaction one can make a precise determination of $G_V \equiv G \cos\theta$ and thus, by comparison with mu-decay, of $\cos\theta$, provided all electromagnetic effects are ignored. The analysis of nuclear $0 \to 0$ beta transitions yields a value of θ between 0.16 and 0.24 allowing for about $\pm 1\%$ uncertainty in radiative corrections and about $\pm 1\%$ uncertainty in the experimental ft values or in Z-dependent effects.[2]

2.   PARTIALLY-CONSERVED AXIAL CURRENT

The assumption about the axial vector current corresponding to CVC is much harder to state precisely. It is usually called partial conservation or PCAC. Two of the simplest consequences of this are the Goldberger-Treiman relation and the value of the induced pseudoscalar coupling. Let me summarize the derivation of these. We start with the assumption that the divergence of the axial vector current satisfies a once-subtracted dispersion relation. (This is weaker than the usual assumption of an unsubtracted dispersion relation.) For nucleon states we write the general form

$$\langle p | A_\lambda | n \rangle = \bar{U}_p \left[-g_A (q^2) \, \gamma_\lambda \, \gamma_5 \right.$$
$$\left. + \, g_P (q^2) \, \gamma_5 \, q_\lambda \right] U_n \tag{6}$$

so that

$$i \langle p | \frac{\partial A_\mu}{\partial x_\mu} | n \rangle = (\bar{U}_p \, \gamma_5 \, U_n) \left[-2 M g_A (q^2) \right.$$
$$\left. - \, q^2 \, g_P \, (q^2) \right] \equiv \bar{U}_p \, \gamma_5 \, U_n \, F(q^2) \tag{7}$$

The once-subtracted dispersion relation may be written[3]

$$F(q^2) = \frac{c}{q^2 - m_\pi^2} + P + I(q^2) \tag{8}$$

where the first term is the pole term with c expressed in terms of
the pion-decay constant and the pion-nucleon coupling constant, P
is the subtraction constant, and $I(q^2)$ is the dispersion integral
with a threshold at $(3m_\pi)^2$. We assume that, apart from the pion
pole which is associated with g_P but not g_A, the form factors are
smoothly varying so that we can write

$$g_A(q^2) = g_A(0) + g_A'(0) q^2 \tag{9a}$$

$$g_P(q^2) = \frac{d}{q^2 - m_\pi^2} + c \tag{9b}$$

Equating the two expressions for $F(q^2)$ with $I(q^2) = I(0) + I'(0)q^2$
we find

$$2Mg_A(0) = c/m_\pi^2 - (P + I(0)) \tag{10}$$

$$d = -c/m_\pi^2 \tag{11a}$$

$$e = -I'(0) - 2Mg_A'(0) \tag{11b}$$

The following comments are of interest:

1. Equation (10) is the Goldberger-Treiman relation with a
correction given by $P + I(0)$. It is of great interest to know how
large this correction term is. Present empirical data with
$\left|\dfrac{g_A}{g_V}\right| = 1.23 \pm 0.02$ and the pion-nucleon coupling constant
$|g_{\pi N}| = 13.65 \pm 0.15$ give a correction

$$\Delta = 1 - \frac{2M\, g_A}{c/m_\pi^2} = 0.08 \pm 0.02.$$

It has been emphasized recently by Pagels[4] that if Δ is indeed of
the order of 10% it is impossible to explain this in terms of $I(0)$
unless there is a 0^- isovector meson (heavy pion) with a mass not
too much greater than $3m_\pi$. On the other hand the error on Δ may be
underestimated so that Δ may indeed be closer to 1% in which case
it could be explained easily by $I(0)$. Therefore, it is important
to confirm the values for g_A and $g_{\pi N}$ by new experiments. It has
been suggested by Lovelace[5] that the accepted value for $g_{\pi N}$ may
be too large by 3 or 4% which could reduce Δ to 0.04; the most
accurate value up to now has been obtained from the pion-nucleon
dispersion relations, but perhaps in the future nucleon-nucleon
experiments may be precise enough to give a value of comparable
accuracy. Assuming Δ is of the order of 10%, Pagels concludes that
the subtraction constant P is probably needed. He assumes that it
is small because it is related to the breaking of SU(2) x SU(2)

symmetry.

2. The induced pseudoscalar constant for mu-meson capture in hydrogen is given by (using Eqs. (9) and (11))

$$g_P^\mu \equiv m_\mu\, g_P(q^2)$$

$$= -\frac{cm_\mu}{m_\pi^2(q^2-m_\pi^2)} - m_\mu\left[I'(0) + 2M\, g_A'(0)\right] \tag{12}$$

The first term gives the standard value based on evaluating the pion pole in terms of the pion-decay constant and the pion-nucleon coupling constant and not by expressing g_P in terms of g_A using the Goldberger-Treiman relation. The result setting $q^2 = 0.9m_\mu^2$ is $g_P^\mu \simeq 7.3\, g_A$. We can now consider the corrections to this value. The only correction that is related to the failure of the Goldberger-Treiman relation is the term in $I'(0)$. It follows from the arguments of Pagels that all the contributions to $I(0)$ are small so that $I'(0)$ is bound to be very small. Even if we assume a 0^- meson of mass μ to give the entire value of $I(0)$ and let $I(0)$ explain a value of $\Delta = 0.1$ we find

$$\left|m_\mu I'(0)\right| = \frac{m_\mu I(0)}{\mu^2} \simeq \frac{m_\mu}{\mu^2}\, 2Mg_A\, \Delta \lesssim 0.1g_A$$

assuming a lower limit of $\mu \simeq 3m_\pi$. The other correction can be evaluated once $g_A(q^2)$ has been determined from experiment. If one assumes the rather arbitrary form

$$g_A(q^2) = g_A(0)\, \frac{m_A^4}{(q^2-m_A^2)^2}$$

the correction is given by

$$- 2m_\mu M\, g_A'(0) = -\frac{4m_\mu M\, g_A(0)}{m_A^2} \simeq -0.7g_A \tag{13}$$

if $m_A = m_\rho$. This is probably an upper limit. When it becomes possible this correction should be included in the analysis of muon capture.

The important point to note is that the theory, even with the fairly weak assumption of a once-subtracted dispersion relation[3], predicts g_P to within 1 or 2% in terms of empirical quantities even though the Goldberger-Treiman relation may fail by 10%. The failure of this prediction for g_P would shed doubt on our understanding of the Goldberger-Treiman relation.

3.   SOME PROBLEMS IN WEAK INTERACTION THEORY

The theory we have just discussed is wrong (or perhaps incomplete) for at least two reasons: (1) calculations beyond the lowest order in perturbation theory give infinite results and (2) the theory is CP-invariant whereas CP violation has been observed in the decays of the long-lived K^0 meson.

It is generally believed that higher-order weak interaction effects are small but finite; the one clear-cut example of a second-order weak interaction effect is the small mass difference between the neutral K_L and K_S mesons, $\delta m \sim 10^{-14} m_K$. It used to be believed that the smallness of this result (in contrast to the divergent result of the theory) was related to the cutoffs due to strong-interaction form factors rather than to a cutoff associated with the weak interaction. It was first pointed out by Ioffe[6] that if the hadronic currents satisfy exactly the Gell-Mann current algebra then there is no strong-interaction cutoff and there needs to be a weak-interaction cutoff of the order of 4 GeV. There is no evidence, however, that the current algebra correctly describes the high-momentum behavior of the weak currents.

One of the most common variations on the theory presented above involves the introduction of the intermediate vector boson (IVB) as a mediating field of the weak interaction in analogy to the photon of electrodynamics. While this does not solve the divergence problems, it provides a somewhat more familiar theoretical setting for attacking them. Experiments attempting to produce the IVB in high-energy neutrino experiments and by other methods set a lower limit of at least 2 GeV on its mass. An interesting possibility not ruled out by present experiments is that the IVB may have some type of strong interaction as well as weak and electromagnetic ones.[7] One theory that solves the divergence difficulties in principle has been elaborated by Christ[8]; in this theory the usual weak interaction involves two intermediate spin-0 heavy bosons and two intermediate spin-1/2 heavy fermions. Unfortunately, in order to avoid disagreement with experiment, the theory must take a very ugly form involving many different intermediate particles. The most important point of these various speculations lies in the fact that their equivalence to the usual theory is only approximate so that they may predict small but highly significant deviations in experimental quantities.

Present experiments still leave it uncertain whether CP violation and T violation is (a) a general feature of all processes including strong and electromagnetic, (b) a special feature of weak interactions, or (c) effectively limited to K^0 processes which have a peculiar sensitivity to very weak interactions because of the near degeneracy of K_L and K_S. The search for possible CP and

T violation in processes other than K^0 decay is important in distinguishing between these models.

An interesting possibility is that CP violation is one of those small effects that distingiush the exact (or more exact) theory of weak interactions from the phenomenological approximation described above. Thus in one theory[9] involving a set of IVB's, CP violation shows up characteristically in order g^3 where g is the weak IVB coupling and the usual weak interactions are of order $g^2 \simeq G m_W^2$. Another possibility is that CP-violating effects are large at large momentum transfers but that the usual CP-invariant theory is a good approximation at the low momentum transfers involved in the usual weak processes.

4. MUON DECAY

Muon decay is the only purely leptonic weak interaction that has been studied experimentally; thus, it is the only place where we may study the weak interaction without the complication of unknown effects of strong interactions. Present experiments on $\mu \to e + \nu + \bar{\nu}$ are in complete agreement with the standard theory; in particular, precision experiments on the electron spectrum[10] give the Michel parameter $\rho = 0.752 \pm 0.003$ in agreement with the theoretical prediction $\rho = 3/4$. It was early pointed out by Lee and Yang[11] that small deviations from $\rho = 3/4$ might reveal the non-point-like nature of the four-fermion vertex. In particular, an intermediate vector boson of mass m_W yields an effective ρ

$$\rho - \frac{3}{4} \simeq \frac{1}{3} (m_\mu/m_W)^2$$

Similar deviations occur for the more complicated structure discussed by Christ[8]. Unfortunately, with the present limit $m_W > 2$ GeV, this corresponds to a deviation of ρ from 3/4 of one part in a thousand or less. Obsiously an experimental search for such small deviations is extremely difficult. Since the radiative corrections give about a 5% effect in the determination of ρ, any attempt to determine ρ to better than 0.1% would require a careful reevaluation of the radiative corrections including the question of fourth-order corrections.

While the ρ parameter has been measured to a high degree of accuracy other parameters in muon decay are known with a much smaller accuracy. The shape of the spectrum for low-energy electrons is sensitive to the parameter η, which has recently been measured by Hildebrand and collaborators[10]

$$\eta = -0.12 \pm 0.21$$

to be compared with the theoretical prediction $\eta = 0$. The asymmetry of the electrons with respect to the muon spin averaged over the spectrum determines the parameter ξ; the most recent result from a Soviet group[12] gives

$$|\xi| = 0.975 \pm .015$$

This is not significantly different from unity, the theoretical prediction; we shall use this result in the form

$$|\xi| \geq 0.95$$

The average helicity $|h|$ of the electrons has been measured to be 1.00 ± 0.13. To my knowledge no experiment has measured the electron polarization transverse to the plane defined by the muon spin and the electron direction, which would be a measure of time-reversal violation.

To put these experiments in a theoretical context, we note that the value $\rho = 3/4$ (and also the value $\delta = 3/4$ which gives the energy dependence of the asymmetry) follows entirely from the assumption that one left-handed and one right-handed neutrino are emitted in muon decay, in which case the most general local form for the muon decay interaction may be written

$$\bar{\nu}_\mu \gamma_\lambda (1 + \gamma_5) \nu_e \; \bar{e} \, \gamma_\lambda \left[(1 + \gamma_5) \, \alpha + (1 - \gamma_5) \beta \right] \mu \tag{14}$$

where we set $|\alpha|^2 + |\beta|^2 = 1$. In the usual theory $\alpha = 1$; the only free parameter β is seen to be the admixture of electrons with the wrong helicity. It should, of course, be noted that if $\beta \neq 0$ and we rewrite Eq. (14) in the usual order (charge-exchange) it involves couplings other than V or A. In terms of β we have for the observables discussed above

$$\eta = - \, \mathrm{Re} \, \alpha^* \beta$$

$$1 - |\xi| \;\; = \;\; 1 - |h| \;\; = \;\; 2|\beta|^2$$

$$\text{Transverse polarization} = 2 \, \mathrm{Im} \, \alpha^* \beta$$

The lower limit on $|\xi|$ provides the best limit on β: $|\beta| \leq 0.16$. This, however, would still allow a transverse polarization as large as 30%.

Searches for possible rare decay modes of muons are related to the question of the accuracy of the lepton number conservation laws of the standard theory. It has been particularly emphasized by Pontecorvo[13] that the present accuracy is no better than was our

knowledge of the accuracy of CP-invariance before the experiment
that discovered the rare decay mode $K_L \rightarrow 2\pi$. In particular, the
present theory implies a separate conservation of muons and elec-
trons and so forbids the decays $\mu \rightarrow e + \gamma$ and $\mu \rightarrow 3e$, for which
the present limits on the branching ratios are 2×10^{-8} and
1.5×10^{-7}, respectively. These correspond to limits on the amplitude ratio of lepton non-conservation of the order of 10^{-3}. Related questions are the rates for $\mu^- + Z \rightarrow e^- + Z$,
$\mu^- + Z \rightarrow e^+ + (Z-2)$, and neutrinoless double β-decay.

5. PION DECAY

The equality of lifetimes of π^+ and π^- is the best test for
the validity of CPT for strangeness-conserving non-diagonal processes. Any CPT-violating amplitude must be less than 10^{-3}. For
diagonal matrix elements by far the best test comes from the mass
difference between K^0 and $\bar{K}^0$, which places a limit of 10^{-8} to 10^{-9}
on CPT violation. Other tests of CPT involving diagonal matrix
elements (as mass differences or magnetic moment differences) are
therefore of little value but further measurements on possible
lifetime differences are valuable.[14]

The rare decays of pions into electrons play a major role in
confirming the fundamental hypotheses of weak interaction theory.
The rate ratio

$$R = \frac{\Gamma(\pi \rightarrow e + \nu)}{\Gamma(\pi \rightarrow \mu + \nu)}$$

provides the best test of muon-electron universality that has so
far been performed. The experimental result[15]
$R = 1.247 \pm 0.028 \times 10^{-4}$ agrees with the theoretical prediction
(including radiative corrections) of 1.23×10^{-4} within the experimental error of about 2%. This, of course, tests only the
universality of the coupling to the axial vector current. Other
tests of universality (comparisons between muon capture and nuclear
β-decay, the ratio of $K \rightarrow e + \nu$ to $K \rightarrow \mu + \nu$, and the ratio[16] of
$\Sigma^- \rightarrow n + \mu^- + \nu$ to $\Sigma^- \rightarrow n + e^- + \nu$) are much less accurate. A
large number of expensive experiments on the decays $K \rightarrow \pi + e + \nu$
and $K \rightarrow \pi + \mu + \nu$ still leave the question of the universality of
the strangeness-changing vector current very confused.

The rate of decay for $\pi^+ \rightarrow \pi^0 + e^+ + \nu$ as compared with the
rate for nuclear beta-decay for 0^{14} provides a test for the conserved vector current theory to the extent that electromagnetic
corrections can be calculated. The experimental branching ratio[17]
of $1.00^{+.08}_{-.10} \times 10^{-8}$ is in perfect agreement with the theoretical

result 1.035×10^{-8}. Alternatively, this rate, if determined more
accurately, would provide a method for determining $\cos\theta$ independent of any uncertainties of nuclear physics involved in the use
of $0 \to 0$ beta-decays. On the other hand there still remains the
uncertainty ($\sim 1\%$) in the radiative correction as well as a 0.5%
uncertainty due to the present error in the π^+ - π^0 mass difference.

Detailed observations on the decay $\pi^+ \to e^+ + \nu + \gamma$ allow the
detection of the direct emission of the gamma ray as well as the
standard internal bremsstrahlung (which is essentially the radiation from the emerging electron in $\pi \to e + \nu$.) The direct emission
has a contribution from the weak vector current that can be related to the decay $\pi^0 \to \gamma + \gamma$ by CVC and a contribution from the
weak axial-vector current that can be calculated under various
assumptions using current algebra. Present experimental results[18]
allow the determination of only a single number, from which the
axial-vector contribution can be deduced if CVC and the π^0 lifetime are assumed. There are two solutions for the axial contribution, one of which is in general agreement in sign and magnitude
with the current algebra predictions. More detailed observations
might make possible a check of the application of CVC, although
this requires a reliable value for the π^0 lifetime as well. Furthermore, the theory that there exist second-class axial vector
currents that are T-violating[19] leads to the prediction that the
electrons may be polarized transversely to the decay plane by about
10%.[20] Alternatively, one might look for an up-down asymmetry of
electrons relative to the plane of the Dalitz pair in the process
$\pi^+ \to e^+ \nu e^+ e^-$.[21]

6. MUON CAPTURE

The reaction of muon capture in hydrogen

$$\mu^- + p \to n + \nu_\mu$$

plays the same basic role for the semi-leptonic processes of muons
as neutron β-decay does for electrons. From general invariance
arguments the nucleon matrix elements of the weak V and A currents
can be described in terms of six form factors. For the V current
these are

$$g_V(q^2) \equiv g_V(o) \, F_V(q^2)$$

$$g_M(q^2) \equiv g_M(o) \, F_M(q^2)$$

$$g_S(q^2)$$

For the A current these are the two form factors defined in Eq.(6):
$g_A(q^2) \equiv g_A(o) F_A(q^2)$ and $g_p(q^2)$ plus the "pseudo-weak-magnetism"
$g_T(q^2)$. The basic theory described above allows for almost a complete prediction of these as follows:

(a) $g_S(q^2)$ and $g_T(q^2)$ are zero as a consequence of Eq.(5).

(b) The CVC hypothesis tells us that the weak-magnetism coupling $g_M(o) = (\mu_p - \mu_n) g_V(o)$ where $(\mu_p - \mu_n) \approx 3.7$ is the difference of the anomalous moments of neutron and proton. Furthermore $F_V(q^2)$ and $F_M(q^2)$ are determined by the charge and magnetic isovector electromagnetic form factors of the nucleon.

(c) $g_p(q^2)$ is determined from PCAC as discussed in Section 2.

(d) Muon-electron universality allows us to determine $g_A(o)$ and $g_V(o)$ from the corresponding values for beta-decay.

Outside of the problem of radiative corrections the main uncertainty in the theoretical prediction is the value of $F_A(q^2)$ at the value $q^2 \approx .9m_\mu^2$. Since this is a small value of q^2 it may be appropriate to write

$$F_A(q^2) = 1 - \frac{1}{6} q^2 \left\langle r^2 \right\rangle_A$$

and state the major unknown as the axial form-factor radius. Thus in principle one precision experiment on muon capture in hydrogen would have as its main consequence the determination of $\left\langle r^2 \right\rangle_A$. A variation of $\left\langle r^2 \right\rangle_A$ between zero and the value $\left\langle r^2 \right\rangle_V$ has, however, only about a 3% effect on the capture rate in the singlet state,[22] which is the hyperfine state from which muons are expected to be captured in gaseous hydrogen. Thus the uncertainties in $g_A(o)$ and $g_V(o)$ deduced from β-decay as well as the uncertainty in the radiative corrections make a determination of $\left\langle r^2 \right\rangle_A$ in this manner extremely difficult. Rather a measurement of the singlet capture rate to a precision of the order of 2 or 3% would serve to verify the four basic assumptions (a through d) to this level of accuracy. At present there exists one measurement of the capture rate in hydrogen gas, the recent beautiful experiment of Zavattini and coworkers,[23] which gives a singlet capture rate of 651 ± 57 sec^{-1} in complete agreement at this level of accuracy with the theoretical prediction of 660 ± 20 sec^{-1}.[22]

It would clearly be desirable to have more than a single number to test these assumptions. For example, the singlet capture rate is almost unaffected if both g_M and g_p are set to zero; thus the singlet capture rate serves as a check on either CVC or PCAC only if the other is assumed correct. An example of another number of great interest is the capture in the triplet state. In the simple V-A theory ($g_V = -g_A$, $g_P = g_M = 0$) this rate is zero, so that in the standard theory it is very sensitive to the small in- duced couplings g_P and g_M. In fact the triplet capture rate is cut in half if either g_P or g_M is reduced from its theoretical value to zero.[22] Unfortunately the predicted triplet capture rate is already so small (approximately 1/50 of the singlet rate) that a measurement of it may not be possible in the foreseeable future, although limits can already be set on it[23] by comparing the capture rates in gaseous and liquid hydrogen. A still smaller rate but one that has a distinctive signal is the radiative captive $\mu^- + p \rightarrow n + \nu + \gamma$. This has a particular sensitivity to g_P be- cause for high-energy photons (due to diagrams in which the photon is emitted by the proton) it is possible to go much closer to the pion pole. Of course, the theoretical uncertainty in g_P really lies in the correction terms in Eq.(12) and not in the pole term. The theoretical calculation of the radiative capture rate by standard Feynman techniques[24] gives a result entirely in terms of the effective couplings g_V, g_M, g_A, and g_P, but excludes possibly significant contributions associated with the structure of the weak hadronic vertex.[25]

Other muon capture processes that are of interest in con- firming the theory are $\mu^- + d \rightarrow n + n + \nu$ and $\mu^- + He^3 \rightarrow H^3 + \nu$. The capture in deuterium involves problems associated with the structure of the deuteron and the final state wave function of the two neutrons but allows measurements on the outgoing neutrons as a check on theoretical calculations. The capture in He^3 has the ad- vantage that it can be directly compared to the β-decay of H^3 and at the moment this comparison provides a confirmation of the theory comparable in accuracy to that of capture in hydrogen. However, significant problems exist in extrapolating matrix elements from β-decay to muon capture particularly with respect to pion exchange effects. In spite of the problems mentioned these capture pro- cesses, as well as others such as $\mu^- + C^{12} \rightarrow B^{12} + \nu$ discussed by Professor Walecka, may serve to put significant constraints on possible values of some of the coupling constants when combined with an accurate measurement of the capture rate in hydrogen.

A great deal of effort is now being put in a study of the "elastic" neutrino processes

$$\bar{\nu}_\mu + p \rightarrow \mu^+ + n$$

$$\nu_\mu + n \rightarrow \mu^- + p$$

For example, this is the primary goal of the huge 12-ft. bubble chamber now being installed at Argonne. Here again the theory makes a complete prediction except for the unknown form factor $F_A(q^2)$. Thus the first results of these experiments will be a more accurate determination of $F_A(q^2)$. These results may prove of some use in determining the value of $F_A(0.9m_\mu^2)$ needed for analyzing muon capture results, although the values of q^2 will in general be much higher so that the extrapolation to low q^2 will be dependent on a theoretical formula. At the energies at which the neutrino experiments will be done the induced pseudoscalar coupling g_P becomes unimportant; on the other hand, a comparison of neutrino and antineutrino cross-sections (which differ in theory by the sign of the interference term between A and V) should allow a check of CVC. The neutrino experiments (in contrast to the antineutrino ones) must be done on deuterium and thus require an understanding of the effect of the spectator proton; possibly a comparison with muon capture in deuterium may help elucidate this problem.

No tests of time-reversal invariance have been made so far in the strangeness-conserving semi-leptonic muon processes. In muon capture such a test would require looking for a correlation of the form $\langle \vec{S} \rangle \cdot \vec{\nu} \times \langle \vec{S'} \rangle$ where $\langle \vec{S} \rangle$ is the initial muon or nucleus spin, $\langle \vec{S'} \rangle$ is the spin of the final nucleus, and $\vec{\nu}$ is the neutrino or recoil momentum. Because of the hyperfine coupling, the muon spin and the nuclear spin cannot be independently oriented so that both the polarization and direction of the recoil are required. In the "elastic" neutrino processes the test requires looking for a muon polarization normal to the scattering plane.

7.   CONCLUSION

In spite of extensive experiments with high-energy accelerators, much of our basic information on weak interactions is seen to be the result of low-energy nuclear physics experiments and studies of pions and muons at low energies or at rest. Many of these experiments were performed in the last few years. With the advent of "meson factories" many of these experiments with pions and muons can be done to a higher level of accuracy and certain new experiments may become possible. Thus it remains likely that in the future many of the most accurate tests of weak interaction theory will come from this area with the ever-present possibility that small but very significant failures of the theory may be found.

I wish to thank the Lawrence Radiation Laboratory of the University of California, where much of this talk was prepared, for its hospitality. This work has been partially supported by the U. S. Atomic Energy Commission.

REFERENCES

No attempt is made here at a comprehensive bibliography. Details and references concerning most of this paper may be found in T. D. Lee and C. S. Wu, Ann. Rev. Nuc. Sci. $\underline{15}$, 381 (1965).

1. S. Weinberg, Phys. Rev. $\underline{112}$, 1375 (1958).
2. For a detailed review see R. J. Blinstoyle, Proceedings, Topical Conference on Weak Interactions, p. 495 (CERN, 1969). For a discussion of radiative correction uncertainties see E. S. Abers, D. A. Dicus, R. E. Norton, and H. R. Quinn, Phys. Rev. $\underline{167}$, 1461 (1968).
    Also B. Beg et al, Phys. Rev. Letters $\underline{23}$, 270 (1969).
3. This assumes $F(q^2)$ approaches a constant as $q^2 \to \infty$. A still weaker assumption would be $F(q^2)/q^2 \to 0$ as $q^2 \to \infty$. I am grateful to Dr. Pagels for comments on this point.
4. H. Pagels, Phys. Rev. $\underline{179}$, 1337 (1969), and to be published.
5. C. Lovelace, invited paper at the Conference on πN Scattering, University of California, Irvine, Dec. 1967.
6. B. L. Ioffe and E. P. Shabalin, Jour. of Nucl. Physics (USSR) $\underline{6}$, 828 (1967).
7. For a summary of these ideas see R. E. Marshak et al in Proceedings of the Topical Conference on Weak Interactions, CERN 1969, p. 371.
8. N. Christ, Phys. Rev. $\underline{176}$, 2086 (1968).
    W. Kummer and G. Segré, Nuclear Physics $\underline{64}$, 585 (1965).
9. S. Okubo, Nuov. Cim. $\underline{54A}$, 491 (1968).
10. Our summary of muon decay experimental results is from S. E. Derenzo, R. H. Hildebrand, and C. Vossler, Physics Letters $\underline{28B}$, 401 (1969).
    See also S. E. Derenzo, Phys. Rev. $\underline{181}$, 1854 (1969).
11. T. D. Lee and C. N. Yang, Phys. Rev. $\underline{108}$, 1611 (1957).
12. Akhmanov et al, Soviet Journal of Nuclear Physics $\underline{6}$, 230 (1968) [Russian original 6, 316 (1967)]
13. B. Pontecorvo in Old and New Problems in Elementary Particles, p. 251 (Academic Press, 1968).
14. For a further discussion see L. Wolfenstein, Nuovo Cimento, to be published.
15. Di Capua et al, Phys. Rev. $\underline{133B}$, 1333 (1964).
16. N. V. Baggett et al, Phys. Rev. Letters $\underline{23}$, 249 (1969).
17. P. De Pommier et al, Nucl. Phys. B4, 189 (1968).
18. P. De Pommier et al, Phys. Letters 7, 285 (1963).
19. L. Maiani, Physics Letters 26B, 538 (1968).
20. B. Holstein, private communication.
21. W. Flagg, Phys. Rev. $\underline{178}$, 2387 (1969).
22. A. Fujii, Nuovo Cimento 27, 1025 (1963). A. Fujii, M. Morita, and H. Ohtsubo, Supplem. to Prog. Theor. Physics, 1968, p.303. The value of $\langle r^2 \rangle_A$ affects not only $g_A(q^2)$ but also $g_P(q^2)$

through the correction in Eq.(13). The singlet capture rate
in the second reference is in agreement with that given by
P. K. Kabir, Z. Physik 191, 447 (1966).

23. A. Quarenta et al, Phys. Rev. 177, 2118 (1969).

24. G. A. Lobov, Nucl. Phys. 43, 430 (1963), G. I. Opat, Phys. Rev.
    134 B 428 (1964), and references therein.

25. Some of these have been calculated by S. L. Adler and Y. Dothan,
    Phys. Rev. 151, 1267 (1966).

MEASUREMENTS OF THE VACUUM POLARIZATION IN MUONIC ATOMS

G. Backenstoss, S. Charalambus, H. Daniel, H. Koch,
Ch. v.d. Malsburg, G. Poelz, H. Schmitt and L. Tauscher

CERN, Geneva, Switzerland
Max-Planck-Institut für Kernphysik, Heidelberg, Germany
Institut für Experimentelle Kernphysik, Karlsruhe,
Germany
Physik-Department der Technischen Hochschule, Münich,
Germany

The energy levels in atoms are shifted by radiative corrections. While in electron atoms the self-energy is dominating compared with the vacuum polarization, the opposite holds for muonic
atoms. Muonic transitions, where other corrections caused by the
finite size and nuclear polarization are small compared to the
vacuum polarization, provide an excellent tool for the measurement
of the vacuum polarization.

In the present experiment the 5g-4f and the 4f-3d transitions
in Bi and natural Pb were chosen. The experiment was performed at
the CERN muon channel. Conventional telescope techniques in connection with two Ge detectors were used. In order to reduce calibration errors, the spectra of the calibration sources ^{54}Mn, ^{88}Y,
^{192}Ir, and ^{207}Bi were measured simultaneously with the muonic X-ray
spectra by random coincidences with telescope signals.

For the evaluation of the spectra, muonic satellite lines were
calculated and taken into account, as well as the HFS-splitting of
the Bi lines and the isotopic shift of the three Pb isotopes. Theoretical values of the transition energies were obtained by numerical
integration of the Dirac equation, using a Fermi-type charge distribution. They were corrected for the self-energy, the shift due
to the anomalous magnetic moment of the muon, the screening by the
K- and L-electrons, and the shift due to nuclear polarization. The
difference between these calculated values and the measured energies
is the measured vacuum polarization. The values obtained are listed

Table 1

Element	(1) Transition	(2) $E_{Calc.}^{Vac.pol.}$ (keV)	(3) $E_{Calc.}^{Hi.Ord.}$ (keV)	(4) $E_{meas.}^{Vac.pol.}$ (keV)
$^{209}_{83}Bi$	$4f_{7/2} - 3d_{5/2}$	6.49	0.12	6.24 ± 0.26
	$4f_{5/2} - 3d_{3/2}$	7.07	0.13	7.11 ± 0.27
	$5g_{9/2} - 4f_{7/2}$	2.202	0.036	2.196 ± 0.050
	$5g_{7/2} - 4f_{5/2}$	2.293	0.039	2.271 ± 0.050
$^{208}_{82}Pb$	$4f_{7/2} - 3d_{5/2}$	6.29	0.11	6.18 ± 0.20
	$4f_{5/2} - 3d_{3/2}$	6.84	0.12	7.06 ± 0.22
	$5g_{9/2} - 4f_{7/2}$	2.134	0.035	2.140 ± 0.040
	$5g_{7/2} - 4f_{5/2}$	2.219	0.037	2.212 ± 0.040

in Table 1, column 4. The errors contain the experimental errors as well as errors in the theoretical transition energies caused by uncertainties in the charge distribution. Comparison is made with calculated values of the vacuum polarization, including corrections up to the order $(\alpha Z)^3$, according to calculations recently performed by Fricke[1]. Column 2 of Table 1 shows the theoretical values of the vacuum polarization, including the higher order corrections, whereas in column 3 these terms are listed separately. The accuracy of the measured values for Bi and Pb is 4.5 and 3.5 per cent for the 4f-3d transitions, and 2.2 and 1.8 per cent for the 5g-4f transitions, respectively. Within these errors the agreement between measured and calculated vacuum polarization is excellent. Furthermore, the weighted mean of the relative deviations of the experimental vacuum polarization values from the calculated values amounts to 0.2 ± 0.9 per cent. By omitting the higher order terms in the calculated values, this value changes to -1.5 ± 0.9 per cent. Hence, the agreement between theory and experiment is improved significantly by including these higher order terms.

1) B. Fricke: Zeitschrift für Physik 218, 495 (1969).

RECENT MUONIUM HYPERFINE STRUCTURE MEASUREMENTS*

P. Crane, J.J. Amato, V.W. Hughes, D.M. Lazarus,
G. zu Putlitz, and P.A. Thompson

Gibbs Laboratory, Yale University

New Haven, Connecticut

Further data on the hyperfine structure interval $\Delta\nu$ of muonium (μ^+e^-) have recently been obtained through the observation of hfs transitions at very weak magnetic field over a wider range of gas pressures than was studied in our previous work.[1,2,3] In particular, we have now observed hfs transitions with krypton as the stopping gas at 6 different pressures in the range from 6 atm to 73 atm, and also with argon at 6 different pressures in the range from 30 atm to 108 atm. The new data allow an increase in the precision of the determination of $\Delta\nu$, principally because of the reduction in the uncertainty in the extrapolation to zero gas pressure.

The unresolved transitions $(F,M_F) = (1,1) \leftrightarrow (0,0)$ and $(1,-1) \leftrightarrow (0,0)$ were observed at a magnetic field of 10mG, and the line center frequency is equal to $\Delta\nu$. The basic experimental technique has been described.[2,3] However, in order to extend our measurements to the low pressure of 6 atm, we used some plastic scintillation counters with thicknesses between 1/16 in. and 1/8 in., which were placed inside the pressure vessel close to the microwave cavity. The scintillators were viewed through quartz pressure windows by photo-tubes fastened directly to the pressure vessel. These counters allowed rejection of muons which stop in the end walls of the pressure vessel and in the gas between the end walls and the microwave cavity, and hence reduced the background. The stopping rate for muons was about 200 per sec in the 6 atm of krypton, and 100 per sec in the microwave cavity end walls and the scintillators themselves. These rates represent a considerable improvement in signal to background compared with that obtained without the use of inside counters.

In order to study the functional dependence of $\Delta\nu$ on pressure, we have fit our data to a function of the form:

$$\Delta\nu \ (P) \ = \ \Delta\nu(0) \ (1 \ + \ a \ P \ + \ b \ P^2).$$

For one case we use only the linear pressure coefficient a and set b=0, and for the second case we use both a and b as parameters in the fit. The results of these two fits to the data for argon and krypton are given in Table 1. The quality of the fits as given by the χ^2 values is about the same for the two cases, and there is no compelling evidence for the existence of a quadratic pressure term. The upper limit for b for muonium in argon from the present experiment is about one order of magnitude smaller than the value of b for rubidium in argon[4]; it is reasonable to expect that the value of b for muonium in argon will be less than that for rubidium in argon. The values of $\Delta\nu$ for the two cases for each gas are in good agreement. Hence we consider only the fits with the linear pressure term alone.

The value of a = (-4.09 ± 0.08) x 10^{-9}/torr for muonium in argon is significantly lower than the measured value of a = (-4.78 ± 0.03) x 10^{-9}/torr for hydrogen in argon from optical pumping experiments.[5] The value of a = (-9.98 ± 0.08) x 10^{-9}/torr for muonium in krypton is also somewhat lower than the measured value of a = (-10.4 ± 0.2) x 10^{-9}/torr for hydrogen in krypton.[6] Our measurements were made in the range of temperature from 15°C to 20°C and the optical pumping measurements were typically made at temperatures greater than 30°C. Present evidence[5] indicates that the temperature dependence of the pressure shift is too small to account for the difference from the hydrogen values. These differences could be due to an isotope dependence of the fractional pressure shift.[7]

The values for $\Delta\nu$ at zero pressure are:

$\Delta\nu$ = 4463.262 $\pm$ 0.012 MHz (from krypton data)

$\Delta\nu$ = 4463.220 $\pm$ 0.020 MHz (from argon data)

$\Delta\nu_{average}$ = 4463.249 $\pm$ 0.031 MHz

The indicated 1 standard deviation errors in the individual measurements are due to counting statistics, which is the dominant source of error. The error in the average value for $\Delta\nu$ has been increased to account for the variation of the two measurements about their weighted average. The value $\Delta\nu_{average}$ is in very good agreement with our previously quoted value[1] of

$\Delta\nu_{1968}$ = 4463.260 $\pm$ 0.040 MHz

It differs by about 2 std. dev. from a value recently reported from Chicago.[8]

Our latest value for $\Delta\nu$ can be used to determine the ratio of the muon magnetic moment to the proton magnetic moment, or the ratio of the muon mass to the electron mass, using a value of α determined from the ac Josephson measurement.[9]   The values are:

$$\frac{\mu_\mu}{\mu_p} = 3.183306 \pm 0.000028 \quad (9\text{ppm}, 1 \text{ std. dev.})$$

$$\frac{m_\mu}{m_e} = 206.771 \pm 0.002 \quad (9\text{ppm}, 1 \text{ std. dev.})$$

Table 1:   Results of Fits to Data on Muonium HFS Interval.

$$\Delta\nu(P) = \Delta\nu(0) \, [1 + aP + bP^2]$$

Gas	a $\times 10^9$/torr	b $\times 10^{15}$/(torr)^2	χ^2/degrees of freedom	$\Delta\nu(0)$MHz
Argon a free, b≡0	−4.09±0.08	0	3.35/4	4463.220±0.020
Argon a,b free	−4.06±0.58	1.4±5.3	3.34/3	4463.218±0.058
Krypton a free, b≡0	−9.98±0.08	0	5.17/4	4463.262±0.012
Krypton a,b free	−10.4±0.4	6.9±5.9	3.8/3	4463.285±0.028

*Research supported in part by AFOSR (at Yale) and in part by NSF (at Columbia).

[1]P.A. Thompson, J.J. Amato, P. Crane, V.W. Hughes, R.M. Mobley, G. zu Putlitz, and J.E. Rothberg, Phys. Rev. Letts. 22, 163 (1969).
[2]W.E. Cleland, J.M. Bailey, M. Eckhause, V.W. Hughes, R.M. Mobley, R. Prepost, and J.E. Rothberg, Phys. Rev. Letts. 13, 202 (1965).
[3]V.W. Hughes, Ann. Rev. Nucl. Sci. 16, 445 (1966).
[4]E.S. Ensberg and G. zu Putlitz, Phys. Rev. Letts. 22, 1349 (1969).
[5]R.A. Brown and F.M. Pipkin, Phys. Rev. 174, 48 (1968).
[6]E.S. Ensberg and C.L. Morgan, Phys. Letts. 28A, 106 (1968).
[7]G.A. Clarke, J. Chem. Phys. 36, 2211 (1962).
[8]R.D. Ehrlich, H. Hofer, A. Magnon, D. Stowell, R.A. Swanson, and V.L. Telegdi, Phys. Rev. Letts. 23, 513 (1969).
[9]W.H. Parker, B.N. Taylor and D.N. Langenberg, Phys. Rev. Letts. 18, 287 (1967).

SEARCH FOR THE $\pi^+ \rightarrow e^+ + e^+ + e^- + V$ DECAY

S.M. Korenchenko, B.F. Kostin, A.G. Morozov, G.V.
Micelmacher, K.G. Nekrasov

Joint Institute for Nuclear Research
Laboratory of Nuclear Problems
Dubna, U S S R

Hitherto nobody has searched experimentally for the
$\pi^+ \rightarrow e^+ + e^+ + e^- + V$ decay. Apart from ordinary electromagnetic interaction /1/, this process could be caused by some
"exotic" interactions (for instance, such as 6-fermion interaction /2/ or four leptons anomalous interaction /3/).

In our investigation we used a cylindrical spark chamber
placed into the magnetic field having 9300 Oe intensity. The
chamber was 39 cm in diameter and 28 cm high. The total amount of
gaps was 18 and the size of each gap was 5 mm. In the centre of
the chamber a target made of plastic scintillator was placed. Its
thickness was 3.5 g/cm^2 (in the plane normal to the beam axis).
As many as 18 scintillation counters were placed inside and
outside the chamber. This hodoscope system was used both for
chamber triggering and for the subsequent analysis of events.
Pulses were displayed on the five-trace oscilloscope the screen
of which was photographed at the same time the chamber was
triggered. Triggering was possible if there were at least two
particles passing through the chamber volume. Pulses from the
counters had to pass through a 70 nsec time gate, opened by any
pion hitting the target. The geometry of the counters imposed
some restrictions on event kinematics.

The total amount of stopped pions was $5.9 \cdot 10^9$ and 80 thousand
pictures were taken. Photographed events were selected for the
analysis by using the following criteria:
1) there should be two positrons and one electron passing
   from one point (x_0, y_0) in the photograph plane (x, y)
   and matching the chosen scheme of triggering;

2) there should be all proper pulses on the oscillograph
    sweeps;

3) an event is treated as background if an angle between
    e^+ and e^- momenta is equal to $180° \pm 10°$ and their
    energies differ not greater tahn 10 MeV, because such
    two traces (e^+ and e^-) may be imitated by a single e^+.

Twenty-eight events were selected for the further analysis.
Then by using the χ^2-criterion with a 90% confidence we expected
events where tracks were not intersected at one point M (x,y,z)
of the target.

After such a procedure we had only two events. Their total
energies were 92 ± 12 MeV and 225 ± 45 MeV instead of the value
140 MeV, expected.

So there were no single event which one could interprete as
the $\pi^+ \rightarrow e^+ + e^+ + e^- + V$ decay.

Detection efficiency related to decay kinematics, chamber
geometry and the logics of triggering was determined by the
Monte-Carlo method. The matrix element of the decay was assumed
to be constant. After all necessary corrections (gate width,
camera dead time, etc.) detection efficiency was found to be
$1.2 \pm 0.2\%$.

The expected value of the background was evaluated from the
28 selected events and was about 1 event.

As a result of the experiment we give the upper limit for
the branching ratio with a 90% confidence
$$\frac{W(\pi \rightarrow e + e + e + V)}{W(\pi \rightarrow \mu + V)} \leq 3.4 \cdot 10^{-8}$$
and with a 50% confidence
$$\frac{W(\pi \rightarrow e + e + e + V)}{W(\pi \rightarrow \mu + V)} \leq 10^{-8}$$

Authors are grateful to B. Pontecorvo, L.I. Lapidus and D.Y.
Bardin for valuable discussions, to V.S. Smirnov for the adjustment of fast electronics.

References

1. W. Flagg, Phys. Rev., 178 (1969) 2387.
2. T. Ericson, S.L. Glashow, Phys. Rev., 133 (1964) B130.
3. L. Okun, B. Pontecorvo, K. Rubia, Preprint JINR, D-2768, 1966.

P AND T VIOLATION IN NUCLEAR PHYSICS[*]

Bruce H. J. McKellar

Department of Theoretical Physics

The University of Sydney, Sydney, N.S.W., Australia

1. INTRODUCTION

The classic paper of Feynman and Gell Mann [1] on the current × current theory of the weak interaction pointed out that on this picture one would expect self interaction terms like

$$H^W_{np} = \frac{1}{\sqrt{2}} \; G \; \{\bar{\psi}_n \; \gamma_\mu \; (1 + i\gamma_5)\psi_p\}\{\bar{\psi}_p \; \gamma_\mu \; (1 + i\gamma_5)\psi_n\}$$

to appear in the weak interaction. Such a term would give rise to a weak force between neutrons and protons, which would be completely swamped by the strong force, were it not for the fact H^W_{np} does not conserve parity.

This would therefore predict that the states of nuclei are not eigenstates of parity. As well as the dominant state of spin J and parity $\pi, |J^\pi\rangle$, one has a mixture of the state $|J^{-\pi}\rangle$ of opposite parity. The mixing amplitude we call F, following the notation of Lee and Yang [2] and so we write the nuclear state $|J\rangle$ as

$$|J\rangle = |J^\pi\rangle + F|J^{-\pi}\rangle .$$

The current × current hypothesis predicts

$$F \sim \frac{Gm_\rho^2}{g^2/4\pi} \sim 10^{-6}$$

where $g^2/4\pi = 14\cdot6$ measures the strength of the strong interaction.

There are essentially two ways to detect a non zero F, as was
pointed out by Wilkinson [3].

One, is to look for violation of a selection rule which is good
if F = 0. This type of experiment measures F^2 and thus has to be
very sensitive. The classic such experiment is the search for α
decay of the 2^- state at $8 \cdot 88$ MeV in ^{16}O to the 0^+ ground state of
^{12}C [4]. The present results indicate $|F|^2 \lesssim 3 \times 10^{-13}$, in the
region where one would expect a positive result to show up.

The other looks for a correlation which would be absent if
F = 0. The simplest such correlation indicating parity non conservation is of the form $J \cdot p$ where J is the spin of one of the particles
and p the momentum of one of the particles. Circular polarisation
of γ rays emitted by nuclei, and asymmetry in the angular distribution of γ rays emitted after capture of polarised neutrons by a
nucleus have been looked for.

The amplitude for processes of this type will be of the form

$$a + bF \; J \cdot p$$

where a and b are factors depending on the details of the nuclear
transitions involved. The parity violating correlation is then of
order F b/a = FR. In general R $\sim$ 1 and the correlation is of
order 10^{-6}. However one can find situations in which a << b, when
the normal transition is forbidden by some selection rule. R may
then be several hundred. The results of experiments on these
enhanced transitions are summarised in table 1, and will be described
in more detail by Dr. Hamilton in the next talk. As you can see
the experimental situation is somewhat confused. Perhaps it will
be cleared up at this conference. Table 1 also shows the results
predicted by the Cabibbo Hamiltonian [14].

The first calculations of a parity violating force between
nucleons were made using the non-strange current J_{NS} only in the
Hamiltonian

$$H^w(NS) = 2^{-\frac{1}{2}} G \{J_{NS}^{\;\;l}, J_{NS}^{\;\;-}\}.$$

This includes strong interaction effects which smear the contact
interaction H^w_{np}. The simplest model [15] takes this smearing to be
dominated by the ρ coupling to the vector current. Blin-Stoyle [16],
and later Lacaze [17], have considered two pion exchange contributions.

It was noticed by Blin-Stoyle [16] and proved in general by
Barton [18], that the non-strange currents cannot contribute to a
π exchange force. This is because only the isovector part of H^w

TABLE 1: Parity Violating Effects in Nuclei.

Circular polarisations and asymmetries are given in units of 10^{-4}.

Transition	Experiment Value	Ref.	Cabibbo Theory fg < 0	Cabibbo Theory fg > 0
^{175}Lu 396 kev polarisation	0·4 ± 0·1	5	± (0·3 ± 0·2)	± (0·45 ± 0·3)
^{175}Lu 343 kev polarisation	0·2 ± 0·3	6	− (0·1 ± 0·05)	− (0·15 ± 0·1)
^{181}Ta 482 kev polarisation	− (0·1 ± 0·4)	6	− (0·2 ± 0·1)	− (0·3 ± 0·1)
	− (0·9 ± 0·6)	7		
	+ (0·3 ± 2·1)	8		
	− (0·32 ± 0·08)	9		
	− (0·06 ± 0·01)	10		
^{203}Tℓ 273 kev polarisation	− (0·2 ± 0·3)	6	− (0·3 ± 0·1)	− 0·45 ± 0·2
^{114}Cd 9·05 mev asymmetry	− (3·7 ± 0·9)	11	± 2	± 3
	− (3·5 ± 1·2)	12		
	− (2·5 ± 2·2)	13		

can contribute to the weak NNπ vertex, and one cannot form an isovector from the symmetrised product of two isovector currents.
Other terms in the Hamiltonian, such as the product of strange currents $J(k^+)$ $J(k^-)$, can give a pion exchange force. The relative
strength of the isovector and isoscalar part of H^W varies greatly
from theory to theory—for example the isovector part of the
Cabibbo Hamiltonian is 1/16th as strong as the isoscalar part.
Dashen et al [20] therefore suggested that experiments sensitive to
the isovector part of the weak nucleon-nucleon force be carried out
to discriminate between weak interaction theories. The π exchange
force was subsequently calculated [21] and it was suggested by
McKellar [22, 23] and Tadic [24] that the results of existing experiments may suffice to eliminate some weak interaction theories.
This conclusion may depend on a too simplified picture of the
nuclear physics involved, as has been pointed out by Vinh Mau and
Bruneau [25], who obtained an effect from the one pion force 1/5
times that of McKellar and Tadic. However, they did not calculate
the parity mixing of the ρ exchange force, so we do not know if it is

similarly reduced. We therefore continue to use the earlier calculations.

The choice of heavy nuclei as the most convenient for experiments has as an unfortunate consequence the impossibility of a
precise theoretical calculation of the transition matrices involved.
The standard procedure [15, 26] has been to replace the nucleon
nucleon force by an effective single particle force defined in the
Hartree-Fock manner

$$< \alpha | W | \beta > = \sum_{\gamma} < \alpha\gamma | V | \beta\gamma - \gamma\beta > .$$

The average has been performed not in the nucleus concerned but in
nuclear matter. Adams [27] pointed out that Michel's [15] original
estimate that the hard core correlation had a negligible effect on
the average was in error, and that in fact the use of plane waves
for the nuclear states overestimated the averaged ρ exchange
potential by a factor of 2. This was confirmed by McKellar [14]
who calculated the average using Bethe-Goldstone wave functions in
nuclear matter.

Having obtained an effective potential W in this way, one then
used it with a single particle wave function to obtain the transition matrix elements [15, 26].

There have recently been some calculations which work directly
with the two body potential [25, 28]. These give rather different
answers, and suggest that the nuclear physics will have to be done
more carefully in the future.

Two excellent reviews of these developments have recently
appeared [29,30] so I intend to concentrate on some points of particular interest.

1.   The development of the Dashen et al idea that one can use the
      parity violating nucleon-nucleon force as a probe to investi-
      gate the structure of the weak interaction. This work has
      proceeded rapidly in the last year by the efforts of Fishbach,
      Trabert, Tadic, McKellar and Pick [22, 23, 24, 31, 32, 33].

2.   The possibility of turning the emphasis around and using the
      observation of parity violation in nuclei to obtain informat-
      ion about the nuclear structure.

3.   The analogous work on prediction and detection of a breakdown
      of time reversal symmetry in nuclear physics.

2. USE OF PARITY VIOLATION IN NUCLEI AS

A PROBE OF THE WEAK INTERACTION HAMILTONIAN

Our knowledge of the leptonic and semi-leptonic weak interaction can be summarised by the current × current Hamiltonian of Cabibbo [19]

$$H(C) = \frac{G}{\sqrt{2}} \{ J_T^* , J_T \}$$

where J_T is the total current of V-A form. It is written as the sum of a leptonic current ℓ and a hadronic current

$$J = J_{1 - i2} \cos \theta + J_{4-i5} \sin \theta$$

which is a member of an SU3 octet.

It would be simplest if this Hamiltonian were also to describe the non leptonic weak interactions of the hadrons. Here one has the difficulty of explaining the $\Delta T = \frac{1}{2}$ rule observed in kaon and hyperon decays [34]. One possible explanation is to retain the Cabibbo Hamiltonian and require that the strong interactions enhance the octet matrix elements leading to the $\Delta T = \frac{1}{2}$ rule [35]. Another is to modify the Hamiltonian so that no $\Delta T = 3/2$ terms appear [36 - 39]. Other modifications of H(C) have been proposed to introduce weak bosons [36, 38, 40], to include CP violation [41, 42], to make the theory renormalisable [43].

Naturally if we extend the range of experimental data to include the $\Delta S = 0$ non-leptonic weak interactions—which is what we do in studying parity violation in nuclei—we impose more stringent tests upon the weak interaction theories.

As we remarked above H(C; $\Delta S = 0$, $\Delta T = 1$) is weaker than H(C; $\Delta S = 0$, $\Delta T = 0$) by a factor $\tan^2 \theta \sim 1/16$. Experiments which can sort out the isotopic structure of the nuclear parity violation are thus highly significant.

The classic case is

$$n + p \to d + \gamma .$$

Here the circular polarisation of the γ is determined by the isoscalar part of the weak force, and the asymmetry of the γ with respect to the neutron spin by the isovector part [44]. The parity violating effects in this reaction are of order 10^{-6}, so that the experiments are extremely difficult.

Henley [45] suggested that certain γ transitions in light self-conjugate nuclei would be sensitive to the isovector interaction. The transitions

2^- (5·105 mev) → 1^+ (0·717 mev) in ^{10}B

0^- (1·080 mev) → 1^+ (0 mev)   in ^{18}F

appear to be good candidates, with polarisations of order 10^{-5}, on the Cabibbo Theory.

It was even suggested that the present data may suffice to eliminate some of the theories. This suggestion rests on Barton's theorem [18, 21] that the π exchange part of the parity violating force is derived from the isovector part of the Hamiltonian in a CP conserving theory. This suggests that in any theory in which the isovector and isoscalar parts of the ΔS = 0 weak Hamiltonian are comparable in magnitude the π exchange force will give the dominant effect because of its long range. This statement is verified for the deuteron and in effective potential calculations. It is how-ever of great importance to test it in a calculation of the type carried out by Vinh Mau and Bruneau.

A general formula for the π exchange force V_π is readily obtained using current algebra [21], SU3 symmetry [24] or a pole model for parity violating hyperon decays [46] to determine the weak NNπ vertex.

A similar general formula for the ρ exchange potential V_ρ has been given on the basis of vector meson dominance of the vector form factors [33], while a similar formula has been derived using field algebra to determine the weak NNρ vertex [32].

In CP conserving theories, the π exchange potential has the structure

$$V_\pi = \frac{gf}{4\pi\sqrt{2}\ M_N} \left[\sigma^{(1)} + \sigma^{(2)}\right] \cdot [p_{12}, \ e^{-m_\pi r_{12}}/r_{12}]\ T_{12}^{(-)}$$

where $p_{12} = \frac{1}{2}(p_1 - p_2)$, $r_{12} = |r_1 - r_2|$ and

$$T_{12}^{(\pm)} = \tau_-^{(1)}\ \tau_+^{(2)} \pm \tau_+^{(1)}\ \tau_-^{(2)} \ .$$

f is the parity violating amplitude for n → p + π⁻ . It differs from theory to theory, depending on the isotopic structure of the

Hamiltonian. In the Cabibbo Theory $f = f_c$,

$$|f_c| = 5 \cdot 8 \times 10^{-8} .$$

(It is interesting to note that the magnitude of f_c is that of $Gm_\pi^2 \sin^2\theta = 1 \cdot 2 \times 10^{-8}$!)

V$_\rho$ has a more complicated dependence on the particular theory. In general

$$V_\rho = \frac{- G\, m_\rho^2}{4\pi\sqrt{2}\ m_N} \left[\Gamma \cdot \{p_{12}, \ e^{-m_\rho\ r_{12}}/r_{12}\} \right.$$

$$\left. + i\,(\sigma_1 \times \sigma_2) \cdot [p_{12}, \ e^{-m_\rho\ r_{12}}/r_{12}]\ T \right)$$

Here Γ is a spin isospin operator and T an isospin operator. In the Cabibbo theory

$$\Gamma_c = (\sigma_1 - \sigma_2)\ T_{12}^{(+)} ; \quad T_c = (1 + \mu^V)\ T_{12}^{(+)} .$$

To crudely estimate the relative effect of V_ρ and V_π, we note the well known result that the low energy behaviour of a potential depends on Vr^2, and thus estimate (in Cabibbo theory)

$$\frac{F_\pi}{F_\rho} \sim \frac{g\, G\, m_\pi^2 \sin^2\theta\ m_\pi^{-2}}{\frac{1}{2}(1 + \mu^V)G\, m_\rho^2\ m_\rho^{-2}} \sim 0 \cdot 4$$

where the factor 1/2 takes into account the suppression of F_ρ by the hard cores. We see that it is not possible to neglect the effect of V_π even where it is suppressed by the factor $\sin^2\theta$. A calculation of the effective potential shows that, in ^{181}Ta, $F_\pi/F_\rho = 0 \cdot 25$, so that the above crude estimate is quite good. The results of this effective potential calculation for H(C) are shown in table 1, with the observed values. As the sign of f is not known we obtain two predictions. The sign of G is taken as positive, according to vector boson theory. We can conclude that the Cabibbo theory is not inconsistent with the observations.

In table 2 we show the effective potentials W_ρ and W_π appropriate to ^{181}Ta which one obtains from several weak interaction theories. W_ρ is divided into two parts, W_ρ^d which comes from the anticommutator term and W_ρ^e which comes from the commutator term.

TABLE 2:   Comparison of the parity violating potentials of some
weak interaction theories.

Theory	Ref.	W_ρ^d	W_ρ^e	W_π		W	P_γ Units of 10^4	
Cabibbo	19	0	0·4	0·1	0·3	0·5	− 0·2	−0·3
d'Espagnat	36	− 0·63	0·49	1·5	−1·4	1·6	0·0	−1·1
Brunet	37	0·04	0·46	0·07	0·4	0·5	− 0·2	−0·3
Lee and Yang	38	0·07	0·73	0	0·8	0·8	− 0·5	−0·5
Michel	39	0·05	0·53	0·9	−0·3	1·5	+ 0·2	−0·9
Segré	40							
(a) γ_5 invariant		− 0·01	0·41	0	0·4	0·4	− 0·2	−0·2
(b) γ_5 non invariant		0·05	0·35	1·9	−1·5	2·3	+ 0·9	−1·4
Oakes	41	− 0·13	0·37	3·2	−3·0	3·4	+ 1·8	−2·2
Glashow	42	0	0·4	0·1	0·3	0·5	− 0·2	−0·3
Lee	43	− 0·23	0·43	0·75	−0·5	0·9	+ 0·4	−0·5

The effective potentials are expressed as a ratio to Michel's effective potential W_M, which is obtained by averaging V_ρ between plane-
wave states and letting $m_\rho \to \infty$.

$$< \underset{\sim}{k}' S_z' t_z' |W_M| \underset{\sim}{k} S_z t_z > \; = \; \delta_{S_z S_z'} \; \delta_{t_z t_z'} \; \delta(\underset{\sim}{k} - \underset{\sim}{k}')$$

$$\times \; (\mu^V + 1) \; \frac{G}{\sqrt{2}M_N} \; \rho \; [\tfrac{1}{2} + t_z \frac{N - Z}{A}] \; <S_z'|\sigma|S_z> \cdot \underset{\sim}{k}$$

where ρ is the density of nuclear matter.

Also shown are the predictions on the effective potential model
for the γ polarisation in the 482 kev transition in ^{181}Ta.

The errors in the calculated values are difficult to estimate, and the numbers are best regarded as indicating the order of magnitude of the effect. Some of the theories, notably the Oakes Theory of CP violation, the d'Espagnat and Michel theories of the $\Delta T = \frac{1}{2}$ rule, and the γ_5 non invariant W boson model of Segré are distinguished by very large pion potentials. It may be possible to improve the calculations and the experiments to the point where one can discriminate against such theories.

Equally striking are the zero pion potential of the γ_5 invariant model of Segré and the schizon model of Lee and Yang, but it would be difficult to experimentally distinguish between a zero pion potential and a weak pion potential.

There is another feature of the weak interaction Hamiltonian which has implications for the parity violating nucleon nucleon force. It has been suggested from time to time that the weak currents may contain terms with the wrong G parity [47]. In particular some investigations suggest there is an induced pseudo tensor term in the axial vector form factor [48], which is then

$$< N' |A_\mu| N > = \bar{u}' (f_A \gamma_\mu \gamma_5 + f_T \sigma_{\mu\nu} \frac{k^\nu}{2M_N} \gamma_5 + f_p k_\mu \gamma_5)u.$$

$f_T(0)$ was thought to be in the range $(-1, -10)$ but its existence is now considered doubtful [49]. Such a term in the A form factor gives rise to additional terms in the ρ exchange potential. They have been discussed by Blin Stoyle and Herczeg [50] and Tadic [24].

If we compute the equivalent effective potential, W_T, we find that

$$W_T = \frac{f_T(0)}{2f_A(0)} \tau_z W_\rho (C)$$

We then obtain the restriction that $-2 < f_T(0)/f_A(0)$, because if this condition were violated P_γ would have the wrong sign. This conclusion depends only on the observations that $W_\rho(C)$ dominates $W_\pi(C)$ and that the sign of the parity violation given by $W_\rho(C)$ is correct. These observations are dependent on the model used to evaluate the effective potentials, but one hopes they are characteristic of all such models. The relation between W_T and $W_\rho(C)$ is independent of the details of the model and one would thus expect the restriction on $f_T(0)/f_A(0)$ to survive improvements in nuclear physics calculations.

3.   PARITY VIOLATION   AS A NUCLEAR STRUCTURE PROBE

In the future when we have sorted out the experimental difficulties and can do the nuclear physics calculations better, would
it be possible to use the observationsof parity violations as a
probe of nuclear structure effects? The most obvious suggestion,
if we accept the Cabibbo theory so that the ρ force dominates, is
that we have a potentially useful technique for observing the rather
elusive short range correlation in nuclei.

Strong short range correlations are predicted by the hard core
nucleon-nucleon potential. Direct experimental observation of
such correlations has been attempted [51]. However the results of
these experiments are not conclusive, being obscured by final state
interaction effects.

As pointed out by Adams, the hard core induced correlations in
nuclei reduce W_ρ by a factor of about 2. If we could be sure of
the rest of our calculation to this accuracy we could thus obtain
evidence for hard core correlations.

But could we? The radial dependence of V_ρ is determined by
our assumption that the vector form factor is dominated by the ρ
meson. This gives the radial factor $\exp(- m_\rho r)/r$ in V_ρ. Experimentally however a dipole form is preferred for the vector form
factor. This would alter V_ρ to V_ρ', in which the radial factor is
$\tfrac{1}{2} m_\rho \exp(- m_\rho r)$.

This potential could be averaged to produce a new effective
potential W_ρ'. W_ρ' again is reduced by a factor of about 2 when hard
core correlations are included, so that this alteration of the form-
factor does not change the sensitivity to the presence of the hard
core.

One can make more sophisticated assumptions about the form-
factors—for example introduce a varying axial vector form-factor.
In the present uncertain circumstances this does not seem to be
warranted. We can conclude from the above example that, provided
the radial factor does not vanish at the origin the effective potential will still be sensitive to the presence of hard cores.

If we can refine the nuclear structure calculations to the
point where we can believe them to within a factor of two we can
use the observations of parity violation in nuclei to probe the
short range correlations.

4. T VIOLATION IN NUCLEAR PHYSICS

The study of T violation has not proceeded as far as that of P violation.

From a theoretical point of view one has no clear idea of the origin of CP violation, and hence T violation if CPT holds, in the kaon system. The source of T violation has been variously proposed to be in the medium-strong, electromagnetic, weak or some new super-weak interaction. Crudely these predict a T violating part of the nuclear wave function which has an amplitude of 10^{-3}, 10^{-3}, 10^{-6} or 10^{-9} respectively. The situation for observing T violation in the weak interaction is not as grim as this suggests, as one may expect effects of order 10^{-3} in β decay.

Detailed calculations of the T violating amplitudes are few because one has a wide variety of theories to deal with [50, 53]. There is a further difficulty in that higher order electromagnetic effects also lead to correlations of the type which are looked for to establish T violation [54]. For example one can look for corre-lations in a βγγ transition to establish a phase difference between the components of a mixed transition, or one can look for correlat-ions of the type $J_n \cdot (p_e \times p_\nu)$ in β decay which establish a phase difference between the vector and axial vector currents. Detailed balance experiments do not suffer from these difficulties, but in reactions it is much harder to predict the T violation from the T odd Hamiltonian.

Present low energy experiments [55] suggest that the amplitude of these T odd terms in the matrix element is less than about 10^{-2}. While no T violation has yet been seen the sensitivity of the experiments has not yet embarrassed any of the contending theories. This is a very active field and we should soon see experiments pushed to the stage where they could decide the existence of T vio-lation of medium-strong or electromagnetic origin.

5. CONCLUSION

In such a fluid situation, our conclusions are tentative. We do not have sufficient agreement between the observations of parity violation to unequivocally conclude that it exists, but it seems likely that it does. If the effect is established, it would be very useful to do experiments in light nuclei to unravel the isotopic structure of the parity violating force. This would permit firm con-clusions to be drawn about the weak Hamiltonian. T violation experi-ments seem to be on the threshold of detecting T odd medium-strong and electromagnetic forces if they exist.

As is often the case, if we can do more difficult experiments and more difficult calculations, we should be able to obtain a lot more useful information from nuclear physics studies about the nat-ure of the weak interaction.

REFERENCES

*Supported in part by the Science Foundation for Physics within The University of Sydney.

1. R.P. Feynman & M. Gell Mann, Phys. Rev. 109, 193 (1968).
2. T. D. Lee and C. N. Yang, Phys. Rev. 104, 254 (1956).
3. D. H. Wilkinson, Phys. Rev. 109, 1603 (1958).
4. R. E. Segel, J. W. Olness and E. L. Sprenkel, Phil. Mag. 6, 163 (1961) and Phys. Rev. 123, 1382 (1961)
   D. P. Boyd, P. F. Donovan, B. Marsh, D. E. Alburger, D. H. Wilkinson, P. Assimakopoulos and E. Beardsworth, Bull. Am. Phys. Soc. 13, 1424 (1968)
   R. E. Segel, Bull. Am. Phys. Soc. 14, 565 (1969)
   and private communication.
   M. Gari and H. Kummel, Phys. Rev. Letters 23, 26 (1969).
5. V. M. Lobashov, V. L. Nazarenko, L. F. Saenko, L. M. Smotritskii and G. T. Kharkevich, Zh. Eksp. Teor. Fiz. Pisma 3, 268 (1966) [translation in JETP Letters 3, 173 (1966)].
6. F. Boehm and E. Kankeleit, Nucl. Phys. A109, 457 (1968).
7. D. W. Cruse and W. D. Hamilton, Nucl. Phys. A125, 241 (1969).
8. P. Bock and H. Schopper, Phys. Letters 6, 284 (1965).
9. E. Bodenstedt, L. Ley, H. O. Schlenz and U. Wehman, Phys. Letters 29B, 165 (1969).
10. V. M. Lobashov, V. A. Nazarenko, L. F. Saenko, L. M. Smotritskii and G. Kharkevich, Zh. Eksp. Teor. Fiz. Pis'ma 5, 73 (1967) [translation in JETP Letters 5, 59 (1967)] and Phys. Letters 25B, 104 (1967).
11. Yu.G. Abov, P. A. Krupchitsky, and Yu.A. Oratovskii, Yadern Fizika 1, 479 (1965) [translation in Soviet J. Nuclear Physics 1, 341 (1965)].
12. Yu.G. Abov, P. A. Krupchitsky, M. I. Bulgakov, O. N. Yermakov and I. L. Karpikhin, Phys. Letters 27B, 16 (1968).
13. E. Warming, F. Stecher-Rasmussen, W. Ratymski and J. Kopecky, Phys. Letters 25B 200 (1967).
14. B.H.J. McKellar, Phys. Rev. Letters 20, 1542 (1969).
15. F. C. Michel, Phys. Rev. 133, B329 (1964).
16. R. J. Blin Stoyle, Phys. Rev. 118, 1605 (1960).
17. R. Lacaze, Nucl. Phys. B4, 657 (1968).
18. G. Barton, Nuovo Cimento 49, 512 (1961).
19. N. Cabibbo, Phys. Rev. Letters 10, 531 (1963).
20. R. F. Dashen, S. C. Frautschi, M. Gell Mann and Y. Hara, The Eightfold Way, edited by M. Gell Mann and Y. Ne'eman (W. A. Benjamin Inc., New York, 1964) p. 254.

21. B.H.J. McKellar, Phys. Letters 26B, 107 (1967).
22. B.H.J. McKellar, Phys. Rev. Letters 21, 1822 (1968).
23. B.H.J. McKellar, Phys. Rev. 178, 2160 (1969).
24. D. Tadic, Phys. Rev. 174, 1694 (1968).
25. N. Vinh Mau and A. M. Bruncau, to be published.
26. S. Wahlborn, Phys. Rev. 138, B530 (1965).

27. J. B. Adams, Phys. Rev. 156, 1611 (1967).
28. E. Manqueda and R. J. Blin Stoyle, Nucl. Phys. A91, 460 (1967).
29. R. J. Blin Stoyle, Proc. Topical Conference on Weak Interactions (CERN report 69 - 7, 1969) p. 495.
30. E. M. Henley, Parity and Time Reversal Invariance in Nuclear Physics, to be published in Annual Reviews of Nuclear Science.
31. E. Fischbach, contribution to this conference.
32. E. Fischbach, D. Tadic and K Trabert, to be published.
33. B.H.J. McKellar and P. Pick to be published.
34. See for example, J. P. Berge, Proc. XIII Intern. Conf. on High Energy Physics (University of Calif. Press, 1967).
35. N. Cabibbo, Phys. Rev. Letters 12, 62 (1964)
    M. Gell Mann, Phys. Rev. Letters 12, 155 (1964).
36. B. d'Espagnat, Phys. Letters 7, 209 (1963).
37. R. C. Brunet, Nuovo Cimento 50A, 562 (1967).
38. T. D. Lee and C. N. Yang, Phys. Rev. 119, 1410 (1960).
39. F. C. Michel, Phys. Rev. 156, 1698 (1967).
40. G. Segre, Phys. Rev. 173, 1730 (1968).
41. R. J. Oakes, Phys. Rev. Letters 20, 1539 (1968).
42. S. L. Glashow, Phys. Rev. Letters 14, 35 (1965)
    B. R. Holstein, Phys. Rev. 171, 1668 (1968)
    J. C. Pati, Phys. Rev. Letters 20, 812 (1968)
    The calculations of table 2 are based on the model in which the phase of the axial vector current is the same for both the π^+ and K^+ components.
43. T. D. Lee, Phys. Rev. 171, 1731 (1968).
44. G. S. Danilov, Phys. Letters 18, 40 (1965).
45. E. M. Henley, Phys. Letters 28B, 1 (1968).
46. L. R. Ram Mohan, to be published.
47. S. Weinberg, Phys. Rev. 112, 1375 (1958).
48. A. Fujii and H. Ohtsubo, Nuovo Cimento 42, 109 (1966).
    H. Ohtsubo, Phys. Letters 22, 480 (1966).
    F. Krmpotic and D. Tadic, Phys. Letters 21, 680 (1966).
49. M. Rho, Phys. Rev. Letters 18, 671 (1967); 19 248 (1967)
    B. Eman, F. Krmpotic, D. Tadic and A. Nielsen, Nucl. Phys. A104, 386 (1967).
50. R. J. Blin Stoyle and P. Herczeg, Nucl. Phys. B5, 291 (1968).
51. See the review by G. E. Brown, Comments on Nuclear and Particle Physics 3, 48 (1969) and 3, 78 (1969).
52. cf refs 49 and 20.
53. E. M. Henley and B. A. Jacobsohn, Phys. Rev. 113, 225 and 234 (1959)
    P. A. Moldauer, Phys. Rev. 165, 1136 (1968)
    Z. Szymanski, Nucl. Phys. A113, 385 (1968)
    C. W. Kim and H. Primakoff, to be published
    A. H. Huffman, to be published
54. E. M. Henley & B. A. Jacobsohn, Phys. Rev. Letters 16, 706 (1966)
    J. P. Hannon & G. T. Trammell, Phys. Rev. Letters 21, 726 (1968)
    C. G. Callen, Jr. and S. B. Treiman, Phys. Rev. 162, 1494 (1967).

55.   W. G. Weitkamp, D. W. Storm, D. C. Shreve, W. J. Braithwaite
      and D. Bodansky, Phys. Rev. 165, 1233 (1968).
      W. von Witsch, A. Richter and P. von Bretano, Phys. Rev. 169,
      923 (1968).
      S. T. Thornton, C. M. Jones, J. K. Bair, M. D. Mancusi, and
      H. B. willard, Phys. Rev. Letters 21, 447 (1968)
      R. Handler, S. C. Wright, L. Pondrom, P. Limon, S. Olsen and
      P. Kloepel, Phys. Rev. Letters 19, 933 (1967)
      E. E. Gross, J. J. Malanify, A. van der Woude and A. Zucker,
      Phys. Rev. Letters 21, 1476 (1968)
      O. C. Kistner, Phys. Rev. Letters 19, 872 (1967)
      M. Atac, B. Christman, P. Debrunner and H. Frauenfelder, Phys.
      Rev. Letters 20, 691 (1968)
      J. Eichler, Nucl. Phys. A120, 535 (1968)
      F. D. Calaprice, E. D. Commins, H. M. Gibbs, and G. L. Wick,
      Phys. Rev. Letters 18, 918 (1967) and Phys. Rev., to be
      published.
      B. G. Erozolimsky, L. N. Bondarenko, Yu. A. Mostovoy, B. A.
      Obinyakov, Z. P. Zacharova and V. A. Titov, Phys. Letters 27B,
      557 (1968).

DISCUSSION

R. Segel: Do the nuclear-matter calculations you mention allow for
the properties of nearby states of opposite parity?   B.H.J.M.:
Yes, they do.

V.M. Lobashov: The difference between Danilov's and Tadic's calculations of the γ asymmetry in polarized neutron-proton capture was
carefully checked by Danilov and Moskalev in Leningrad.   It was
found that a necessary part of the derivation was omitted in
Tadic's work, but no error was found in Danilov's original paper.

EXPERIMENTAL INVESTIGATIONS OF P- AND T-VIOLATIONS

D. Hamilton

University of Sussex, Brighton, England
CERN, Geneva, Switzerland

1.   INTRODUCTION

Experiments to test parity and time violations are difficult. Theoretical estimates of the expected effects are uncertain and it appears that accurate experimental results will be necessary in order to clarify the theoretical position.

The theoretical estimates as have been discussed by Dr. McKellar [1] indicate that the parity mixing of nuclear states could have an amplitude in the range $10^{-6} \gtrsim F \gtrsim 10^{-7}$. The situation in the case of time reversal is not so clear since the source of a possible time reversal interaction has not been identified. For the case when the interaction is due to a $T_{odd} P_{even}$ term in the Hamiltonian then the most favourable estimates arise from strong and electromagnetic interactions and have an amplitude of $\gtrsim 10^{-3}$, while for a $T_{odd} P_{odd}$ interaction the amplitude is probably not greater than $\sim 10^{-9}$ times the regular interaction.

2.   EXPERIMENTAL TESTS OF PARITY VIOLATION

The basis of these tests was laid by Wilkinson [2]. One may search for a transition which could only occur if parity mixing of nuclear states occurred. The intensity of the transition will be of the order F^2, i.e. $\sim 10^{-12}$ to 10^{-14} compared with the intensity that a similar allowed transition would have. The alternative and more direct test is to establish the presence of a pseudoscalar term. This test falls into two categories; a measurement of a forward-backward asymmetry in the radiation emitted by polarized nuclei, or the detection of the circular polarization of radiation from a

random assembly of nuclei. The extent of these pseudoscalar interactions is proportional to the amplitude of the parity admixture, F,
since they arise from an interference between the normal electromagnetic decay and that occurring as a result of the parity mixing of
the nuclear states. Thus one is searching for an effect which may
be as small as one part in 10^6 compared to the regular process.

Let us now consider in some detail the present position in
these different types of experiment.

2.1 Parity Violating Transition

The most sensitive conditions for observing a parity violating
transition occur in the α-decay of the 8.88 MeV 2^- state in ^{16}O to
the 0^+ ground state of ^{12}C. Normal α-decay is absolutely forbidden.
A good resolution experiment is necessary in order to identify the
weak parity violating transition in the tails of neighbouring intense regular transitions and also to distinguish the possibility
of decay of the 2^- state by an allowed higher order process. One
must also be confident that the decay scheme could not allow a weak
normal transition of comparable energy.

The 8.88 MeV 2^- state is populated following the β^- decay of
the 2^- ground state of ^{16}N, and this ensures that the decay to the
even parity states in ^{16}O is relatively small. From the systematics
of 2^+ states one can estimate that expected α-decay width of an

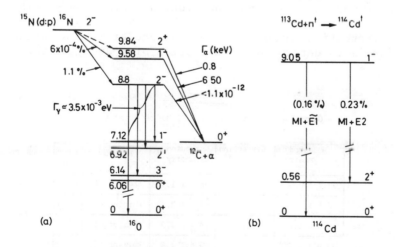

Fig. 1   a) The β^- decay of ^{16}N to levels in ^{16}O which may subsequently
decay by α-emission to ^{12}C.
    b) The decay of the neutron capture state in ^{114}Cd to the ground
and first excited state. The majority of decays occur to higher energy
levels.

8.88 MeV 2^+ state is about 6.7 keV. Measurements by Boyd et al. [3] indicate that the α-decay width of the equivalent 2^- state is less than 1.1×10^{-12} keV. It can be concluded that $|F| \gtrsim 4 \times 10^{-7}$.

2.2 Gamma-Ray Asymmetry from Polarized Nuclei

The measurement of pseudoscalar quantities which can be associated with parity violation has been the most popular way of trying to establish the strength of the parity violating interaction.

The first measurement which reported a non negative result was carried out by Abov et al. [4]. They measured an asymmetry in neutron capture γ-rays from ^{114}Cd with respect to the neutron polarization. The transitions in the 9 MeV region were selected since the M1 ground state transition from the 1^- level at 9.05 MeV might be expected to contain an irregular E1 admixture. These measurements have been repeated by Warming et al. [5] and also by Abov et al. [6]. The most recent measurements by Warming [7] used a Ge-detector and with the improved resolution it was possible to separate easily the 9.05 MeV and 8.48 MeV transitions. Since it is not expected that the E2 + M1 8.48 MeV transition contains any significant parity admixture she corrected previous data for the contribution of this transition in the 9.05 MeV region. The results are given in Table 1.

The original Abov data indicate a very strong asymmetry; there is less than 0.1% probability that it is zero. However both data by Warming are consistent with a small or zero asymmetry. Theoretical values have poor accuracy and most include zero as a possible answer.

An alternative method of selecting nuclei with a particular polarization is to determine the direction of β-decay of these nuclei.

Table 1

Gamma-ray asymmetry of the 9.05 MeV M1 Transition emitted by polarized ^{114}Cd nuclei

Measured asymmetry	Correction factor[a]	Corrected asymmetry	Reference
-3.7 ± 0.9	40	-6.1 ± 1.5	Abov et al. [4]
-2.5 ± 2.2	0	-2.5 ± 2.2	Warming et al. [5]
-3.5 ± 1.2	20	-4.4 ± 1.5	Abov et al. [6]
-0.6 ± 1.7[b]	5	-0.6 ± 1.8	Warming [7]

a) The correction factor represents the contribution of the unresolved 8.48 MeV γ-ray in the detected spectrum. This is an M1 + E2 transition and may not be expected to show an asymmetry.

b) Measured using a Ge-detector, all other results were obtained using NaI-detectors.

The asymmetry of a subsequent γ-ray can then be measured with respect to the β-direction.

An experiment of this type was first carried out by Boehm and
Hauser [8] who search for the presence of an odd Legendre polynomial
term in a β-γ correlation in which the β-transition was allowed.
They used a conventional correlation technique with a moving detector and their accuracy was limited. A novel method capable of an
inherently high degree of accuracy is being used by Baker and
Hamilton [9]. It is based on a symmetric four detector arrangement
which can simultaneously measure the forward (0 - 0) and backward
(θ = π) β-γ correlations.

Uncertainties in detector efficiency and solid angle corrections
are removed by interchanging one pair of detectors since by combining the results of both measurements these factors cancel. In addition this combination can be arranged such that the measurements
give the value 1 - 8A where A is the asymmetry. A detailed analysis also shows that no correction is required for accidental coincidences; higher order terms are negligible if $A \lesssim 10^{-3}$. Electronic
stability is largely achieved by having one master coincidence circuit for the four possible coincidence combinations and using routing
units to sort these combinations. The more important effects due to
electronic drifts can be removed by choosing a short counting period
between each rotation of the γ-detectors and using data sampling
techniques.

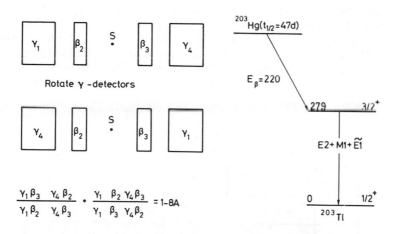

Fig. 2 A schematic representation of the detector system is shown together with an expression indicating how the coincidence counts may be
combined. Also shown is the decay scheme of ^{203}Hg.

They chose to measure the β-γ correlation in ^{203}Hg for which
they confirmed that the coefficient of the even Legendre polynomial,
P_2, was zero, i.e. the normal β-γ correlation is isotropic. It is
a particularly suitable case since the 279 keV d$\frac{3}{2}$ state in ^{203}Tl
is fed 100% by the β-decay. This state then decays by an E2-M1
transition to the s$\frac{1}{2}$ ground state. The M1 component is thus ℓ-
forbidden which results in a relative enhancement of a parity
violating $\widetilde{E}1$ admixture. The decay scheme is also favourable since
the energies of the β- and γ-transitions allow discrimination against
possible sources of error such as bremsstrahlung and multiple scattering.

A similar method has been used by Bock [10] who used a correlation angle of 45° for which any contribution from a P_2 term will be
zero. The results are given in Table 2.

2.3 Gamma-Ray Circular Polarization

The most popular method of searching for parity violation is
to determine the degree of helicity of γ-rays emitted by a random
assembly of nuclei. At first sight it is an attractive experiment
since no coincidence or correlation measurements are necessary.
However, it suffers from the serious disadvantage that the only
method of analysing circular polarization available to low-energy
nuclear physics is by Compton scattering from polarized electrons.
In magnetized iron only 7% of the electrons are polarized and the
polarization efficiency of an analyser is seldom more than a few
per cent. This poor analysing power of the detection system to
some extent may be compensated for by choosing a suitable transition.

The 482 keV E2 + M1 transition in ^{181}Ta has been studied most
often. The M1 component is hindered by the selection rules associated with the Nilsson asymptotic quantum numbers and has a hindrance
of ∿ 3×10^6, while the possible $\widetilde{E}1$ admixture is unhindered. Thus
the circular polarization, which is proportional to the
$<||\widetilde{EL}||> : <||ML||>$ matrix element ratio is large. This ratio is

Table 2

Gamma-ray asymmetries in β-γ correlations

Nucleus	Asymmetry × 10^{-4}	F	Reference
^{133}Xe	15 ± 15	⩽7.5 × 10^{-6}	Boehm and Hauser [8]
^{203}Hg	1.5 ± 0.9	∿10^{-6}	Baker and Hamilton [9]
^{203}Hg	⩽10	⩽10^{-5}	Bock [10]

commonly written as the product FR where R contains the nuclear structure information. In the case of ^{181}Ta, special circumstances and certain assumptions make it possible to evaluate R with some confidence [11] and one obtains for the circular polarization

$$P_\gamma = - (1.6^{+1.8}_{-0.9}) \times 10^2 \, F \, .$$

Thus one may expect a circular polarization of $P_\gamma \sim -10^{-4}$, and because of the low efficiency of the analyser the effect searched for is no larger than one part in 10^6.

The most serious limitation in the past has been that set by pulse counting rate. It is difficult to achieve much more than several MHz without pulse pile-up limitations. This requires the experiment to be run for many months in order to achieve the required statistical accuracy and to carry out the various control experiments. A very significant improvement in experimental technique by Lobashov et al. [12] removes this limitation. They use an integral counting method and measure the change in the scintillator light output when the sensitivity of the circular polarimeter is reversed. The source strength may be increased by several orders of magnitude so that an effective counting rate of GHz may be used, which significantly shortens the time required to obtain adequate statistical accuracy. The method is now being used in other laboratories. The more recent results obtained for circular polarization measurements are given in Table 3.

Until recently only the result obtained in 1967 by Lobashov et al. [12] excluded a zero effect with any great confidence. It is also unlikely that an effect as small as that reported by Lobashov could be seen using a conventional pulse-counting system. However, using just such a system but with an array of detectors, Bodenstedt et al. [15] have obtained a result which is about five times larger than that of Lobashov and differs from it by more than three standard deviations. The preliminary results of Vanderleeden and Boehm [17] and Bock and Jentschke [18] support Lobashov. Indeed the result of Bock and Jentschke, which contains no correction for bremsstrahlung, indicates that P_γ may be much smaller since they expect their bremsstrahlung correction to be comparable to the observed effect. It should be stressed that some of the data in Table 3 are preliminary and very recent.

The integral counting technique offers the only way so far of measuring such small effects in a comparatively short experimental time. It does suffer from the disadvantage that energy discrimination is impossible and one must rely on lead filters to remove unwanted low-energy radiation such as strongly polarized bremsstrahlung.

Table 3

The circular polarization of γ-rays emitted in nuclear transitions
by a random assembly of nuclei. The results marked thus *
should only be considered as preliminary values.

Nucleus	Method		Polarization × 10^{-5}	Theory[a] × 10^{-5}	Ref.
	Magnet	Counting			
[181]Ta	Fwd. sc	integral	-0.6 ± 0.1	-2 ± 1	Lobashov et al. [12]
	Trans.	pulse	-1.0 ± 4.0		Boehm and Kankeleit [13]
	Fwd. sc.	pulse	-2.0 ± 4.0		Bock [10]
	Trans.	pulse	-9.0 ± 6.0		Cruse and Hamilton [14]
	Back sc.	pulse	-3.2 ± 0.8		Bodenstedt et al. [15]
	Trans.	integral	*-1.3 ± 0.7		Diehl et al. [16]
	Fwd. sc.	integral	*-0.38± 0.13		Vanderleeden and Boehm [17]
	Fwd. sc.	integral	*-0.6 ± 0.2[b]		Bock and Jentschke [18]
[175]Lu 396 keV	Fwd. sc.	integral	+4 ± 1	±(3 ± 2)	Lobashov et al [19]
343 keV	Trans.	pulse	+2 ± 3	-1.0 ± 0.5	Boehm and Kankelcit [13]
[203]Tl	Trans.	pulse	-2 ± 3	-3 ± 1	Boehm and Kankeleit [13]
	-	pulse	-1.3 ± 0.8[c]		Baker and Hamilton [9]

a) The theoretical values are those of McKellar [20] and are based on the conventional Cabibbo
model.

b) Obtained using only 1 mm of lead as filter and uncorrected for bremsstrahlung, which is expected to be comparable to the size of the measured effect.

c) Deduced from β-γ anisotropy, cf. Table 2.

2.4 Mössbauer Experiments Sensitive to Parity Mixing

The remaining method which has been used in low-energy nuclear
physics has searched for small differences in the intensities of
the hyperfine components of a Mössbauer spectrum. The intensities
of the Δm = ±1 components of an M1 transition with respect to the
direction of nuclear polarization will be unequal if an Ê1 admixture
is present.

No recent results have been reported and the most accurate
value is that of Kankeleit [21], who showed that the $+\frac{1}{2} \rightarrow -\frac{1}{2}$ and
$-\frac{1}{2} \rightarrow +\frac{1}{2}$ components in the 14.4 keV transition in [57]Fe were equal
to at least one part in 10^5.

2.5 High-Energy Tests

There have been very few specific high-energy experiments to
test parity violation. Several were made about ten years ago and
in addition to possible instrumental limitations the most serious
difficulty was lack of statistics.

Garwin et al. [22] and Heer et al. [23] searched for pseudo-scalar terms associated with pion decay. The lowest limit which it was possible to set was $F \leq 10^{-3}$. Jones et al. [24] attempted to measure the longitudinal polarization of a 350 MeV neutron beam produced by the interaction of unpolarized protons with an unpolarized target. Again only a limit, $|F| \leq 2 \times 10^{-3}$, could be set.

As we shall see when discussing time-reversal experiments, the accuracy has been improved by about an order of magnitude, but low-energy experiments remain the most accurate way of obtaining evidence for parity violation.

2.6 Future Experiments

There still remains confusion about the extent to which parity is violated in strong and electromagnetic interaction processes. The range of results which are available serve to show how difficult these experiments are, rather than providing a quantitative estimate on which theoretical evaluations may be based. Thus although the latest results for ^{181}Ta indicate that the extent of parity violation may be as much as an order of magnitude less than predictions based on the conventional Cabibbo model, it is unlikely that the nuclear model aspects of the problem could be evaluated with sufficient accuracy to allow a decision to be made about the terms present in the parity violating potential.

It is thus important to have accurate experimental results for less complex systems. Henley [25] has proposed several cases in light nuclei where the nuclear aspects of the problem appear to be well understood and one may also see the effects of the isobaric spin selection rule. An alternative approach is to study reactions of the type

$$n + p \rightarrow d + \gamma .$$

Using polarized neutrons one could search for an asymmetry in the γ-ray emission. However, since there is no large retardation of the normal radiative process, as may occur in complex nuclei, the effect is not expected [26] to be bigger than $\simeq 10^{-7}$ and may be as small as 10^{-8}. Experiments on the reaction ^{2}H(n,γ) are in progress but as yet no results of a statistical significance have been reported.

3. EXPERIMENTAL TESTS OF T-VIOLATION

In the same year in which two papers, presented as the Paris Conference (1964), reported evidence for P-violation, Christenson et al. [27] published their results showing CP-violation. And while the first parity results now appear to have been an order of magnitude too large, Christenson's results are well established.

However the complementary problem of determining the extent of T-violation has made little progress over this period and the accuracy of any experiment is not more than $\sim 0.1\%$.

This large difference of 10^3 in the limit of accuracy that may be obtained between P- and T-violation experiments largely represents the difference in the degree of difficulty between the two types of measurement. Time reversal is insensitive to selection rules and, since both angular and linear momentum change sign under T-reversal, one requires experiments capable of measuring quantities of the type[*])

$$\vec{\sigma} \cdot \vec{p}(1) \times \vec{p}(2) .$$

This correlation between two momentum vectors in the plane perpendicular to the polarization axis of the interaction system should be zero for T-invariance. Such a quantity is obviously several orders of magnitude more difficult to measure than say a γ-ray asymmetry from a polarized nucleus, i.e. $\vec{\sigma} \cdot \vec{p}$, which will test P-violation.

Although it is generally accepted that CP is violated, and consequently T, there is no experimental evidence or good theoretical reasons which allow the source of the violation to be identified. This allows the experimentalist a wide range of tests.

The present experimental position will be summarized from the viewpoint of the interaction which is tested, or the type of test carried out.

3.1 Weak Interaction Tests

Time invariance implies that there should be no phase difference between the V and A interactions. The electron–neutrino angular correlation following β-decay of a polarized nucleus, i.e. $\vec{\sigma}(N) \cdot \vec{p}(e) \times \vec{p}(\nu)$, will be sensitive to this interference. Two accurate experiments have been carried out: Erozalimsky et al. [30] measured the asymmetry in the electron–proton correlation on reversing the spin of the neutron, and Calaprice et al. [31] used an atomic beam experiment to polarize ^{19}N (ground state spin $\frac{1}{2}$) from which they measured the positron – ^{19}F correlation. The results of these measurements are given in Table 4.

Time invariance implies that θ is $0°$ or $180°$ and the experiments indicate that T-violation is less than $\sim 1\%$. Final state electromagnetic interactions are negligible compared with the present

---

[*]) Comprehensive surveys of these types of measurements are given in Refs. [28] and [29].

Table 4

Weak interaction tests of T-invariance

Decay process	Phase shift θ	Reference
$n \to p + e^- + \bar{\nu}$	$178.7° \pm 1.3°$	Erozolimsky et al. [30]
$^{19}Ne \to {}^{19}F + e^+ + \nu$	$180.2° \pm 1.6°$	Calaprice et al. [31]
$K^0 \to \pi + \mu + \nu$	$180.5° \pm 2.2°$	Helland et al. [33]
$\Lambda^0 \to \pi^- + p$ a)	$9.0° \pm 5.5°$	Overseth and Roth [34]
	$9.6° \pm 5.3°$	Anderson et al. [35]

a) Final state interactions are important and give a phase
shift of $6.4° \pm 1.7°$, i.e. the result is consistent with
T-invariance.

accuracy[*]). It has also been pointed out by Marchioro et al. [32]
that if second class axial currents are a source of T-violation
then one might expect null results for n and ^{19}Ne decays.

The other tests of this type are high-energy experiments and
are less accurate. The correlation $\vec{\sigma}_\mu \cdot \vec{p}_\pi \times \vec{p}_\mu$ has been studied
in the decay $K \to \pi + \mu + \nu$. The correlations give the value Im ξ,
where ξ is a ratio of form factors describing the decay. From a
knowledge of Re ξ one can construct $\theta = \arg \xi$. There exists un
certainty at different stages of the analysis but Helland et al.
[33] have analysed the K^0 decay and obtain $\theta = 180.5° \pm 2.2°$. Again
final state interactions are negligible.

In the decay $\Lambda^0 \to \pi^- + p$ there can be no term of the type
$\vec{\sigma}_p \cdot \vec{\sigma}_\Lambda \times \vec{p}_p$ unless there is a phase difference between the s- and
p-wave amplitudes. Experimentally the average value of the correla-
tion is $\theta = 9.3° \pm 3.8°$ [34, 35]. However, final state interactions
are important and the π-N scattering phase shift gives $\theta = 6.7° \pm 1.7°$.

3.2 Electromagnetic Interaction Tests

A T_{odd} term will allow a phase difference, η, between components
of a mixed cascade which, for T-invariance, should be 0° or 180°. If
in addition a P_{odd} term is present multipoles of opposite parity
will have a phase difference of $\pi/2 \pm \eta$ and the phase difference

[*) It   ould also be added that strong and electromagnetic interactions are also present in the decays, and if a non-zero effect
was found then it would be necessary to locate its source.

between similar parity multipoles is a second order effect, $\lesssim 10^{-9}$, and negligible. Jacobsohn and Henley [28] and Boehm [29] have set out the principles on which these tests can be made.

The tests are often low-energy experiments using radioactive sources and, in principle, it should be possible to obtain good statistical accuracy. The correlation $\vec{\sigma} \cdot \vec{k}_1 \times \vec{k}_2$ is measured using an initially polarized nucleus which decays by a $\gamma-\gamma$ cascade. The first member is mixed and the two components should have approximately equal amplitudes since the magnitude of the interference term, $\sin \eta$, in the correlation is proportional to the multipole mixing ratio.

The initial polarization may be provided by a β-decay and Garrell et al. [36] examined two $\beta-\gamma-\gamma$ triple correlations in ^{56}Mn. They have opposite signs for δ. By combining the two results, inherent asymmetries in the system should be removed. They found that the over-all difference in phase between the two transitions was $(4 \pm 26) \times 10^{-3}$. The individual results are listed in Table 5 together with the result for a similar correlation in ^{106}Pd [37].

An alternative method of polarizing the initial state is by the capture of polarized neutrons. The lowest limit of a phase difference so far reported is by Eichler [38].

The Mössbauer effect also provides a sensitive method of testing for T_{odd} terms. These experiments are sophisticated in concept

Table 5

Angular correlation tests of T-invariance in the electromagnetic interaction. The measured phase difference between the E2 and M1 components in the mixed cascade is denoted by η.

Experiment	Source	$\eta \times 10^{-3}$		Reference
Mössbauer absorption	^{99}Ru	1.1 ± 1.7 [a]		Kistner [40]
Mössbauer emission	^{193}Ir	1.1 ± 3.8 [a]		Atac et al. [42]
$\beta-\gamma(E2 + M1) - \gamma(E2)$ correlation	^{56}Mn	-26 ± 14 -45 ± 27	$(\delta = -0.28)$ $(\delta = 0.18)$	Garrell [36]
" "	^{106}Pd	4 ± 18		Perkins and Ritter [37]
Pol. n capture $\gamma(E2 + M1) - \gamma$	^{36}Cl	0.8 ± 2.3		Eichler [38]
	^{49}Ti	17 ± 25		Kajfoz et al. [39]

a) After correction for a phase shift introduced by the internal conversion process, these results are $(0 \pm 1.7) \times 10^{-3}$ [^{99}Ru] and $(0.2 \pm 3.8) \times 10^{-3}$ [^{193}Ir].

and complex in operation. The experiment by Kistner [40] seeks to
measure an asymmetry in linearly polarized Mössbauer components following their resonant absorption by polarized nuclei. He used a
^{99}Ru source and chose the $\frac{5}{2} \rightarrow \frac{3}{2}$ and $-\frac{5}{2} \rightarrow -\frac{3}{2}$ components. The
interaction of the radiation with a magnetized absorber produces
linearly polarized radiation and a second similar absorber may be
used to analyse the change in intensities of the two selected components when the field direction in this second absorber is reversed.
The experiment is complicated by the fact that the analysing absorber, and its field, were placed at an angle with respect to the
photon direction in order to make the P_{odd} term maximum. This in
turn caused the radiation resonantly emitted by the analysing absorber
have a different polarization to the incident radiation so that the
subsequent behaviour of the resonant radiation is different. Thus
the polarization of the radiation changes as it passes through this
absorber – Faraday rotation. The accurate assessment of this
Faraday rotation [41] is necessary for the analysis of the asymmetry.

A similar experiment has been carried out by Atac et al. [42]
using the 73 keV transition in ^{193}Ir. In their case they were able
to remove the Faraday rotation effect by maintaining the analysing
field perpendicular to the photon direction.

An important limitation to the sensitivity of these experiments
arises as a result of the interaction of the radiation field with
the atomic electrons by the internal conversion process [43]. There
exists a phase difference between the internal conversion currents
for the M1 and E2 processes which is not associated with T-violation
and is important for these accurate Mössbauer experiments. The
corrections have been calculated [44] and the corrected result for
^{193}Ir is

$$\eta = (0.2 \pm 3.8) \times 10^{-3}$$

and like Kistner's result, for which the correction is less certain
because of Faraday rotation, does not show any evidence of T-violation.

The high-energy tests of the electromagnetic interaction are
still at a preliminary stage. The detailed balance test involving
the emission or absorption of a photon in the reaction

$$\gamma + d \leftrightarrow n + p$$

has been studied using 300 MeV γ-ray and the differential cross-
section was measured by Anderson et al. [45]. The reverse reaction
has been measured by Friedberg et al. [49] using neutrons in the 200
to 600 MeV energy range but the final data analysis has not yet been
published.

An alternative method is to observe the inelastic scattering
of electrons by polarized protons. Again the quantity
$\vec{\sigma}(p) \cdot \vec{p}(e_1) \times \vec{p}(e_2)$ can be constructed and virtual photon emission
serves as the T-violation test. The analysis of the data is very
dependent on assumption about the reaction and the accuracy obtained
by Chen et al. [50] was only about 5%.

3.3 Detailed Balance

Besides the preceding detailed balance experiments, other
reactions, as discussed by Henley and Jacobsohn [28], may be used
in sensitive tests. However, some care must be taken since detailed
balance does not necessarily imply T-invariance. This is most
easily seen for reactions which are described by the Born approximation, since in this case the transition matrix elements are
Hermitian.

Although it appears that the best tests involve reactions which
proceed through a many channel compound nucleus, uncertainties in
nuclear reaction theory limit this method. The few very accurate
experiments that have been carried out are limited to direct interactions for which the T_{odd} part of the interaction potential
could be identified.

The experimental situation is simplest when the first excited
state is relatively high as in light nuclei. In addition the analysis is greatly simplified if the number of spin channels available
for the reaction is small. This may be achieved by having low spin
values for the nuclei involved in the reaction and choosing suitable
reaction angles.

The results of recent accurate experiments are given in Table **6**.

In the case of the $^{24}Mg(\alpha;p)^{25}Mg$ reaction the angles were 30°
and 120° with respect to the beam and many spin channels were possible, while only one spin channel was possible for the $^{24}Mg(\alpha;p)^{27}Al$
reaction since the particles were detected in the backward direction.

Table 6

Detailed balance tests of T-violation. The ratio of the
T_{odd} to T_{even} matrix elements is denoted by A.

Reaction	$A \times 10^{-3}$	Reference
$^{24}Mg + d \leftrightarrow {}^{25}Mg + p$	≤ 3	Weitkamp et al. [46]
$^{24}Mg + \alpha \leftrightarrow {}^{27}Al + p$	≤ 3	Von Witsch et al. [47]
$^{16}O + d \leftrightarrow {}^{14}N + \alpha$	≤ 3	Thornton et al. [48]

4.   TESTS FOR P_{odd} T_{odd} TERMS

The simultaneous violation of P and T may be expected to give rise to an effect having a magnitude equal to the product of the separate P- and T-violating processes, i.e. not greater than $\sim 10^{-9}$. And, although one may choose a γ-ray transition in which selection rules enhance the P-violating component, the angular correlation tests as outlined by Boehm [29] appear beyond the range of presently available techniques.

However another aspect of the problem shows that P- and T-invariance of the electromagnetic field implies that a non-degenerate state cannot have odd electric or even magnetic multipoles.   The most suitable system in which one might search for these forbidden multipoles would be a neutral atom of spin $\frac{1}{2}$. And the simplest of these is the neutron.

Several very accurate experiments [51,52] have searched for the electric dipole moment of the neutron. They owe their high degree of accuracy to the precision with which small changes of frequency in the RF region can be measured. A combination of electric and magnetic fields can be set up in an RF resonance spectrometer and the frequency set to the precessional frequency of the neutrons. On reversing the electric field the presence of an electric dipole moment would be observed as a shift in magnetic resonance frequency. The most recent measurements by Baird et al. [53] set the limit on the electric dipole moment of the neutron at

$$|\mu_e| < 5 \times 10^{-23} \text{ e cm} .$$

This value is about an order of magnitude less than the upper limit of $\mu_e \lesssim 6 \times 10^{-22}$ e cm for an electromagnetic violation of CP invariance [54]. However it is still larger than the range of values expected from a milliweak theory, 10^{-24} e cm $\lesssim \mu_e \lesssim 10^{-23}$ e cm.

In conclusion, the evidence for P-violation now appears to be well-established following the most recent experiments and the results are consistent with the current-current interaction of the non-leptonic part of the weak interaction. The experimental accuracy for the T-violation experiments is barely adequate to test the origin of T-violation, although only a small improvement would bring the accuracy limit to within the predictions based on T-violation stemming from the electromagnetic interaction of the hadrons.

REFERENCES

1. B.H.J. McKellar, preceding contribution.
2. D.H. Wilkinson, Phys.Rev. 109, 1603 (1958).
3. D.P. Boyd, P.F. Donovan, B. Marsh, D.E. Alburger, D.H. Wilkinson, P. Assimakopoulos and E. Beardsworth, Bull.Am.Phys.Soc. 13, 1424 (1968).
4. Y.G. Abov, P.A. Krupchitsky and Y.A. Oratovsky, Phys.Letters 12, 25 (1964).
5. E. Warming, F. Stecher-Rasmussen, W. Ratynski and J. Kopecky, Phys.Letters 25B, 200 (1967).
6. Y.G. Abov, P.A. Krupchitsky, M.I. Bulgakov, O.N. Yermakov and I.L. Karpikhin, Phys.Letters 27B, 16 (1968).
7. E. Warming, Phys.Letters 29B, 564 (1969).
8. F. Boehm and U. Hauser, Nucl.Phys. 14, 615 (1959).
9. K.D. Baker and W.D. Hamilton, to be published.
10. P. Bock, Thesis, Karlsruhe (1969).
11. S. Whalborn, Phys.Rev. 138, B530 (1965).
12. V.M. Lobashov, V.A. Nazarenko, L.F. Saenko and L.M. Smotritskii, JETP Letters 5, 59 (1967).
13. F. Boehm and E. Kankeleit, Nucl.Phys. A109, 457 (1968).
14. D.W. Cruse and W.D. Hamilton, Nucl.Phys. A125, 241 (1969).
15. E. Bodenstedt, L. Ley, H.O. Schlenz and U. Wehmann, Phys. Letters 29B, 165 (1969).
16. H. Diehl, G. Hopfensitz, E. Kankeleit and E. Kuphal, contributed paper to this conference.
17. J.C. Vanderleeden and F. Boehm, contributed paper to this conference.
18. P. Bock and H. Jentschke, Private communication, September 1969.
19. V.M. Lobashov, V.A. Nazarenko, L.F. Saenko and L.M. Smotritskii, JETP Letters 3, 173 (1966).
20. B.H.J. McKellar, Phys.Rev.Letters 21, 1822 (1968).
21. E. Kankeleit, Comptes Rendus du Congrès Int. de Phys.Nucl., Paris 1964, paper 5/C293.
22. R.L. Garwin, G. Gidal, L.M. Lederman and M. Winrich, Phys. Rev. 108, 1589 (1957).
23. E. Heer, A. Roberts and J. Tinlot, Phys.Rev. 111, 645 (1958).
24. D.P. Jones, P.G. Murphy and P.L. O'Neill, Proc.Phys.Soc. 72, 429 (1958).
25. E.M. Henley, Phys.Letters 28B, 1 (1968).
26. G.S. Danilov, Phys.Letters 18, 40 (1965).
27. J.H. Christenson, J.W. Cronin, V.L. Fitch and R. Turlay, Phys. Rev.Letters 13, 138 (1964).
28. E.M. Henley and B.A. Jacobsohn, Phys.Rev. 113, 225 (1959).
29. F. Boehm in Hyperfine Structure and Nuclear Radiations, North-Holland, Amsterdam (1968).
30. B.G. Erozolimsky, L.N. Bondarenko, Y.A. Mostovoy, B.A. Obinyakov, V.P. Zacharova and V.A. Titov, Phys.Letters 27B, 557 (1968).

31. F.P. Calaprice, E.D. Commins, H.M. Gibbs, G.L. Wick and
    D.A. Dobson, UCRL-18647 (1969) and to be published in
    Phys.Rev.
32. C. Marchioro, A. Prosperetti and A. Pugliese, preprint (1969).
33. J.A. Helland, M.J. Longo and K.K. Young, Phys.Rev.Letters 21,
    257 (1968).
34. O.E. Overseth and R.F. Roth, Phys.Rev.Letters 19, 391 (1967).
35. T. Anderson, J.K. Bienlein, W.E. Cleland, G. Conforto,
    G.H. Eaton, H.J. Gerber, M. Reinharz, M. Veltman, A. Gautschi,
    E. Heer, J.F. Renevey, C. Revillard and G. von Dardel,
    Proc. 14th Int.Conf. on High-Energy Physics, Vienna (1968).
36. M.H. Garrell, Thesis, University of Illinois (1966).
37. R.B. Perkins and E.T. Ritter, Phys.Rev. 174, 1426 (1968).
38. J. Eichler, Nucl.Phys. A120, 535 (1968).
39. J. Kajfosz, J. Kopecky and J. Honzatko, Nucl.Phys. A120, 225
    (1968).
40. O.C. Kistner, Phys.Rev.Letters 19, 872 (1967).
41. M. Blume and O.C. Kistner, Phys.Rev. 171, 417 (1968).
42. M. Atac, B. Chrisman, P. Debrunner and H. Frauenfelder, Phys.
    Rev.Letters 20, 691 (1968).
43. E.M. Henley and B.A. Jacobsohn, Phys.Rev.Letters 16, 706 (1966).
44. J.P. Hannon and G.T. Trammell, Phys.Rev.Letters 21, 726 (1968).
45. R.L. Anderson, R. Prepost and B.H. Wiick, Phys.Rev.Letters 22,
    651 (1969).
46. W.G. Weitkamp, D.W. Storm, D.C. Shreve, W.J. Braithwaite and
    D. Bodansky, Phys.Rev. 165, 1233 (1968).
47. W. von Witsch, A. Richter and P. von Brentano, Phys.Rev. 169,
    923 (1968).
48. S.T. Thornton, C.M. Jones, J.K. Blair, M.D. Mancusi and
    H.B. Willard, Phys.Rev.Letters 21, 447 (1968).
49. C.E. Friedberg, D.F. Bartlett, K. Goulianos, I.S. Hammerman
    and D.P. Hutchinson, Bull.Am.Phys.Soc. 14, 76 (1969).
50. J.R. Chen, J. Sanderson, J.A. Appel, G. Gladding, M. Goiten,
    K. Hanson, D.C. Imrie, T. Kirk, R. Madaras, R.V. Pound,
    L. Price, R. Wilson and C. Zajde, Phys.Rev.Letters 21,
    1279 (1968).
51. C.G. Shull and R. Nathans, Phys.Rev.Letters 19, 384 (1967).
52. W.B. Dress, J.K. Baird, P.D. Miller and N.F. Ramsey, Phys.
    Rev. 170, 1200 (1968).
53. J.K. Baird, P.D. Miller, W.B. Dress and N.F. Ramsey, Phys.Rev.
    179, 1285 (1969).
54. G. Barton and E.D. White, Preprint (1969).

DISCUSSION

H. Primakoff: In the experimental investigation of time-reversal
asymmetries in the semi-leptonic decays of hadrons one should, if
at all possible, avoid those transitions where the initial and final
hadrons are members of the same isodoublet or SU3 octet [e.g.,

(n,p), (Λ,p), $(Ne^{19}F^{19})$]. This follows because, in a class of theories where the time-reversal violation is ascribed to the presence of second-class weak currents with normal behavior under charge symmetry, or its SU3 generalization, such asymmetries vanish in the limit of SU2, or SU3, symmetry and zero momentum transfer. More specifically, one should search for time-reversal asymmetries in beta decays between initial and final nuclei with isospins differing by unity and with anomalously small Gamow-Teller matrix elements. (e.g. $S^{32} \rightarrow P^{32} + e^- + \bar{\nu}_e$).

RECENT WORK ON PARITY MIXING IN NUCLEAR STATES

V.M. Lobashov

A.F. Ioffe Physicotechnical Institute

Leningrad, USSR

The basic problem in the present state of the studies on the weak interaction between nucleons is to obtain data which can be interpreted with sufficient reliability, from the point of view of parameters depending on nuclear properties.

Of the greatest interest from this point of view are the reactions n+p=D+γ, and n⏐D=T+γ, in which the relation between the amplitudes of the weak nucleon-nucleon interaction and the value of the effect measured may be calculated in the existing theory with good certainty.

The investigation of parity non-conservation effects in complex nuclei appears to be also desirable. The main problem in this case should be, apparently, the determination of the parameters of the PNC-potential. The most essential thing for solving this problem is to increase the range of the nuclei investigated.

The present report concerns two experiments on the measurements of the circular polarization of the γ-quanta emitted by the non-polarized nuclei.

In the first experiment, performed in the Physico-Technical Institute Academy of Sciences of the USSR by V.M. Lobashov, N.A. Lozovoy, V.A. Nazarenko, L.M. Smotritsky and G.I. Kharakevitch, the study was made of the circular polarization of the 1290 Kev M2 γ-transition exited in ^{41}K after ^{41}A β-decay. The circular polarization of this transition was found to be $P_\gamma = + (1.9 \pm 0.3)10^{-5}$.

In the second experiment, carried out in the same institute by V.M. Lobashov, D.M. Kaminker, V.A. Nazarenko, L.F. Sayenko,

L.M. Smotritsky, G.I. Kharkevitch and V.A. Kniazkov, the circular
polarization of γ-quanta in the reaction n+p=D+γ was measured.   The
report describes the experimental set-up and measurement procedure.
The preliminary result will be also presented.

In these experiments use was made of the γ-quanta integral
detection method and the resonance separation of the periodical
signal.   In the both experiments the polarimeter, based on the
transmission of γ-quanta through magnetized absorber, was used.

The most important problem, arising in such experiments are
as follows: to obtain the most powerful source of γ-quanta and
to avoid the background of β-decay or K-capture bremsstrahlung
γ-quanta imitating the small circular polarization.

^{41}K Experiment

In this experiment the source of ^{41}A was used with the activity
of about 700 Curie.

In Fig. 1 the experimental set-up is shown.   The 0,6 litre
tank placed in the active zone of the reactor was connected by two
pipes with the source volume.   The source was filled with liquid
argon from the liquefier placed at the input-pipe.   At the output-
pipe the heater, evaporating the liquid argon, was attached.   The
gas circulation in the loop, arising when the argon is liquefied
was sufficient to provide the required constancy of the ^{41}A source
activity.

The use of the "transmission" polarimeter in this experiment
was on account of the fact that the "transmission" polarimeter
in spite of its smaller efficiency for 1290 kev γ-quanta yields
an effect from the bremsstrahlung radiation of ^{41}A β-decay by an
order smaller than that in the case of a "forward scattering" method.

Besides, the calculation shows, that for the 60 mm magnetized
iron absorber the contribution from the bremsstrahlung radiation
is effectively zero.

The design of the polarimeter is shown in fig. 1.   The effective
thickness of the absorber is 60 mm.

The magnetization of the polarimeter was changed every second.
The switching frequency was kept constant to about 1 part in 10^7.
The variable component of the voltage at the load resistance of
the photo-diode after the resonance amplifier, tuned to the frequency
0,5 c/s, was fed to the pendulum filter, tuned to the switching
frequency with the accuracy of about 1 in 10^7.   The measurement
procedure was similar to that described in ref. /1/.

Zero control experiments were carried out with sources of

^{46}Sc, the γ-quanta of which (1120 kev and 880 kev E2) can be considered as non-polarized.

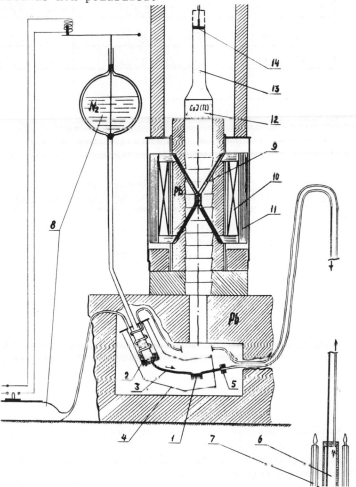

Fig. 1. Experimental set-up in ^{41}K experiment. 1-source, 2-liquefier, 3-pipes, 4-vacuum jacket, 5-heater, 6-tank with argon, 7-active zone, 8-liquid nitrogen feed, 9-magnetized absorber, 10-coils, 11-magnetic yoke, 12-scintillator, 13-light-pipe, 14-photo-diode.

In order to check the correctness of the calculations of the bremsstrahlung effect measurements were made with a source of ^{188}Re (β-spectrum end-point: 2120 kev.) The effect measured proved to be in good agreement with the calculated one.

The results of the control experiments with ^{46}Sc and the measurements with ^{41}A are presented in Table 1. The results of control experiments show, within the limits of the error absence of the false effects.

Source	Run 1, $\delta . 10^7$	Run 2, $\delta . 10^7$	Run 3, $\delta . 10^7$
^{46}Sc	$- 0.4 \pm 0.6$	$+ 0.4 \pm 0.7$	
	No Pb filter	8mm Pb filter	4mm Pb filter
^{41}A	$+ 6.5 \pm 1.0$	$+ 10 \pm 3$	$+ 4.6 \pm 2.2$

Table 1 - Experimental results.

The results of all runs with ^{41}A show the presence of the effect, the weighted average of which is:
$$\delta = + (6.5 \pm 0.85) \ 10^{-7}.$$
The polarization efficiency of the polarimeter for the energy 1290 Kev is 3.3%, therefore the value of the c. p. corresponding to this effect is:
$$P_\gamma = + (1.9 \pm 0.25) \ 10^{-5}.$$
Allowing for the control experiment error the resultant error of this figure must be increased up to $\pm 0.3. \ 10^{-5}$.

The corresponding analysis shows that this effect cannot be caused by admixture of any β-activity. The absence of activities decaying by K-capture was also verified.

The c. p. of the 1290 Kev transition must arise as a result of the interference of the regular M2 and irregular E2 transitions.

The calculations of the amplitude of the E2 admixture transition can be carried out on the basis of universal weak interaction theory within the framework of the finite size Fermi-system models. For ^{41}K such estimate was made by Yu. Gaponov and Yu. Fursov /2/
$$P_\gamma \simeq (0.6 - 2.0) \ 10^{-5}.$$
The estimate is preliminary and not all calculations are completed, but it does not contradict our result.
                         (n,p)-Capture Experiment.

The measurement of the c. p. of γ-quanta in the reaction n+p=D+γ enables one to measure directly the isoscalar part of the amplitude of the weak nucleon-nucleon interaction. As seen from the estimates, performed in a number of papers /3/, one can expect that the value of c. p. may be of order 10^{-6}. As a consequence of this a very powerful γ-quanta source is needed to obtain the statistical accuracy required.

As such source in our experiment use was made of the water neutron trap in the active zone of the WWRM reactor of the Physico-Technical Institute. The experimental set-up is shown in fig. 2.

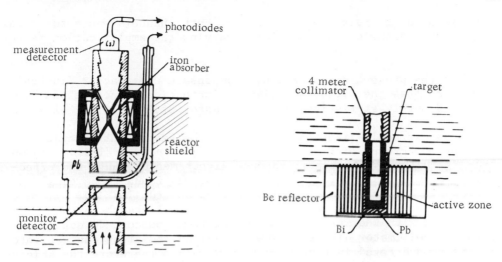

Fig. 2 - An experimental set-up in the (n,p)-capture experiment.

Light water, which is the moderator and coolant for the given type of reactors, served as a hydrogen target. The trap being 3 litres in volume and the neutron flux in the centre of the trap $\sim 3.10^{14}$ n/cm$^2\cdot$ sec, the effective activity of the source was $\sim 10^{16}$ /sec.

γ-quanta from the trap passed outside to the polarimeter through the channel-collimator. In the beam of γ-quanta, in the front of the absorber a plate CsJ(Tl) with a photo diode was placed serving as monitor.

The effective number of γ-quanta, registered in the measurement detector, was $\sim 5.10^{10}$ /sec.

The need for the monitor detector was due to the fact that the relative amplitude of the neutron flux fluctuations in the rector was $1-2.10^{-3}$ at frequencies of order 1-10 cps. This exceeds the statistical fluctuation of the measurement detector current almost by two orders, so that the resultant error must have deteriorated by the same factor. In order to avoid this, compensation of the fluctuation was applied by using the monitor signal.

Fig. 3 shows block diagram of the electronics used. The variable components of the signals from the measurement detector and the monitor were amplified, the signal of the monitor passing through the variable gain amplifier, and then the both signals were fed to the differential amplifier. By proper variation of the amplification factor of the v.g.a. one can achieve almost complete compensation of the reactor fluctuations. A deterioration of the com-

pensation due to relative drift of the both channels was automatically
restored by means of the correlator varying the amplification factor
of the v.g.a.

At the present stage we could compensate the reactor fluctuations, so that the difference signal fluctuations exceed those
expected purely statistically by the factor of 1.5 - 2.

The extremely important problem in the experiment described
was the background of bremsstrahlung γ-quanta from fission fragment
β-decay in the active zone. Through direct measurements the effective c.p. of γ-quanta in the reactor core was found to be equal to
$1-2.10^{-3}$. The corresponding calculations showed that this can almost
fully account for the polarization measured. Although in the
experiment with the water trap γ-quanta from the core do not get
into the collimator directly, bremsstrahlung radiation can get into
it, being scattered in the water trap volume. Therefore the shielding of the water cavity proved to be quite necessary.

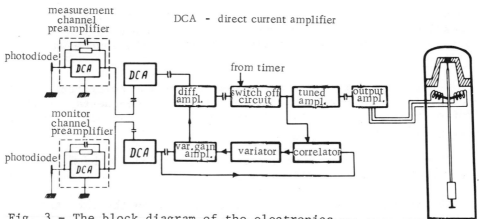

Fig. 3 - The block diagram of the electronics.

A number of experiments and calculations were carried out in
order to determine the dependence of the value of the bremsstrahlung
radiation effect on the thickness of the lead screen surrounding
the water trap.

It is possible to measure the effect from the external background if the water in the trap is replaced with the material which
has similar density (in order to provide the same scattering characteristics) and a small neutron capture cross-section. For this purpose
Be and carbon targets were used.

Table 2 shows the experimental results for targets consisting
of H_2O, Be, Al, and carbon (graphite). A good coincidence of the

calculated internal bremsstrahlung effect with the experimental
values obtained for the case of absence of shield and for the 1 cm
lead shield thickness, allowed us to determine with sufficient
reliability the minimum thickness of lead screen which enables one
to reduce the bremsstrahlung effect down to the value corresponding
to the effect $\sim 10^{-8}$. This thickness proved to be 60 mm.

Pb shield thickness	$\delta \cdot 10^6$			
	0	1 cm	3 cm	6 cm
Exper. H_2O	-5.5 ± 0.5	-1.2 ± 0.3	-	0.0 ± 0.025[x]
Exper. Be	-	-2.0 ± 0.4[xx]	-	-
Calc. H_2O	-5.0	-1.7	-0.17	-0.01

Table 2. The experimental effect with different targets and the
calculated internal bremsstrahlung effect.

[x] - The result with carbon target from table 3.
[xx] - For Be and carbon targets as a constant component of intensity
(voltage) when calculating δ, voltage measured for H_2O target was
taken.

Zero control experiments for checking up the absence of false
effects for this polarimeter were associated with some difficulties.
It is desirable to use for the control experiments γ-quanta with
the similar energy. Unfortunately, for all the $(n\gamma)$-reactions
except (np)-capture, the γ-quanta energy is not below 4 Mev. As a
consequence of this some of the control experiments were performed
with radioactive sources of ^{24}Na emitting 2.7 Mev E2 γ-quanta.
Others were carried out with γ-quanta from the reaction ^{48}Ti (n,γ)
^{49}Ti, in which 7 Mev E1 γ-quanta are emitted. In the both cases
γ-quanta are expected to be non-polarized. γ-quanta of ^{24}Na are
closer in energy; however, the limited activity of the sources of
^{24}Na (only 10000 Curie) did not permit us to obtain good accuracy.

As a result of several runs with ^{24}Na we obtained:
$$\delta = - (0.3 \pm 0.4) \, 10^{-7}.$$

A few runs were performed with the Ti target in the trap which
showed, that for high energy γ-quanta there exist a number of
phenomena causing false effects. These apparently are connected
with the influence of the magnetic field of the polarimeter on
electron-positron pairs generated in the absorber and detector.
For example, in one of the runs with Ti we observed the effect:
$$\gamma = - (5.0 \pm 0.7) \, 10^{-7}.$$

Immediately before this, the run with water in the same conditions yielded $\delta = - (0.9 \pm 0.8) 10^{-7}$

After some improvements of the symmetry of the system, switching the current of the polarimeter and the polarimeter position, the effect disappeared.

This is the evidence of the fact that γ-quanta of ^{49}Ti are not quite a suitable source for the control experiments, although probably more critical.

Allowing for these remarks it is possible to give the results of several measurements, all performed under the same conditions, with different targets in the trap. The runs are arranged in the chronological order.

Target	H_2O	Ti	H_2O	H_2O	C	Ti
$\delta.10+7$	-0,8 ±0,4	+0,5 ±0,5	-0,88 ±0,35	-1,5 ±0,55	0,0 ±0,25	+0,6 ±0,5

Table 3.   The experimental results.

As seen from the table, for H_2O target an effect was observed, the weighted average of which is:
$$\delta = - (1.0 \pm 0.23) 10^{-7}.$$

At the same time the background measurements with a graphite target gave approximately a zero effect and in the case of the Titanium target the average effect was:
$$\delta = + (0.55 \pm 0.35) 10^{-7}.$$

Allowing for the error in control measurements with the Ti target only, for γ-quanta emitted in the reaction $n+p = D+\gamma$ the c.p. can be written as: $P_\gamma = - (1.8 \pm 0.9) 10^{-6}$.

Taking into account the previous remark concerning control experiments, this figure can be considered only as a preliminary one. One can believe with a greater certainty, that
$$0 \leq - P_\gamma < 3. \quad 10^{-6}.$$

This estimate appears sufficiently reliable.

The main problem henceforth is zero control experiments. The improvement of their accuracy will allow one to give the value of c.p. in the np-reaction more reliably.

REFERENCES

[1] V.M. Lobashov, V.A. Nazarenko, L.F. Sayenko, L.M. Smotritsky, and
G.I. Kharkevitch; Phys. Lett. 25B, (1967), 104.
[2] Y. Gaponov and Y. Fursov; Report on XIX conference on nuclear
spectroscopy, Erevan, USSR, (1969).
[3] G.S. Danilov; Phys. Lett. 18, (1965), 40. D. Tadic; Phys. Rev.
174, (1968), 1694.

Discussion

V. Telegdi: If the sign of the parity-violating effect in the (np)
system is uncertain, how can you give a sign for the effect in the
case of finite nuclei?   V.M.L.: If you suppose a definite sign
for the nucleon case, - the positive sign for parity mixing, you
can obtain the sign as it is written here. It is of course a
problem in all the experiments.   B. McKellar: The sign of the
ρ contribution can be found in the weak vector boson model, provided
one knows missing parameters of the transition as one does for
the ^{181}Ta case.   M. Goldhaber: Did you correct for bremsstrahlung
in the ^{41}A case? V.M.L.: No we didn't make any corrections because
the conditions of the experiment were arranged so that the expected
bremsstrahlung contribution would be negligible.   E. Fischbach: A
new calculation of the parity-violating effects in p(n,γ)d is
presently under way by Hadjimichael, Tadic and myself. This reaction
is most attractive from a theoretical point of view since the
nuclear physics can be done in an essentially exact manner. We
have solved the Schrodinger equation for the parity admixed states
using the Yale strong interaction potential and the most general
weak interaction potential derived from π, ρ, ω and ϕ exchange. We
hope that final results will be available shortly.

FIRST RESULTS WITH A NEW CIRCULAR POLARIZATION ANALYZER ON THE

PARITY MIXING IN ^{181}Ta

H. Diehl, G. Hopfensitz, E. Kankeleit and E. Kuphal

Institut fur Technische Kernphysik
Technische Hochschule Darmstadt, Germany

(Presented by E. Kankeleit)

The parity admixture in nuclear states, expected according to the current-current hypothesis (1) of weak interaction, has been tested by several groups (2-5) by the detection of the circular polarization of nuclear gamma rays from unpolarized sources. The best evidence for the existence of this polarization has been obtained up to now by Lobashov et al. (2) with a value of $P = -(6\pm1)\cdot10^{-6}$ for the 482 keV transition in ^{181}Ta. Because of the importance and need for further independent confirmation of this effect we have set up a new analyzer system which in main parts deviates from the usual construction and takes in particular care of some of the serious error sources.

The general layout of the system is shown in Fig. 1. The source is positioned in the center of a magnet in which a saturated magnetization of the iron is produced in radial direction by a pair of coils. Gamma rays penetrating this iron are detected in ten plastic scintillators surrounding the magnet. Two mu-metal containers between magnet and scintillator reduce the small stray field further. Via a long conical plexiglas light pipe the scintillation is detected by a multiplier. From the anode current only the averaged low frequency part is passing a band filter consisting of one low pass and two (because of leakage current) high pass filters for which high quality operational amplifiers are used. This is followed by an 400 channel ADC digitizing the current integrated over two seconds. With an offset current the average channel number is adjusted to be at 200. This part of the electronics is enclosed in the multiplier head. Particular

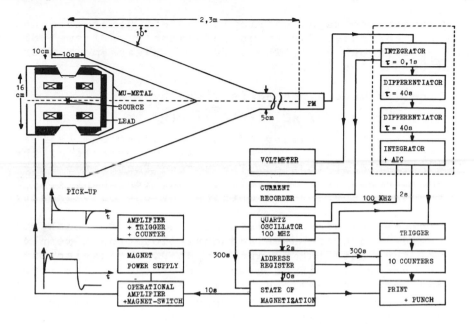

Figure 1

care has been taken to avoid correlations to other parts
of the electronics. The pulse train is incremented and
stored in one of ten counters which during one period are
gated open in succession. After each halfperiod of 10 s
the magnet current is reversed. This is done by an operational power amplifier. The current impressed on to the
magnet with a stability of about 10^{-5} has a square wave
form with a 30 % overshoot in the beginning of each halfperiod. The magnet is thereby brought further into saturation
for a short time, symmetrizing this way the hysteresis
loop.   After a "run" of 15 periods the sign sequence is
reversed. The state of magnetization is indicated on the
print-out of the counters taking place then.

The solid angle transmission of the magnet is 24 %.
With the plastic scintillators the overall detection efficiency is 0.5 % at 482 keV. The polarization efficiency
of the magnet according to a definition of the relative
anode current difference
$$\delta = 2(I^+ - I^-)/(I^+ + I^-) = \varepsilon \cdot P$$
(plus sign indicating magnetization in direction of gamma
ray) has been determined by the quadratic compton scatte-
ring effect to be $\varepsilon(482 \text{ keV}) = -(3.2 \pm 0.3)$ %. In this

measurement the magnetostriction effect being of equal
size was corrected for by use of gamma rays from ^{46}Sc for
which ε is about zero. The value is in agreement with that
given by Chesler (6).

The variance expected is given by

$$\frac{\Delta \delta}{\delta} = (\sqrt{2}/\sqrt{\text{å}T}) \ (\sqrt{E(Q^2)}/ \ E(Q))$$

åT being the total number of counts during one halfperiod.
E(Q) and E(Q^2) are the first and second moments of the
pulse height distribution. The second factor for this
system is about 1.4. The relation between count rate å
and the anode current was obtained by use of weak sources.

In the computer analysis of the data for each run of
15 periods δ and the standard deviation from a mean δ of
several days data taking are computed. A gaussian fit to
the frequency distribution was performed and the total
standard error determined. It turned out that the data
showed pure statistical behaviour and that agreement with
the above formula was achieved within 3o %. The dark
current fluctuation of the system has been tested to be
of negligible influence on the variance.

The main advantages of this system may be summarized
as follows: 1. The system has a high detection efficiency.
The absorption in the 5 cm of iron is useful in reducing
the Bremsstrahl contribution. 2. Due to the mirror symmetry of the magnet, stray fields are small and drop off
with fourth power along the axis. No stray field effects
were detected, even though the scintillators showed a
rather large field dependence in their light output. 3.
Very important is the nearly axial symmetry. Homogenous
magnetic (earth) fields cancel in their effect on the
magnet and scintillators. An other error source was put
forward by Bock (7). This is supposed to be due to a term
$(\vec{K} \times \vec{K'}) \cdot \vec{S}$ in the scattering cross section. Again due
to the symmetry this effect should cancel. In contrast
to conventional arrangements high reproducibility in radial
location of sources is not required. 4. The quadratic
polarization effect can be used to determine the efficiency with good accuracy. 5. By digitizing the anode
current five times each halfperiod effects resulting from
the switching of the magnet can be corrected for.

In the parity experiment performed so far a 34 Ci
source of ^{181}Hf was used. This corresponds to a count
rate of $4 \cdot 10^9$ c/s in the analyzer. The source was pre-

pared by irradiation in the FR 2 reactor Karlsruhe of
1 g HfO_2 enriched to 98.2 % in [180]Hf. The oxide was mixed
with 2 g of graphite for reduction of Bremsstrahlung,
pressed to a cylinder and surrounded by an Al-can. The
purity of the material used except the Hf was checked by
a dummy irradiation.

The experiments with the Hf source were interrupted
by control experiments using [46]Sc and 103Ru sources. All
data taken showed a purely statistical behaviour. The
results are as follows:

[181]Hf-Ta   482 keV $\delta = +(5.9 \pm 1.4) \cdot 10^{-7}$
[46]Sc-Ti   889,11205 keV, $\delta = +(1.0 \pm 1.0) \cdot 10^{-7}$
[103]Ru-Rh   497 keV $\delta = -(2.2 \pm 3.2) \cdot 10^{-7}$

The average of the two control experiments $\delta = +(0.7 \pm 1.0) \cdot 10^{-7}$ is considered as an information on a
possible bias of our system. Taking the difference we
obtain for the [181]Ta measurement

$$\delta = +(5.2 \pm 1.7) \cdot 10^{-7}$$

With $\epsilon = -(3.2 \pm 0.3) \cdot 10^{-2}$ for the efficiency we obtain

for the [181]Hf source

$$P(^{181}Ta) = -(16 \pm 7) \cdot 10^{-6}$$

This value includes a contribution from the Bremsstrahlung.
With an effective average charge $Z_{eff} = 20$ for our mixed
source we calculate a contribution of $P = -6 \cdot 10^{-6}$. This
value is after correcting for this transmission geometry
considerably smaller than that derived by other authors
(4,5). But experiments on the [170]Tm Bremsspectrum and
also an evaluation of the experiments by Lobashov et al.
(2) indicate that this calculated value must be too large
by about a factor of two. As a final result for the circular polarization of the 482 keV gamma ray we obtain

$$P(482 \text{ keV}) = -(13 \pm 7) \cdot 10^{-6}.$$

This result is in rough agreement with the value given by
Lobashov et al. (2), but obviously bears less significance.
Up to this conference in total only 30 days of data taking
were (unexpectedly) available. This paper is mainly meant
to demonstrate the sensitivy of the instrument. Experiments
with stronger sources under various conditions are in progress.

The help of J. Foh in constructing the electronics
and of Dr. Frank and D. Türck in handling the strong sources
we like to acknowledge.

REFERENCES

(1) R.P. Feynmann and M. Gell-Mann, Phys. Rev. 109 (1958)
    193.
(2) V.M. Lobashov, V.A. Nazarenko, L.F. Saenko, L.M. Smo-
    tritskii and G.I. Kharkevich, Phys. Lett. 25 B
    (1967) 104.
(3) P. Bock and H. Schopper, Phys. Lett. 16 (1965) 284;
    F. Boehm and E. Kankeleit, Nucl. Phys. A 109 (1968)
    457; J.J. Van Rooijen, P. Pronk, S.U. Otterangers
    and J. Blok, Physica 37 (1967) 32.
(4) E. Bodenstedt, L. Ley, H.O. Schlenz and U. Wehmann,
    Phys. Lett. 29 B (1969) 165.
(5) D.W. Cruse and W.D. Hamilton, Nucl. Phys. A 125 (1969)
    241.
(6) R.B. Chesler, Nucl. Instr. and Meth., 37 (1965) 185.
(7) P. Bock, private communication.

DISCUSSION

V. M. Lobashov: It has been pointed out by Tolhoek that because the
electron spin is 1/2 and the proton spin 1, the spurious term
$(\vec{K} \times \vec{K}') . \vec{S}$ must vanish.   E.K., and others (unidentified): There
appear to be contrary suggestions from both theoretical and experi-
mental sources.

AN EXPERIMENTAL TEST OF TIME-REVERSAL INVARIANCE

IN THE PHOTODISINTEGRATION OF THE DEUTERON[*]

B. L. Schrock, J. -F. Detoeuf,[†] R. P. Haddock, and
J. A. Helland[‡]

University of Californa, Los Angeles, California 90024

M. J. Longo, K. K. Young,[§] and S. S. Wilson

University of Michigan, Ann Arbor, Michigan 48014

D. Cheng,[||] J. Sperinde, and V. Perez-Mendez

Lawrence Radiation Laboratory
University of California, Berkeley, California 94720

(Presented by R. P. Haddock)

We present data from a measurement of angular distributions
of the reaction $n + p \rightarrow \gamma + d$ (1a) for neutrons with energies between 300 and 700 MeV. These data are compared with recent
data for the inverse reaction $\gamma + d \rightarrow n + p$ (1b). Time-reversal
invariance implies that apart from a normalization factor the angular distributions should be the same. We find an intriguing discrepancy in the vicinity of the well-known peak in the total cross
section of reaction (1b) and apparent agreement elsewhere. The
discrepancy occurs in the shape of the angular distributions and

---

[*] This work was supported jointly by the U. S. Atomic Energy
Commission Contract AT(11-1)-34 and the U. S. Office of Naval
Research Contract NONR 1224(23).

[†] N.S.F. Senior Foreign Scientist Fellow, on leave from C. E. N. S.
Saclay, France.

[‡] Present address: University of Notre Dame, Notre Dame,
Indiana.

[§] Present address: University of Washington, Seattle, Washington.

[||] Present address: Brookhaven National Laboratory, Upton,
New York.

does not depend on the normalization of our data or those of the inverse reaction experiments.

Bernstein, Feinberg, and Lee[1] noted the absence of experimental evidence to exclude the possibility that the violation of CP in $K_L^0 \to \pi^+ + \pi^-$ (2) decay was due to a possible failure of C and T in the electromagnetic interactions of hadrons. A number of possible experimental[1,3] tests appear now to be expected to yield only relatively small effects even if there is a maximal violation. Other tests, such as reciprocity in photopion production from nucleons are complicated by the neutron, $\gamma + \binom{P}{N} \rightleftharpoons \binom{N}{P} + \pi^{(\pm)}$ (2), involved in the comparison. However, Christ and Lee[3] concluded from their study of reaction (2) that a failure of C and T should manifest itself most likely near a pion-nucleon resonance and only for processes which actually correspond to a photopion production with the pion subsequently reabsorbed for reaction (1). Reaction (1) has a peak in the cross section near $k_L = 300$ MeV as shown in Fig. 1, which is attributed to the excitation of a nucleon to the $N^*(1236)$ in an intermediate state. It is then worth while to test reciprocity in the vicinity of this peak.

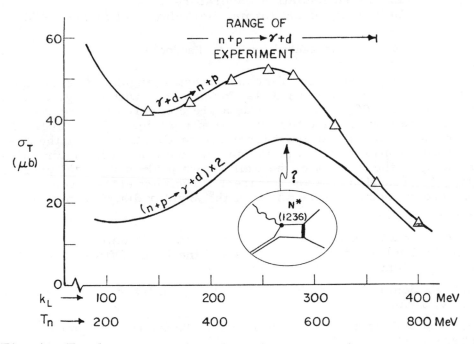

Fig. 1.   Total cross section of $\gamma + d \to n + p$ and $n + p \to \gamma + d$ as a function of lab kinetic energy of the incident photon and neutron, respectively.

Barshay[4] has used a specific model due to Austern[5] to estimate the possible magnitude of the effect to the expected. The matrix element for the excitation of $J = I = 3/2$ isobar by absorption of a magnetic dipole (M1) photon leading to a $^1D_2$ state with $I = 1$ for the final two-nucleon system is calculated by evaluating Feynman graphs like that shown in Fig. 1. The hypothetical failure of T invariance is introduced by giving the γ-N-N* vertex a phase, $e^{i\Delta}$, different from 0 or π. The matrix element for the reverse reaction is shown to have a phase $e^{-i\Delta}$. If, e.g., $\Delta = \pi/2$ and we are at resonance so that the resonance-energy denominator of the matrix element is purely imaginary, then $M(1a) = - M(1b)$. And if there is another relatively real amplitude with which the isobar amplitude interferes, reciprocity may be grossly violated. The $M1 \to {}^1S_0$ transition that exists at lower energy can interfere while the $E1 \to {}^3P_0$ transition cannot. After an estimate of the relative weights of each amplitude, the angular distributions for reactions (1a) and (1b) at $k \approx 290$ MeV are given

as $\dfrac{d\sigma}{d\Omega}\begin{pmatrix}1a\\1b\end{pmatrix} \propto 10.66 - 2\, P_2(\cos\theta)\,[\,1 + 0.94\cos(\delta_r \mp \Delta)\,]$

$$= A_0 + A_2\, P_2(\cos\theta),$$

where P_2 is the second-order Legendre polynomial, θ is the CMS angle between photon and neutron, and δ_r is the phase of the resonance denominator which may at resonance be unequal to $\pi/2$ because of the deuteron's internal momentum. The essential features are that the total cross sections of reactions (1a,b) are equal, and that the effect of a T violation is expected to show up only in the normalization-free ratio A_2/A_0. A complete description of the experimental data requires addition of odd-order Legendre polynomials to account for the fore-aft asymmetry attributed to interference between E1-E2 transitions. An estimate of the failure of reciprocity at resonance may be expressed as $A_2/A_0(1b)-A_2/A_0(1a) \approx \sin\Delta/3$.

The coefficient of $\sin\Delta$ is proportional to the imaginary part of the $M(1) \to {}^1D_2$ amplitude and should vanish at lower and higher energies at least as fast as δ_r, which is scaled by $\gamma^* \approx 120$ MeV, the isobar width.

The neutron's lab kinetic energy, T_N equals $2\, k_L$, the photon's lab energy at the same CMS total energy as shown in Fig. 1. So neutrons with energies up to 700 MeV could test the energy dependence of any possible failure of reciprocity. We decided a test (to $\Delta \approx \pm 15°$) by using the UCLRL 184" Synchrocyclotron was reasonable. A similar test has been made by Bartlett et al.[6] at PPA. The inverse reaction (1b) has recently been remeasured at Cornell,[7] Orsay,[8] and Stanford.[9]

A presumably unpolarized neutron beam was produced at $0°$ from a Be target in the internal cyclotron beam at an energy of

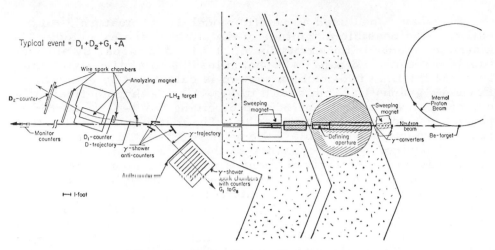

Fig. 2.   Experimental arrangement.

$\approx$ 715 MeV (see Fig. 2).   The beam passed through a Pb filter to
remove γ rays, passed through a sweeping magnet to remove
charged particles, then through a carefully constructed collimator,
another sweeping magnet, and finally appeared at a 3" diameter
by 2" long liquid hydrogen target about 40' from the Be target.   At
this point the beam had a 2" diameter with very little halo.   The
neutron flux was estimated to be $\approx 5 \times 10^6$/sec with a peak in the
spectrum near 670 MeV due to charge exchange, a flat valley between 500-600 MeV, and a broad hump at lower energies due to
pion production.   After passing through the target the beam was
monitored by three sets of monitor telescopes containing polyethylene converters.   These tracked one another to within 0.1%
over the time of the data presented.   Two side-by-side counters
connected to rate meters permitted continuous monitoring of the
beam-centering and target setting.

Charged particles were detected by counters D_1 and D_2 before and after passing through a C magnet.   Two sets of wire
spark chambers before and two sets after the magnet provided
the particle's three-momentum.   Each set had four planes having
1 mm resolution per plane with magnetostrictive readout.   The
gammas were converted in an array of ten one-radiation-length
30"$\times$30" optical spark chambers with a total of 40 gaps.   A set of
8 counters—G_1, $\cdots G_8$—were placed between the first nine chambers to detect the converted pair.   These counters were 24" high
by 27" wide and established the fiducial area of the chamber.   An
anti-counter in front of the gamma chambers insured that the incoming particle was neutral.   Three lead scintillation sandwich
anti-counters were placed near the LH target to reduce the princi-
pal source of background due to the reaction $n + p \rightarrow d + (\pi^0 \rightarrow 2\gamma)$(3),
which occurs $\approx$ 50 times more frequently than reaction (1a).

The spark chambers were triggered by a coincidence between D_1D_2 (with the timing appropriate for a deuteron) and a pulse from any one of the counters, G_1, $\cdots$, G_8, without a pulse from any of antis. Because of the good duty cycle of the cyclotron, accidental rates of all kinds were found to be very low. Those between D_1 and D_2 were always $< 1\%$. Accidental vetos from antis were monitored continuously and were $\leq 7\%$. Scalers recorded singles and coincidences as well as time. Some of these were gated on only during the live time of the equipment.

The following information was recorded for each trigger. The gamma chamber with its fiducials and the run and event was photographed. A PDP-5 computer recorded the event and run number, and the 16 coordinates from the wire chambers (four planes for each of four chambers. This redundancy allowed us to determine the deuteron's trajectory with high accuracy and an efficiency $> 98\%$.) The computer also recorded the D_1-D_2 time-of-flight, the gamma counters that fired, and other less essential information to be used to monitor the experiment on line. In addition, at every twelfth event the scalers were automatically recorded on the magnetic tape.

A total of 10^6 good pictures, of which $\approx 5\%$ are due to reaction 1a (events due to reactions 1a and 3 are called γds and π^0ds respectively), were taken at five angles ranging from $\approx 25°$ to $150°$ in the CMS. Both the magnet and gamma chamber were mounted on rolling carriages to facilitate the changes. The gamma chamber distance was varied to keep the angular interval of accepted events about $\pm 10°$ (CMS). The deuteron spectrometer's position and field were adjusted to accept events corresponding to $200 \lesssim T_N(MeV) \lesssim 750$ at each angle. Target-full runs were alternated with target-empty runs. The latter showed a negligible background due to target walls.

The data were processed as follows: The gamma chamber film was digitized by a Vidicon scanning system followed by a pattern-recognition program to yield one or more vertices and the corresponding spark count. The efficiency for this process is $\gtrsim 99\%$ at all angles. Less than $1/2\%$ of the events with two distinct vertices is rejected from the analysis. Other events where two vertices are found in the same shower are resolved by choosing the vertex closest to the target. Scanning shows this is the correct choice more than 99% of the time. The overall reproducibility of the vertex is accurate to $\pm 1/2$ mm. The vector momentum of the charged particle was determined from the wire chamber data, three-dimensional magnetic field maps, and a particle-tracking program to $\Delta p/p \approx 1\%$. The mass of the particle is then calculated from its momentum and D_1-D_2 time of flight. A small fraction of protons is included in the final analysis to insure that no deuterons are lost even though the separation is quite clean at all angles. The deuteron's direction is projected back to the center plane of the LH target that served as the origin from which the gamma's angle is determined. A χ^2 of the event according to reaction (1a) was calculated. The fit is over determined (a 2C fit) and the χ^2 is essentially a measure of the coplanarity (assuming the neutrons come from the Be target) and the changed particle's missing mass. Figure 3 shows a χ^2 distribution of events binned according to the reconstructed neutron energy for a best and worst case situation, respectively.

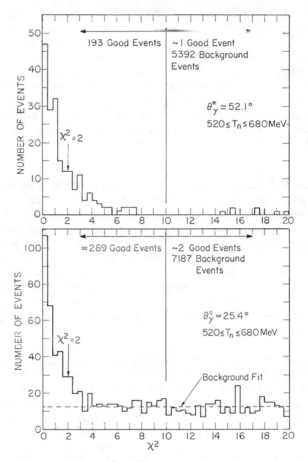

Fig. 3a, b. χ^2 plots of typical best and worst case of background subtraction due to $n + p \rightarrow d + (\pi^0 \rightarrow 2\gamma)$. The χ^2 is calculated according to reaction 1a for each event.

It is seen that there is a nearly complete separation in one case while in the other a significant subtraction of $\pi^0 d$ events must be made. Independent checks on the alignment of the equipment were made by examining the coplanarity and missing-mass plots; e. g. , the mass corresponding to the center of the deuteron's peak agreed with the accepted mass of the deuteron to within ± 0.05% for most data points.

A Monte Carlo program was constructed to simulate $\pi^0 d$ and γd events. The purpose of the program is (1) to test the entire data reduction scheme, (2) to check that the experimental and theoretical resolutions agree, (3) to yield the neutron energy resolution of the reconstructed events, (4) to yield the solid angle as a function of T_N for the fiducial areas of the equipment, and (5) to assure that the $\pi^0 d$ background extrapolates smoothly under the γd peak in the χ^2 plots. Good agreement was found between the Monte Carlo and experimental χ^2 plots at small χ^2 for the γd events and at large χ^2 for the $\pi^0 d$ events. This gives us considerable confidence in the background subtraction, which is made by a smooth extrapolation of the χ^2 distribution at high χ^2 under the peak at low χ^2 to obtain the net number of γd events, $N(\theta, \overline{T}_N)$, and the overall statistical error of this method.

The data presented here are based on the final analysis of a sample of 20% of our data. The differential cross section, $d\sigma/d\Omega$, aside from an unknown normalization factor, F, is $d\sigma/d\Omega$ = $F\, N(\theta, T_N)/M\,\Delta\Omega\,\epsilon$, where M is the corresponding number of monitor counts, $\Delta\Omega$ is the CMS solid angle subtended by the system, and ϵ is the efficiency. Our data are divided into four neutron energy bins (300-400, 400-500, 500-600, and 600-700 MeV) for comparison with the inverse data.

Results of three recent measurements of reaction (1b) are shown in Fig. 4. The coefficients A_0, A_1, A_2 are from a least-squares fit of the data to a Legendre polynomial series. Three curves are shown, of which the center one corresponds to average values over the interval of our T_N bins and the outer ones correspond to either the standard deviation of the points or their statistical uncertainty, whichever is larger. The agreement is quite good for the Orsay and Cornell data while the Stanford data (which is statistically more accurate) shows a slight disagreement.

Our angular distributions, $d\sigma(\theta)/\sigma_{total}$, are shown in Fig. 5. The errors are only statistical and include the uncertainty in background subtraction. The curve is that obtained from the average values of A_1/A_0 and A_2/A_0 obtained from Fig. 4. Agreement is quite good except in the 500-600 MeV bin, where the 26° point is above the line and the 51° and 88° points are below it. The effect of this reversal is most clearly seen in Fig. 6. Good agreement is obtained in the A_1/A_0 ratio but the A_2/A_0 ratios do not agree with the inverse to $\approx$ 3 standard deviations for the 500±50 MeV bin and $\approx$ 1 standard deviation for the 450±50 MeV bin. We have had a result of this nature in this data sample for over a

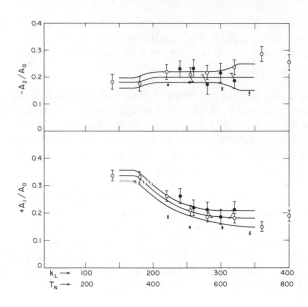

Fig. 4. Results of recent measurements of reaction 1b as a function of photon energy, k_L. The CMS scattering data is θ. The three lines represent the average value and its uncertainty in these ratios.
Legend:
- Stanford-SLAC ($\simeq$ 40 MeV wide)
- Orsay (40 MeV wide)
- Cornell-Ithaca-Princeton (20 MeV wide)

year and have found no systematic errors in our treatment that could lead to this discrepancy. The ratio of wire chamber tracks to triggers was 98% after subtracting randoms. Scattering from the pole-tips is easily rejected in the momentum fitting program. Hence, we trust our estimates of $\overline{\Delta\Omega\epsilon}$ for this part of the system and our Monte Carlo results based on uniform chamber efficiency. Our data do not include a small correction for the efficiency of detecting a gamma. The spark count on low-χ^2 events shows $\geq$ 15 sparks/event at all angles. This represents 2-3 chambers involved in a detected shower. We estimate the detection efficiency to be > 98% for photon energies > 140 MeV, the lowest energy involved in the 500 < T_N < 600 MeV data. Furthermore, this correction should be most important for the lower energy neutrons. Its effect will be to increase the discrepancy slightly since our cross sections are high in the region where the γ energy is lowest. We have not found any reason to change the method of background subtraction, which could easily lead to this result. Data over the entire neutron energy spectrum was taken simultaneously to reduce the time-dependent efficiency of the apparatus. Systematic

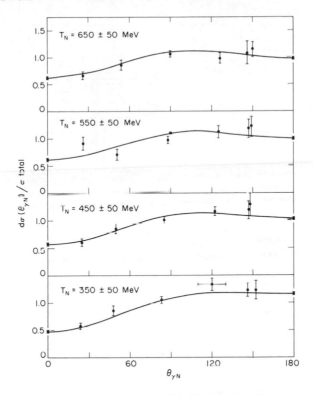

Fig. 5. Angular distributions of our data for four intervals of
neutron energy where $\theta_{\gamma N}$ is the CMS angle. The error bars
shown at 0° and 180° for all energies and at 90° for
T_N = 550 MeV reflect the uncertainty in these curves as
shown in Fig. 4.

errors in the reconstructed neutron energy have been extensively
investigated to remove the possibility that the rapidly changing
structure of the neutron energy spectrum could produce our result. The resolution of T_N varies from 3% - 1% at low and
high energy, respectively. Neutron energy spectra at each angle
agree well in shape and to ± 1.5% at the high energy cutoff at
≈ 720 MeV.

The vertical bar in Fig. 6 represents Barshay's estimate of
the possible consequence of T violation from no violation
(Δ =0°) to maximal violation (Δ =-90°). The photodisintegration of the
deuteron has been studied intently for many years. It is interesting to
ask whether there is any direct information which a priori precludes a
large T violation and, if not, what the energy dependence of an observed
violation might be. The horizontal bar is an estimate of the
FWHM of the effect of the isobar amplitude taken from the analysis
of Pearlstein and Klein. [10] Also shown is their result for A_2/A_0,
which agrees surprisingly well with recent data. They disagree

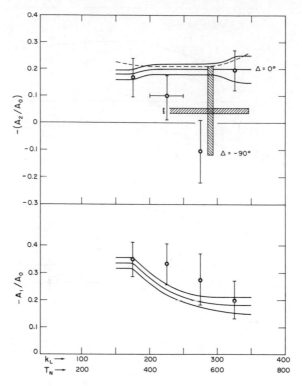

Fig. 6. Comparison of our results with the inverse reaction.
Legend:
      O   UCLA-UM-LRL (50 MeV bins)
     — — — Theory (Pearlstein and Klein)

      Barshay

      $\sqrt{\sigma}\ (MD) > 0.5\ \sqrt{\sigma}\ (MD)_{max}$

with Barshay as to the contribution of E1, M1 transitions to the
total cross section. A possible T violation should show up in
the value of $\delta_2 - \delta_0$ (the $^1D_2$ and $^1S_0$ n-p phase shifts) required to
obtain a fit to data. They required a value that can be interpreted
to be $\approx 90°$ different from recent analyses of nucleon-nucleon
scattering.[11] We interpret this as a restatement of the observation of Ref. 1.

    In view of the importance of our results it is essential that
further studies be made of reactions (1a) and (1b) to remove the uncertainties in the experimental data and to investigate theoretically
the magnitude of a possible violation allowed by all existing data.

REFERENCES

1. J. Bernstein, G. Feinberg, and T. D. Lee, Phys. Rev. 139, B1650, (1965).

2. J. H. Christenson, J. W. Cronin, V. L. Fitch, and R. Turlay, Phys. Rev. Letters 13, 138 (1964).

3. N. Christ, and T. D. Lee, Phys. Rev. 148, 1520 (1966) and Phys. Rev. 143, 1310 (1966).

4. S. Barshay, Phys. Rev. Letters 17, 49 (1966).

5. N. Austern, Phys. Rev. 100, 1522 (1955).

6. D. F. Bartlett, C. E. Friedberg, K. Goulianos, I. S. Hammerman, and D. P. Hutchinson (private communication).

7. D. I. Sober, D. G. Cassel, A. J. Sadoff, K. W. Chen and P. A. Cream, Phys. Rev. Letters 22, 430 (1969).

8. J. Buon, V. Grocco, J. Lefrancois, P. Lehrmann, B. Merkel, and Ph. Roy, Phys. Letters 26B, No. 9, 595 (1968).

9. R. L. Anderson, P. Prepost, and B. H. Wiik, Phys. Rev. Letters 22, 651 (1969).

10. L. D. Pearlstein, and A. Klein, Phys. Rev. 118, 193 (1960).

11. M. H. MacGregor, R. A. Arndt, and R. M. Wright, Phys. Rev. 169, 1149 (1968).

DISCUSSION

G.E. Brown: The Barshay mechanism might also give observable T-violation effects in low energy nuclear phenomena. A promising place to look would be the γ-γ correlations involving ℓ-forbidden M1 transitions. Here the main, allowed, nuclear effects are suppressed and the effect from isobar admixture should be particularly large, giving phases of 20°-30° or so in favourable cases.

R.P. Haddock (in reply to question by J. Levinger): The error limits shown are strictly statistical. The statistical errors introduced by the background subtraction are also included. I might say we know of no mechanism whereby one would get exactly this energy dependence in the ratio a_2/a_1.

<u>C. Friedberg</u>: We find generally similar results in our experiment
(at P.P.A.). If we fit the differential cross section to $\frac{d\sigma}{d\Omega} =$
$A + B \cos \theta + C \cos^2 \theta + D \cos^3 \theta$, we find $C/A = 0.5 \pm 0.3$
at a neutron energy of 580 ± 60 MeV. The inverse reaction $(\gamma + d \rightarrow n + p)$,
yields $C/A = -0.2 \pm .02$ (Stanford-Slac group). (These results will
be published in Phys. Rev. Letters.)

TEST OF TIME-REVERSAL INVARIANCE IN ELECTROMAGNETIC INTERACTIONS*

L. W. Mo†

Stanford Linear Accelerator Center, Stanford University

Stanford, California

In this short report, a brief description is given of a recently completed experiment, carried out by an UC/LRL-SLAC collaboration[1], which tests the time-reversal invariance in electromagnetic interactions of hadrons[2] by scattering high energy electrons inelastically from a polarized proton target.[3]   An experiment of this kind was done before at CEA and the reported result was consistent with no violations in time-reversal invariance.   However, with the availability of the SLAC high energy electron and positron beams, high-resolution magnetic spectrometers, and also the improved polarized proton target, we felt that it is of great urgency to re-examine this important problem once again as the experiment can be done at SLAC with much higher accuracy and sensitivity.

The high energy electrons (or positrons) from the two-mile long linear accelerator were momentum analyzed and sent through a polarized proton target made of butanol alcohol.   In order to reduce or totally eliminate the depolarization effect due to local beam heating, a pair of air-core Helmholtz coils were installed at $\sim$100 ft upstream from the target to sweep the beam across the target on a pulse-to-pulse basis.   The size of the beam spot on the target was typically 3 mm in diameter and the target was uniformly illuminated in a square of $\sim$1" × 1".   Normally the beam intensity was held at $\sim$5 × 10^8 electrons per pulse and the experiment was conducted at $\sim$180 pulse/sec. The scattered electrons were momentum analyzed and detected using the 20-GeV/c magnetic spectrometer at SLAC.   The beam intensity was monitored by two induction toroids and a secondary emission quantameter

*Work supported by U.S. Atomic Energy Commission.
†Permanent address: Dept. of Physics, The University of Chicago.

which served also as a beam dump.   Asymmetries in scattered electrons, as weighted by the target polarization along the direction
perpendicular to the scattering plane, were measured continuously
from the elastic peak to a missing mass value of 2 GeV.

The depolarization effects due to local beam heating and
radiation damage were also investigated in great detail in this
experiment.   Relaxation of the target polarization was measured by
turning off the rf-power which polarizes the target.   As no
difference in the relaxation of polarization was observed with the
beam off and on, it provided the evidence that the target was not
instantaneously depolarized during the scattering.   The maximum
polarization achieved in this experiment was ∿38% for a ∿1" cube
target.

Possible asymmetry arising from the interference between one-
photon and two-photon exchange processes can be checked by using
positrons.   The asymmetry due to this interference will change
sign as one changes from electrons to positrons, while that due to
T-violation will not.

Experiments were performed with electrons of incident energies
of 12, 15, and 18 GeV; and positrons of incident energy 12 GeV.
Typically the target polarization was reversed every 2 minutes.
True asymmetry as well as various possible false asymmetries were
evaluated by an on-line computer.   The four-momentum transfer
squared, q^2, were covered from 0.3 to 1.0 $(GeV/c)^2$.   No T-violation
was found, and only a small two-photon exchange effect was observed
in the (33) resonance region.   Detail numbers have to be decided
by further analysis.

REFERENCES

[1]The participants of this experiment are: M. Borghini, O.Chamberlain,
 C. Morehouse, T. Powell, S. Rock, G. Shapiro, and H. Weisberg from
 UC/LRL; and R. Cottrell, J. Litt, L.W. Mo, and R.E. Taylor from SLAC.
[2]J. Bernstein, G. Feinberg and T.D. Lee, Phys. Rev. 139, B1650(1965).
[3]N. Christ and T.D. Lee, Phys. Rev. 143, 1310 (1965).
[4]J.R. Chen, J. Sanderson, J.A. Appel, G. Gladding, M. Goitein,
 K. Hanson, D.C. Imrie, T. Kirk, R. Madaras, R.V. Pound, L. Price,
 Richard Wilson, and Zajde, Phys. Rev. Letters 21, 1279 (1968).

DISCUSSION

R. Wilson: From Stanford and CEA there is evidence that the longitudinal cross section σ_L is different from zero at 0.3 $(GeV/c)^2$ and
below.   Above this point the only evidence comes from a doubtful
joining of two experiments.   At CEA we ran at 0.3 $(GeV/c)^2$ deli-
berately.   You run at 0.37 $(GeV/c)^2$ and above where σ_L could be

zero, and hence no asymmetry.   L. Mo: As measured by Ritson et al. and also by a DESY group, the longitudinal cross section at the (33) resonance is non-zero when q^2 is less than 0.7 $(GeV/c)^2$.

L. Mo (in response to question): We cannot at this time quote an estimate for the ratio of 2 photon and 1 photon amplitudes. Before we can, we must, inter alia, determine the ratio $\sigma(C+H)/\sigma(H)$.

Experimental Tests of Models of the Weak Hamiltonian

Ephraim Fischbach[*†]

Institute for Theoretical Physics, State University of
New York, Stony Brook, New York, U.S.A.

Dubravko Tadić

Institute Ruder Bosković, Zagreb, Yugoslavia

Kenneth Trabert[§]

Department of Physics, University of Pennsylvania,
Philadelphia, Pennsylvania, U.S.A.

Models of the weak interaction Hamiltonian density $\mathcal{H}_w(x)$ are currently the subject of intensive investigation for a variety of reasons having to do with various shortcomings of the conventional current-current model. These shortcomings include the failure of the conventional model to naturally account for either the nonleptonic $|\Delta \vec{I}| = \frac{1}{2}$ rule or for the CP violation observed in weak hadronic processes and also the well known divergence difficulties which arise in attempts to apply the conventional model to higher order weak

[*] Work supported under Contract No. AT(30-1)-3668B of the United States Atomic Energy Commission. [†]Address as of September, 1969: Niels Bohr Institute, Copenhagen Ø, Denmark.

[§] Work supported by the Life Insurance Medical Research Fund and Contract No. GP-5524 of the National Science Foundation.

interaction processes. In the conventional model $\mathcal{H}_w(x)$ is given by

$$\mathcal{H}_w(x) = \frac{G}{\sqrt{2}} \left[J_\lambda(x)J_\lambda^*(x) \right]_s \equiv \frac{G}{2\sqrt{2}} \left[J_\lambda(x)J_\lambda^*(x) + J_\lambda^*(x)J_\lambda(x) \right]$$

$$J_\lambda^*(x) = (-J_i^\dagger(x),\ J_4^\dagger(x)) \qquad i = 1,2,3$$

$$J_\lambda(x) = \ell_\lambda^{(e)}(x) + \ell_\lambda^{(\mu)}(x) + \cos\theta\ h_\lambda^{\Delta S=0}(x) + \sin\theta\ h_\lambda^{\Delta S=1}(x), \quad (1)$$

where $\ell_\lambda^{(\ell)}(x) = i\bar{\psi}_\ell(x)\gamma_\lambda(1+\gamma_5)\psi_\nu(x)$ (ℓ=e or μ) is the lepton current, $h_\lambda^{\Delta S=0}(x)(h_\lambda^{\Delta S=1}(x))$ is the strangeness-conserving (strangeness-changing) hadron current, $\theta \cong 0.21$ is the Cabibbo angle and $G \cong 1.02 \times 10^{-5}/m_p^2$. $h_\lambda(x)$ may be decomposed into its polar-vector and axial vector pieces, $V_\lambda(x)$ and $A_\lambda(x)$ respectively, and these in turn are given by

$$V_\lambda^{\Delta S=0}(x) = \mathcal{F}_\lambda^1(x) - i\,\mathcal{F}_\lambda^2(x) \qquad A_\lambda^{\Delta S=0}(x) = \mathcal{F}_{5\lambda}^1(x) - i\,\mathcal{F}_{5\lambda}^2(x)$$

$$V_\lambda^{\Delta S=1}(x) = \mathcal{F}_\lambda^4(x) - i\,\mathcal{F}_\lambda^5(x) \qquad A_\lambda^{\Delta S=1}(x) = \mathcal{F}_{5\lambda}^4(x) - i\,\mathcal{F}_{5\lambda}^5(x). \quad (2)$$

The F-spin currents in Eq. (2) are assumed to obey the usual SU(3)⊗ SU(3) equal time commutation relations. The model of Eqs. (1)-(2), when augmented by the conserved vector current (CVC) and partially conserved axial-vector current (PCAC) hypotheses, provides a generally good description of a variety of leptonic and semileptonic phenomena and hence the alternatives to the conventional model are generally constructed to reproduce its predictions for these classes of processes. One is thus motivated to examine nonleptonic weak interactions in an effort to discriminate among the various models of $\mathcal{H}_w$. In principle the differences in various models should be manifested in the absolute magnitudes of the amplitudes for the $\Delta S=1$ nonleptonic processes such as $\Lambda \rightarrow p\pi^-$. In practice, however, the absolute magnitudes of these amplitudes are difficult to calculate with the result that the dependence of these amplitudes on $\mathcal{H}_w$ cannot be reliably extracted. The problem is that while principles such as CVC and PCAC enable us to evaluate the one hadron $\rightarrow$ one hadron matrix elements of a single current there are as yet no general principles which permit us to evaluate the matrix elements of a product of currents such as occurs in nonleptonic hyperon decays. The ratios of $\Delta S=1$ S-wave (parity-violating) B $\rightarrow$ B'π amplitudes (where B and B' are baryons), or alternatively, sum rules among these amplitudes, can be calculated with some confidence but are rather insensitive to models of $\mathcal{H}_w$. In particular the $|\Delta \vec{I}| = \frac{1}{2}$ relations[1]

$$\Lambda_-^0 + \sqrt{2}\ \Lambda_0^0 = 0$$

$$\Xi_-^- - \sqrt{2}\ \Xi_0^0 = 0$$

$$\Sigma_-^- + \Sigma_+^+ - \sqrt{2}\ \Sigma_0^+ = 0 \tag{3}$$

and the Lee-Sugawara relation

$$2\Xi_-^- - \Lambda_-^0 - \sqrt{3}\ \Sigma_0^+ = 0 \tag{4}$$

follow from any CP conserving model of $\mathcal{H}_w$ in which the nonleptonic Hamiltonian density $H_w(x)$ transforms predominantly as the sixth component of an SU(3) octet. As this transformation property holds true for all of the models under consideration, we conclude that sum rules involving only $\Delta S=1$ amplitudes are not useful for our purposes. It turns out, however, that ratios of the $\Delta S=1$ S-wave amplitudes to the $\Delta S=0$ S-wave amplitude for the weak process $n \to p\pi^-$ (denoted by n_-^0) are extremely sensitive to models of $\mathcal{H}_w$ so much so that $|n_-^0/\Lambda_-^0|$ can vary by more than an order of magnitude from one model to another.[2-5] This observation forms the basis of the current attempts to discriminate among models of $\mathcal{H}_w$.

Although it is not directly observable for physical pions, the amplitude n_-^0 may be studied for virtual pions through its effects on the nucleon-nucleon potential as shown in Fig. 1.

The weak parity-violating (p-v) $np\pi$ vertex contributes to a parity-violating piece V_{12} of the total nucleon-nucleon potential $\mathcal{V}_{12}$ and through V_{12} effects a variety of parity-violating phenomena in nuclear physics some of which have been discussed previously by other speakers. V_{12}^π, the one pion exchange contribution to V_{12}, is given by[6]

$$V_{12}^\pi = \frac{ig_{\pi NN}\ m_\pi^{-1/2}\ |n_-^0|}{16\pi\sqrt{2}\ m_N} (\vec{\sigma}_1 + \vec{\sigma}_2) \cdot [\vec{p}_{12},\ \exp(-m_\pi r)/r] \otimes$$
$$(\vec{\tau}_1 \times \vec{\tau}_2)z, \tag{5}$$

where $\vec{p}_{12} = \vec{p}_1 - \vec{p}_2$ is the relative momentum of the two nucleons and $g_{\pi NN}^2/4\pi \simeq 14.6$. The dependence of n_-^0 on models of $\mathcal{H}_w$ is contained in the triangular relation[3]

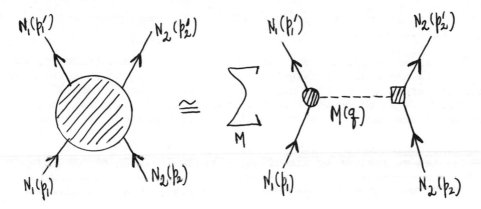

Fig. 1. One meson exchange contributions to the parity-violating internucleon potential V_{12}. Shaded circle indicates a weak parity-violating interaction while shaded square denotes a purely strong vertex. From the low-lying 0- and 1- nonets only M = $\pi^\pm$, $\rho^\pm$, ρ^0, ω and ϕ contribute to V_{12}.

$$\varepsilon^- - 2\Lambda n^0_- + (\sqrt{3}/2A)n^0_- = 0, \tag{6}$$

where $\Lambda = A(H^{pv}_w)$ is a dimensionless parameter defined by

$$H^{pv}_w = A\, T^{(8)}_{0,1,0} + T^{(8)}_{1,1/2,-1/2} . \tag{7}$$

In Eq. (7) H^{pv}_w is the parity-violating part of H_w and $T^{(8)}_\nu (\nu=Y,I,I_3)$ is an octet tensor operator. We note that any H^{pv}_w whose $\Delta S=0$ and $\Delta S=1$ pieces belong to the same octet necessarily assumes the form given in Eq. (7). From Eq. (6) the magnitude of n^0_- may be deduced from the data for the $\Delta S=1$ S-wave B → B'π amplitudes and is

$$|n^0_-| = (1.25 \times 10^5 \text{ sec}^{-1/2})A = (0.037\, \frac{G}{\sqrt{2}}\, m^2_p\, m^{1/2}_\pi)A . \tag{8}$$

Evidently the viability of the present program depends on the sensitive dependence of A on H_w, which in turn is a consequence of the $|\Delta \vec{I}| = 1$ isospin transformation property of V^π_{12}. A detailed explanation of how the dependence of A on H_w comes about will be given elsewhere.

In order to extract the contribution of V_{12}^π (and hence n_-^0) to V_{12}, it is necessary to take into account as well the contributions from the exchange of other mesons in the low-lying (0^- and 1^-) nonets. It can be shown[6,7] that in the 0^- nonet π^0, η, and $\eta'(X^0)$ exchanges are forbidden so that only $\pi^\pm$ exchanges need be considered.

The contributions from vector meson (1^-) exchanges have been calculated recently[10] by use of the current-field identity (CFI) and field algebra. For the weak p-v $NN\rho$ vertex the field algebra treatment was shown to reproduce the results of the factorization approximation

$$<N\rho\,|V_\lambda(0)A_\lambda^*(0)|N> \;=\; <\rho\,|V_\lambda(0)|0><N|A_\lambda^*(0)|N>\;. \tag{9}$$

If the __strong__ $NN\rho$ vertex is assumed to be described by the usual phenomenological Lagrangian density

$$\mathcal{L}_{\rho NN} = f_\rho\,\vec{\rho}_\mu\cdot i\bar{N}(\gamma_\mu + (\frac{i\mu_N}{2m_N})\,\sigma_{\mu\nu}\partial_\nu)\frac{\vec{\tau}}{2}N\;, \tag{10}$$

where μ_N is the nucleon anomolous magnetic moment and $f_\rho^2/4\pi \simeq 2.4$, then the contribution from the ρ-exchange diagram of Fig. 1 is identical to that obtained from the factorization approximation[11]

$$<NN|H_w^{pv}(0) \sim V_\lambda(0)A_\lambda^*(0)|NN> \;\simeq\; <N|V_\lambda(0)|N><N|A_\lambda^*(0)|N> \tag{11}$$

under the assumption that the ρ-meson dominates the polar vector form factors. The contributions from ω and ϕ exchanges may be treated in the same fashion if we neglect the effects of ω-ϕ mixing. The vector meson exchange contribution (V_{12}^V) to V_{12} is then given by

$$V_{12}^V = \frac{-h_A f_\rho}{8\pi\sqrt{2}\, m_N} \left\{ i\vec{\sigma}_1 \times \vec{\sigma}_2 \cdot [\vec{p}_{12}, \exp(-m_\rho r_{12})/r_{12}]_- \left\{ (1+\mu_p-\mu_n) \otimes \right. \right.$$

$$\left[T_{12}^{(+)} + \frac{B}{4} \tau_1^{(z)} \tau_2^{(z)} + \frac{C\xi}{4} \left(\tau_1^{(z)} \tau_2^{(0)} + \tau_2^{(z)} \tau_1^{(0)} \right) \right]$$

$$+ (1+\mu_p+\mu_n) \frac{\sqrt{3}}{8} C' (\tau_1^{(z)} \tau_2^{(0)} + \tau_2^{(z)} \tau_1^{(0)}) + \frac{\sqrt{3}}{2} D\xi \tau_1^{(0)} \tau_2^{(0)} \right\}$$

$$+ (\vec{\sigma}_1 - \vec{\sigma}_2) \cdot [\vec{p}_{12}, \exp(-m_\rho r_{12}/r_{12}]_+ \cdot \left\{ T_{12}^{(+)} + \frac{B}{4} \tau_1^{(z)} \tau_2^{(z)} \right.$$

$$+ \frac{\sqrt{3}}{2} D\xi \tau_1^{(0)} \tau_2^{(0)} \right\} + [\vec{p}_{12}, \exp(-m_\rho r_{12})/r_{12}]_+ \cdot \left\{ \frac{\sqrt{3}}{4} C' \right. \qquad \otimes$$

$$\left. \left. \left(\vec{\sigma}_1 \tau_1^{(z)} \tau_2^{(0)} - \vec{\sigma}_2 \tau_2^{(z)} \tau_1^{(0)} \right) + \frac{C\xi}{2} \left(\vec{\sigma}_1 \tau_2^{(z)} \tau_1^{(0)} - \vec{\sigma}_2 \tau_1^{(z)} \tau_2^{(0)} \right) \right\} \right\} \cdot$$

$$\tag{12}$$

In Eq. (12) $h_A = \frac{G}{\sqrt{2}} \cos^2\theta \, (\sqrt{2}\, m_\rho^2 \, G_A/f_\rho)$, with $G_A \cong 1.2$, and $\tau^{(0)}$ is the 2×2 isospin identity matrix. The constants, B, C, C' and D define the contribution to $H_W^{PV}(\Delta S=0)$ from the non-strange F-spin currents (which are the only currents that contribute to V^V):

$$H_W^{PV}(\Delta S=0) \equiv \frac{G}{\sqrt{2}} \cos^2\theta \left\{ (V_\lambda^{\pi+} A_\lambda^{\pi-} + V_\lambda^{\pi-} A_\lambda^{\pi+}) + B V_\lambda^{\pi0} A_\lambda^{\pi0} \right\}$$

$$+ C V_\lambda^{\pi0} A_\lambda^\eta + C' V_\lambda^\eta A_\lambda^{\pi0} + D V_\lambda^\eta A_\lambda^\eta + \cdots \quad,$$

$$V_\lambda^{\pi0} = \mathcal{J}_\lambda^{(3)}, \quad V_\lambda^\eta = \mathcal{J}_\lambda^{(8)} \text{ etc.} \tag{13}$$

Finally, the parameter ξ in Eq. (13) is defined by

$$\langle N(\nu_3) | \mathcal{J}_{5\lambda}(\nu_2) | N(\nu_1) \rangle = \left\{ (\nu_1^8 \nu_2^8 \nu_3^8 \,{}^s) D_A + (\nu_1^8 \nu_2^8 \nu_3^8 \,{}^a) F_A \right\} \otimes$$

$$\bar{u}(p')\gamma_\lambda\gamma_5 u(p) ; \quad \xi \equiv \frac{1}{\sqrt{12}} (1 - 2(1-F_A/D_A)/(1+F_A/D_A)). \tag{14}$$

A recent analysis by Brene et al.[12] suggests the value $F_A/D_A \cong \frac{1}{2}$ and hence $\xi \cong 1/(6\sqrt{3})$.

We wish to stress that although the field algebra treatment of vector meson exchanges and the factorization approximation of Eq. (11)

give identical results, it is only through the former approach that
we can assess the validity of the final result. In other words the
usefulness of studying parity-violating nuclear interactions in an
effort to discriminate among models of $\mathcal{H}_w$ would be lost if we were
unable to estimate the errors incurred in computing V_{12}^V given a
particular model of $\mathcal{H}_w$. While the factorization approximation is
just an ansatz whose validity cannot be independently checked, the
field algebra formalism has been applied to numerous weak and electromagnetic processes with good results. Similar remarks apply as well
to the calculation of n^0 using current algebra and PCAC. Our experience indicates that the effective coupling constants in V_{12}^π and
V_{12}^V may be expected to be correct to within 20 per cent or so.
Since different models of $\mathcal{H}_w$ lead to order of magnitude changes in
the effective coupling constant in V_{12}, we can expect that the
differences among various models of $\mathcal{H}_w$ will still be quite pronounced, even after the uncertainties of nuclear physics are added
to those inherent in the calculation of V_{12}.

We have attempted in the above discussion to clarify some
of the ideas underlying the current attempts to discriminate among
models of $\mathcal{H}_w$ through the study of parity-violating nuclear interactions.[13] Detailed numerical results will be discussed elsewhere.

REFERENCES

[1]We denote the S-wave amplitude for $B^a \to B'+\pi^b$ (where a and b are
the charges of the respective particles) by B_b^a.

[2]D. Tadić, Phys. Rev. 174, 1694 (1968).

[3]E. Fischbach and K. Trabert, Phys. Rev. 174, 1843 (1968).

[4]B.H.J. McKellar, Phys. Rev. 178, 2160 (1969); Phys. Rev. Letters
21, 1822 (1968); ibid 20, 1542 (1968).

[5]R.F. Dashen, S.C. Frautschi, M. Gell-Mann and Y. Hara, in The
Eightfold Way, edited by M. Gell-Mann and Y. Ne'eman (W.A. Benjamin,
Inc., New York, 1964), p. 254.

[6]E. Fischbach, Phys. Rev. 170, 1398 (1968).

[7]B.H.J. McKellar, Phys. Letters 26B, 107 (1967).

[8]J.P. Berge, in Proceedings of the Thirteenth Annual International
Conference on High-Energy Physics, Berkeley (University of
California Press, Berkeley, 1967), p. 46.

[9]G. Barton, Nuovo Cimento 19, 512 (1961).

[10]E. Fischbach, D. Tadić, and K. Trabert, Phys. Rev. (to be published).

[11]F.C. Michel, Phys. Rev. 133, B329 (1964); R.J. Blin-Stoyle, ibid 118, 1605 (1960).

[12]N. Brene, M. Roos, and A. Sirlin, Nucl. Phys. B6, 255 (1968).

[13]For further discussion see R.J. Blin-Stoyle, Proceedings of the Topical Conference on Weak Interactions, CERN Report 69-7 (1969), p. 495; R.H.J. McKellar, contribution to this Conference.

Acknowledgment

Two of us (E.F. and K.T.) would like to thank Professor Henry Primakoff for introducing us to the problem of computing parity-violating potentials and for his constant, essential advice throughout the development of our work. We also wish to thank Dr. W. K. Cheng for his vital and fundamental help particularly in the initial phases of our calculations.

DISCUSSION

K. Trabert: Olesen and Rao (Phys. Lett. 29B, 233 (1969)) have recently suggested that the entire vector meson exchange contribution to the parity-violating potential V_{12} should be zero. They argue that the Schwinger terms, which we have shown are the dominant contribution to the weak-parity-violating NNρ vertex, should be cancelled by 'seagull' terms, and, therefore the weak NNρ vertex should vanish. It is not easy to see, however, how such a cancellation could occur because the factorization approximation contribution to V_{12} is finite and equivalent to contribution from a W meson pole diagram. It should also be mentioned that the proposed cancellation of 'seagull' and Schwinger terms is usually required by a condition such as gauge invariance or current conservation. Since the parity-violating weak interaction Hamiltonian in a current-current model couples the vector current to a non-conserved axial current, it is not clear, when one is using the assumptions of the current-field identity and field algebra in the presence of weak interactions, that gauge invariance with respect to the ρ meson can be allowed to first order in the weak interaction Fermi constant G_F. In addition, we do not expect the Ademollo-Gato theorem to apply to our calculation, since we work to first order in the momentum transfer q_μ.

A. Goldhaber: How much does the present uncertainty about vector

meson exchange affect the sensitivity of parity-violating effects
to the pion exchange term whose importance you emphasize?
E.F.: In certain models, for instance the Cabibbo model, the vector
meson contribution to the parity violating potential is larger by
a factor of 18.   How much this will be changed by the nuclear
physics is not definitely known.

G.E. Brown: I would like to correct what I think is a misconception
-- that the main ambiguity in the problem lies in the nuclear
physics part of the calculation.   Given a parity violating potential one ought to be able to calculate the effects of parity violation to something like 20% -- not a factor of twenty -- by following
a procedure like that used by Vin Mau & Bruno, which puts in short
range correlations.   I say this because the problems that enter
are similar to those one encounters in evaluating effective nucleon-
nucleon forces.   I would like also to point out that the connection
between calculations carried out in nuclear matter as that done by
McKellar and those done assuming finite nuclei is non-trivial.

H. Primakoff: The parity-violating α-decay for which there may be
some indication (c.f. contributed paper by Sprenkel-Segel, Segel
and Siemssen), requires an isoscalar component in the parity-
violating weak nuclear force.   As we have seen, such an isoscalar
component cannot arise from π-exchange and must therefore be
largely due to ρ exchange.   Thus, if the α-decay in question does
occur, the weak vertex $N \rightarrow N + \rho$ does not vanish, which supports
your conclusions.   E.F.: Just so.   B. McKellar: Likewise, if
the ρ contribution does vanish, Lobashov should not see any parity-
violating effect in $n + p \rightarrow d + \gamma$.

ANGULAR MOMENTUM NONCONSERVATION SCHEME

Franco Selleri

Istituto di Fisica dell'Università di Bari

INFN-Sottosezione di Bari

As I have shown in two recent papers[I], the phenomenology of elementary particles can be interpreted and, in some cases, better understood by attributing to the strange particles spin values higher than the traditional ones. Thus the K-meson has spin $\frac{1}{2}$, Λ and Σ become spin I bosons, $K^*(890)$ has spin 3/2, and so on.

This scheme is possible because the spin of the strange particles has never been measured directly, but always inferred from the total angular momentum of their decay products, assuming angular momentum conservation in the weak decay. Direct measurements of the spin would be, for instance, the study of fine splitting in K-mesic atoms and the application of detailed balancing to the reactions $K^-D \leftrightarrow \Sigma^- P$. If angular momentum is instead non-conserved during the decay process and the violation shows some kind of regularity (for instance by taking place according to a $\Delta J = I/2$ rule) a new attribution of spin-values is possible.

A very important question which has to be answered is whether angular momentum nonconservation can be compatible with Lorentz-invariance. This question was answered positively in the first paper by studying the spurion model according to which the missing angular momentum is carried away by an unobserved spin-I/2 "particle" with zero four-momentum. In the present note I will try and answer again the same question in a more complete manner by using a noninvariant vacuum and the indefinite metric. The latter ingredient is important in order to be able to work with the finite-dimensional representations of the Lorentz group, which are non-unitary.

Indefinite metric. The scalar product of two elements η_1 and η_2 of the vector space is defined to be[2]

$$P_{\eta_1 \eta_2} \equiv (\Omega \eta_1 , \eta_2)$$ (I)

where Ω is an operator satisfying

$$\Omega^2 = I$$ (2)

$$\Omega^+ = \Omega$$ (3)

It follows from (I) and (2) that the squared norm of a vector η, $P_{\eta\eta}$, is not necessarily positive. The mean value of an observable A is defined to be

$$\langle A \rangle = (\Omega \eta , A \eta)$$ (4)

We define two types of observables:

Class I : $A^+ = A$; $[A, \Omega]_- = 0$ (5)

Class II: $A^+ = -A$; $[A, \Omega]_+ = 0$ (6)

Both types satisfy

$$A^+ \Omega = \Omega A$$ (7)

It follows from (7) that both Class I and Class II operators have real mean values. In fact

$$\begin{aligned}
\langle A \rangle &\equiv (\Omega \eta , A \eta) \\
&= (A^+ \Omega \eta , \eta) \\
&= (\Omega A \eta , \eta) \\
&= (A \eta , \Omega \eta) \\
&= (\Omega \eta , A \eta)^* \equiv \langle A \rangle^*
\end{aligned}$$ (8)

In particular, if $A\eta = a\eta$, a being a number, it follows that a is real, unless $(\Omega\eta, \eta) = 0$.

Poincaré invariance.   Correspondingly to an arbitrary proper Poincaré transformation

$$x'_\mu = \Lambda_{\mu\nu} x_\nu + a_\mu$$ (9)

we assume the existence of a non-singular operator $U(\Lambda, a)$ transforming the elements of the vector space according to

$$\eta \longrightarrow \eta' \equiv U(\Lambda, a)\eta$$ (IO)

The invariance postulate of the scalar product

$$P_{\eta_1 \eta_2} = P_{\eta'_1 \eta'_2} \tag{II}$$

for arbitrary η_1, η_2 can easily be shown to imply

$$U^+(\Lambda, a) \, \Omega \, U(\Lambda, a) = \Omega \tag{I2}$$

whence

$$U^+(\Lambda, a) \, \Omega = \Omega \, U^{-1}(\Lambda, a). \tag{I3}$$

The latter condition substitutes the unitarity condition $U^+ = U^{-1}$ obtained with the usual treatment in the Hilbert space. Therefore our operator $U(\Lambda, a)$ needs not being unitary.

The invariance of the S-matrix elements[3)]

$$(\Omega \eta_1, S \eta_2) = (\Omega \eta'_1, S \eta'_2) \tag{I4}$$

can easily be shown to lead to

$$[S, U(\Lambda, a)]_- = 0 \tag{I5}$$

because of (I3).

If we write

$$U(\Lambda, a) = \exp\left\{ i P_\mu a_\mu + \frac{i}{2} M_{\mu\nu} \Lambda_{\mu\nu} \right\} \tag{I6}$$

it does not follow anymore that P_μ and $M_{\mu\nu}$ are hermitean. It follows however from (I3) that

$$P_\mu^+ \Omega = \Omega P_\mu \quad ; \quad M_{\mu\nu}^+ \Omega = \Omega M_{\mu\nu} \tag{I7}$$

Therefore P_μ and $M_{\mu\nu}$ are either of Class I or of Class II (or sums of pieces of Class I and Class II operators) and can be assumed to represent observables. Defining the operators

$$J_\ell = -\frac{1}{2} \varepsilon_{\ell mn} M_{mn} \quad ; \quad K_\ell \equiv M_{\ell o} \tag{I8}$$

(latin indeces run from I to 3), from the group properties of the U-operators

$$U(\Lambda', a') \, U(\Lambda, a) = U(\Lambda'\Lambda, \Lambda'a + a') \tag{I9}$$

the following commutation relations can be deduced[4)]

$$[P_\mu, P_\nu] = 0, \tag{20}$$

$$\begin{cases} [J_1 , P_0] = 0 \\ [J_1 , P_m] = i\varepsilon_{1mn}P_n \\ [K_1 , P_0] = -i\,P_1 \\ [K_1 , P_m] = i\,g_{1m}P_0 \end{cases} \qquad (21)$$

$$\begin{cases} [J_1 , J_m] = i\,\varepsilon_{1mn}J_n \\ [J_1 , K_m] = i\,\varepsilon_{1mn}K_n \\ [K_1 , K_m] = -i\,\varepsilon_{1mn}J_n \end{cases} \qquad (22)$$

Notice that (20), (21) and (22) follow uniquely from (19). The unitarity condition is never used. Therefore these commutation rules are exactly the same as in the Hilbert space. It follows from (15) that all the generators of the Poincarè group commute with S:

$$[S, P_\mu] = [S, K_1] = [S, J_1] = 0 \qquad (23)$$

As for the class properties of the generators only three possibilities are compatible with the commutation rules (20)-(22). These are the following:

	Class I	Class II
i)	P_0, P_m, K_1, J_1	---
ii)	P_0, J_1	P_m, K_1
iii)	P_m, J_1	P_0, K_1

Other possibilities are excluded because they would imply that some of the generators are identically zero. In general we can envisage the possibility that different field operators give rise to contributions to P_μ, J_1, K_1 of different type. Thus a boson field could fall in the case i), while some fermion field could fall in the case ii). The alternative between ii) and iii) seems more or less arbitrary and we will therefore disregard the case iii).

    <u>Energy-momentum conservation.</u>   Consider two eigenstates of P_μ

$$P_\mu \, \eta_i = P_\mu^{(i)} \eta_i \qquad (i=1,2) \qquad (24)$$

where $P_\mu^{(i)}$ are the eigenvalues. From $[S, P_\mu] = 0$ it follows

$$0 = (\Omega \eta_1, [S, P_\mu] \eta_2) = (P_\mu^{(I)} - P_\mu^{(2)}) (\Omega \eta_1, S \eta_2) \tag{25}$$

by virtue of (I7). Therefore $(\Omega \eta_1, S \eta_2) = 0$ unless $P_\mu^{(I)} = P_\mu^{(2)}$, that is unless energy and momentum are conserved.

Angular momentum conservation. In the same way it can be shown that angular momentum is conserved. This is however not necessarily the same as the physical conservation law, if the vacuum is not invariant, as we will show next.

Non-invariant vacuum. We assume the existence of five states[5] $\phi_0, \phi_1, \phi_2, \phi_3, \phi_4$ satisfying

$$P_\mu \phi_0 = P_\mu \phi_1 = P_\mu \phi_2 = P_\mu \phi_3 = P_\mu \phi_4 = 0 \tag{26}$$

$$(\Omega \phi_0, \phi_0) = (\Omega \phi_1, \phi_1) = (\Omega \phi_2, \phi_2) = 1 \tag{27}$$

$$(\Omega \phi_3, \phi_3) = (\Omega \phi_4, \phi_4) = -1 \tag{28}$$

$$(\Omega \phi_A, \phi_B) = 0 \quad ; \text{ if } A \neq B ; (A, B = 0, 1, 2, 3, 4) \tag{29}$$

The physical vacuum ϕ is defined by

$$\phi = \phi_0 + \sum_{n=1}^{4} \varepsilon_n \phi_n \tag{30}$$

with

$$|\varepsilon_1| = \ldots = |\varepsilon_4| \equiv \varepsilon \tag{3I}$$

It follows from the above definitions

$$(\Omega \phi, \phi) = 1 \tag{32}$$

The generators of the Lorentz group are assumed to have the following action on ϕ_0 :

$$J_\ell \phi_0 = K_\ell \phi_0 = 0 \tag{33}$$

Therefore ϕ_0 is the invariant part of the vacuum. Finally the action of J_1 and K_1 on $\phi_1, \phi_2, \phi_3, \phi_4$ can be expressed algebraically in a way obtainable from the following representation:

$$\phi_1 = \begin{vmatrix} 1 \\ 0 \\ 0 \\ 0 \end{vmatrix} , \quad \phi_2 = \begin{vmatrix} 0 \\ 1 \\ 0 \\ 0 \end{vmatrix} , \quad \phi_3 = \begin{vmatrix} 0 \\ 0 \\ 1 \\ 0 \end{vmatrix} , \quad \phi_4 = \begin{vmatrix} 0 \\ 0 \\ 0 \\ 1 \end{vmatrix} \tag{34}$$

$$\Omega = \begin{vmatrix} I & 0 \\ 0 & -I \end{vmatrix} \quad ; \quad J_\ell = \begin{vmatrix} \frac{1}{2}\sigma_\ell & 0 \\ 0 & \frac{1}{2}\sigma_\ell \end{vmatrix} \quad ; \quad K_\ell = \begin{vmatrix} 0 & \frac{i}{2}\sigma_\ell \\ \frac{i}{2}\sigma_\ell & 0 \end{vmatrix} \quad (35)$$

where all the elements are two by two matrices and σ_ℓ are the Pauli matrices. It is straightforward to check that J_1 and K_1 so defined satisfy the commutation relations (22), that J_1 is hermitean and K_1 antihermitean and that the relations $J_1^+ \Omega = \Omega J_1$, $K_1^+ \Omega = \Omega K_1$ are satisfied as well. Therefore J_1 and K_1 as given in (35) give a realization of the possibility ii) defined above. Furthermore Ω satisfies (2) and (3). It follows from (33) and from (34)-(35)

$$\vec{J}^2 \phi_o = 0 , \qquad\qquad (36)$$

$$\vec{J}^2 \phi_r = \frac{1}{2}\left(\frac{1}{2} + 1\right) \phi_r . \qquad (r=I,2,3,4) \qquad (37)$$

Therefore the vacuum state can be divided into two parts, ϕ_o having spin 0 and $\phi' = \sum_r \varepsilon_r \phi_r$ having spin $\frac{1}{2}$.

Goldstone theorem.   The non-invariance of the vacuum can be understood by assuming the existence of an universal field Ψ such that[6]

$$(\Omega \phi_o , \Psi \phi_o) = 0 ,$$
$$(\Omega \phi , \Psi \phi) \neq 0 . \qquad\qquad (38)$$

Those interactions whose lagrangian does not contain Ψ proceed on-ly between equal vacuum states (strong interactions) and angular mo-mentum is conserved also for the particles partecipating in the in-teraction. Instead the interactions with a lagrangian containing linearly Ψ (which we identify with the weak interactions) to the first order proceed necessarily between ϕ_o and one of the states ϕ_r . They give rise to observed transitions apparently violating angular momentum according to the rule $\Delta J = I/2$. We demonstrate this explicitly in the final section.

The universal field Ψ.   We assume that the field Ψ introdu-ced above is a constant (with respect to space-time) four-component field operator, transforming under an infinitesimal Lorentz trans-formation according to

$$U^{-1}(\Lambda, a) \Psi U(\Lambda, a) = \left(1 + \Lambda_{\mu\nu} \Sigma_{\mu\nu}\right) \Psi \qquad (39)$$

where

$$\Sigma_{\mu\nu} = -\frac{1}{4i} [\gamma_\mu , \gamma_\nu]_- . \qquad (40)$$

We assume furthermore that Ψ obeys the following anticommutation rules

$$[\Psi, \Psi^+]_+ = I \quad ; \qquad [\Psi, \Psi]_+ = [\Psi^+, \Psi^+]_+ = 0 . \qquad (41)$$

We write, in analogy with the usual fermion fields

$$\Psi = \sum_{\imath=1}^{2} \left(u_{\imath} a_{\imath} + v_{\imath} b_{\imath}^{+} \right) \qquad (42)$$

and assume that

$$\sum_{\imath=1}^{2} \left(u_{\imath} u_{\imath}^{+} + v_{\imath} v_{\imath}^{+} \right) = I \qquad (43)$$

$$[a_{\imath}, a_{\jmath}^{+}]_+ = \delta_{\imath\jmath} = [b_{\imath}, b_{\jmath}^{+}]_+ \qquad (44)$$

all other anticommutators being identically zero. Obviously (43) and the anticommutation rules of a_r, b_r, a_r^+, b_r^+ imply (41). Notice that we do not have a Dirac equation in the case of a constant spinor. Therefore the spinors u_r and v_r are assumed to be the eigenstates of the γ_5-matrix

$$\gamma_5 u_r = u_r \quad ; \quad \gamma_5 v_r = -v_r \qquad (r = I, 2) \qquad (45)$$

Finally we assume that the vacuum states $\phi_1, \ldots \phi_4$ are obtained from ϕ_0 upon application of the creation operators a_r^+ and b_r^+

$$\phi_{\imath} = a_{\imath}^{+} \phi_0 \quad ; \quad \phi_{\imath+2} = b_{\imath}^{+} \phi_0 . \qquad (r = I, 2) \qquad (46)$$

In order to satisfy (27) and (28) we require that

$$\Omega \phi_0 = \phi_0 \qquad (47)$$

and that

$$\Omega a_{\imath}^{+} \Omega = a_{\imath}^{+} \quad ; \quad \Omega b_{\imath}^{+} \Omega = -b_{\imath}^{+} . \qquad (48)$$

It follows then

$$\Omega \Psi \Omega = \gamma_5 \Psi . \qquad (49)$$

Interestingly, the latter condition requires that the lagrangians in which Ψ enters linearly contain the $I - \gamma_5$ combination.

Particle states. We assume that all the field operators of the observable particles commute with the metric operator Ω. That is the field Ψ is the only one not commuting with Ω. Obviously all the destruction and creation operators for physical particles commute with Ω as well. Therefore the algebra of the field operators can remain exactly the same as in the traditional theories. It follows from this that the stable particle states, obtained by applying creation operators to the physical vacuum ϕ , are all of positive norm. The relation (32) is needed in the proof, which is straightforward and will be omitted. Our theory satisfies then the Lee-Wick condition for having an unitary S-matrix.[3]

Angular momentum nonconservation. We will now illustrate the possibility of angular momentum nonconserving transitions by studying the decay $K \to \pi\pi$ where the K is assumed to have spin I/2. The invariant lagrangian for this decay is assumed to be

$$\mathcal{L}_{int} = G \bar{\Psi} \mathcal{L}' + h.c. \quad ,\text{where} \quad \mathcal{L}' = (1-\gamma_5) \psi_k(x) \, \phi_\pi^+(x) \, \phi_\pi(x) \tag{50}$$

To the first order in the coupling-constant G the matrix element for $K \to \pi\pi$ is given by

$$M(K \to \pi\pi) \sim (\Omega \, a_\pi^+(\vec{q}) \, a_\pi^+(-\vec{q}) \, \phi, \; \mathcal{L}_{int} \, a_k^+(0) \, \phi) \tag{51}$$

where $a_A^+(\vec{p})$ is the creation operator for the particle A with momentum $\vec{p}$. Using (50) with ψ given in (42) one obtains

$$M(K \to \pi\pi) \sim \sum_{n=1}^{2} \bar{u}_n (\Omega \, a_\pi^+(\vec{q}) \, a_\pi^+(-\vec{q}) \, a_n \phi, \; \mathcal{L}' a_k^+(0) \, \phi)$$

$$+ \sum_{n=1}^{2} \bar{v}_n (\Omega \, a_\pi^+(\vec{q}) a_\pi^+(-\vec{q}) \, \phi, \; \mathcal{L}' a_k^+(0) \, b_n \, \phi) \tag{52}$$

It follows from (44) that $a_r \phi = \varepsilon_r \phi_0$, $b_r \phi = \varepsilon_{r+2} \phi_0$. Therefore, reducing the field operators contained in $\mathcal{L}'$ with the creation operators of the states, one obtains in momentum space

$$M(k \to \pi\pi) \sim \sum_{n=1}^{2} \left[\varepsilon_n \bar{u}_n (1-\gamma_5) u_k(0) + \varepsilon_{n+2} \bar{v}_n (1-\gamma_5) u_k(0) \right] \tag{53}$$

since $(\Omega\phi_0, \phi) = (\Omega\phi_0, \phi_0) = I$. Now $\bar{u}_r(I-\gamma_5) = 0$ and $\bar{v}_r(I-\gamma_5) = 2\bar{v}_r$ Therefore

$$M(K \to \pi\pi) \sim \sum_{n=1}^{2} \varepsilon_{n+2} \bar{v}_n \, u_{k,s}(0) = \sum_{n=1}^{2} \varepsilon_{n+2} \, \delta_{n,s} = \varepsilon_{s+2} \tag{54}$$

One sees then that $M(K \to \pi\pi) \neq 0$ and the transition can take place.

References

I) F. Selleri, Nuovo Cimento 57A,678(68); ibid. 60A,291(69).

2) K. Bleuler, Helv. Phis. Acta 23,567(50); S. Gupta, Proc. Phys. Soc. (London) A63,681(5I).

3) T. D. Lee and G. C. Wick, Nucl. Phys. B9,209(69); ibid. BIO,I(69).

4) E. g. see S. Gasiorowicz, Elementary Particle Physics, John Wiley & Sons, Inc., New York/London/Sidney, I966.

5) Four states in addition to ϕ_0 is the minimum possible. In fact there are no two and three-dimensional representations of the algebra of the operators J_1 , K_1 , Ω .

6) J. Goldstone, A. Salam and S. Weinberg, Phys. Rev. I27,965(62).

THE CIRCULAR POLARIZATION OF γ RAYS FROM Ta^{181} AND THE

PARITY VIOLATING ONE PION EXCHANGE POTENTIAL

N. Vinh Mau and A.M. Bruneau

Division de Physique Théorique*

Institut de Physique Nucléaire -91 ORSAY(FRANCE)

The circular polarization of γ rays emitted in the 482 kev transition of Ta^{181} has been measured first by Boehm and Kankeleit[1] who reported a polarization $P = -(2.0 \pm 0.4)10^{-4}$. More recently, new experiments found a smaller effect, with $P = -(6. \pm 1.)10^{-6}$[2a] or $P = -(3.9 \pm 0.8)10^{-6}$[2b]. This effect, although very small, yields evidence for a parity non conserving component in nuclear forces due to the weak interaction.

In the present paper, the circular polarization of γ rays emitted by the 482 kev transition in Ta^{181} is calculated assuming that the dominant contribution to the PNC nucleon-nucleon potential is due to the one pion exchange because of its long range. Such potential arises in a current-current weak interaction scheme, from the $\Delta S = 1$ non leptonic current. The contribution from the $\Delta S = 0$ current comes mainly from the vector meson exchange and therefore has a shorter range. Actually, it will be shown that the polarization decreases rapidly with the range of the potential. We use the unified collective model for describing the Ta^{181} nucleus, since this model has been successful for describing the excited states and the γ.transitions in the region of Tantalum[3][4].

The one pion exchange, parity violating nucleon-nucleon interaction has been calculated by K. Mc Kellar[5].

The sequence of the lowest level of Ta^{181} and neighbouring nuclei are well reproduced by the Nilsson scheme.

Hence we consider

$$H = H_o + \sum_{i<j} V_\pi(i,j)$$

where H_o is the collective model hamiltonian including the intrinsic motion of nucleons in a Nilsson deformed field. Following ref. 6, we take for Ta^{181} a deformation corresponding to $\delta = 0.36$ then the single particle eigenvectors denoted by $|N\Omega\varepsilon>$ where ε labels the states $(N\ell)$ with different energies, are known by their expansion on spherical states $|N\ell\Lambda\Sigma>$ [6]. In this model, the regular $5/2^+$ and $7/2^+$ states correspond to 108 neutrons and 72 protons filling the lowest Nilsson orbitals with an odd proton in the second $/4, 5/2>$ or the second $/4, 7/2>$ Nilsson orbitals respectively. In our calculation, they are mixed through V_π with all the negative parity $|5, 5/2, \varepsilon_k>$ and $|5, 7/2, \varepsilon_\ell>$ Nilsson orbitals.

Then the initial and final perturbed states can be written as

$$\psi_o(5/2) = \phi_o(5/2^+) + \sum_k \alpha_k \phi_k (5/2^-)$$

$$\psi_f(7/2) = \phi_j(7/2^+) + \sum_\ell \beta_\ell \phi_\ell (7/2^-)$$

(2)

where

$$\alpha_k = \sum_n \frac{<5,5/2,\varepsilon_k; N_n\Omega_n\varepsilon_n|V_\pi|4,5/2; N_n\Omega_n\varepsilon_n>-\text{exch.}}{E_k - E_o}$$

$$\beta_\ell = \sum_n \frac{<5,7/2,\varepsilon_\ell; N_n\Omega_n\varepsilon_n|V_\pi|4,7/2; N_n\Omega_n\varepsilon_n>-\text{exch.}}{E - E_f}$$

As the odd particle is a proton which interacts through V_π only with the neutrons, the sum runs over the core neutrons.

The coefficients α_k and β_ℓ have been calculated as functions of the $n \rightarrow p + \pi^-$ weak amplitude f and the polarization is: $P = -18.3$ f. The values of P calculated with the different non-leptonic weak interaction models are given in Table I.

	Cabbibo model	Extra-current (d'Espagnat)	Schizon model (T.D. Lee)
$f^{(8)}(10^{-7})$	0.43	8.45	4.3
$P(10^{-5})$	-0.08	-1.55	-0.8

Table I

The weak $n \rightarrow p + \pi^-$ amplitude calculated from different non leptonic weak interaction models and the polarization of 482 kev-γ rays from Ta^{181}.

One might argue that the inclusion of vector meson exchange, such as the ρ meson, could modify the value of f. We have estimated the influence of the ρ meson exchange on the polarization by substituting the ρ mass for the π mass in the expression of PNC interaction. The contribution from ρ meson is found to be inhibited by a factor $\simeq 8$. This estimate is rather crude, nevertheless it shows that the ρ exchange would not change drastically our conclusions. Furthermore, recently, Olesen [7] has shown that with the assumption of field-current identity and of cancellation of Schwinger's and Seagul's terms, the vector meson exchange must be negligible.

Then our result, different from previous ones[9][10] seems to be in disagreement with the standard model including only charged currents and in favour of extra-current models.

The hard core effects of the strong N-N interaction could be appreciable[9] in the nuclear matrix elements. We have estimated such corrections on radial integrals involving S-states of relative motion. The error is always less than 10% and is certainly in the range of uncertainty of nuclear models.

We take pleasure in thanking Profs. G.E. Brown and T.N. Truong for useful comments and stimulating discussions and Prof. F. Boehm for communicating to us his last measurement before publication.

REFERENCES

(1) F. Boehm and E. Kankeleit, Phys. Rev. Lett. <u>14</u>, 312 (1965).

(2) a) V.M. Lobashov, V.A. Nazareno, L.F. Saenko, L.M. Smotritsku and G.I. Kharevitch, J.E.T.P. Letters <u>3</u>, 173 (1966 and <u>5</u>, 59 (1967).

   b) F. Boehm, Private communication.

(3) B.R. Mottelson and S.G. Nilsson, Dan. Vid. Mat. Fis. Skryfter <u>1</u>, n° 8(1959).

(4) C.J. Gallagher, Phys. Rev. <u>126</u>, 1525 (1962).

(5) B.H.J. Mc Kellar, Phys. Letters <u>26B</u>, 107 (1967).

(6) S.G. Nilsson, Dan. Mat. Fys. Medd. <u>29</u>, n°16(1955).

(7) P. Olesen and J.S. Rao, Phys. Letters <u>29B</u>, 233(1969).

(8) E. Fishbach and K. Trabert, Phys. Rev. <u>174</u>, 1843(1968)

(9) B.H.J. Mc Kellar, Phys. Rev. Lett. <u>20</u>, 1542(1968).

(10) E. Maqueda and R.J. Blin-Stoyle, Nucl. Phys. <u>A91</u>, 460(1967).

----

* Laboratoire associé au C.N.R.S.

E. L. Sprenkel-Segel, Illinois Institute of Technology
and Argonne National Laboratory; R. E. Segel, Argonne
National Laboratory* and Northwestern University†; and
R. H. Siemssen, Argonne National Laboratory*

It was pointed out about ten years ago[1] that a very sensitive
way to search for parity nonconservation in nuclei is to look for
an alpha branch from the 8.88-MeV state in ^{16}O in the delayed alpha
spectrum following the β decay of ^{16}N. Figure 1 shows the energy
level diagram. Previous attempts[2—4] to detect this decay mode
found $\Gamma_\alpha \leq 10^{-8}$ eV, which implies an upper limit $F^2 \lesssim 2 \times 10^{-12}$
for the intensity of the opposite-parity component in the wave
function. We have repeated this experiment, using new techniques
which have improved the sensitivity by at least an order of
magnitude.

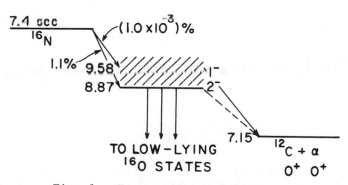

Fig. 1 - Energy level diagram.

* Work supported by the U.S. Atomic Energy Commission.
† Work supported by the National Science Foundation.

The previous experiments have been limited by counting statistics. The 400-keV-wide 9.58-MeV state, which decays solely by alpha emission, is fed by a 10^{-5} branch, and the alphas from the 8.88-MeV state lie on the low-energy tail of the broad group. Thus, the search is for a sharp peak superimposed on a continuum. Two major improvements in technique have led to a much higher counting rate. First, the ^{16}N was separated out from the target material, thus greatly increasing the specific activity of the source. Since the total source thickness is limited by the requirements of good alpha-energy resolution (and therefore small loss of alpha energy) increasing the specific activity increases the counting rate. The second major innovation was that the detection of electrons was prevented by sweeping them out with a magnetic field. There are 10^5 electrons produced for every alpha particle in the decay of ^{16}N, and these electrons would smear the energy resolution of the detector if the counting rate gets too high. However, by substantially eliminating these betas it was possible to take full advantage of the higher yield occasioned by the increased specific activity.

The ^{16}N was formed in the ^{19}F(n, α) reaction, Q = -1.50 MeV, produced by fast neutrons from the Argonne CP-5 research reactor. The ^{16}N recoils emerged from thin layers of CaF coated on aluminum plates, and were thermalized in a carrier gas. A trace of NO in the nitrogen carrier was found to pick up a substantial fraction of the active nitrogen. After being carried about 50 ft, the ^{16}NO was frozen out on a surface viewed by a silicon detector about 1.5 cm away. The source and detector were in a 50 kG field generated by a superconducting magnet.

The running procedure was to flow gas for 10 sec and then count for 10 sec with the chamber evacuated. Data were taken over a 10-day period. Figure 2 shows the spectrum obtained by combining the data from the many individuals runs taken during this period. A total of about 10^6 alpha particles are in the spectrum— at least 20 times the yield of any of the previous experiments.[2—4] An arrow points to the expected position of the parity-violating group and, indeed, there is a small bump in the spectrum at the right place.

The broad group was fitted with a single-level Breit-Wigner formula, suitably modified for extranuclear effects and for transitions directly through the continuum. In the difference spectrum obtained by subtracting off the broad groups, a sharp peak with a magnitude of $2\frac{1}{2}$ standard deviations was found at 1.287 ± 0.015 MeV—in good agreement with the expected position of 1.295 ± 0.010 MeV. No other peak of comparable statistical significance was found in the difference spectrum.

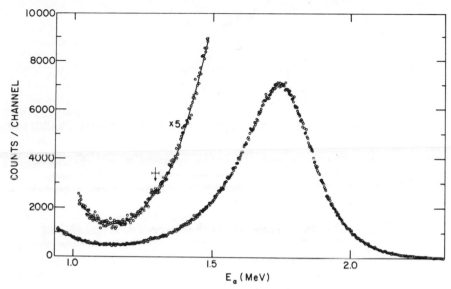

Fig. 2. Delayed alpha-particle spectrum following the beta decay of ^{16}N. The rise at low pulse heights is due to electrons. The arrow points to the expected position of the parity violating group.

It was not possible to further improve the counting statistics or to check for systematic error because of a scheduled 1-yr shut-down of the reactor. Until the results can be checked further, they must be considered as preliminary.

If the branching ratio to the 9.58-MeV state is taken to be 1×10^{-5} and the lifetime of the 8.88 MeV state to be 1.5×10^{-13} sec, the tentative effect found here corresponds to an alpha width of 1×10^{-9} eV which, in turn, corresponds to a parity impurity $F^2 \approx 2 \times 10^{-13}$. An impurity of this magnitude is consistent with the parity-violating component to be expected in the nucleon-nucleon potential on the basis of the conserved-vector-current theory of the weak interactions, although the theoretical uncertainties are quite large.

[1]R. Segel, J. Olness, and E. Sprenkel, Phil. Mag. 6, 163 (1961).
[2]R. Segel, J. Olness, and E. Sprenkel, Phys. Rev. 123, 1382 (1961).
[3]P. F. Donovan, D. E. Alburger, and D. H. Wilkinson, Proc. Rutherford Jubilee Conf., Manchester, 1961, J. B. Birks, Ed. (Heywood and Co., London, 1961) p. 827.
[4]W. Kaufmann and H. Waffler, Nucl. Phys. 24, 62 (1961).

EXPERIMENTS ON PARITY NON-CONSERVATION IN NUCLEAR GAMMA TRANSITIONS*

J. C. Vanderleeden and F. Boehm

California Institute of Technology

Pasadena, California

The existence of a strangeness conserving non-leptonic weak interaction has long been postulated.[1]   The resulting parity admixtures in nuclear states and nuclear gamma transitions have been the subject of several examinations[2-6] of which the recent work of Lobashov[3] has been the most outstanding.   In this work the limitation due to counting statistics are overcome by the use of very strong sources and an ingeneous current-integrating detector scheme. We have been conducting experiments with a current integration technique whereby we integrate and digitize the signal-current from a germanium counter.   While a full account of our work is in preparation we present here a brief description of our method and results.

We have reinvestigated the circular polarization of the 482 keV transition in Ta^{181} (regular M1 interfering with irregular E1) and the 396 keV transition in Lu^{175} (regular E1 admixed with irregular M1) using unpolarized sources of Hf^{181} and Yb^{175}.   Both cases have been studied previously[2-5], and some of the results as quoted below have been conflicting.

The Hf^{181} and Yb^{175} sources were produced by neutron capture in 1.0 gram and 0.85 gram, respectively, of the isotopically enriched oxide at a flux of 5×10^{14} cm^{-2} sec^{-1} at the MTR.   To reduce the intensity of the external bremsstrahlung the oxide samples were diluted with 2 gram of 1/4 micron diamond powder.   For experiments designed to study the bremsstrahlung we used Lu^{177} sources obtained from neutron capture of 0.3 gram of natural abundance Lu_2O_3, mixed with 0.6 gram of diamond powder.   An Ru^{103} source, needed in a control experiment, was made by neutron capture in 10 gram of natural abundance Ru metal sponge.

Gamma rays from Hf^{181} and Yb^{175} sources of strength 200 Ci and 800 Ci, respectively, were forward scattered on the inside surface of a hollow cylindrical Armco iron magnet. Pb filters of various thicknesses were used to control the contribution from low energy circularly polarized bremsstrahlung quanta originating in the preceding beta decay. The scattered photons are detected with a 40 cm^3 Ge(Li) counter. The detector current, of about 40 μA for a 200 Ci Hf^{181} source (equivalent to ~5 × 10^9 counts/sec.), is integrated in a charge sensitive circuit. At the end of preset time intervals the voltage, V, across a capacitor in the feedback loop of an operational amplifier is recorded with a digital voltmeter.

The magnetization of the polarimeter was reversed at unequal time intervals constituted in such a fashion as to eliminate slow drifts in the signal (such as from source decay) up to the 5th order term of an expansion of the signal in powers of t. Twelve unequal time intervals formed a sequence lasting 120.8 sec., which was repeated over a three-day measuring period. It is possible with this technique[7] to reduce the undesired part of the observed asymmetry stemming from the exponential decay of the source by two orders of magnitude, compared with a sequence of equal time intervals of the same average length. Experimentally the magnitude of this asymmetry can easily be determined from reversing the magnetization without changing the pattern.

We have calculated as a function of Pb absorber thickness the asymmetry due to the circular polarization of bremsstrahlung (internal and external) originating in the Hf^{181}, Yb^{175}, and Lu^{177} beta decay. The accuracy of the method of calculation has been checked by comparing with measurements on Lu^{177} sources, where the endpoint energy is close to that of Hf^{181} and Yb^{175}, and where the asymmetry due to bremsstrahlung is conveniently large. The calculation agrees with a curve drawn through the experimental points, to about 5% for the 2 and 3 mm Pb absorbers and to 30% for smaller and larger absorber thicknesses as illustrated in Fig. 1a.

In the control experiment with an Ru^{103} source we have measured the composite of the switching asymmetries due to the exponential decay as well as other possible instrumental asymmetries that cannot easily be calculated (such as those due to magnetostriction effects). The gamma ray of 498 keV in Rh^{103} is a fast E2 transition, thus is not expected to show a circular polarization. The resulting average values of the asymmetry, A_{Fe} , are listed for Yb^{175} and Ru^{103} for a 3 mm Pb absorber in Table 1, column 2. Figure 1b gives additional results for different absorber thicknesses. A_{Fe} is defined by $A_{Fe} = (V^+ - V^-)/(V^+ - V^-)$, where V^+ is the digitized voltage for one direction of magnetization of the polarimeter.

Another control experiment consisted of covering the scattering surface of the polarimeter with a 10 mm Pb sleeve. We have used

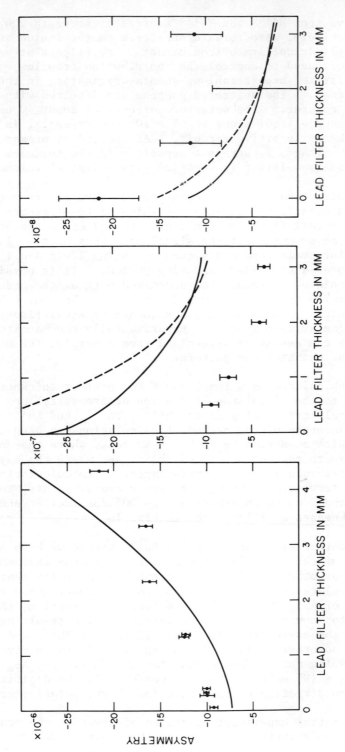

Fig. 1 - Results for Lu^{177} (a), Yb^{175} (b), and Hf^{181} (c). The solid line is the calculated asymmetry due to the circular polarization of bremsstrahlung. The dashed line is a corrected calculation.

this control for the measurements on Hf^{181} and Ru^{103} which lasted over a period of several months. In both cases data were taken successively with and without Pb sleeve in intervals of three days, and the observed asymmetries A_{Pb} and A_{Fe} were averaged. The quantities $<A_{Fe} - A_{Pb}>$ are given in Table 1, column 2, for Hf^{181} and Ru^{103}, and in Fig. 1c. The average values for A_{Pb} in the case of Hf^{181} were $(+0.3 \pm 2.3) \times 10^{-8}$ and $(-2.1 \pm 3.6) \times 10^{-8}$ for 3 mm and 2 mm absorbers, respectively, and $(-4.6 \pm 3.3) \times 10^{-8}$ and Ru^{103}.

In Figs. 1b and c, the solid line is the calculated asymmetry due to circular polarization of bremsstrahlung, and the dashed line is a correction to this calculation based on the deviation between experiment and theory in the Lu^{177} test case. We take the difference between these two curves to be a measure of the uncertainty to be attached to the calculation. We then calculate the net asymmetry by subtracting from each of the four experimental points the bremsstrahlung contribution to the asymmetry, A_{brem} (Table 1, col. 3), correct the result for dilution of the asymmetry due to other gamma rays in the Hf^{181} and Yb^{175} decay, and take a weighted average of the four values. This net asymmetry divided by the polarimeter efficiency of 1.8%(Hf^{181}) and 1.6%(Yb^{175}) gives the circular polarization P_γ, listed in column 4 of Table 1. The values of the circular polarization are subject to an additional uncertainty estimated at 15% arising from the efficiency calibration of the polarimeter.

The result for Hf^{181} is consistent with that of Lobashov, et al.[3], $P_\gamma = (-6.1 \pm 0.7) \times 10^{-6}$, as well as with that of Boehm and Kankeleit[2], $P_\gamma = (-10 \pm 40) \times 10^{-6}$, and that of Cruse and Hamilton[5], $P_\gamma = (-90 \pm 60) \times 10^{-6}$. It conflicts with the work of Bodenstedt, et al.[4], $P_\gamma = (-32 \pm 8) \times 10^{-6}$. Our result for Yb^{175} is in reasonable agreement with that of Lobashov, et al.[3], $P_\gamma = (40 \pm 10) \times 10^{-6}$.

The present findings confirm the existence of a parity-nonconserving nuclear force of the magnitude predicted by the current-current theory of the weak interaction. (For a recent review we refer the reader to ref. 8).

REFERENCES

[1] R.P.Feynmann and M. Gell-Mann, Phys.Rev. 109 (1958) 193.
[2] F. Boehm and E. Kankeleit, Nucl. Phys. A109 (1968) 457.
[3] V.M.Lobashov, V.A.Nazarenko, L.F.Sayenko, L.M.Smotritskii, and G.I.Kharkevich, JETP Letters 5 (1967) 59; Proc. of International Conf. on Nucl. Structure, Tokyo. Physical Soc. of Japan (1968) 443; Proc. of Conf. on Electron Capture and Higher Order Processes in Nucl. Decays, Debrecen. Eötvös Lóránd Phys. Soc. (1968) 480.
[4] E.Bodenstedt, L.Ley, H.O.Schlenz, & L.I.Wehman, Phys.Let. 29B (1969) 165.
[5] D.W. Cruse and W. D. Hamilton, to appear in Nuclear Physics; and references cited.

[6]Yu. G. Abov, P.A. Krupchitsky, M.I. Bulgakov, O.N. Yermakov, and
I.L. Karpikhin, Phys. Letters 27B (1968) 16; E. Warming, F. Stecher-
Rasmussen, W. Ratynski and J. Kopecký, Phys. Letters 25B (1967) 200.
[7]J.D. Bowman and J.C. Vanderleeden, to be submitted to Nuclear
Instruments and Methods.
[8]R.J. Blin-Stoyle, Proc. of Conf. on Weak Interactions, CERN,
Geneva (1969) 495; E.M. Henley, to be published in Annual Review
of Nuclear Science (1969).

TABLE I

Observed Asymmetry, A, and Circular Polarization, P_γ

Source	$\langle A_{Fe} \rangle^{a)}$ $(\times 10^{-8})$	A_{brem} $(\times 10^{-8})$ (Calculated)	P_γ $(\times 10^{-6})$ (All Data)
Yb^{175} (3 mm Pb)	-41 ± 7	-106	62 ± 8
Ru^{103} (3 mm Pb)	-8.5 ± 2.1	-3.7	
Source	$\langle A_{Fe} - A_{Pb} \rangle (\times 10^{-8})$	A_{brem} $(\times 10^{-8})$ (Calculated)	P $(\times 10^{-6})$ (All Data)
Hf^{181} (3 mm Pb)	-11.3 ± 2.8	-2.8	-3.8 ± 1.3
Ru^{103} (3 mm Pb)	-2.9 ± 4.2	-3.7	0.5 ± 2.3

a)Corrected for asymmetries due to exponential decay, which
amounted to 5.1×10^{-8} and 1.9×10^{-8} for Yb^{175} and Ru^{103}
respectively.

T-INVARIANCE CHECK IN 635 MEV ELASTIC PROTON-PROTON SCATTERING

R.Ya.Zulkarneyev,V.S.Nadezhdin,V.I.Satarov

Laboratory of Nuclear Problems
Joint Institute for Nuclear Research
Dubna, U S S R

Interaction invariance with respect to time-inversal results in the well-known "polarization asymmetry" equality $(P=\mathcal{Q})$ for nucleon-nucleon elastic scattering /1,2/. The experimental check of this conclusion performed earlier at 142-210 MeV /3-5/ is also important even at higher energies both due to the fundamental principles on which the equality under study is based and the feasible dependence of T-invariance violation upon energy. For this purpose the polarization $P(\theta)$ was measured in the angle region from 34° to 117° c.m.s. The results were compared with the appropriate values of $\mathcal{Q}(\theta)$ obtained in ref./6/. The polarization $P(\theta)$ was determined in the triple proton scattering on parallel planes when the first scattering occurred inside the accelerator vacuum chamber. The obtained proton beam having the polarization as large as 0.425 ± 0.013 /7/ was scattered for a second time on a hydrogeneous (CH_2-C) target at an angle θ_2. The polarization was analysed by proton scattering on carbon at an angle of $\theta_3 = 8^{\circ}+1^{\circ}30'$. In the case when the planes of all scatterings coincided, the left-right asymmetries e_+ and e_- which are observed in the analysing scattering and correspond to reciprocally opposite normal directions to the second scattering plane make is possible to find polarization by the known formular /1,3/ $P(\theta)=\frac{1}{2P_3}\{e_+(1+\mathcal{Q}P_1)-e_-(1-\mathcal{Q}P_1)\}.$

Here P_3 is the analysing power in the third scattering; P_1 is the beam polarization after the first scattering; $\mathcal{Q}$ is asymmetry arising in scattering of the totally polarized proton beam on hydrogen.

Special attention was paid to the determination of the effective zero reference on the angle scale in the third scattering. This was achieved by a method similar to that described in /8/. In our case this allowed the determination of the position of the incident beam axis to an accuracy better than $\pm 0°2'$.

The analysing power of P_3 of the scatterers was defined in a special run. The method and the developed apparatus were preliminary tested in the scattering experiment at an angle of $90°$ c.m.s. The Pauli principle and P-parity requirements lead to the fact that $P(90°)$ should be zero for pp-systems. The results of the test experiment and those of measuring $P(\theta_2)$ are summarized in Table 1. As follows from the Table, $P(90°) = 0.004 \pm 0.020$, which does not contradict the equality $P(90°)=0$. This result can be considered as an experimental proof of the absence of essential systematic errors in the measurements.

Table 1

$\theta_2°$ c.m.s.	$P \pm \Delta P$	$Q \pm \Delta Q$	$P-Q$	$\dfrac{ReT}{\sqrt{I_0}}$;% **
34.5	0.558±0.039	0.496±0.024	0.062±0.046	−4.0±3.0
41	0.563±0.041	0.518±0.014	0.045±0.044	−2.6±2.5
48	0.473±0.020	0.485±0.009	−0.012±0.022*	0.8±1.6
61	0.378±0.020	0.395±0.014	−0.017±0.024	−1.5±2.1
72	0.274±0.034	0.296±0.014	−0.022±0.037	−3.2±5.4
90	0.004±0.020	0.012±0.009	−0.008±0.022*	3.5±9.7
106	−0.358±0.028	−0.268±0.015	−0.090±0.037	13.2±4.9
117	−0.338±0.039	−0.400±0.020	0.062±0.044	8.2±5.8

* Preliminary results have been published in /9/ .
** In our calculations it has been assumed that $ReT = ImT$.

The values of $P(\theta)$ are given without the account of corrections due to the existence of angle-energy correlations, in the third scattering. These corrections are about 0.004 ± 0.002 and 0.002 ± 0.002 in the angle region of $106°$ and $41°$ c.m.s., respectively. The above errors are complete and, in contrast to the results of other authors (refs./3,4,5,10/) they include the errors of primary beam polarization measurements and the target analysing powers.

By taking into account both the obtained values of $P-Q$, the contribution of the matrix term T of elastic pp-scattering (see, Table 1) as well as the noninvariance phase λ_2 /2/ were evaluated. For the former it was suggested that Re T = Im T, and for the latter $\lambda_4 = 0$.

By averaging over all the angles we have

$$Sin\lambda_2 = -0.11 \pm 0.10 \; ; \; \frac{ReT}{\sqrt{I_o}} = \frac{ImT}{\sqrt{I_o}} = (-0.8 \pm 1.1) \cdot 10^{-2}$$

If one assumes that the results of S.Wright et al. /10/ obtained for 435 MeV are valid also for 635 MeV, their treatment together with our experimental data allows one to find ReT and ImT . Calculations have shown that in our case

$$ReT = (-0.008 \pm 0.021) \cdot \sqrt{10^{-27}} cm; \; ImT = (0.015 \pm 0.022)\sqrt{10^{-27}} cm$$

$$|T| = (0.017 \pm 0.030) \cdot \sqrt{10^{-27}} cm; \; \sqrt{I(30^\circ_{c.m.s})} = 2\sqrt{10^{-27}} cm \; .$$

Thus, as is seen, even if there is any effect of T-invariance violation in the angle and energy regions, it occurs at the level not higher than $(2-3) \times 10^{-2}$ of the total amplitude of elastic pp-scattering.

References

1. L.Wolfenstein. Ann.Rev.Nucl.Sci. 6 (1956) 43.
   R.I.N.Phillips. Nuovo Cim., 8, (1958) 265.
   I.Bell, F.Mandl. Proc.Phys.Soc. 71(1958)272.
2. A.E.Woodruf. Ann.Phys. 7 (1959) 65.
3. C.F.Hwang, T.R.Ophel, E.H.Thorndike, R.Wilson. Phys.Rev. 119 (1960) 352.
4. P.Hillman, A.Johansson, G.Tibbel. Phys.Rev.110, (1958) 1218.
5. A.Abashian, E.M.Hafner. Phys.Rev.Lett.1(1958)255 .
6. R.Ya.Zulkarneyev, V.S.Kiselev, V.S.Nadezhdin, V.I.Satarov. Yad.Phys. 6 (1967) 995.
7. R.Ya.Zulkarneyev, V.S.Nadezhdin, V.I.Satarov. Preprint P1-3189, Dubna, 1967.
8. O.Chamberlain, E.Segre, R.Tripp et al. Phys.Rev. 105 (1957) 288.
9. R.Zulkarneyev, V.Nadezhdin, V.Satarov. Rev.Mod.Phys. 39 (1967) 509 (see, Yu.M.Kazarinov's paper).
10. R.Handler, S.Wright, L.Pondrom et al. Phys.Rev.Lett. 19 (1967) 933.

AN UNDERGROUND EXPERIMENT ON NEUTRINOLESS DOUBLE BETA DECAY

E. Fiorini and A. Pullia
Istituto di Fisica and I.N.F.N. - Milano

G. Bertolini, F. Cappellani and G. Restelli
Euratom - C.C.R. - Ispra

Various theoretical and experimental investigations have been carried out on double beta decay, and particularly on neutrinoless $\beta\beta$ decay as a very sensible way of testing lepton non conservation. This decay has been mostly considered, and its lifetime calculated, as the product of two ordinary beta decays (1-3). Recently however B. Pontecorvo (4) has considered the possibility of a "direct" super-weak interaction with $\Delta L=2$. Experiments based on direct methods of detection of the electrons (nuclear emulsions, cloud and spark chambers, counters) have given up to now no evidence for any type of decay (3,5). On the contrary evidence for $\beta\beta$ decay of ^{128}Te and ^{130}Te has been obtained with chemical and mass spectroscopic analysis of geological samples (6). This technique cannot however discriminate between neutrinoless and two neutrino decay; and possible background effects simulating $\beta\beta$ decay are hard to be excluded.

Preliminary results are reported of an experiment presently carried out at the Monte dei Cappuccini Laboratory, as a depth of about 70 meters water equivalent, on the decay ^{76}Ge $\rightarrow$ ^{76}Se + 2e$^-$, which, if found, would indicate violation of the lepton conservation law. The negative results of a previous experiment at sea level have been published already (5). The experiment is based on the use of a Ge(Li) crystal both as source and detector of the decay. Isotope 76, whose percentage is 6.67% in natural germanium, could undergo $\beta\beta$ decay with a transition energy of (2.048 ± 0.007) MeV. The possibility of two successive single beta decays is energetically forbidden since the ground state of the intermediate isotope (As), lies well above that of the parent nucleus. If no neutrino is emitted the transition peak should appear in the spectrum of pulses from the detector for this experiment due to its low intrinsic radioactivity and high resolution.

The detector used in this experiment, a 24 cm^3 coaxial litium compensated germanium crystal, is shielded against local activity by layers of mercury (3cm), copper (4cm), low activity lead (10 cm) and normal lead (10 cm). Neutrons are partially slowed down and absorbed by paraffin and cadmium layers of 20 and 0.2 cm thickness, respectively. During part of the experiment a further reduction of background due to cosmic rays was accomplished by an anticoincidence plastic scintillator of 70 x 70 cm^2 area placed above the shielding. An external antenna, twisted around the shielding and connected in anticoincidence with the Ge(Li), was used to reduce the background due to spurious signals.

The signals from the chain are sent to a multi-channel analyser with a scale of 4 keV and 2 keV per channel for the regions 0-1.8 MeV and 1.8-2.7 MeV, respectively. The stability and resolution of the chain was routinely tested every three days by extracting the detector and exposing it to Ra and Th sources.

The spectrum, obtained in about 2,400 hours of running time, shows a peak at 511 keV, which is due to annihilation of positrons generated by radioactivity and electromagnetic radiation of cosmic rays muons. The high energy part of the spectrum is smooth and the counting rate in the region of neutrinoless decay is of (5.5 ± 0.5) x 10^{-3} counts hour^{-1} keV^{-1}. We have investigated the spectrum up to 36 MeV to understand the origin of background counting, and attribute the counting rate in the region of neutrinoless $\beta\beta$ decay to be due in good part to cosmic rays. It is not yet clear which is the role played by cosmic ray neutrons leaking through the absorber and by electromagnetic radiation generated by muons leaking through the anticoincidence plastic scintillator or inclined with respect to the vertical. To determine a limit for the rate of neutrinoless $\beta\beta$ decay we have applied a maximum likelihood procedure assuming a continuous background spectrum and a counting rate in the region of interest distributed according to the Poisson law. We can exclude with 68% confidence level an half lifetime lower than $1.1x10^{21}$ years, thus improving by a factor of three our previous results (5). The theoretical predictions for the lifetime of neutrinoless and two neutrino $\beta\beta$ decay are 8 x 10^{16} and 10^{23} years, respectively.

It is a pleasure to thank Prof. C. Castagnoli and his group for the generous hospitality at the Monte dei Cappuccini Laboratory, and the Societa Tonolli - Milano for the gracious loan of the lead.

REFERENCES

(1) G.F. Dell'Antonio and E. Fiorini - Suppl. Nuovo Cim. <u>17</u>, 132 (1960) also for previous references.
(2) S. Rosen and H. Primakoff - Alpha, betta and gamma spectroscopy - North Holland Publ. Co. (1965).

(3)   R.K. Bardin, P.J. Gollon, J.D. Ullman and C.S. Wu - Phys.
      Letters 26 B, 602 (1967), also for previous references.
(4)   B. Pontecorvo - Phys. Letters 26 B, 630 (1968)
(5)   E. Fiorini, A. Pullia, G. Bertolini, F. Cappellani and G.
      Restelli - Phys. Letters 26 B, 602 (1967)
(6)   T. Kirsten, W. Gentner and O.A. Shaeffer - Z. Physic, 202,
      273 (1967) also for previous referneces.

PHENOMENOLOGICAL INVESTIGATIONS OF CHARGE SYMMETRY

BREAKING IN THE Λ-N INTERACTION

K. Hartt and E. Sullivan

University of Rhode Island

Kingston, Rhode Island

Although a charge-symmetry-breaking (CSB) component is known to occur in the Λ-nucleon (Λ-N) interaction,[1,2,3] there is not any general agreement as to the strength or spin-dependence of CSB. We are engaged in a series of calculations which attempt to delimit the strength and spin-dependence required of the CSB S-wave interaction to explain low-energy Λ-p elastic scattering cross sections ($\sigma(\Lambda p)$) at c.m. energies E < 18 Mev[4] as well as light hyperfragment binding energies.[1]

Our assumptions are: (a) Two-body, central, non-local separable (NLS) interactions of the Yamaguchi form describe Λ-N and (b) N-N interactions. This facilitates a numerical solution for the binding energy, $B_\Lambda(_\Lambda H^3)$. (c) There is a common intrinsic range, b, for all terms in the Λ-N interaction. Such a simplifying assumption is common in phenomenological analyses,[5] and to some extent might parallel the present degree of refinement of experimental knowledge. (d) Information about the mirror hypernuclei $_\Lambda He^4$ and $_\Lambda H^4$ is provided by a dimensionless parameter, Δ, representing the relative strength of the perturbation in the Λ-N potential causing the splitting of these 4-baryon binding energies.

Let λ_{npp} = spin-averaged Λ-N interaction strength in $_\Lambda He^4$, and let λ_{nnp} = spin-averaged Λ-N interaction strength in $_\Lambda H^4$. Then we define

$$\Delta = (\lambda_{npp} - \lambda_{nnp})/(\lambda_{npp} + \lambda_{nnp}).$$

Potentials obtained by Herndon and Tang[5] yield values of Δ in the range .0091 to .014. Using the values .01 and .015 for Δ, and making no assumptions concerning the CSB spin-dependence,[6] we obtain the following results:

(i) Good fits to $\sigma(\Lambda p)$, in the range $\chi^2 \leqslant 3.2$, are found for 131 Λ-p potentials. The best fits occur at b = 2.1 F and 1.8 F.

(ii) Requiring consistency of the Λ-N interaction with B$_\Lambda$($_\Lambda H^3$), Δ, and $\sigma(\Lambda p)$ forces a rejection of more than two-thirds of our Λ-p potentials, and yields a strong, completely attractive Λ-p CSB interaction ($\sim$15-39% of the total Λ-p singlet and $\sim$6-23% of the total Λ-p triplet S-wave interaction). Our best χ^2 fits satisfying consistency occur at b = 1.8 F. For a number of our Λ-N potentials, the CSB interaction appears to lie within estimated requirements for the existence of a particle-stable $_\Lambda Li^6$ state.[7] There is still controversy over whether such a state has actually been observed.[7,8] Our present results provide independent evidence for such a large CSB interaction, which apparently cannot be explained on the basis of electromagnetic mechanisms.[7]

(iii) As the spin of $_\Lambda He^4$ (which we label I) is experimentally undetermined, the values I = 0 and I = 1 are separately considered throughout these investigations. A spin-difference (singlet minus triplet) CSB interaction strength is obtained which is larger for I = 1 than for I = 0, by as much as 33% for a Λ-N intrinsic range of 1.8 F.

(iv) CSB creates isospin mixing in $_\Lambda H^3$ whose effects we determine precisely in our solutions of the three-body problem. Allowing for an admixture of a T = 1 state to the dominant T = 0 state causes the CSB spin-difference interaction to increase by 8%. All other CS and CSB Λ-N interaction parameters are also modified, but to a smaller extent.

(v) Our main results are insensitive to Δ. Even the choice Δ = 0 leads to quantitatively similar predictions for the Λ-N interaction. Our spin-independent CSB term (α) makes large numerical cancellations with the $\sigma \cdot \sigma$ term in its contribution to the 4-baryon splitting parameter, Δ. Thus the 4-baryon splitting is seen to involve only a small fraction of the full CSB potential.

Calculations are in progress which employ short-range repulsions and which treat the 4-baryon systems more completely. Allowing a longer intrinsic range for the CSB potential might reflect the role of π^o-exchange, and is also under investigation.

REFERENCES

1) G. Bohm et al., Nucl Phys. B4, 511 (1968).
2) R.H. Dalitz and F. von Hippel, Phys. Letters 10, 153 (1964).
3) B.W. Downs, Nuovo Cimento 43A, 454 (1966).
4) G. Alexander et al., Phys. Rev. 173, 1452 (1968); B. Sechi-Zorn et al., Phys. Rev. 175, 1735 (1968).
5) R.C. Herndon and Y.C. Tang, Phys. Rev. 159, 853 (1966).
6) We employ a general spin-dependence of the form ($\sigma(\Lambda) \cdot \sigma(N) + \alpha$). Ref. 5 specializes to the case where $\alpha = 0$, necessarily excluding the possibility that $T = 1$ for $_\Lambda He^4$.
7) L. Lovitch, S. Rosati and R. Dalitz, Nuovo Cimento 53A, 301, 1060 (1968).
8) D.-M. Harmsen, Phys. Rev. Letters 19, 1186 (1967); W.A. Barletta and D.-M. Harmsen, to be published.

GAUGE TRANSFORMATIONS AND GENERALISED

MULTIPOLE MOMENT OPERATORS

J.Leite Lopes and Mário Novello

Centro Brasileiro de Pesquisas Físicas

Rio de Janeiro, Brasil

In this note we show that the unitary implementation in Hilbert space corresponding to a space-time dependent gauge transformation:

$$\psi'(x) = e^{i\Lambda(x)} \psi(x)$$

is of the form:

$$e^{-iJ} \psi(x) e^{iJ} = e^{i\Lambda(x)} \psi(x) \tag{1}$$

The following assumptions and definitions are made:
1. The function $\Lambda(x)$ is a test function in the sense of L.Schwartz; 2. A vector current $j^\mu(x)$ is defined; 3. The operator J is taken to be a linear functional of Λ as follows:

$$J\left[x^0; \Lambda\right] = \int \Lambda(x) j^0(x) d^3x \tag{2}$$

4. $\Lambda(x)$ vanishes outside of a small set around the origin in three-dimensional space:

$$\Lambda(\underline{x}, x^0) = 0 \qquad \text{for} \quad |\underline{x}| > a \tag{3}$$

so that $\Lambda(x)$ can be developed around the origin:

$$\Lambda(\underline{x}, x^0) = \Lambda(0, x^0) + x^k \Lambda_{,k}(0, x^0) +$$

$$+ \frac{x^k x^\ell}{2!} \Lambda_{,k,\ell}(0, x^0) + \ldots \tag{4}$$

If $j^0(x)$ makes the integral (2) such that in the exponential in equation (1) we may neglect terms of higher order, we shall have for an infinitesimal transformation:

$$\{I-iJ\}\psi(x)\{I+iJ\} = \{i+i\Lambda(x)\}\ \psi(x)$$

The series (4) in this equation will lead to the following commutation rules:

$$[\psi(x),Q] = \psi(x); [\psi(x),Q^k] =$$

$$= x^k\psi(x); [\psi(x),Q^{k\ell}] = x^kx^\ell\psi(x), \quad \text{etc.}$$

where:

$$Q = \int j^0(x)d^3x; \quad Q^k = \int x^kj^0(x)d^3x; \quad Q^{k\ell} =$$

$$= \int x^kx^\ell j^0(x)d^3x, \text{ etc.}$$

are the charge and multipole moment operators of the field.

This may be generalized to a set of non-hermitian fields $\psi_\alpha(x)$ where the label α refers to space-time components of tensor or spinor character as well as to internal symmetry variables.

The infinitesimal gauge transformation

$$\psi'_\alpha(x) = \{\delta_{\alpha\lambda} + iC_{\alpha\beta\lambda}\ \Lambda_\beta(x)\}\ \psi_\lambda(x)$$

will induce the following unitary transformation in Hilbert space:

$$[I-i\int\Lambda_\beta(x)j^0_\beta(x)d^3x]\psi_\alpha(x)\ [I+i\int\Lambda_\beta(x)j^0_\beta(x)d^3x] =$$

$$= \{\delta_{\alpha\lambda} + iC_{\alpha\beta\lambda}\ \Lambda_\beta(x)\}\ \psi_\lambda(x)$$

summation being understood over repeated indices.

One then deduces the commutation rules:

$$[\psi_\alpha(x),\ Q_\beta] = C_{\alpha\beta\lambda}\ \psi_\lambda(x);$$

$$[\psi_\alpha(x),\ Q^k_\beta] = C_{\alpha\beta\lambda}\ x^k\ \psi_\lambda(x);$$

$$[\psi_\alpha(x),\ Q^{k\ell}_\beta] = C_{\alpha\beta\lambda}\ x^kx^\ell\ \psi_\lambda(x); \text{ etc.}$$

where the Q operators are the generalized multipole moment operators. These may be associated to systems with electric charge but also to systems with isospin, baryon, lepton currents etc. Thus a positive pion which is formed, according to the quark model, of a p-quark and an anti-n_o-quark could have a baryonic dipole moment although its total baryon number vanishes.

If the generalised charges form a Lie Algebra as a result of the commutation rules:

$$[j_\alpha^0(x), j_\beta^0(y)]_{x^0=y^0} = if_{\alpha\beta\lambda}j_\lambda^0(x)\ \delta(\underline{x}-\underline{y})$$

then the set of multipole moment operators is closed under the commutation rule:

$$[Q_\alpha, Q_\beta] = if_{\alpha\beta\lambda}\ Q_\lambda;$$

$$[Q_\alpha, Q_\beta^{k\ell\ldots n}] = if_{\alpha\beta\lambda}\ Q_\lambda^{k\ell\ldots n};$$

$$[Q_\alpha^k, Q_\beta^\ell] = if_{\alpha\beta\lambda}\ Q_\lambda^{k\ell};$$

$$[Q_\alpha^k, Q_\beta^{\ell\ldots n}] = if_{\alpha\beta\lambda}\ Q_\lambda^{k\ell\ldots n};\ \text{etc.}$$

THE OPTICAL PROPERTIES OF NUCLEI AT VERY SHORT WAVELENGTHS

Kurt Gottfried

Cornell University

If you will allow me to use some rather archaic language, I would describe the topic of this conference as the passage of highly energetic radiation through the atomic nucleus. Surely electromagnetic radiation is the most familiar of these, and it is therefore fitting that we examine what is now known about the optical properties of nuclei in the regime of very short wavelengths. According to current theoretical ideas these properties are actually quite strange and unexpected. Several elegant experiments have been completed in recent weeks to test these startling predictions of the theory. Here I shall sketch this theory and describe the present experimental situation.

PHOTON INTERACTIONS WITH INDIVIDUAL NUCLEONS

Before I go to the question of nuclear optics, I shall briefly summarize certain salient aspects of what is now known about photon interactions with individual nucleons.

The total cross section on protons, $\sigma_{\gamma p}$, has just been measured in tour-de-force experiments at SLAC[1] and DESY[2]. The dependence of $\sigma_{\gamma p}$ on the photon energy k is as follows:

k (GeV)	5.1-6.3	7.4-9.4	12-15	14.4-18.3
$\sigma_{\gamma p}$ (μb)	121±4	120.2±2.2	112.6±2.5	113.5±2.3

The DESY experiment has also measured the total cross section on deuterium, and from this deduces that $\sigma_{\gamma n} \approx \sigma_{\gamma p}$.

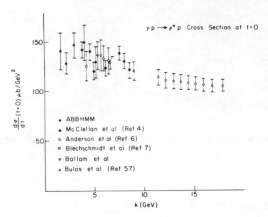

Fig. 1. $d\sigma_{\gamma\rho}/dt$ at t=o on hydrogen.

From these results we infer that the photon's mean-free-path in nuclear matter, ℓ_γ, is about 700 fermis (F) in the multi-GeV region.   Please bear this in mind.

The other important fact is that photons copiously produce vector mesons (V), especially ρ. If one writes the photoproduction differential cross section as $d\sigma_{\gamma V}/dt = a_V \exp(b_V t)$, one finds that a is almost independent of energy, and that the slope b has roughly the value characteristic of elastic processes such as πN and NN. The data on a_ρ is compiled in Fig. 1: for[3] 8 < k < 9 GeV, $a\phi \approx 5\mu b/GeV^2$. The slopes in 7-9 GeV region are[3,4] $b_\rho \approx 8.5$ and $b_\phi \approx 5.2$ GeV^{-2}.

The spin and iso-spin dependence of the V-production amplitudes $f_{\gamma V}$ can be partially unravelled by measuring the production from deuterium, by using polarized photon beams, and by studying the decay distribution of the V's. From experiments of this type[3,4] we have learned that $f_{\gamma V}$ is only weakly dependent on spin and isospin. The overall picture that emerges is then that V-photo-production at small angles has all the aspects of high energy elastic scattering--it is diffractive. This is not entirely sur-prising as no change in quantum numbers is involved in the transition $\gamma \to V^0$. The lack of energy and iso-spin dependence of the total γN cross section is also reminiscent of hadronic diffraction scattering.

COHERENT PHOTOPRODUCTION OF VECTOR MESONS

The transformation $\gamma \to V$ can take place when a photon impinges on a nucleus while that nucleus remains in its ground state. The production amplitude is therefore coherent over the nucleus, and would provide an A^2 dependence of the cross section were it not for two effects: (1) The mass change $\gamma \rightarrow V$ implies a momentum transfer $\Delta_V = m_V^2/2k$ even in forward scattering, and therefore brings with it a form factor which decreases rapidly if $R\Delta_V \gg 1$; (2) V-mesons are absorbed in nuclear matter, and the cross section would therefore grow less rapidly than A^2 even if Δ_V were zero. By varying V one can measure the V's attenuation, and thereby the total VN cross section σ_{VN}.

An optical model that accounts for these features was first formulated by Drell and Trefil. I shall summarize briefly a somewhat more refined treatment[5] that also incorporates effects due to the finite lifetime of the produced particles. These effects are, in part, akin to collision broadening, and become increasingly noticeable as k decreases below 5 GeV. The $\pi^+\pi^-$ production amplitude is

$$F = 4\pi \, \gamma_{\rho\pi\pi} \, f_{\gamma\rho} \, \hat{\varepsilon} \cdot (\vec{q}_1 - \vec{q}_2) \int dz \, d^2b \, G(\vec{pm};bz) n(b,z) e^{ikz} \qquad (1)$$

Here $\gamma_{\rho\pi\pi}$ is the $\rho\pi\pi$ coupling constant, $\hat{\varepsilon}$ the photon's polarization vector and n the nuclear density. G is the probability amplitude for finding two pions of total momentum $\vec{p} = \vec{q}_1 + \vec{q}_2$ and invariant mass m if a ρ is produced at (b,z); it satisfies the equation

$$\left[\nabla^2 + k^2 - m^2 + i\Gamma_\rho m_\rho + i p\sigma_{\rho N}(1+i\alpha_{\rho N})n(r)\right] G(\vec{p};\vec{r}) = \Phi_1^*(\vec{r})\Phi_2^*(\vec{r}) \qquad (2)$$

where $\alpha_{\rho N} = \text{Re } f_{\rho N}/\text{Im } f_{\rho N}$, and Φ_i are the pion wave functions including final state interactions. This equation therefore describes the collision broadening effects, when the ρ is inside the nucleus, and the familiar vacuum propagation when it is outside; it also incorporates the absorption of the pions in case the ρ decays inside the nucleus. For the extremely small production angles of interest to us, (2) can be integrated analytically, and F is thereby reduced to quadratures. The variation of the minimum momentum transfer with $\pi\pi$-mass is automatically incorporated in this method of calculation.

The finite ρ-width corrections incorporated in (1) and (2) shift the $\pi\pi$-mass peak towards lower mass, but this effect is no longer significant at the high energies used in the measurements of $\sigma_{\rho N}$ that I shall discuss in a moment. Further details concerning these corrections can be found in ref. 5.

There are several effects which are neglected in (1): (a) there are corrections to the optical model due to ground state correlations; (b) there is a background of "incoherent" reactions involving nuclear excitation, which does not vanish in the diffraction peak; (c) the optical model incorporates intrinsic inaccuracies of order $1/A$.

The pair correlation corrections have recently been considered by several groups[6], and by common agreement the conclusion is that the true $\sigma_{\rho N}$ differs from that appearing in (2) by approximately -2 mb if $\sigma_{\rho N}$ is in the range of 25-40 mb. For various reasons this correction is itself probably uncertain by ±1 mb. There is no doubt, however, that higher order corrections due to triplet correlations are entirely negligible. The incoherent background has been estimated by various authors[7], while the 1/A troubles can be avoided by using Glauber's detailed multiple scattering theory. Nevertheless, it is safer to avoid these necessarily more complicated theories: The most reliable approach to the determination of $\sigma_{\rho N}$ is undoubtedly to delete nuclei in the range $2 < A \lesssim 25$ from the analysis.

Despite all this fine theoretical work, we are at present in an unpleasant quagmire which has already been described here by Prof. Leith: The DESY group finds $\sigma_{\rho N}$ = 25±2 mb, Cornell[8] finds $\sigma_{\rho N}$ = 38±4 mb, while SLAC obtains $\sigma_{\rho N}$ = 37±4 mb if it excludes all light nuclei, and ≈45 mb if hydrogen is included in the analysis.

Insofar as ϕ-production is concerned, there have been measurements at DESY and Cornell[9]. The Cornell data yields $\sigma_{\phi N}$ = 20±3 mb, that from DESY, $\sigma_{\phi N}$ = 12±4 mb.

Some feeling for the sensitivity of these measurements can be attained by examining Fig. 2, which is based on the Cornell data, and from Fig. 3.

To preclude any misconceptions I should like to stress most emphatically that nuclear physics cannot be blamed for this mess. As I have indicated, the corrections to the optical model used here are small and in good control. Furthermore, if the nuclear theory is wrong it cannot resolve the dilemma because all groups use essentially the same theoretical model. The discrepancy must stem

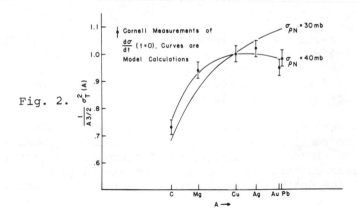

Fig. 2.

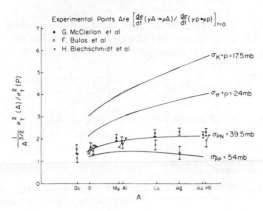

Fig. 3.   The experimental points are from Cornell (McClellan et
          al.) and SLAC (Bulos et al.). For comparison the ratios
          $\sigma_T^2(A)/\sigma_T^2(H)$ for p, $\bar{p}$, π^+ and K^+ are also shown, to-
          gether with the known cross sections on hydrogen. (These
          ratios equal the forward differential cross section ratios
          if the amplitudes are pure imaginary.) From this plot
          one sees that the Cornell data implies $\sigma_{\rho N} \simeq \sigma_{pN}$, while
          the SLAC data gives a somewhat larger value of $\sigma_{\rho N}$.
          Furthermore, it is clear that they are inconsistent
          with $\sigma_{\pi N}$, which approximately equals $\sigma_{\rho N}$ as measured
          at DESY.

from the experiments themselves, or possibly from ambiguities such
as background subtractions that are present even when hydrogen is
the target. In fact, large nuclear targets are superior to hydrogen
in many respects because they produce a coherent peak that singles
out the 1$^-$ channel of primary interest to us.

                          VECTOR MESON DOMINANCE

     I now come to the heart of this review. At this time the most
popular and attractive idea for unifying the electromagnetic properties of hadrons is vector meson dominance (VMD). According to this
notion the only hadronic degrees of freedom that are coupled to the
electromagnetic field are associated with the known neutral vector
mesons ρ^0, ω and ϕ.

     There are two ways of stating this notion with some precision.
Consider a hadronic transition a$\rightarrow$b induced by a photon of momentum
k. One formulation of VMD states that the amplitudes satisfy
unsubtracted dispersion relations in k^2 which are saturated by the
contributions from ρ, ω and ϕ when $0 \lesssim |k^2| \lesssim m_V^2$. Another statement of
VMD is field-theoretic. In the case of the isovector current j_μ^V it

would claim that: (a) the hadronic amplitude M for the process $\rho_\lambda a \rightarrow b$, where ρ_λ is a ρ with helicity λ, is related to the electromagnetic transition amplitude by

$$\langle b | j_\mu^V | a \rangle = \frac{e m_\rho^2}{2\gamma_\rho} \Delta_{\mu\nu}(k^2) \epsilon_\lambda^\nu M_{\rho_\lambda a \rightarrow b}(k^2), \tag{3}$$

where $\Delta_{\mu\nu}$ is the ρ's propagator; and (b) that $M(k^2)$ varies slowly when $0 \lesssim |k^2| \lesssim m_\rho^2$. Obviously assumption (a) is equivalent to that of an unsubtracted dispersion relation, while (b) amounts to its saturation by the ρ-pole.

The most direct determination of the coupling constants is afforded by the leptonic decay as measured in storage rings. These yield the values $\gamma_\rho^2/4\pi = 0.52 \pm 0.03$, $\gamma_\omega^2/4\pi = 3.70 \pm 0.65$, and $\gamma_\phi^2/4\pi = 2.75 \pm 0.40$. Note that these determinations are at $k^2 = m_V^2$.

If VMD is valid, it provides several powerful relations between various amplitudes. Let f_{ab} be the amplitude for the transition $a \rightarrow b$ on a single nucleon. The photoproduction amplitude for a particle d can then be expressed as

$$f_{\gamma d} = \sum_V C_V f_{Vd}, \tag{4}$$

while for Compton scattering

$$f_{\gamma\gamma} = \sum_{VV'} C_V C_{V'} f_{VV'}. \tag{5}$$

Henceforth we shall only use (4) and (5) for light-like photons ($k^2 = 0$). A strict interpretation of VMD would require $C_V = e/2\gamma_V$, with γ_V being given by the quoted measurements. Indeed, an unthinking application of VMD would require (4) and (5) to hold if the target where a complex nucleus and not just a nucleon. However, it is intuitively clear that VMD cannot be applied blindly: we must be well above threshold for V-production, and the mass change involved in the transformation $\gamma \rightarrow V$ must be small so that "form factor" effects are negligible. In the case of ρ's and nucleons the latter effects are already insignificant when $k \gtrsim 1$ GeV, while in the nuclear case these effects can be calculated with considerable confidence, as we already saw in the problem of coherent V-production.

There is both indirect and direct evidence that one must be cautious in identifying the C_V's in (4) and (5) with the constants $e/2\gamma_V$ as measured at $k^2 = m_V^2$. As very indirect evidence I remind you that in the analogous case of pion-dominance of the weak axial current (PCAC) there is an error of $\sim 20\%$ in extrapolating over the tiny k^2 interval $m_\rho^2/30$. Another indirect piece of evidence is that the nucleon form factor agrees with the famous dipole formula for surprisingly small values of k^2. But far more direct evidence is available: the amplitudes for $\gamma_\perp N \to \pi N$ and $\pi N \to \rho_\perp N$, where $\perp$ indicates polarization perpendicular to the reaction plane, are related to each other by a factor which disagrees by 30-60% from the value expected from VMD.[9] For polarization in the production plane there are even greater discrepancies, but there is no Lorentz-invariant definition of what one is talking about, and this opening has resulted in a host of unlikely explanations based on various prejudices as to what is an esthetic coordinate frame. Schmidt[10] has recently used fixed-t dispersion relations to show that significant k^2-variation of $M_{\rho N \to \pi N}(k^2)$ is probably responsible for these discrepancies. His model does not contradict the more basic assumption of VMD, eqn. 3, but it does show that in processes where one-particle exchange diagrams are significant the hadronic amplitude is sensitive to k^2.

In view of these observations it would be wise to adopt a cautious attitude towards VMD. In particular, the relationships between amplitudes, eqns. 4 and 5, are probably far more reliable for small-angle diffractive processes where Born terms are insignificant, and the detailed kinematics of the reaction are (hopefully) irrelevant, than they are for reactions where quantum numbers are exchanged. Even in these diffractive processes we should not be surprised if the C_V's differ from $e/2\gamma_V$; the essential point is that the same C_V's should appear in forward V-production and Compton scattering. If the experiments show that this last statement is also incorrect, VMD will have become a rather sterile concept.

The V-production amplitude is proportional to $f_{\gamma V}$ (see eqn. 1), and this means that the absolute value of the coherent V-production cross section provides a measure of C_V. Here the disagreement between SLAC-Cornell and DESY plagues us again. The values of $\gamma_\rho^2/4\pi$ found by these laboratories are ~ 1.1, ~ 1.1, and ~ 0.5, respectively. Cornell[9] has also measured the γ-ϕ coupling constant, and finds $\gamma_\phi^2/4\pi = 8.5 \pm 0.3$, which is roughly three times larger than that obtained in the storage ring experiment.

NUCLEAR OPTICS AND TESTS OF VECTOR MESON DOMINANCE

The groundwork having been laid, I can now turn to the topic mentioned in the title, nuclear optics.

Coherent Photoproduction and Babinet's Principle[12]

At very high energies, where the minimum momentum transfer Δ_V becomes negligible in comparison to $1/R$, the proportionality relation (4) states that the angular distribution in elastic V-scattering and in V-photoproduction are identical. At first sight this is astonishing, because one thinks of the nucleus as being rather opaque to V's, and transparent to photons.

A similar phenomenon actually occurs in the diffraction of light, where it goes under the name of Babinet's Principle. Let ψ^A be the wave function that describes diffraction by an aperture A in an opaque screen S, and ψ^S that for diffraction by the complimentary arrangement where A is opaque and S is removed. These functions are related by

$$\psi^S + \psi^A = \psi_{inc}. \tag{6}$$

If ψ_{inc} is a plane wave incident normally, it does not contribute to (6) when A is viewed along a direction off the normal. For such an observer $|\psi^A|^2 = |\psi^S|^2$: the diffraction pattern of the aperture and complimentary screen are identical.

In photoproduction the mesons all emanate from the nucleus, which therefore plays the role of a small aperture in an infinite opaque screen. But in elastic V-scattering only those V's that "miss" the nucleus are observed, and here the nucleus acts as the complementary opaque screen. The equation that replaces (6) in our problem is

$$\psi_{\gamma \to V} - (C_V)^{-1} \psi_{V \to V} = \psi_{inc} + 0(\Delta_V R). \tag{7}$$

At high energy the last term is negligible, and $|\psi_{\gamma \to V}|^2 \propto |\psi_{V \to V}|^2$ off the exact forward direction.

It should be noted that VMD was **not** involved in the foregoing discussion. Only the one-photon amplitudes are involved here--only eqn. 4. Furthermore, in a large nucleus only the $t \approx 0$ part of the one-nucleon amplitudes enter, and so we have only relied on (4) at one value of the kinematic variables. The proportionality assumption (4) is therefore empty in this particular case; all that it does is to fix the relative magnitude of the diffraction patterns in the two reactions $\gamma \to V$ and $V \to V$.

Transmission of Photons Through Nuclear Matter

I now turn to a problem where VMD plays an essential role. Consider forward photon scattering off a complex nucleus at energies so high that all minimum momentum transfers for production of the known vector mesons are negligible. If $\sigma_\gamma(A)$ and $\sigma_V(A)$ are the total nuclear cross sections for photons and V's, we can use (5) to obtain[*]

---

[*]Here I have ignored $V \to V'$ reactions in using (5). These processes are believed to be unimportant.[11] It is actually not difficult to include them but I shall not do so because none of the qualitative features I shall describe here are affected significantly thereby.[12]

$$\sigma_\gamma(A) = \sum_V c_V^2 \sigma_V(A). \tag{8}$$

All the evidence points to the ρ as the dominant contributor to (8)--at least for masses below 2 GeV. But ρ has a nuclear mean-free-path, ℓ_ρ, of order 3F, and a ρ-wave is therefore depleted rapidly as it traverses the nucleus. To a first approximation the total cross section $\sigma_\rho(A)$ for a heavy nucleus will therefore be given by the cross section of an opaque disc. More precisely,

$$\sigma_V(A) = 2\pi R^2 \, G(\infty, \ell_V/R), \tag{9}$$

where the function G will be evaluated in a moment. At this point its only property of concern to us is that for $\ell_V/R \ll 1$, $G \simeq 1 - 2(\ell_V/R)$. The VDM prediction for the high energy limit of total photoproduction cross sections on heavy nuclei is therefore

$$\sigma_\gamma(A) = 2\pi R^2 \sum_V c_V^2 \left(1 - \frac{2\ell_V}{R}\right). \tag{10}$$

Let me now remind you of a fact that I stated near the outset: The measured total photo-absorption cross section on isolated nucleons implies that the photon's mean-free-path in nuclear matter, ℓ_γ, is of order 700 F. Taken at face value, this would mean that every nucleon in the nucleus is exposed to a virtually undiminished incident wave, and should therefore contribute its full measure, $\sigma_{\gamma p}$ or $\sigma_{\gamma n}$, to the total cross section.[13] Common sense therefore seems to dictate

$$\sigma_\gamma(A) = A\sigma_{\gamma N} . \tag{11}$$

Having gone through the elementary--and convincing--argument leading to (11) really underscores the paradoxical nature of the VMD result, eqn. 10. Indeed, we are now faced with several questions:

1. How can one reconcile the immense mean-free-path ℓ_γ with the shadow-like cross section (10)?

2. Assuming that an explanation of the first question exists, how does the total cross section $\sigma_\gamma(A)$ behave as the energy is raised from threshold for V-production (where (8) does not hold) towards the high-energy regime where (10) is applicable?

3. How is this presumed energy dependence of $\sigma_\gamma(A)$ reflected in the cross section for specific inelastic processes?

The first question was asked and answered by Stodolsky several years ago.[14] The other questions have been examined by several groups during the past year.[12,15-18] This work has

stimulated several experiments which I shall discuss after I sketch
the theoretical developments.

The essential clue towards a reconcilation of a huge ℓ_γ with
the formation of a shadow $(\sigma_\gamma(A) \propto R^2)$ is that we are dealing with a
coupled channel problem: As we have seen, not only $\gamma \to \gamma$, but also
$\gamma \to V$ is coherent at high energy. Hence a photon can be scattered
into the forward direction in two distinct ways: (1) by Compton
scattering from a nucleon, or (2) by the two-step process $\gamma \to V \to \gamma$
(see Fig. 4). As you know from many examples in all branches of

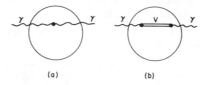

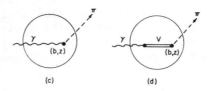

Fig. 4.   (a) Compton scattering a nucleon.   (b) Coherent
          V-production, followed by V-propagation, and the
          inverse process V→γ.   (c) Direct pion-photoproduction
          from a nucleon at (b,z).   (d) π-production from a
          nucleon at (b,z) by a V that was photoproduced
          coherently.

physics, the propagation constants of an eigenmode in a coupled-
channel system can differ markedly from those that characterize
the individual modes. This being said, it is clear that the mere
existence of coherent V-production--which has nothing to do with
VMD--must lead to a modification in the way that light propagates
through nuclear matter. This modification is particularly drastic
if VMD is correct, and as $k \to \infty$ it leads to the paradoxical result
(10).

The problem can be analyzed quantitatively by introducing a
set of optical potentials $U_{\gamma\gamma}$, $U_{\gamma V}$, and U_{VV}, which describe the
indicated transformations. The coupled system of equations is[11]

$$(\nabla^2 + k^2 - U_{\gamma\gamma})\psi_\gamma = \sum_V U_{\gamma V}\psi_V \quad ,$$

$$(\nabla^2 + k^2 - m_V^2 - U_{VV})\psi_V = U_{V\gamma}\psi_\gamma \quad . \tag{12}$$

Consider the photon wave-function $\psi_\gamma(z)$ at a depth z into a slab of nuclear matter. To lowest order in α it is convenient to write $\psi_\gamma = \psi_\gamma^1 + \psi_\gamma^2$, where ψ_γ^1 and ψ_γ^2 are the portions of the amplitude corresponding to the one- and two-step processes depicted in Fig. 4 (a) and (b), respectively. Their explicit forms are

$$e^{-ikz}\psi_\gamma^1 = 1 - \frac{iz}{2k} U_{\gamma\gamma}, \tag{13}$$

$$e^{-ikz}\psi_\gamma^2 = \frac{iz}{2k}\sum_V \frac{U_{\gamma V}U_{V\gamma}}{\varepsilon_V} + \sum_V \frac{U_{\gamma V}}{\varepsilon_V}\psi_V(z) \quad , \tag{14}$$

where $\varepsilon_V = 2k_V(k - k_V + U_{VV}/2k)$, $k_V^2 = k^2 - m_V^2$. The V-meson wave function is

$$\psi_V = \frac{U_{V\gamma}}{\varepsilon_V} e^{ikz}\left(1 - e^{i\varepsilon_V z/2k}\right). \tag{15}$$

These wave functions are sketched in Fig. 5. As we see, ψ_γ^1 falls

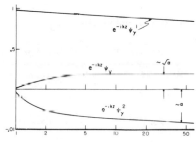

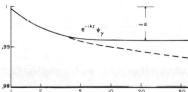

Fig. 5.   Various components of the photon wave function as a function of depth z in fermis. The solid and dashed curves show the large-k limits of ψ_γ when VMD is valid or invalid, respectively.

off linearly with z (provided z<<700 F!!). By itself it would therefore lead to the "common sense" result $\sigma_\gamma(A) \propto A$. ψ_γ^2 contains a linear term, and also a term that rises exponentially to its asymptotic value within a distance $\approx \ell_V$. The complete coefficient of the term linear in z is

$$U_{\gamma\gamma} - \sum_V \frac{U_{\gamma V} U_{V\gamma}}{\varepsilon_V} \equiv K. \tag{16}$$

The portion of ψ_γ that is proportional to z provides a contribution to $\sigma_\gamma(A)$ proportional to volume, or equivalently to A. For sufficiently large A this will always dominate because the remainder of ψ_γ, which describes the shadow cast by the absorption of the _real_ intermediate V's, gives a contribution roughly proportional to the cross sectional area of the target, or $A^{2/3}$. As $k\to\infty$, $\varepsilon_V \to U_{VV}$, because $k\to k_V$ and $U_{VV} \propto k$. Thus as $k\to\infty$,

$$K \to U_{\gamma\gamma} - \sum_V \frac{U_{\gamma V} U_{V\gamma}}{U_{VV}}. \tag{17}$$

But the VMD relations (4) and (5) say this expression vanishes, that is to say, the portion of $\sigma_\gamma(A) \propto A$ disappears as $k\to\infty$. Thus we see that the puzzling VMD result (10) occurs because of a delicate destructive interference between two waves: one which has been Compton scattered, the other having undergone the $\gamma\to V\to\gamma$ transformations. As we see from (17) this destructive interference is always present, but it is only complete if VMD holds.

    Now a few bookkeeping details. For finite k, and assuming the VMD relations (4) and (5) with Re f_{ij} = 0, we find

$$\sigma_\gamma(A) = \sum_V c_V^2 \left[A\sigma_{VN} g(\xi_V) + 2\pi R^2 G(\xi_V, \ell_V/R) \right] \tag{18}$$

where

$$g(\xi) = \xi^2 |\eta|^{-2}$$

$$G(\xi,x) = \text{Re}\left\{ \eta^{-2} - 2x\eta^{-4}\left[1 - (1 + \eta x^{-1}) \right] e^{-\eta/x} \right\} \tag{19}$$

and $x = \ell/R, \eta = 1 + i\xi$, while

$$\xi_V = \frac{m_V \ell_V}{k} = 2\Delta_V \ell_V. \tag{20}$$

ξ_V is the dimensionless parameter which tells us whether we are approaching the VMD limit ($\xi_V \ll 1$) with $\sigma_\gamma(A) \propto A^{2/3}$, or whether we are in the "common sense regime" ($\xi_V \gg 1$) where $\sigma_\gamma(A) \propto A$. Note that ξ_V involves the minimum momentum transfer and the mean-free-path ℓ_V. The transition therefore occurs at a rather high energy ($\xi_\rho \approx 1$ when $k \approx 7.5$ GeV if $\ell_\rho \approx 3$ F), far higher than if one had thought that the VMD limit is approached as soon as one is well above threshold for all the strongly produced V's. The limiting behavior of the functions in (18) is: $g\to1$ as $\xi\to\infty$, $g\to0$ as $\xi\to0$; $G\to0$ as $\xi\to\infty$, and if $\xi\to0$ and $x\to0$, G attains the limiting form already given after eqn. 9. The detailed behavior of g and G is shown in Fig. 6. In Fig. 7 the k-dependence of $\sigma_\gamma(A)$ is shown for several nuclei, assuming not just VMD, but ρMD. From what we

now know about C_ϕ, C_ω, and $f_{\gamma\phi}$, $f_{\gamma\omega}$, ρMD is almost certainly valid
if ρ, ω and ϕ are the only V's significantly coupled to photons.
As we see from Fig. 7, the effect we are considering is really
quite large, especially for heavy nuclei.

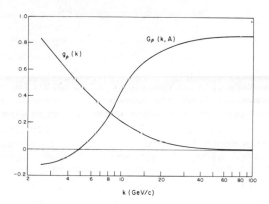

Fig. 6.   The functions g and G; $\sigma_{\rho N}$ = 40 mb, and
          G is shown for A = 208, with R = 1.15
          $A^{1/3}$F.

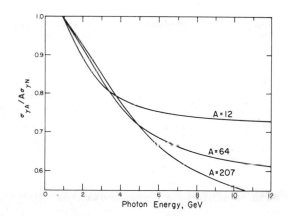

Fig. 7.   Variation of σ_γ(A) with photon energy, taken from
          ref. 16. The nuclear parameters are $\sigma_{\rho N}$ = 30 mb,
          R = 1.3 $A^{1/3}$F. Observe that the transition energy
          is not solely determined by ξ_0; it also depends
          weakly on R, growing roughly like $\sqrt{R}$.

    Brodsky and Pumplin[16] have pointed out that these results are
easily extended to inelastic electron scattering by nuclei. If
the virtual photon has a mass q^2 ($q^2<0$ for electron scattering), we

need merely replace ε_V by $\varepsilon_V - q^2$ throughout. The cross sections $\sigma_\gamma(A, q^2)$ for the absorption of virtual photons can be extracted from measured inelastic electron scattering cross section.

Inelastic Processes[12]

The total cross section is merely the sum of its parts, and therefore the cross sections for specific photo-processes (e.g. π-production) must also show a volume to surface transition. That is to say, when $\xi_V \ll 1$, the photo-produced hadrons should appear to emanate only from the surface, whereas when $\zeta_V \gtrsim 1$ their effective source distribution should be the whole nucleus. (The attenuation of the outgoing particles by nuclear processes must be taken into account throughout of course). To calculate the cross section, we consider a nucleon at (b,z). It will be struck by the incident photon wave, which gives rise to an amplitude $A^1(b,z)$ for directly producing the final state in question (see Fig. 4 (c)). But it is also struck by a variety of vector mesons produced coherently downstream by the γ; these V's then produce the final state with an amplitude $A_V^2(b,z)$. If the momentum transfer is sufficiently large (>>250 MeV/c) the contributions of the various nucleons is incoherent, and the cross section is proportional to

$$\int dz\, d^2b\; n(b,z)\; \left| A^1(b,z) + \sum_V A_V^2(b,z) \right|^2. \qquad (21)$$

Comparison with Experiment

Total cross sections $\sigma_\gamma(A)$ have just been measured at SLAC[1] and DESY[2] for $1 \leq A \leq 208$, $1.5 \leq k \leq 18$ GeV. The two experiments do not quite overlap in energy, and DESY has only reported results up to $A = 64$. Nevertheless, the measurements appear to be consistent with each other.

The most accurate and extensive measurements available are for C, and are shown in Fig. 8. A compilation of the SLAC data is shown in Fig. 9. The data unambiguously demonstrates that $\sigma_\gamma(A)$ is _not_ proportional to A--that some shadowing is taking place. It is likely that a tolerable fit can be attained with $\sigma_{\rho N} \sim 25$ mb and a moderate real part for $f_{\gamma\gamma}$ and $f_{\gamma\rho}$. If the Cornell-SLAC determination of $\sigma_{\rho N}$ is correct, however. This data is in contradiction with ρMD. Further vector mesons (beyond ϕ and ω) will have to be introduced, and if $\sigma_{\rho N}$ is as large as 38 mb, their mean mass would have to be well in excess of 2 GeV.

There are two recent studies of incoherent processes: a Cornell experiment[19] on ρ^0 production, and a SLAC measurement[20] of $\pi^\pm$ production. Both disagree strongly with ρMD if $\sigma_{\rho N} \gtrsim 30$ mb. The

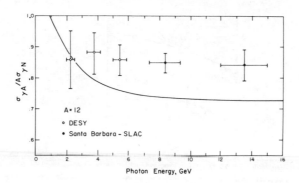

Fig. 8.   Data for C compared with theoretical prediction
          taken from Fig. 7.

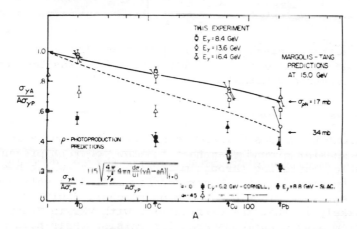

Fig. 9.   SLAC measurements of $\sigma_\gamma(A)$.   The curves are computed
          from eqn. 18, but taking a realistic density distribu-
          tion[18]. Some ρMD-predictions based on Cornell and SLAC
          measurements of ρ-photoproduction are also shown.

more violent disagreement concerns π-production, and I therefore
show some of the SLAC results in Fig. 10. No energy dependence of
the cross section is discernible. A quantitative measure of the
disagreement with theory can be obtained by inserting a factor w
into the two-step term of eqn. 21. If VMD is correct, w = 1. The
data[20] requires w = 0.32±0.12, where the error is not mainly experimental, but largely due to uncertainties in $f_{\rho N}$; the quoted error
covers the range $26 < \sigma_{\rho N} < 38$, and $|\alpha_{\rho N}| < 0.3$. Hence there is a
striking disagreement even if the DESY value for $\sigma_{\rho N}$ is correct.

There has been some criticism[21] of the theory[12], eqn. 21, for

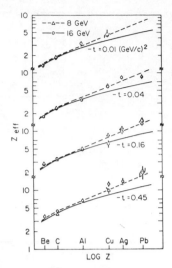

Fig. 10.   Measurements[20] of Z_{eff}, defined so that
            $d\sigma(\gamma A \rightarrow \pi^{+}A')/dt = Z_{eff}\ d\sigma(\gamma p \rightarrow \pi^{+}n)/dt$.   The theoretical
            curves are normalized to the data at C.

these incoherent processes. There is no doubt that the existing
theory is on shakier ground than that for the coherent processes,
and that double scattering through moderate angles, which has been
ignored, may be significant. Nevertheless, I find it unlikely that
the disagreement shown in Fig. 10 can be reconciled with VMD. What
seems more likely is that the difficulties with VMD in π-photo-
production from single nucleons are also raising their head here.

CONCLUSIONS

    I shall confine myself to several essential concluding
remarks.

1. The theory of coherent V-production and forward Compton
   scattering by nuclei is reliable, and cannot be used as a
   scapegoat for discrepancies between different measurements
   of $\sigma_{\rho N}$ and γ_{ρ}.

2. The data on total photo cross sections is in very serious
   conflict with ρ-dominance if $\sigma_{\rho N} \gtrsim 30$ mb.

3. Incoherent π-production appears to disagree strongly with VMD
   even if $\sigma_{\rho N}$ has the smallest value compatible with existing
   measurements, i.e. $\sigma_{\rho N} = 23$ mb.

4. It is quite conceivable that as a result of these (and other) experiments we shall have to adopt a more conservative view of VMD, to wit, that it only applies to small angle diffractive processes, and that important mass extrapolation effects are present in other types of reactions.

References

1. D. O. Caldwell et al., Univ. of Calif. at Santa Barbara preprint, unpublished. See the contribution by R. J. Morrison to this Conference.

2. H. Meyer et al., Contributions to Int. Symp. Electron and Photon Interactions, Liverpool, September 1969.

3. G. McClellan et al., Cornell preprint CLNS-69.

4. G. McClellan et al., Phys. Rev. Letters $\underline{22}$, 374 (1969); G. Diambrini-Palazzi et al., Cornell preprint CLNS-83; F. Bulos et al., Contribution to High Energy Physics Conference, Boulder, Colorado, August 1969; K. Gottfried, A. Lebedev and D. Julius, to be published.

5. K. Gottfried and D. I. Julius, Cornell preprint CLNS-66.

6. G. von Bochmann, B. Margolis, and C. L. Tang, DESY preprint; E. J. Moniz, G. D. Nixon, and J. D. Walecka, contribution to this conference; and also ref. 5.

7   K. S. Kölbig and B. Margolis, Nuc. Phys. $\underline{B6}$, 85 (1968); J. S. Trefil, University of Illinois preprint.

8. J. Swartz and R. Talman (Cornell preprint CLNS-79) have found a good fit to the Cornell data with $\sigma_{\rho N} = 27$ mb and $\sigma_{\rho N} = -0.45$. According to D. Leith's report to this Conference, however, such a large real part is probably inconsistent with the energy dependence of the ρ-cross section as measured very recently at SLAC.

9. S. C. C. Ting, Proc. 14th Int. Conf. High Energy Physics, Vienna, Sept. 1968; G. McClellan et al., Cornell preprint CLNS-70. The Cornell data is also consistent with $\sigma_{\phi N} = 12$ mb, $\alpha_{\phi N} = -0.35$; in this case $\gamma_{\phi}^2/4\pi \simeq 3.4$, in tolerable agreement with the storage ring result.

10. W. Schmidt, Cornell preprint CLNS-72.

11. When applying (4) and (5) we shall usually delete the terms
     $f_{VV'}$. In the case of $f_{\rho\phi}$ and $f_{\rho\omega}$, this is justified on general
     grounds because these reactions require $\Delta I = 1$ in the t-channel,
     are therefore non-diffractive, and small. The amplitude $f_{\omega\phi}$
     requires the exchange of the $I = 0$ member of an octet; it is
     therefore "less" diffractive than elastic scattering, but
     "more so" than, say, $f_{\rho\phi}$. (See M. Ross and L. Stodolsky. Phys.
     Rev. Letters 17, 563 (1966)). The data at present[9] of $\phi-\omega$
     transformation.
12. K. Gottfried and D.R. Yennie, Phys. Rev. 182, 1595 (1969).
13. This argument also requires that the wavelength be short com-
     pared to the inter-nucleon spacing, a condition which is
     admirably satisfied for all energies of concern to us.
14. L. Stodolsky, Phys. Rev. Letters 18, 135 (1967). The conjecture
     that a weak process could provide a cross section $\propto R^2$ first
     arose in the far more startling case of neutrino scattering by
     nuclei: J.S. Bell, Phys. Rev. Letters 13, 57(1964). Here it
     is pion dominance of the axial current (PCAC) that plays the
     role of VMD.
15. J.S. Bell, CERN Report TH 877 (unpublished).
16. S.J. Brodsky and J. Pumplin, Phys. Rev. 182, 1595 (1969).
17. M. Nauenberg, Phys. Rev. Letters 22, 556 (1969).
18. B. Margolis and C.L. Tang, Nuc. Phys. B10, 329 (1969).
19. G.M. McCellan et al., Phys. Rev. Letters 23, 554 (1969).
20. A.M. Boyarski et al., Contribution to Int. Symp. on Electron
     and Photon Interactions, Liverpool, Sept. 1969.
21. B. Margolis, private communication.

DISCUSSION

A. Goldhaber: In the case of pion photo-production there are, presumably, data on hydrogen at the appropriate energies. Can one make
phenomenal predictions for complex nuclei using such data?
K.G.: Yes, there has been some preliminary work of this sort by
Yennie and by Schmidt. There are difficulties in using vector-meson
dominance with π - photo-production. Schmidt used a very conservative
model - essentially a dispersion-theory model where the Born terms
are calculated using V.M. dominance but also including empirical
information about the 3-3 resonance. With this one can calculate
the k^2 dependence of the process - with remarkably good results;
removing much of the discrepancy between the photo-production
process, and the process of ρ-production by pions. This theory
gives also predictions about the polarization structure of the
process.

POSSIBILITIES OF USING NUCLEAR TARGETS IN INVESTIGATIONS OF

HADRONIC INTERACTIONS AT HIGH ENERGIES

W. Czyż

Institute of Nuclear Physics, Kraków, Bronowice

Poland

1. Introduction

From the extensive literature on high energy hadron-nucleus scattering we can conclude that we do understand these processes, provided we limit ourselves to the small angle, purely elastic scattering and to the small angle inelastic scattering - in which only target nuclei are excited and no new particles are produced (one may call it almost elastic scattering). Our understanding is, however, rather crude; there are large uncertainties in the experimental data and we can explain the data using some quite crude models for the incident hadron-nucleon amplitudes and for the ground state wave functions of the target nucleus.

It was already stressed (see e.g. [1]) that if one uses the simplest possible target nuclei, one may hope to learn something about the hadron-nucleon amplitude. By "simplest possible" we mean: nuclei with spin zero, spherical, with the first excited state as high as possible and, if possible, with the isobaric spin zero. The examples of such targets are He^4 and O^{16} nuclei. The Pb^{208} nucleus is also an attractive possibility in spite of its isobaric spin different from zero. We shall discuss this last case in more detail later.

We are limiting ourselves to such "simplest possible" targets because we want to discuss the processes with <u>only one</u> amplitude. For instance, if a beam of high energy pions scatters elastically from He^4, O^{16} or Pb^{208} targets, only one (complex) amplitude describes the process. The fact that these targets are spherical further simplifies the analysis (e.g. C^{12} nucleus has spin zero but is probably deformed in its ground state - which

fact introduces very large uncertainties in the analysis - the
deformation is poorly known - and, in addition, makes the calculation extremely complicated[2]. If the scattering angles are
small, even a spin 1/2 incident hadron-nucleus amplitude is virtually without its spin-flip part (which goes to zero with angle
like $\sin\theta$). This is one of the reasons why the Pb[208] target is an
interesting possibility: because of its large radius the diffractive structure appears at small angles (of the order of a few
miliradians at $\sim$ 20 GeV incident energy). Hence even proton -
Pb[208] scattering can be described with good accuracy, by only one
amplitude.

2. Elastic scattering

We shall discuss here elastic scattering of a spin zero projectile from a spin zero nuclear target and illustrate it mostly
on the case of Pb[208] target.

First, let us treat the target nucleus as a spin zero, charged particle, without going into its structure. We shall assume
that, in the high energy limit, the total phase shift is the sum
of the Coulomb phase shift (χ_c - the phase shift we would get if
the strong interactions were switched off) and the strong interaction phase shift (χ_s - the phase shift we would get if the
Coulomb interaction were switched off). In the impact parameter
representation we have:

$$\mathcal{M} = ik \int_0^\infty db\, b\, J_0(\Delta b)\left\{1 - e^{-i(\chi_c(b) + \chi_s(b))}\right\} \tag{1}$$

where Δ is the momentum transfer. χ_c is purely real, χ_s is not,
and we shall split it into its real and imaginary parts $\chi_s = i\xi + \tilde{\xi}$.
From (1) we get the following expressions for the real and imagi-
nary parts of $\mathcal{M}$:

$$i\, \partial m\, \mathcal{M} = ik \int_0^\infty db\, b\, J_0(\Delta b)\left\{1 - e^{-\xi(b)}\cos\left[\chi_c(b) + \tilde{\xi}(b)\right]\right\} \tag{2a}$$

$$\mathcal{R}e\, \mathcal{M} = k \int_0^\infty db\, b\, J_0(\Delta b)\, e^{-\xi(b)}\sin\left[\chi_c(b) + \tilde{\xi}(b)\right] \tag{2b}$$

and $d\sigma_{el}/d\Omega = |\partial m\, \mathcal{M}|^2 + |\mathcal{R}e\, \mathcal{M}|^2$. From (2) we see that the
Coulomb interaction may help us in learning about the real part
of the strong interaction phase shift χ_s. For instance, if the
Coulomb interaction is absent ($\chi_c = 0$) the elastic cross section
is invariant against the change of sign of $\tilde{\xi}$. If, however, χ_c
is present, the change of sign of $\tilde{\xi}$ introduces drastic changes,
especially at the diffractive minima (compare examples below).
The change of sign of χ_c (which can be achieved by changing the
sign of the charge of the projectile) introduces identical chan-
ges in $d\sigma_{el}/d\Omega$ as the change of sign of $\tilde{\xi}$. That means that if

the target has isospin zero (like e.g. He^4 or O^{16}) and one finds
no difference between the elastic cross-sections of π^+ and π^- this
implies that there is no real part in the $\pi^\pm$ -nucleus elastic
scattering strong interaction phase-shift χ_s. That leads us to
a question: is there any point in getting information on e.g. the
strong interaction $\pi^\pm$ - nucleus phase shift ? The answer seems
to be yes, because we believe in a straightforward connection between the real and imaginary parts of χ_s and the real and imaginary parts of the forward scattering amplitudes of the incident
hadron and the nucleons of the target nucleus. In order to see
it more clearly we have to employ a model and we choose the model
proposed some time ago by Glauber [3], and we shall use its "optical limit" (see [4]), which is presumably an excellent approximation in the case of Pb^{208} nucleus which is going to serve us as
an illustration.

In this model the phase shift χ_s is given as follows [4]:

$$i\chi_s(b) = -\frac{N}{2}(1-i\alpha_n)\,\sigma_n\,\tilde{\rho}_n(b) - \frac{Z}{2}(1-i\alpha_p)\sigma_p\,\tilde{\rho}_p(b) \Rightarrow -\xi(b)(1-i\alpha) \qquad (3)$$

where the indices n and p refer to the neutrons and protons
in the target nucleus, N and Z are the number of neutrons
and protons, respectively, $\tilde{\rho}(b)$ are the single particle densities
at the impact parameter b into which the size of the incident
particle was folded [4] , σ -the total incident baryon - target
nucleon cross-section, and α the ratio of the real to the imaginary part of the forward incident baryon - target nucleon scattering amplitude. We obtain the last relation of (3) if we accept
the same densities and cross-sections for protons and neutrons
(then $\alpha = (N\alpha_n + Z\alpha_p)A^{-1}$) .

We are of the opinion that the relation (3) stands a good
chance to be very accurate for large nuclei. A more detailed discussion of this point is included in [5] from which all the numerical results shown below are taken. Here, let us only stress
three points. The large size of the target nucleus implies that
only the forward scattering amplitude α is significant. If this
ratio does depend on the momentum transfer, it is extremely improbable that this dependence is comparable or stronger than the
target nucleus form factor dependence. If it were so, the range
of the forces which produce the real part of the incident hadron
- nucleon amplitude would be of the order of magnitude of e.g.
6 fm in the case of Pb^{208} target, which is virtually impossible.
We believe that this range is, in fact, much smaller than the
nuclear radius, hence the stabilizing influence of the large nuclear size makes only its forward value to play any role. The second point is that, in the case of incident pions, and a large
target nucleus only the non spin-flip part of the pion - nucleon
amplitude comes into play because the angles involved are of the
order of a few miliradians. Besides even if the angles were not

so small the spin effects of the pion - nucleon amplitude would
average out as the calculations of [6] show. The third point is
a very serious objection to the validity of (3) , based on the
estimates of the so called "inelastic shadow" effect [7]. At very
high incident energies the probability of producing particles at
certain points inside the nucleus and re-absorbing them at some
other points inside the nucleus seems to be considerable. Were
such effects important they would introduce serious corrections
to (3). However we do not really know whether such effects exist
and if so, how and under which circumstances they are important
(see also [8]). So, we shall proceed assuming that there exist
cases where the simple formula (3) is quite accurate. From this
assumption and from the general properties of the amplitude (2)
it follows that what was said before about e.g. π - nucleus amplitude applies to π - nucleon amplitude. For instance: if the
target has zero isospin (e.g. He^4 , O^{16}), and one finds no difference between the elastic cross-sections of π^+ and π^- , α
of (3) is zero, hence $\alpha_n + \alpha_p = 0$. So, either $\alpha_n = \alpha_p = 0$, or
$\alpha_n = -\alpha_p$; assuming charge symmetry α_n belongs to $\pi^- p$ or $\pi^+ n$
and α_p to $\pi^- n$ or $\pi^+ p$ scattering amplitudes.

Let us show how the formulae (2) and (3) work in the case
of a charged hadron scattering from Pb^{208} nucleus. The single particle density was taken, identical for protons and neutrons, in
the form of a two parameter Fermi distribution

$$\varrho(r) = \varrho_0 \left(1 + \exp\left(\frac{r-R}{c}\right)\right)^{-1} \tag{4}$$

where ϱ_0 is the normalization constant and, for Pb^{208}, we took
R = 6.5 fm, c = 0.523 fm. The $\tilde{\varrho}_n(b)$ and $\tilde{\varrho}_p(b)$ densities of (3)
were obtained by integrating [4] along the axis perpendicular to
the impact parameter vector $\vec{b}$ and then the "profile" of the
neutron or proton [1], [4]

$$\gamma_{n,p}(b) = \frac{(1-i\alpha_{n,p})\sigma_{n,p}}{4\pi\alpha_{n,p}} e^{-b^2/2a} \tag{5}$$

with a = 10(GeV/c)^2, which is appropriate for the incident hadron
energy about 20 GeV, was folded into it. The Coulomb phase $\chi_c(b)$
was calculated from the Coulomb potential folded into the density (4).

$$\chi_c(b) = -\frac{1}{v} \int_{-\infty}^{+\infty} V_c\left(\sqrt{b^2+z^2}\right) \tag{6}$$

where

$$V_c(r) = Ze^2 \int d^3r' d^3r'' d^3r''' \frac{\varrho(r') h_p(r'') h_\pi(r''')}{|\vec{r}+\vec{r}''' -\vec{r}' -\vec{r}''|} \tag{7}$$

h_p and h_π being the charge density distributions of the protons in the target nucleus and the incident hadron, respectively. The densities h_p and h_π introduce only negligible modification and can be safely replaced by the δ-function distributions. There are, of course, many points concerning reliability of such evaluation of $\chi_c(b)$ which should be discussed. For instance, one can also calculate χ_c, similarly like χ_s, from the individual Coulomb amplitudes. One can show that the result of such calculations is approximately the same as given by (6) and (7). More details concerning this problem will be given in [5].

The results of calculating $\mathrm{Im}\,\mathcal{M}$ and $\mathrm{Re}\,\mathcal{M}$ and $d\sigma_{el}/d\Omega$ are shown in Figs 1 and 2. Let us first discuss Fig. 1. First of all we see that the Coulomb interaction profoundly changes the structure of the elastic scattering amplitude. When we compare the complete real and imaginary parts of the amplitude with the real and imaginary parts of purely strong interaction amplitude (Coulomb interaction switched off, subscript s in Fig. 1) we see that the Coulomb interaction managed to interchange their roles: in the absence of χ_c the imaginary part of the amplitude produces maxima and the minima are filled by the real part, however, the result of introducing χ_c into (2) is that the real part produces maxima which are filled by the imaginary part. One can see this easily from (2). First let us observe that $\chi_c + \tilde{\xi}$ is very nearly a constant if α of (3) is equal to about -0.3 and, as b becomes slightly larger than the nuclear radius $R \approx 6.5$ fm, it decreases very slowly to $\frac{1}{2}\pi$, hence $\cos(\chi_c + \tilde{\xi}) \approx 0$ and $\sin(\chi_c + \tilde{\xi}) \approx 1$ at the nuclear surface (and, approximately, inside of the nucleus too). The effect of all this is that at $b \approx R$ the integrand of the real part (2b) changes rapidly, instead the rapid change of the integrand of the imaginary part (2a) occurs for b's definitely larger than R. This results in oscillation which are shifted in phase (see Fig. 1). If we put $\chi_c = 0$, the roles of (2a) and (2b) interchange, because $\tilde{\xi}$ is always small at high energies, hence $\cos\tilde{\xi} \approx 1$, $\sin\tilde{\xi} \approx 0$ at the nucleus boundary. Consequently, the roles of the imaginary and the real parts of the amplitudes with

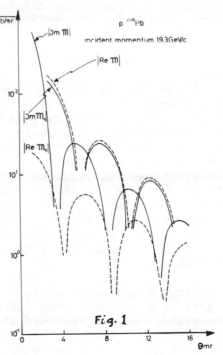

Fig. 1

and without Coulomb interaction get interchanged – provided x_c is large enough (and
α is small enough) which is
the case in the scattering from
Pb ^{208}. Of course the whole
spectrum of intermediate cases
is possible for smaller target
nuclei. In order to estimate
whether the influence of the
Coulomb interaction is important, it is enough to compute
the point charge ($z e^2$) phase
shift at the target nucleus
radius. If it is comparable
to $\tilde{\xi}$ – it should be taken
into account. For instance
even for such light nuclei as
O^{16}, the Coulomb interaction
changes the diffractive minima depths quite appreciably
[2], [5].

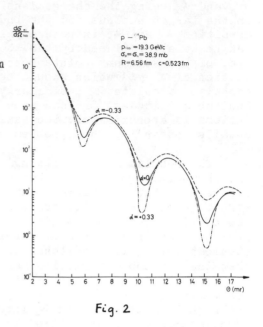

Fig. 2

In Fig. 2 the changes introduced by changing the sign of α
are shown. Remember that the changes are identical when we change the sign of the incident particle charge (hence x_c).

3. Almost elastic scattering

It is very difficult to measure the nuclear elastic scattering cross-sections at very high energies because of the very high
energy resolution requirements (the first nuclear levels are separated from the ground state by the excitation energy of the order of 1 MeV). It is much easier to measure cross-sections with
poor energy resolution where the experiment sums over all nuclear excitations but excludes meson production. Is it possible to
extract α's for π – nucleon amplitudes from such "poor energy
resolution" cross-sections measured at some multi-GeV machine ?
The answer depends very much on how reliably one can calculate
the corrections to $d\sigma_{el}/d\Omega$ coming from all the inelastic nuclear processes: $d\sigma_{inel}/d\Omega$ (the sum of these two let us denote by
$d\sigma_{sc}/d\Omega$).

Since there exist a series of poor energy resolution measu-
rements of $d\sigma_{sc}/d\Omega$ at ~20 GeV [9], we calculated $d\sigma_{inel}/d\Omega$
for several nuclei, then computed $d\sigma_{sc}/d\Omega$ and compared with [9].
The results are shown in Fig. 3. Before we comment on them let
us make a few general remarks.

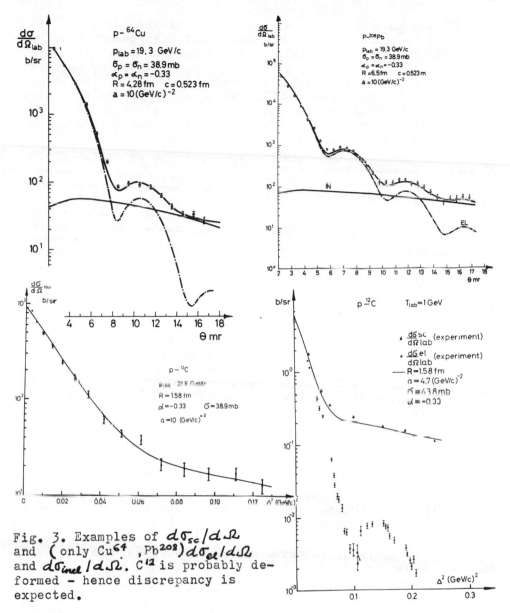

Fig. 3. Examples of $d\sigma_{sc}/d\Omega$ and (only Cu^{64}, Pb^{208}) $d\sigma_{el}/d\Omega$ and $d\sigma_{inel}/d\Omega$. C^{12} is probably deformed – hence discrepancy is expected.

First of all it is very convenient to split $d\sigma_{sc}/d\Omega = d\sigma_{el}/d\Omega + d\sigma_{inel}/d\Omega$. This is so, because some factors important in $d\sigma_{inel}/d\Omega$ are unimportant in $d\sigma_{el}/d\Omega$ and vice versa. In order to analyze $d\sigma_{inel}/d\Omega$ we used the formula

$$\frac{d\sigma_{inel}}{d\Omega} = \left(\frac{k}{2\pi}\right)^2 \int d^2b\, d^2b'\, e^{i\vec{\Delta}\cdot(\vec{b}-\vec{b}')} \left[\langle\psi_o|\Gamma^*(b)\Gamma(b')|\psi_o\rangle - \langle\psi_o|\Gamma^*(b)|\psi_o\rangle\langle\psi_o|\Gamma(b')|\psi_o\rangle\right] \tag{8}$$

with $\Gamma(\vec{b}) = 1 - \prod_{j=1}^{A}(1 - \gamma_j(\vec{b} - \vec{s}_j))$, where γ_j is the profile of the j -th nucleon and includes the Coulomb contribution in the case of protons, ψ_0 is the ground state of the target nucleus. The expression (8) follows naturally from the considerations of unitarity relations [1] in the high energy scattering model [3]. We found the following properties of $d\sigma_{inel}/d\Omega$: i) in contrast to elastic scattering the Coulomb interactions virtually do not play any role, ii) in contrast to elastic scattering, correlations, (hence Pauli exclusion principle) play, at small momentum transfers, an important role and, percentagewise, amount to similar corrections as in single scattering processes. Its role can be estimated from correlation functions, as it was suggested in [10]. Consequently, the difference between e.g. $d\sigma_{sc}/d\Omega$ for π^+ and π^- does not depend on $d\sigma_{inel}/d\Omega$. This gives us a method of measuring α's from $d\sigma_{sc}/d\Omega$ which does not depend on $d\sigma_{inel}/d\Omega$.

Now back to Fig. 3. The curves shown there were computed from (2), (3) and (8) and the ground state density in (8) was taken in the form $\psi_0^* \psi_0 = \prod_{j=1}^{A} \rho_j(r_j)$, where the densities ρ_j were computed from the oscillator potential wave functions for Li, Bi, C and for Al, Cu, Pb and U nuclei the Fermi density (4) was used. No other approximations were made, and all the integrals were computed numerically. The results are preliminary in the sense that some more work to improve on the approximation $\psi_0^* \psi_0 = \prod_{j=1}^{A} \rho_j(r_j)$ is under way. More details will be included in [5]. Fig. 3 shows also $d\sigma_{sc}/d\Omega$ for p - C^{12} scattering at 1 GeV [11] incident kinetic energy. A different experimental technique was used in [11] than in [9]. We can see that the "low energy" (1 GeV) data can be consistently explained together with the "high energy" data (20 GeV).

4. Conclusions

First of all one should remember that only one target nucleus (Pb^{208}) among the ones shown in Fig. 3 satisfies the criteria, stated at the beginning of the paper (but even in this case the target was not pure Pb^{208}), hence some nuclei with spin different from zero and perhaps deformed were admixed . So, one should look at the results as very encouraging indeed, more so because all the parameters used were not free but had to be consistent with the results of some other experiments.

There is one very clear result that if we go over to very large nuclei (especially Pb^{208} is worth recommendation) the elastic diffractive structure, which was completely destroyed by the inelastic processes in light nuclei, becomes visible again. That strongly suggests that the "background" $d\sigma_{inel}/d\Omega$ is small enough and smooth enough to be reasonably reliably subtracted. The situation obviously improves when we go to pion beams. Fig. 4 shows $d\sigma_{sc}/d\Omega$ for $\pi^{\pm}$ - Pb^{208} at 40 GeV.

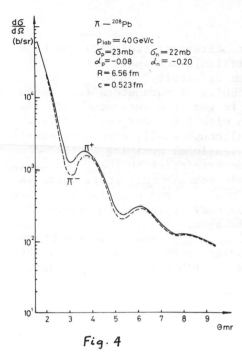

Fig. 4

By comparing it with the scattering cross section of p - Pb^{208} we see that they do not differ very much. The pion beam has however this great advantage of containing spin zero particles, which makes the interpretation much more safe.

We can conclude by saying that if the incident particles are spin zero hadrons (pions) and the targets are large, spin zero, spherical nuclei, even poor energy resolution cross-sections stand a good chance to provide us with the α parameters. The only very serious obstacle could be the "inelastic shadow scattering" [7] about which, at the moment, we know very little.

REFERENCES

[1] W.Czyż, High Energy Collisions of Hadrons and Nuclei, Proceedings of the Inaugural Conference of the European Physical Society, Florence, 8 - 12 April 1969, preprint INP 678/PL/PH.

[2] L.Leśniak, Ph.D. dissertation (unpublished).

[3] R.J.Glauber, High Energy Collision Theory, Lectures in Theoretical Physics, Vol. I Interscience Publishers, Inc., New York 1959 .

[4] W.Czyż, L.Maximon, Annals of Physics, 52, 59 (1969).

[5] W.Czyz, L. Lesniak, H. Wolek, to be published, see also L. Lesniak, Proceedings of the IX Cracow School of Theoretical Physics, preprint INP No. 682/PL/PH, Vol. II, June 1969.

[6] J.Formanek, J.S.Trefil, Nuclear Phys., B3, 155 (1967).

[7] J.Pumplin, M.Ross, Phys.Rev.Letters, 21, 1778 (1968).

[8] E.Abers, H.Burkhardt, V.Teplitz, C.Wilkin, Nuovo Cimento, 42A, 365 (1966).

[9] G.Bellettini, G.Cocconi, A.Diddens, E.Lillethun, G.Matthiae, J.Scanlon and A.Wetherell, Nuclear Phys., 79, 609 (1961).

[10] A.S.Goldhaber, C.J.Joachain, Phys.Rev., 171, 1566 (1968).

[11] J.F.Friedes, H.Palevsky, R.J.Sutter, G.W.Bennett, G.J.Igo, W.D.Simpson, D.M.Corley, Nuclear Phys., A104, 294 (1967). See also: D.M.Corley, Quasi-free scattering of 1 BeV protons from C^{12} and Ca^{40} , Thesis, University of Maryland.

DISCUSSION

H.J. Lubatti: In earlier attempts to interpret (ref. 10) the CERN
results (ref. 9), there was some difficulty at small angles,
especially in Pb.   Do you have such difficulty?   W.C.: No.
A. Goldhaber: With any reasonable choice of parameters for the Pb
nucleus, it does not seem possible to get good agreement with the
rate of change of the cross-section with t at very small angles.
Discrepancies of this sort (which although small, seem persistent)
are not readily revealed in the conventional semi-log plots!
W.C.: There are also considerable uncertainties in the experimental
results.   S.J. Lindenbaum: In these measurements where there is
an extremely sharp t-dependence at small t, and uncertainty in the
measurement of t itself, it would be easy to introduce errors of
the sort (∿10%) mentioned by A. Goldhaber.

J. Trefil: Was the nucleon distribution in Pb^{208} you assumed, the
same as the electric charge density distribution?   W.C.: Yes.

BARYON RESONANCES IN THE NUCLEAR GROUND STATE: HOW ABOUT DOUBLE COUNTING?

Michael Danos

National Bureau of Standards

Washington D.C. 20234

It is very easy to state that the nucleus is made up of all kinds of particles, the most prominent of which are protons and neutrons, with π-mesons and other Bosons floating around and with occasional heavier particles, viz., baryon resonances and hyperons also making an appearance. It is more difficult to give a precise meaning to this statement. The reason for this is obvious: one would need a workable theory. Namely, one would like to talk exclusively in terms of dressed particles. Since the dressed particles have structure, this structure must be described either in terms of bare particles (field theory) or in terms of dressed particles (boot strapping). As is well known, neither of these approaches has as yet been successful.

Lately, first order perturbation theory has again become fashionable as a means of avoiding the convergence problems of the above theories. As long as one deals with scattering and production processes one always can define effective coupling constants and form factors which reproduce the experiments. It is not clear that one can do it consistently in that the same vertex in all processes can have the same characteristics. It is, in fact, extremely unlikely that this will be so; c.f. quantum electrodynamics, where the radiative corrections are different in different states and different processes. At any rate, by ignoring the particle structure one in essence eliminates the possibility of computing the dynamics, e.g., the form factors, from more basic principles. This procedure thus is useless for our purposes. If one wants to go beyond calling a nucleus, say, ^{16}O, an elementary particle, one is compelled to go to higher than first order perturbation theory. Then one is back in the thicket of the convergence problems, and one is back

at the question of the structure of the dressed particles, and
one is again stuck at the question: what is, say, the difference between having a meson in the nucleus, or having A-1
nucleons and one baryon resonance? This question, in fact, is
the title of the talk expressed in different words.

As for the present experimental evidence for the need to
go beyond protons and neutrons in nuclei, let me just mention
the exchange current contribution to the magnetic properties of
nuclei,[1,2,3] and the Kerman-Kisslinger interpretation of the
(p,d) elastic backward scattering.[4] I shall not discuss either
of them here.

Let me now turn to the question on how to give a consistent
and precise meaning to the concept of the particles. This will
then also allow a precise counting procedure. The central
problem is to define a convergent approximation scheme, since an
exact theory does not exist. This can be attempted in the
following manner.[5]

It has always been the belief of the physicist that one
can understand nature "a piece at a time", that one can cut up
physics into separate parts. That means that one can do classical physics without having to do atomic physics; that one
can do atomic and molecular physics without having to do nuclear
physics; that one can do nuclear physics without having to do
nucleon dynamics. The meaning of these cuts is the following.
One decides which phenomena one wants to compute. This tells
which range of energies will be needed in the calculation. From
that one then can tell how large a function space will be needed.
Finally, this determines what kind of Hamiltonian one will have
to employ. For example, one may want to do atomic and molecular
physics. The relevant energies then are 10-100 eV. One thus
can safely cut the Hilbert space at, say, 100 keV. Therefore
the nuclear phenomena are eliminated from the picture, and the
nuclear properties enter the problem via certain phenomenological parameters. The Hamiltonian then must contain explicitly terms for the participating nuclei, say, ^{1}H, ^{2}H, ^{16}O and
^{17}O in addition to the terms describing the electrons; and the
diverse electric and magnetic moments of the nuclei in the
ground states must be given as input parameters: one cannot
compute them in the chosen limited function space. To treat
the nuclear ground state properties one would have to truncate
at a higher energy, say 10 MeV.

Similarly, in treating nuclear problems, the truncation of
the Hilbert space at, say, 100 MeV does not allow for a calculation of the ground state properties of the nucleons. Anyway,
now the Hamiltonian is different than the Hamiltonian of atomic
physics: one has protons and neutrons in the place of the
intact nuclei.

To proceed further one must continue to approach the problem with humility. One has to decide at the outset on the limits one is going to impose on the calculation. One cannot solve everything. One thus must begin by choosing a truncation energy, say 1 GeV above the ground configuration. This then determines the form of the Lagrangian. It will have to contain explicitly that high-energy information which cannot be treated in the limited function space. For example, it may be possible to compute in this space the proton-neutron mass difference, and the anomalous magnetic moments; it is possible but not likely that one will be able to compute the $\pi^+ - \pi^0$ mass difference; it is very unlikely that one can compute the intermediate-range behaviour of the nuclear forces; the repulsive core description is out of the question. The humility-Lagrangian thus will have to contain nucleons with a certain semi-undressed mass, π -mesons, ρ -mesons, etc. also more or less completely dressed, and practically completely dressed baryon resonances, together with the different coupling terms. When talking about mesons in nuclei, or baryon resonances in nuclei, one has to do it in the context of a definite humility-Lagrangian. Only this way is it possible to give a precise meaning to the terms of the discussion. As compared with atomic physics the situation is less clear-cut: there seems to be no energy gap which would give a natural criterion on the choice of the truncation energy. It is at present only a matter of conjecture and hope that the low energy results will be insensitive to the choice of the truncation energy.

With this we are ready to define an asymptotic particle. We choose, say, a complete set of harmonic oscillator basis functions, and do a shell-model type calculation using the following configurations: (1) a single bare nucleon; (2) states of one bare nucleon and one bare meson with up to such kinetic energy in the relative motion that the total energy is $\leq$ 2 GeV; (3) states of one bare nucleon and two bare mesons with a set of relative kinetic energies, etc. Next, (4) similar states where the bare nucleon has been replaced by a bare baryon resonance and finally, at a cut at 2 GeV barely, a Λ - K meson configuration. The different particles interact with appropriate Chew-Low type interactions. Upon diagonalization of this finite-dimensional Hamiltonian matrix, one obtains the dressed nucleon as the lowest eigenstate. Similarly, from a Hamiltonian matrix built with configurations having the quantum numbers of a π -meson one obtains the dressed π -meson. No essential difficulties arise since all matrices are finite-dimensional. One must choose the input parameters such that these dressed particles have the experimentally observed properties.

In precisely the same way one treats the deuteron. Here

one has to include those configurations which have a total energy
(bare masses and kinetic energies) of up to 3 GeV. Now one can
project out from the dressed deuteron ground state the previously
computed dressed proton and dressed neutron. The excess then
contains all other particles whose participation amplitudes can
be obtained by projection. Evidently both dressed mesons and
dressed baryon resonances are going to be present. Finally,
the binding energy then is given by the difference between the
mass of the dressed deuteron and the sum of the dressed proton
and neutron masses. Note than in such a treatment no ambiguity
arises in the treatment of the many-meson exchange. Also, if
one so wishes, one can generate a non-relativistic nucleon-
nucleon potential by the Born-Oppenheimer procedure.

The three-body problem can be treated, in principle,
albeit not yet in practice, by the same procedure as discussed
above for the two-body problem.

I now address myself to the problem of how to separate,
e.g., the meson + two-nucleon state from the nucleon + baryon
resonance state. A baryon resonance, say the Fermi resonance,
has the following constituents in terms of bare particles: the
3/2, 3/2 resonance; a nucleon and a π -meson in a relative
p-state, a nucleon and two π -mesons, coupled to 3/2, 3/2, etc.
In a two-nucleon system the configuration two bare nucleons +
two bare π -mesons contributes both to the two dressed nucleon
configuration and to the two dressed 3/2, 3/2 resonance configuration. In a complete consistent treatment clearly both
these configurations must be considered. Also, without doing
a complete treatment, these two configurations cannot be disentangled, and it then definitely would be very dangerous to
consider the admixture of both π -mesons and baryon resonances.

The next question is whether it is possible to go one step
beyond the standard non-relativistic treatment without having to
do the horribly involved consistent treatment. This would be
very easy if the amplitude of the dressed baryon resonance
configuration were much larger than the amplitude of the dressed
nucleon + dressed meson configuration. One then could justi-
fiably disregard the latter configuration and enlarge the
nuclear physics simply by admitting baryon resonances to the
nuclear wave function, while continuing to keep the mesons in
exile and replacing them by a two-body potential.[6] Naturally,
the potential would have to contain a term which transforms a
nucleon into a baryon resonance. In that model no overcounting
will occur. However, an undercounting is inevitable. It is
associated with the non-vanishing of the amplitude of the
dressed nucleon + dressed meson configuration. It will have to
be determined by detailed work how serious the undercounting is.
If the indications taken from the computations of the magnetic
moment exchange currents[1,2,3] are correct then the undercounting

is, in fact, not serious. Anyway, a very important aspect of
relativity, viz., particle creation, in this manner has been
successfully incorporated into the non-relativistic nuclear
structure theory, so to say, by the back door. If the smallness
of the undercounting error should withstand further scrutiny
this quasi-relativistic model could carry nuclear physics a
reasonable distance beyond the present theory.

REFERENCES

1. H. Arenhövel and M. Danos, Phys. Lett. 28D, 200 (1968).

2. H. Arenhövel, M. Danos and H.T. Williams, to be published.

3. L. Kisslinger, to be published.

4. A. Kerman and L. Kisslinger, to be published; L. Kisslinger, this Conference.

5. M. Danos, Lectures at the Les Hoches Summer School, 1968 (to be published).

6. H. Sugawara and F. von Hippel, Phys. Rev. 172, 1764 (1968).

DISCUSSION

G.E. Brown, M. Rho and K. Gottfried expressed, each in turn,
surprise and dismay at what they interpreted to be the unwarranted
gloomy implications of Danos' argument. They felt that such
items and theories as duality, current algebra, PCAC, provide the
possibility of obviating just those problems of double-counting
which are Danos' concern. M. Danos in reply, contended that
these were low energy theorems, and applicable to low-energy and
long-range phenomena. The approach he indicated was more appropriate to high energy and for many meson exchange.

BARYON RESONANCES IN NUCLEI[*]

Leonard S. Kisslinger

Department of Physics, Carnegie-Mellon University

Pittsburgh, Pennsylvania   15213

In the traditional picture, the nucleus is treated as a system of neutrons and protons. This picture is successful because the neutron and proton constituents are rather weakly bound in the nucleus, and the explicit presence of other objects such as mesons apparently can be neglected with appropriate renormalization of the properties of the nucleons. It is the purpose of today's discussion to suggest that at higher energies such a picture may not be possible. Arguments will be given suggesting that the degrees of freedom associated with the baryonic resonances must be explicitly included for these particular properties. The first part of the talk will be concerned with the forces which bind the baryonic resonances to nucleons while the second part will review the present evidence for the role played by these resonances in high energy back scattering of protons on deuterons.

I.   NUCLEON-BARYON RESONANCE INTERACTION

First let us review the properties of the objects which will enter into today's discussion. Fig. 1 gives the spectroscopic information for the proton nucleus - the mass 1 nucleus. The ground state of this nucleus is the nucleon while the excited states are referred to as the baryon resonances. Note that the characteristic excitation energies of the levels in this system are hundreds of MeV which is the main reason why the excited states have been neglected in studying nuclear processes with their characteristic energies of the order of an MeV.

An important property which has not been included in Fig. 1 is the width of the resonances. The characteristic widths are 100 or

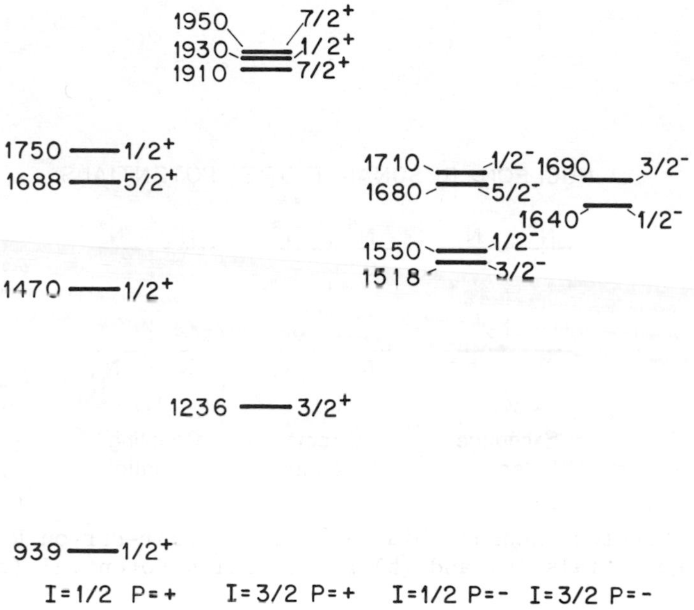

```
                          1950 ——— 7/2⁺
                          1930 ═══ 1/2⁺
                          1910 ——— 7/2⁺


         1750 ——— 1/2⁺                  1710 ——— 1/2⁻  1690 ——— 3/2⁻
         1688 ——— 5/2⁺                  1680 ——— 5/2⁻
                                                       1640 ——— 1/2⁻
                                        1550 ═══ 1/2⁻
         1470 ——— 1/2⁺                  1518 ——— 3/2⁻



                          1236 ——— 3/2⁺




                939 ——— 1/2⁺
            I=1/2  P=+    I=3/2 P=+   I=1/2 P=-   I=3/2 P=-
```

Fig.1. Proton Nucleus. States Grouped by Isospin(I) and parity(P).

more MeV since these resonances decay strongly into the nucleons and mesons. From these widths we can determine one component of the long-range part of the interaction between these resonances and nucleons. This component is the exchange force illustrated in Fig. 2a in which the needed coupling constant g^* can be determined from a suitable model for the decay of the resonance and from the experimental width. Unfortunately, one of the coupling constants needed for the direct component of the interaction illustrated in Fig. 2b is not known for it involves information about the meson resonance coupling.

We have adopted simple static models for the pion-baryon coupling from which we have derived interactions corresponding to the one-pion exchange forces illustrated in Fig. 2a) and b). Using these forces, we have made estimates of the resonance-nucleon binding in various states. These are very simple, crude, calculations but should give us some idea of the magnitude of the binding energies which one might expect. The results are given in Table I. Presumably, this perturbation theory calculation can underestimate the binding in some states, and the binding might be 10 or more MeV in such a case.

Because of the many uncertainties in this type of calculation, the quantitative studies of these baryon resonance nucleon interactions must await experiments. The result of a "typical" experiment is illustrated in Fig. 3. The figure gives the cross section

NUCLEON-RESONANCE O.P.E. POTENTIALS

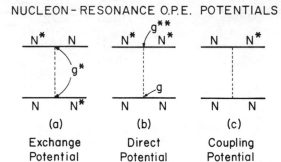

Fig.2. One-Pion Exchange Diagrams for Nuclear-Baryon Resonance
Potentials (a) and (b) and Coupling Potential (c).

TABLE I

Baryon Resonance-Nucleon Binding

Resonance(or Particle) (I,J)	S	L	J	I	ΔE(MeV)
Nucleon (Deuteron)	1	0	1	0	-1.0
	0	0	0	1	-1.0
1236 $(3/2,3/2^+)$	1	0	1	2	-2.3
	2	1	1	2	-4.1
1470 $(1/2,1/2^+)$	0	0	0	1	-0.6
	1	0	1	0	-0.6
1518 $(1/2,3/2^-)$	2	0	2	0	-1.5
	2	1	2	0	-3.1
1688 $(1/2,5/2^+)$	2	0	2	1	-1.2
	3	0	3	0	-2.6
1710 $(1/2,1/2^-)$	0	0	0	1	-1.3
	1	1	1	0	-3.6

as a function of pion energy for one of a number of processes in
which a π meson is incident upon a nucleus. At the energy E_R one
sees more or less the free baryon resonance. The other bump in the
figure occurs at a lower energy corresponding to the binding of the
baryon resonance. This part of the reaction takes place through a
specific bound state of the resonance to the rest of the nucleus.
If the binding is strong enough, by studying angular distributions
and perhaps other things, one can learn the quantum numbers of the
bound state and from this can extract valuable information about
the resonance-nucleon interactions. Next we take up the question
of whether the presence of these resonances as components of ordinary nuclei is ever an important feature or indeed can even be detected.

II.   BARYON RESONANCES IN NUCLEI AND PICKUP REACTIONS

From the simplest considerations one expects baryon resonance
components of several percent in the nuclear wave function, for the
excitation energies of the resonances are hundreds of MeV and nuclear matrix elements of several MeV or more are expected.[1] Recent
calculations indicate that some or even most of the so-called meson
current corrections to the magnetic moments can be taken into account through these components,[1,2] although it will probably be
rather difficult to make this quantitative. From symmetry considerations the 1236 state is important for double beta decay.[3] We would
like to suggest that pickup reactions will perhaps be the best place
to test the presence of these components for as we have learned at
lower energies the pickup reactions are most useful for testing the
particle aspects of nuclear structure. Here we shall discuss only
the pickup reactions on the deuteron or, in other words, high energy proton-deuteron backward elastic scattering. (At high energy
the impulsive mechanism should give a rather small amount of the
backward scattering compared to the pickup mechanism.[1,3]) Nuclear
physicists can most easily understand why we feel this pickup process will be sensitive to the presence of small components of baryon resonances by recalling the results of the shell model calculations of (p,d) reactions at lower energies

$$\left(\frac{d\sigma}{d\Omega}\right)_{pickup} \quad \text{Const}(\kappa^2 + \Delta^2)^2 |\psi_{Deut}(\Delta)|^2 |\psi_{j\ell}(\Delta')|^2 \tag{1}$$

where Δ and Δ' are momentum transfer variables ($|\Delta| = |\Delta'|$ for
(p,d) elastic scattering). The major angular or momentum transfer
dependence of the cross section is given by the Fourier transform
of the single particle wave function called $\psi_{j\ell}(\Delta')$ in Eq.(1).
From this follows the well-known result that the pickup peaks
occur at larger and larger angles corresponding to higher orbital
angular momenta of the shell model particles. This result follows
from the fact that for higher and higher orbital angular momenta
the Fourier transform of the wave function peaks at higher and
higher values of the momentum.

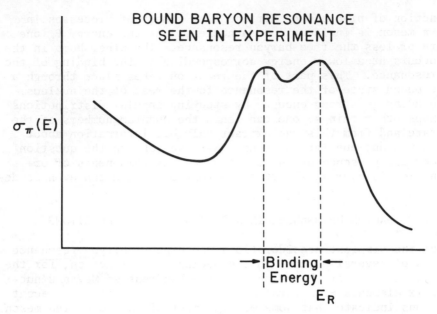

Fig. 3.

From this we can see first why we would expect the presence of
baryon resonances in the nucleus to become more and more important
for pickup reactions at high energy, and second, we can guess which
resonances will be important at any particular energy. To be explicit, let us consider the pickup from the deuteron. Since the
total angular momentum of the deuteron is 1, the highest angular orbital momentum component of the deuteron is 2. We now know a good
deal about the wave functions of the deuteron and that the S state
has a node at about 2 f^{-1} in momentum space while the D state peaks
at approximately that same momentum and has a node at about 4 f^{-1}.
Components with orbital angular momentum of 4 will tend to peak at
even higher values of momentum. In order to have these, of course,
one must have higher spin particles which suggests that one looks
for the highest spin projectile with lowest mass as possible candidates for giving important contributions to these reactions.

In fact, one is led to this conslusion in another way which
historically was the approach Kerman and I used in studying the
backward proton-deuteron scattering.[1] In the Regge theory only the
nucleon Regge trajectory should contribute to the backward proton-
deuteron scattering if one can follow the models that have been
used for nucleon scattering. In this model one is led to consider
first the 1688 MeV spin $5/2^{+}$ resonance which is the first excited
baryonic state on the nucleon Regge trajectory. The next state to
be considered would be the spin $9/2^{+}$ baryonic resonance which in
fact is not very well established. For energies between 500 MeV

and 1 GeV to which we will restrict most of the rest of the discussion the D states are most important. Thus we are led to consider the component of the deuteron wave function which explicitly contains the 1688 resonance. The Born approximation expression for the differential cross section at back angles which results from the pickup mechanism including the 1688 particle in an orbital momentum 2 state is given in Ref. 1. Among the many unknowns is the amplitude for this component of the wave function. We have estimated this amplitude in perturbation theory using the coupling potential which follows from the one π meson exchange diagram, Fig.2c. The explicit form for this potential is given in Ref. 1, and a crude but probably reasonable estimate is that there is a probability of something like 1% or so of this resonance in the deuteron.

The results at 1 GeV and 590 MeV are shown in Fig. 4. The differential cross section is plotted as a function of the variable $\Delta=|\underline{p}-\underline{d}'/2|$, which we shall discuss below. This variable corresponds to the momentum transfer of the protons. As seen from Eq.(1) it is the natural variable to use if the basic mechanism is that of pickup. Note that the experimental[5] points at 590 MeV and at 1 GeV run smoothly together which helps confirm that the pickup mechanism is the basic process. As was mentioned earlier, the zero of the Fourier component of the S-state part of the wave function occurs at about $2f^{-1}$ which is in the physical region of the pickup peak at 590 MeV, but which occurs outside of the pickup region at 1 GeV. Thus at both of these energies the S state contribution is very small or negligible which in fact is why one can see the effect of the smaller components. With the D-state alone, the cross section is too small, while with about a 2% admixture of a component with a 1688 resonance one can obtain a fit to the 1 GeV cross section with somewhat of an underestimate of the 590 MeV cross section. The neglected contribution from the impulsive scattering is probably somewhat larger at the lower energy. Also, the corrections to the Born approximation have not yet been concluded, and they might well be somewhat different at the two energies.

The nature of the Δ variable is explained in Fig.5. At 180° Δ is equal to 1/2 the center of mass momentum. Over the angular distribution in the region of the pickup peak the Δ variable increases a rather small amount. But by changing the incident momentum one has contributions from different Fourier components of the wave function. This gives us the picture shown in Fig.6. At low energy the S-state is dominant. These are the energies at which the distorted wave Born approximation calculations are being done to get nucleon transfer information which has been so valuable. At the energy of the NASA cyclotron, 590 MeV, the S-state has disappeared and the D state is dominant, but already the contribution from the D* state (the component of the wave function with a 1688 baryonic resonance) is quite significant. At 1 GeV the contribution from the D* component is perhaps more than 1/2 of the total

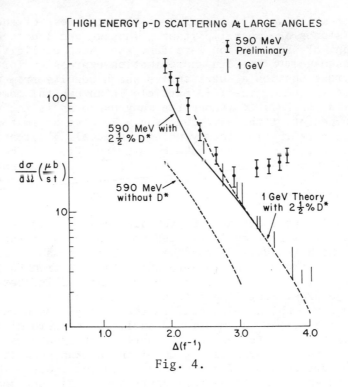

Fig. 4.

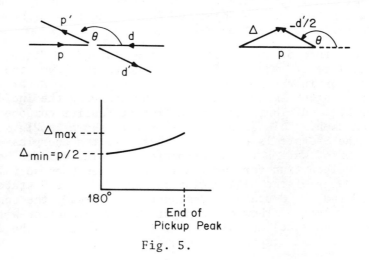

Fig. 5.

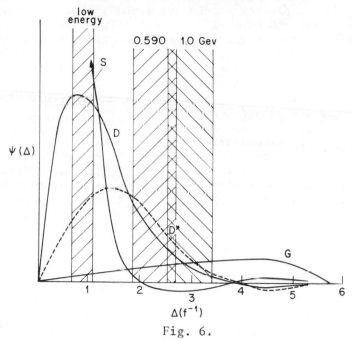

FOURIER TRANSFORM OF, S,D,D,* AND G
COMPONENTS OF DEUTERON WAVE FUNCTION

Fig. 6.

pickup cross section. At the higher energies, one can look forward
to major or even dominant contributions from the L = 4 states
corresponding to a spin 9/2 half particle, the next resonance on
the nucleon Regge trajectory. This is depicted in Fig.6. However,
at energies above 1 GeV inelastic effects become very important
and one must do a more sophisticated calculation. Still, these
high-spin baryonic resonances, even though they are present only as
minor components of the nuclear wave functions, can play a dominant
part in the pickup process at very high energy.

* Supported in part by the National Science Foundation. Work done
in collaboration with A. K. Kerman.
1. A. K. Kerman and L. S. Kisslinger, Phys. Rev. 180, 1483 (1969);
    MIT Summer Study (1967), p. 399.
2. H. Arenhövel and M. Danos, Phys. Letters 28B (1968) 299;
    L. S. Kisslinger, Phys. Letters 29B, 211 (1969).
3. H. Primakoff et al, to be published.
4. L. Bertocchi and A. Capella, Nuovo Cim. 51, 369 (1967).
5. Experimental points at 1 GeV are taken from Bennett et al,
    Phys. Rev. Letters 19, 387 (1967). Experiments of Coleman
    et al, Phys. Rev. 16, 761 (1966) give slightly smaller re-
    sults. Experiments at 590 MeV by the NASA group, to be
    published.

DISCUSSION

P. Minkowski: Has a dip been found corresponding to the Regge
trajectory's passage through -1/2?   L.K.: This would only manifest
itself at energies much higher than those studied here.

A.H. Cromer: How sensitive are results of the calculations you report
to the amount of D-state deuteron wavefunction?   L.K.: At any one
point, a factor 2 at most.   With 6.5% maximum D-state contribution,
wavefunctions do not change very much.

T.E.O. Ericson: The value 2-1/2% for the baryon-resonance in the
nuclear wavefunction should be regarded as an upper limit.
Evidence from an analysis (by Feld) of the π-d system indicates
1 to 1-1/2% as a more likely figure.   A larger quantity would destroy the agreement in the experiment.   L.K.: Not necessarily:
the influence of the D* may be greatly reduced in this case.

SOFT AND REAL PIONS IN NUCLEI[*)]

M. Ericson

Institut de Physique Nucléaire, Lyon, France

In the last few years some very interesting studies have been
made of the interactions of pions of variable mass. As you may
imagine, these investigations have been carried out by theoreticians,
although it would of course be very interesting if the experimentalists also could vary the mass of the pion continuously.

As the mass of the pion is pushed down towards zero some very
remarkable theorems emerge: one of them says that the elastic
scattering of such pions at low energies becomes universal; it
depends only on the isotopic spin of the target and not on its size,
shape or structure. The universal formula that results is the
following: the scattering length a is related by an isospin factor to a universal constant L which depends on the decay constant
of the pion f_π

$$a = L \ 2\underline{t}\cdot\underline{T} \qquad L = \mu/8\pi f_\pi^{\ 2}$$

This formula, that was derived by Weinberg and Tomozawa, has been
applied to elementary particles and it works very well. But it is
a universal expression and it should apply to any target, to nuclei
and to bigger systems like a water melon or the moon! Now in these
big systems we do not expect the scattering length to depend only
on the number of protons and neutrons. For these systems we would
rather expect the scattering to be a measure of the size. Already
light nuclei show important deviations from this simple expression.

Of course the real mass is not zero and we should perhaps not
be surprised by deviations. But the simple expression works for

*) This talk was presented by T. Ericson, CERN, Geneva, Switzerland.

elementary particles and we may well wonder by what mechanism
exactly the mass causes the deviations. Many arguments have been
given: in particular, the many low-lying excited states in these
systems have been blamed. There are millions of states of energy
well below the pion mass and as we extrapolate to zero mass we en-
counter successively all these states. As this did not look harm-
less, a common reaction of elementary particle physicists has been
to forget about nuclei.

But nuclear physicists are not allowed to forget about nuclei
and it is a very interesting problem to find the right way to look
at these soft pion theorems in order to apply them. The secret of
this can be guessed   quite easily just by looking at the previous
expression for the scattering length. This is proportional to the
pion mass and as this mass goes to zero the amplitude, and therefore
the threshold interaction, also goes to zero. If there is no interaction the pion , like a neutrino, penetrates fully the scattering
system. The scattering is then given by the Born approximation.
Nobody is surprised that in the Born approximation the scattering
is proportional to the number of constituents, even for the moon!

Why then should the Born approximation be good for the nucleon
and bad for nuclei and the moon? In the nucleon case we have only
one scale, the ratio μ/m of the pion to the nucleon mass, which is
a small quantity. In nuclei we have other scales than the ratio
of the masses:

One is the ratio of the radial extent to the pion Compton wavelength μR. This is far from small and one therefore expects
important size effects. This is indeed the case.

Another scale characterizes the granular structure of the nucleus.
The typical excitation energies representative of this structure,
ε_N are much smaller than the pion mass, $\varepsilon_N/\mu \ll 1$. However, the
scale associated with this structure is not the huge ratio
$\mu/\varepsilon_N \gg 1$, but instead $\mu d \approx 1$, where d is the correlation distance, essentially the internucleon distance. These excitations
are then not disastrous, as they were thought to be.

Before giving the quantitative arguments to support these
statements, I first want to remind you of the ingredients that enter
into the composition of these soft pion theorems. They are essentially three:

i)   the partially conserved axial current hypothesis, PCAC, which
     relates the divergence of the axial current to the pion field
     Φ: $\partial_\nu J_\nu^{\ A} = f_\pi \mu^2 \ \Phi$;

ii) the current algebra which provides us with a certain number of
     basic commutators which are used in the derivation of these
     low-energy theorems;

iii)   the soft pion assumption $q_\lambda = 0$ which takes all the components
       of the pion four momentum equal to zero.

   The question is now how to generalize the expressions obtained
with this soft pion assumption to the real pion? How should we relate the physical amplitude to the universal constant L? This has
been done in the nucleon case by de Alfaro-Fubini-Furlan-Rosetti
[1,2] and we apply their method to the nuclear case [3].

   We take the equal time commutator of two axial charges Q_A^+ and
Q_A^-. Its expectation value between the two nuclear ground states
(at rest) is from the algebra of currents proportional to the third
component of the isospin.

$$\langle N | [Q_A^+ \ Q_A^-] | N \rangle = 2M \ \tau_3 \ .$$

This particular commutator will lead us to the isospin antisymmetric
amplitude a^-, which describes isospin effect and charge exchange.
The symmetric amplitude will be discussed later. This commutator
is then expanded by introducing a complete set of states; and let us
see which ones are of interest to us. They are:

1)   The nuclear states alone. They correspond to the nucleonic
     degrees of freedom without pion. The axial charge has a nega-
     tive parity and can carry only to the states of negative parity
     relative to the ground state. The ground state itself gives
     no contribution.

2)   The second important states are the rescattering states where
     the nucleus is left in its ground state and a pion is also
     present.

3)   There can be inelastic excitation of a nucleus with a pion
     present.

There are also many other states which could contribute like the
states with more than one meson. However, it is known from the
study of the corresponding problem in the nucleus case that these
states are unimportant and we will neglect them also in the nuclear
case.

   We proceed then by replacing the matrix element of the axial
charge by that of the pion field using PCAC. And doing this we get
the following relation

$$a^- = L + \text{corrections} \ .$$

   The important point is that the threshold amplitude is taken
for the real pions and not for the synthetical pions of zero mass,
so that we have an expression for the corrections to the soft pion
limit.

What are these corrections? They are what is left of the expansion of the commutator when we have extracted the physical threshold amplitude. I do not want to enter into the technical details but what remains is essentially the three contributions that we mentioned. Only a piece of the second one has been used explicitly in the threshold amplitude, the so-called semi-disconnected part.

I discuss now these corrections starting with the most important one -- the coherent rescattering. That is the nuclear size effect. If we keep only these terms we can write a dispersion relation which involves the matrix elements f_{0q} for the off-shell scattering of a pion of momentum zero to a momentum q

$$a^- = L + \frac{1}{2\pi^2} \int d^3q \; \frac{\left|f_{0q}\right|^2}{q^2\left(1 + \frac{q^2}{\mu^2}\right)^{3/2}} \; .$$

For simplicity we have not written the isospin indices in f_{0q}. Since we have to treat off-shell effects, it is rather natural to introduce an optical potential for the pion nucleus interaction. For potential scattering the corresponding dispersion relation can be written and it is very similar indeed. The amplitude a is the sum of the Born amplitude (the characteristic constant in a non-relativistic dispersion relation for potential scattering) and of an integral over off-shell matrix elements

$$a = a_{Born} + \frac{1}{2\pi^2} \int d^3q \; \frac{\left|f_{0q}\right|^2}{q^2} \; .$$

The integrals in the two relations have the same structure so that if we take the pion to be non-relativistic in the first one ($q \ll \mu$) they look precisely the same. [The convergence of the integrand is rapid enough in moderately light nuclei ($A \lesssim 30$) to take the pion non-relativistic.]

In view of the similarity of the two equations, we can then try to compare the Born term to the universal constant L. The deviation of the amplitude from the soft pion limit is then the distortion effect. This is of course approximate since we did not include the other contributions to the sum rule. But we can see how this interpretation stands   comparison with the experiments.

From π mesic atom data the charge exchange integral is found to have a volume integral so that

$$\frac{1}{2\pi} \int V \, d^3x = a_{Born} = 0.094 \; \mu^{-1} \; .$$

In our optics this should be the universal constant L = 0.09 μ^{-1}.
The test is quite successful. It is a sensitive test; it is not
at all indifferent to interpret the soft pion value as the physical
amplitude or as the Born value. These two quantities are very different. For a nucleon number A = 18 for instance, the amplitude
$a^{(-)}$ is 0.047 μ^{-1}, only one half of the Born value.

We have now to check that the remaining contributions to the
sum rule do not destroy this beautiful agreement. The first group
of intermediate states to consider are the excited states of the
nucleus with no pion present. If the nucleons were at rest in the
nucleus there would be no contribution, for an axial charge has no
matrix elements between nucleons at rest. Because of the fermi
motion there is an effect but it is a small one ($\sim$ 4%) [4]. This
effect is associated with the scale μd.

However, this is not the whole story for these states because
there exists also pion absorption. For excitation energies of the
order of the pion mass, pion absorption occurs by a many-body process
and this has not been included in the previous description. This
effect is somewhat hard to estimate and although no good argument
can be given for its detailed magnitude the absorption rate of pions
in nuclei indicates that this effect is small, probably of a few
per cent only.

The next group of states to be discussed are the inelastic
excitations of the nucleus by the pion. They can be related to the
correlations in the nucleus. They give two kinds of terms:

- First, the correlation that exists for the free nucleon, the self
  correlation. It is the rescattering effect on the nucleon itself
  and it increases the amplitude by 6 to 9%.

- A second correction arises by the pair correlations. This is
  the effective field correction in a multiple scattering theory.
  There is an absence of matter around one particle which produces
  a distortion of the pion wave. Its effect is opposite to that
  of the self correlation (they are complementary since they correspond to the absence or presence of nuclear matter). Its value
  depends on the nucleus. In the region 10 < A < 20 and for isospin
  $\frac{1}{2}$ nuclei the pair correlation nearly compensates the self correlation effect so that no net effect results from the correlations.

We summarize now the main steps which have improved the approximations starting from the soft pion limit to reach the physical
amplitude. Only the rescattering states are included and we have
indicated the pion wave which corresponds to every step.

soft pion   +   coherent rescattering   +   inelastic $\longrightarrow$ physical
                                                excitations   amplitude

incident wave   average wave in   effective wave
(Born approximation)   the medium   at the scatterer

Can something similar be done for the isospin independent part of the interaction which is responsible for the bulk of the energy shifts in π mesic atoms? In principle it can, but instead of considering the commutator of two axial charges we have to take the commutator of an axial charge with the divergence of an axial current. This is not as well defined an object as the previous one. It is not given by the fundamental identities of the current algebra and its introduction necessitates other hypotheses.

However, we know from the nucleon case that this commutator produces a small scattering length ($\frac{1}{2}|a_n + a_p| < 0.02 \; \mu^{-1}$) and indeed, theoretically, we expect this quantity to be of order μ^2. There is then no term in the expansion of first order in the pion mass. In order to have a consistent expansion it is essential to include all the terms of order μ^2, in particular the correlation effects. The soft pion limit plus the inelastic excitations gives a scattering length

$$a^+ \propto \frac{a_n + a_p}{2} - \frac{(a_n + a_p)^2 + 2(a_n - a_p)^2}{4} \left\langle \frac{1}{r} \right\rangle_{corr}$$

where $\left\langle 1/r \right\rangle_{corr}$ is the average value of $1/r$ over the correlation function. In a multiple scattering theory the normal parameter of the expansion is the single scattering amplitude and there is no stringent reason why the second term should be included. Here instead the parameter of the expansion is the pion mass. To be consistent the second term has to be included since both are of order μ^2. In fact it even dominates the first one (it accounts for about 70% of the potential).

The coherent rescattering represents, as previously, the distortion of the pion wave.

The interpretation that we propose for the soft pion limit as a Born approximation is not restricted to the elastic scattering. Another example is the radiative capture of the pion $\pi^- + A \rightarrow \gamma + B$.

In the soft pion limit the amplitude is proportional to the matrix element, $\left\langle A|J_A^\mu \; \epsilon_\mu|B \right\rangle$, of the product of the axial current with the photon polarization vector. This relates this process to the axial part of β decay and μ capture. However, the finite pion mass introduces deviations from this expression. A similar method as before shows that the most important deviation is the distortion of the pion wave and that the soft pion limit is essentially the

Born approximation for the interaction [5]. This gives the effective Hamiltonian for this transition $\mathcal{H}_{eff} = (e/2f_\pi)J_A^\mu \, \varepsilon_\mu$. The inelastic excitation here also represent the effective field correction.

To conclude my talk I want to summarize what we have learned. First we have a reasonable answer to a problem interesting in itself: what do the soft pion theorems become in nuclei? What is the meaning of the soft pion limit? The fact that it is such a simple thing as the Born approximation for the interaction should rehabilitate the nucleus in the eyes of the high-energy physicists. The nucleus is not such a complicated thing that these theories cannot be handled; and we can make the link with the multiple scattering theory and the impulse approximation.

These methods also provide the effective Hamiltonian for the interaction, independently of the impulse approximation.

Finally, we learn something about the applicability of PCAC to nuclei. The rapid variation of nuclear matrix elements of the axial currents due to anomalous thresholds has cast doubt about this applicability. It is certainly true that the soft pion limits are not to be interpreted as the ultimate predictions for the physical amplitudes. These deviate appreciably from that limit and these deviations reflect the variations of the matrix element of the axial current. But these deviations have a calculable physical interpretation.

REFERENCES

1. S. Fubini and G. Furlan, Ann. Phys. (N.Y.) 48, 322 (1968).

2. V. de Alfaro and C. Rossetti, Nuovo Cimento Suppl. 6, 515 (1968).

3. M. Ericson, A. Figureau and A. Molinari, Nuclear Phys. B10, 501 (1969).

4. R. d'Auria, A. Molinari, E. Napolitano and G. Boliani, Phys. Letters 29B, 285 (1969).

5. M. Ericson and A. Figureau, Nuclear Phys. B11, 621 (1969).

DISCUSSION

M. Rho: Your prediction that the shift goes from attractive to repulsive depends on the magnitude of $\bar{a}_s$ and $\bar{a}_p$. How accurately can these quantities, which are presumably A dependent, be determined? T.E.O.E.: These quantities, which depend on neutron excess rather than A, can be calculated from optical parameters, but this calculation does not, of course, make their origin very transparent.

ON A NEW PERIPHERAL MODEL

A. Dar

Deutsches Elektronen-Synchrotron DESY, Hamburg, &

Technion, Israel Institute of Technology, Haifa*

In a recent publication[1] we have proposed a new peripheral model for exchange reactions in high energy hadron collisions, and demonstrated its success in reproducing experimental data on meson exchange reactions. Here I shall only review it briefly and concentrate on one interesting aspect, namely, the origin of dips and spikes in the differential cross section for these reactions.

The model is based on two main observations:

a) The experimental data on exchange reactions (Those reactions where non-vanishing intrinsic quantum numbers are exchanged in the t channel, where we call intrinsic any quantum number other than angular momentum and associated parity) strongly suggest a reaction mechanism based on single particle exchanges in the t channel. This can be deduced from the following characteristic features:

   i) The cross-sections for reactions with t channel quanta that do not correspond to any known particle are extremely small.

   ii) The differential cross-sections for reactions with exchange of t channel quanta that correspond to known particles are appreciable and are peaked at small momentum transfers.

b) Total reaction cross-sections for hadron collisions are

* Permanent address.

large, implying that the unitarity of the S matrix should
play a major role in hadron collisions.

The way that these two elements, exchange mechanism and unitarity, are incorporated in our new peripheral model is described below.

Consider the differential cross-section for a two body exchange
reaction $a + b \rightarrow c + d$ with unpolarized incident beam

$$\frac{d\sigma}{dt} = \frac{1}{64\pi sk^2} \frac{1}{(2s_a + 1)(2s_b + 1)} \sum_{\lambda} |<\lambda_c\lambda_d|M|\lambda_a\lambda_b>|^2 , \qquad (1)$$

where s_i and λ_i are the spin and helicity respectively, of particle
i. s and k are the square of the c.m. energy, and the c.m.
momentum. $M_{[\lambda]}$ is the scattering amplitude for the specified heli-
city situation $[\lambda] \equiv \lambda_a\lambda_b\lambda_c\lambda_d$. Its partial wave expansion is
given by[2]

$$<\lambda_c\lambda_d|M|\lambda_a\lambda_b> = \sum_j (j + \tfrac{1}{2})<\lambda_c\lambda_d|M^j|\lambda_a\lambda_b> d^j_{\mu\nu}(\cos\theta) , \qquad (2)$$

where $d^j_{\mu\nu}(\cos\theta)$ is a rotation function, $\mu = \lambda_a - \lambda_b$, $\nu = \lambda_c - \lambda_d$.
In order to make our model physically transparent we will use the
impact parameter representation for the amplitudes: If many values
of j contribute importantly to M one may replace the sum over j by
an integral over the impact parameter b

$$M_{[\lambda]} \overset{\sim}{=} \int b\,db\, M_{[\lambda]}(b)\, J_{\Delta\lambda}(\sqrt{-t'}b) , \qquad (3)$$

according to the relations

$$j + \tfrac{1}{2} \approx kb , \quad \sum_j \approx k \int db , \quad d^j_{\mu\nu}(\cos\theta) \approx J_{\Delta\lambda}(\sqrt{-t'}b) . \qquad (4)$$

$\Delta\lambda$ is the total helicity change in the reaction $\Delta\lambda = \mu - \nu =$
$(\lambda_a - \lambda_b) - (\lambda_c - \lambda_d)$, and $J_{\Delta\lambda}$, the cylindrical Bessel function
of $\Delta\lambda$ order, is a small angle and large j approximation to the
rotation functions.

In the short wave length limit of potential theory one can
construct the S-matrix, or equivalent T-matrix, for colliding particles directly from the Born matrix, via the simple eikonal prescription:[3]

$$S(b) = \exp(i\tilde{B}(b)) ; \quad T(b) = i(I - S(b)) \qquad (5), (6)$$

where $\tilde{B}_{ij}(b) = (\sqrt{k_ik_j}/8\pi\sqrt{s})\, B_{ij}(b) \qquad (7)$

TABLE I

Comparison between predicted structure and experimental results for $d\sigma/dt'$ for various meson baryon collisions.

a) Reactions dominated by ρ and ω exchange

Reaction	Exchanged particles	Dominant amplitudes	Dip at t = 0			Dip at $-t \sim .6$ GeV^2/c^2		
			Theory	Exp.	Ref.	Theory	Exp.	Ref.
$\pi^- p \to \pi^0 n$	ρ	$\Delta\lambda \doteq 1$	+	+	10	+	+	10
$\pi^+ n \to \omega p$	ρ	none	−	−	11	−	−	11
$\gamma p \to \pi^0 p$	ω, ρ	$\Delta\lambda = 1$	+	+	12	+	+	12
$\gamma p \to \eta^0 p$	ω, ρ	$\Delta\lambda = 1,2$	+	+	13	−	−	13
$\pi^+ p \to \pi^0 N^*$	ρ	$\Delta\lambda = 1$	+	Indication	14	+	+	14
$\gamma p \to \eta^0 N^*$	ρ	$\Delta\lambda = 2$	+	No data		−	−	
$\gamma p \to \pi^0 N^*$	ρ	$\Delta\lambda = 2$	+	No data		−	−	
$\pi^0 p \to \rho^0 p$	ω	$\Delta\lambda = 1$	+	+	15	+	+	15
$\pi^+ p \to \rho^+ p$	π, ω	$\Delta\lambda = 1$	+	+	14	+	+	14
$\pi^+ p \to \omega N^*$	ρ	Depend on energy	Depend on energy	not conclusive	14	−	−	14

b) Reactions dominated by π exchange

Reaction	Exchanged particles	Dominant amplitude	Forward Spike			Forward Dip		
			Theory	Exp.	Ref.	Theory	Exp.	Ref.
$\pi^- p \to \rho^0 n$	π, A_2	$\Delta\lambda = 1$	−	−	16	+	+	16
$\gamma p \to \pi^+ n$	π, A_2, ρ	$\Delta\lambda = 0$	+	+	17	−	−	17
$\pi^+ p \to \rho^+ p$	π, ω	$\Delta\lambda = 1$	−	−	14	+	+	14
$\pi p \to \rho^0 N^*$	π, A_2	$\Delta\lambda = 0$	+	+	14	−	−	14
$\gamma p \to \pi^- N^*$	π, A_2, ρ	$\Delta\lambda = 1$	−	−	18	+	+	18
$\pi^+ p \to f^0 N^*$	π	$\Delta\lambda = 0$	+	+	14	−	−	14
$\pi^- p \to \rho^0_{tr} n$	π, A_2	$\Delta\lambda = 0$	+	not conclusive		−	not conclusive	
$\pi^+ p \to \rho^0_{tr} N^*$	π, A_2	$\Delta\lambda = 0$	−	not conclusive		+	not conclusive	
$K^\pm p \to K^{*\pm} p$	π, ω, ρ	$\Delta\lambda = 1$	−	−	19	+	+	19
$K^- p \to K^* n$	π, ρ	$\Delta\lambda = 1$	−	−	20	+	+	20
$K^\pm p \to K^{*0} N^*$	$\pi \, \rho$	$\Delta\lambda = 0$	+	+	21	−	−	21

*The subscript t_r stands for helicity states ±1 in the <u>helicity frame</u>.

TABLE II

Coupling constants

Coupling Constants	Source and Remarks	Coupling Constants	Source and Remarks
$\dfrac{g^2_{\rho\pi\pi}}{4\pi} = 2.6$	$\Gamma(\rho \rightarrow 2\pi) = 130$ MeV[22]	$g_{\rho\eta\gamma} = \dfrac{-g_{\omega\pi\gamma}}{\sqrt{3}}$	Quark Model[24]
$g^V_{\rho\rho\rho} = \dfrac{1}{2}g_{\rho\pi\pi}$	Universality[23]	$g_{\omega\eta\gamma} = \dfrac{-g_{\omega\pi\gamma}}{3\sqrt{3}}$	(In the calculations we used ηx mixing angle of
		$g_{x\rho\gamma} = -\sqrt{2}\,g_{\rho\eta\gamma}$	$-.19$.[25]
$\dfrac{g^T_{\rho\rho\rho}}{g^V_{\rho\rho\rho}} = 3.7$	Vector meson-photon analogy[23]	$g_{x\omega\gamma} = -\sqrt{2}\,g_{\omega\eta\gamma}$	
$\dfrac{g^2_{\omega\rho\pi}}{4\pi} = 14$	$\Gamma(\omega \rightarrow 3\pi) = \Gamma(\omega \rightarrow \rho\pi \rightarrow 3\pi)$ $= 12$ MeV[22]	$\dfrac{G^2_{N^*\rho\rho}}{4\pi} = 36$	SU(6) or vector dominance applied to $N^* \rightarrow p\gamma$
$g^V_{\rho\rho\pi} = \sqrt{2}\,g^V_{\rho\rho\rho}$	Isospin Invariance	$(G_1 = G_2, G_3 = 0)$	Stodolsky Sakurai Model[23]
$\dfrac{g^2_{\omega\pi\gamma}}{4\pi} = .038$	$\Gamma(\omega \rightarrow \pi^0\gamma) = 1.2$ MeV[22]	$\dfrac{g^2_{\rho\pi p}}{4\pi} = 14.7$	Dispersion Theory applied to πN scattering.[26]
$g_{\rho\pi\gamma} = \dfrac{1}{3}\,g_{\omega\pi\gamma}$	SU(6)	$g_{p\pi n} \quad \sqrt{2}\,g_{p\pi p}$	Isospin Invariance
$g^V_{\omega\rho\rho} = 3\,g^V_{\rho\rho\rho}$	Universality[23]	$\dfrac{G^2_{N^*p\pi}}{4\pi} = .37$	$\Gamma(N^* \rightarrow p\pi) = 120$ MeV[22]
$\dfrac{g^T_{\omega\rho\rho}}{g^V_{\omega\rho\rho}} = -.12$	Vector meson-photon analogy[23]	$\dfrac{g^2_{f\pi\pi}}{4\pi} = 2.5$	$\Gamma(f \rightarrow 2\pi) = 95$ MeV[22]
		$\dfrac{g^2_{K^*K\pi^0}}{4\pi} = .80$	$\Gamma(K^* \rightarrow K\pi^0) = 16$ MeV[22]
		$\dfrac{g^2_{K^*K\pi^+}}{4\pi} = 1.61$	$\Gamma(K^* \rightarrow K\pi^+) = 32$ MeV[22]

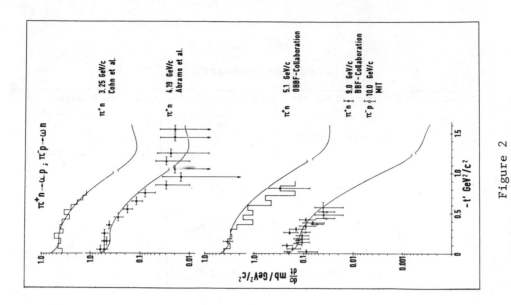

Figure 2

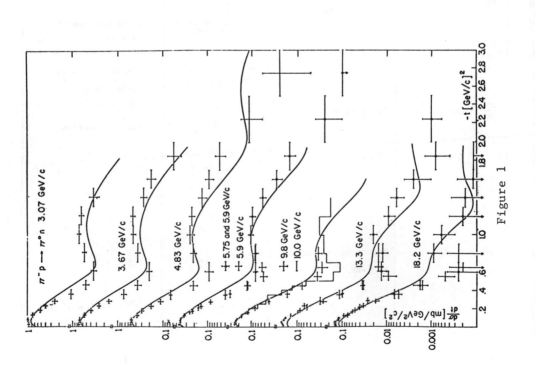

Figure 1

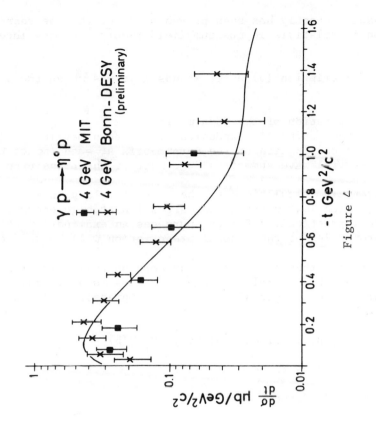

Figure 4

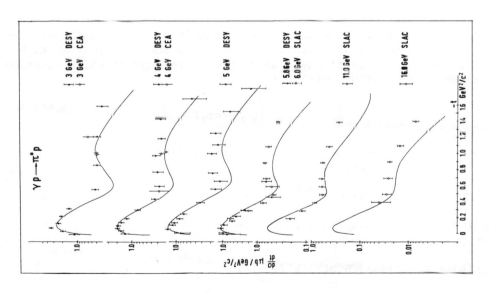

Figure 3

(This result recently has been proven also for the one channel situation in relativistic quantum field theory, by many independent authors.[4])

The prescription (6) as it stands involves an enormous amount of labour:

    a)   Computation of the Born matrix.
    b)   Impact parameter expansion of the Born matrix.
    c)   Diagonalization of the Born matrix at each impact parameter.
    d)   Partial wave summation to construct the transition amplitude.

Straightforward calculations in the case of a large number of open channels, which is the prevailing situation at high energy, are therefore impractical. For our purposes an examination of a few simple properties of the eikonal prescription will be sufficient.

First note that for large b the Born approximation for exchange of one or perhaps several exchange particles is very small (see below for explicit expressions). From expression (6) one obtains then, for $b \gg R$

$$T(b) \cong \tilde{B}(b), \text{ or equivalently } M_{[\lambda]}(b) \cong B_{[\lambda]}(b) \qquad (8),(9)$$

where R is a radius to be determined.

One may also expect that

$$M_{[\lambda]}(b) \ll B_{[\lambda]}(b) \quad \text{for} \quad b \ll R , \qquad (10)$$

on the grounds that, at small impact parameters, there is competition among a large number of reaction channels, so the amplitude in each particular channel is strongly depressed. We express this situation by means of a function $\eta(b)$:

$$M_{[\lambda]}(b) = \eta(b) \, B_{[\lambda]}(b) \; ; \; \eta(b) = \begin{cases} 1 & b \gg R \\ 0 & b \ll R \end{cases} . \qquad (11)$$

However, if $\eta(b)$ is a universal function, independent on the specific final channel, the requirement

$$i(I - e^{i\tilde{B}(b)}) = \eta(b) \, \tilde{B}(b) \qquad (12)$$

implies a simple relation between the phase and magnitude of $\eta(b)$,[5] namely

$$\eta(b) = \exp[\tfrac{1}{2}\lambda(b)] \, Sin(\tfrac{1}{2}\lambda(b)/(\tfrac{1}{2}\lambda(b) \qquad (13)$$

where $\lambda(b)$ is a real function of b. Our method consists then in

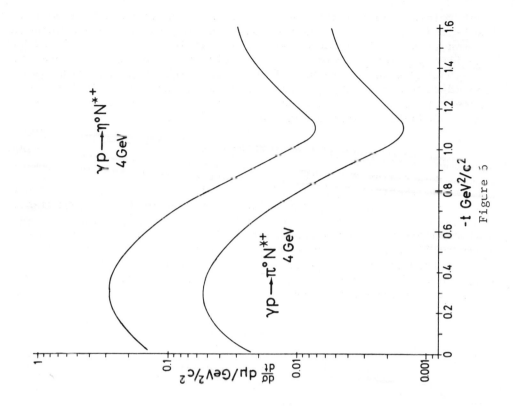

Figure 5

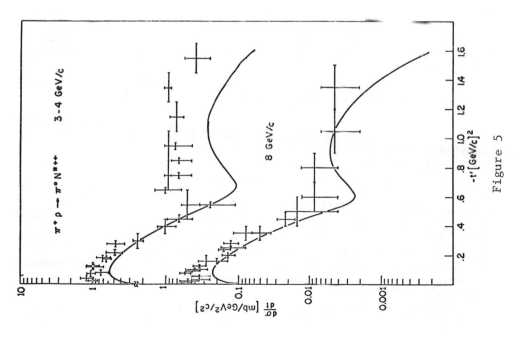

Figure 5

determining the function $\eta(b)$ from the experimental data on a given
selected reaction.   Once the function $\eta(b)$ is determined (also as
a function of s) any other exchange reaction can be calculated with
no free parameters, providing the relevant coupling constants are
reasonably known.   In Ref. 1 as a zero order approximation we proposed a simple assumption regarding the behaviour of $\eta(b)$ for $b \sim R$,
namely that it is a real function and that it goes from zero to one
as a Woods-Saxon expression does:

$$\eta(b) = [(1 + \exp(R-b)/d]^{-1} , \tag{14}$$

where the radius R and the width d are energy dependent.   (We differ
from the Absorption Models of Dar and Tobocman[6] and Gottfried and
Jackson[7] only in our prescription for the absorption coefficients).

     In order to get a quantitative idea about the dependence of
our amplitudes on the relevant quantities we make use of the asymptotic form of the Born approximation at large b   ($\mu_e b > \Delta\lambda$) for the
exchange of a particle with mass m_e and spin J_e in a reaction where
the minimum momentum transfer is

$$t_{min} \cdot \mu_e^2 = m_e^2 - t_{min} ;   t_{min} \rightarrow 0 \text{ for } k \rightarrow \infty .$$

$$\sim c_{[\lambda]} (s/s_o)^{J_e - 1} [\exp(-\mu_e b)]/\sqrt{(\mu_e b)} \tag{15}$$

where $c_{[\lambda]}$ is a constant depending on the helicity situation and the
coupling constants for the exchange diagram.   A look at (14) and
(15) shows that because of the exponential fall off of $B_{[\lambda]}(b)$ and
the step nature of $\eta(b)$, only a narrow region of $b \sim R$ contributes
to the integral (3) and we obtain approximately (see appendix of
Ref. 1)

$$M_{[\lambda]} \approx c_{[\lambda]} (s/s_o)^{J_e} \sqrt{(\mu_e R)} \exp(-\mu_e R) J_{\Delta\lambda}(\sqrt{-t'}R)/(\mu_e^2-t') \tag{16}$$

where $t' = t - t_{min}$.   From (6) and (15) one may deduce that R
has to increase at least logarithmically with s.   Physically, the
increase of R with s reflects the fact that at higher energies the
exchange of higher angular momentum become more important (see
Ref. 1).   The Born approximation using known mesons only as
exchange particles may then be valid only beyond a radius which is
larger at high energies than at low ones.   A slow increase of the
radius R has a strong effect on the energy dependence of the reaction cross section because of the exponential dependence of the
amplitude on R in (16).   In particular, the following observations
regarding the energy behaviour of the cross sections can be made:

     a)   The exponential dependence on the mass of the exchanged

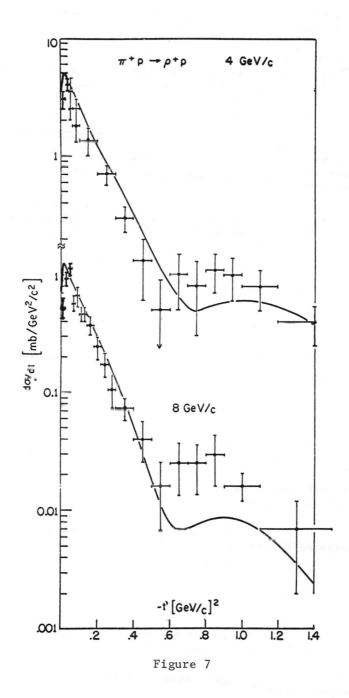

Figure 7

particles depresses the exchange of particles with the
same spin but with higher mass.

b) In the case of pion exchange reactions the Absorption
    Model predictions of Gottfried and Jackson[7] are not
    appreciably altered because of the small mass of the pion.

c) If R depends logarithmically on s one obtains simple power
    laws for the energy dependence of the cross sections.

Let us now turn to the t dependence of the cross sections and
let us first discuss the intermediate and large t region
$(-t \gg m_\pi^2)$. $M_{[\lambda]}$ as given by (16) can be recognized as a diffraction pattern of a radiating ring with radius R and with angular
momentum change $\Delta\lambda$.[8]   The Bessel functions which appear in (16)
have simple properties as function of their argument and their
order: Beyond their first maximum (around $x \sim \Delta\lambda$) the Bessel
functions approach rather quickly their asymptotic oscillatory form:

$$J_{\Delta\lambda}(x) \rightarrow \sqrt{(2/\pi x)} \, \cos(x - \tfrac{1}{2}\pi[\Delta\lambda + \tfrac{1}{2}]) \; . \tag{17}$$

Consequently, helicity amplitudes with even $\Delta\lambda$ and with odd $\Delta\lambda$
oscillate exactly out of phase. (16) and (14) lead then to the
following general predictions for the large t behaviour of the
cross sections:

a) Individual helicity amplitudes should exhibit a diffrac-
    tion pattern of maxima and minima.   The position of
    these maxima and minima do not depend on the specific
    reactions.

b) If helicity amplitudes with even $\Delta\lambda$ and with odd $\Delta\lambda$ con-
    tribute about equally to the unpolarized cross section all
    the diffraction minima and maxima should be smeared.

c) If a single helicity amplitude dominates the reaction, the
    unpolarized cross section too, should exhibit a diffrac-
    tion pattern of maxima and minima (follows from a)

d) Polarized cross sections should exhibit more structure
    than unpolarized cross sections (follows from a)

e) The diffraction patterns should shrink with increasing
    energy since R increases with s .

We would like to stress, however, that all these predictions
are only approximate. In general, both finite width effects and
the imaginary part of the amplitude tend to wash out the diffraction pattern. In particular we do not expect any significant

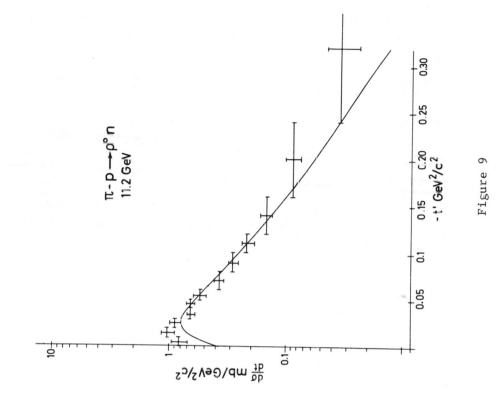

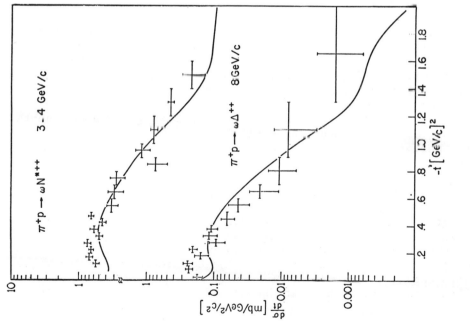

Figure 9

Figure 8

structure in dσ/dt for large t, for pion exchange reactions, since
for pion exchange the reaction is poorly localized at a well
defined radius.

Next let us examine the region $-t \lesssim m_\pi^2$. From (16) one may
make the following predictions:

a) a forward peak is expected in reactions dominated by
    helicity amplitudes with $\Delta\lambda \doteq 0$.   For exchange of parti-
    cles <u>other than the pion</u> one may neglect the slowly vary-
    ing propagator $(m_e^2 \quad L)^{-1}$ in (16).   The slope of the for-
    ward peak is then given by R and it is thus independent on
    the specific reaction.   However, <u>for pion exchange reac-
    tions</u> one may not neglect the pion propagator in (16).
    This propagator turns the forward peak into a <u>sharp spike</u>
    with a width of about $\mu_e^2/2$.

b) A forward dip followed by a maximum around $-t' \sim <\Delta\lambda>^2/R^2$
    is expected in reactions dominated by helicity amplitudes
    with $\Delta\lambda \neq 0$, where $<\Delta\lambda>$ is the average helicity change
    in the reaction.   However, for pion exchange reactions,
    due to the pion propagator, the maximum should occur
    around $-t' \sim \mu_e^2$.

From (9) we see also that for π exchange reactions the forward
structure becomes narrower and more pronounced, and the forward
cross section tends to increase when μ_e^2, the distance of the physical region from the pion pole, decreases (i.e., when s increases).

By now one may have realized that the key to the understanding
of the structure in the angular distribution of an exchange reaction
is the relative magnitudes of all helicity amplitudes for that particular reaction.   This however, in general involves a) knowledge of
all relevant coupling constants, b) laborious Feynman algebra.   In
this paper we present a comparison between theory and experiment
only for a selected set of reactions for which the Feynman algebra
is relatively simple and the coupling constants are known either
from experimental decay rates or from reasonably successful symmetry
schemes.

Table I summarizes a comparison between our theoretical predictions and experimental data.   The coupling constants that were used
are listed in Table II.   We conclude from Table I that agreement
between theory and experiment is remarkably good in those cases
where the experimental data is conclusive.   Figures 1-10 illustrate
some of our detailed predictions.   The calculations are described
in more detail in Ref.(1).   As is evident from these figures good
agreement is obtained between theory and experiment.

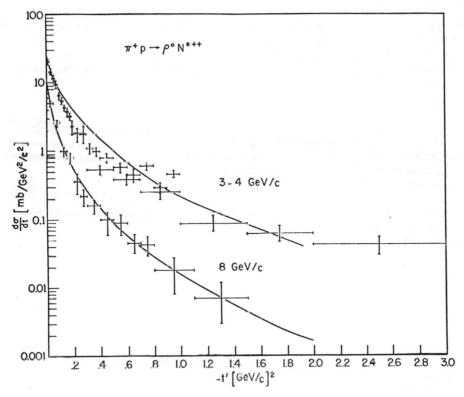

Figure 10a

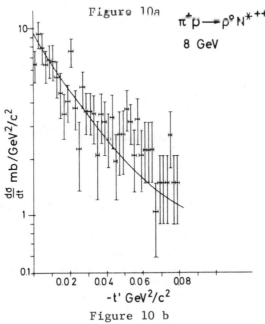

Figure 10 b

We would like to conclude by drawing attention to the fact that any model that has the long range particle exchange pole built in together with <u>strong</u> absorption of low partial waves is bound to yield the same qualitative results.   In particular the strong absorptive Regge pole exchange models[9] should result in the same qualitative predictions.

Acknowledgement

The author would like to thank Profs. W. Jentschke, E. Lohrmann, S.C.C. Ting and M.W. Teucher for their hospitality during the author's visit at DESY.

*Figures 1-10: Comparison of calculated $d\sigma/dt$ with experimental data for various reactions: 1) $\pi^- p \to \pi^\circ n$, 2) $\pi^+ n \to \omega p$, 3) $\gamma p \to \pi^\circ p$, 4) $\gamma p \to \eta^\circ p$, 5) $\pi^+ p \to \pi^\circ N^*$, 6) $\gamma p \to \eta^\circ N^*$, 7) $\pi^+ p \to \rho^+ p$, 8) $\pi^+ p \to \omega N^*$, 9) $\pi^- p \to \rho^\circ n$, 10) $\pi^+ p \to \rho^\circ N^*$ a) large t behaviour, b) small t behaviour.

References & Footnotes

[1] A. Dar, T.L. Watts and V.F. Weisskopf, "Towards a New Peripheral Model for Hadron Collisions", to be published in Nucl. Phys. B. See also A. Dar, Proceedings of the Montreal Summer School on Diffractive Processes, July, 1969.

[2] M. Jacob and G.C. Wick, Ann. Phys. N.Y. 7, 404 (1959).

[3] A. Dar, Ann. Phys. to be published, and references therein.

[4] V.N. Gribov and A.A. Migdal, preprint, and Communication to the 1968 Vienna Conference on High Energy Physics by R.A. Ter-Martirosyan. H.D.I. Abarbanel and C. Itzykson, Phys. Rev. Lett. 23, 53 (1969). M. Levy and J. Sucher, Preprint. R. Sugar and R. Blankenbecler to be published.

[5] G. Berlad, Ph.D. Thesis to be published (Technion Haifa).

[6] A. Dar and W. Tobocman, Phys. Rev. Lett. 12, 511 (1964).

[7] K. Gottfried and D. Jackson, Nuovo Cimento 34, 735 (1965).

[8] A. Dar, M. Kugler, Y. Dothan and S. Nussinov, Phys. Rev. Lett. 11, 265 (1964).

[9] F. Henyey, G.L. Kane, J. Pumpkin and M. Ross, Phys. Rev. Lett. 23, 946 (1969).   Earlier references for the Absorptive Regge Pole Exchange Model are: R.C. Arnold, Phys.Rev. 153, 1523 (1967); A. Morel and H. Navelet, Nuovo Cimento 48A, 1075 (1967), J.N.J. White, Phys.Lett. 26B, 461 (1968); F. Schrempp, Nucl. Phys. B6, 487 (1968).   For detailed discussion and other references see J.D. Jackson, Proceedings of the Lund Conference, June 1969.

[10] M. Wahling et al. (MIT-PISA), Proc. Int. Conf. on High Energy Physics, Dubna 1964. A.V. Stirling et al. (Saclay-Orsay), Phys.Rev.Lett.14, 763 (1965). P. Sonдеregger et al. (CEN-Saclay-Orsay), Phys.Lett.20, 75 (1966).

[11] A. Cohn et al., Phys.Rev. 178, 2061 (1969); G. Benson et al., Univ. of Mich. Report No. COO-1112-4 (1966); G.S. Abrams et al. Univ. of Ill. Preprint; N. Armenise et al. (OBBF Collaboration) Phys.Lett. 26B, 336 (1968) and (BBF Collaboration) Submitted to the 1969 Lund Conference on High Energy Physics.

[12] G.C. Bolon et al. (MIT) Phys.Rev.Lett. 18, 926 (1967); M. Braunschweig et al. (BONN-DESY) Phys.Lett. 22, 705 (1966); Phys. Lett. 26B, 405 (1968); R. Anderson et al. (SLAC) Phys.Rev. Lett. 21, 334 (1968).

[13] D. Bellenger et al. (MIT) Phys.Rev.Lett. 21, 1205 (1968). W. Braunschweig et al. (Bonn-DESY) preliminary results.

[14] D.G. Brown, Ph. D. Thesis UCRL-18254, May 1968; G. Gidal et al. UCRL-17984, December 1967; A.B.C. Collaboration, Nucl. Phys. B8, 45, (1968); D.R.O. Morrison, private communication.

[15] A.P. Contogouris et al. Phys. Rev. Lett. 19, 1352 (1967).

[16] B.D. Hyams et al., Nucl. Phys. B7, 1 (1968).

[17] A.M. Boyarski et al. (SLAC) Phys.Rev.Lett. 20, 300 (1968); 21, 1767 (1968) G. Buschhorn et al. (DESY) Phys.Rev.Lett. 17, 1027 (1966); 18, 571 (1967).

[18] A.M. Boyarski et al. (SLAC) Phys.Rev.Lett, 22, 1131 (1969).

[19] ABCLV Collaboration Nucl. Phys. B7, 111 (1968). S.C. Berlinghieri et al. (Rochester) Nucl. Phys. B8, 333 (1968). D.D. Carmony et al., Nucl. Phys. B12, 9 (1968).

[20] ABCLV Collaboration Nucl. Phys. B7, 111 (1968); B5, 567 (1968).

[21] J.C. Berlinghieri et al. Ref. 19. Y. Goldshmidt-Clermont, private communication. B. Bassompierre et al. (CERN-Bruxelles) to be published.

[22] Particle Data Group, Rev. Mod. Phys 40, 77 (1968).

[23] J.J. Sakurai, Ann. Phys., (N.Y.) 11, 1 (1960). M. Gell-Mann and F. Zachariasen, Phys. Rev. 124, 953 (1961). L. Stodolsky, Phys. Rev. 134, B 1099 (1964). J.J. Sakurai, Phys. Rev. Lett. 17, 1021 (1966). M.K. Banergee and C.A. Levinson, Phys. Rev. 176, 2140 (1968).

[24] See for instance, A. Dar and V.F. Weisskopf, Phys. Lett. 26B, 670 (1968).

[25] R.H. Dalitz and G. Sutherland, Nuovo Cimento 37, 1777 (1965); 38, 1945 (1965).

[26] V.K. Samaranayake and W.S. Woolcock, Phys. Rev. Lett. 15, 936 (1965).

EFFECTIVE NUCLEON NUMBER IN HIGH ENERGY REACTIONS

Daphne F. Jackson

Department of Physics, University of Surrey

Guildford, England

High energy reactions such as incoherent scattering of 20 GeV protons and pion production by 600 MeV protons have recently been analysed[1,2] using impulse approximation, the semi-classical approximation and closure. This treatment introduces the effective nucleon number

$$N(A) = \int T(\underset{\sim}{b}) \, e^{-\sigma T(\underset{\sim}{b})} \, d^2b$$

where σ is the total cross-section for two-body scattering and

$$T(\underset{\sim}{b}) = \int_{-\infty}^{\infty} A \, \rho(r) \, dz \quad .$$

In plane wave approximation, $N(A) = A$, so that $N(A)/A$ yields a rough estimate of the reduction in magnitude of the cross-section due to absorption. An alternative derivation of these expressions has been given[3] in terms of the distorted wave impulse approximation for nuclear reactions.

Calculations of $N(A)$ have been carried out[3] for nucleon-nucleus scattering at 1 GeV ($\sigma \approx 44$ mb) and pion-nucleus scattering at 2 GeV ($\sigma \approx 27$ mb). The nuclear density distribution has been taken to be spherically symmetric and of fermi or parabolic fermi shape. The region of the density to which the process is most sensitive depends on the degree of absorption of the projectile. The behaviour of the integrand $T(b) \, e^{-\sigma T(b)}$ for pion scattering from ^{40}Ca and ^{209}Bi is shown in the figure for parameters of the density which give agreement with the elastic electron scattering data; it can be seen that the integrand peaks at the halfway radius and the most important contribution comes from the 90%-10% transition region. It has already been shown[1] that for high

energy nucleon scattering the integrand peaks approximately 0.5 fm
beyond the halfway radius and our calculations confirm this.   Thus
pion-nucleus scattering in the 1-2 GeV region is sensitive to the
transition region which is the region most accurately determined
from electromagnetic measurements, while nucleon-nucleus scattering
in the GeV region is sensitive to the outer part of the nuclear surface.   We may conclude that both processes provide information
about the nucleus which is complementary to that obtained from
electron scattering.

Further calculations with density distributions which also fit
the electron scattering data yield variations in $N(A)$ of 3-5%, but
use of a different distribution for ^{209}Bi which fits the muonic
X-ray data but does not fit the electron data yields a variation in
$N(A)$ of 11%.   A more serious variation is caused by the difference
between the proton and neutron distribution, and estimates for Ag
yield a change in $N(A)$ of 23% for nucleon-nucleus scattering and
18% for pion-nucleus scattering.

(1)   R. J. Glauber, Proceedings of the Conference on
      High Energy Physics and Nuclear Structure,
      Rehovoth, 1967 (ed. Alexander, North-Holland,
      1967).

(2)   B. Margolis, Nucl. Phys. B4, 433 (1968).
      W. Hirt, Nucl. Phys. B9, 447 (1969).

(3)   D. F. Jackson, Nuo. Cim., to be published.

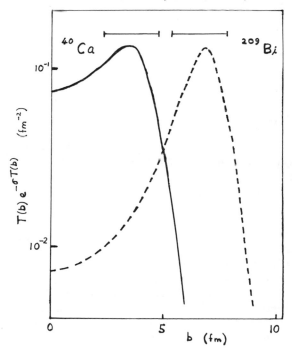

Fig. 1 - The integrands
$T(b)^{-\sigma T(b)}$ for pion
scattering from ^{40}Ca and
^{209}Bi.   The bar indicates
the region of the 90%-10%
transition region for each
nucleus.

THE OPTICAL POTENTIAL FOR HIGH ENERGY PION-NUCLEUS AND NUCLEON-NUCLEUS SCATTERING

V. K. Kembhavi and Daphne F. Jackson

Department of Physics, University of Surrey

Guildford, England

The optical potential for high energy scattering is given in the impulse approximation by

$$V(r) = -A(2\pi)^{-2} K \int e^{-i\underset{\sim}{q}\cdot\underset{\sim}{r}} M(q) F(q) d\underset{\sim}{q} \qquad (1)$$

where A is the number of nucleons in the target nucleus, K is a kinematic factor, $M(q)$ is the free two-body scattering amplitude, and $F(q)$ is the nuclear form factor. In the limit of large A, the scattering amplitude varies slowly with q compared with the form factor so that the potential is given approximately by

$$V(r) \simeq -A(2\pi) K M(0) \rho(r) \qquad (2)$$

where $\rho(r)$ is the nuclear matter distribution. This second form is the one normally used in calculations at high energies.

We are studying the potentials given by equation (2) (Potential A) and equation (1) (Potential B) for pion-nucleus and nucleon-nucleus scattering in the GeV region, using a matter distribution constructed from single-particle wavefunctions which fit electron scattering (SP) and also a fermi distribution (F). Results show that the difference between the potentials is greater for the single-particle distribution than for the fermi distribution, and that the difference is not negligible even for ^{40}Ca. This can be seen from the figure, which shows the imaginary part of the potential for nucleon-nucleus scattering for ^{40}Ca at 1 GeV. (The real part is repulsive with a central strength of 20-30 MeV.) Calculations of the effective nucleon number using these potentials give a difference of about 1%, and calculations of the differential cross-section and total absorption cross-section are in progress.

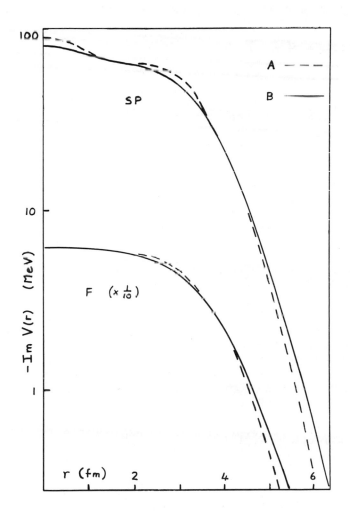

COMPATIBILITY BETWEEN DIFFRACTIVE AND STATISTICAL MECHANISMS

E. Predazzi

University of Torino (Italy) and Indiana University

Bloomington, Indiana

It has become clear, in recent times, that there is a large
similarity between the high energy scattering of nucleons on composite objects (i.e. nuclei) and on "elementary" particles. For
this connection we refer to the large literature on the subject[1]
and we merely recall here that the implication is that "elementary"
particles are, in fact, far from being elementary; a conclusion to
which one comes from many different points of view. It has also
been shown[2] that one can reverse the usual approach to the problem
by first deriving an empirical formula that fits the data and then
analyzing its theoretical implications. This analysis has been
completely performed for p-p elastic scattering[2] and results are
in surprisingly close agreement with the conclusions of Ref. 1.
Here, however, we want to focus our attention on another point.
The very good agreement with experiments of the formula proposed
in Ref. 2 holds throughout the entire angular interval. This requires the small angle data to be in the form of a sum of Gaussians
in the momentum transfer (or equivalent variable in this domain).
In the following we refer to such an expansion as a "diffractive
expansion". On the other hand, the large angle data are well known[3]
to be very well reproduced by a sum of exponentials. This sort of
expansion will be referred to as a "statistical expansion" because
of its obvious analogies with a statistical model.[4] Clearly, there
is no a priori reason why the two mechanisms of a diffractive and
a statistical model, acting in different angular regions should have
anything to do one with the other. One of the results of Ref. 2,
however, is that, starting from a diffractive expansion (i.e. a
series of Gaussians) in which all the parameters are determined
through i) experimental p-p data in the small angle region and ii)
the break points in the same angular distribution; one can exactly
convert it into a statistical expansion (i.e. a series of exponentials)

whose aymptotic behavior agrees (within 10%) with Orear's formula.[3]
Notice that the above is an exact result which follows for a large
class of possible parametrizations of the diffractive expansion
provided only the dependence on the summation index of the exponent
of the Gaussians be of the form $(cn + 1)^{-1}$ with c = constant (as
suggested also by other considerations[5]).

The picture that emerges (see Ref. 2 for details) is the
following. At very small angles just one or a few terms of the
diffractive expansion are sufficient to account for the data but
more and more terms become necessary as we go to larger momentum
transfers since the convergence of the series becomes slower and
slower. Conversely, one term in the expansion in exponentials is
an excellent approximation to the data in the large angle domain
whereas more and more terms become necessary as we move toward the
forward direction. This is, of course, exactly what one expects
from a diffractive and a statistical model respectively.

The situation outlined previously is very interesting since
it opens the problem of whether or not diffractive effects which
are responsible for the small angle data (and which are usually
understood as the shadow of inelastic processes through the mechanism of unitarity), are indeed to be considered as inherently different from statistical effects that are conventionally attributed to
"central collisions" and which are supposed to dominate the large
angle data. Contrary to a rather widespread belief, the results of
Ref. 2 suggest that there is a smooth transition between the two
mechanisms. Thus, the use of one or another of the two formalisms
is merely dictated by the convenience of choosing the expansion that
converges more rapidly. The sets of coefficients of the two expansions, however, are strictly related one to the other.

It is clear that this problem needs further clarification in
order to understand what are exactly the physical relevance and the
dynamical implications of the mathematical developments which are
given in Ref. 2. It is, however, also clear that if the situation
described before could be shown to apply to processes other than p-p,
this would give rise to the very interesting possibility that one
should ultimately be able to correlate large angle phenomena to small
angle effects in an essentially one to one way.

REFERENCES

[1]See for instance: R. J. Glauber, Proccedings of the Conference on High Energy Physics and Nuclear Structure, ed. G. Alexander
(North Holland Pub. Co., Amsterdam 1967) p. 311; T. T. Chou and
C. N. Yang (ibidem) p. 348 and Phys. Rev. 175 1832 (1968); N. Byers
and S. Frautschi, Calt. 68-191 and references therein.

[2]H. Fleming, A. Giovannini and E. Predazzi, Nuovo Cimento 56A 1131 (1968) and Annals of Physics 54 6 2 (1969).

[3]J. Orear, Phys. Rev. Letters 12 112 (1964); Phys. Rev. Letters 13 190 (1964).

[4]See for instance: R. Hagedorn, Nuovo Cimento Suppl. 3 147 (1965).

[5]G. Cocconi, Nuovo Cimento 57A 837 (1968), see also M. M. Nieto, Nuovo Cimento 61A 105 (1969).

INDEX